# POLYMER ELECTROLYTE FUEL CELLS

## Physical Principles of Materials and Operation

# POLYMER ELECTROLYTE FUEL CELLS

## Physical Principles of Materials and Operation

Michael Eikerling • Andrei Kulikovsky

CRC Press
Taylor & Francis Group
Boca Raton London New York

CRC Press is an imprint of the
Taylor & Francis Group, an **informa** business

CRC Press
Taylor & Francis Group
6000 Broken Sound Parkway NW, Suite 300
Boca Raton, FL 33487-2742

First issued in paperback 2017

© 2015 by Taylor & Francis Group, LLC
CRC Press is an imprint of Taylor & Francis Group, an Informa business

No claim to original U.S. Government works
Version Date: 20140812

ISBN 13: 978-1-138-07744-7 (pbk)
ISBN 13: 978-1-4398-5405-1 (hbk)

---

### Library of Congress Cataloging-in-Publication Data

---

Eikerling, Michael.
   Polymer electrolyte fuel cells : physical principles of materials and operation / Michael Eikerling, Andrei Kulikovsky.
      pages cm
   Includes bibliographical references and index.
   ISBN 978-1-4398-5405-1 (hardback)
   1. Proton exchange membrane fuel cells.  I. Kulikovsky, Andrei A. II. Title.

TK2933.P76E39 2014
621.31'2429--dc23                                                          2014024439

---

**Visit the Taylor & Francis Web site at**
**http://www.taylorandfrancis.com**

**and the CRC Press Web site at**
**http://www.crcpress.com**

# Contents

**v**

# Foreword

## PHYSICS OF POLYMER ELECTROLYTE FUEL CELLS: THE STORY OF THE MOST IMPORTANT ISSUES

It is my great pleasure and privilege to write a foreword to this encyclopedic book (a comprehensive monograph or a textbook, depending on how you look at it) written by my friends and former coworkers.

The book is, in many respects, unique. First of all, everywhere it starts from physics, oriented on physical understanding of how all components of fuel cells work and, most importantly, how they work together. Indeed, fuel cells are very interesting objects for physics—and both authors are physicists (with all of their degrees received in physics).

Conventionally speaking, a fuel cell is composed of a membrane, a membrane–electrode assembly (MEA), current collectors and gas diffusion layers, and bipolar plates containing gas supply channels.

If we speak specifically about polymer electrolyte fuel cells (PEFCs), to which this book is devoted, the theory of the polymer electrolyte membrane is the theory of proton transport in a complex water-containing porous environment; it is also the story of water in the membrane, of its sorption and distribution coupled with the proton transport.

In the MEA, key things happen. At the anode catalyst layer, the fuel reacts via a hydrogen oxidation reaction and it splits into protons and electrons at the interface of the catalyst and the membrane material (the two phases must interpenetrate). The catalyst must be a part of a continuous percolating network of an electronically conducting medium leading the newly born electrons to the current collector, whereas the membrane phase inside the catalyst layer must provide a continuous path for the protons to the bulk of the membrane. At the cathode, oxygen reduction takes place by recombination of oxygen molecules with protons and electrons to form water. The phase boundary between the catalyst and membrane fragments must be accessible to oxygen, and the formation of water must not hamper other processes, such as gas transport. Thus, the story of each MEA is the story of multicomponent percolation, transport of gases, molecular electrochemistry, nanoscale structures, and processes in them—all together in one place.

The problems with gas diffusion layers are less crucial but there are also some related to potential inhomogeneity of distribution of gases and electronic current flows, depending on the design of bipolar plates. The latter are related with the design of stacks, and in the case of a wrong design, they may cause fuel cell starvation or flooding.

This is an oversimplified, bird's-eye view of the types of problems encountered in the theory of polymer electrolyte fuel cells, but all of them are covered or at least touched in this book. It is essentially a first comprehensive book on the modern theory of these cells.

Before going into the details of the book chapters, I want to write about the *authors* and their contributions to fuel cell research:

Michael Eikerling joined my Theory Group in Jülich in 1995 as a PhD student. At that time, the group was doing all sorts of challenging electrochemical physics projects. Embedded in the Institute of Materials and Processes in Energy Systems, it had to work on fuel cells. A couple of years before Michael joined the group, I actually entered that field working on some percolation models of high-temperature *solid oxide fuel cells* and found it to be extremely interesting. The theme *polymer electrolyte fuel cells* was a logical continuation. I offered Michael three "chemical physics" topics to choose from: one, in which I knew everything what to do, one more challenging, and one topic where I almost knew nothing—and this was in the PEFC area. Unmistakably, Michael has chosen the last one, the most challenging one—on the polymer electrolyte membrane structure and performance. The work resulted in a series of papers, in which we modeled the membrane at different levels of complexity—from phenomenological percolation models to the analysis of proton transport mechanisms.[1] Only then we proceeded to modeling the porous composite electrodes.[2] The work was so successful that we decided to describe the impedance of the electrode, and because of that Michael did something unusual for a career-oriented person—he stayed for 1 year as a postdoc in my group to complete that work.[3] It appeared to be a good time investment, as all those works became highly cited and widely known.

Next, Michael undertook a "real postdoc" in the group of Tom Zawodzinski and Shimshon Gottesfeld at Los Alamos National Laboratory moving to molecular level modeling of the membrane structure and properties, including quantum-chemical computations.[4] After that he returned to Germany, but now to the Department of Physics at the Technical University of Munich, to join the Institute of Interfaces and Energy Conversion headed by Ulrich Stimming. There he worked on the models of molecular electrocatalysis in nanostructured electrodes.[5]

Close to the end of his term there, I moved from Jülich to Imperial College London, and was about to initiate a prestigious fellowship for Michael to bring him to our Department of Chemistry. However, at that time, Michael received a very attractive offer of assistant professorship at Simon Fraser University, combined with a senior researcher position at the newly created Institute for Fuel Cell Innovation of the Canadian National Research Council (NRC). I strongly recommended him to accept this position straightaway.

Having settled in Vancouver, Michael has built his vibrant and successful research group, working on various fuel cell-related problems spanning from macro- to nanoscale using all sorts of theoretical methods including molecular dynamics and quantum chemistry. During his work in my group, Michael was never afraid of unknown territories, bravely cracking any problem he wanted to solve. Over the years at Simon Fraser University and NRC, this intrepid spirit led to a series of breakthrough findings.

A concerted program of quantum mechanical simulations and theory development he pursued elucidated the effects of molecular structure and packing density of acid-functionalized surface groups on spontaneous ordering, acid dissociation, water binding, and proton dynamics at highly charged interfaces.[6,7,8] The key finding, which his group made, was regarding a mechanism of soliton-like proton transport, which was explored all the way up to calculating a proton mobility. The editors of *Journal of Physics: Condensed Matter* chose the 2013 paper on this topic as an IOP select article "for novelty, significance and potential impact on future research" (http://iopscience.iop.org/0953-8984/labtalk-article/52010).

In membrane research, Michael has developed a poroelectroelastic theory of water sorption and swelling together with Peter Berg (NTNU Trondheim). It rationalizes the impact of external conditions, statistical distribution of anionic head groups, and microscopic elastic properties of the polymer on water sorption and swelling.[9] This work has opened up an intriguing research area: the study of internal mechanical stresses in charged elastic media, induced by water sorption.

In nanoparticle electrocatalysis, the area that Michael entered just some time ago in Munich, he and his coworkers rationalized the sensitivity of electrocatalytic processes to the structure of nanoparticles and interfaces. Studies of catalytic effects of metal oxide support materials revealed intriguing electronic structure effects on thin films of Pt, metal oxides, and graphene.[10,11] In the realm of nanoparticle dissolution and degradation modeling, Michael's group has developed a comprehensive theory of Pt mass balance in catalyst layers.[12,13] This theory relates surface tension, surface oxidation state, and dissolution kinetics of Pt.

Moving up the scale to the level of flooded nanoporous electrodes, Michael's group has developed the first theoretical model of ionomer-free ultrathin catalyst layers—a type of layer that promises drastic savings in catalyst loading.[14,15,16] Based on the Poisson–Nernst–Planck theory, the model rationalized the impact of interfacial charging effects at pore walls and nanoporosity on electrochemical performance. In the end, this model links fundamental material properties, kinetic parameters, and transport properties with current generation in nanoporous electrodes.

Together with Kourosh Malek at the NRC, Michael started pioneering coarse-grained molecular dynamics studies of self-organization in catalyst layer inks.[17,18] Simulations revealed, among other interesting findings, the formation of a skin-type ionomer morphology, which is nowadays widely explored experimentally.

Along these lines, Michael's group continually integrates findings from nanoparticle studies, pore-level modeling and molecular modeling into models of catalyst layer operation. The current version represents the most complete catalyst layer model to date. It accounts for statistical effects in catalyst utilization and nonuniform reaction conditions across the full range of scales.[19] Perhaps the most important result from a practical perspective is the calculation of the effectiveness factor of Pt utilization. His group has established an extensive network of collaborations with academia, government institutions, and industry, where theoretical findings are tested experimentally and integrated into diagnostic approaches.

Michael's progress was steady, the key achievements opening new horizons in understanding of fuel cells, and highly cited. He was elected as chairman of the Division of Physical Electrochemistry of the International Society of Electrochemistry, and was promoted in 2012 to full professorship at Simon Fraser University. Over the years, we maintained good contact, having published a number of joint follow-up papers on various aspects of fuel cells, including review articles.[20]

The story of my relationship with Andrei Kuikovsky is different. We graduated from the same elite Technical University of Nuclear Sciences (MEPHI) in Moscow, he being several years younger than me. We met, however, first on ski slopes, being members of our university alpine ski team. Andrei joined that team, being an excellent slalom racer, but best in giant slalom. We continued training and racing together over many years after graduation, within the racing team of the USSR Academy of Sciences and became close personal friends, who have experienced together very tough situations on the slopes—as in those days, paradoxically for adults, racing was no less important for us than science. But as those years passed, our sport activities reduced to less exciting veteran racings, and science fully conquered our hearts and minds.

Years later, just around the time when Michael decided to stay 1 year more with me in Jülich, I got a senior scientist slot in my group and convinced the Jülich administration to offer it to Andrei, who by that time became an authority in the theory of low-temperature plasma discharge. He knew nothing about fuel cells then, but was a computational champion in solving nonlinear equations of a class not that different from those emerging in the fuel cell theory. He started with computational three-dimensional modeling of physical (not engineering) models of fuel cells with the most important effects taken into account, which opened a road to what I called *a functional map of a fuel cell*, the term later used by many. Such maps have revealed that some parts of the cell, depending on the geometry of current collectors, do not perform at all and just take space and add weight.[21]

Several models of more efficient architectures came out of it.[22] But interestingly, inspired by these findings, we started to do analytical modeling of some of these processes which resulted in simple laws of "starving" fuel cells and the recipes one needs to follow to feed them properly.[23] The important turn for Andrei was that he got himself keenly interested in analytical theory, and in his later work, after I moved to Imperial College London, he split his time between computation and analytics. After my leave, we published several follow-up papers together, including a monographic chapter written together with Michael in *Encyclopedia of Electrochemistry*,[24] a modest, pale prototype of this book.

During his further years in Jülich, Andrei has shown enormous creativity and productivity. Inspired by the paper by Perry, Newman, and Cairns,[25] and our paper on the catalyst layer,[2] Andrei has extended the theory to a more general class of electrodes, suitable for fuel cells of various types, having obtained a number of novel asymptotic solutions of the problem,[26] which was later applied to hydrogen fuel cells (PEFCs), direct methanol fuel cells (DMFCs), and even solid oxide fuel cells (SOFCs).[27] Having combined the model of catalytic layers with the description of the processes in the diffusion layers and the gas-supplying channels, he has obtained analytical models of half cells, which have already been used by engineers in the design of PEFC stacks.[28] Whereas in our earlier work, Michael and I proposed a basic model of impedance of the catalyst layer,[3] Andrei developed a generalized physical model for the whole fuel cell impedance, including oxygen transport in gas channels and the gas diffusion layer.[29]

He has published a series of seminal papers on the mechanisms of fuel cell degradation and possible scenarios of "catastrophic" worsening of the fuel cell performance, having described, for example, effects such as the degradation wave along the air channel in PEFCs,[30] and Ru corrosion due to methanol depletion in DMFCs.[31] He contributed to the explanation of exotic "direct" and "reversed" performance domains in PEFCs under the lack of hydrogen at the anode[32] (in the reverse domain, carbon in the cathode catalyst layer is utilized as a fuel, while oxygen on the anode side is reduced; the model reveals the formation of a local hydrogen–oxygen fuel cell at the direct/reverse domain interface, which supports large in-plane proton current in the membrane).

He also made an important contribution to the background of novel diagnostic methods, such as description of the local current distribution around a Pt-free spot in the PEFC anode, a key problem for transmission x-ray adsorption spectroscopy of the fuel cell cathode.[33] He filed many patents based on his theoretical work.

Many more of Andrei's findings could be named, but most of them found their place in this book. A lone hiker, he publishes on average 10 papers per year, 9 of it alone! These are not repetitions of the same theme: there are always new ideas behind practically each of his works. On average, they are less cited and known than Michael's, but this does not make them less important. For his contributions to understanding fuel cells, he recently received the prestigious Alexander M. Kuznetsov Prize for Theoretical Electrochemistry of the International Society of Electrochemistry.

Andrei has published in Elsevier his first book on fuel cells,[34] heavily biased on analytical models of their performance. Very elegantly written, it covers the findings and methods of his in total 40 papers and the key works of many other authors, giving theorists who want to work in the area of fuel cells full guidance and a powerful tool for their research. By its nature, it is, however, oriented to a more specialized audience than the current book, although generally it has greatly influenced its content.

Having met each other 15 years ago in my group, Andrei and Michael had no joint publications except for a couple of reviews that we wrote together and one beautiful paper, very useful for experimentalists, on the development of the model of impedance of the catalyst layer, which is about how to get its parameters from the impedance spectrum "fitting-free."[35] After that milestone in their cooperation, they seemed to have found the right "chemistry" for working together. Over the years, each of them had done so many different things on the theory of fuel cells that it was logical that they joined forces and completed this voluminous project covering all aspects of the fundamental theory of PEFC. I am glad that they were able to complete it, and—amazingly—almost without impeding the flow of their original studies.

I wrote here so much about the various contributions of the authors of this book to the science of fuel cells in order to make it clear that the book is written by professionals, of mostly complementary knowledge (Michael focussed more on microscopic theory and Andrei more on continuum models), but both have worked at the cutting edge of the theory of fuel cells, contributing a lot toward their understanding.

## The Book

Chapter 1 is the "sweetest" part of the book, masterfully written. In less than 60 pages, it rolls out the basic science of PEFCs, from the main principles of thermodynamics and electrochemistry to the structure and function of key materials within the MEA. Chapter 1 is meant to bolster physical approaches for ushering in future advances in the field.

Chapter 2 dwells on all aspects of the structure and functioning of polymer electrolyte membranes. The detailed treatment is limited to water-based proton conductors, as, arguably, water is nature's favorite medium for the purpose. A central concept in this chapter is the spontaneous formation of ionomer bundles. It is a linchpin between polymer physics, macromolecular self-assembly, phase separation, elasticity of ionomer walls, water sorption behavior, proton density distribution, coupled transport of protons and water, and membrane performance.

The structure-based modeling of fuel cell catalyst layers, described in Chapter 3, dates back to work that Michael and I had started in Jülich in 1998. At that time, we had proposed an intuitive picture of the catalyst layer structure (reproduced in Figure 3.1), which was only supported by sparse data from porosimetry studies. As of today, it has gone through numerous refinements using molecular modeling and massively evolved experimental characterization. Remarkably, key features of this picture have outlasted these developments: agglomeration of Pt/C particles, bimodal pore size distribution, and formation of an ionomer skin layer at agglomerates. An important consequence of this structure is that the ionomer skin and water-filled pores should not be treated as a mixed effective electrolyte phase. Water in pores plays a pivotal role in keeping the catalyst "alive," albeit with a potentially markedly reduced "effectiveness." The effectiveness factor concept described in this chapter is vital for assessing different catalyst layer designs.

The full complexity of catalyst layer performance modeling is dealt with in Chapter 4, in which the general modeling framework is presented. The chapter comprehensively covers all relevant approximations and limiting cases. It demonstrates predictive capabilities for the optimization of composition, porous structure, and thickness. Moreover, it displays the unprecedented level of accuracy of the current generation of catalyst layer models. Models relate polarization curves with shapes in local potentials, concentrations, and species fluxes. These tools, namely, the "shapes" (or functional maps of the catalyst layers), will be highly useful for applied scientists who are interested in performance evaluation and in optimizing the structural design of the catalyst layers.

The performance modeling theme is expanded in Chapter 5 to deal with effects at the MEA level. Practical capabilities of physical modeling are demonstrated by deriving a number of fitting equations for polarization curves and impedance spectra. For instance, Figures 5.9 and 5.10 reveal the remarkable power of this approach, showing a fitted polarization curve and the breakdown of voltage losses into different contributions.

Sprawling "multiscale" and "multiphysics" approaches promise to rapidly reproduce any kind of fuel cell response function if only sufficient input information is provided—the more complex the multiscale model, the more phenomena it can describe, finding answers without asking questions. The approaches that Michael and Andrei overview here follow a different guiding principle that drove our very first steps in Jülich: start with formulating a problem or question of scientific interest; using appropriate scientific concepts, build a consistent model that then is developed and solved, and answer the relevant questions. While a number of new questions arise in the course of the study, select those which should be given priority to be solved and so forth. But the key rule is: the simpler the model that provides consistent answers at any stage of this iterative process, the better. The leitmotif for this approach was formulated allegedly by none other than Albert Einstein: "Everything should be as simple as possible, but not simpler!"

The literature covered in this book is indeed huge but it is not all inclusive (which would be impossible to achieve). Some existing approaches and particular papers are not covered owing to the tastes, preferences, and knowledge of the two authors, as well as space limitations. But I believe that this book will sustain a second edition. After the first one, the authors will receive the necessary feedback from many readers and take on board what they may have omitted in the first attempt, as well as correct any noticed inaccuracies and misprints, inevitable in the first edition of a book of that scale. As Andrei knows this very well, in giant slalom there is always the first and second leg.

Generally, the book is nicely written, simple at the beginning but increasing in complexity on the way. The authors try to keep mathematical formalism as compact as possible, but still comprehensive. Sections are short and informative, not too expansive in experimental data, but containing the conceptually important ones. It will be a must desk-book for PhD students and postdocs; theorists and experimentalists; electrochemists who want to learn more about the physics of fuel cells; physicists who wish to enter the field, at the beginning not knowing much about chemistry (!); and of course also electrochemical engineers working in industry. The latter usually employ the alternative, engineering approach, and work with very complicated sets of equations having many parameters, the values of some of which are not even known. This book can navigate them in their complex numerical modeling studies. All in all, I believe in the success of this ambitious project, but only the future will show if I am right.

**Alexei A. Kornyshev**
*Professor of Chemical Physics*
*Imperial College London*
*London, June 2014*

## References

1. For a review, see M. Eikerling, A.A. Kornyshev, and E. Spohr, Proton-conducting polymer electrolyte membranes: Water and structure in charge. In *Fuel Cells*, Ed. G.G. Scherer, *Advances in Polymer Science* **215**, 15–54, 2008; M. Eikerling, A.A. Kornyshev, and A.R. Kucernak, Water in polymer electrolyte fuel cells: Friend or foe? *Phys. Today* **59**, 38–44, 2006.
2. M. Eikerling and A.A. Kornyshev, Modelling the cathode catalyst layer of polymer electrolyte fuel cell. *J. Electroanal. Chem.* **453**, 89–106, 1998. For later developments, see M. Eikerling, A.S. Ioselevich, and A.A. Kornyshev, How good are the electrodes we use in PEFC? *Fuel Cells* **4**, 131–140, 2004.
3. A.A. Kornyshev and M. Eikerling, Electrochemical impedance of the catalyst layer in polymer electrolyte fuel cells. *J. Electroanal. Chem.* **475**, 107–123, 1999.
4. M. Eikerling, S.J. Paddison, L.R. Pratt, and T.A. Zawodzinski, Defect structure for proton transport in a triflic acid monohydrate solid. *Chem. Phys. Lett.* **368**, 108–114, 2003.

5. See, for example, a champion paper of that period—F. Maillard, M. Eikerling, O.V. Cherstiouk, S. Schreier, E. Savinova, and U. Stimming, Size effects on reactivity of Pt nanoparticles in CO monolayer oxidation: The role of surface mobility. *Faraday Discuss.* **125**, 357–377, 2004.

6. A. Golovnev and M. Eikerling, Theoretical calculation of proton mobility for collective surface proton transport. *Phys. Rev. E* **87**, 062908, 2013.

7. A. Golovnev and M. Eikerling, Soliton theory of interfacial proton transport in polymer electrolyte membranes. *J. Phys.: Cond. Matter* **25**, 045010, 2013.

8. S. Vartak, A. Roudgar, A. Golovnev, and M. Eikerling, Collective proton dynamics at highly charged interfaces studied by *ab initio* metadynamics. *J. Phys. Chem. B* **117**, 583–588, 2013.

9. M. Eikerling and P. Berg, Poroelectroelastic theory of water sorption and swelling in polymer electrolyte membranes. *Soft Matter* **7**, 5976–5990, 2011.

10. L. Zhang, L. Wang, T. Navessin, K. Malek, M. Eikerling, and D. Mitlin, Oxygen reduction activity of thin-film bilayer systems of platinum and niobium oxides. *J. Phys. Chem. C* **114**, 16463–16474, 2010.

11. L. Zhang, L. Wang, C.M.B. Holt, B. Zahiri, K. Malek, T. Navessin, M.H. Eikerling, and D. Mitlin, Highly corrosion resistant platinum-niobium oxide–carbon nanotube electrodes for PEFC oxygen reduction reaction. *Energy Environ. Sci.* **5**, 6156–6172, 2012.

12. S.G. Rinaldo, W. Lee, J. Stumper, and M. Eikerling, Theory of platinum mass balance in supported nanoparticle catalysts. *Phys. Rev. E* **86**, 041601, 2012.

13. S.G. Rinaldo, J. Stumper, and M. Eikerling, Physical theory of platinum nanoparticle dissolution in polymer. *Electrolyte Fuel Cells. J. Phys. Chem. C* **114**, 5775–5783, 2010.

14. K. Chan and M. Eikerling, Water balance model for polymer electrolyte fuel cells with ultrathin catalyst layers. *Phys. Chem. Chem. Phys.* **16**, 2106–2117, 2014.

15. K. Chan and M. Eikerling, Impedance model of oxygen reduction in water-flooded pores of ionomer-free PEFC catalyst layers. *J. Electrochem Soc.* **159**, B155–B164, 2012.

16. K. Chan and M. Eikerling, Model of a water-filled nanopore in an ultrathin PEFC cathode catalyst layer. *J. Electrochem. Soc.* **158**, B18–B28, 2011.

17. K. Malek, T. Mashio, and M. Eikerling, Microstructure of catalyst layers in polymer electrolyte fuel cells redefined: A computational approach. *Electrocatalysis* **2**, 141–157, 2011.

18. K. Malek, M. Eikerling, Q. Wang, T. Navessin, and Z. Liu, Self-organization in catalyst layers of polymer electrolyte fuel cells. *J. Phys. Chem. C* **111**, 13627–13634, 2007.

19. E. Sadeghi, A. Putz, and M. Eikerling, Hierarchical model of reaction rate distributions and effectiveness factors in catalyst layers of polymer electrolyte fuel cells. *J. Electrochem. Soc.* **160**, F1159–F1169, 2013.

20. See, for example, a pedestrian oriented review—M. Eikerling, A.A. Kornyshev, and A.A. Kulikovsky, Can theory help to improve fuel cells? *Fuel Cell Rev. (IOP)* **1**, 15–24, 2005.

21. A.A. Kulikovsky, J. Divisek, and A.A. Kornyshev, Modeling of the cathode compartment of polymer electrolyte fuel cell: Dead and active reaction zones. *J. Electrochem. Soc.* **146**, 3981–3991, 1999.

22. A.A. Kulikovsky, J. Divisek, A.A. Kornyshev, Two dimensional simulation of direct methanol fuel cell. A new (embedded) type of current collectors. *J. Electrochem. Soc.* **147**, 953–959, 2000.

23. A.A. Kulikovsky, A. Kucernak, and A.A. Kornyshev, Feeding PEM fuel cells. *Electrochim. Acta* **50**, 1323–1333, 2005.

24. M. Eikerling, A.A. Kornyshev, and A.A. Kulikovsky, Physical modeling of fuel cells and their components. In *Encyclopedia of Electrochemistry* **5**, Eds. A. Bard et al., Wiley-VCH, New York, 429–543, 2007.

25. M.L. Perry, J. Newman, and E.J. Cairns, Mass transport in gas diffusion-electrodes: A diagnostic tool for fuel cell cathodes. *J. Electrochem. Soc.* **145**, 5–15, 1998.

26. A.A. Kulikovsky, The regimes of catalyst layer operation in a fuel cell. *Electrochim. Acta* **55**, 6391, 2010; Catalyst layer performance in PEM fuel cell: Analytical solutions. *Electrocatalysis* **3**, 132–138, 2012.

27. A.A. Kulikovsky, A model for DMFC cathode performance. *J. Electrochem. Soc.* **159**, F644–F649, 2012; A model for Cr poisoning of SOFC cathode. *J. Electrochem. Soc.*, **158**, B253–B258, 2011.

28. A.A. Kulikovsky. The effect of stoichiometric ratio $\lambda$ on the performance of a polymer electrolyte fuel cell. *Electrochim. Acta* **49**, 617–625, 2004.

29. A.A. Kulikovsky, A model for local impedance of the cathode side of PEM fuel cell with segmented electrodes. *J. Electrochem. Soc.* **159**, F294–F300, 2012.

30. A.A. Kulikovsky, H. Scharmann, and K. Wippermann, Dynamics of fuel cell performance degradation. *Electrochem. Comm.* **6**, 75–82, 2004.

31. A.A. Kulikovsky, A model for carbon and Ru corrosion due to methanol depletion in DMFC. *Electrochim. Acta* **56**, 9846–9850, 2011.

32. A.A. Kulikovsky, A simple model for carbon corrosion in PEM fuel cell. *J. Electrochem. Soc.* **158**, B957–B962, 2011.

33. A.A. Kulikovsky, Dead spot in the PEM fuel cell anode. *J. Electrochem. Soc.* **160**, F1–F5, 2013.

34. A.A. Kulikovsky, *Analytical Modelling of Fuel Cells*, Elsevier, Amsterdam, 2010.

35. A.A. Kulikovsky and M. Eikerling, Analytical solutions for impedance of the cathode catalyst layer in PEM fuel cell: Layer parameters from impedance spectrum without fitting. *J. Electroanal. Chem.* **691**, 13–17, 2013.

# Preface

More than 175 years have passed since the discovery and first demonstrations of the fuel cell principle by Christian Friedrich Schoenbein and Sir William Grove. However, in spite of many years of research, fuel cells still are somewhat exotic and expensive power sources. The main reasons are materials cost and a lack of fundamental knowledge of fuel cell operation.

In this book, we discuss low-temperature fuel cells with a polymer electrolyte membrane (PEM). Two main representatives of the family of low-T cells are hydrogen-fed polymer electrolyte fuel cells (PEFCs) and liquid-fed direct methanol fuel cells (DMFCs). Though the major part of this book is devoted to materials and performance modeling of PEFCs, some features of DMFCs will also be discussed due to a great potential of these cells for small-scale mobile applications.

Noting that PEFCs are highly efficient and environmentally friendly power sources is a commonplace. The only chemical product of the fuel cell reaction is water; the exhaust contains neither carbon dioxide nor toxic oxides, water's evil twin products from the combustion of fossil fuels. In contrast to internal combustion engines (ICEs), fuel cells produce no noise. Another advantage of fuel cells is simplicity; they are much simpler than ICEs. When opening the cowling of an ICE car, a lot of pipes are seen. In an FC system, pipes exist as well but they are invisible, as they are of nanoscale size.

Nanoscale feature sizes are the distinguishing structural characteristic of electrochemical energy conversion systems. The inherent performance gains due to the small dimension pose formidable challenges for scientists and engineers in materials and performance research. Electrochemistry requires the active surface area of the electrocatalyst material in anode and, predominantly, cathode to be as high as possible; this implies that fuel cell electrodes must be designed as porous composite structures. The electrodes, together with gas diffusion layers and proton-conducting membrane, form a membrane–electrode assembly (MEA), a layered porous material, where molecular chemistry, physics, electrochemistry, and power engineering meet. Fuel cell performance crucially depends on MEA kinetic and transport properties, which are not well understood. The largest part of this book is devoted to the physics of materials used in MEAs and related performance problems.

Fuel cells are, undoubtedly, the best way to convert the free energy of the hydrogen–oxygen reaction into electricity. Once we have hydrogen, there is no better means to utilize it but a fuel cell. Hydrogen production, storage, and distribution is a separate bunch of scientific and engineering challenges that are beyond the scope of this book. As any naturally abundant power source (like direct sun light or wind) could be used for $H_2$ production, fuel cells represent an absolutely vital element of any future energy economy that is highly efficient, carbon–neutral, and emission–free.

Like any potentially revolutionary technology transition, the hydrogen economy arouses tremendous expectations and apprehensions as well as a fair share of scepticism. So far, many of the expectations have not been realized. However, seen in a sober light, scientific and technological progress has been amazing and there are strong indications that this economy is the only viable solution to the global energy challenge, and that it is actually emerging.

Over the past several decades, we have witnessed an exponential growth of computing power (Moore's law). Computers had no analogs in the beginning of the twentieth century; however, as hundred years ago, we still rely on ICEs in our cars and use heat machines (coal-, gas-, and oil-based power plants) for large-scale electricity production. The nuclear catastrophes in Chernobyl and Fukushima have undermined the future of nuclear power generation. Rapid development of electrochemical power sources has seemingly no alternatives, and hydrogen fuel cells take one of the leading positions in the transition. Importantly, one of the key parameters in PEFCs—the mass activity of Pt catalyst—also exhibits exponential growth over the last 50 years, a distinct sign of an emerging technology!

Fuel cell science has grown into a huge field at overlapping frontiers of physics, chemistry, and engineering. It is impossible to equally address all aspects of the field. We had to cut short many interesting discussions of fundamental, albeit specific, materials aspects as well as of interesting engineering aspects. Instead, we tried to keep our focus on generic aspects of materials structure and function, for example, flow, transport, and reaction in porous media, as well as vital fuel cell phenomena, for example, water-based proton transport. The selection of material that is covered in this book reflects research priorities of the authors and we apologize for missing some aspects of PEFC materials and operation that other researchers deem important.

It is virtually impossible to thank all of our colleagues for teaching us at seminars, conferences, and in private discussions. First and foremost, both of us are deeply indebted and grateful to Alexei Kornyshev, who engaged us in fuel cell studies. He has been an inspirational force and continues to serve to us as a role model of a principled, creative, and spirited scientist. We are thankful to numerous colleagues, coworkers, and students in Jülich, Los Alamos, and München whom we had a pleasure and a privilege to work with along various stretches of our fuel cell journeys. M.E. wishes to express his gratitude to an unparalleled force of experts and students from industry, National Research Council, and universities, who are about to transform *The Lower Mainland* into *Fuel Cell Valley*.

This book would have never been written without the steadfast support from our wives Silke and Maria. We highly appreciate the time and encouragement that they have given us to bring this work to completion. M.E. wishes to dedicate his contributions of this book to Finn, Liv, and Edda for their unblemished inspiration to work on something that they will hopefully benefit from in the future.

# Authors

**Michael Eikerling** is a Professor of theoretical chemical physics and electrochemical materials science at Simon Fraser University in Burnaby, British Columbia, Canada. From 2003 to 2013, he also led activities in physical modeling of fuel cells at the National Research Council of Canada. Prof. Eikerling received his diploma in theoretical condensed matter physics from RWTH Aachen (Germany) in 1995. He obtained his doctoral degree from the Technical University of Munich in 1999 for a dissertation on theoretical chemical physics of polymer electrolyte fuel cells. Prior to joining Simon Fraser University in 2003, he spent periods as a research associate at Research Center Jülich, Los Alamos National Laboratory, and TU München. His main scientific contributions are in the theory and modeling of electrochemical systems, with a focus on materials and processes in polymer electrolyte fuel cells. This prioritization is owed to the fuel cell cluster in the Vancouver region, which holds a globally recognized reputation as a center of excellence. Moreover, he has a general interest in energy science and its implications for societal development. Research interests in the Eikerling Research Group span a diverse range of fundamental as well as applied topics, encompassing transport phenomena at interfaces and in polymeric materials, theory and modeling of electrocatalytic phenomena, self-organization in electrochemical materials, statistical physics of heterogeneous media, porous electrode theory, advanced electrochemical diagnostics, fuel cell design, and performance modeling. The group has extensive expertise in molecular modeling, physical theory, and continuum modeling. It interacts extensively with experimental groups in academia and industry. In addition to leadership and executive roles in pan-Canadian and international research networks, Prof. Eikerling engages strongly in activities of the international scientific community.

**Dr. Andrei Kulikovsky** graduated from the Faculty of Theoretical and Experimental Physics of the Moscow Engineering–Physical Institute, one of the leading schools in physical sciences in the former USSR. In 1984, he obtained his PhD from the Institute for High Temperatures of the USSR Academy of Sciences. In 1998, he received the Doctor of Sciences (Research Professor) degree in physics and mathematics from the M.V. Lomonosov Moscow State University. While working in Russia, his main research interests were in the field of modeling of gas discharge plasmas.

In 1998, Dr. Kulikovsky moved to the Forschungszentrum Juelich (Research Centre Juelich), Germany, where Alexei Kornyshev engaged him into modeling of fuel cells, cell components, and stacks. Since 1998, Dr. Kulikovsky has published more than 80 research papers in high-ranked electrochemical journals; most of these papers have a sole author. In 2010, Dr. Kulikovsky published a book *Analytical Modeling of Fuel Cells* (Elsevier), which was the first monograph on theory and modeling of fuel cells. His current research interests include modeling of low-, intermediate-, and high-temperature fuel cells, and catalyst layers, macroscopic modeling of aging processes and defects in cells, analytical study of the transport and kinetic processes in cells and stacks, and impedance spectroscopy of cells. Dr. Kulikovsky's work has always been focused on the development of simple analytical models, aiming at understanding the phenomena of interest.

# Introduction

Late in the year 1838, the Swiss professor of chemistry Christian Friedrich Schoenbein performed experiments on water electrolysis in his laboratory. Though Schoenbein did not publish details of his experiments, he reported results and we could reconstruct the details based on statements in his paper. To collect hydrogen and oxygen, two Pt wire electrodes were mounted inside glass vessels and both vessels were immersed into a water–acid solution (Figure I.1a). Schoenbein connected a battery (that had been invented around 1800 by Alessandro Volta) to the electrodes to split water into oxygen and hydrogen. The gases were collected inside the glass vessels. After disconnecting the battery, Schoenbein noticed a small potential between the electrodes (Mogensen, 2012).

Schoenbein reported this effect in a paper to *Philosophical Magazine*, which was published in January 1839 (Schoenbein, 1839). He wrote: "... The most delicate test to ascertain that electrolyzation has taken place, is the polarized state of the electrodes." This phrase undoubtedly means that Schoenbein related the source of potential to the presence of hydrogen inside the anode and oxygen inside the cathode vessels.

The key is that after water electrolysis one of the electrodes appeared to be in a hydrogen and the other in an oxygen atmosphere *surrounding the* electrode/electrolyte interface. Schoenbein had made the first observation of the *fuel cell effect* and his system could be considered the first fuel cell.

It was seemingly this work that urged Welsh-born Sir William Robert Grove, who resided in London as a barrister and physical scientist, to verify this effect in his home laboratory (Bossel, 2000). Grove checked the effect, and the potential was there. The schematic of a later experiment of Grove with several fuel cells connected in series is shown in his famous drawing (Figure I.1b).

Over almost two centuries of research, the idea of a fuel cell-powered world went through ups and downs. The unrivaled thermodynamic efficiencies and low environmental impact of fuel cells have been famed ever since the publication of a visionary article by the Baltic-German chemist Friedrich Wilhelm Ostwald (1909 Nobel Laureate in Chemistry) from 1894 on the topic *Die Wissenschaftliche Elektrochemie der Gegenwart und die Technische der Zukunft** published in the *Zeitschrift für Elektrotechnik und Elektrochemie* (Ostwald, 1894). Ostwald starts by exclaiming what sort of atrocious energy conversion devices steam engines are, stating further that this miserable outcome is a consequence of principal thermodynamic limitations of combustion processes that are conducted at temperatures of 1000°C. He further wrote that thermodynamics allows the reaction energy to be converted into mechanical energy via an alternative pathway, which he referred to as a *cold combustion*. As he reasons further in his educational pamphlet" *Der Weg nun, auf welchem diese grösste aller technischen Fragen, die Beschaffung billiger Energie, zu lösen ist, dieser Weg muss*

---

* On the scientific electrochemistry of today and the technical electrochemistry of tomorrow.

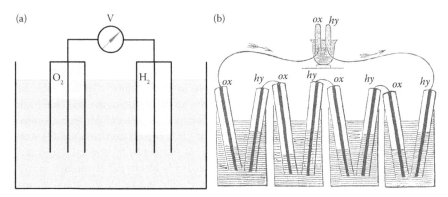

**FIGURE I.1**   (a) Presumable schematic of Schoenbein's experiment (1838). (b) Grove's drawing of his experiment (1839). Interestingly, Grove's picture contains a flaw: in fact, the hydrogen volume must have been twice the amount of the oxygen volume. (Adapted from Mogensen, 2012. Private Communication.)

*von der Elektrochemie gefunden werden.*"* This sentence encompasses two remarkable revelations: Ostwald considered securing cheap energy the biggest challenge of his (!) time and he asserted that electrochemistry would be the only solution. Whereas heat engines produce unacceptable levels of atmospheric pollution, electrochemical energy conversion would offer a direct route to electricity generation, which would be highly efficient, silent, and pollution-free. In Ostwald's vision, this transition would mark a technological revolution but he was intelligent and responsible enough to dun his audience that in practice the transition could take a long time!

Almost two centuries after Schoenbein's discovery and more than a century after Ostwald's stunning discourse, polymer electrolyte fuel cells have become the most promising members of a highly diverse family of fuel cells (Eikerling et al., 2006, 2007b). Owing to unrivaled thermodynamic efficiencies, high energy densities, and ideal compatibility with hydrogen as a fuel, PEFCs have emerged as a key technology in ever-escalating efforts to address the global energy challenge (Smalley, 2005). These cells could replace internal combustion engines in vehicles and supply energy to power-hungry portable devices and stationary systems. The successful introduction of PEFCs hinges, however, on breakthrough progress on two scientific–technical frontiers: (i) the development of materials that are inexpensive, earth-abundant, sufficiently stable, and optimized in view of their function, as well as (ii) the engineering of cells and fuel cell stacks that optimize power densities, voltage efficiencies, and heat and water management. Moreover, fuel cells must be competitive with other energy technologies in terms of operating conditions, durability, lifetime, and costs.

---

\* The way now, on which all these great technical challenges could be solved, this way has to be found in electrochemistry.

## GLOBAL ENERGY CHALLENGE

It is almost universally recognized that the global energy infrastructure must embark on a dramatic transformation (Kümmel, 2011, Smalley, 2005). The demand for power, as energy input per time, is predicted to double by 2050 and triple by 2100 (Nocera, 2009). An era of abundant and cheap fossil fuels, which stifled Ostwald's vision during the twentieth century, is coming to an end. Within a timescale of about 50 years, production rates of the major sources of fossil fuels, which account for more than 80% of the current global energy use, will pass a peak and adopt a steeply declining trajectory.

Figure I.2 compares the trend in global refined oil production and consumption to the rate of crude oil discoveries. Similar curves abound in the literature for the global production of crude oil, coal, and natural gas (Berg and Boland, 2013, Nashawi et al., 2010). The plots vary somewhat in shape and peak positions due to varying scenarios of growth in power demand, strategic manipulation in resource estimation, and technical advances in resource extraction technology. However, the underlying message remains uncontested: the fossil fuel age will be winding down, with sharp declines to be felt within the span of one or two human generations. The diverging trends of increasing demand for energy input and decreasing rate of production from fossil fuel resources create a growing demand–supply gap, looming on the horizon with shortages and soaring costs of energy.

As depicted in Figure I.3, the energy use per capita and year shows a correlation to the human development index (HDI) (Smil, 2010, Chapter 35). The HDI is a composite metric established by the United Nations Development Office to evaluate and compare indicators of life expectancy, education, and income. As can be seen, for a per capita energy use above 100 GJ per year, the correlation with the HDI is

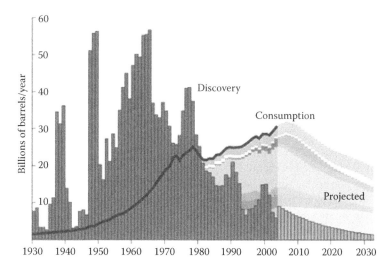

**FIGURE I.2** The growing gap between petroleum crude oil discoveries and refined oil production/consumption. (From Rep. Roscoe Bartlett, Maryland.)

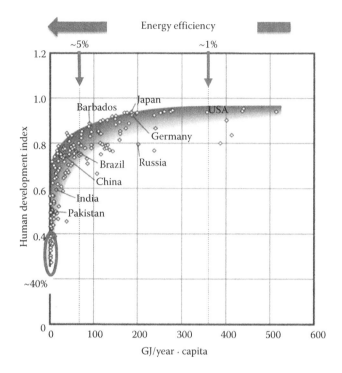

**FIGURE I.3** Human development index (HDI) versus energy input per year and per capita. Energy input was converted to a "societal" energy efficiency based on an estimate that the average daily energy intake of an adult is 2500 kcal. (Adapted with permission from Smil, V. Vision of Discovery: New Light on Physics, Cosmology, and Consciousness. Cambridge University Press (2010).)

weak and it exhibits strong fluctuations. These trends are indicative of an increasing inefficiency or "lavishness" of energy use.

To better illustrate this point, the per capita energy use per day can be converted to a societal energy efficiency, which is calculated as the recommended dietary energy intake of a single adult per day (assumed as 2500 kcal) divided by the average daily energy use per capita. With this definition living on the breadline corresponds to a "societal" energy efficiency of about 40%. Several values of this "societal" energy efficiency are indicated in Figure I.3. A "societal" energy efficiency above 5% seems undesirable due to the strong anticorrelation with the HDI in this range. On the other hand, for values below ~5%, a decrease in energy efficiency does not seem to have any predictable impact on the HDI. Moreover, the "societal" energy efficiency exhibits huge fluctuations among countries with similar HDI. Countries with an excessively high annual per capita energy use have an energy efficiency of about 1–2%. If evaluated as thermodynamic entities, for which high "societal" energy efficiency were considered a viable metric of performance and healthy habits of energy use, highly developed countries would fare miserably in this comparison.

The trinity of growing power demand, dependence on diminishing fossil fuel resources, and miserable energy efficiencies of developed countries let alone fatal implications for environment and climate (IPC, 2011) poses an unprecedented threat for the socioeconomic prospects of the human society.

## TOWARD AN AGE OF ELECTROCHEMISTRY

The inevitable transformation in the global energy infrastructure entails formidable socioeconomic challenges. At the same time, it bears tremendous opportunities for widespread use of abundant and renewable sources of energy, and the development of innovative energy technologies and infrastructures.

Concepts and technologies of energy storage and conversion are evaluated and compared in terms of (i) tangible performance metrics, viz., efficiency, power density, and energy density, (ii) input of natural resources during fabrication and operation, (iii) infrastructure requirements for widespread deployment, (iv) lifetime and durability, and (v) adaptability to prevailing environmental conditions. Electrochemical energy technologies provide a benign combination of these factors. Favorable evaluations according to these criteria draw technology innovators toward electrochemical technologies.

Unique modularity renders electrochemical technologies suitable for handheld electronics ($\sim$1–10 W), transportation ($\sim$100 kW), and stationary applications (>1 MW). Their diversity is a further asset: photoelectrochemical cells, batteries, supercapacitors, and fuel cells can be combined in autonomous energy systems to supply varying power demands, overcome issues due to cost and abundance of materials, and fulfill requirements of infrastructural integration and availability. Correspondingly, massive clusters forging the development of electrochemical technologies and infrastructures are emerging. Worldwide investments in clean energy projects, with electrochemical technologies as the main drivers, amounted to $211 billion in 2010. Investment in small-scale government and industry R&D projects had increased significantly from $26 billion in 2008, to $37 billion in 2009, and to $68 billion in 2010 (UNEP, 2011).

The combination of urgent demand and unique prospects for innovation in materials and systems render electrochemical energy technologies a scientifically stimulating and dynamic field of research. For the curious-minded, there is a lot of science to be discovered!

However, as known to anybody working in the field, from fresh university student to scientific or technological expert, the challenge of bringing breakthroughs in understanding of electrochemical phenomena and in design of electrochemical materials to technological fruition is complex.

## ENERGY CONVERSION IN CHEMISTRY, BIOLOGY, AND ELECTROCHEMISTRY

The public perception of electrochemical energy conversion and storage technologies has experienced seismic shifts, particularly in the last decade. As the demand for electrochemical devices is becoming more apparent and technological readiness is getting closer to the level deemed critical for commercialization, the debate is becoming more polarized. We skip this discussion and focus on one aspect that is not up for debate: electrochemical energy technologies exhibit an undeniable advantage in thermodynamic efficiency. In this context, we could reiterate Ostwald's statements and with a bit of polishing his historical article would still pass off as a contemporary perspective paper (Ostwald, 1894).

The primary question to address is thus: Why are electrochemical energy technologies so extraordinarily efficient? The answer has to do with fundamental principles of thermodynamics and electrostatics and with the spatial separation of partial reactions that are needed to complete a chemical redox process.

*Reaction enthalpy.* Let us consider a chemical reaction that releases a certain amount of enthalpy. Any spontaneous chemical reaction, during which reactants are in direct contact, transforms the reaction enthalpy completely in heat. Combustion, a specific type of exothermic reactions between a fuel and an oxidant, is arguably the first chemical process harnessed by humans. Among its many other uses, heat generated by direct combustion can be converted into mechanical work in a heat engine, which could be used in an electric generator to produce electricity. Now, consider an electrochemical reaction that transforms the same reactants into the same product species: forcing the reaction to follow the electrochemical pathway converts a major portion of the released enthalpy directly into electricity.

The most general way to think of a chemical process is a redistribution of electronic charge among available energy levels of atomic or molecular species or in condensed matter. Processes include the hybridization and shifting of electronic orbitals in atoms and molecules or the filling, depletion, formation, or shifting of electronic energy levels in condensed matter. However, in a direct chemical process, transferred electronic charge cannot perform any work due to the microscopic distance of the transfer.

*Biological energy transduction.* Next, let us separate the process of electron loss/oxidation of one species from electron gain/reduction of another species; this separation is the basis of energy transduction in biological organisms. The separation distance in this case is of the order of $\sim 100$ nm; it exists across inner and outer membranes of mitochondria; the short separation is the reason for the high efficiency of biological energy conversion processes.

*Electrochemical energy conversion.* Finally, let us consider further separation of redox species and partial reactions to macroscopically large distances ($\sim 1$ mm). Now we are dealing with a technological electrochemical process. Completing this process requires a few adjustments due to the increase in separation distance of oxidized and reduced species. The net reaction might be the same as or similar to a process in direct combustion or in biological energy conversion. Thermodynamics tells us that it will have the same reaction enthalpy and Gibbs energy, if conducted reversibly, that is, infinitely slow or close to equilibrium: any electrochemical process is theoretically as efficient as the same process conducted under chemical or biological conditions.

Figure I.4, taken from Hambourger et al. (2009), compares spontaneous oxidation of a fuel to produce water in a biological organism and in a hydrogen–oxygen fuel cell. The sophisticated machinery of biological energy conversion is based on

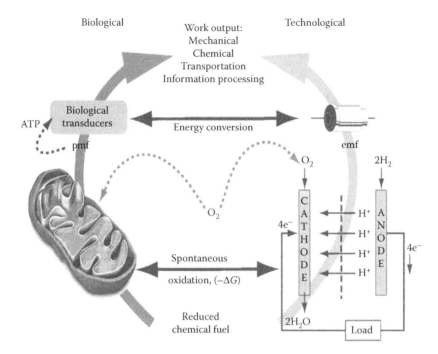

**FIGURE I.4** Comparison of biological and electrochemical energy conversion. (M. Hambourger et al., 2009. *Chem. Soc. Rev.* **38**, 25–35, Fig. 1(b), 1999. Reproduced by permission of The Royal Society of Chemistry.)

microscopic charge separation and it is built from earth-abundant elements. In biological organisms, the oxidation of a *fuel* with $O_2$, supplied by the respiratory system, produces a certain amount of work. In some organisms, the ratio of work to enthalpy change, defined as *energy conversion efficiency*, could reach values of 90% or higher. In comparison, the practical efficiency of a polymer electrolyte fuel cell is significantly lower at 60%, even though it involves similar processes and is operated under similar conditions.

Why is the efficiency of a PEFC worse? Charge separation at the macroscopic scale in PEFCs brings about additional requirements and needs in terms of components and processes: reactant gases must be supplied through flow fields and porous electrodes; electrons and protons must be transported over macroscopic distances through conduction media to complete the net reaction; partial redox reactions proceed at interfaces, where they must overcome significant activation barriers. Macroscale transport processes cause significant losses in efficiency, since the effective resistance of any transport process scales with the transfer distance.

As argued above, biology as a blueprint for fuel cells is seemingly better, viz., more efficient at converting energy. However, this success comes at a price. What is it? Car manufacturers will tell us that they need fuel cell stacks with a power density of $1–2$ kW $L^{-1}$. Bioorganisms work at power densities of $1–2$ W $L^{-1}$; they operate at much lower current densities and correspondingly produce much less water and heat per unit volume and per time. Thus, the higher efficiency of biological energy conversion is paid for by a significantly lower power density that would not be feasible for a scale-up to technological applications.

## PRINCIPLES OF ELECTROCHEMICAL ENERGY CONVERSION

The core of any electrochemical system is the electrified interface between the metal electrode and the electrolyte, as depicted in Figure I.5. Charge storage and transfer proceed at this interface. Correspondingly, the *interfacial area per unit volume* of the electrochemical medium is the key structural parameter; it can be directly related to *charge storage capacity, energy density*, and *power density*. Maximization of the interfacial area per unit volume enforces the use of *nanocomposite* or *nanoporous media*. In such media, an intricate interplay unfolds between interfacial processes, which involve electrostatic charge separation, adsorbate formation/removal, electrochemical charge transfer, as well as the transport of electrons, protons, ions, solvent molecules, reactants, and reaction products in interpenetrating percolating phases.

Electrochemical cells, for example, photoelectrochemical cells, batteries, fuel cells, or supercapacitors, consist of assemblies of functional electrode and electrolyte layers. Individual cells are linked together electrically to form stacks, whereas stacks form part of energy systems, that involve additional units for fuel storage, thermal management, and so on. The development of energy systems is thus a hierarchical and cross-disciplinary exercise, with strongly coupled phenomena arising across multiply connected structural levels.

The science of electrochemical materials and processes plays a key role in the design of functionally optimized component materials that meet requirements in (i) operational range flexibility (e.g., temperature range of operation; refuelling range of vehicles), (ii) affordability (cost of materials and manufacturing), (iii) performance (e.g., voltage efficiency of fuel cells; charge storage capacity of batteries; power density), and (iv) durability and lifetime.

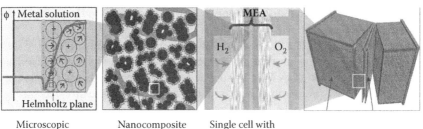

| Microscopic electrochemical interface | Nanocomposite multifunctional materials | Single cell with membrane–electrode assembly (MEA) | Fuel cell stack |

**FIGURE I.5** Structural hierarchy involved in understanding, designing, and optimizing a polymer electrolyte fuel cell system. Figures illustrate the transition from fundamental electrochemical science and materials science to single cell and stack level design. The membrane–electrode assembly is pivotal; it incorporates scientific and engineering challenges from all other scales.

The outlined general design principles illustrate that the development of functional materials needs to be informed by comprehensive and concise knowledge of system-level requirements. Successful design of materials with desired physical or electrochemical properties hinges, furthermore, on advances in understanding quantum mechanics of condensed matter, molecular and surface chemistry, electrostatics and statistical mechanics at the electrified interface, thermodynamics and kinetics of electrochemical reactions, and flow and transport in nanocomposite and porous media.

Science-driven strategies in the development of future energy technologies demand long-term research on fundamental physical and chemical phenomena and in materials science. The timescale that Whitesides and Crabtree (2007) proposed in their perspective *Science* article from 2007 is 50–100 years. This estimate is sobering; yet, it is in accord with the history of research on key–fuel cell materials, viz., Pt-based catalysts and proton-conducting polymer electrolytes that have been in use for over 170 and 50 years, respectively. Engineering could provide immediate solutions to improve upon existing materials and technologies, but it does not solve fundamental challenges. Real groundbreaking progress needs a well-balanced approach that merges science- and engineering-driven strategies.

## SLEEPING BEAUTY: 100 YEARS IS NOT ENOUGH!

Schoenbein made his discovery in 1838 and Ostwald wrote his visionary essay in 1894. Where have fuel cells been so long? The development of electrochemical technologies was held in check by the fossil fuel age, spring-fed by the fairytale-like ease of availability and seemingly unlimited supply of fossil fuels. Throughout the bigger portion of this age, forseeable issues like resource depletion, accelerated by inefficient energy use, as well as detrimental impact of energy use on climate and environment have been put aside for the sake of short-lived economic prosperity.

The development of fuel cell technology was slow until the 1980s. The term "fuel cell" was coined in 1889 by Charles Langer and Ludwig Mond. It took until 1959 before Francis Bacon, professor of engineering at Cambridge, demonstrated the first practical 5 kW fuel cell system, an alkaline fuel cell. The development of proton exchange membrane fuel cells, nowadays predominantly referred to as polymer electrolyte fuel cells, started at General Electric in the late 1950s. Willard Thomas Grubb at General Electric is credited with their invention; the first patent on solid polymer electrolytes in 1959 was issued in his name (Grubb, 1959). He continued this development together with Leonard Niedrach, another researcher at General Electric. The Grubb–Niedrach cell was developed in collaboration with NASA for the Gemini space program. Similar programs in fuel cell development for space missions were pursued in the former Soviet Union.

From the mid-1960s onwards, General Motors and Shell experimented with hydrogen fuel cells and direct methanol fuel cells for vehicle applications. By the mid-1970s, other vehicle manufacturers in Germany, Japan, and the United States launched programs to develop fuel cell electric vehicles. In 1983, the Canadian company Ballard, located first in North Vancouver and nowadays at the Fraser River in

Burnaby, British Columbia, became the first company to predominantly focus on fuel cell research and development. Since then, Ballard Power Systems has been a frontrunner in developing PEFC and DMFC technology. The company has trained a huge cohort of experts and it has been a seed for the creation of other fuel cell companies.

Fuel cells used in the Gemini space program in the early 1960s were extremely expensive and they had short lifetimes of less than 500 h at 60°C. The sulfonated polystyrene–divinylbenzene copolymer membranes used were prone to oxidative degradation, rendering these cells too costly and short-lived for commercial use. A milestone that launched the current era in the development of polymer electrolyte fuel cell technology was the invention of Nafion by Walter R. Grot at DuPont du Nemours in the late 1960s (Grot, 2011). *Nafion* that found immediate use in the chloralkali industry helped to extend the lifetime of PEFCs significantly beyond 1000 h. Until today, it has remained as the prototypical material for the prevailing class of poly(perfluorosulfonic acid) ionomer membranes.

Amazingly, it took about 25 years to seize the great potential of Nafion as an electrode separator for fuel cells. In the beginning of the 1990s, a group at the Los Alamos National Laboratory (Springer, Zawodzinski, and Gottesfeld) and a group from General Motors (Bernardi and Verbrugge) extensively studied hydrogen PEFCs. Their famous papers (Bernardi and Verbrugge, 1992, Springer et al., 1991) crashed down an exponentially growing avalanche of publications on the topic. Nowadays, fuel cells based on PEFC technology are found in applications ranging from several watts in portable applications to several hundred watts (residential systems designed by United Technologies) to hundred–kilowatt stacks developed by Siemens for submarines. The exceptional scalability, demonstrated by these examples, is among the greatest assets of PEM fuel cells.

## POLARIZATION CURVES AND "MOORE'S LAW" OF FUEL CELLS

The quintessential question of practical PEFC research is: How to evaluate and compare the electrochemical performance of cells that incorporate different materials that employ different design layouts or that operate under different conditions?

The single characteristic that encompasses phenomena at all scales and in all fuel cell components is the fuel cell *polarization curve*. The polarization curve of a membrane–electrode assembly (MEA), a single cell, or a fuel cell stack furnishes the link between microscopic structure and physicochemical properties of distinct cell components on the one hand and macroscopic cell engineering on the other. It thus condenses an exuberant number of parameters, which lies in the 50s to 100s, into a single response function. Analysis of parametric dependencies in the polariztion curve could be extremely powerful; at the same time, it could as well be highly misleading if applied "blindly."

For MEA designers or fuel cell stack engineers, a polarization curve is an immensely useful practical analysis tool. It allows for a comparative assessment of sources of voltage losses in the cell, fuel cell failure modes, critical or limiting current densities, as well as impacts of degradation and water management. For materials scientists, the polarization curve entails useful information on performance effects

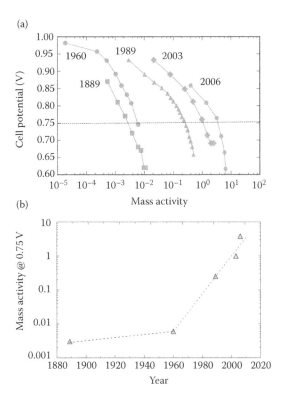

**FIGURE I.6** Polarization curves of fuel cells, extracted from the literature over the last 125 years (Debe et al., 2006, Gore and Associates, 2003, Grubb and Niedrach, 1960, Mond and Langer, 1889, Raistrick, 1989). The fuel cell voltage, $E_{cell}$ ($j_0$), is shown as a function of the logarithm of the mass activity $j_0/m_{Pt}$ (in units of kW $g_{Pt}^{-1}$) in (a). The plot in (b) shows $\log (j_0/m_{Pt})$ at $E_{cell} = 0.75$ V over time, revealing roughly an exponential growth in PEFC performance over the last 50 years.

of materials modifications. To be of value for these purposes, analyses of polarizaton curves must be accompanied by physicochemical characterization of structure and composition as well as measurements of electrocatalytic and transport properties of fuel cell materials. In this way, voltage loss phenomena could be deconvoluted into separate components and processes and distinct voltage loss contributions could be quantified. This assessment corresponds to a so-called voltage-loss-breakdown, as described in Baghalha et al. (2010). It is, however, an inadequate and misleading practice to employ polarization curves as a sole validation for multiscale and multiparameter models of fuel cell systems and stacks.

Figure I.6a displays a selection of fuel cell polarization curves extracted from publications that span a period of 125 years. The ordinate depicts the fuel cell voltage $E_{cell}$ ($j_0$), as given in Equation 1.21. The abscissa represents, on a logarithmic scale (to the base of 10), the fuel cell current density that has been normalized to the surface area-specific mass loading of Pt, $m_{Pt}$.

Fuel cell engineers refer to $j_0/m_{Pt}$ as the *mass activity*. From an electrochemical perspective, Figure I.6a represents a Tafel plot. However, the voltage $E_{cell}$ ($j_0$) does not only account for a single plain electrified interface, but also depends on electron flux, ion flux, and mass transport in all components and across all interfaces of the MEA.

Essential performance metrics can be deduced from Figure I.6a. The mass activity at a fixed value of $E_{cell}$, usually chosen arbitrarily as 0.9 V, is widely employed as a comparative metric of catalyst activity, although it is not a genuine kinetic parameter of the catalytic system, as the exchange current density or rate constant would be. The fuel cell voltage is directly related to the *voltage efficiency*, $\epsilon_v = E_{cell}/E_{O_2,H_2}^{eq}$, and the volumetric *energy density* $W_{cell} = FE_{cell}/l_{CL}$ (in J L$^{-1}$). The volumetric *power density* is $P_{cell} = j_0 E_{cell}/l_{CL}$ (in W L$^{-1}$) and the *specific power* is $P_{cell}^s = j_0 E_{cell}/m_{Pt}$ (in W g$_{Pt}^{-1}$).

Figure I.6a also reveals the timeline of milestones in fuel cell design. The left-most curve is the performance curve of the first practical $H_2/O_2$ fuel cell, built by Mond and Langer in 1889 (Mond and Langer, 1889). The electrodes consisted of thin porous leafs of Pt covered with Pt black particles with sizes of 0.1 μm. The electrolyte was a porous ceramic material, earthenware, that was soaked in sulfuric acid. The Pt loading was 2 mg cm$^{-2}$ and the current density achieved was about 0.02 A cm$^{-2}$ at a fuel cell voltage of 0.6 V. The next curve in Figure I.6a marks the birth of the PEFC, conceived by Grubb and Niedrach (Grubb and Niedrach, 1960). In this cell, a sulfonated cross-linked polystyrene membrane served as gas separator and proton conductor. However, the proton conductivity of the polystyrene PEM was too low and the membrane lifetime was too short for a wider use of this cell. It needed the invention of a new class of polymer electrolytes in the form of Nafion PFSA-type PEMs to overcome these limitations.

The polarization curve from 1989 exhibits the most significant performance leap in Figure I.6a. The innovative cell design was proposed by Ian D. Raistrick at the Los Alamos National Laboratory (Raistrick, 1989), the hotbed for frontier fuel cell research of the time. This MEA design utilizes nanoporous carbon-based electrodes that are loaded with nanoparticles of Pt and impregnated with ionomer. These two modifications combined allowed for improved performance with an order of magnitude lower Pt loading, reduced from 4 to 0.4 mg cm$^{-2}$. The latter value still represents the typical loading in modern PEFCs. Optimization of catalyst layer composition, in part enabled by systematic modeling studies, together with advances in structural design of diffusion media and progress in MEA fabrication led to the performance gain exhibited by the Gore-select® MEA from 2003 (Gore and Associates, 2003).

Another milestone in performance was achieved with the nanostructured thin-film technology of catalyst layer fabrication, developed by Mark K. Debe at the company 3M (Debe et al., 2006). With this innovative approach, $m_{Pt}$ could be reduced by another order of magnitude at the critical cathode side, while performance is maintained and catalyst durability and lifetime are in fact significantly enhanced.

The staggering progress in fuel cell performance is illustrated in Figure I.6b. It shows the logarithm of the specific power (or, equivalently, of the mass activity) at $E_{cell} = 0.75$ V as a function of the year of publication. Since the advent of PEFCs in 1960, the specific power has roughly followed an exponential growth law; during this period, it has doubled approximately every 5.4 years, resulting in a total increase in mass activity and specific power by a factor $10^3$. Remarkably, all improvements that have contributed to this increase represent preindustrial developments. Another intriguing observation is that all of the advances have been achieved with Pt-based catalysts, in spite of tremendous research on new catalyst materials and frequent reports on new materials with enhanced activity. Decisive steps involved a scaling down of the catalyst size to the nanoscale, densification of the catalyst on a nanoporous support material, improvement of local reaction conditions at the catalyst by impregnation of catalyst layers with ionomer, systematic model-based optimization of porous electrodes for selective transport and reaction, and, in the case of the 3M technology, a fundamentally different design of ionomer-free ultrathin catalyst layers.

In general, the plots in Figure I.6a do not exhibit linear slopes as would be expected for Tafel curves of simple faradaic electrode processes. These curves embody the convoluted impacts of phenomena at a hierarchy of scales, from surface processes at molecular scale to the stack level. At the molecular scale, pathways and mechanisms of surface reactions at the catalyst determine effective exchange current densities and electronic transfer coefficients. These parameters are functions of the electrode potential. The nonlinear interplay of transport and reaction in nanoporous electrodes leads to nonuniform reaction conditions, which transpire as changes in the effective Tafel slope. Further averaging of reaction rates over spatially varying conditions in "thick" porous electrodes leads to new effective parameters of a catalyst layer. Ohmic losses in PEM and transport losses in porous diffusion media cause additional departures from the equilibrium fuel cell voltage. Finally, water distribution and fluxes at MEA, cell, and stack levels lead to nonlinear effects in fuel cell voltage versus current density relations. The main chapters of this book will address these phenomena and explain their contributions to the effects seen in Figure I.6a.

## ABOUT THIS BOOK

The history of research on polymer electrolyte fuel cells spans about 50 years. PEFCs appeared in the focus of scientific interest toward the end of the 1980s. Generally, PEFC design is simple and all the needed components are available on the market. Take two gas-diffusion electrodes separated by a polymer electrolyte membrane and clamp this membrane–electrode assembly between two graphite plates with channels for hydrogen and air supply—the cell is ready.

However, this apparent simplicity hides tremendous complexity of the cell components' structure and function. Obviously, the first working PEFC prototypes were constructed by the trial–and–error method. However, a significant market penetration requires that the cells must be cheap, efficient, and long-living. Nowadays, it is evident that the solution of these problems demands concerted efforts of specialists in electrochemistry, quantum chemistry, physics, fluid mechanics, mechanical, and chemical engineering.

This list of disciplines alone demonstrates the scale of the problem. It is scarcely an exaggeration to say that commercialization of fuel cells is one of the greatest challenges in the history of civilization. The expected changes in people's everyday life due to fuel cells are comparable to the discovery of electricity. No more power lines, transformers, grids, and so on; the required electricity is produced locally, at the place of use: in homes, cars, wearable gadgets, and so on. This futuristic picture gets to reality closer and closer.

This book represents the authors' view of the current situation and trends in PEFC studies. The book does not pretend to be complete and unbiased: the selection of material reflects the authors' scientific interests and priorities. The focus is on generic aspects of structure, properties, and function of key materials in PEFCs: principles of electrocatalysis; flow, transport, and reaction in porous electrodes; equilibration and transport of water in nanoporous media; and water-based proton conduction.

We begin with the discussion of cell thermodynamics and electrochemistry basics (Chapter 1). This chapter may serve as an introduction to the field and we hope it would be useful for the general reader interested in the problem. Chapter 2 is devoted to basic principles of structure and operation of the polymer electrolyte membrane. Chapter 3 discusses micro- and mesoscale phenomena in catalyst layers. Chapter 4 presents recent results in performance modeling of catalyst layers, and in Chapter 5 the reader will find several applications of the modeling approaches developed in the preceding chapters.

The models considered in this book cover space scales that differ by many orders of magnitude, from several nanometers (e.g., the scale of proton-conducting channels in the membrane) to tens of centimeters (a typical length of the oxygen channel in PEFCs). We, however, avoid using the terms *multiscale modeling*, or *multiphysics*, which are nowadays commonly found in the fuel cell literature. These terms have been introduced by software companies to promote their products as "universal multiscale and multiphysics" numerical codes for fuel cell modeling. Using these terms is nothing but marketing, which might be misleading for those who begin working in the field. We prefer to speak about *hierarchical modeling*, which means that information obtained at the lower-scale level is "compressed" into force fields, rate constants, transport coefficients and so on, which are then used at the larger–scale level. This is a standard approach established over many years in physics and chemistry.

This book is mainly on theory and modeling; however, these terms have been deliberately removed from the title to emphasize the focus on *understanding*. This cannot be achieved by theory and modeling alone, but requires close interaction with experimentalists on development of characterization and diagnostics tools, with materials scientists on understanding of modification strategies and limitations, and with engineers on rationalizing the environmental conditions that will translate into boundary conditions of physical models.

The book is intended to reach a diverse readership:

- Theory and modeling experts, interested in problems and approaches.
- Experimentalists, interested in modeling results and, in particular, in theoretical limitations.
- Fuel cell developers and engineers, interested in the status of understanding, capabilities of modeling and diagnostics, obstacles, and shortcomings.
- Chapter 1 is intended to serve as an introduction into the field for a general reader.

# 1 Basic Concepts

This chapter gives an overview of basic concepts in polymer electrolyte fuel cells (PEFCs). The intent is to provide the reader with an intuitive understanding of the processes that underlie fuel cell operation. General and engineering aspects of fuel cell design and operation are treated in greater detail in recently published books (Bagotsky, 2012; Barbir, 2012). Please refer to these books for further discussions of different types of fuel cells and specific aspects of their operation.

## FUEL CELL PRINCIPLE AND BASIC LAYOUT

### NATURE'S BLUEPRINT FOR FUEL CELLS

Energy transduction in a biological cell proceeds via *redox reactions* in aqueous media that involve electron and proton transfer between molecular species (Kurzynski, 2006; Kuznetsov and Ultrup, 1999). During photosynthesis, absorption of a photon creates an excited molecular state. Cascades of electron transfer processes, coupled to protein-driven proton pumping across membranes of mitochondria, transform this molecular excitation into a stable charge-separated state. Thereby, photon energy is absorbed and converted into a gradient of the electrochemical potential of protons, known as a *protonmotive force* in the jargon of bioenergetics. The protonmotive force consists of a transmembrane pH difference and a *Galvani potential difference*. It corresponds to energy that is stored in the biological cell to perform chemical, osmotic or mechanical work upon release (Hambourger et al., 2009). The reverse spontaneous mitochondrial process, oxidative phosphorylation, is a highly efficient way of releasing stored chemical energy. The respiratory system supplies oxygen that is consumed in this process so as to transform the stored energy into muscular work.

### ELECTROMOTIVE FORCE

Electrochemical energy conversion in a PEFC follows similar principles. It involves the creation of a gradient in the *electrochemical potential*, referred to as an *electromotive force* (EMF), across a highly proton-selective membrane, the PEM. In the PEFC configuration, illustrated schematically in Figure 1.1, the PEM provides a low-resistive pathway for proton passage while being highly effective at blocking the flows of electrons and gaseous reactants.

The electromotive force is maintained by keeping electrode compartments on anode and cathode sides of the *polymer electrolyte membrane* (PEM) at different chemical compositions. One electrode, the anode, is supplied with a fuel while the other electrode, the cathode, is supplied with an oxidant. Unless otherwise stated, this book assumes operation with hydrogen as the fuel and oxygen (air) as the oxidant.

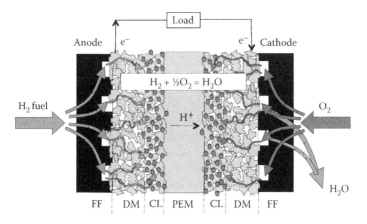

**FIGURE 1.1**  Layout of a polymer electrolyte fuel cell, showing functional components and processes. FF, DM, and CL abbreviate flow field, diffusion media, and catalyst layer, respectively. Note that throughout this book, DM will also be referred to as a GDL (gas diffusion layer).

Different chemical compositions of reactant gas mixtures at electrodes, give rise to a difference in *electrode potentials*. The anode is immersed in a hydrogen-rich atmosphere, which is the feedstock of the *hydrogen oxidation reaction* (HOR)

$$2H_2 \rightleftarrows 4H^+ + 4e^-. \tag{1.1}$$

The *equilibrium electrode potential* of this electrode configuration, the *hydrogen electrode*, is given by

$$E^{eq}_{H^+,H_2} = E^{0,M}_{H^+,H_2} + \frac{R_g T}{F} \ln\left(\frac{a_{H^+}}{p_{H_2}^{1/2}}\right), \tag{1.2}$$

where $p_{H_2} = P_{H_2}/P^0$ is the dimensionless hydrogen partial pressure, normalized to the standard pressure $P^0 = 1$ bar, and $a_{H^+}$ the activity of protons in electrolyte solution. By writing Equation 1.2, the standard electrochemical convention is followed: electrode potentials are given for the reaction proceeding in the cathodic direction, that is from right to left in Equation 1.1.

The first term on the right-hand side of Equation 1.2 is the *standard equilibrium potential* of the HOR for a given metal material (superscript M)

$$E^{0,M}_{H^+,H_2} = \frac{2\mu^{0,s}_{H^+} + 2\mu^{0,M}_{e^-} - \mu^{0,g}_{H_2}}{2F}. \tag{1.3}$$

This expression entails standard chemical potentials for molecular hydrogen in the gas phase, $\mu^{0,g}_{H_2}$, protons in electrolyte solution, $\mu^{0,s}_{H^+}$, and electrons in the metal phase of the electrode, $\mu_{e^-}^{0,M}$, respectively. The latter term incorporates the dependence of $E_{H^+,H_2}^{0,M}$ on the metal material.

The most widely used convention in the electrochemical literature is to define the standard equilibrium potential of the HOR at platinum as the reference or zero point of the electrochemical potential scale, $E_{SHE} = E_{H^+,H_2}^{0,M} \equiv 0$. This specific electrode configuration is known as the *standard hydrogen electrode* (SHE).

At the cathode, protons supplied from the membrane and electrons arriving from the external electric circuit have to be converted into a neutral molecule. A natural choice for the cathode reaction is the *oxygen reduction reaction* (ORR), since oxygen is readily available from the atmosphere. By immersing the cathode into an oxygen-rich atmosphere, the ORR converts proton and electron fluxes into a flux of water according to the following reaction:

$$O_2 + 4H^+ + 4e^- \rightleftarrows 2H_2O(l). \tag{1.4}$$

The equilibrium electrode potential of the *oxygen electrode* is given by

$$E_{O_2,H^+}^{eq} = E_{O_2,H^+}^{0,M} + \frac{R_g T}{F} \ln \left( \frac{a_{H^+} p_{O_2}^{1/4}}{a_{H_2O}^{1/2}} \right), \tag{1.5}$$

where $p_{O_2} = P_{O_2}/P^0$ is the normalized oxygen partial pressure. Note that water in the ORR is produced in liquid form and, hence, its activity may be set as $a_{H_2O} = 1$. The standard equilibrium potential of the oxygen cathode is

$$E_{O_2,H^+}^{0,M} = \frac{\mu_{O_2}^{0,g} + 4\mu_{H^+}^{0,s} + 4\mu_{e^-}^{0,M} - 2\mu_{H_2O}^{0,l}}{4F}, \tag{1.6}$$

where $\mu_{O_2}^{0,g}$ and $\mu_{H_2O}^{0,l}$ are the standard chemical potentials of gaseous oxygen and liquid water, respectively. Relative to the SHE, the standard equilibrium potential of the ORR is $E_{O_2,H^+}^{0,M} = 1.23$ V.

The overall fuel cell reaction

$$2H_2 + O_2 \rightleftarrows 2H_2O(l) \tag{1.7}$$

creates an equilibrium potential difference, referred to as *electromotive force* (EMF), given by

$$E_{O_2,H_2}^{eq} = E_{O_2,H^+}^{eq} - E_{H^+,H_2}^{eq} = E_{O_2,H_2}^0 + \frac{R_g T}{4F} \ln \left( p_{H_2}^2 p_{O_2} \right), \tag{1.8}$$

with the standard EMF

$$E_{O_2,H_2}^0 = \frac{\mu_{O_2}^{0,g} + 2\mu_{H_2}^{0,g} - 2\mu_{H_2O}^{0,l}}{4F}. \tag{1.9}$$

Equation 1.8 together with Equation 1.9 form the famous *Nernst equation*, which relates the equilibrium potential difference of an electrochemical cell to standard equilibrium potential, cell composition and temperature.

In deriving Equation 1.8, it was assumed that both electrodes are made of identical metals and that they are in contact with electrolyte with identical composition and Galvani potential, $\Phi$. Therefore, specific terms, depending on chemical potentials of electrolyte protons or metal electrons will cancel out of this equation. Indeed, the difference in electrochemical potentials $\tilde{\mu}_{e^-}$ of metal electrons at cathode and anode is proportional to the EMF:

$$\tilde{\mu}_{e^-}^c - \tilde{\mu}_{e^-}^a = \mu_{e^-}^c - \mu_{e^-}^a - F\left(\phi^{c,eq} - \phi^{a,eq}\right) = FE_{O_2,H_2}^{eq}. \tag{1.10}$$

If both electrodes are made of the same material, corresponding to the case considered above, electrons must have the same chemical potential at anode and cathode, $\mu_{e^-}^c = \mu_{e^-}^a$. Therefore, the difference in electrochemical potential or EMF is equal to the difference in electrostatic Galvani potentials $\phi$ between the metal electrodes at equilibrium:

$$E_{O_2,H_2}^{eq} = \phi^{c,eq} - \phi^{a,eq}. \tag{1.11}$$

This Galvani potential difference can be measured with a voltmeter. It represents the maximal driving force of the electron flux from anode to cathode. Since it is reasonable to assume that electron transport in metal wires occurs under negligible ohmic resistance losses, the potential difference between metal ports at anode and cathode is almost completely available to perform electrical work in electrical loads or appliances, indicated in Figure 1.1. The standard EMF of the $H_2/O_2$ fuel cell is $E_{O_2,H_2}^0 = 1.23$ V.

In order for a constant electron current to flow through the external wiring, in the direction indicated in Figure 1.1, the difference in reactant gas compositions at the electrodes must be maintained in a steady state. Otherwise, the cell would relax to global thermodynamic equilibrium, as signified by a decreasing cell current. During this relaxation, hydrogen and oxygen would be used up until a configuration with identical compositions in electrode compartments and equal electrode potentials is attained.*

Under conditions of constant current, namely, under steady-state operation, reactants must be supplied continuously and at a constant rate to precisely balance the rate of reactant consumption in electrode reactions. The coupled fluxes of electrons, protons, and gaseous reactants are subdued to two fundamental conservation laws, which complete the description of the fuel cell principle: conservation of charge and mass. These laws allow balance or continuity equations to be written among all involved

---

* This is the relaxation phenomenon that Schoenbein must have seen in his experiment, given the finite amounts of hydrogen and oxygen available in the closed electrode chambers. Indeed, this mode of operation resembles the principle of battery operation, where a finite amount of reactants is used up during battery discharge; in the course of this discharge, the composition of the electrodes relaxes toward global thermodynamic equilibrium; equilibrium is attained when the difference of Galvani potentials between anode and cathode goes to zero. In this sense, Schoenbein discovered the fuel cell principle, but he observed it for a system that operated like a battery!

species. The rate of electron and proton generation in the HOR at the anode, Equation 1.1, must match the electron flux through the external metal wires, as well as the proton flux through the PEM. At the cathode, these fluxes must exactly balance the rate of consumption of electrons and protons in the ORR, Equation 1.4.

All in all, the fuel cell principle explains how an electrostatic potential gradient or electromotive force is created and maintained by controlling the unequal composition of feed components. The current flowing through the load is uniquely determined by the coupled and balanced rates of reactant supply through diffusion media, rates of anode and cathode reactions at electrodes, and electron and proton fluxes through their respective conduction media.

## BASIC LAYOUT OF A SINGLE CELL

The simplest fuel cell can be constructed using two planar platinum electrodes separated by an electrolyte, shown in Figure 1.2a. Bringing hydrogen molecules to the left Pt/electrolyte interface and oxygen molecules to the right electrolyte/Pt interface polarizes the cell and creates a Galvani potential difference between the electrodes. Physically, the potential corresponds to a separation of opposite charges, deposited at the left and right interface, respectively. In the electrochemical literature, electrode polarization is called *double layer charging*. The term "double layer" reflects the shape of the dipolar charge density profile with the maximum and minimum just near the interface (Schmickler and Santos, 2010).

However, such a simple cell generates insufficient current due to the small catalyst surface area available for reactions. For the cell to provide a useful current in the order of 1 A cm$^{-2}$, the electrode/electrolyte interface area must be expanded by a sizable factor. This brings us to the multilayered design of a single PEFC, depicted in Figure 1.1.

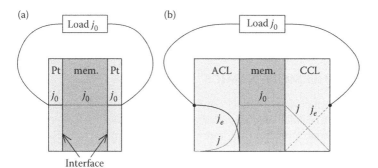

**FIGURE 1.2** (a) Schematic of currents in the cell with two planar Pt electrodes. (b) Schematic of currents in a PEM fuel cell with porous electrodes. ACL and CCL stand for the anode and cathode catalyst layer, $j$ and $j_e$ are the local proton and electron current densities in the catalyst layers, and $j_0$ is the cell current density. Note that the shapes of the local current in the electrodes correspond to a small cell current (see the section "Regimes of Electrode Operation").

The general layout of a cell includes a proton-conducting *polymer electrolyte membrane* (PEM), sandwiched between the anode and the cathode. Each electrode compartment is composed of (i) an active catalyst layer (CL), which accommodates finely dispersed nanoparticles of Pt that are attached to the surface of a highly porous and electronically conductive support, (ii) a gas diffusion layer (GDL), and (iii) a flow field (FF) plate that serves at the same time as a current collector (CC) and a bipolar plate (BP). This plate conducts current between neighboring cells in a fuel cell stack. At the cathode side, usually a strongly hydrophobic microporous layer (MPL) is inserted between CL and GDL, which facilitates the removal of product water from the cathode CL. The central unit including PEM and porous electrode layers, excluding the bipolar plates, is called the *membrane electrode assembly* (MEA).

The main structural components in modern PEFCs are the porous composite electrodes. The primary purpose of utilizing porous electrodes is to enhance the *active surface area* of the catalyst by several orders of magnitude in comparison to planar electrodes with the same in-plane geometrical area. In the CLs, the fluxes of reactant gases, protons, and electrons meet at the catalyst particle surface. Active catalyst nanoparticles are located at spots that are connected simultaneously to the percolating phases of proton, electron, and gas transport media. An important implication of the electrode's finite thickness is the necessity to provide transport of neutral molecules and protons through the depth of the porous electrode. Additional overhead is caused by the transport of neutral reactants through FF, GDL, and MPL. This leads to specific potential losses in the electrodes, which will be considered in detail in Chapter 4.*

A schematic graph of currents in the cell with porous electrodes is shown in Figure 1.2b. As can be seen, proton and electron currents in catalyst layers are continuously converted into one another, so that on the opposite sides of each layer, the currents are either completely protonic or electronic in nature.

Fuel cells that consume (oxidize) other hydrocarbon-based fuels, for example, methanol, ethanol, or formic acid, have the same basic layout. They follow the same thermodynamic principles as $H_2/O_2$ fuel cells, whereas the redox couples could be $(CH_3OH/O_2)$, $(C_2H_5OH/O_2)$, and so on, with the product state involving $H_2O$ and $CO_2$.

## FUEL CELL THERMODYNAMICS

As discussed in the section "Electromotive Force," the equilibrium or open-circuit cell potential can be calculated from thermodynamics. Indeed, thermodynamics states that for any reversible reaction, the following relation holds:

$$\Delta H = T\Delta S + \Delta G, \qquad (1.12)$$

---

* Note that in the fuel cell literature, the term *electrode* may have two meanings: either a catalyst layer only, or a combination of catalyst and gas diffusion layers. The actual meaning must be inferred from the context.

---

**TABLE 1.1**

**Standard Thermodynamic Values for the Oxygen–Hydrogen Reaction (1.7) with Liquid Water as a Product**

| $\Delta H^0$ | $\Delta G^0$ | $\Delta S^0$ |
|---|---|---|
| $-142.9$ kJ mol$^{-1}$ | $-118.6$ kJ mol$^{-1}$ | $-81.7$ J mol$^{-1}$ K$^{-1}$ |

Note that all values are given per mole of protons exchanged in the reaction.

---

where $\Delta H$ is the enthalpy change, $T$ the absolute temperature, $\Delta S$ the entropy change, and $\Delta G$ the change in Gibbs energy between product state ($H_2O$) and reactant state of the redox couple ($H_2/O_2$). For convenience in this chapter, all thermodynamic potentials, work, and heat terms are specified in units of joules per mole of protons exchanged in the fuel cell reaction. With this convention, standard thermodynamic parameters for the fuel cell reaction of the redox couple ($H_2/O_2$) to liquid water in Equation 1.7 are listed in Table 1.1.

The direct combustion of hydrogen in an oxygen atmosphere follows the same reaction as in Equation 1.7. In this process, $\Delta H$ is transformed completely into thermal energy (heat), which can be converted into mechanical work using a steam turbine. Thereafter, it can be transformed into electrical work in an electric generator. The upper limit of the thermodynamic efficiency for any heat or steam cycle corresponds to the efficiency of the hypothetical Carnot heat engine:

$$\epsilon_{rev}^{heat} = -\frac{W_{rev}}{\Delta H} = 1 - \frac{T_2}{T_1}, \tag{1.13}$$

where $W_{rev,m}$ is the reversible mechanical work extracted from the process, and $T_1$ and $T_2$ are the temperatures of the upper and lower heat reservoirs, between which the engine is operated.

In a fuel cell, the difference in reactant gas compositions at the two electrodes leads to the formation of a difference in Galvani potential between anode and cathode, as discussed in the section "Electromotive Force." Thereby, the Gibbs energy $\Delta G$ of the net fuel cell reaction is transformed directly into electrical work. Under ideal operation, with no parasitic heat loss of kinetic and transport processes involved, the reaction Gibbs energy can be converted completely into electrical energy, leading to the *theoretical thermodynamic efficiency* of the cell,

$$\epsilon_{rev} = \frac{\Delta G}{\Delta H} = 1 - \frac{T\Delta S}{\Delta H}. \tag{1.14}$$

If the isothermal entropy change of the reaction is negative, as is the case for the $H_2/O_2$ fuel cell as well as for most other technically relevant fuel cell configurations, the thermodynamic efficiency will be smaller than one, $\epsilon_{rev} < 1$, and it will decrease linearly

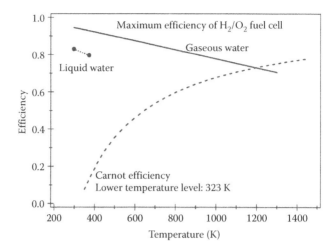

**FIGURE 1.3** Thermodynamic efficiencies of a fuel cell and a hypothetical Carnot heat engine, plotted as a function of temperature. A fuel cell that forms water in a gaseous phase has a higher efficiency compared to a fuel cell that produces liquid water. The efficiency difference corresponds to the enthalpy of vaporization.

with increasing temperature. The trends in theoretical thermodynamic efficiency versus temperature for heat engine and $H_2/O_2$ fuel cell are illustrated in Figure 1.3. The fuel cell efficiency decreases with temperature, while the Carnot efficiency increases. In the temperature range well below 1000°C, fuel cells have a decisive advantage in thermodynamic efficiency over heat engines.

To a good approximation, the temperature dependence of the equilibrium fuel cell potential $E^{eq}_{O_2,H_2}$ is linear (Kulikovsky, 2010a):

$$E^{eq}_{O_2,H_2} = E^0_{O_2,H_2} + \frac{\Delta S^0}{F}\left(T - T^0\right). \tag{1.15}$$

The entropy change for the $O_2$–$H_2$ reaction, with the product water in the liquid state, is negative (Table 1.1). Thus, the cell potential decreases linearly with the growth of temperature. The factor $\Delta S^0/F \simeq 0.85 \cdot 10^{-3}$ V K$^{-1}$ implies that the cell potential at 100°C is $\simeq 63$ mV lower than at standard conditions.

The property

$$Q_{rev} = -T\Delta S \tag{1.16}$$

is the reversible heat that is exchanged with the environment during the fuel cell reaction. The amount of heat transferred to the environment is smaller if the reaction product is water in gaseous form, as opposed to liquid water. Therefore, the former process exhibits the larger thermodynamic efficiency (Figure 1.3). At standard conditions, thermodynamic efficiencies of $H_2/O_2$ fuel cells are 96% for product water in gaseous form, and 83% for product water in liquid form. This reduction in thermodynamic efficiency for liquid water production is an intrinsic disadvantage of fuel cells operated with aqueous electrolytes.

The maximum amount of electrical work corresponding to idealized reversible (or infinitely slow) progress of the fuel cell reaction in Equation 1.7, is given by

$$W_{rev,el} = -\Delta G = FE^{eq}_{O_2,H_2}. \tag{1.17}$$

The reversible thermodynamic efficiency of the fuel cell can thus be expressed as

$$\epsilon_{rev} = \frac{FE^{eq}_{O_2,H_2}}{FE^{eq}_{O_2,H_2} + Q_{rev}}. \tag{1.18}$$

A cell that is operated infinitesimally close to electrochemical equilibrium (or open circuit conditions) will not produce any useful power output. To produce a significant power output, sufficient to propel a vehicle, for instance, the cell must be operated at a current density on the order of 1 A cm$^{-2}$. Under load, the value of the current density $j_0$ of fuel cell operation determines the power output. The current density is directly related to reaction rates at catalyst layers, as well as flows of electrons, protons, reactants, and product species in the cell components. Each of these processes contributes to irreversible heat losses in the cell. These losses diminish the amount of electrical work that the cell could perform.

The main contributions to *irreversible heat loss*, listed in the order of decreasing significance, are due to (i) kinetic losses in the ORR at the cathode ($Q_{ORR}$), including losses due to proton transport in the cathode catalyst layer, (ii) resistive losses due to proton transport in the PEM ($Q_{PEM}$), (iii) losses due to mass transport by diffusion and convection in porous transport layers ($Q_{MT}$), (iv) kinetic losses in the HOR at the anode ($Q_{HOR}$), and (v) resistive losses due to electron transport in electrode and metal wires ($Q_M$). Some of these losses are indicated in Figure 1.4. Energy (heat) loss terms are related to overpotentials by $|\eta_i| = Q_i/F$, which will be discussed in the section "Potentials."

The resulting amount of work that the cell performs is

$$W_{out} = W_{rev,el} - \sum_i Q_i. \tag{1.19}$$

The *cell efficiency* is defined as the product of $\epsilon_{rev}$ and the *voltage efficiency* $\epsilon_v = W_{out}/W_{rev,el}$,

$$\epsilon_{cell} = \epsilon_{rev}\epsilon_v = \frac{\Delta G}{\Delta H}\frac{W_{out}}{W_{rev,el}} = \frac{FE^{eq}_{O_2,H_2} - \sum_i Q_i}{FE^{eq}_{O_2,H_2} + Q_{rev}}. \tag{1.20}$$

Nowadays, efficiencies for automotive PEFC stacks, reported by car manufacturers, are in the range of 60% or above. In order to determine a precise value of the voltage efficiency, the irreversible heat losses $Q_i$ must be quantified. Knowing these values, we obtain the terms $|\eta_i|$ and the corresponding values of $W_{out}$ and $\epsilon_{cell}$. Therefore,

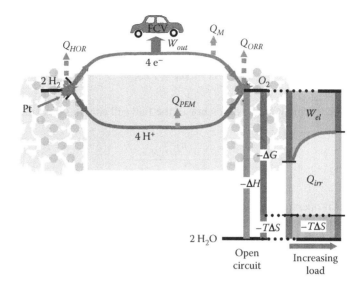

**FIGURE 1.4** Illustration of basic fuel cell processes and their relation to the thermodynamic properties of a cell. The electrical work performed by the cell, $W_{el}$, corresponds to the reaction enthalpy, $-\Delta H$, minus the reversible heat due to entropy production, $-T\Delta S$, and minus the sum of irreversible heat losses at finite load, $\sum_i Q_i$. These losses are caused by kinetic processes at electrochemical interfaces as well as by transport processes in diffusion and conduction media.

the cell voltage at any current density can be found,

$$E_{\text{cell}}(j_0) = E^{eq}_{O_2,H_2} - \sum_i |\eta_i|. \tag{1.21}$$

The power density of the cell can be calculated from

$$P_{\text{cell}}(j_0) = j_0 E_{\text{cell}}(j_0) = j_0 \left( E^{eq}_{O_2,H_2} - \sum_i |\eta_i| \right). \tag{1.22}$$

Typical data for $E_{\text{cell}}(j_0)$ and $P_{\text{cell}}(j_0)$ are depicted in Figure 1.5.

$Q_i$ or $\eta_i$ are complex functions of cell design, specific properties of cell materials, cell composition, thermodynamic conditions, and current density of operation. All efforts in fundamental fuel cell science are geared toward unraveling these functional dependencies. Applied research and development focuses on finding ways to minimize contributions of $Q_i$ and $|\eta_i|$, whereby optimal values of $\epsilon_{\text{cell}}$ and $P_{\text{cell}}$ could be obtained. The target values and constraints on performance, cost, durability, and lifetime of this optimization problem are set by end-user requirements.

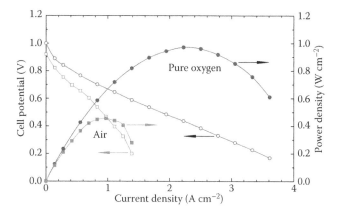

**FIGURE 1.5** Polarization and power density curves of a typical PEFC.

## MASS TRANSPORT PROCESSES

### OVERVIEW OF TRANSPORT PROCESSES

A schematic drawing of a typical PEM fuel cell is shown in Figure 1.6. One of the greatest assets of PEFCs is the small thickness of the MEA in combination with its large geometric surface area. A typical MEA is about 1 mm thick, while the cell active area is usually about $10 \times 10 \text{ cm}^2$. The ratio of the MEA thickness to its characteristic in-plane length, is about $10^{-2}$–$10^{-3}$. This is the prerequisite for constructing simple transport models for the fuel cell.

To distribute oxygen over the whole active area of the cell, a system of channels is machined into the *current collector* (Figure 1.6). The channels on either side

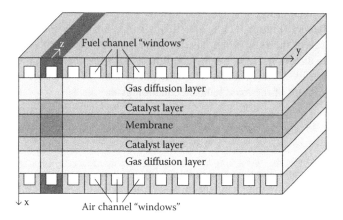

**FIGURE 1.6** Schematic of a typical PEFC with the meandering flow field. Note that the figure is strongly not to scale: the MEA thickness is $10^2$–$10^3$ times less, than the in-plane size of the current collector plate. The color indicates an elementary cell fragment, which is often considered in cell modeling.

of the cell usually are of meandering form covering the whole cell surface, while some exotic fractal geometries have also been suggested (Senn and Poulikakos, 2004, 2006). The system of channels covering the active area is called *flow field*.

The first functional porous electrode layer under the channel is the GDL (Figure 1.6). The role of the GDL is to minimize the in-plane inhomogeneity of the oxygen concentration caused by alternating channels and current collector ribs (Figure 1.6). Another important role of the GDL is to prevent fast oxygen consumption: in a cell without the GDL, oxygen is rapidly consumed at the channel inlet, leaving most of the cell area in oxygen starvation conditions. From this point of view, the role of the GDL is to *retard* the transport of oxygen to the catalyst layer. Last but not least, the GDL must provide a proper balance of liquid water on the cathode side, by facilitating the removal of water produced in the ORR. Oftentimes, a strongly hydrophobized microporous layer is inserted between cathode GDL and catalyst layer. This layer facilitates water removal at high currents and it keeps the membrane well hydrated at small currents.

Owing to the finite thickness of the catalyst layer, oxygen must be transported through the CL depth. Thicknesses of modern catalyst layers are usually in the range of 5–10 μm. The electrode porosity is typically large enough to minimize the oxygen transport loss. However, at high current densities, the electrode tends to be flooded, which might dramatically increase its resistance to oxygen transportation.

## Air Flow in the Channel

Generally, owing to the presence of liquid water produced in the ORR, the flow in the cathode channel is two-phased. However, at low and moderate current densities, the mass fraction of liquid water is small and does not strongly affect the flow's velocity. To a good approximation, it can be considered as a plug flow (well-mixed flow with constant velocity).*

Under this assumption, the mass balance for oxygen concentration $c_h$ in a small box in the channel (Figure 1.7), is $hwv^0(c_h(z+dz) - c_h(z)) = -w\,dzj_0/(4F)$, that is, the variation of the flow along $z$ is due to the stoichiometric oxygen flux in the through-plane direction. Here, $z$ is the coordinate along the channel, $j_0(z)$ is the local cell current density, $h$ is the channel height (depth), and $w$ is the channel width. Taking the continuum limit of the mass balance equation gives

$$v^0\frac{\partial c_h}{\partial z} = -\frac{j_0}{4Fh}. \tag{1.23}$$

Equation 1.23 means that the oxygen molar flux $v^0 c_h$ decreases toward the channel outlet at a rate given by the local current $j_0$.

A fundamental parameter in chemical engineering of fuel cells is the *oxygen stoichiometry* $\lambda$. By definition, $\lambda$ is the ratio of the total inlet oxygen flux $4Fhwv^0 c_h^0$

---

* Note that the flow in the cathode channel is typically laminar, so "well-mixed" simply means that the diffusion transport across the channel is fast. The estimate, which demonstrates validity of this assumption, is given at the end of this section.

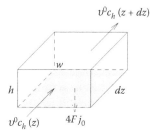

**FIGURE 1.7**   Elementary box in the fuel cell channel.

(expressed in electric units) to the total current $L_h w J$ produced in the cell:

$$\lambda = \frac{4 F h v^0 c_h^0}{L_h J}. \tag{1.24}$$

Here, $c_h^0$ is the inlet oxygen concentration, $L_h$ is the channel length, and $J$ is the mean current density in the cell. Note that the presented treatment ignores details of the channel/rib geometry and it assumes that the channel covers the whole cell active area.*

Equation 1.23 can be expressed in terms of $\lambda$:

$$\lambda J \frac{\partial \tilde{c}_h}{\partial \tilde{z}} = -j_0, \quad \tilde{c}(0) = 1, \tag{1.25}$$

with dimensionless variables

$$\tilde{z} = \frac{z}{L_h}, \quad \tilde{c} = \frac{c_h}{c_h^0}. \tag{1.26}$$

Integrating Equation 1.25 over $\tilde{z}$ and taking into account that $\int_0^1 j_0 \, d\tilde{z} = J$ results in the simple but useful relation

$$\tilde{c}_1 = 1 - \frac{1}{\lambda}. \tag{1.27}$$

This equation can be rewritten as $\tilde{c}_1 = 1 - u$, where $u$ is the *oxygen utilization*, which is equal to the inverse stoichiometry

$$u = \frac{1}{\lambda}. \tag{1.28}$$

For example, $\lambda = 2$ means that $u = 0.5$, that is, 50% of the oxygen flow is consumed and 50% leaves the cell. Using this property, a total *fuel efficiency* of the fuel cell can

---

* In other words, the cell in Figure 1.6 is replaced by an equivalent cell with a long straight channel covering the whole cell active area. The length of the straight channel equals the length of the meandering channel.

be defined by

$$\epsilon_{fuel} = u\epsilon_{rev}\epsilon_v. \tag{1.29}$$

The assumption of a "well-mixed" flow is valid provided that oxygen diffusion across the cathode channel is fast. The variation of the oxygen flux along the channel is $v^0 \partial c_h / \partial z$. The variation of the flux across the channel due to diffusion is $D_{\text{free}} \partial^2 c_h / \partial x^2$, where $D_{\text{free}}$ is the oxygen diffusion coefficient in free space and $x$ is the coordinate across the channel. Estimating these variations as $v^0 c_h / L_h$ and $D_{\text{free}} c_h / (h/2)^2$, respectively, and dividing the first by the second, it can be seen that diffusion transport across the channel is fast, if the dimensionless Péclet number is small:

$$\text{Pe} = \frac{v^0 h^2}{4 D_{\text{free}} L_h} \ll 1. \tag{1.30}$$

Estimated parameter values $v^0 = 10^2$ cm s$^{-1}$, $h = 0.1$ cm, $L_h = 10$ cm, and $D_{\text{free}} \simeq 0.2$ cm$^2$ s$^{-1}$ give Pe $\simeq 0.13$, which means that the required criterion is fulfilled.

## TRANSPORT IN GAS DIFFUSION AND CATALYST LAYERS

The mean free path of molecules in air at atmospheric pressure is $l_{\text{free}} \simeq 1/(N_L \sigma_g)$, where $N_L \simeq 2.69 \cdot 10^{19}$ cm$^{-3}$ is the number density of gas molecules and $\sigma_g \simeq 10^{-14}$ cm$^2$ is the cross section for elastic collisions of molecules. These numbers result in $l_{\text{free}} \simeq 3.7 \cdot 10^{-6}$ cm, or 37 nm. The mean pore radius of the GDL is in the order of 10 µm, which means that the flow in the GDL pores occurs in a continuum regime. Thus, pressure-driven oxygen transport in a dry porous GDL can be modeled as a viscous Hagen-Poiseuille flow in an equivalent duct. However, determination of the equivalent duct radius and the dependence of this radius on the GDL porosity is a nontrivial task (Tamayol et al., 2012). Much work has recently been done to develop statistical models of porous GDLs and to calculate viscous gas flows in these systems using Navier–Stokes equations (Thiedmann et al., 2012).

In the absence of an external pressure gradient, the dominating mechanism of oxygen transport in a GDL is diffusion in a concentration gradient (Benziger et al., 2011; Thiedmann et al., 2012). This situation is realized in PEM fuel cells operating on air and hydrogen at atmospheric pressure. Owing to the large average pore radius in the GDL, diffusion proceeds in a molecular regime and the oxygen diffusion coefficient in the porous medium may be expected to closely resemble that in free space. However, oxygen diffusivity in the GDL is affected by the tortuous nature of the media. Physically, tortuosity increases the average diffusion path between two points in the GDL. Let the straight distance between two points $a$ and $b$ be $l_s$; the path length the oxygen molecule must travel from $a$ to $b$ is $\tau l_s$, where $\tau$ is the GDL tortuosity. Equating the expressions for the effective and real mass transfer coefficients, $D_b/l_s = D_{\text{free}}/(\tau l_s)$, it can be seen that the following relation holds:

$$D_b = \frac{D_{\text{free}}}{\tau},$$

where $D_{\text{free}}$ is the diffusion coefficient in free space. A more general expression includes correction for the GDL porosity $\epsilon$

$$D_b = \frac{\epsilon}{\tau} D_{\text{free}}$$

[see Fishman and Bazylak (2011) and the literature cited therein].

Rationalizing a relation between the GDL structure and its effective hydraulic permeability is a more difficult task. Over the past decade, numerous efforts have attempted to establish the structure–function relations for transport in GDLs. See the recent reviews on this subject (Thiedmann et al., 2012; Zamel and Li, 2013); see also a paper of Becker et al. (2009) and the references therein.

Oxygen transport in the catalyst layer may differ from transport in the GDL due to the nonnegligible impact of its small pores with radii of $\sim$0.01 μm. In these pores, oxygen transport is controlled by the Knudsen diffusion mechanism (see below). For a detailed discussion of transport mechanisms in the porous CL and GDL, see Hinebaugh et al. (2012), Wang (2004), and Weber and Newman (2004a).

In a mixture of gases of comparable concentrations, diffusion of any component due to a concentration gradient causes the counter-diffusion of other components. Physically, redistribution of one component must be compensated for by the diffusion of other components in order to preserve constant pressure in the system. The fluxes of gaseous components in a multicomponent mixture are related by Stefan–Maxwell equations (Bird et al., 1960):

$$\sum_i \frac{\xi_i N_k - \xi_k N_i}{D_{ik}} = -c_{\text{tot}} \nabla \xi_k, \tag{1.31}$$

where $\xi_k$ and $N_k$ are the molar fraction and flux, respectively, of the $k$th component, $c_{\text{tot}}$ is the mixture total molar concentration, and $D_{ik}$ is the binary diffusion coefficient of component $i$ in a mixture with component $k$.

Summing up Equations 1.31 over $k$, the left side becomes zero, and we obtain an identity

$$c_{\text{tot}} \sum_k \nabla \xi_k = 0.$$

Indeed, as $\sum_k \xi_k = 1$, we have $\sum_k \nabla \xi_k = \nabla \left( \sum_k \xi_k \right) = 0$. This means, that Equations 1.31 are not independent: any of these equations can be expressed as a linear combination of the others. Physically, Equations 1.31 determine the relative fluxes only; to establish the reference frame for the fluxes, an independent closing relation has to be prescribed. In PEFC modeling, nitrogen flux at the cathode side is zero, and the condition $N_{N_2} = 0$ is usually used as the closing relation.

In small pores of radius less than 10 nm (0.01 μm), the frequency of molecular collisions is much less than the frequency of molecule collisions with pore walls. The flow is no longer of a continuum-type; nonetheless, the flux of species in a pore can still be expressed by Fick's laws of diffusion, Equation 1.34, with the Knudsen diffusion coefficient $D_K$ proportional to the product of the pore radius $r_p$ and the mean thermal velocity of the molecules

$$D_K = \psi r_p \sqrt{\frac{8R_g T}{\pi M}}, \tag{1.32}$$

where $M$ is the molar mass of the molecules, and $\psi$ is a correction factor in the order of unity. Note that in pores of an intermediate radius 10 nm $< r_p <$ 100 nm, both molecular and Knudsen mechanisms contribute to the flux. In that case, Fick's diffusion equation still can be used, with the average effective diffusion coefficient $D_{\text{eff}}$ given by

$$\frac{1}{D_{\text{eff}}} = \frac{1}{D} + \frac{1}{D_K}. \tag{1.33}$$

This highly simplified approach can be used in analytical modeling, aiming to understand the trends; more accurate numerical models are based on the idea of volume averaging (Whitaker, 1998).

In this book, pressure-driven species transport in porous electrode layers is not considered and we will always assume diffusion transport caused by a concentration gradient. Moreover, this transport will always be described by Fick's diffusion equation:

$$N_i = -D\nabla c_i, \tag{1.34}$$

that is, the molar flux $N_i$ is assumed to be proportional to the gradient of the species concentration. Equation 1.34 is exact for the molecular diffusion of gases constituting a small fraction in the mixture with other components. The content of oxygen in air is 0.21, which can be considered a small parameter for most practical aspects studied in this book. Equation 1.34 also formally describes Knudsen diffusion and, hence, we may expect that in all the cases considered, this equation leads to physically consistent results.

## POTENTIALS

### POTENTIALS AND OVERPOTENTIALS IN A FUEL CELL

#### Planar Electrodes

Consider again the simplest fuel cell configuration with a proton-conducting membrane sandwiched between two planar electrodes, cf. Figure 1.8a. Keeping the electrodes at a different composition with hydrogen at the anode/electrolyte interface (left) and oxygen at the cathode/electrolyte interface (right) gives rise to the Galvani potential difference between the two metal sides. This difference defines the equilibrium cell potential, given in Equation 1.11.

Under normal conditions, the anode surface is negatively charged and the cathode surface is positively charged owing to electrochemical double layers established at both metal/electrolyte interfaces. The value of the Galvani potential in the electrolyte will have a value intermediate between Galvani potentials at anode and cathode. Note that the potential drop arises at the metal/electrolyte interface without any external "feeding," simply because electrons in the metal attract cations and repulse anions in the electrolyte, leading to the formation of a double layer at the interface. However, the presence of feed molecules changes the potential drop at the interface in order to facilitate continuous current production. This will be discussed in the section "Electrochemical Kinetics."

The left interface is charged as the HOR splits hydrogen into protons and electrons. The protons charge up the electrolyte close to the metal surface, while excess electrons form a charged skin layer of thickness $\sim 1$ Å on the metal surface. On the right interface, the ORR consumes protons from the electrolyte, leaving the nearest electrolyte domain negatively charged. ORR also consumes electrons from the metal, positively charging the skin layer at the metal interface (Figure 1.8a). Thereby, states with defined electrode potential and surface charge density, $\sigma$, are established independently on each electrode.

The resulting distribution of charges leads to the shape of electric potential schematically shown in Figure 1.8a. At electrochemical equilibrium, the bulk phases of electrodes and electrolyte are equipotential; strong variations in potential exist in the double layer regions only (Figure 1.8a). Note that in Figure 1.8, the potential is continuous and, hence, $\phi$ and $\Phi$ denote the same physical property. However, here we keep different notations to emphasize the transition to the membrane potential $\Phi$, upon replacing the smooth variation of potential at the interface by the step jump from $\phi$ to $\Phi$ (see below).

If an electrical current $j_0 > 0$ is drawn from the cell, the shape of the potential profile changes. Figure 1.8b is obtained under the assumptions of (i) ideal kinetics of the anode reaction and (ii) ideal proton conductivity of the electrolyte. The first assumption implies that the deviation of the potential distribution from electrochemical equilibrium is negligible in the anode double layer region. The second assumption implies a constant electrolyte phase potential. The only change observed in this case is a shift in the metal phase potential on the cathode side away from equilibrium, caused by the poor kinetics of the cathode reaction; it corresponds to a decrease of

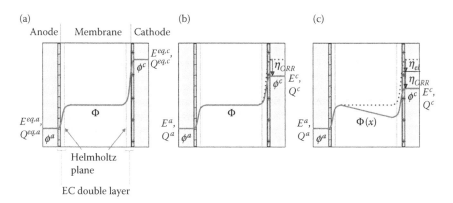

**FIGURE 1.8** Potential distribution in a fuel cell with planar Pt electrodes. (a) Equilibrium (open-circuit) conditions; (b) under load, assuming infinite membrane conductivity, and (c) under load, assuming finite membrane conductivity. In cases (b) and (c), the anode polarization is assumed to remain constant, implying a negligible shift of the anode electrode potential.

the potential drop across the double layer region at the cathode/electrolyte interface. This negative potential shift at the oxygen cathode is defined as the *overpotential* $\eta_{ORR}$, which can be written as

$$\eta_{ORR} = E_{O_2,H^+} - E_{O_2,H^+}^{eq} = \phi^c - \Phi^c - \left(\phi^{c,eq} - \Phi^{c,eq}\right). \qquad (1.35)$$

Here, $\Phi^c$ is the potential on the cathode side of the membrane just outside the interfacial double layer.

When the assumption of ideal proton conductivity of the electrolyte is relaxed, the situation depicted in Figure 1.8c is encountered. The interfacial potential drop at the cathode, and thus the value of $\eta_{ORR}$, remain unchanged but the nonideal proton conductivity in the membrane incurs an additional potential drop, an overpotential $\eta_{PEM}$. Assuming that the electrolyte behaves like an ohmic resistor gives

$$\eta_{PEM} = R_{PEM} j_0, \qquad (1.36)$$

where $R_{PEM}$ is the electrolyte (membrane) resistance.

In another step, the assumption of ideal reaction kinetics at the hydrogen anode can be relaxed. This modification leads to a positive deviation of the anode potential from equilibrium, referred to as the anode overpotential $\eta_{HOR}$, defined by

$$\eta_{HOR} = E_{H^+,H_2} - E_{H^+,H_2}^{eq} = \phi^a - \Phi^a - \left(\phi^{a,eq} - \Phi^{a,eq}\right). \qquad (1.37)$$

It is left to the reader to draw the profile of the potential distribution for this case. Here, $\Phi^a$ is the potential on the anode side of the membrane just outside the double layer.

In general, overpotentials describe the departure of the potential from equilibrium, caused by running the electrochemical reaction. For the simple cell configuration, all three potential losses sum up to lower the potential difference between anode, $\phi^a$, and cathode, $\phi^c$, metal phases by the value

$$\eta_{tot} = E_{O_2,H_2}^{eq} - E_{cell}(j_0) = \eta_{HOR} - \eta_{ORR} + R_{PEM} j_0$$
$$= \eta_{HOR} + |\eta_{ORR}| + R_{PEM} j_0. \qquad (1.38)$$

The characteristic length for the separation of charges at both interfaces is in the order of 10 Å. In modeling of fuel cells, the continuous potential distribution in this region is usually not resolved and two potentials are introduced instead: the metal phase potential, which drives electrons, and the electrolyte phase potential, which controls the proton current in the ionomer phase. The two potentials are separated by finite jumps at the metal/electrolyte interface (Figure 1.8).

In the context of classical electrochemistry, this convention implies that Frumkin corrections (Schmickler, 1996) to the electrochemical kinetics of interfacial faradaic reactions are either negligible or constant. How good is this simplifying assumption? It is justified at the hydrogen anode, where the overpotential is generally small. At the cathode, where the interfacial potential drop varies significantly with the value of the current density $j_0$ drawn, it is disputable. It would only be justified if the dominant double layer potential drop occurs in a narrow region confined by metal surface and position of the reaction or Helmholtz plane on the electrolyte side. In other words, the potential drop over the diffuse part of the double layer (zeta potential) should be small as compared to the drop over the Stern layer, which is warranted at sufficiently large equilibrium proton concentration in the electrolyte. The assumption of negligible potential drop in the diffuse double layer is generally adopted in classical porous electrode theory or catalyst layer modeling, where highly concentrated electrolyte is provided either in liquid form or via ionomer impregnation. In cases where this assumption cannot be made, for example, encountered in ionomer-free ultrathin catalyst layers (UTCLs), distributions of protons and electrolyte potential must be determined explicitly using Poisson–Nernst–Planck theory, as discussed in the section "ORR in Water-Filled Nanopores: Electrostatic Effects" in Chapter 3.

## Porous Electrodes

Going from planar to *porous electrode* introduces another length scale, the electrode thickness. In the case of a PEM fuel cell catalyst layer, the thickness lies in the range of $l_{CL} \simeq$ 5–10 µm. The objective of *porous electrode theory* is to describe distributions of electrostatic potentials, concentrations of reactant and product species, and rates of electrochemical reactions at this scale. An accurate description of a potential distribution that accounts explicitly for the potential drop at the metal/electrolyte interface would require spatial resolution in the order of 1 Å. This resolution is hardly feasible (and in most cases not necessary) in electrode modeling because of the huge disparity of length scales. The simplified description of a porous electrode as an effective medium with two continuous potential distributions for the metal and electrolyte phases appears to be a consistent and practicable option for modeling these structures.

More specifically, the conventional approach to describe this situation is to introduce a so-called *representative elementary volume* (REV).* The size of an REV has to be small on the scale of the electrode thickness, and large on the scale of microscopic variations in electrode structure and composition. Fluctuations on the scale of the double layer thickness, that is, below ~1 nm, are averaged out. The electrochemical properties of an REV are defined by local values of metal and electrolyte phase potentials. These potentials are continuous functions of spatial coordinates.

In one-dimensional electrode modeling, $\phi(x)$ denotes the metal phase potential and $\Phi(x)$ the electrolyte phase potential. The gradient of the metal (carbon) phase potential drives the electron flux, while protons move along the potential gradient of the electrolyte (ionomer) phase. At equilibrium, these gradients are zero and the potentials in the distinct phases are constant, $\phi(x) = \phi^{eq}$ and $\Phi(x) = \Phi^{eq}$. The potential distribution of a working PEFC with porous electrodes of finite thickness is shown in Figure 1.9. For illustrative purposes, a simple assembly of anode catalyst layer, PEM and cathode catalyst layer is displayed.

---

* The representative elementary volume approach in continuum modeling of random heterogeneous media is similar to unit cell approaches in the theory of solids with crystalline ordering.

The *local overpotential* $\eta(x)$ is a measure of electrode departure from equilibrium in an REV at position $x$ in the porous electrode. By definition, $\eta$ is given as

$$\eta(x) = \phi(x) - \Phi(x) - \left(\phi^{eq} - \Phi^{eq}\right). \tag{1.39}$$

Noting that $\phi^{eq} - \Phi^{eq} = E_{1/2}^{eq}$, where $E_{1/2}^{eq}$ is the equilibrium potential of the half-cell reaction, Equation 1.39 takes the form

$$\eta(x) = \phi(x) - \Phi(x) - E_{1/2}^{eq}. \tag{1.40}$$

In the cathode catalyst layer, the metal phase potential $\phi$ must be lowered relative to its value at equilibrium in order to enhance the rate of the ORR. The true cathode overpotential $\eta_{ORR}$ is thus negative (Figure 1.9). Note that in many cases, it is convenient to work with the positive ORR overpotential, which is $|\eta_{ORR}| = -\eta_{ORR}$. The anode overpotential has a positive value. For example, if the anode is grounded ($\phi^a = 0$), $\Phi$ is negative, while $\eta_{HOR} = 0 - \Phi$ is positive.

A simple but intuitive way to illustrate potential distributions and current fluxes in a porous electrode is the *transmission line model* (TLM) that was developed by R. De Levie in the 1960s (de Levie, 1964; Levie, 1963). Figure 1.10 shows the transmission-line equivalent circuit for the CCL under steady-state current flux. Resistances due to electron transport in the metal phase, $R_M$, proton transport in

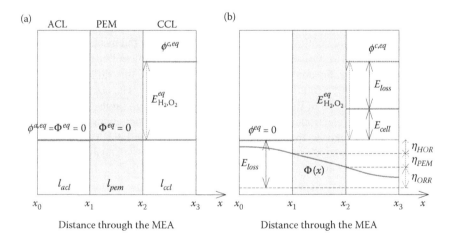

**FIGURE 1.9** Potential distribution of a PEFC with porous electrodes of finite thickness at (a) equilibrium and (b) under load. The metal phase potentials in the electrodes (horizontal lines below the label $\phi^{a,eq}$ in ACL and $\phi^{c,eq}$ in CCL) are constant along $x$, but shifted as a function of current density. The electrolyte phase potential (continuous line in (b) labeled with $\Phi(x)$) exhibits a continuous decrease from anode to cathode; the shape of this profile depends on the proton conductivity in electrodes and PEM. The total potential loss $\eta_{tot} = \eta_{HOR} - \eta_{ORR} + R_{PEM}j_0$ is the sum of the overpotentials in the anode and cathode, plus the resistive potential loss in the membrane.

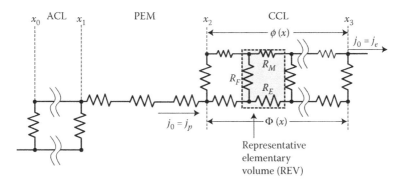

**FIGURE 1.10**  Cathode catalyst layer as a transmission line (see text). (Adapted from Levie, R. De. 1963. *Electrochim. Acta*, **8**(10), 751–780.)

ionomer phase, $R_E$, and faradaic processes at interfaces between these phases, $R_F$, determine the potential distribution in the electrode.

Three of these resistive elements define an REV (or unit cell) of the electrode, as indicated by the dotted box. Current arriving at the CCL from the PEM side is entirely due to proton flux, $j_0 = j_p(x_2)$. At the right boundary of the CCL, this current has been converted completely into an electron flux, $j_0 = j_e(x_3)$. Current conversion takes place in the REVs along the electrode thickness. Owing to opposite signs of proton and electron charges, both electron and proton currents flow from left to right, while potentials in electrolyte and metal phases decrease from left to right. Usually, high metal phase conductivity can be represented by $\phi(x) = $ const, corresponding to $R_M \approx 0$.

For the discrete TLM, current and potential distributions can be obtained from solving algebraically the set of Kirchhoff equations for the given equivalent circuit. If the thickness of an REV can be chosen as very small compared to the electrode thickness, properties of the porous electrode can be studied with high accuracy by taking the continuum limit of the TLM. In this limit, the thickness of an REV goes to zero and the number of REVs to infinity in such a way that the electrode thickness remains finite; the discrete Kirchhoff equations are replaced by a set of first-order differential equations, the solution of which will give potential and current distributions, see the section "Physical Modeling of Catalylst Layer Impedance" in Chapter 5.

The total overpotential of the CCL corresponds to the potential drop between positions $x_3$ and $x_2$, in Figure 1.10, taken at finite current density, minus the same difference taken at equilibrium:

$$\eta_{ORR} = \phi(x_3) - \Phi(x_2) - \left(\phi^{c,eq} - \Phi^{eq}\right)$$
$$= \phi(x_3) - \Phi(x_2) - E_{ORR}^{eq}. \tag{1.41}$$

The overpotential contributions incurred by PEM and ACL in Figure 1.9 are

$$\eta_{PEM} = \Phi(x_1) - \Phi(x_2) \tag{1.42}$$

$$\eta_{HOR} = \phi(x_0) - \Phi(x_1) - \left(\phi^{a,eq} - \Phi^{eq}\right)$$

$$= \phi(x_0) - \Phi(x_1) - E_{HOR}^{eq}, \tag{1.43}$$

respectively. It shall be left to the reader to verify that the following intuitive relation holds:

$$E_{\text{cell}}(j_0) = E_{O_2,H_2}^{eq} - \eta_{HOR} - \eta_{PEM} - |\eta_{ORR}|$$

$$= \phi(x_3) - \phi(x_0). \tag{1.44}$$

Absolute values of potentials in metal and electrolyte phases do not matter; besides, they cannot be measured. For the determination of catalyst layer local overpotentials, it only matters by how much the local values of $\phi$ and $\Phi$ deviate from their equilibrium values.

To complete this section, it should be noted that any transport loss in the fuel cell translates into the growth of the electrode overpotential. This statement can be illustrated using the Tafel equation for the rate of electrochemical conversion. Suppose that the reaction rate is uniform through the electrode depth; then, the current density produced in the electrode $j_0$ is simply a product of the Tafel reaction rate by the electrode thickness $l_{CL}$:

$$j_0 = l_{CL} i^0 \left(\frac{c_t}{c_h^0}\right) \exp\left(\frac{|\eta_0|}{b}\right).$$

Oxygen transport loss in a cell lowers the oxygen concentration $c_t$ in the catalyst layer. From the equation above it is obvious that if $c_t$ decreases, $|\eta_0|$ must increase to maintain $j_0$.

## HEAT PRODUCTION AND TRANSPORT

### HEAT PRODUCTION IN THE CATHODE CATALYST LAYER

The largest source of *heat generation* in PEFCs is the ORR in the cathode CL. The volumetric rate of heat production in this reaction $\mathcal{Q}_{ORR}$ (W cm$^{-3}$) is given by

$$\mathcal{Q}_{ORR} = \left(\frac{T|\Delta S_{ORR}|}{F} + |\eta_{ORR}|\right) R_{ORR}, \tag{1.45}$$

where $T$ is the cell temperature, $\Delta S_{ORR}$ is the entropy change in the ORR per mole of electrons,[*] and $R_{ORR}$ is the ORR rate (total charge of protons converted in the reaction in unit volume per second, A cm$^{-3}$).

---

[*] Note that this value is four times less than the usually used entropy change per mole of oxygen.

The first term in Equation 1.45 represents the reversible thermodynamic heat of the ORR. At equilibrium, the reaction runs spontaneously in both reduction and oxidation directions, and the heat $T\Delta S_{ORR}/F$ is released or consumed in every reaction event, depending upon the event direction. The change in the reaction direction changes the sign of $\Delta S_{ORR}$. In an operating fuel cell (under load), $\Delta S$ is negative, and $|\Delta S_{ORR}|$ will be used in the expressions below, keeping in mind that this is the source of heat. The second term in Equation 1.45 is the rate of irreversible heating associated with the charge transfer in the ORR.

It is insightful to estimate $Q_{ORR}$ in Equation 1.45. $|\Delta S_{ORR}|$ can be taken from Table 1.1; the characteristic temperature of PEFC operation is 350 K. These numbers provide an estimate $T|\Delta S_{ORR}|/F \simeq 0.3$ V. A typical value of $\eta_{ORR}$ in a working cell is between 0.3 and 0.5 V, so that both terms in Equation 1.45 contribute amounts of the same order to the total rate of heat generation.

Another source of heat is Joule heating generated by current passage through the cell. The electronic conductivity of the catalyst layer is two orders of magnitude larger than the proton conductivity, and the heat released in the electron-conducting phase can be ignored. The local rate of the *Joule heating*, $Q_J$, in the CCL is given by

$$Q_J = \frac{j^2}{\sigma_p}, \tag{1.46}$$

where $j$ is the local proton current density and $\sigma_p$ is the proton conductivity of the electrolyte phase.

The parameters $\eta_{ORR}$, $j$, and $R_{ORR}$ in Equations 1.45 and 1.46 are functions of the distance through the catalyst layer. The total heat flux from the CCL should be calculated based on the solution of the CCL performance problem; this will be done in the section "Heat Flux from the Catalyst Layer."

## HEAT PRODUCTION IN THE MEMBRANE

The only source of heat in the membrane is the Joule heating produced by the proton current. The volumetric rate of heat production is given by $j_0^2/\sigma_{PEM}$, where $j_0$ is the cell current density, and $\sigma_{PEM}$ is the bulk membrane proton conductivity. Assuming that the membrane is uniformly hydrated, the potential drop in the membrane is $\eta_{PEM} = R_{PEM}j_0 = l_{PEM}j_0/\sigma_{PEM}$, where $l_{PEM}$ is the membrane thickness. Taking $j_0 = 1$ A cm$^{-2}$, $\sigma_{PEM} = 0.1$ $\Omega$ cm$^{-1}$ (the conductivity of a fully hydrated Nafion® membrane) and the membrane thickness of $2.5 \cdot 10^{-3}$ cm (25 µm), we get $\eta_{PEM} = 0.025$ V. In the section "Heat Flux from the Catalyst Layer" in Chapter 4, we will see that $R_{ORR}$ in Equation 1.45 can be estimated as $j_0/l_{CL}$. On the other hand, the rate of the Joule heating in the membrane is $\eta_{PEM}j_0/l_{PEM}$. As $l_{CL} \simeq l_{PEM}$, we can simply compare $\eta_{PEM}$ with the potentials in Equation 1.45. Under normal operating conditions, $\eta_{PEM} \simeq 0.1\eta_{ORR}$, and, hence, the heat released in the membrane is small. Note that this estimate is correct if the membrane is fully hydrated. Under strong drying, heat production in the membrane can be an order of magnitude larger and it cannot be ignored.

## WATER EVAPORATION

Evaporation of liquid water produced in the ORR contributes to cell cooling. The rate of heat consumption $Q_{vap}$ (W cm$^{-3}$) by *liquid water evaporation* is obtained if we multiply the molar transfer rate related to evaporation by the respective enthalpy change $\Delta H_{vap}$ (Natarajan and Nguyen, 2001). This yields

$$Q_{vap} = -K_{vap}\xi_{CL}^{lv}\left(\frac{\Delta H_{vap}\rho_w}{M_w}\right)(P^s - P^v),\qquad(1.47)$$

where $K_{vap}$ is the evaporation rate constant, $\xi_{CL}^{lv}$ the total liquid/vapor interfacial area per total geometric electrode surface area, $\Delta H_{vap}$ the enthalpy of water evaporation, $\rho_w$ the liquid water density, $M_w$ the molecular weight of water, $P^s$ the pressure of saturated water vapor, and $P^v$ the partial pressure of water vapor in the CL. Note that if $P^v > P^s$, Equation 1.47 changes the sign and it gives the rate of heat production during water condensation.

Physically, $\xi_{CL}^{lv}$ is a heterogeneity factor for vaporization exchange in the partially saturated porous electrode, while $\Delta H_{vap}\rho_w/M_w$ is the energy consumed for the evaporation of a unit volume of liquid water; $(P^s - P^v)$ is the "driving force" for evaporation or condensation (depending on the sign), which is proportional to the difference of saturated and actual water vapor pressures. The product $K_{vap}\xi_{CL}^{lv}$ describes the kinetics of water phase transformation in the porous media, averaged over the REV volume.

The pressure of saturated water vapor is given by

$$P^v = P_0\tilde{p}(T),\qquad(1.48)$$

where $P_0 = 1$ atm, and

$$\tilde{p}(T) = \exp\left(a_0 + a_1(T - 273K) + a_2(T - 273K)^2 + a_3(T - 273K)^3\right)\qquad(1.49)$$

accurately fits the Arrhenius-like temperature dependence (the coefficients $a_0, \ldots, a_3$ are given in Table 1.2).

Written in the form of Equation 1.47, the rate $Q_{vap}$ is determined by the liquid/vapor interfacial area, the temperature, and the density of water vapor in the CL. It does not explicitly depend on the cell current density. However, the dependence on $j_0$ is "hidden" in the structural parameter $\xi_{CL}^{lv}$, which is a function of the current-dependent water accumulation in the CCL.

---

## TABLE 1.2
## The Coefficients in Equation 1.49

| $a_0$ | $a_1$ | $a_2$ | $a_3$ |
|---|---|---|---|
| $-2.1794\ln(10)$ | $0.02953\ln(10)$ | $-9.1837 \times 10^{-5}\ln(10)$ | $1.4454 \times 10^{-7}\ln(10)$ |

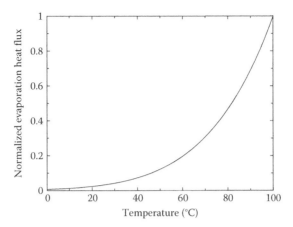

**FIGURE 1.11** The dimensionless heat flux due to liquid water evaporation in the catalyst layer. The curve is obtained with the data from Table 1.3 and $P_v = 0$, which corresponds to zero pressure of water vapor in the CCL (the limit of fast vapor removal). The heat flux is normalized to that flux at 100°C.

The *heat flux* related to water evaporation in the CCL, $q_{vap}$, is obtained by multiplying $Q_{vap}$ by the CCL thickness $l_{CL}$. Unfortunately, the evaporation constant $K_{vap}$ is poorly known, which makes numerical evaluation of $q_{vap}$ rather unreliable. Figure 1.11 shows the dimensionless heat flux due to water evaporation in the CCL, normalized with respect to this flux at 100°C, assuming $P_v = 0$, that is, fast water vapor removal. As can be seen, the heat flux $q_{vap}$ is a strong function of the cell temperature. Evaporation is an important mechanism of cell/stack cooling, and Figure 1.11 provides another argument in favor of cell operation at an elevated temperature.

The order-of-magnitude estimate of the heat flux $q_{vap}$ can be done using the following arguments. Suppose that all water is vaporized in the CCL; the heat consumed

---

**TABLE 1.3**

**Heat Transport Parameters $\lambda_T$ Is Taken From Khandelwal and Mench (2006); the Product $K_{vap}\xi_{CL}^{lv}$ Corresponds to the Rate Constant Used in Natarajan and Nguyen (2001)**

| | |
|---|---|
| Thermal conductivity of CL, $\lambda_T$ (W m$^{-1}$ K$^{-1}$) | 0.27 |
| Liquid water density $\rho_w$ (kg m$^{-3}$) | $10^3$ |
| Molecular weight of water $M_w$ (kg mol$^{-1}$) | 0.018 |
| Heat capacity of liquid water $c_{pw}$ (J kg$^{-1}$K$^{-1}$) | 4190 |
| Enthalpy of water evaporation $\Delta H_{vap}$ (J mol$^{-1}$) | $41.7 \times 10^3$ |
| Evaporation rate constant $K_{vap}$ (atm$^{-1}$ s$^{-1}$) | 0.01–1 |
| Liquid/vapor interfacial area factor $\xi_{CL}^{lv}$ | 10–100 |

by vaporization is $\frac{1}{2} \cdot 42$ kJ mol$^{-1}$ = 21 kJ mol$^{-1}$ of electrons (Table 1.3). Assuming that the total overpotential of the cathode side is $\eta_0 \simeq 0.3$ V, the amount of heat produced in the CCL according to Equation 1.45 is $\simeq 60$ kJ per mol of electrons at 350 K. Thus, complete evaporation of liquid water produced in the ORR removes about one-third of reaction heat.

## HEAT TRANSPORT EQUATION

Generally, two mechanisms contribute to heat transport in the PEFC: heat conduction in solid media and convection due to the transport of liquid water. The estimate in the section "Remarks" in Chapter 4 shows that the heat flux associated with the transport of liquid water is not large. It is less than 20% of the total heat flux leaving the CCL. Typically, the nature of the heat flux from the CCL is of minor interest, as only a total value of this flux appears in the heat balance equations for the whole cell or cell components. Note also that owing to the large liquid/metal interfacial area and sufficient residence time of liquid in the porous media, thermal equilibrium between liquid water and solid phase is usually established. Hence, separate equations will not be formulated for the heat transport in solid and liquid phases. Instead, a single equation for the single solid/liquid phase temperature $T$ will be written, taking into account all the heat sources and sinks.

According to Fourier law, the heat flux in the solid media is proportional to the temperature gradient, $q = -\lambda_T \nabla T$, where $\lambda_T$ is the media thermal conductivity. Divergence of the conductive heat flux, $\nabla \cdot (-\lambda_T \nabla T)$, equals the sum of volumetric sources. In the catalyst layer through-plane direction, this equation reads

$$-\lambda_T \frac{\partial^2 T}{\partial x^2} = Q_{ORR} + Q_J + Q_{vap}, \tag{1.50}$$

where the terms on the right-hand side have been discussed above.

In general, Equation 1.50 is nonlinear, as $R_{ORR}$ and $j$, which appear in the source terms (1.45) and (1.46), exponentially depend on temperature. However, the CCL is thin, and the temperature variation along $x$ is usually very small. To a good approximation, the CCL temperature *in the source terms* can be taken to be constant. This leads to a simple general expression for the heat flux produced in the CCL, which will be derived in the section "Heat Flux from the Catalyst Layer" in Chapter 4. This expression can be used in the modeling of application-relevant cells and stacks, in which quite significant in-plane temperature gradients may arise based on a nonuniform distribution of reactants.

The thermal conductivity of a porous medium depends on the material composition and structure. There has been a number of works on modeling the thermal conductivity of porous GDLs (Sadeghi et al., 2008, 2011). Cell and stack modeling usually employ experimental data on $\lambda_T$, which can be found in Khandelwal and Mench (2006).

## BRIEF DISCOURSE ON FUEL CELL ELECTROCATALYSIS

The first primitive cells of Schoenbein and Grove utilized platinum wire electrodes. Today, high-performance PEFC stacks are approaching commercial readiness in transportation applications, yet Pt remains as their critical materials component. Pt exhibits a benign combination of an exceptional activity for hydrogen oxidation and oxygen reduction reactions, and an unusual corrosion resistance under strongly acidic conditions in the fuel cell (Debe, 2012).

In spite of an endless research history, basic questions pertaining to Pt's exceptional properties remain

- What physical and chemical properties make Pt such an excellent electrocatalyst?
- What is the relationship between the size, shape, composition, and atomistic structure of Pt-based nanomaterials and their electrocatalytic activity?
- What interfacial properties and reaction conditions cause intermediates of the ORR to act as promoters or inhibitors?
- How do Pt oxide layers affect the mechanism and kinetics of the ORR?
- What is the rate of Pt dissolution, what factors catalyze or inhibit the dissolution of Pt and/or Pt oxides, and what happens to the dissolution products?
- How do applied potential, oxide layers, or adsorbed oxygen intermediates modify the wettability of electrodes?
- How do chemical composition and electronic properties of metallic electrodes interrelate and how do they determine the reaction mechanism and kinetics?
- What is the synergy between support material and Pt electrocatalysts?
- How does the composition of the electrolyte, and the presence of specific additives and impurities, influence the direction of electrochemical processes?
- Can nonnoble electrocatalysts compete with Pt in terms of the low-activation-energy reaction path, as well as the density of active catalyst sites?

Fuel cell engineers and entrepreneurs might prefer to wink at this thought, but success or failure of fuel cell technology might come down to the scientific community's ability of, first, finding answers to these questions and, second, transforming them into markedly improved materials and fuel cell designs.

Fuel cell modeling is particularly concerned with the ORR that occurs at the cathode. The ORR and its inverse process, the water-splitting or oxygen evolution reaction, are among the most important reactions in electrochemistry. The overall ORR reaction requires the exchange of four electrons between catalyst and protons in solution. It involves formation and reduction of several adsorbed *oxygen intermediates*, which compete for free catalyst surface sites with adsorbed oxygen-containing species that form during oxidation of water.

An overarching requirement for any Pt-based catalyst is that it must be able to successfully navigate the delicate balance between ORR activity and durability. The "catch-22" is that ORR activity and Pt dissolution cannot be approached separately. Both processes proceed via formation of surface oxide species at Pt (Rinaldo et al., 2010, 2012, 2014).

Specific processes involved in the ORR will be dealt with in the section "Electrocatalysis of the Oxygen Reduction Reaction at Platinum" in Chapter 3. Here, the focus is on basic phenomenological concepts that are required to incorporate electrocatalytic reactions in device-level modeling of PEFCs. For a more detailed treatment of fundamental concepts in electrocatalysis, see the recent edition of the textbook by Schmickler and Santos (2010), as well as the classical textbook by Bard and Faulkner (2000).

Concepts will be explained for a generic electrode process of the type

$$O + e^- \rightleftarrows R, \qquad (1.51)$$

where O and R represent oxidized and reduced electroactive species. These electron acceptor and donor species are available in the electrolyte with concentrations $c_O$ and $c_R$.

## BASIC CONCEPTS OF ELECTROCATALYSIS

The term "electrocatalysis" refers to the acceleration of the rate of an electrochemical reaction taking place at a solid electrode. The main role of the catalyst material is to lower the activation energy for electron transfer between energy levels of electroactive species on the electrolyte side and electronic states at the Fermi level in the metal. Although the electrode itself does not undergo any chemical transformation, it participates in the reaction indirectly by acting as a reservoir of electrons. Moreover, the catalyst surface provides active sites for the adsorption of reaction intermediates.

A good electrocatalyst should possess the following characteristics: (i) high electrical conductivity, (ii) chemical stability, (iii) mechanical stability, (iv) large surface area, (v) long lifetime, (vi) low production cost, and (vii) abundant availability. The last point is a major concern in commercialization scenarios for PEFC technology since Pt ranks highly among the rarest elements in the Earth's continental crust.

Two electrochemical parameters define the electrochemical properties of an electrocatalyst material: the *intrinsic exchange current density* $j_*^0$ and the *Tafel parameter* $b$. These parameters are determined from current–potential relationships measured by voltammetric methods of classical electrochemistry (Bard and Faulkner, 2000). One could write an expression for $b$ of the form

$$b = \frac{R_g T}{\alpha_{\text{eff}} F} \qquad (1.52)$$

with an effective *electron transfer coefficient* $\alpha_{\text{eff}}$ that will be explained below.

For simple electrode processes, which involve the exchange of a single electron between electroactive redox species in solution and metal surface, $\alpha_{\text{eff}}$ and $b$ have a simple interpretation. For so-called outer-sphere electron transfer reactions, which do not involve adsorption and desorption processes and leave the chemical structure of electroactive species, including their solvation shell, unaffected, values $\alpha = 1/2$ and $b \approx 50$ mV (at room temperature) are expected. For multielectron transfer processes involving adsorbed intermediates, like the four-electron pathway of the ORR, $\alpha_{\text{eff}}$ assumes a sequence of discrete values between 1 and $1/2$ as the overpotential $\eta$ increases. The values of $\alpha_{\text{eff}}$ can be found from an analysis of reaction pathways and mechanisms, as discussed in the section "Deciphering the ORR" in Chapter 3. As the reaction pathway evolves in discrete ranges of $\eta$, not only must one find a value of $b$ for a given potential range, but also the corresponding value of $j_*^0$.

A good catalyst should exhibit high value of $j_*^0$ and a small value of $b$. A low value of $b$ is beneficial, since it allows for a reduction of the overpotential $\eta$, which drives an electrochemical process. An effective electrocatalyst can deliver a large current density at low $\eta$. In energy-generating electrochemical devices, such as PEFCs, a large $\eta$ translates into significant power losses. A low value of $j_*^0$ demands catalyst layer designs that enhance the catalyst surface area per unit of the fuel cell active area.

## ELECTROCHEMICAL KINETICS

The main function of a fuel cell electrode is to convert a chemical flux of reactants into fluxes of charged particles, or vice versa, at the electrochemical interface. Electrochemical kinetics relates the local interfacial current density $j$ to the local interfacial potential drop between metal and electrolyte phases, illustrated in Figure 1.8. A deviation of the potential drop from equilibrium corresponds to a local overpotential $\eta$ at the interface, which is the driving force for the interfacial reaction. The reaction rate depends on overpotential, concentrations of active species, and temperature. For the remainder of this section, it is assumed that the metal electrode material is an ideal catalyst, that is, it does not undergo chemical transformation and serves as a sink or source of electrons. The basic question of electrochemical kinetics is: how does the rate of interfacial electron transfer depend on the metal phase potential?

A simple one-electron transfer reaction proceeds isoenergetically between electronic states at the Fermi level of the metal electrode and donor or acceptor states of the species R and O in Equation 1.51. Figure 1.12 shows an intuitive explanation of a simple outer-sphere one-electron transfer process. The vertical axis is the single-electron energy scale. On the metal side, electronic states are filled up to the Fermi level, $\varepsilon_F$.[*] The density of metal electrons with energy $\varepsilon_F$ is given by the electron density of states, which can be calculated using electronic structure theory (Ashcroft and Mermin, 1976).

On the electrolyte side, Figure 1.12 shows the probability density distributions $W$ for finding a single R or O species in the electrolyte with energy $\varepsilon$ of its donor or acceptor state. The maxima of these distributions are shifted by $2\lambda$, where $\lambda$ is the solvent reorganization energy. Probability density distributions of finite width arise because of thermal fluctuations of the solvation state of R and O in solution. They

---

[*] Strictly speaking, the Fermi energy is defined at zero temperature. However, at room temperature, deviations of the electrochemical potential from the Fermi energy are small (Ashcroft and Mermin, 1976).

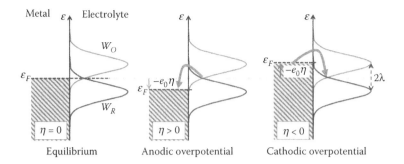

**FIGURE 1.12** Schematic of a simple outer-sphere one-electron transfer process at the metal/electrolyte interface.

could be represented by Gaussian distribution functions (Figure 1.12). The energy, at which the probability densities for R and O are equal, is often referred to as a pseudo-Fermi energy of the electrolyte, $\varepsilon_F^{el}$. As should be clear from this explanation, $\varepsilon_F^{el}$ does not represent an eigenenergy of a quantum mechanical electron state; it is a thermal average over the combined probability density distributions for energies of donor and acceptor states in R and O species, and depends thus on thermal fluctuations.

Equilibrium for the single-electrode system implies that reactive electrons on metal side and on the species R in solution possess equal total energy (electrochemical potential), which means that $\varepsilon_F = \varepsilon_F^{el}$. Therefore, $\varepsilon_F^{el}$ can be used to define the equilibrium electrode potential $\varepsilon_F^{el} = -e_0 E^{eq}$, where $e_0$ is the electron charge. Now, let us consider a potential shift at the electrode, which will result in a new metal phase potential $\phi = \phi^{eq} + \Delta\phi$. This potential change shifts the Fermi level of the metal by $\Delta\varepsilon_F = -e_0\Delta\phi$. On the electrolyte side, $\varepsilon_F^{el}$ is considered independent of $\phi$; in practice, this assumption implies that a concentration of electrolyte is high and it effectively screens the electrode potential. Under these conditions, the interfacial potential drop on the electrolyte side will occur in the Stern layer. The potential drop in the diffuse double layer is negligible, and the Galvani potential in solution is not affected by a variation of the metal phase potential. Thermodynamically, with the application of $\phi$, the reaction Gibbs energy of the forward reaction in Equation 1.51 changes by $\Delta G = -F\Delta\phi$. A positive value $\Delta\phi > 0$ corresponds to $\Delta G < 0$, meaning that it will accelerate the process to proceed in anodic direction, with an electron transferred from R in solution to the metal, leaving an O species in the solution. Electrochemically, the electrode potential shifts to a new value $E = E^{eq} + \Delta\phi$. This means that $\Delta\phi = \eta$, that is, the shift in the Galvani potential of the metal phase, is equal to the overpotential of the electrode reaction. An anodic process requires $\Delta\phi = \eta = E - E^{eq} > 0$, while for a cathodic process, $\Delta\phi = \eta = E - E^{eq} < 0$.

At equilibrium, the rates of anodic and cathodic charge transfer are exactly balanced and the net rate of interfacial electron transfer processes is zero. Out of equilibrium, at $\eta \neq 0$, the net rate $v_{net}$ is given by

$$v_{net} = K_R c_O^S - K_O c_R^S. \tag{1.53}$$

Here, $c_O^s$ and $c_R^s$ are surface concentrations of oxidized and reduced species, and $K_O$ and $K_R$ are the oxidation and reduction rate constants, respectively. By definition, the net rate or current of a cathodic reaction (left to right in Equation 1.51) will be positive.

In order to understand how fast the reaction proceeds, one needs to determine the rate constants in Equation 1.53. Using transition state theory, these rate constants can be written as

$$K_R = A \exp\left(-\frac{\Delta G_R^{\ddagger}(E)}{R_g T}\right) \text{ and } K_O = A \exp\left(-\frac{\Delta G_O^{\ddagger}(E)}{R_g T}\right). \tag{1.54}$$

In the next step, one must determine how the molar Gibbs energies of activation, viz., $\Delta G_R^{\ddagger}(E)$ and $\Delta G_O^{\ddagger}(E)$, depend on $E$. It is convenient to choose the standard equilibrium potential $E^0$ as a reference. As illustrated in Figure 1.12, the reaction Gibbs energy changes by an amount $\Delta G_r(E) = F\eta = F\left(E - E^0\right)$. Validity of the Brønsted–Evans–Polanyi relation implies a linear relation between reaction Gibbs energy and activation Gibbs energy:

$$\Delta G_R^{\ddagger}(E) = \Delta G_R^{\ddagger}\left(E^0\right) + \alpha F\left(E - E^0\right),$$

$$\Delta G_O^{\ddagger}(E) = \Delta G_O^{\ddagger}\left(E^0\right) - \beta F\left(E - E^0\right), \tag{1.55}$$

with dimensionless cathodic transfer coefficient $\alpha$ and anodic transfer coefficient $\beta$. In general, both of these transfer coefficients are functions of $E$. They could be rationalized as coefficients that arise from linear terms in Taylor expansions of $\Delta G_{O,R}(E)$ around $E = E^0$, which means that

$$\alpha = \frac{1}{F} \left.\frac{\partial \Delta G_R^{\ddagger}}{\partial E}\right|_{E=E^0} \text{ and } \beta = -\frac{1}{F} \left.\frac{\partial \Delta G_O^{\ddagger}}{\partial E}\right|_{E=E^0}. \tag{1.56}$$

Higher-order contributions to expressions in Equation 1.55 become significant at larger deviations of $E$ from $E^0$. Typical values of $\alpha$ and $\beta$ are between 0 and 1. For simple outer-sphere one-electron transfer reactions that do not involve adsorbate formation at the metal, the relation $\alpha + \beta = 1$, thus $\beta = 1 - \alpha$, is valid and both values lie close to 1/2. More rigorous derivations of values of $\alpha$ and $\beta$ utilize concepts of electron transfer theory that explicitly treat effects of solvent reorganization during charge transfer. Basic concepts underlying Marcus–Hush theory, as well as intuitive concepts of Gerischer theory are discussed in Schmickler and Santos (2010).

The trends of kinetic rates of electron transfer processes due to changes in electrode potential $E$ are now obvious. Increasing $E$ relative to $E^o$ decreases the value of $\Delta G_O^{\ddagger}$ and accelerates the anodic reaction; decreasing $E$ relative to $E^o$ accelerates the cathodic reaction. Using this formalism, it is straightforward to determine the current density of the generic electrode reaction in Equation 1.51, in units of A cm$^{-2}$,

$$j = F\nu_{net} = F\left[K_R c_O^s - K_O c_R^s\right].\tag{1.57}$$

Using $\beta = 1 - \alpha$ and

$$k^0 = A\exp\left(-\frac{\Delta G^{\ddagger,0}}{R_g T}\right), \text{ with } \Delta G^{\ddagger,0} = \Delta G_O^{\ddagger,0} = \Delta G_R^{\ddagger,0},\tag{1.58}$$

one finds

$$j = Fk^0\left\{c_O^s\exp\left(-\frac{\alpha F\left(E - E^0\right)}{R_g T}\right) - c_R^s\exp\left(\frac{(1 - \alpha)F\left(E - E^0\right)}{R_g T}\right)\right\}.\tag{1.59}$$

This is the famous Butler–Volmer equation. Incorporating the Nernst equation, which relates the equilibrium electrode potential to the standard equilibrium potential and to the equilibrium composition of the bulk electrolyte (concentrations with superscript $b$) via

$$E^{eq} = E^0 + \frac{R_g T}{F}\ln\left(\frac{c_O^b}{c_R^b}\right)\tag{1.60}$$

gives

$$j = j_*^0\left\{\frac{c_O^s}{c_O^b}\exp\left(-\frac{\alpha F\eta}{R_g T}\right) - \frac{c_R^s}{c_R^b}\exp\left(\frac{(1 - \alpha)F\eta}{R_g T}\right)\right\}\tag{1.61}$$

with the exchange current density

$$j_*^0 = Fk^0\left(c_O^b\right)^{1-\alpha}\left(c_R^b\right)^{\alpha}.$$

In the special case that mass transport from the bulk electrolyte to the surface of the electrode is fast, implying that the surface concentrations deviate only insignificantly from the bulk concentrations, we obtain the simplest form of the Butler–Volmer equation,

$$j = j_*^0\left\{\exp\left(-\frac{\alpha F\eta}{R_g T}\right) - \exp\left(\frac{(1 - \alpha)F\eta}{R_g T}\right)\right\}.\tag{1.62}$$

Which of these equations is the appropriate form to use for the modeling of reactions in electrochemical devices? It depends on electrode configuration and conditions. Equation 1.62, is the least general form. It can only be used, if all sorts of mass transport limitations could be neglected, that is, for uniform distributions of reactants, simple non-porous electrode geometries, or well-stirred electrolyte solutions. Equations 1.61 and 1.59 are equivalent, provided that the bulk concentrations in Equation 1.61 are those at equilibrium conditions.

For a specific reaction, a standard exchange current density can be defined by $j^{00} = Fk^0 \cdot 1$ mol L$^{-1}$. Values of $j^{00}$ span a range of many orders of magnitude for different reactions. One-step single-electron transfer processes have large values of $j^{00}$, so that even a small deviation of the electrode potential from equilibrium results in a large current density. Multistep electron transfer processes that involve strongly adsorbed reaction intermediates, like the ORR, exhibit small values of $j^{00}$ and they possess different kinetic regimes with different effective values of $\alpha_{eff}$ and $j^{00}_{eff}$. In this case, the relation $\alpha + \beta = 1$ is not valid anymore.

The *limit of small overpotential* is encountered when $|\eta| < R_g T/F$. This condition usually coincides with negligible *mass transport limitations*, that is, uniform reaction conditions. A Taylor expansion of Equation 1.62 gives

$$j = R_{ct}\eta, \text{ with } R_{ct} = \frac{j^0_* F}{R_g T}, \tag{1.63}$$

which resembles the Ohm's law with a *charge transfer resistance $R_{ct}$*.

The case $|\eta| \gtrsim 3R_g T/F$ is the so-called *Tafel regime*. In this regime, one of the partial currents in Equation 1.59 dominates. Assuming for simplicity that mass transport limitations are still negligible, the Tafel relation is obtained for the cathodic current at $\eta < 0$,

$$\log_{10}|j| = \log_{10}(j^0_*) - \frac{\alpha F}{2.3R_g T}\eta \tag{1.64}$$

and for the anodic current at $\eta > 0$,

$$\log_{10}(j) = \log_{10}(j^0_*) + \frac{(1-\alpha)F}{2.3R_g T}\eta. \tag{1.65}$$

A Tafel plot shows $\log_{10}|j|$ as a function of $\eta$. The slope of a Tafel plot gives $\alpha$ and the crossing of this linear fit with the ordinate gives $j^0_*$.

The oxygen reduction reaction is notoriously sluggish. The faradaic current density for this reaction is given by the cathodic branch of the Butler–Volmer equation:

$$j = j^0_* \left(\frac{c_{O_2}}{c^0_{O_2}}\right)^{\gamma_{O_2}} \left(\frac{c_{H^+}}{c^0_{H^+}}\right)^{\gamma_{H^+}} \exp\left(-\frac{\alpha_{eff}F}{R_g T}\eta\right), \tag{1.66}$$

where $j^0_*$ is the exchange current density corresponding to reference concentrations $c^0_{O_2}$ and $c^0_{H^+}$. The value of the exchange current density reported in the literature is in the order of $j^0_* \sim 10^{-8}$–$10^{-9}$ A $cm^{-2}_{Pt}$, which is 5–6 orders of magnitude lower than that of the HOR (Neyerlin et al., 2006). The reaction order of oxygen is usually considered as $\gamma_{O_2} = 1$. The reaction order of protons, $\gamma_{H^+}$, depends on the potential range and the corresponding adsorption conditions. For conventional ionomer-impregnated catalyst layers, the proton concentration is determined by the amount of ionomer. If the proton concentration is sufficiently high, its variation in the diffuse double layer is negligible. In this limit, a constant proton concentration can be assumed at the reaction plane. In this case, the second fraction in Equation 1.66 is constant, which is usually included in the exchange current density.

The assumption of constant $c_{H^+}$ fails in water-filled agglomerates, considered in the section "Hierarchical Model of CCL Operation" in Chapter 4 and in water-filled pores of ionomer-free UTCLs layers, studied in the section "ORR in Water-Filled Nanopores: Electrostatic Effects" in Chapter 3. In these cases, proton concentration must be considered explicitly as a variable, and it exerts a strong impact on the effectiveness factor of Pt utilization.

## KEY MATERIALS IN PEFC: POLYMER ELECTROLYTE MEMBRANE

To a large degree, the PEM determines the operational range of a fuel cell, that is, temperature range, pressures, and humidification requirements. Peculiar properties of the PEM, namely, being gas tight, highly proton conductive, and electronically insulating, are of fundamental importance for the fuel cell principle. This rests upon the spatial separation of partial redox reactions on anode and cathode sides. The membrane should be a medium with high concentration and mobility of protons. Moreover, the membrane should be mechanically and chemically stable over desired operation times.

Fuel cell lifetime requirements range from 3000 to 5000 h for passenger cars, up to 20,000 h for buses, and up to 40,000 h for stationary power generation. Notwithstanding the wide range of lifetime requirements, they have to be accomplished under widely varying operating conditions. Whereas fuel cells for automotive applications need to be compatible with temperatures from $-40°C$ to $100°C$, stationary applications are much less demanding, in this respect, but require the longest lifetimes.

PEMs can contribute to a significant fraction of irreversible voltage losses during fuel cell operation due to their resistance to proton transport and the crossover of reactant gases. The crossover of unreacted fuel from anode to cathode is a major problem of direct alcohol fuel cells, since PEMs that contain large volume fractions of liquid water easily dissolve and transport polar alcohol molecules like methanol or ethanol. The voltage losses in PEMs can be particularly harmful if PEFCs are operated outside of their benign range of operation, that is, under conditions that are too hot ($T > 90°C$) or too dry. In addition to voltage losses incurred directly in PEMs, structure and processes in the polymer electrolyte affect water management in all components and at all scales of PEFCs.

### MEMBRANE RESEARCH

PEM research is a multidisciplinary, hierarchical exercise that spans scales from Ångstrom to meters. It needs to address challenges related to (i) to ionomer chemistry, (ii) physics of self-organization in ionomer solution, (iii) water sorption equilibria in nanoporous media, (iv) proton transport phenomena in aqueous media and at charged interfaces, (v) percolation effects in random heterogeneous media, and (vi) engineering optimization of coupled water and proton fluxes under operation. Figure 1.13 illustrates the three main levels of the hierarchical structure and phenomena in PEMs.

The key objective of PEM research is to rationalize and eventually predict the impact of variations in thermodynamic conditions and fuel cell current density on (i) chemical structure and composition, (ii) water sorption properties, (iii) proton conductivity, (iv) water flux mechanisms and parameters, (v) electrochemical

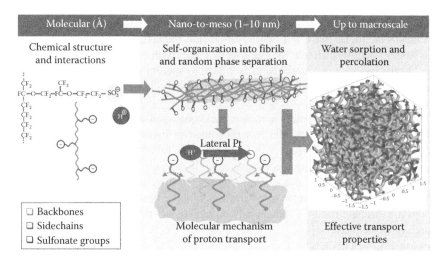

**FIGURE 1.13** The three main levels of the structural hierarchy and associated phenomena in PEMs.

performance, and (vi) degradation mechanisms and rates of the PEM. Specific structural functionalities that steer physical properties and electrochemical performance could be programmed into a material at the fabrication stage; they could as well evolve reversibly during operation in response to changing conditions; moreover, they could deteriorate irreversibly as a consequence of membrane degradation.

Comprehensive and consistent knowledge of these multivariate response functions is a prerequisite for tuning chemical structure and nanoscale morphology of PEMs in view of optimum performance and durability. Striving for complete knowledge of these relations is illusional. It is, however, of utmost importance to rationalize the main principles and limitations of membrane structure and function. This understanding is a prerequisite for tangible predictions of performance and lifetime. Moreover, it provides the basis for the exploration of systematic approaches in membrane design.

## BASIC STRUCTURAL PICTURE

Excellent reviews on chemical structure, morphology and properties of acid-bearing polymers can be found in Mauritz and Moore (2004), Peckham and Holdcroft (2010) and Yang et al. (2008). The base polymer of the prototypical DuPont$^{TM}$ Nafion *perfluorosulfonic acid (PFSA) ionomer*, shown in Figure 1.13(left), consists of a tetrafluoroethylene (TFE) backbone with randomly attached pendant sidechains of perfluorinated vinyl ethers. Sulfonic acid groups are fixed at the sidechain heads (Kreuer et al., 2004; Tanimura and Matsuoka, 2004; Yang et al., 2008; Yoshitake and Watakabe, 2008).

The *equivalent weight* (EW) of Nafion, defined as the mass of dry polymer per mole of ion exchange sites, can be calculated from the formula $EW = (100 \cdot n + 444)$ g mol$^{-1}$. The first number in brackets, 100 g mol$^{-1}$, represents the molar mass of the TFE backbone monomer (CF$_2$CF$_2$), and $n$ the average number of TFE groups between neighboring sidechains. The second additive term in brackets is the molar mass of a single Nafion sidechain ($C_7F_{13}O_5SH$), 444 g mol$^{-1}$. This number will be different for ionomer materials with a different sidechain chemistry. Nowadays, the preferred parameter used to specify the ion content of a PEM is the *ion exchange capacity*, defined as the inverse of the equivalent weight, that is, $IEC = EW^{-1}$.

Upon immersion in water, ionomer backbones self-assemble into fibrillar or ribbon-like aggregates or bundles, as illustrated in Figure 1.13 (middle) (Rubatat et al., 2002). The driving force for this process is the minimization of the free energy of the ionomer–water system, as discussed in the section "Aggregation Phenomena in Solutions of Charged Polymers" in Chapter 2. Sidechains, end-grafted to backbones, form a dense array at hydrated surfaces of these fibrils. The sulfonic acid head groups at sidechain heads spontaneously dissociate and release protons into the surrounding water phase, where they move almost as freely as in bulk water, except at very low levels of hydration. The hydrated array of $SO_3^-$ anions forms a neutralizing environment at the fibril–water interface, relative to which protons move. Protons are distributed according to the rules of hydrogen bonding in water and electrostatic interactions with pore walls; they acquire mobilities that are determined to some extent by dynamic properties of flexible surface groups and predominantly by the aqueous environment.

Hydrophilic domains contain water, hydrated mobile protons, and flexible anionic head groups; they form a continuous percolation network of pores or channels, depicted on the right-hand side of Figure 1.13. Geometric properties of ionomer aggregates and the morphology of the percolating network of hydrophilic pores determine the transport properties of the PEM. These transport properties determine a voltage loss response function, $\eta_{PEM} = f(j_0)$, and they play a key role in water management and degradation phenomena in PEFCs.

## WHO IS A PROTON'S BEST FRIEND?

A proton never travels alone. Proton motion is strongly correlated with its environment, that is usually composed of an electrolyte and a solid or soft host material.

Figure 1.14 displays proton conductivities of various solid-state compounds and liquid electrolytes as well as composite materials like PEM that inherit properties of both solid state and liquid electrolytes. Classifications of *proton-conducting materials* can be made with respect to temperature range of operation, attainable proton conductivity, and transitions in conductivity upon temperature variation. The main groups of proton-conducting materials are:

- *Proton-conducting oxides* acquire proton conductivities $> 10^{-3}$ S cm$^{-1}$ in the temperature range from 400°C to 1000°C. The mechanism of proton conduction rests upon the chemistry of defects in the crystalline configuration.
- *Solid acid proton conductors* exhibit a sharp transition at a critical temperature in the range of 100–200°C to a high-temperature superprotonic state. Conductivities attained in this structurally disordered high-temperature state could approach values of $10^{-1}$ S cm$^{-1}$.

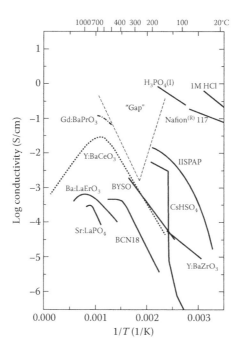

**FIGURE 1.14** Proton conductivities of various solid-state compounds, liquid electrolytes, and composite materials. (Reprinted from Solid State Ionics, 125 (1-4), T. Norby, Solid-state protonic conductors: Principles, properties, progress and prospects, 1–11, Copyright (1999), with permission from Elsevier.)

- *Phosphoric acid-doped (PA) poly-benzimidazole (PBI) films* attain proton conductivities >0.25 S cm$^{-1}$ at temperatures above 150°C (Xiao et al., 2005). The shift to higher temperatures of fuel cell operation, enabled by the use of these membranes, improves the fuel tolerance to carbon monoxide impurities, enhances the electrode kinetics, and alleviates humidification requirements. The problem is that the higher temperature not only accelerates the kinetics of desired electrode reactions, but also those of degradation processes. Moreover, PBI membranes function poorly under ambient conditions. PEFCs using PBI PEMs are suitable for residential applications, but they are currently not envisaged for automotive applications.

- *Water-based PEMs* exhibit proton transport mechanisms and mobilities similar to those in liquid electrolytes like hydrochloric acid; proton conductivities could reach up to 0.1 S cm$^{-1}$ in the case of PFSA-type ionomers, and up to 0.5 S cm$^{-1}$ in the case of block copolymer systems. The temperature range of operation of PEMs stretches from −30°C to 90°C, the lower bound being determined by the freezing point of water, which is suppressed because of the high surface energy of water in nanopores. The upper limit is determined by evaporation of water; only a few water-based PEMs have been demonstrated that could maintain a sufficient conductivity above 100°C.

Water is nature's favorite proton solvent and shuttle. In liquid form, it enables the highest mobility of protons of any known material. The simple explanation is that any excess proton entering the membrane from the anode side of the PEFC could easily switch its role, namely, the privilege to migrate, with any of the protons that water consists of. Thus the concentration of potential charge carriers in water is 110 mol $L^{-1}$. For comparison, phosphoric acid ($H_3PO_4$) offers a total proton concentration of 58 mol $L^{-1}$.

Water is the lifeblood of all living organisms and it works best under benign living conditions. There is thus a perfect adaptation between properties of water and the requirements of natural processes. Nature's primer provides a compelling argument for using water-based proton conductors in technological energy applications, considering that they should operate under the same conditions, at least most of the time.

For water, insights from *ab initio* molecular dynamics methods, generated over the last two decades, have furnished a precise picture of the molecular mechanism of proton migration. Excess proton movement in water consists of sequences of transitions between hydrated proton complexes known as Eigen and Zundel ions. Elementary transitions between these complexes are triggered by rotational fluctuations, as well as hydrogen bond breaking-and-making in the surrounding water network. Nowadays, there is little debate about the validity of this mechanism in water; it has replaced the mechanistic picture of proton motion in water that was put forth by Baron C.J.T. de Grotthuss in 1806 (de Grotthuss, 1806). However, essential elements of Grotthuss' intuitive mechanism have survived until today: for long-range transport, a protonic defect is passed along a line-up of water molecules without the proton ever existing as a separate species.

## COUPLED TRANSPORT OF PROTONS AND WATER

The section "PEM Conductivity: Simply a Function of Composition" in Chapter 2 presents a rudimentary conductivity model, based on Equation 2.1. Local variations of proton density and mobility in pores render the simple factorized form of Equation 2.1 inapplicable at the pore level. In the case of a cylindrical pore of length $L_p$, radius $R_p$, with uniform surface charge density at the pore walls, the conductance $\Sigma_{\text{pore}}$ is given by an integral expression

$$\Sigma_{\text{pore}} = \frac{2\pi}{L_p} \int_0^{R_p} r\, \rho_{H^+}(r) \mu_{H^+}(r)\, dr. \tag{1.67}$$

It accounts for variations in density and mobility of protons along the radial direction. A similar expression including distributions in proton charge density and proton mobility relative to pore walls can be written for slab-like pores:

$$\Sigma_{\text{pore}} = \frac{L_x}{L_y} \int_{-z_0}^{z_0} \mu^+(z) \rho^+(z)\, dz. \tag{1.68}$$

Here, $L_y$ is the length of the pore in the direction of transport, $L_x$ is the in-plane width of the pore perpendicular to the direction of transport, and $\pm z_0$ denote the positions of charged interfacial layers in thickness direction $z$. These expressions assume that proton density and mobility are uncorrelated.

To scale up the single pore conductance to the conductivity of a random network of water-swollen pores in the hydrated PEM, percolation concepts, field theory, and homogenization approaches are employed. These approaches are rooted in the theory of random heterogeneous media (Torquato, 2002). A PEM is, however, an unconventional porous medium. Water sorption incurs continuous changes in pore sizes and shapes as well as in the pore network topology. Because of this reorganization, the total number of pores is not conserved during swelling.

The coupling between mobilities of protons and water by the electroosmotic effect is normally an undesired effect, as the resulting electroosmotic flux shuffles water molecules from the anode to the cathode. Backflux of water toward the anode side caused by diffusion or hydraulic permeation partly balances this flux. However, this flux requires establishment of internal gradients in water concentration or hydraulic pressure, which implies depletion of water at the anode side of the PEM. The drying of the PEM on the anode side by electroosmosis could drastically diminish its conductive abilities, since the conductivity of presently available PEMs is a strong function of local water content.

On the other hand, water traffic toward the cathode side could cause flooding of the porous catalyst layer and gas diffusion media. Pore blockage by liquid water could drastically diminish the rate of oxygen supply toward the electrocatalyst. This effect could cause sizable voltage losses and lead to the observation of limiting current behavior (Berg et al., 2004; Eikerling and Kornyshev, 1998; Ihonen et al., 2004; Kulikovsky, 2002a; Mosdale and Srinivasan, 1995; Paganin et al., 1996; Weber and Newman, 2004b).

Overall, the coupled fluxes of protons and water create a problem of *water management in the membrane*, catalyst layers, and the whole cell (Eikerling, 2006). Moreover, dimensional PEM instabilities related to excessive swelling and deswelling (or shrinking) of different PEM parts compromise the durability of MEAs. This is often the origin of delamination between membrane and electrode layers (Mathias et al., 2005). Mathematical models by Weber and Newman (2004b) as well as by Nazarov and Promislow (2007), examined the effect of mechanical compression on water uptake and swelling in PEMs. Mechanical compression leads to decreased water content, more uniform water distribution, and significantly reduced water transport.

For all of these matters, it is of utmost importance to understand (i) ionomer self-organization in PEMs, (ii) the thermodynamic state of water as a function of external conditions, (iii) pore swelling and network reorganization upon water uptake, and (iv) associated transport properties of protons and water. These relations will be developed in Chapter 2.

## KEY MATERIALS IN PEFC: POROUS COMPOSITE ELECTRODES

Electrochemical electrode reactions occur at metal/electrolyte interfaces. The overall interfacial reaction involves mass transport of reactants to the electroactive interface. It also involves formation of adsorbed reaction intermediates at the metal surface, electron transfer between electroactive species in solution or adsorbed intermediates and metal, desorption of product species, and mass transport of the product species away from the interface. All of these processes, that is, interfacial charge transfer and transport of electrons, protons, reactant gases, and water, take place in *porous electrodes*.

Porous electrodes are ubiquitous in electrochemical energy conversion and storage. They are needed in photoelectrochemical cells, electrochemical double layer capacitors, supercapacitors with pseudocapacitive effect, batteries, and fuel cells.

The main intrinsic parameter of the electrocatalytically active electrode material is the *exchange current density*, $j_*^0$. It tells us how rapid electron exchange can take place across the electrified interface under a dynamic equilibrium, with balanced rates of anodic and cathodic interfacial charge transfer, see the section "Electrochemcial Kinetics." This parameter represents a reaction rate per unit of the metal surface real area.

The primary optimization parameter of porous electrodes is the ideal *electrochemically active surface area* per unit volume $S_{ECSA}^{id}$. In rough approximation, the value of $S_{ECSA}^{id}$ is proportional to the amount of the electrocatalytically active material Pt, in the case of PEFC electrodes. It is inversely proportional to the feature size $d$, which could represent diameters of catalyst particles, of pores in a porous catalytic medium, or of rod-like structures (nanotubes or nanorods), onto which a thin film of catalyst is deposited. On the other hand, $S_{ECSA}^{id}$ is also roughly proportional to the energy density and current density that the electrode is capable of generating:

$$S_{ECSA}^{id} \sim \frac{1}{d} \sim \begin{cases} \text{energy density,} \\ \text{power density.} \end{cases} \tag{1.69}$$

Using these basic parameters, we can define the intrinsic electrocatalytic activity per unit volume of the porous electrode, $i^{0,id} = j_*^0 S_{ECSA}^{id}$, and the ideal exchange current density per unit of geometrical surface area of the electrode, $j^{0,id} = j_*^0 S_{ECSA}^{id} l_{CL}$. This form reveals the explicit dependence of the electrode activity on the thickness, $l_{CL}$.

Further requirements of porous electrode design and fabrication are to ensure optimal surface accessibility of the catalyst and to provide uniform reaction conditions. If a fraction of the surface of the active electrode material is inaccessible for one of the species consumed in the reaction, the electrochemically active surface area will be reduced, relative to the ideal surface area. In general, this results in $S_{ECSA} < S_{ECSA}^{id}$; the ratio

$$\Gamma_{stat} = \frac{S_{ECSA}}{S_{ECSA}^{id}} \tag{1.70}$$

is a *statistical utilization factor*. Uniform reaction conditions demand rapid transport of species. Too low rates of transport of electrons, protons, or gaseous reactants will lead to nonuniform reaction rate distribution. The total *effectiveness factor* $\Gamma_{CL}$, defined in the section "Nonuniform Reaction Rate Distributions: Effectiveness Factor" in Chapter 3, quantifies the compounded impact of statistical utilization and transport effects.

## CATALYST LAYER MORPHOLOGY

Improvements in design and fabrication of catalyst layers have led to the typical structure that is depicted in Figure 1.15. Nanoparticles of Pt with size distributions ranging from 2–5 nm are attached to primary particles of the catalyst support material, which have typical sizes of 5–20 nm. The most widely used catalyst support materials are carbon black or graphitized carbon black (Soboleva et al., 2010, 2011; Zhang, 2008). Over the last decade, nanostructured carbon supports employing, for instance, carbon nanotubes have been tested (Zhang et al., 2012). As in many other energy applications that require substrates with high surface area and high electronic conductivity, surface-treated or doped graphene is emerging as a promising new catalyst support material in fuel cell electrochemistry (Brownson et al., 2012; Shao et al., 2010). Moreover, significant efforts have exploited the unique properties of metal oxide materials, most prominently $TiO_2$ and $Nb_xO_y$, as support materials (Zhang et al., 2010).

The Pt-decorated C particles form *agglomerates* with sizes in the range of 100 nm, which could assemble further into aggregates with sizes of ~1 µm (Soboleva et al., 2010). Different stages of the aggregation process create a pore space with three types of pores: (1) micropores (<2 nm) between crystalline domains inside primary carbon particles, (2) mesopores (2–20 nm) between primary Pt/C particles inside agglomerates, and (3) macropores (>20 nm) between agglomerates. Meso- and macropores in CLs are also commonly referred to as primary and secondary pores, respectively. The pore space is partially filled with liquid water, which could accumulate predominantly in hydrophilic primary pores, whereas secondary pores, which are typically larger and more hydrophobic, are usually filled with gas.

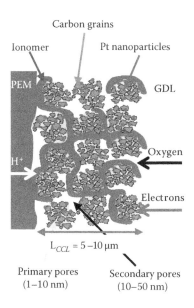

**FIGURE 1.15**  Schematic of a typical catalyst layer structure.

A *catalyst ink* is prepared by mixing the catalyst powder that consists of primary Pt/C particles in dispersion media that are mixtures of water, alcohols, or other organic compounds (Xie et al., 2008). Ionomer is added until a desired ionomer-to-carbon mass ratio is reached. For a review of *catalyst layer fabrication* approaches, see Chapter 19 in Zhang (2008). The added ionomer self-assembles into a separate interconnected phase in the pore space, primarily in secondary pores. The final CL structure depends on materials used, ink composition, dispersion medium, fabrication conditions, and the protocol of MEA fabrication and drying.

The gravimetric composition of the CL is defined by the contents of Pt, C, and ionomer, namely, $m_{Pt}$, $m_C$, and $m_I$, given in units of mg cm$^{-2}$, referring masses to the geometric electrode surface area. Physical properties of the layer depend on the volumetric composition, connectivity, and tortuosity of the different phase domains (solid Pt/C, ionomer, and pore space). Of vital relevance are the specific areas of different types of interfaces that are formed in the composite. The volumetric composition could be determined from the following relation:

$$1 - X_p = \frac{1}{l_{CL}} \left( \frac{m_{Pt}}{\rho_{Pt}} + \frac{m_C}{\rho_C} + \frac{m_I}{\rho_I} \right), \tag{1.71}$$

where $\rho_{Pt}$, $\rho_C$, and $\rho_I$ stand for the mass densities of components. Using a measurement of either thickness, $l_{CL}$, or total porosity, $X_p$, the remaining property in Equation 1.71 could be solved for and, thereafter, volume fractions of all components $i$ could be calculated from $X_i = m_i/l_{CL}\rho_i$.

Higher-order structural information, beyond volume fractions, requires more sophisticated experimental methodologies. *Pore size distribution* functions can be obtained from porosimetry studies using nitrogen physisorption (Soboleva et al., 2011) or the versatile method of standard porosimetry (Vol'fkovich et al., 2010). Differential measurements with standard porosimetry using water and octane allow the wetting angle of pores to be calculated as a function of pore radius. Thereby, amounts of hydrophilic and hydrophobic pores could be quantified. These studies were employed to analyze the impact of Pt deposition and ionomer loading on the hydrophilic–hydrophobic properties of CLs formed from different carbon support materials.

Moreover, isopiestic water sorption data and dynamic vapor sorption (Soboleva et al., 2011) provide information on water uptake as a function of relative humidity in the surrounding gas phase. This information, if combined with pore size distribution data, could help determine the wetting angle in pores or the Gibbs energy of water sorption as functions of pore size.

Particle size distributions of C and Pt, as well as the structure and morphology of individual particles, could be visualized directly using high-resolution transmission electron microscopy (TEM) (Mayrhofer et al., 2008; Meier et al., 2012; Shao-Horn et al., 2007). Tomographic approaches are under development, which can combine information from TEM and focused ion beam scanning electron microscopy (FIB-SEM), to reconstruct a 3D image of the catalyst layer (Thiele et al., 2013). At present, the minimal feature size that can be resolved with tomographic imaging of catalyst layers is approximately 10 nm.

A challenge for any of the structural characterization methods mentioned so far is the determination of the ionomer structure. Coarse-grained molecular dynamics simulations, discussed in detail in the section "Mesocale Model of Self-Organization in Catalyst Layer Inks" in Chapter 3, suggest that ionomer forms an adhesive skin layer with thickness in the range of 3–10 nm at the surface of agglomerates of Pt/C particles (Malek et al., 2008, 2011). The formation of this thin-film morphology is supported by analyses of experimental data on CL structure and composition. The properties of the Pt/C surface determine the ionomer structure, that is, the thickness of the ionomer film and the ionomer coverage on agglomerates. Moreover, the *ionomer skin layer* is characterized by a preferential orientation of dissociated sulfonic acid head groups at ionomer sidechains, which could be oriented toward or away from the agglomerate surface, depending on substrate properties.

The peculiar ionomer morphology influences the formation of a thin film of liquid water between Pt/C surface and ionomer. Resulting distributions of ionomer and water at the agglomerate surface and in agglomerate pores are essential structural properties. These distributions determine the real value of the electrochemically active surface area of the catalyst, $S_{ECSA}$, the proton concentration (or pH) at the catalyst surface, and the proton conductivity of the layer.

Generally, catalyst layer modeling has to account for transport phenomena in interconnected and interpenetrating phases of Pt/C, ionomer, liquid water, and gas pores as well as for processes at interfaces between these components. Thus, catalyst layers in PEFCs represent unconventional four-phase composite media. If we ignore the distinction of water in pores and at ionomer thin films, meaning that we treat these components as one electrolyte phase with effective properties, structural models reduce to those of conventional *three-phase gas diffusion electrodes*. This effective electrolyte approach has been the standard in CL modeling until recently. However, it fails in explaining the impact of the substitution of Nafion by other ionomers in CLs, and in rationalizing water management in CCLs and MEAs, for which we must explicitly consider water distribution and fluxes.

An important group of structural properties are the interfacial areas formed between Pt and support on the one hand, as well as ionomer and liquid water or gas phase, on the other. These interfacial areas will determine Pt utilization and electrochemical properties of the layer. The *electrocatalytically active surface area* in the catalyst powder can be estimated from the charge under either the H-adsorption or CO-stripping waves measured by thin-film rotating disk electrode (TF-RDE) voltammetry in $H_2SO_4$ solution (Easton and Pickup, 2005; Schmidt et al., 1998; Shan and Pickup, 2000). The *total catalyst surface area* can be calculated from the catalyst loading and the particle size distribution, obtained from x-ray diffraction patterns, or from TEM images. This analysis is based on the assumption that catalyst particles are spheres (Rudi et al., 2012). Cyclic voltammetry can be used to deduce the values of the surface area for hydrogen adsorption (at the catalyst) and double layer formation (at catalyst and support). So far, it has remained challenging to deconvolute information from these measurements and unambiguously determine contributions of different interfacial areas.

## THE PLATINUM DILEMMA

Progress in bringing down Pt loading while increasing cell performance has been impressive. *Pt nanoparticles* supported on carbon provide a factor 100 increase in mass-specific surface area over bulk polycrystalline Pt. Dispersed nanoparticle catalysts of Pt on highly porous carbon materials were introduced at the end of the 1980s by Ian D. Raistrick at Los Alamos National Laboratory. Simultaneously, he proposed impregnation of porous Pt/C fuel cell electrodes with *ionomer electrolyte* (Raistrick, 1986, 1989). These steps represented a major breakthrough in CL design for PEFC. They enabled a reduction of the Pt loading from approximately 4 mg cm$^{-2}$ to 0.4 mg cm$^{-2}$ at the cathode. More recently, researchers and material developers have looked for alternatives to state-of-the-art Pt/C catalyst materials in an effort to reduce the Pt loading, although even the most promising candidates still contain Pt as the base catalytic material (Debe, 2012; van der Vliet et al., 2012).

In spite of the progress, the ORR in the cathode still incurs about 40% of all irreversible energy losses in the cell, as well as a proportional fraction of voltage losses. Moreover, at the current mass loadings required for high cell performance, Pt is responsible for 30–70% of the total cost of a fuel cell stack, although it only amounts to about 0.1% of the stack volume. The foremost challenge in PEFC research remains to maximize performance with a minimal amount of Pt.

Any significant reduction in Pt loading could translate into huge cost savings of PEFC stacks, if performance and lifetime requirements could still be fulfilled. The importance of a further Pt loading reduction is self-evident. In charts with the abundances of elements in the upper crust of the Earth, Pt is found close to the bottom. The ratio of demand to abundance of a material determines its price.

Let us do a simple estimate of the Pt loading requirement in automotive PEFC stacks. Assuming a current Pt loading of 0.5 mg cm$^{-2}$ (including anode and cathode) and a power density of 1.5 W cm$^{-2}$, the Pt requirement is 0.3 g kW$^{-1}$. The number of cars in the world is a little over 1 billion and it exhibits rapid growth rates. Assuming there is an average power of 50 kW per car, generating the accumulated power of 1 billion cars with PEFCs at the current technology standard would require 15,000 tons of Pt. The estimated world reserves of Pt are 66,000 tons (with 70–80% of these reserves found in South Africa).

What rate of car production would the current world production of Pt sustain? The world production of Pt was 200 tons per year in 2012. If 50% of the Pt production were to be diverted for the manufacturing of fuel cell electric vehicles, it would suffice to produce 6 M vehicles. The current annual production of cars is 60 M. Thus, even under bold commercialization scenarios, the Pt requirement would drastically curtail the contribution of PEFC technology to the world car production. A further reduction of the Pt requirement by a factor 10 would be a game changer.

Visual inspection of Figure 1.15 suggests that the statistical utilization of Pt is expected to be far below optimal, and reaction conditions are expected as being highly nonuniform. The interplay of mass transport and interfacial electrochemical kinetics is inherent in the operation of porous electrodes with finite thickness. It leads to a distinction of several regimes of CCL operation that will be rationalized below. The foremost questions are: How low is the utilization of the catalyst and what is the impact of Pt under-utilization on performance?

Before diving deeper into the secrets of catalyst layer design, it is insightful to ponder a few basic questions and simple estimates. To establish a baseline for the evaluation of catalyst loading, 1 mg of Pt is considered to be spread out as a perfect monolayer with the crystalline configuration of a Pt(111) surface and the lattice constant of bulk Pt ($a = 3.92$ Å). What will be the total surface area obtained? The simple calculation gives a surface area of 2000 cm$^2$. Nowadays, the typical Pt mass loading of ionomer-impregnated CCLs, as shown in Figure 1.15, is $m_{Pt} = 0.4$ mg cm$^{-2}$, which would correspond to a *surface area enhancement factor* of $\xi_{Pt} = 800$, for ideal monolayer spreading. The parameter $\xi_{pt}$ is defined as the ratio of the total (real) catalyst surface area to the geometric (apparent) electrode surface area.

What is the surface area enhancement factor that is required for achieving performance targets of PEFCs? The best-in-class catalyst layers in terms of electrochemical performance (leaving aside problems with water management) are UTCLs. UTCLs are fabricated by the *nanostructured thin-film (NSTF) technology*, invented at 3M by Mark K. Debe (Debe, 2012, 2013). The catalyst in UTCLs consists of continuous layers of Pt generated by physical vapor deposition at the surface of whiskers with a core of perylene red. Assuming cylindrical geometry of whiskers, the surface area enhancement factor can be calculated from

$$\xi_{NSTF} = \left(1 - X_p\right) \frac{4 l_{UTCL}}{d}. \tag{1.72}$$

Estimating the thickness as $l_{UTCL} = 300$ nm, porosity as $X_p = 20\%$, and whisker diameter as $d = 50$ nm, one obtain's $\xi_{NSTF} = 20$. Under the monolayer constraint, this surface area enhancement factor would be achieved with a Pt loading of 0.01 mg cm$^{-2}$. Although based on crude estimates, this number gives an idea what the room for further Pt loading reduction in CCLs might be. The ratio of minimal Pt loading (under ideal conditions) to actual Pt loading in the state-of-the-art CLs could be considered as an estimate of the *effectiveness factor of Pt utilization*. According to the numbers provided, this effectiveness is expected to lie in the range of $\sim$2–3%. As demonstrated in the section "Hierarchical Model of CCL Operation" in Chapter 4, these values agree with effectiveness factors obtained from physical models of structure and function of conventional ionomer-impregnated CLs. They are, moreover, consistent with an experimental assessment of the effectiveness factor of Pt utilization.

As for 3M-type UTCLs, actual Pt loadings in them are in the range of 0.05–0.1 mg cm$^{-2}$. This loading deviates from the ideal monolayer loading for several reasons: Pt layers in UTCLs are about 5–10 monolayers thick, leading to a low surface atom ratio of Pt. Moreover, the electrostatic effectiveness factor in UTCLs is significantly below one.

## CATALYST LAYER DESIGNS

The merit function of a catalyst layer is the current density that it could generate at a given fuel cell voltage and time of operation, $j\left(E_{cell}, t\right)$, divided by the mass loading of Pt at the beginning of life, $m_{Pt}^0$ (in mg cm$^{-2}$). This function can be defined through

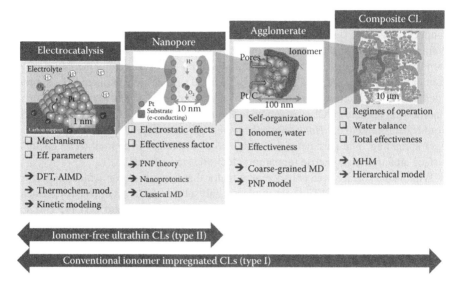

**FIGURE 1.16**  Scales in the cathode catalyst layer structure and function.

the following proportionalities:

$$\frac{j(E_{cell}, t)}{m_{Pt}^0} \propto j_*^0 \Gamma_{np} \Gamma_{stat} f \text{ (structure, conditions)}. \tag{1.73}$$

In this expression, $\Gamma_{np}$ is the *surface atom ratio of catalyst nanoparticles*, $\Gamma_{stat}$ accounts for percolation and wettability effects that diminish the *statistical utilization* of the catalyst, and the function $f$ (structure, conditions) incorporates all effects due to nonuniform reaction rate distributions.

Understanding catalyst layer structure and function is a hierarchical challenge. The dependencies in Equation 1.73 can be rationalized by distinguishing four scales, illustrated in Figure 1.16. At the lowest scale, size, shape, and surface atom arrangement of catalyst nanoparticles as well as interactions between catalyst and support material, determine the intrinsic electrocatalytic activity. These influences are embodied in $j_*^0$. Since only surface atoms participate in reactions, the surface atom ratio was introduced in Equation 1.73. This factor is roughly inversely proportional to particle radius, $\Gamma_{np} \sim 1/r$, and it is sensitive to the particle shape. Particles with more spherical shapes, like cuboctahedral particles, possess the smallest values of $\Gamma_{np}$, whereas nonspherical shapes, like tetrahedral particles, exhibit significantly higher surface atom fractions.

The next level in Figure 1.16 shows a single water-filled pore with Pt deposits at pore walls. Typical pore sizes are in the range of 2–20 nm. Ultimately, all electrochemical reactions in PEFC must proceed at Pt/water interfaces, and it can be assumed that a large portion of these interfaces exists in water-filled nanopores. Gas-filled nanopores will not make a contribution to current generation. Single pore effects will be accounted for in the function $f$ (structure, conditions), predominantly as an *electrostatic effectiveness factor*. The rate of current conversion at the pore level depends on the amount and distribution of Pt at the interface, the pore radius, and the surface charge density of the metal walls, $\sigma_M$, which varies with metal-phase potential, $\phi_M$.

Water-filled nanopores exist inside of *agglomerates of Pt/C*, which represent the next structural level. Typical agglomerate sizes are in the range of 100–300 nm. Agglomerates are partially covered by a thin ionomer film with a thickness of 3–10 nm. Thickness and coverage of the ionomer film on agglomerates are important structural characteristics at this level. Furthermore, it is essential that a large fraction of sulfonic acid head groups at the ionomer film are oriented toward the agglomerate surface. This favorable orientation renders the interfacial region between agglomerate surface and ionomer film hydrophilic and it confers high proton concentration to the water film at the agglomerate surface. Therefore, a high interfacial density of sulfonic acid head groups with orientation toward the agglomerate surface enhances the electrocatalytic activity and the proton conductivity at the agglomerate level. Using oxygen concentration, electrolyte phase potential, and proton concentration at the agglomerate surface, a model can be developed to rationalize the reaction rate distribution and *effectiveness factor of agglomerates*. The agglomerate effectiveness factor $\Gamma_{agg}$ represents an important contribution to $f$ (structure, conditions). Owing to partial ionomer coverage on the agglomerate surface and partial liquid saturation of pores in agglomerates, a fraction of Pt particles may form contacts with the gas phase and thus remain inactive. The inactive portion of the Pt surface is accounted for in $\Gamma_{stat}$.

At the macroscopic scale, transport properties of the porous composite layer depend on percolation effects in the interpenetrating functional phases. These transport properties determine the reactant distribution at the macroscale. Nonuniform distribution of oxygen due to poor oxygen diffusivity will diminish the value of $f$ (structure, conditions). Moreover, parts of the Pt surface may not be covered by water, rendering them inactive. Agglomerate and macroscopic scale must be coupled in modeling studies in order to determine the reaction rate distribution across the layer and inside agglomerates. The coupled model will give the agglomerate effectiveness factor as a function of the thickness coordinate $x$, $\Gamma_{agg}(x)$. Averaging over $\Gamma_{agg}(x)$ will give an effective agglomerate effectiveness factor.

Using the introduced levels of the structural hierarchy, we can define the two basic design types of catalyst layers in PEFCs, as indicated in Figure 1.16: *three-phase gas diffusion electrodes* (type I electrodes) and *two-phase flooded electrodes* (type II electrodes). Type I electrodes represent the prevailing catalyst layer design. These layers, fabricated by conventional ink-based approaches, incorporate all functional phases (metal: Pt/C; electrolyte: ionomer and liquid water; gas pores) and they span all four levels of the structural hierarchy. The requirement for the active catalyst surface area enhancement factor and volumetric requirements for the interpenetrating phases, determine the viable thickness range of this electrode design. The optimal thickness range is 5–10 μm. At lower thickness, catalyst layers of this type will not provide a sufficient $S_{ECSA}$. They will be too dense, critically affecting their gas diffusion capability. Thickness and composition must be optimized concertedly. For the optimal thickness range given above, the optimal ionomer volume fraction is $X_{PEM} \approx 35\%$. This volume fraction maximizes the fuel cell power density at a target current density of 1 A cm$^{-2}$.

Incorporating statistical factors, which define $\Gamma_{np}$ and $\Gamma_{stat}$, as well as transport processes at all four scales in Figure 1.16 into model studies gives an estimate of the total effectiveness of Pt utilization. It lies in the range of ~3%. This value is consistent with the result of an experimental assessment of the *effectiveness* factor *of Pt utilization* (Lee et al., 2010).*

The main distinguishing characteristic of type II electrodes is that the fabrication process neither involves mixing with ionomer nor an ionomer impregnation step. These structures are ionomer-free.[†] The thickness of these electrodes is in the range of ~200 nm. They are referred to as UTCLs. At or below the nanopore scale, processes in UTCLs are similar to those in conventional CLs. A UTCL could be considered as an agglomerate with planar geometry. Since their thickness is about two orders of magnitude lower than that of conventional CLs, they operate in a different transport regime. Diffusion rates of reactant gases dissolved in water are sufficient to ensure rapid transport to the catalyst surface. Therefore, it is beneficial if the pore space of UTCLs is fully saturated with liquid water in order to maximize $\Gamma_{stat}$. While mass transport effects of gaseous reactants are less pronounced in UTCLs, their electrochemical performance is strongly affected by electrostatic interaction of protons with charged metal walls. These *interfacial charging effects* determine the pH in water-filled pores and thereby the activity of the catalyst.

Two subcategories of type II electrodes exist, corresponding to UTCLs with (a) conductive nanoporous or nanostructured catalyst support, which could accommodate nanoparticles or nanoislands of Pt at pore walls, or (b) insulating catalyst support, which requires a continuous film of Pt at pore walls in order to ensure continuous electron flux. The first design option, using conductive supports, could be realized with nanoporous metal foils or foams, fabricated by dealloying or template-based approaches (Zeis et al., 2007). The second design option corresponds to 3M's NSTF technology. Realizations of both designs exist, and these have demonstrated excellent performance at markedly reduced Pt loading in comparison to type I electrodes. UTCLs represent a significant design simplification; since these layers are ultrathin and ionomer-free, they do not require a gas porosity. The effectiveness factor of Pt utilization of 3M-type NSTF layers was found to lie in the range of ~15%; this estimate does not factor in a surface atom ratio to account for the thickness of the sputter-deposited Pt film. Notably, recently tested NSTF layers with Pt alloy catalysts exhibit the highest electrocatalytic activities of all fuel cell CLs for the ORR, hitherto tested (Debe, 2013).

---

* When comparing the theoretical estimate of the Pt effectiveness factor with the experimental values in Lee et al. (2010), it must be taken into account that the experimental study did not include the impact of $\Gamma_{np}$.

[†] Although ionomer penetration could occur at the interface with the PEM.

## PERFORMANCE OF TYPE I ELECTRODES

### IDEAL ELECTRODE OPERATION

In the prevalent type I electrodes, large $S_{ECSA}$ can only be achieved at the cost of a significant electrode thickness. This complicates transport of reactant and product species in the layer. The CL function in the ideal regime with minor transport losses is illustrated schematically in Figure 1.17, which shows the shapes of currents and concentrations through the catalyst layer of a finite thickness. The proton current density $j$ linearly decreases through the electrode depth, while the electron current density $j_e$ linearly increases. The variation of the overpotential $\eta$ along $x$ is small, as in the ideal case, the CL proton conductivity is assumed to be large. The variation of the oxygen concentration $c$ is also small, as the oxygen transport through the layer is assumed to be fast.

Far from equilibrium, the rate of the electrochemical conversion (the ORR rate) on the cathode side, $R_{ORR}$, can be described by the Tafel equation

$$R_{ORR} = i^0 \left( \frac{c}{c_h^0} \right) \exp\left( \frac{\eta}{b} \right), \tag{1.74}$$

where $i^0$ is the volumetric exchange current density (in A per $cm^3$ of electrode volume), and $c$ and $c_h^0$ are local and the reference oxygen concentration, respectively.

In the situation depicted in Figure 1.17, $\eta \simeq \eta_0$ and $c \simeq c_1$ hold true. The ORR rate is thus constant along $x$, and the conversion current density in the CCL is simply the product $l_{CL}R_{ORR}$, or

$$j_0 = l_{CL}i^0 \left( \frac{c_1}{c_h^0} \right) \exp\left( \frac{\eta_0}{b} \right). \tag{1.75}$$

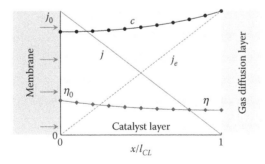

**FIGURE 1.17** Schematic of the ideal regime of the type I cathode catalyst layer operation. The proton current density $j$ linearly decreases through the electrode depth, while the electron current density $j_e$ linearly increases. The variation of the overpotential $\eta$ along $x$ is small, as the CL proton conductivity is assumed as large. The variation of the oxygen concentration $c$ is small, as the oxygen transport through the layer is fast.

This equation is the simplest polarization curve of the CCL, demonstrating a typical exponential dependence of the converted current density on the cathode overpotential. Solving Equation 1.75 for $\eta_0$ results in

$$\eta_0 = b \ln \left( \frac{j_0}{j^0 c_1 / c_h^0} \right), \quad \text{where} \quad j^0 = l_{CL} i^0. \tag{1.76}$$

Note that $j^0$ is the exchange current density *per surface area of the porous electrode*. It is related to the intrinsic exchange current density $j_*^0$ of the catalyst as $j^0 = \xi_{Pt} j_*^0$, where $\xi_{Pt}$ is the specific catalyst surface in the electrode per $cm^2$ of electrode area.

At small current densities, $j_0$ and $\eta_0/b$ are small, the reverse reaction cannot be neglected, the exponent in Equation 1.75 should be replaced by $2 \sinh(\eta_0/b)$, and Equation 1.76 transforms to

$$\eta_0 = b \operatorname{arcsinh} \left( \frac{j_0}{2 j^0 c_1 / c_h^0} \right). \tag{1.77}$$

Finally, at small cell current densities, expansion of the arcsinh function simplifies Equation 1.77 to

$$\eta_0 = R_{ct} j_0, \tag{1.78}$$

where

$$R_{ct} = \frac{b}{2 l_{CL} i^0 c_1 / c_h^0} \tag{1.79}$$

is the catalyst layer *charge transfer resistance* ($\Omega \ cm^2$). Note that this parameter is defined at small current densities only. Note also that increasing $l_{CL}$ at a constant $i^0$ means higher catalyst loading in the CL, which decreases $R_{ct}$.

## REGIMES OF ELECTRODE OPERATION

The *continuity equation for proton flux* in the CCL reads

$$\frac{\partial j}{\partial x} = -R_{ORR}, \quad j(0) = j_0. \tag{1.80}$$

The variation of proton current density along $x$ is due to the proton consumption in the ORR at a rate given by the right-hand side.

## Small-Cell Current Density

At small cell currents, transport losses are negligible, $R_{ORR}$ is constant along $x$ and the solution to Equation 1.80 describes the linear shape

$$j(x) = j_0 \left( 1 - \frac{x}{l_{CL}} \right), \tag{1.81}$$

where $j_0$ can be expressed through $\eta_0$ by Equation 1.75. This linear shape of the proton current density in the CCL, depicted in Figure 1.17, means that all parts of the layer contribute equally to the current conversion; the catalyst is uniformly utilized.

This ideal situation corresponds to *a priori* assumption of ideal reactant transport, which is realized in any electrode at sufficiently small currents. In this case, the local proton current is linear, while the overpotential and the reaction rate are constant along $x$.

In the nonideal case, one has to evaluate the interplay of three effective electrode parameters, viz. volumetric exchange current density $i^0$, proton conductivity $\sigma_p$, and oxygen diffusivity $D$. Instead of considering these parameters, it is more insightful to evaluate and compare three corresponding characteristic current densities. These include the current conversion capability $j^0 = i^0 l_{CL}$, defined above, the *characteristic current density due to proton transport*

$$j^p = \frac{\sigma_p b}{l_{CL}}, \tag{1.82}$$

and the *characteristic current density due to oxygen diffusion*

$$j^d = \frac{4FDc_1}{l_{CL}} = \frac{4DP_1}{b_* l_{CL}}. \tag{1.83}$$

Here, $c_1$ is the oxygen concentration at the CCL/GDL interface, $P_1$ the corresponding oxygen partial pressure (assuming ideal gas behavior), and

$$b_* = \frac{R_g T}{F}.$$

The regime of CCL operation depends on the values of these characteristic current densities relative to each other and relative to $j_0$. Ideal operation with uniformly distributed reaction rates corresponds to two simultaneously fulfilled conditions

$$\frac{\sigma_p b}{l_{CL}} \gg j_0 \quad \text{and} \quad \frac{4FDc_1}{l_{CL}} \gg j_0. \tag{1.84}$$

Typical parameters for a conventional cathode CL are $l_{CL} = 10$ µm, $\sigma_p = 0.01$ S cm$^{-1}$, $b = 25$ mV, $D = 10^{-3}$ cm$^2$ s$^{-1}$, and $c_1 = 7.36 \cdot 10^{-6}$ mol cm$^{-3}$. Using these parameters, one finds $j^0_{ORR} \sim 10^{-6}$–$10^{-7}$ A cm$^{-2}$ for the cathode superficial exchange current density, $j^p \sim 0.3$ A cm$^{-2}$ and $j^d \sim 3$ A cm$^{-2}$. As can be seen, proton transport limitations are expected to be more severe than oxygen diffusion limitations, except at high liquid water saturation in the CCL. Note that for hydrogen diffusion in the anode CL, $j^d$ will be an order of magnitude larger than the estimate above.

If the current density is small, that is in the linear regime of the Butler–Volmer equation, the shape of the proton current and of the local overpotential is, in general, exponential in $x$. The characteristic scale of the exponential current decay law is given by the reaction penetration depth, an important intrinsic electrode parameter

$$l_N = \sqrt{\frac{\sigma_p b}{2i^0}} = \sqrt{\frac{j^p}{2j^0}}\, l_{CL}. \tag{1.85}$$

The term "penetration depth" reflects the idea that the electrochemical reaction penetrates to the porous electrode from the membrane/electrode interface. This idea stems from classic studies of reaction kinetics at planar electrodes, as replacement of a planar by a porous electrode extends the reaction domain into the electrode depth.

Equation 1.85 shows that $l_N$ describes the competition of the electrode conversion ability and the proton conductivity. If $l_N \gg l_{CL}$, the shape of local proton current along $x$ appears to be linear. If, however, $l_N \lesssim l_{CL}$, the shape of local $j$ along $x$ is exponential.

In the anode catalyst layer of a PEFC, hydrogen transport can be assumed ideal. Moreover, the electrode conversion ability is very high, while the rate of proton transport is typically not perfect. The shape of the proton current density through the electrode depth is exponential (Figure 1.2b). Formally, the exponential shape arises because the exchange current density of the HOR is many orders of magnitude larger than that of the ORR, so that the reaction penetration depth on the anode side is $l_N \lesssim l_{CL}$. In contrast, in the cathode, the intrinsic reaction penetration depth is much larger than the electrode thickness $l_N \gg l_{CL}$, with the above estimates giving $l_N \sim 10^3 l_{CL}$. At small current density, the shape of proton current in the CCL is linear, Equation 1.81, which is independent of $\sigma_p$.

## Large Cell Current Density

The parameter $l_N$ is independent of the cell current density $j_0$. If the cell current is large, two new current-dependent space scales arise in the problem. If the transport of neutral reactant is still fast, while the rate of proton transport is finite, the local proton current and the reaction rate exhibit exponential-like decay in a thin sublayer of a finite thickness at the membrane interface. The characteristic space scale of the corresponding exponential function is given by the reaction penetration depth

$$l_p = \frac{\sigma_p b}{j_0} = \frac{j^p}{j_0}\, l_{CL}. \tag{1.86}$$

The shapes of proton and electron currents in this case are illustrated in Figure 1.18a. As can be seen, the reaction runs predominantly in a conversion domain at the membrane interface. In contrast to the low current density regime, the penetration depth $l_p$ is inversely proportional to the cell current density $j_0$, that is, with the growth of the cell current density the conversion domain shrinks.

In the opposite case, with fast proton transport and severely impaired diffusion transport of neutral molecules, caused for instance by flooding of pores with liquid water, the conversion domain shifts to the CL/GDL interface and proton current and concentration exhibit exponential decay at this interface (Figure 1.18b). The respective *reaction penetration depth* (the width of the conversion domain) is given by

$$l_d = \frac{4FDc_1}{j_0} = \frac{j^d}{j_0} l_{CL}.$$

$$(1.87)$$

Note that in either case, the reaction penetration depth is proportional to the "poor" transport coefficient, and it is inversely proportional to the cell current density (cf. Equations 1.86 and 1.87).

Thus, at large currents, electrode operation is characterized by the parameters $l_p$ and $l_d$, which play a key role in electrode design. Ideally, $l_p$ and $l_d$ should be large compared to $l_{CL}$. The effect of these parameters on electrode performance is discussed in Chapter 4.

Note that both $l_p$ and $l_d$ are independent of the exchange current density $j^0$. Physically, the dependence on $j^0$ appears in the problem if both $l_p$ and $l_d$ are much larger than $l_{CL}$. This case corresponds to a small current regime discussed in the section "Small Cell Current Density." If one of $l_d$ and $l_p$ is less than $l_{CL}$, the "rate-determining" process is the transport of the respective reactant, and the reaction penetration depth is independent of the catalyst active surface, which is incorporated in $j^0$.

Dividing $l_p$ by $l_d$, gives a composite parameter

$$\kappa = \frac{\sigma_p b}{4FDc_1}.$$

$$(1.88)$$

If $\kappa \ll 1$, with the growth of cell current, the CCL enters the proton-transport-limiting regime, with the conversion domain of thickness $l_p$ residing at the membrane/CCL interface. In the opposite limit of $\kappa \gg 1$, at sufficiently high current, the CCL enters the oxygen-transport-limiting regime with the conversion domain of thickness $l_d$ at

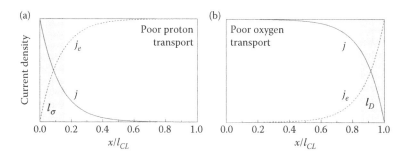

**FIGURE 1.18** Schematic of the two large-current regimes of CCL operation for the cases of (a) poor proton transport and (b) poor oxygen transport. The membrane is at the left side, while the GDL is on the right. In either case, the electrochemical conversion runs in a small conversion domain $l_\sigma$ or $l_D$ at the CCL interface, through which poorly transported particles arrive. Note that in both the cases, the thickness of the conversion domain is inversely proportional to the cell current density.

the CL/GDL interface. In either of these two regimes, the effect on polarization curves is a doubling of the apparent Tafel slope

$$j_0 \propto \exp\left(\frac{\eta_0}{2b}\right), \tag{1.89}$$

as is evident from comparison to the simple Tafel law in Equation 1.75. Note that neither of the conditions $\kappa \ll 1$ or $\kappa \gg 1$ itself guarantees the onset of the corresponding transport limitation regime. The value of $\kappa$ only shows what type of limitation will be encountered first upon increasing the cell current density. In contrast, the condition $l_d \lesssim l_{CL}$ or $l_p \lesssim l_{CL}$ manifests as the onset of the respective transport limitation.

One performance regime remains to be discussed. At large current density, a situation with $j_0 \gg \{j^p, j^d\}$ or, equivalently, $l_{CL} \gg \{l_p, l_d\}$, might arise. This case reflects poorly designed electrodes with insufficient ionomer loading and pore space flooded by liquid. What is the concerted impact of severely restricted proton and oxygen transport? Oxygen starvation will prevail and restrict the active part of the CCL to the domain of thickness $l_d$ at the CCL/GDL interface. The inactive part of the CCL closer to the PEM will merely act like an ohmic proton resistor, albeit with poor conductivity. The signature of this regime is a "knee" in polarization curves, which indicates severely impeded proton transport. This regime often materializes in aged catalyst layers.

## WHAT IS PERFORMANCE MODELING?

Chapters 4 and 5 are devoted to various aspects of electrode and fuel cell *performance modeling*. The type of modeling discussed in these chapters deals with the question of how to improve *fuel cell performance*.

The term "performance" depends on the context and it may have different meanings. In one case, the cell's electric efficiency might be the key optimization parameter. In another case, cell designers might strive to achieve the highest cell power density. Other target parameters of *design optimization* may be indicated by lowest thermal loss, minimal cell volume (for mobile applications), largest cell lifetime, or any combination of these. In this book, "performance" is related to the electric power of a fuel cell.

The performance portrait of the fuel cell is the polarization curve illustrating how much of the open-circuit potential has to be spent in order to generate a given load current. The product of the cell current density by the available cell potential gives the cell power density, that is, power generated by a unit cell active area. Understanding the contribution of every transport and kinetic process in the cell to the potential loss, is a key task of performance modeling.

There is a great variety of approaches to *fuel cell performance modeling*. The simplest approach used in system simulations deals with the semiempirical polarization curves of the cell or stack under investigation. Such curves are obtained by fitting a simple analytical model equation to measured data. This philosophy is very useful in the optimization of FC systems with numerous peripheral components (blowers, heaters, recycling units, etc.) However, this approach is limited to a certain type, or even implementation of a cell or stack—thus it lacks generality.

A hierarchy of cell models is often represented as a chain of models of various dimensionality (from 1D to 3D). This picture is inherited from fluid dynamics, where 3D models traditionally have been considered as being superior over 1D or 2D models. However, in fuel cell studies, the situation is more complicated.

Classical computational fluid dynamics (CFD) deals with a well-established system of equations. Typically, transport parameters and kinetic constants (if any) are also well defined. Scientific CFD problems are complicated by the presence of turbulence in the system. The spectrum of turbulence covers many orders of magnitude, which requires exhausting computing resources to resolve all the space scale and timescales.

In FC modeling, *the equations themselves* are not well defined. What is the best way to describe the species transport through the membrane? What is the most adequate model for the two-phase transport in porous layers? How transport properties of porous layers depend on their structure? All of these issues remain, to a large extent, unresolved.

Essentially, each FC model consists of several building blocks: models for the catalyst layers (CLs) and membrane, and models for the GDLs and for the flow fields (FFs). The largest gradients in concentrations and potentials in a cell are directed through-plane. This renders 1D through-plane models of primary interest and importance for FC technology. Owing to their low dimensionality, models of that type can incorporate very sophisticated and detailed submodels of MEA layers. These models can easily handle nonlinearities, and last but not least, they provide fast responses to user's requests. In the following chapters, the main focus will be on 1D models.

At each level (1D, 2D, or 3D), there is a great variety of approaches to modeling of the cell components' performance. For example, 1D models typically resolve catalyst layers, while 3D models usually treat CLs as thin interfaces.

Modeling of large-size application-relevant cells requires multidimensional approaches. Fortunately, the fuel cell is a two-scale system: its in-plane size is two to three orders of magnitude larger than the MEA thickness. This gives an opportunity to separate the in-plane and the through-plane transport processes. The idea of separation is simple: inside the MEA, all processes are assumed to be directed through-plane, while in bipolar plates, mass and heat transport occur mainly in-plane. This enables to construct fast 1D+2D hybrid models. An obvious simplification is a 1D+1D model for the cell, with a single straight channels on either side. This 1D+1D, or quasi-2D model, is very useful for understanding the effects related to feed molecule depletion along the channel (Chapter 5). For more detailed discussions of 3D modeling, see original papers reviewed by Wang (2004).

## SPACE SCALES IN FUEL CELL MODELING

The space scale of events determining fuel cell operation spans the range from nanometers (the elementary steps in reaction kinetics, proton transport in the nanopores) to meters (the length of a cathode channel). Over the past years, in fuel cell literature, the term "multiscale modeling" has emerged, and applied to models incorporating various space scales.*

To illustrate this term, consider the following chain of information transfer between the space scales shown schematically in Figure I.5. State-of-the-art small-scale modeling of reaction steps is based on quantum mechanical calculations using density functional theory (DFT). The next level are surface phenomena at the electrochemical interface. These phenomena are described by a set of mass balance and kinetic equations for concentrations and surface coverages of reactants, products, and adsorbed reaction intermediates on the catalyst (Pt) surface. The largest space scale provides performance modeling of a catalyst layer or fuel cell, which employs the laws of reaction kinetics in mass, charge, and heat transport equations.

The goal of DFT modeling is to understand the chain of elementary reaction events in the electrochemical conversion and to calculate the rate constants for these steps. The reaction mechanism and the rate constants, obtained from DFT, are then used to establish and parameterize time-dependent mass balance equations for the adsorbed/desorbed species. The steady-state solution of the surface coverage equations provides the conversion function, which can be used in the simplified current conservation equation in the CL model. The solution of the CL performance model yields the CL polarization curve, which can be used in the fuel cell or stack model. The chain of information transfer looks schematically like this: $\boxed{\text{DFT}}$ → $\boxed{\text{Kinetic model of surface processes}}$ → $\boxed{\text{CL performance model}}$ → $\boxed{\text{Fuel cell model}}$. Note that at every level, the results of modeling are usually verified by experiments. Moreover, more complicated verification schemes are also applied. For example, the results of CL performance modeling allow for fitting the electrode polarization curves, developing impedance models, and so on. Comparison of measured and predicted electrode properties enables the estimation of the quality of the underlying reaction scheme, which emerges from the DFT box.

The lower-scale level typically provides the upper-scale level with space- and time-averaged transport parameters, and kinetic rate constants. This information transfer between the levels is typical for fuel cell modeling. Each level has a long history of research.

In essence, the term "multiscale modeling" does not possess any new content: it is something that researchers in the field have done over many decades. What could be perceived as misleading in the use of the term, is that the multiscale model does not replace the need for a justification of assumptions made, or validation of parameters incorporated at different levels of the space-scale hierarchy. Simply put, the accuracy of a multiscale model could not be better than the accuracy of its weakest part. Moreover, occasional error cancelation could lead to fortuitous, unreliable, and misleading results.

---

* Science is not free from the tricks commonly used in promoting goods. A scientific result or a model is often considered as a merchandise that needs to be advertised. The term "multiscale modeling" is an example. The term is repeatedly employed to promote new models, often of very different sophistication and fidelity.

The approaches presented in this book mainly focus on simple solvable models on levels two and three (the macroscopic reaction kinetics and the CL/cell performance). Such models rarely give the exact numbers; however, they explicitly show the basic features and parametric dependencies in the system. Analytical studies of that type aim at understanding rather than giving a complete description. They ignore many secondary details in order to capture the essential features of the system. The nearest analogy is *avantgarde portraying*, which fixes the characteristic features of a human face. Being far from a detailed classical portrait, an avantgarde image often tells more about the person's character.

# 2 Polymer Electrolyte Membranes

## INTRODUCTION

The polymer electrolyte membrane (PEM) is the heart of the polymer electrolyte fuel cell (PEFC). It separates the partial redox reactions at anode and cathode and, thereby, enables the fuel cell principle.

### BASIC PRINCIPLES OF PEM STRUCTURE AND OPERATION

#### Principle 2.1: Stable Separation and Highly Selective Transport

The PEM must suppress the crossover of gaseous reactants, thereby, forcing hydrogen and oxygen to react separately at anode and cathode. If a PEM would fail to fulfill this fundamental function, *reactant crossover* would cause parasitic voltage losses due to the formation of a mixed potential. Moreover, reactant crossover accelerates the degradation processes. The PEM must exhibit electronic insulation, with electronic conductivities $\ll 10^{-4}$ S cm$^{-1}$, thereby forcing electrons, produced at the anode, to flow to the cathode through the electrical circuit. The driving force for electron flux is the Gibbs energy difference between anode and cathode. At the same time, the PEM must allow for rapid—low-resistive—proton passage, from anode to cathode.

#### Principle 2.2: Water as Nature's Benign Proton Shuttle

The prevailing class of PEMs exploits the unique properties of water as a proton solvent and shuttle. As a result of the high concentration of protons and the peculiar nature of hydrogen bonding, liquid water is an ideal medium for *proton transport*. As a blueprint for this fundamental principle, nature relies entirely on liquid water to facilitate proton transfer in intracellular energy transduction (Kuznetsov and Ultrup, 1999).

#### Principle 2.3: Proton Conductivity

The main physical property of a PEM is its *proton conductivity* $\sigma_{PEM}$ or the proton resistance $R_{PEM} \propto 1/\sigma_{PEM}$. This property determines the heat loss, $Q_{PEM}$, from Joule heating and the associated voltage loss, $\eta_{PEM}$.

#### Principle 2.4: Ion Exchange Capacity

The key material characteristic that is implanted into the material during chemical design and fabrication is the *ion exchange capacity* (IEC), defined as the number of moles of ion exchange sites per mass of dry polymer. It determines the volumetric charge density $\rho_{H^+}$, resulting from excess protons.

**Principle 2.5: Water Content as the State Variable of the PEM**

Water content is the variable that defines the thermodynamic state of a PEM. The water content controls the formation of the phase-segregated PEM morphology and it determines its transport properties for protons and water. The crucial challenge of PEM research is to understand water sorption, membrane swelling, and water retention as functions of thermodynamic conditions and current density of operation.

**Principle 2.6: The Operational Range of a PEM**

PEFCs and, thus, PEMs for automotive applications should remain fully functional under a wide range of conditions, from $-40°C$ to $100°C$ and from rather dry to fully wetted conditions. The operation above $120°C$ would enhance *electrode kinetics*, improve catalyst tolerance to CO poisoning and contamination, and facilitate heat removal from the cell. The prevailing type of water-based proton conductors suffers from extensive evaporative water loss at elevated temperatures, rendering their operation above $90°C$ virtually impossible. The materials design challenge that derives from the dependence on liquid water is to replace water by a different proton solvent or to retain a sufficient amount of water above $100°C$ in the PEM, by means of strong hydrogen bonding to the polymeric host.

**Principle 2.7: Water Management**

Water management is a key issue in view of optimizing PEFC operation. Attributable to the role of the PEM as a medium for proton transport and the sensitivity of its properties to water content, the membrane determines the feasible operational range and water management in all other components of the fuel cell.

**Principle 2.8: PEM Degradation and Lifetime**

A PEM should be mechanically and chemically robust. It has to survive in an electrochemically aggressive environment, over demanded duty cycles and lifetimes of envisaged applications ($>5000$ h in automotive applications, $>20,000$ h in heavy duty bus applications, and up to 40,000 h in stationary systems). Excessive cycling of voltage loads, humidification conditions, and temperature causes premature and irreversible aging of the PEM.

## CONDUCTIVITY ESTIMATES

Let us assume that a fuel cell manufacturer has specified a tolerated voltage loss $\eta_{PEM}$ at a target current density $j_0$ of fuel cell operation. The corresponding minimum conductivity is $\sigma_{PEM}^{min} = l_{PEM} j_0 / \eta_{PEM}$. In this way, the conductivity requirement is linked to the membrane thickness $l_{PEM}$. For a basic estimate, using typical values $j_0 = 1$ A cm$^{-2}$ and $\eta_{PEM} = 50$ mV, the *conductivity requirement* is $\sigma_{PEM}^{min} \simeq 0.05$ S cm$^{-1}$ at $l_{PEM} = 25$ μm and $\sigma_{PEM}^{min} \simeq 0.1$ S cm$^{-1}$ at $l_{PEM} = 50$ μm, which represent the thickness range of current PEMs (Adachi et al., 2010; Peron et al., 2010). PFSA-type PEMs like Nafion, Aquivion$^{®}$, 3Ms PEM and similar materials meet this conductivity requirement.

In this estimate, the thickness emerges as the key determinant of $\sigma_{PEM}^{min}$. Indeed, the most significant progress in PEM development over the last 10 years has been a thickness reduction from $l_{PEM} \approx 200\,\mu m$, the state of the art in the year 2000 (DuPont Nafion 117) to $l_{PEM} \approx 25\,\mu m$, as the industry standard in 2012 (DuPont Nafion PFSA NRE211 membrane). However, the thickness decrease leads to concomitant increases in water fluxes and permeation of unreacted fuel and oxygen. Reactant crossover diminishes the voltage efficiency resulting from the formation of a mixed electrode potential and accelerated PEM degradation.

Does a PEM behave like an ohmic resistor, with a linear relation between voltage drop and current density, as suggested by the above estimate of $\sigma_{PEM}^{min}$? The conductivity $\sigma_{PEM}$ is a function of membrane water content, temperature, and time. The temperature dependence is owed mainly to the fact that proton transport is an activated molecular process that essentially follows an Arrhenius-type dependence under normal conditions (Cappadonia et al., 1994, 1995; Kreuer et al., 2008). The dependence on water content involves complex effects of molecular state of water, ionomer aggregation, nanophase separation, and formation of percolating water-containing pathways. Partial PEM dehydration under PEFC operation can lead to nonlinearity in the relation between $\eta_{PEM}$ and $j_0$. The "nonohmic" resistance regime will be discussed in the section "Membrane in Performance Modeling" of Chapter 5 (Eikerling et al., 1998). Time dependence of $\sigma_{PEM}$ derives from the fact that water content, microstructure, and transport properties undergo reversible changes in response to varying external conditions, as well as irreversible changes resulting from long-term impacts of chemical and mechanical degradation. In general, proton transport properties of a PEM are more complex than those of an ohmic conductor.

## PEM CONDUCTIVITY: SIMPLY A FUNCTION OF COMPOSITION?

In well-hydrated PEMs, structural and dynamic properties of water and protons *closely* approach those of free bulk water. A combination of high proton concentration owing to *nearly* perfect acid dissociation of the highly acid-loaded medium, *almost* bulk water-like mobility of protons and *good percolation* of the random network of water-filled pores, guarantees high proton conductivity. These statements embody the main principles of PEM operation. The challenges lie in clarifying the precise meaning of "nearly," "almost," and "good."

It is possible to derive a simple expression for the *PEM conductivity* based on these essential principles of PEM structure and functioning. As demonstrated below, this simple treatment is possible within the limits of high water content. Assuming 100% dissociation of sulfonic acid sites and assuming further that all protons acquire a bulk-like proton mobility as in free liquid water, $\mu_{H^+}^b$, the proton conductivity is

$$\sigma_{PEM}^b = \rho_{H^+}^{eff} \cdot \mu_{H^+}^b \cdot f(X_w). \tag{2.1}$$

This expression radically simplifies structural effects on proton charge density and proton mobility. It serves as a springboard for detailed treatments of these effects in subsequent sections.

The proton charge density in the ideal case of complete acid dissociation is

$$\rho_{H^+}^{\text{eff}} = F \cdot IEC \cdot \rho_p^{\text{dry}} \cdot (1 - X_w), \tag{2.2}$$

with $F$ being Faraday's constant, $\rho_p^{\text{dry}}$ the density of the dry polymer, and $X_w$ the water volume fraction. The factor $(1 - X_w)$ accounts for the proton dilution effect upon increasing water uptake. Parameters that define $\rho_{H^+}^{\text{eff}}$ are effective membrane properties that can be either controlled at the fabrication stage or obtained from macroscopic measurements.

In a random porous network with fixed geometry and topology, the structure-based factor $f(X_w)$ exhibits a percolation-type dependence on $X_w$:

$$f(X_w) \simeq \left(\frac{X_w - X_c}{1 - X_c}\right)^\nu \Theta (X_w - X_c), \tag{2.3}$$

where $\Theta (X_w - X_c)$ is the Heaviside step function. For three-dimensional random porous networks, $\nu = 2$ was proposed as a universal critical percolation exponent of the conductivity (Sahimi, 1994; Stauffer and Aharony, 1994; Zallen, 1983). The percolation threshold $X_c$ corresponds to a critical value of the water volume fraction, at which a sample-spanning continuous pathway of water-filled pores emerges that connects to the opposite membrane surfaces. It depends on the topology of the random network. In the case of PEMs, one is dealing with a quasi-percolation phenomenon (Eikerling et al., 1997; Hsu et al., 1980). Below $X_c$, a network of minimally hydrated and, thus, poorly conductive pores remains. Pores in this low hydration state sustain a mechanism of surface proton conductivity $\sigma_{PEM}^s$, since sulfonate anions retain a small number of strongly bound water molecules in their hydration shell. Above $X_c$, percolation in water-filled pores gives the high bulk-like conductivity $\sigma_{PEM}^b$. Conductivity increases strongly with hydration, that is, $\sigma_{PEM}^s \ll \sigma_{PEM}^b$. The percolation threshold of Nafion-type membranes was determined as $X_c = 0.1$ or smaller, indicative of a highly connected network of water-filled domains.

The mobility of protons in bulk water at infinite dilution and 25°C is $\mu_{H^+}^b = 3.63 \cdot 10^{-3}$ cm$^2$ V$^{-1}$ s$^{-1}$ (Adamson, 1979; Meiboom, 1961). DuPont Nafion PFSA-type membranes (NRE211 and NRE 212) have $IEC \approx 1$ mmol g$^{-1}$ and $\rho_p^{\text{dry}} \approx 2$g cm$^{-3}$, corresponding to a maximum proton charge density of $\rho_{H^+}^{\max} = F \cdot IEC \cdot \rho_p^{\text{dry}} \approx 2 \cdot 10^8$ C m$^{-3}$. The volume-averaged proton concentration of the hydrated sample is approximately 2 mol L$^{-1}$.

The main unknown in Equation 2.1 is the structure-based factor $f(X_w)$. Assuming normal percolation behavior with $X_w = 0.4$, $X_c = 0.1$ and $\nu = 2$, gives $f(X_w) \approx 0.1$ and $\sigma_{PEM}^b \simeq 0.05$ S cm$^{-1}$ (at 25°C). This conductivity estimate is close to experimental values for Nafion® and similar PEMs. Pore network reorganization as a result of swelling, including the merging of pores, could result in larger values of $f(X_w)$.

Equation 2.1 reveals several options for achieving higher values of $\sigma_{PEM}^b$, by increasing (i) the effective proton charge density $\rho_{H^+}^{\text{eff}}$, (ii) the factor $f(X_w)$, and (iii) the intrinsic proton mobility in water $\mu_{H^+}^b$. As concerns option (i), the obvious way to obtain larger $\rho_{H^+}^{\text{eff}}$ is to increase the IEC; however, the increase in ion loading competes with the dilution effect due to a larger water uptake by PEMs with higher IEC (Tsang et al., 2007, 2009). As seen in Figure 2.1a, the water uptake of a saturated PEM could increase dramatically at high IEC. The compounded impact on conductivity owing to proton dilution and $f(X_w)$, is given by the proportionality

$$\sigma_{PEM}^b \propto (1 - X_w) \cdot \left(\frac{X_w - X_c}{1 - X_c}\right)^\nu \Theta(X_w - X_c). \tag{2.4}$$

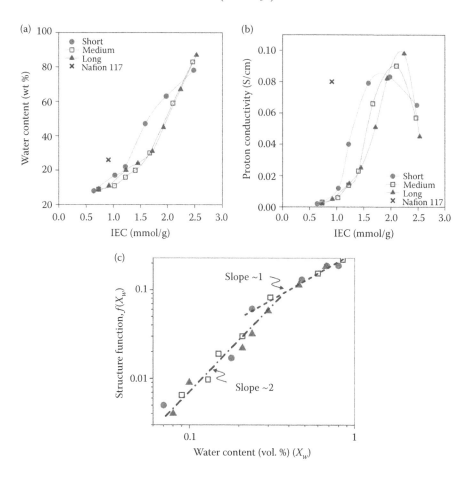

**FIGURE 2.1** Relations between (a) water content $X_w$ and IEC, (b) proton conductivity $\sigma_{PEM}^b$ and IEC, and (c) structure-based function $f(X_w)$ and water content $X_w$, for a series of graft-copolymer ion exchange membranes with varying ion exchange capacity and length of sidechains. The function $f(X_w)$ is obtained by dividing proton conductivity by the effective proton charge density and the proton mobility in bulk water. This function exhibits two distinct scaling regimes, namely, a percolation regime with $f(X_w) \sim (X_w - X_c)^2$ for $0 < X_w < 0.4$ and a regime with $f(X_w) \sim X_w$ at $0.4 < X_w < 1.0$ that is characteristic of a hydrogel-like state of the ionomer. ((a) and (b) are reprinted from Tsang, E. M. W. et al. 2009. *Macromolecules*, **42**(24), 9467–9480. Copyright (2009) American Chemical Society. With permission.)

This functional dependence, which accounts for the interplay of increasing IEC, proton dilution, and random network percolation, gives rise to a nonmonotonic dependence of $\sigma^b_{PEM}$ on IEC and $X_w$. Upon increasing IEC, $\sigma^b_{PEM}$ is observed to pass through a maximum, as seen in Figure 2.1b.

Application of Equations 2.1 and 2.2 to experimental data, under the assumption of perfect acid dissociation and bulk-like proton mobility of all protons, provides a means for analyzing the structural factor $f(X_w)$ and assessing whether it is consistent with structural models of random heterogeneous media. Figure 2.1c shows the function $f(X_w)$ extracted from the conductivity data in Figure 2.1b. PEMs displayed in this plot are based on graft-copolymer chemistries. The evaluated materials possess varying lengths and grafting densities of sidechains, and they span a wide range of IEC values, as indicated, and water contents from $X_w = 0.07$ to $X_w \approx 0.90$. The double logarithmic plot reveals the existence of two scaling regimes of the function $f(X_w) \sim (X_w - X_c)^\nu \Theta (X_w - X_c)$. At $X_w \leq 0.4$, the exponent is $\nu \approx 2$, resembling the percolation behavior of a typical porous medium. The percolation threshold obtained in this regime is low $X_c < 0.05$. At $X_w > 0.4$, scaling with $\nu \approx 1$ is observed, which is indicative of an ionomer hydrogel formed by interconnected ionomer fibrils.

The best compromise of proton conductivity and stability is achieved at intermediate IEC and moderate water uptake. Insufficient water content extinguishes the bulk-water-like proton mobility. Excessive swelling is a problem, as well. It reduces the steady-state performance as a result of proton dilution. Moreover, excessive swelling increases the microscopic stress on polymer fibrils, rendering PEMs with high IEC more prone to mechanical degradation.

As a conclusion from these basic considerations, high proton charge density in PEMs, achieved through high IEC, is beneficial. Sufficient hydration of sulfonic acid groups is needed to accomplish (almost) complete acid dissociation and bulk water-like mobility of hydrated protons. An increase in water volume fraction improves random network percolation but causes proton dilution, as well. Excessive swelling of PEMs is, therefore, not beneficial. PEMs with high IEC and well-connected pore network but constrained swelling of pores are desirable.

What is the maximum proton conductivity that could be obtained? An upper estimate for proton conductivity that could be achieved with the prevailing class of ionomer materials is $\sigma^{max}_{PEM} \simeq 0.5$ S cm$^{-1}$. This estimate is obtained by considering operation at 80°C, with $\mu^b_{H^+} = 6.75 \cdot 10^{-3}$ cm$^2$ V$^{-1}$s$^{-1}$, assuming $IEC = 2.5$ mmol g$^{-1}$, and making optimistic assumptions that $X_w \approx 0.5$ and $f(X_w) \approx 0.3$.

The treatment presented, so far, does not invoke detailed knowledge of structure formation and transport phenomena in PEMs. It is valid under well-hydrated conditions, when it predicts basic conductivity trends upon variation of effective PEM parameters such as IEC. Moreover, it can be applied to extract information on percolation effects in PEMs from conductivity data.

The main assumptions of the simple bulk conductivity model are valid at high water contents. This wording implies that a lower characteristic value of water content exists, below which the model is invalid. This value emerges as an important membrane characteristic. It will be discussed below, based on an assessment of water sorption properties and dynamics of proton and water transport.

If operated under conditions that are too hot or too dry, $\eta_{PEM}$ shows a significant, sometimes dramatic, increase resulting from a marked decrease in the proton conductivity. Vaporization of weakly bound liquid-like water at $T > 90°C$ disables the highly efficient structural diffusion of protogenic defects in bulk-like water. Moreover, the electro-osmotic water drag in an operating PEFC dehydrates the PEM close to the anode side leading as well to poor conductivity (Eikerling, 2006; Eikerling et al., 1998, 2007a; Springer et al., 1991; Weber and Newman, 2004b).

## CHALLENGES IN UNDERSTANDING PEM STRUCTURE AND PROPERTIES

Unraveling relations among structure, physical properties, and performance of PEMs is a hierarchical challenge, as reflected in Equation 2.1. It evolves around the versatile role of water as pore former, pore filler, proton solvent, and proton shuttle. The two key challenges for in-depth studies of membrane materials are

(1) How does the primary chemical architecture of ionomer molecules determine ionomer aggregation, formation of water containing pathways in the PEM, and water sorption properties of the PEM?
(2) How do morphology, water distribution, and water sorption properties determine transport properties and electrochemical performance of the PEM?

To name a few, here are some of the complicating traits involved, which will be discussed in subsequent sections: (i) The degree of acid dissociation is usually not perfect. This detrimental effect is a result of steric constraints, which obstruct the solvation of acid head groups at ionomer sidechains. (ii) The proton mobility is a microscopic parameter, which depends on the coupled molecular dynamics of protons and water. These dynamics are affected by nanoscale confinement of water in pores and the proximity of protonated water complexes to polymer–water interfaces, which are lined by a dense array of anionic surface groups. Electrostatic interaction of protons with surface groups could slow down proton motion. (iii) At the single pore level, proton concentration and mobility are spatially varying functions. Proton concentration increases in the direction toward the negatively charged pore walls whereas proton mobility exhibits the opposite trend. (iv) The coupling of proton and water motion gives rise to the so-called *electro-osmotic effect*. This effect leads to nonuniform distributions of water and proton conductivity in PEMs under operation.

Assuming that the size and shapes of water-filled domains are known, as well as the structure of polymer/water interfaces, proton distributions at the microscopic scale can be studied with molecular dynamics simulations (Feng and Voth, 2011; Kreuer et al., 2004; Petersen et al., 2005; Seeliger et al., 2005; Spohr, 2004; Spohr et al., 2002) or using the classical electrostatic theory of ions in electrolyte-filled pores with charged walls (Commer et al., 2002; Eikerling and Kornyshev, 2001). An advanced understanding of spatial variations of proton mobility in pores warrants quantum mechanical simulations.

## STATE OF UNDERSTANDING POLYMER ELECTROLYTE MEMBRANES

### CHEMICAL ARCHITECTURE AND DESIGN OF PEMs

The first perfluorinated ionomer was developed in the early 1960s by Walther G. Grot at E.I. DuPont de Nemours (Grot, 2011). It became famous under the tradename Nafion. From the mid-1960s, Nafion found use as an electrochemical separator material in the chlor-alkali industry. Exploration of Nafion as a fuel cell electrolyte started at about the same time.

The success of DuPont's Nafion spurred the development of other polymeric materials with similar chemical architecture. The most notable material developments have been the Dow experimental membrane (Dow Chemicals), Flemion (Asahi Glass), Aciplex (Asahi Kasei), as well as Hyflon Ion and its most recent modification Aquivion (SolviCore). In addition to excellent ionic conductivity, materials of the PFSA family, illustrated in Figure 2.2, exhibit exceptional stability and durability in highly corrosive acidic environments, owing to their Teflon-like backbone (Yang et al., 2008; Yoshitake and Watakabe, 2008).

PFSA ionomers are linear statistical copolymers of hydrophobic polytetrafluoroethylene backbones with randomly grafted vinyl ether sidechains, terminated by sulfonic acid ($-SO_3H$) head groups. Materials of this type vary in the chemical structure of the pendant sidechains. Other materials modifications are done to reduce membrane thickness and increase the IEC, through increasing the grafting density

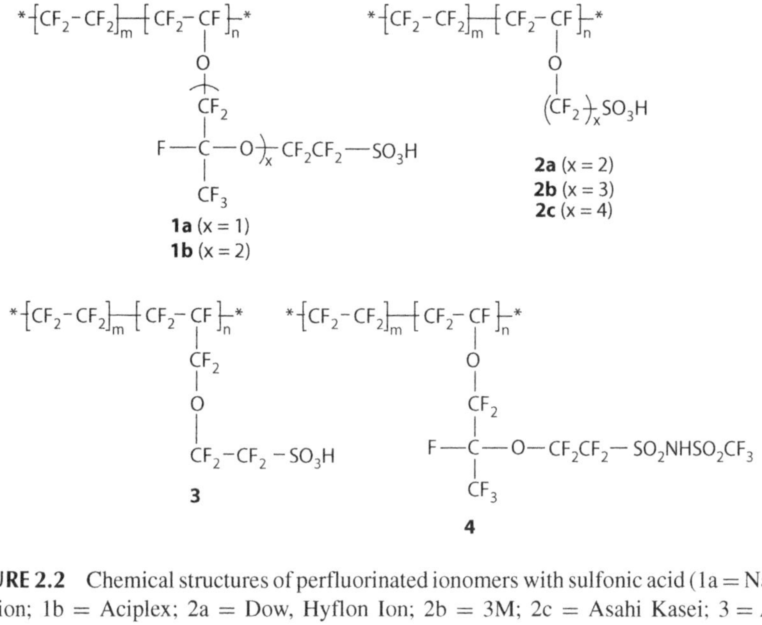

**FIGURE 2.2** Chemical structures of perfluorinated ionomers with sulfonic acid (1a = Nafion, Flemion; 1b = Aciplex; 2a = Dow, Hyflon Ion; 2b = 3M; 2c = Asahi Kasei; 3 = Asahi Glass) and bis[(perfluoro)alkyl sulfonyl] groups (4). (Reprinted with permission from Peckham, T. J., Yang, Y., and Holdcroft, S. et al., *Proton Exchange Membrane Fuel Cells: Materials, Properties and Performance*, Wilkinson, D. P. et al., Eds., Figure 3.16, 138, 2010, CRC Press, Boca Raton. Copyright (2010) CRC Press.)

of sidechains. Recently, PFSA membranes with short sidechains and high volumetric density of ion exchange sites like Aquivion, a modern version of the Dow experimental membrane, have garnered primary attention related to their markedly enhanced proton conductivity, water retention, and thermal stability. These materials are gradually replacing Nafion as the industry standard.

Alternative ionomer materials with controlled polymer architectures are obtained using block and graft copolymers with varying chemical composition of blocks or repeat units (Gubler et al., 2005; Hickner et al., 2004; Peckham and Holdcroft, 2010; Schuster et al., 2005; Smitha et al., 2005; Yang and Holdcroft, 2005). Block copolymer systems offer the possibility to design high proton conductivity and good mechanical strength, which usually represent mutually exclusive properties, into a single material. Besides perfluorinated ionomers, polymer systems that have been designed and studied include partially fluorinated ionomers, polystyrene-based ionomers, poly(arylene ether)s, polyimides, polybenzimidazoles, and polyphosphazenes. Reviews on membrane synthesis and experimental characterization can be found in Smitha et al. (2005), Hickner and Pivovar (2005), and Yang et al. (2008).

Increased thermal, mechanical, and electrochemical stability can be achieved by impregnating Nafion into inorganic matrices of clays, silica, or phosphotungstic acid or into porous Teflon (Yang et al., 2008). Such hybrid membranes exhibit reduced swelling, improved water retention, and an improved aptitude for operation at $T > 120°C$. Higher power densities could, thus, be achieved. Other directions of research pursue the development of water-free systems, by utilizing acid-base complexes or simply by replacing water in Nafion by other proton-conducting groups such as imidazole, which can be immobilized by covalent attachment to the polymer backbone (Herz et al., 2003).

## ROLE OF WATER

Water is the lifeblood of PEFCs and it is the working liquid in PEMs. From an operational point of view, the major challenge in PEFC research is to understand the double-edged role of water for structure and processes in all functional layers of a PEFC.

Water-mediated interactions between fixed anionic groups at polymer sidechains, hydrated protons and hydrophobic ionomer backbones control ionomer self-organization at the mesoscale. The main structural motif emerging from this process corresponds to cylindrical fibrils with radius of a few nanometers and high aspect ratio (Rubatat et al., 2002). Fibrils organize further into cage-like, cylindrical, or lamellar-like superstructures, illustrated in Figure 2.3 (Ioselevich et al., 2004; Schmidt-Rohr and Chen, 2008; Tsang et al., 2009).

The ionomer morphology determines the water uptake. Two definitions of *water content* are commonly used: the total water content, $\lambda = \frac{n_{H_2O}}{n_{SO_3H}}$, is defined as the ratio of the number of moles of water molecules absorbed in the PEM over the number of moles of acid head groups ($-SO_3H$). The *water volume fraction* is defined by $X_w = \frac{V_w}{V_w + V_p^{dry}}$, where $V_w$ is the volume of water and $V_p^{dry}$ is the volume of the dry polymer.

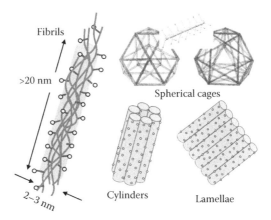

**FIGURE 2.3** Self-organization in ionomer solution into cylindrical fibrils or bundles (Rubatat et al., 2002) and further assembly of bundles into possible superstructures with cage-like (Iose-levich et al., 2004) cylindrical (Rubatat et al., 2004, 2002; Schmidt-Rohr and Chen, 2008) or lamellar-like morphology (Tsang et al., 2009).

The strength of *hydrogen bond* (HB) interactions with anionic head groups defines different types of water in PEMs. A distinction of contributions to $\lambda$ from *surface water* and *bulk-like water*, that is, $\lambda = \lambda_s + \lambda_b$, has proven essential for explaining water sorption and dynamic properties of PEMs, as will be demonstrated below. Surface water is strongly bound to charged protogenic surface groups (SGs) at polymer|water interfaces. The HB strength and fluctuation modes of surface water are strongly influenced by the packing density of SGs (Roudgar et al., 2006). Bulk-like water has weak interactions with SGs. In experimental studies, the amount of surface water is an uncertainty in the definition of water content. It is usually diffi-cult to find out whether surface water is incorporated in values of $\lambda$, or whether it contributes to $V_w$ or $V_p^{\text{dry}}$.

Assuming that only bulk-like water contributes to $V_w$ and, thus, $X_w$, the following relation can be derived:

$$X_w = \frac{\lambda_b}{\lambda_b + \bar{V}_p / \bar{V}_w}, \tag{2.5}$$

where $\bar{V}_p$ and $\bar{V}_w$ are molar volumes of ionomer (per backbone repeat unit that includes one sidechain) and water.

The scheme in Figure 2.4 emphasizes the impact of the different types of water. At $\lambda \leq \lambda_s$, interfacial effects prevail. In this range, interfacial packing density, long-range ordering, and flexibility of sulfonic acid head groups, as well as the distribution and HB structure of interfacial water, control the abundance and mobility of con-ductive protons at the interface. Usually, a strong electrostatic pinning of protons in interfacial water layers would reduce the mobility of protons. However, at a criti-cal density of protogenic surface groups, interfacial protons could acquire high rates of transport owing to a soliton-like mechanism that will be presented in the section "Solitons Coming Alive in Surface Proton Conduction."

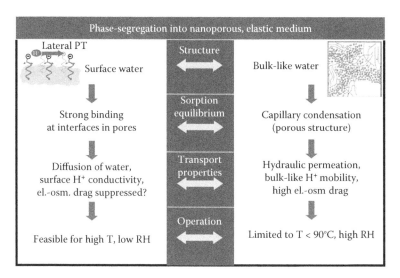

**FIGURE 2.4** The effects of the polymer host material on transport mechanisms and properties. The key distinction required to understand the response of PEM operation to changing external conditions and water uptake is between surface and bulk-like water.

At $\lambda > \lambda_s$, capillary effects control the equilibration of water with the polymer. In this regime, the values of molecular mobilities of protons and water approach the corresponding values in free bulk water, and hydrodynamic effects control transport phenomena. The PEM conductivity is described well by Equation 2.1. Highly functionalized polymer–water interfaces have a minor impact on transport mechanisms in this regime. An important consequence of this picture is that molecular-level studies of proton transport that account for details of ionomer structure are required strictly only for $\lambda \leq \lambda_s$. At $\lambda > \lambda_s$, it is sufficient to employ the well-established mechanism of proton transport in bulk water and incorporate it into structural models of PEMs.

## Membrane Structure: Experimental Studies

Information on PEM structure is mainly obtained from scattering studies. Ultrasmall, small, and wide-angle x-ray scattering (USAXS, SAXS, WAXS), as well as small-angle neutron scattering (SANS), provide insights into structural details at length scales from the micrometer range down to atomic distances (Gebel and Diat, 2005; Rubatat et al., 2004, 2002). Other experimental techniques have been employed to understand the morphology and dynamics of polymer and water in hydrated ionomer membranes at different time and length scales. These include *quasi-elastic neutron scattering* (QENS) (Perrin et al., 2007; Pivovar and Pivovar, 2005; Volino et al., 2006), IR and Raman spectroscopy (Falk, 1980; Gruger et al., 2001), time-dependent FTIR (Wang et al., 2003), NMR (MacMillan et al., 1999; Schlick et al., 1991), electron microscopy (Rieberer and Norian, 1992), positron annihilation spectroscopy (Dlubek et al., 1999), scanning probe microscopy (Lehmani et al., 1998), electrochemical atomic force microscopy (AFM) (Aleksandrova et al., 2007; Hiesgen et al., 2012), scanning electrochemical microscopy (SECM) (Mirkin, 1996), and electrochemical impedance spectroscopy (EIS) (Affoune et al., 2005; Kelly et al., 2005). Reviews on the application of experimental techniques can be found in Kreuer et al. (2004), Hickner and Pivovar (2005), and Mauritz and Moore (2004).

### Insights from Scattering Experiments

The characteristic length scale probed by scattering is $l_s \sim 2\pi/q$, where $q$ is the wave number. A typical x-ray scattering curve for Nafion in water is shown in Figure 2.5, adapted from Rubatat et al. (2004).

Analysis of scattering data, involving form factors and structure factors, provide information on (i) size and shape of scattering objects that could correspond either to hydrophobic domains of ionomer backbones or to hydrophilic domains, which accommodate water, anionic surface groups, and protons, (ii) amount of interfacial area per anionic surface group, (iii) density and size of crystalline domains, (iv) formation of polymer superstructures, as indicated in Figure 2.3, and (v) organization of such objects into pore networks with local or long-range order.

Figure 2.5 shows the scattering intensity, on a log scale, as a function of $q$, also on a log scale. This figure illustrates key features of scattering curves and their structural interpretation. The strong upturn in scattering curves for very low values

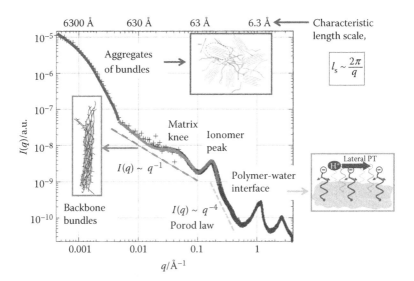

**FIGURE 2.5**  A typical x-ray scattering curve for Nafion swollen in water. The curve shows the scattering intensity as a function of the scattering wave number, $q$, using logarithmic scales on both axes. It reveals different scaling regimes and features. Structural characteristics are illustrated on the corresponding length scales. (Adapted from Rubatat, L., Gebel, G., and Diat, O. 2004, *Macromolecules*, **37**(20), 7772–7783, Figure 4. Copyright (2009) American Chemical Society. With permission.)

$q < 0.004$ Å$^{-1}$ has been attributed to aggregates of bundles of rod-like ionomer backbones with local correlation in orientation that extends over a characteristic length of ~100 nm (Gebel and Diat, 2005; Rubatat et al., 2004). Toward larger $q$, in the range of $0.004$ Å$^{-1} < q < 0.1$ Å$^{-1}$, the prevailing feature is a straight line with scaling behavior $I(q) \propto q^{-1}$, with a superimposed hump referred to as the matrix knee. The matrix knee was found to be relatively insensitive to the hydration state of the membrane. It was, therefore interpreted as scattering from elongated cylindrical bundles of ionomer backbones. The diameter of these bundles lies in the range of ~3 nm and their length was predicted to be about 50 nm. This value of the bundle length provides an estimate of the persistence length of the ionomer.

The main feature in the low-angle region of scattering curves is the ionomer peak that is seen around $q \simeq 0.1 - 0.2$ Å$^{-1}$ ($l_s \simeq 3$ nm). This peak corresponds to the first maximum of the structure factor. It has been interpreted in terms of size and locally ordered distribution of ionic aggregates. The height and position of the ionomer peak has been extensively analyzed as a function of $\lambda$, IEC, and $T$ (Mauritz and Moore, 2004).

Scattering intensities at $q$-values in the range of $0.2$ Å$^{-1} < q < 0.6$ Å$^{-1}$, corresponding to $l_s \simeq 1 - 3$ nm, follow the Porod scaling law, $I(q) \propto q^{-4}$ (Porod, 1982). Observation of this proportionality indicates sharp interfaces between polymeric and water-containing domains. The proportionality factor is given by

$$\lim_{q \to \infty} \left( \frac{Iq^{-4}}{\varphi_p} \right) = 2\pi \Delta\rho^2 \frac{\sigma_{SG}}{v_{SG}}, \tag{2.6}$$

where $\Delta\rho$ is the difference of the scattering length density between polymer and water ($\Delta\rho = 5.27 \cdot 10^{10}$ cm$^{-2}$ for neutrons), $\varphi_p$ is the polymer volume fraction, and $v_{SG}$ is the average polymer volume per ionic head group ($v_{SG} = 0.87$ nm$^3$ for Nafion 117). In the Porod regime, a plot of $Iq^{-4}$ versus $q$ gives a horizontal line, from which the value of specific surface area per ionic group, $\sigma_{SG}$, can be extracted, provided that $\Delta\rho$ and $v_{SG}$ are known. For a Nafion PEM with $IEC = 0.9$ mmol g$^{-1}$, G. Gebel and J. Lambard determined $\sigma_{SG} = 54$ Å$^2$ from this Porod analysis (Gebel and Lambard, 1997). For a range of short-sidechain PFSA PEMs with IECs from 0.8 mmol g$^{-1}$ to 1.5 mmol g$^{-1}$, Porod analysis in the water-swollen state gave an average value $\sigma_{SG} = 61$ Å$^2$ (Gebel, 2000).

Further characteristics in the wide angle part of the spectrum in Figure 2.5 at $q > 0.6$ Å$^{-1}$ correspond to amorphous and crystalline peaks.

Much to the frustration of scientists coming new to this field, lasting controversy ensues with respect to the analysis and interpretation of scattering data. Over a period of research spanning more than 30 years, fitting models have remained ambiguous, rendering structural interpretation of experimental studies still inconclusive.

## Microscopy Studies of the PEM Surface

Direct visualization of domain sizes and random morphology of ionomer membranes can be obtained from microscopy studies (Hiesgen et al., 2010). Information obtained from these approaches is restricted by the need to consider thin films (~100 nm thick films, cast from solution or obtained by microtoming) for *transmission electron microscopy* (TEM) or by the limitation to surface structures in the case of AFM studies.

Holdcroft and coworkers used TEM pictures to compare the morphologies of hydrogen-based graft copolymer membranes with random copolymer membranes with similar ion content (Ding et al., 2001, 2002). For hydrated grafted membranes, a continuous phase-separated network of water-filled channels with diameters of 5–10 nm was observed. Random copolymer membranes exhibit a weaker tendency toward microphase separation with water being dispersed more randomly within polymer domains. The water uptake in the grafted membranes is limited to smaller values, since the hydrophobic nature of channel walls impedes the intrusion of water. Nevertheless, their ionic conductivity is an order of magnitude larger than that of the random copolymer membranes, owing to the increased connectivity of aqueous domains.

Hiesgen and colleagues (Hiesgen et al., 2010, 2012) scan the surface of a PEM with a Pt-covered AFM tip. The back side of the membrane is covered with an extended porous electrode (anode) that contains Pt and generates a proton flux. The AFM tip serves as the cathode, which collects the local proton current that flows out of the PEM. This approach provides current density maps of the Pt surface. The spatial resolution of current measurements is in the nanometer range. In the same experimental setup, the surface can be scanned to probe surface topology and mechanical properties (adhesion force, stiffness). Correlations between conductive and mechanical properties can, thus, be studied. The spatial distribution of the proton current reveals a fine structuring of the PEM surface in terms of conductive and nonconductive areas. The topological model of the heterogeneous PEM surface, reconstructed from the high-resolution conductivity map, is in qualitative agreement with models of locally aligned cylindrical water channels (Gebel, 2000; Rubatat et al., 2002; Schmidt-Rohr and Chen, 2008). It reveals increasing channel sizes and conductivity upon increasing water content.

An interesting finding of the AFM studies is a peculiar restructuring at the PEM surface. At the PEM–gas interface, bundles of ionomer backbones are oriented preferentially parallel to the surface plane, forming a hydrophobic skin layer. This layer at the PEM surface blocks pore entrances and, thereby, inhibits water penetration into the membrane. Contact with a liquid water reservoir induces a rapid reorientation of ionomer bundles at the PEM surface, thereby, facilitating liquid water exchange between PEM interior and reservoir. Moreover, a strong electro-osmotic water flux driven by a high proton current has been seen to revert the pore blocking effect, activating the PEM surface for water exchange with the surroundings. Interfacial effects, thus, play an eminent role in controlling structural relaxation and water flux phenomena in PEMs.

## Local Order

Crystallinity associated with the local ordering of polytetrafluoroethylene (PTFE) backbones is deemed an important PEM property in view of mechanical and thermal stability. The degree of crystallinity has an impact on microscopic stress–strain relationships and water sorption properties of PEMs. High crystallinity improves the stress resistance under excessive swelling and deswelling cycling of the PEM. Nanocrystalline domains act as physical cross-links that enhance the mechanical stability, reduce the swelling, and suppress reactant crossover.

Crystallinity depends on the chemical composition of the ionomer. Moreover, it is affected by the thermal processing history of the PEM, the solvent used during preparation, and the membrane thickness (Kim et al., 2006; Moore and Martin, 1988).

Neutron and x-ray diffractograms can be analyzed in view of size and shape of ordered domains, as well as the distance between crystallites. Scattering from backbone crystallites is believed to be responsible for the matrix knee in SAXS and SANS spectra at about $q \approx 0.04 \text{ Å}^{-1}$. The corresponding distance between crystallites varies in the range from 150 to 500 Å. It increases with increasing PEM thickness.

Information on the morphology and ordering within crystalline domains is obtained from the wide angle part of diffraction spectra (Fujimura et al., 1981). The percentage of crystallinity (Kasai and Kakudo, 2005) is defined as the ratio of scattering intensity due to crystalline domains to the total intensity (Tsang et al., 2009). Nafion with $IEC = 0.9 \text{ mmol g}^{-1}$ exhibits a volume percentage of crystalline domains of 14%, at ambient conditions and 20% volumetric water uptake. This corresponds to a crystallinity of backbone domains of $\sim$28% (Chen and Schmidt-Rohr, 2007). The crystallinity decreases with increasing IEC and vanishes below a critical IEC. Increasing length and irregular distribution of sidechains inhibit the formation of crystalline domains. It is worth noting that the morphology of crystalline domains remains unknown and a crystalline phase is not always observed.

## MEMBRANE MORPHOLOGY: STRUCTURAL MODELS

### Cluster–Network Models of Ion Aggregation

Based on the analysis of WAXS data, Longworth and Vaughan (1968a,b) proposed a *cluster model of ion aggregation* in polyethylene ionomers. A thermodynamic theory of ion aggregation in organic polymers was proposed by Adi Eisenberg in 1970 (Eisenberg, 1970). In the early 1980s, T.D. Gierke and colleagues at DuPont developed the first morphological model of hydrated Nafion membranes, using information derived from SAXS data (Hsu and Gierke, 1982, 1983). The well-known *Gierke model* described the morphology of hydrated ionomers as a network of inverted spherical micelles with nanoscopic dimension, confined by anionic head groups of sidechains. In the dry state, the estimated diameter of micelles was $\sim$2 nm and micelles were assumed to be disconnected from each other. With increasing water uptake, the diameter of inverted micelles was proposed to grow up to 4 nm. In order to form percolating pathways for proton and water transport, it was inferred that narrow aqueous necks with length and diameter of $\sim$1 nm should connect the spherical micelles at intermediate water contents.

In the Gierke model, the formation of a critical number of necks signals the *percolation transition* in proton conductivity, that is, the emergence of an uninterrupted conductive path in the network of micelles and necks. During water uptake, the cluster network continuously reorganizes by swelling and merging of individual clusters and by the formation of additional necks (Eikerling et al., 1997; Hsu and Gierke, 1982, 1983). This pore network evolution involves a reorganization of anionic head groups at polymer–water interfaces. Their density was predicted to decrease with water uptake.

*Random network models*, based upon assumptions of the Gierke model, proved useful for understanding water fluxes and proton transport properties of the PEM in fuel cells. They helped in rationalizing the percolation transition in proton conductivity upon water uptake.

However, the proposal of narrow necks as proton connectors between inverted micelles has spurred debates and caused confusion. As there is no experimental evidence for the formation of such necks, their structure, ion content, and proton-conducting ability have remained elusive. A modern version of the theory of neck formation, proposed by Ioselevich et al. (2004), addressed some of these issues.

The physical attraction of stiff hydrophobic ionomer backbones drives ionomer assembly into bundles with further increased stiffness, consistent with the formation of elongated fibrils, as discussed in the section "Insights from Scattering Experiments." The persistence length of bundles is expected to exceed 10 nm by a significant factor. The presence and high density of sidechains with ionic head groups limits the bundle growth. A bundle of three Nafion backbone chains is shown in Figure 2.3. Thus, fibrillar bundles emerge as a prevailing structural motif at the mesoscopic scale. Bundles could further assemble into a range of ordered or disordered superstructures, for example, with cylindrical or lamellar geometries.

Tetrahedral cage-like superstructures that could accommodate droplet-like water inclusions are depicted in Figure 2.3. In this picture, microscopic swelling during water uptake involves two basic processes: water droplet growth because of bundles sliding along each other and growth of the necks that connect the cages. Ioselevich et al. (2004) speculated that the latter process might be responsible for the percolation transition in proton conductivity.

The different superstructures depicted in Figure 2.3 should be seen as limiting structures. A PEM consists of a random mix of those. However, interpretation of scattering, water sorption, and water transport data suggest that the most likely limiting structure to be found in this mix resembles cylindrical pores, as will be discussed below.

## Fibrillar Structure Model

Deficiencies of the cluster-network model become apparent when it is employed to describe the structural evolution of the PEM from the highly diluted polymer solution $X_W \to 1$, to the dry membrane state $X_W \to 0$. An earlier conceptual model for this morphological reorganization, suggested by Gebel (2000), involved a vague conjecture about the structural inversion at $X_W \approx 0.5$ from a colloidal dispersion of rod-like polymer aggregates in dilute solution, at $X_W > 0.5$, to a cluster network of water-filled ionic domains (or inverted micelles) embedded into a polymer matrix at $X_W \leq 0.5$. This idea evolved into a new structural model of Nafion-type membranes, which produces a continuous transition between dry membrane state and dilute polymer solution (Rubatat et al., 2002).

The currently accepted structural model consists of elongated aggregates of hydrophobic polymer backbones with cylindrical or ribbon-like shape. These elements are lined on the surface by arrays of dissociated sidechains and are surrounded by solvent and mobile counterions (Gebel, 2000; Gebel and Diat, 2005; Gebel and Moore, 2000; Loppinet and Gebel, 1998; Rollet et al., 2002; Rubatat et al., 2004). Simulations of Schmidt-Rohr and Chen of small-angle scattering data of hydrated Nafion support the tubular structure model (Schmidt-Rohr and Chen, 2008). The suggested structure consists of an array of cylindrical, randomly packed ionic water channels, embedded in a locally aligned polymer matrix. Introducing crystallites of hydrophobic polymer as physical cross-links was found to be crucial for reproducing scattering data. By using their empirical fitting approach, Schmidt-Rohr and Chen stipulated that other structural models failed to explain peculiar details of experimental scattering curves.

The persistence length of Teflon (PTFE) lies in the range of $\sim$5–10 nm (Rosi-Schwartz and Mitchell, 1996). Comb-shaped ionomer molecules like Nafion have the same PTFE backbone but they exhibit a markedly enhanced persistence length compared to Teflon. This enhancement is caused by the compounded effect of electrostatic and steric stiffening (Dobrynin, 2005). Further stiffening occurs when rod-like ionomer segments self-assemble into bundles of backbones. These processes will be described in the section "Theory and Modeling of Structure Formation in PEM." A typical bundle contains $\sim$10 rods. It has a diameter of $\sim$2 to 3 nm and a length of stiff segments that exceeds 20 nm.

The formation of elongated fibrous bundles of ionomer backbones is consistent with the majority of scattering and microscopy data, as well as thermodynamic arguments (Gebel and Diat, 2005; Hiesgen et al., 2012; Melchy and Eikerling, 2014; Rubatat et al., 2004; Schmidt-Rohr and Chen, 2008). These bundles are the building blocks of ionomer superstructures as illustrated in Figure 2.3.

Recent models of proton and water transport in PEMs tend to support the notion of cylindrical pore networks. A qualitative distinction between superstructures will be made below, based on the analysis of water sorption data and evaluation of the implications of pore network reorganization upon water uptake.

## Dynamic Properties of Water and Protons in PEMs

Studies of the dynamical behavior of water molecules and protons in PEMs rationalize the influence of random morphology and water uptake on effective physicochemical properties, that is, proton conductivity, water flux, and electro-osmotic drag.

Phase separation and spatial organization of membrane domains determine the state of water, fundamental interactions in the polymer/water/ion system, vibration modes of fixed sulfonate groups, and mobilities of water molecules and protons. Dynamic properties of the membrane can be probed at the microscopic scale with spectroscopic techniques, including Fourier transform infrared spectroscopy (FTIR) and nuclear magnetic resonance (NMR) (Mauritz and Moore, 2004). FTIR

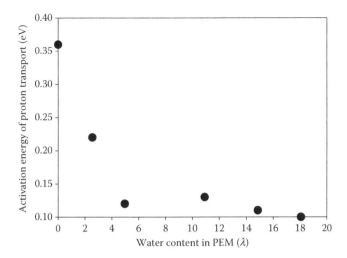

**FIGURE 2.6** Activation energies of proton transport Nafion 117, extracted from the Arrhenius representation of conductivity data at varying water contents. (With kind permission from Springer Science+Business Media: *Fuel Cells I*, Proton-conducting polymer electrolyte membranes: Water and structure in charge, 2008, pp. 15–54, Eikerling, M., Kornyshev, A. A., and Spohr, E.)

measurements provide information about sidechain motions (Cable et al., 1995; Falk, 1980). NMR studies probe mobilities of protons and water molecules at nanometer resolution (Zawodzinski et al., 1991, 1993).

At the macroscopic level, proton transport can be studied with electrochemical impedance spectroscopy (EIS). Cappadonia et al. (1994, 1995) performed EIS studies to explore variations of proton conductivity with water content and temperature for Nafion 117. The Arrhenius representation of conductivity data revealed activation energies between 0.36 eV at lowest hydration and 0.11 eV at highest hydration, as shown in Figure 2.6. The transition occurs at a critical water content of $\lambda_{crit} \sim 3$. At fixed $\lambda$, the transition between low and high activation energies was observed at $\sim$260 K for well-hydrated membranes. This finding was interpreted as a freezing point suppression due to confinement of water in small pores.

Pulsed-field gradient NMR is a powerful tool for studying transport mechanisms and parameters of molecular species in restricted spaces (Stejskal, 1965; Stilbs, 1987). Various groups have adopted this approach to measure the self-diffusivity of water in PEM as a function of water content (Kreuer, 1997; Zawodzinski et al., 1991). QENS has been applied to analyze typical time and length-scales of molecular motions (Perrin et al., 2007; Pivovar and Pivovar, 2005; Volino et al., 2006). The water mobility increases with water content to approach a saturation at almost bulk-like values above $\lambda \sim 10$. In Perrin et al. (2007), QENS data for hydrated Nafion were analyzed with a Gaussian model for localized translational diffusion. Typical sizes of confining domains, local and long-range diffusion coefficients of water molecules, and characteristic times for the elementary jump processes at molecular level, were obtained as functions of $\lambda$. The results were rationalized with respect to PEM structure and sorption characteristics.

NMR relaxometry is a suitable technique for investigating proton motion in the range from 20 ns to 20 $\mu$s, corresponding to three orders of magnitude of Larmor angular frequencies $\omega$ (Perrin et al., 2006). The NMR longitudinal relaxation rate $R_1$ is particularly sensitive to host–water interactions and, thus, well suited to study fluid dynamics in restricted geometries. In polyimide membranes, a strong dispersion of $R_1$ was found that closely followed a power law of the type $R_1 \sim \omega^\alpha$ in the low-frequency range (correlation times from 0.1 to 10 ms). This is indicative of strong attractive interactions of the first 3–4 water molecules with "interfacial" hydrophilic groups at the polymeric matrix. Variations of $R_1$ with $\lambda$ suggest a two-step hydration process; it involves solvation and formation of water clusters centered on interfacial charged head groups, followed by the formation of a continuous hydrogen-bonded network. At low $\lambda$, $R_1$ depends logarithmically on $\omega$, suggesting bidimensional diffusion of protons in interfacial regions at hydrated polymer surfaces in pores.

QENS studies on the dynamics in PEMs suggest that water and protons attain microscopic mobilities that are similar to those in bulk water, with mean jump times of protons in the order of 1 ps. The experimental scattering data include two components, which correspond to "fast" ($\Delta t \sim O\,(1\,\mathrm{ps})$) and "slow" ($\Delta t \sim O\,(100\,\mathrm{ps})$) motions. Observation of the latter type of motion suggests that hydronium ions exist as long-lived entities in Nafion (Perrin et al., 2006).

The local (subscript "t") and long-range (subscript "lr") diffusion coefficients of water, probed by QENS, vary from $D_t = 0.5 \cdot 10^{-5}$ cm$^2$ s$^{-1}$ to $2.0 \cdot 10^{-5}$ cm$^2$ s$^{-1}$ and $D_{lr} = 0.1 \cdot 10^{-5}$ cm$^2$ s$^{-1}$ to $0.6 \cdot 10^{-5}$ cm$^2$ s$^{-1}$, for water contents from $\lambda \approx 3$ to 18 (Perrin et al., 2007). For comparison, the water self-diffusion coefficient in Nafion measured by PFG NMR is $D_s = 0.58 \cdot 10^{-5}$ cm$^2$ s$^{-1}$ at $\lambda = 14$ (Kreuer, 1997; Zawodzinski et al., 1993a,c). As can be seen in Figure 2.7 at $\lambda > 10$, the long-range diffusion coefficient of water in Nafion PEMs measured by QENS, $D_{lr}$, is close to the self-diffusion coefficient from PFG-NMR, $D_s$, that probes mobilities at scales above 0.1 $\mu$m. Long-range diffusion coefficients are reduced by approximately a factor 4 with respect to the water self-diffusion coefficient in bulk water, which is $D_w^b = 2.69 \cdot 10^{-5}$ cm$^2$ s$^{-1}$. This reduction can be attributed to random network effects, that is, pore space connectivity and tortuosity. On the other hand, the local diffusion coefficient of water in PEMs at the subnanometer scale determined from QENS, $D_t$, closely approaches the value of $D_w^b$.

The comparison of local and long-range diffusion demonstrates that in well-hydrated PEMs the major geometric constraints for water mobility unfold at the scales probed by QENS, which is up to several tens of nanometers. This implies that there are no significant restrictions of mobility arising between the nanometric and the micrometric scales, when the membrane is sufficiently hydrated. Similar considerations could be made for proton diffusivities.

These experimental insights bear an important message for the molecular modeling of PEMs: the main effects of PEM structure on transport properties are captured if simulation boxes with size of $\sim$100 nm are used. Multiscale modeling approaches carried out to larger scales will not generate decisive new insights on intrinsic PEM properties. Notably, operation at ultralow hydration emphasizes even more

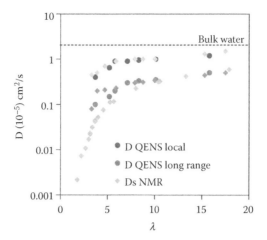

**FIGURE 2.7** Local ($D_t$) and long-range ($D_{lr}$) diffusion coefficients of water in Nafion, probed by QENS, illustrating enhanced water dynamics upon increasing the membrane water content. Self-diffusivity of water in Nafion probed by PFG-NMR and self-diffusivity of bulk water are given for the sake of comparison. (Reprinted with permission from Perrin, J. C., Lyonnard, S., and Volino, F. 2007. Quasielastic neutron scattering study of water dynamics in hydrated Nafion membranes. *J. Phys. Chem. C.*, **111**(8), 3393–3404. Copyright (2007) American Chemical Society.)

the importance of nanoscale phenomena controlled by explicit interactions in the polymer–water–proton system.

## THEORY AND MODELING OF STRUCTURE FORMATION IN PEM

The self-organized structure and mechanical properties of *ionomer aggregates* determine the water sorption properties and the stability in PEMs. These properties in turn are the key to PEM operation in PEFCs: they govern water distribution and transport, proton density and conductivity, and the membrane response to mechanical stressors.

In a broader context, charged polymers control properties and function of a range of key materials in biological and electrochemical systems. They can be classified according to molecular composition and structure of the backbone, length and density of acid-bearing molecular groups at the backbone, and chemical structure of acid groups.

### AGGREGATION PHENOMENA IN SOLUTIONS OF CHARGED POLYMERS

*Aggregation phenomena* are a common characteristic in solutions of charged polymers. They depend on the interplay of long-range electrostatic and short-range hydrophobic interactions, as well as entropic effects (Henle and Pincus, 2005). Aggregation or *bundle formation* occur on limited sections of the polymer, which can be treated as stiff rods. The length of these stiff segments is typically on the order of the polymer persistence length (Rubinstein and Colby, 2003).

The aggregation behavior of biological polyelectrolyte molecules, such as DNA or F-actin, has a critical impact on their biological function. Studies that strive to understand fundamental interactions and aggregation mechanisms in polyelectrolyte solution have a long history (Dobrynin and Rubinstein, 2005; Ha and Liu, 1999; Kornyshev and Leikin, 1997; Manning, 1969, 2011; Oosawa, 1968; Rouzina and Bloomfield, 1996). In salt solutions of highly charged polyelectrolyte chains, such as DNA, condensation of multivalent counterions could generate an effective attraction, which induces aggregation of chains.

Ionomers are sparsely charged polymers, with typically less than 15 mol% of charged groups. The case of ionomers aggregating in the sole presence of monovalent counterions has been studied much less than polyelectrolyte aggregation, in spite of the importance of bundle formation and phase separation for the properties of PEMs. What are the driving forces for bundle formation in ionomer solution? Contrary to polyelectrolytes, ionomers are not charged enough for the electrostatic interaction to be the predominant interaction. Aggregation can only be observed when other forces are at play, which is the case in ionomer systems. Uncharged hydrophobic polymers, like Teflon, exhibit complete phase separation from an aqueous solution. Hence, for an ionomer that is a copolymer of charged sidechains with hydrophobic backbones, the hydrophobicity of backbone segments provides a sufficiently strong complementary interaction. Aggregation in ionomer solution is, therefore, a consequence of the interplay of electrostatic and hydrophobic interactions.

**Theory of Ionomer Bundle Formation**

A physical theory of bundle formation in ionomer solution, specifically geared toward PFSA-type ionomers, was proposed by Melchy and Eikerling (2014). The model system consists of an aqueous solution of charged rigid rods, which represent stiff segments of ionomer molecules. Rods are assumed to be sufficiently diluted and not connected chemically with each other. Moreover, they are considered to be of equal size and length. The basic process explored is the aggregation of rods into close-packed cylindrical bundles, characterized by an aggregation number $k$. Bundles are surrounded by water, which does not contain added electrolyte ions. The dissociation of acid-bearing ionomer groups releases protons into the water phase. The bundle model is illustrated in Figure 2.8.

The calculation of the equilibrium bundle size employs a mean field approach, which can be rationalized in several steps. In the first step, the system is divided into noninteracting cells of equal size, each containing a single bundle in the core and an outer electrolyte shell that accommodates dissociated and hydrated protons. In the second step, the ensemble of cells is replaced by a single effective cell with cylindrical geometry, concentric with the bundle in the core. The length of the cell is assumed to be the same as the rod length $L_R$. The cell radius $r_C$ is a function of the density $\rho$ of rods

$$r_C = \sqrt{\frac{k}{\pi \rho L_R}}. \tag{2.7}$$

Figure 2.9a displays $r_C$ and $r_B$ as functions of $k$, the number of rods per bundle.

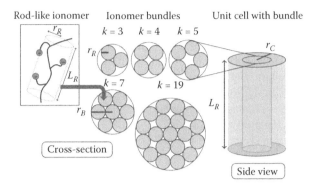

**FIGURE 2.8** A schematic depiction of Nafion ionomer and its representation as a cylindrical rod of radius $r_R$ and length $L_R$. At this level, fixed anionic charges are described by a continuous surface charge density $\sigma$. The middle section illustrates the aggregation of rod-like ionomer molecules into cylindrical bundles with minimal radius $r_B$ for a given aggregation number $k$, assuming close packing of rods into bundles. The mean field approach assumes that a single effective cell (depicted on the right) consists of a bundle in the core, surrounded by a concentric electrolyte shell with radius $r_C$, which accommodates the dissociated protons. Bundle cores are assumed to be electrolyte- and proton-free.

The model assumes that the space between rods inside a bundle is electrolyte-free. Therefore, acid dissociation could only occur at the bundle surface. The degree of acid dissociation in the bundle determines the charge density at the surface of the bundle, assumed to be continuous,

$$\sigma = \eta k \frac{r_R}{r_B} \sigma_R, \tag{2.8}$$

and, thus, the number of protons per unit cell. In Equation 2.8, $\sigma_R$ is the ideal surface charge density of the rod obtained if all ionizable groups were ionized. The fraction $\eta$ of ionized groups depends on the properties of sidechains. Specific dissociation scenarios are illustrated and explained in Figure 2.9b.

In the case of only one type of ionic species in the solution, namely, dissociated protons, electrostatic interactions would drive the system toward a dispersed state, if considered as the sole type of interaction. However, backbone hydrophobicity constitutes a significant driving force for aggregation, reverting this trend.

Protons are represented by a continuous distribution $n_H$. Electrostatic interactions and entropy effects in the proton subsystem are treated implicitly, using the Poisson–Boltzmann approach that is solved in Debye–Hückel approximation. This simplifying treatment is well suited for problems in weak electrolytes, and it was successfully used to describe electrostatic interactions between biological macro-ions, more specifically DNA (Kornyshev and Leikin, 2000). Electrostatic interactions between protons and anionic surface groups at rods cause a net repulsive interaction between these rods. This contribution is given by the electrostatic energy of the proton density $n_H$ in a cylindrical cell that has at its core a bundle with a negatively charged

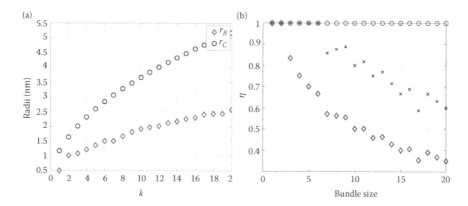

**FIGURE 2.9** (a) Radius $r_B$ of the bundle and $r_C$ of the unit cell as a function of the number $k$ of rods in the bundle. Results are depicted for rod length $L_R = 20$ nm and rod radius $r_R = 0.5$ nm. (b) Fraction of dissociated acid groups per bundle as a function of bundle size. The lower limiting curve corresponds to minimal dissociation for the case of zero length sidechains, $l_{SC} \to 0$. The case with $\eta = 1$, corresponding to perfect dissociation, represents the limit of long sidechains, $l_{SC} \gg r_R$. The intermediate case is $l_{SC} \simeq r_R$, representing a sidechain length that is comparable to the thickness of one layer of rods.

surface. Hydrophobic interactions are accounted for using an interfacial energy determined by the surface tension and by the surface area of the cylinder surface that encapsulates the bundle.

In the treatment sketched above in abbreviated form, the free energy density of a $k$-bundle is given by

$$f(k) = L\rho(\tilde{f}_k - \tilde{f}_1) - T(s_k - s_1),\qquad(2.9)$$

where

$$s_k = -\rho_0 k_B \left( x_k \ln x_k - (1 + x_k)\ln(1 + x_k) \right)\qquad(2.10)$$

is the entropy per $k$-bundle expressed as a function of the molar fraction of $k$-bundles relative to water, $x_k = \rho/(kN_A\rho_w)$, and with $N_A\rho_w = 55$ mol L$^{-1}$. In Equation 2.9, $\tilde{f}_1$ is the free energy in the single-rod limit, which is considered to be a baseline of fixed energy, while $\tilde{f}_k$ is the contribution per rod and per unit length of a $k$-bundle in an isolated cell,

$$\tilde{f}_k = \frac{2\pi}{k} \left\{ \left[ \frac{\gamma}{\beta} - \frac{\sigma}{q}E_{solv} + \int_0^\sigma d\sigma' \varphi_{r_B}(\sigma') \right] r_B + q \int_{r_B}^{r_C} r\, dr\, \varphi(r) n_H(r) \right\}.\qquad(2.11)$$

Here, $1/\beta = k_B T$, $E_{solv}$ is the dissociation and solvation energy of acid head groups, $\gamma$ is the surface tension and $\varphi$ is the potential in the surrounding electrolyte. The first term in square brackets on the right-hand side of Equation 2.11 accounts for hydrophobic interactions. The second term represents the anion solvation energy. The third term expresses the electrostatic repulsion between rods as the energetic cost to build the bundle surface with surface charge $\sigma$. The last term, in curly brackets, represents electrostatic interactions between $k$-bundle and dissociated protons. The proton density $n_H$ is found as the solution of Poisson–Boltzmann equation, evaluated for the cylindrical cell geometry and in Deby–Hückel approximation.

## Stable Bundle Size: Configuration Diagrams

The main outcome of this theory is the stable bundle size $k$ that is obtained by minimization of $f(k)$. The bundle size can be evaluated as a function of geometric parameters of single rods, $r_R$ and $L_R$, rod density, $\rho$, surface charge density, $\sigma$, surface tension, $\gamma$, solvation energy, $E_{solv}$, and the dissociation parameter, $\eta$.

A typical result is depicted in Figure 2.10. This configuration diagram shows the equilibrium bundle size $k$, as a function of $\sigma$ and $\gamma$ in the intermediate dissociation scenario with $l_{SC} \simeq r_R$.

The stable bundle size in Figure 2.10 diverges, $k \to \infty$, as the limit of vanishing surface charge density, $\sigma \to 0$, is approached. This behavior is characteristic of spontaneous phase segregation in uncharged hydrophobic polymers. Another clearly visible limit in the configuration diagram is the dispersed rod region, in which $k = 1$. It corresponds to the regime, where electrostatics dominates. The boundary of this region is parabolic as can be seen explicitly if the free energy is rewritten in the form

$$\frac{k\tilde{f_k}}{2\pi} = \left[ \frac{\gamma}{\beta} + \frac{4\pi}{\epsilon\kappa} \tilde{\Delta}(r_B)\sigma^2 + \left( \frac{1}{q\beta} - \frac{E_{solv}}{q} \right)\sigma \right] r_B + q \int_{r_B}^{r_C} r\, dr\, \varphi(r) n_H(r). \quad (2.12)$$

The terms in square brackets are responsible for the parabolic shape. Between these two limits (complete phase segregation and dispersed rod limit), a cascade of transitions is seen between stable configurations. The main trends are that the aggregation number of stable bundles, $k$, decreases as $\sigma$ increases and $\gamma$ decreases. At sufficiently large $\sigma$, $k$ attains a value of 9. This bundle size allows $\sim$90% of the sidechains to

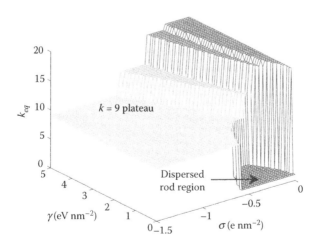

**FIGURE 2.10** The configuration diagram for rods of radius $r_R = 0.5$ nm and length $L = 20$ nm, in the intermediate dissociation scenario with $l_{SC} \simeq r_R$ (cf. Figure 2.9b). The solvation energy is $E_{solv} = -0.2$ eV.

protrude to the bundle surface. The penalty in solvation energy because of incomplete dissociation of ionizable groups is an important parameter in determining the stable size of ionomer bundles.

The sensitivity of bundle formation to dissociation scenarios and to the solvation energy is clearly discernible from a detailed model analysis. It was found that the minimal dissociation scenario, corresponding to zero length sidechains, $l_{SC} \to 0$ (acid groups directly attached to ionomer backbone), results in a strong preference for a nonaggregated state, that is, the dispersed rod limit $k = 1$. The perfect dissociation scenario with $\eta = 1$ favors a strong phase segregation. However, the assumption $l_{SC} > r_B$ underlying this case is bound to fail in the strong segregation limit, since sidechains have a finite length, typically with $l_{SC} \sim r_R$. Essentially, the extent of phase segregation and the size of bundles is determined by the length of sidechains and their flexibility, which must be sufficient for acid head groups to reach to the surface of bundles.

All the major trends discussed in this section are in agreement with experimental findings on the impact of sidechain properties and sidechain grafting density on *ionomer aggregation*.

## MOLECULAR MODELING OF SELF-ORGANIZATION IN PEMS

Basic requirements on feasible systems and approaches for *computational modeling* of fuel cell materials are: (i) the computational approach must be consistent with fundamental physical principles, that is, it must obey the laws of thermodynamics, statistical mechanics, electrodynamics, classical mechanics, and quantum mechanics; (ii) the structural model must provide a sufficiently detailed representation of the real system; it must include the appropriate set of species and represent the composition of interest, specified in terms of mass or volume fractions of components; (iii) asymptotic limits, corresponding to uniform and pure phases of system components, as well as basic thermodynamic and kinetic properties must be reproduced, for example, density, viscosity, dielectric properties, self-diffusion coefficients, and correlation functions; (iv) the simulation must be able to treat systems of sufficient size and simulation time in order to provide meaningful results for properties of interest; and (v) the main results of a simulation must be consistent with experimental findings on structure and transport properties.

Requirement (i) must be enforced in a rigorous fashion. The only possible relief is that in some cases the laws of classical mechanics suffice to establish a consistent approach, rendering computationally intensive quantum mechanical approaches obsolete. Other requirements on the above list offer room for adaptation to peculiar simulation objectives, priorities in terms of physical properties of interest and structural system specifications.

Studies of proton transport in PEMs or at interfaces, as well as studies of processes at the electrified interface, usually demand quantum mechanical simulations to incorporate electronic structure effects and hydrogen bond dynamics. Studies of structure formation and transport properties in heterogeneous media demand computationally efficient algorithms that enable simulations of sufficient length ($>20$ nm) and time

(>20 ns). In *mesoscale simulations*, it is often not sufficient to average out electronic degrees of freedom, by which a quantum mechanical simulation is transformed into a classical all-atom simulation. In many cases, further simplification is required that involves replacement of individual atoms by groups of atoms, employing a so-called coarse graining procedure. This procedure should not only transform the structural representation of the system but it defines a force field that accounts for fundamental interactions and leaves unaltered structural and dynamic properties of interest.

A lot of effort has been recently focused on developing so-called multiscale approaches that incorporate a wide range of scales. The challenge of these simulations is the multiplication of errors occurring by linking simulations that are optimized for distinct length scales. The links involve the transfer of a number of parameters, each subject to uncertainty. Proper calibration of multiscale approaches, thus, requires a good intuition of the researcher and vast experimental data sets for testing and optimization.

A wide range of computer simulations at meso-, micro-, and atomistic levels have been employed to complement experiment and theory in understanding self-organization and structure-related transport properties of PEMs (Cui et al., 2007; Devanathan et al., 2007a,b; Elliott and Paddison, 2007; Elliott et al., 1999; Galperin and Khokhlov, 2006; Goddard et al., 2006; Khalatur et al., 2002; Mologin et al., 2002; Spohr, 2004; Spohr et al., 2002; Venkatnathan et al., 2007; Vishnyakov and Neimark, 2000, 2001; Wescott et al., 2006; Zhou et al., 2007). Similar approaches are being increasingly adopted in studies of catalyst layers, as will be discussed in the section "Mesoscale Model of Self-Organization in Catalyst Layer Links" of Chapter 3.

At least three major scales must be distinguished in *simulations of heterogeneous media*: (i) the atomistic scale, required to account for electronic structure effects in catalytic systems or for molecular and hydrogen bond fluctuations that govern the transport of protons and water; (ii) the scale of the electrochemical double layer, ranging from several Å to a few nm; at this level, simulations should account for potential and ion distributions in the metal–electrolyte interfacial region; and (iii) the scale of about 10 nm to 1 μm, to describe transport and reaction in heterogeneous media as a function of composition and porous structure.

Computational approaches that have been employed to understand the structure and transport properties of water and protons in swollen Nafion membranes include *ab initio* (Eikerling et al., 2003; Roudgar et al., 2006, 2008; Vartak et al., 2013), classical all-atom (Cui et al., 2007; Devanathan et al., 2007a,b; Goddard et al., 2006; Spohr et al., 2002; Vishnyakov and Neimark, 2000, 2001), and coarse-grained descriptions (Galperin and Khokhlov, 2006; Wescott et al., 2006) of the system.

*Ab initio* simulation techniques derive interactions within the system from quantum mechanical principles, computing forces that act on the atomic nuclei in the system by solving the electronic structure problem "on the fly", that is, at each nuclear configuration (Marx and Hutter, 2009). *Ab initio* techniques are highly accurate. However, this accuracy comes at the cost of excessive computational requirements. Their range of applicability is limited to a narrow window of lengths and times that corresponds to the microscopic scale (sizes of ~1 nm and time spans of ~100 ps). The size of systems that is required for meaningful structural studies of hydrated ionomer systems limits the utility of *ab initio* simulations.

*Molecular simulations* of ionomer systems that employ classical force fields to describe interactions between atomic and molecular species are more flexible in terms of system size and simulation time but they must fulfill a number of other requirements: they should account for sufficient details of the chemical ionomer architecture and accurately represent molecular interactions. Moreover, they should be consistent with basic polymer properties like persistence length, aggregation or phase separation behavior, ion distributions around fibrils or bundles of hydrophobic backbones, polymer elastic properties, and microscopic swelling. They should provide insights on transport properties at relevant time and length scales. Classical all-atom molecular dynamics methods are routinely applied to model equilibrium fluctuations in biological systems and condensed matter on length scales of tens of nanometers and timescales of 100 ns.

*Coarse-grained mesoscopic models* offer the largest degree of flexibility in terms of system representation, as well as the feasible range of length and timescales of simulations. In a coarse-grained (CG) model, the fundamental interacting species are not atoms, but groups of atoms, which form beads of the coarse-grained representation. Beads interact via effective forces. The mesoscopic CG approach averages out microscopic degrees of freedom and microscopic fluctuations. Simulations of larger systems and longer time spans can be done, which are relevant to biological matter and complex heterogeneous materials like PEMs or catalyst layers in PEFCs. The reduced resolution of CG models could lead to tremendous gains in computational efficiency. Appropriately constructed and calibrated CG models and force fields can capture major interactions and relevant transitions involved in complex processes such as protein folding. However, the structural representation and force field parameterization are ambiguous. In many cases, the outcome of coarse-grained simulations is limited to qualitative insights and even qualitative conclusions could be misleading. Excessive computational requirements are replaced by the need for excessive calibration of the structural model and effective interaction parameters. This tedious calibration involves comparison to experimental data for known physical properties as well as systematic fine-tuning of interaction parameters by creating consistent structural correlation functions between all-atom and CG models.

Heavy parameterization of CG simulations could undermine their predictive capabilities. There is an underlying danger that simulation results on structure formation or dynamic properties are not predictive in view of the intrinsic properties of the real physical system but have been built into the simulation during parameterization.

## Atomistic Simulations of PEM Fragments and Substructures

Boundaries between atomistic and coarse-grained simulation approaches are floating. Atomistic simulations of ionomer systems typically employ all-atom representations of water molecules, anionic head groups, and protons. For the remaining components, the use of a coarse-grained or united-atom representation for the $CF_x$ groups in both the fluorocarbon backbone and the sidechains could markedly improve the computational efficiency of atomistic simulations. United-atom force fields permit simulations of substantially larger systems compared to all-atom force fields. For instance, Urata et al. (2005) have employed a united-atom representation of $CF_x$

groups for simulations of systems with 12,000–25,000 atoms over 1.3–2.5 ns. A drawback of a mixed atomistic/coarse-grained representation is that results do not accurately account for the impact of ionomer backbones on morphology and transport. In any case, the all-atom approach is computationally demanding, but it is needed to establish rigorous benchmarks and refine force field parameterizations for simulations with coarse-graining.

In general, results of fully atomistic or mixed representations have confirmed the formation of a microphase-separated morphology in PEMs, albeit results on sizes, shapes, and distributions of phase domains have remained inconclusive. In comparison to experiment, molecular models were found to underestimate the sizes of ionic clusters.

Furthermore, atomistic MD studies shed light on the distinction of different types of water. Bound water exhibits suppressed water dynamics as a consequence of having strong electrostatic interactions with charged sulfonate groups. More loosely attached "free" water was found to exhibit bulk-like properties, in agreement with experimental findings on the dynamics of water in PEM (Elliott et al., 2000; James et al., 2000; Urata et al., 2005). A frequent exchange was observed between bound water molecules at interfaces and "free" water molecules.

Dupuis and coworkers (Devanathan et al., 2007a,b; Venkatnathan et al., 2007) simulated the effect of temperature and hydration on membrane nanostructure and mobility of water and hydronium ions. They used classical MD simulations with DREIDING (Mayo et al., 1990) and modified AMBER/GAFF force fields. Their studies showed that interfacial sulfonate groups drift apart with increasing $\lambda$, a finding that is in qualitative agreement with previous MD studies. Simulations suggest that most water molecules and hydronium ions are bound to sulfonate groups at $\lambda < 7$. In comparison with experimental data, these simulations, seemingly overestimate sulfonate–water interactions. Analyzed were the density of the hydrated polymer, radial distribution functions (RDFs) of water, ionomer, and protons, coordination numbers of sidechains, and diffusion coefficients of water and protons. The diffusion coefficient of water agreed well with experimental data, while the diffusion coefficient of hydronium ions was smaller by a factor of 6–10 relative to the value for bulk water.

Jang et al. (2004), using an all-atom MD approach, explored the effect of sidechain distribution at the backbone on structure formation. They showed that blocky Nafion ionomers with highly nonuniform distributions of sidechains at polymer backbones, form larger phase-segregated domains compared to systems with uniform distributions of sidechains at backbones.

Elliott and Paddison (2007) used the ONIOM method of QM/MM calculations (Vreven et al., 2003) to understand the effects of hydration on the local structure of PFSA membranes. Calculations were performed on fragments of a *short sidechain* (SSC) PFSA ionomer with three sidechains affixed to the backbone. Full optimizations of the oligomeric fragment was carried out with six to nine water molecules added. The lowest energetic state was found with six water molecules. Upon addition of further water molecules, the energetic preference for uniform hydration of sidechains via interconnected water clusters disappeared. The optimized structures of a system of two oligomeric fragments at $\lambda = 2.5$ showed that the structure with kinked backbone was energetically preferred by $\sim$37 kJ mol$^{-1}$ over the one, in which the fluorocarbon backbone was fully extended. Paddison and Elliott concluded that conformation of the backbone, sidechain flexibility, and degree of association and aggregation of sidechains under low hydration determine the formation of protonic species in water (Zundel and/or Eigen ions). However, these calculations for single ionomer chains do not account for ionomer aggregation. Therefore, they ignore important correlation effects between backbones, sidechains, protons, and water.

## Mesoscale Simulations

Atomistic molecular dynamic simulations are unable to predict the structure-related properties of PEM at longer time (>10 ns) and length scales (>10 nm). Mesoscale models can bridge the scale gap that exists between the chemical structure of the ionomer and phase-separated morphology of the self-organized membrane. The section "Coarse-Grained Molecular Dynamics Simulations" presents a detailed description of coarse-grained MD studies of structure formation in PEMs. Other simulations that rely on coarse-grained models of the ionomer-solvent system are reviewed in this section.

The first mesoscale simulations of hydrated Nafion were based on a hybrid *Monte Carlo/reference interaction site model* (MC/RISM) (Khalatur et al., 2002). This method uses a combination of an MC routine and *rotational isomeric state* (RIS) theory, developed originally by Flory (Flory, 1969). Khalatur et al. (2002) used a coarse-grained representation, in which each $CF_2$ or $CF_3$ group was represented by a united atom, with a uniform distribution of sidechain groups along the backbone. Results of these calculations suggest that water and polar sulfonic acid groups segregate into a three-layer structure with a central water-rich region and two outer layers of sidechain groups strongly associated with water molecules. In agreement with experiment (Mayo et al., 1990), they found a linear dependence of microscopic swelling on $\lambda$, attributed to the growth of water-filled domains between ionomer fibrils or bundles.

CG models based on dynamic *self-consistent mean field* (SCMF) theory have recently been developed to study the structure of hydrated ionomers at varying $\lambda$. Each sidechain and backbone is constructed of a number of coarse-grained segments (beads), which represent groups of several atoms. The interaction parameters and bead sizes can be obtained by calibration with atomistic MD. In the SCMF approach, the density distribution of beads, $\rho(r)$, evolves under the influence of a slowly varying external potential, $U(r)$, relative to which polymer chains are equilibrated instantaneously. The main assumption of SCMF theory is that the external potential, acting on the ideal system, generates a density distribution, which matches that of the interacting system. The free energy functional consists of terms for the beads in the external potential with the addition of a Gaussian coil stretching term.

Bead–bead interaction parameters can either be generated by force-matching to classical atomistic MD or calculated from Flory–Huggins parameters (Flory, 1969; Galperin and Khokhlov, 2006; Groot, 2003; Groot and Warren, 1997; Wescott et al., 2006). Simulations suggest that at low water content ($\lambda < 6$), isolated hydrophilic domains of anionic surface groups, hydronium ions and water are spherical, while at higher water content ($\lambda > 8$), they deform into elliptical shapes. Since these levels of water content lie significantly above the values required to achieve bulk-like proton conductivity, it was concluded that there may be proton transport through water-depleted regions, either by interfacial diffusion or through a second ionic phase.

Another method that has been applied to predict the mesoscopic structure of hydrated Nafion membranes is dissipative particle dynamics (DPD). This mesoscopic method was introduced in the 1990s for simulations of complex fluids (Hoogerbrugge and Koelman, 1992). It uses a CG model for Nafion and its force-handling scheme is based on the Langevin equation, a stochastic differential equation that describes Brownian motion (Groot, 2003; Groot and Warren, 1997; Yamamoto and Hyodo, 2003). The time evolution of interacting particles is governed by Newton's equations. The total force acting on a particle entails contributions from a conservative force, a dissipative force, a pair-wise random force, and a binding spring force. Conservative interactions are parameterized on the basis of Flory–Huggins parameters.

DPD simulations have been analyzed in view of the microsegregated structure of hydrated Nafion at various $\lambda$ (Hayashi et al., 2003; Vishnyakov and Neimark, 2005; Yamamoto and Hyodo, 2003). A typical structure is depicted in Figure 2.11. Size and separation distance of ionic clusters were found to increase approximately linearly with $\lambda$. Upon increasing $\lambda$, the membrane undergoes a *percolation transition* from isolated hydrophilic clusters to a 3D network of randomly interconnected water channels. Wu et al. (2008) applied extensive DPD simulations to compare the morphologies of Nafion, *short sidechain* (SSC) PFSA PEMs of Solvay–Solexis, and 3M

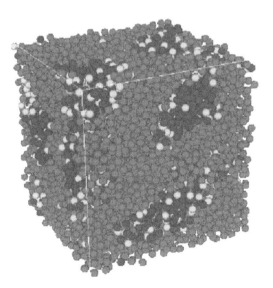

**FIGURE 2.11** A snapshot of a the microstructure of Nafion at $\lambda = 9$ obtained form DPD simulation. Nafion backbone is shown in dark and the first sidechain bead in light. The second sidechain bead, water and hydronium ions are shown in medium grey, defining the ionic cluster region. (Reprinted with permission from Malek, K. et al. 2008. *J. Chem. Phys.*, **129**, 204702, Figures 1,2,5,6,9,10. Copyright (2008), American Institute of Physics.)

PFSA PEMs at various hydration levels and IECs. They found that longer sidechains lead to the formation of larger aggregates of sulfonate groups and consequently to larger water clusters, with cluster sizes varying from 2 to 13 nm for $5 < \lambda < 16$.

In spite of their many assets, DPD and SCMF methods are not able to accurately predict physical properties that rely upon time correlation functions, for example, diffusion coefficients. The more feasible mesoscale approach for hydrated ionomer membranes is *coarse-grained molecular dynamics* (CGMD) simulations.

One should notice an important difference between CGMD and DPD techniques. In comparison to the DPD technique, CGMD is essentially a multiscale method (parameters are directly extracted from classical atomistic MD), and it has a different force field handling scheme, as described below. Angular and dihedral interactions, which are ignored in DPD, are included in CGMD to account for the conformational flexibility of ionomer molecules.

## COARSE-GRAINED MOLECULAR DYNAMICS SIMULATIONS

In recent years, CGMD simulations have found a wide range of applications in soft materials and biomolecular systems. In these systems, the CGMD approach has proven highly insightful and reliable (Voth, 2008).

In order to improve the computational efficiency of molecular simulations, the CGMD approach groups atoms or molecules together into new coarse-grained sites or "beads," which experience redefined effective interactions. The task at hand is to devise a coarse-grained representation of molecular structure and effective interactions that retains essential chemical and physical characteristics of the system. Consistency of the CGMD approach has to be established by benchmarking against atomistic MD and by comparison with experimental data. Force field or effective interaction parameters must be fine-tuned such that an acceptable agreement can be achieved between coarse-grained and all-atom models for structural correlation functions and transport parameters.

In the rigorous formulation of statistical mechanics, the coarse-grained system must represent a reasonable approach in view of the following formula:

$$\exp\left(-F_H/k_BT\right) = C \int d\mathbf{r} \, \exp\left[-V(\mathbf{r})/k_BT\right]$$

$$\approx C' \int d\mathbf{r}_{CG} \, \exp\left[-V_{CG}(\mathbf{r}_{CG})/k_BT\right], \qquad (2.13)$$

where $F_H$ is the Helmholtz energy of the system, $V(\mathbf{r})$ is the system potential energy as a function of all atomistic coordinates, $\mathbf{r}$, and $V_{CG}(\mathbf{r}_{CG})$ is the corresponding potential of the coarse-grained system, which is a function of coordinates of coarse-grained particles. Finding a representation of $V_{CG}(\mathbf{r}_{CG})$ that fulfills Equation 2.13 is the main challenge of CGMD approaches. Coarse-graining schemes relying on "minimalist," "inversion," or "multiscale" techniques are discussed in the compiled volume by Voth (2008).

## Coarse-Grained Model and Simulation Protocol

In the following, a recipe is provided of typical CGMD simulations of the PEM microstructure. In the first step, the atomistic system must be mapped onto a coarse-grained model, in which spherical beads with predefined, subnanoscopic length scale replace groups of atoms. The CGMD model uses polar, nonpolar, and charged beads to represent water, polymer backbones, anionic sidechains, and hydronium ions in a hydrated ionomer system (Marrink et al., 2007).

In the second step, parameters of renormalized interaction energies between beads are specified, defining the force field under which the system trajectory evolves. Interactions between beads could be determined by force matching procedures from atomistic interactions (Izvekov and Violi, 2006; Izvekov et al., 2005) or by fitting of experimental structural correlation functions (Marrink et al., 2007).

After defining the structural representation of the coarse-grained system and inter-action terms, a simulation protocol has to be devised. This protocol involves an equilibration or thermalization phase, as well as production runs. In a production run, the system trajectory is simulated under well-defined thermodynamic conditions, over a specified period of time, in order to generate a statistically meaningful ensemble of uncorrelated system configurations. In the final step, these configurations are analyzed using RDF, and density maps that provide direct information on size, shape, and distribution of phase domains in the composite medium.

Results can be compared to all-atom simulations for calibration and refinement of interaction parameters. Further analysis could involve Monte Carlo procedures in order to obtain shape and size distributions of ionomer and water clusters from the structures formed during CGMD simulations. Moreover, stable structures from CGMD studies could form the basis for mobility studies by atomistic MD simulations, using remapping procedures.

Malek et al. (2008) employed a typical coarse-grained parameterization of hydrated Nafion. Clusters of four water molecules are represented by polar beads. Clusters of three water molecules and a hydronium ion correspond to charged beads. An ionomer sidechain is represented by a single charged bead. A four-monomeric unit of PTFE (–[–CF2–CF2–CF2–CF2–CF2–CF2–CF2–CF2–]– on the backbone, corresponds to a nonpolar bead. All beads are assigned a radius 0.43 nm and a volume of 0.333 nm$^3$.

The interactions between nonbonded uncharged beads in CGMD simulations are modeled by the Lennard–Jones (LJ) potential

$$U_{LJ}\left(r_{ij}\right) = 4\varepsilon_{ij}\left[\left(\frac{\sigma_{ij}}{r_{ij}}\right)^{12} - \left(\frac{\sigma_{ij}}{r_{ij}}\right)^{6}\right], \tag{2.14}$$

where the effective bead radius is assumed as $\sigma_{ij} = 0.43$ nm. The strength of inter-actions $\varepsilon_{ij}$ could assume five possible values ranging from weak (1.8 kJ mol$^{-1}$) to strong (5 kJ mol$^{-1}$). Charged beads $i$ and $j$ interact via coulombic interactions:

$$U_{el}\left(r_{ij}\right) = \sum_{i<j} \frac{q_i q_j}{r_{ij}}. \tag{2.15}$$

**FIGURE 2.12**  Coarse-grained representation of a Nafion chain as a 20-unit oligomer with 40 backbone beads (dark spheres) and 20 sidechain beads (light spheres). This composition corresponds to an equivalent weight of 1100 g of dry polymer per mole of ion exchange sites. (Reprinted with permission from Malek, K. et al. 2008. *J. Chem. Phys.*, **129**, 204702, Figures 1,2,5,6,9,10. Copyright (2008), American Institute of Physics.)

Interactions between chemically bonded beads in ionomer chains are modeled by harmonic potentials for bond length and angle:

$$U_{\text{bond}}\left(r_{ij}\right) = \frac{1}{2}K_{\text{bond}}(r_{ij} - r_0)^2$$

$$U_{\text{angle}}\left(\theta_{ij}\right) = \frac{1}{2}K_{\text{angle}}\left[\cos\theta - \cos\theta_0\right]^2 , \qquad (2.16)$$

where $K_{\text{bond}} = 1250$ kJ mol$^{-1}$nm$^{-2}$ and $K_{\text{angle}} = 25$ kJ mol$^{-1}$ are force constants, and $r_0$ and $\theta_0$ are equilibrium bond length and angle (Marrink et al., 2007).

The simulation box considered in Malek et al. (2008) had a size of $(25 \times 25 \times 25)$ nm$^3$. It contained 72 coarse-grained Nafion chains, as depicted and described in Figure 2.12. 1440 CG hydronium ions were added for electroneutrality. A varying number of water beads was added to simulate systems with varying water contents, corresponding to $\lambda = 4, 9, 15$.

The typical simulation proceeds according to the following sequence of steps, as implemented in Malek et al. (2008):

1. Starting from a random initial configuration, an energy minimization is performed using a steepest descent algorithm.
2. Thermal annealing is conducted (Allen and Tildesley, 1989), where the structure was first expanded over a period of 50 ps by increasing the temperature from 298 to 398 K. This is followed by a short MD simulation for 50 ps in an NVT ensemble,* and further by a cooling procedure down to 298 K.
3. An equilibration is conducted in NVT ensemble with an integration time step of 0.05 ps. The temperature is controlled by a Berendsen thermostat, which mimics a weak coupling to an external heat bath at 298 K (Berendsen et al., 1995; Lindahl et al., 2001). The equilibrium structure is determined by monitoring the total energy. In Malek et al. (2008) the total energy was found to decrease steeply during an equilibration period of 0.05 µs, after which it converged and became stable. The final density of 1.7 g cm$^{-3}$ agreed with the experimental value.

---

* This is the short-hand notation for a canonical ensemble, which represents the possible states of a statistical many particle system that is in thermal equilibrium with a heat bath; in such an ensemble, particle number (N), volume (V) and temperature (T) are constant.

4. Production runs for the statistical sampling of configurations are performed in an NVT ensemble at 298 K for up to 0.7 μs. Structures are saved every 500 steps (25 ps) and used for the analysis.
5. Using structures formed during production runs, size distributions of ionomer and water clusters are obtained by applying a Monte Carlo procedure (Lindahl et al., 2001). Two water beads belong to the same cluster if the distance between them is smaller than 0.43 nm. The radius of a pore was determined by using a procedure based on the CHANNEL algorithm (Kisljuk et al., 1994). In this algorithm, an initial random position is chosen inside the pore network and at any given distance along the pore axis. Pore sizes are evaluated by calculating the maximum size for a spherical probe to still fit in the pore without overlapping with van der Waals radii of beads in the pore wall.

### Analysis of the Coarse-Grained Membrane Structure

The hydrophilic subphase of water, hydronium ions, and charged sidechain beads forms a three-dimensional network of irregular channels. Typical channel sizes are 1, 2, and 4 nm at water contents of $\lambda = 4, 9, 15$, respectively.

Two-dimensional contour plots depicting density distributions of beads corresponding to Nafion backbones, sidechains, hydronium ions, and water provide insights into the self-organized ionomer morphology at different $\lambda$. The effective density of ionomer backbones decreases and it fluctuates less with increasing $\lambda$. This is mainly a consequence of larger, more connected hydrophilic clusters that form at high $\lambda$. At low $\lambda$, below the percolation threshold, ionic clusters are small and isolated.

The analysis of RDFs, $g_{ij}(\mathbf{r}_{ij})$, provides valuable structural information of simulated many-particle systems. It allows rationalizing structural correlations between atoms or beads, aggregation behavior, and phase separation, as well as sizes, shapes, and coordination structure of distinct phase domains. Experimentally, the RDF is obtained from the structure factor $S(\mathbf{q})$, which determines the intensity of x-ray or neutron scattering (Ashcroft and Mermin, 1976). The structure factor for scattering from phase domains formed by a single component, for example, apolar beads in fibrils of ionomer backbones, is given by

$$S(\mathbf{q}) = 1 + \frac{N}{V} \int_V \left[ g(\mathbf{r}) - 1 \right] \exp(i\mathbf{q} \cdot \mathbf{r}) \, d\mathbf{r}. \tag{2.17}$$

Microscopically, the RDF is defined as the conditional probability density of finding a bead $j$ at a position $\mathbf{r}_{ij}$ relative to the position of a reference bead $i$,

$$g_{ij}(\mathbf{r}_{ij}) = \frac{V}{N_i N_j} \langle \delta(\mathbf{r} - \mathbf{r}_{ij}) \rangle, \tag{2.18}$$

where $V$ is the total sample volume, and $N_i$ and $N_j$ refer to the total numbers of beads of type $i$ and $j$ in the volume. The triangle brackets on the right-hand side indicate an ensemble average. For an isotropic system, the RDF is

$$g_{ij}(r) = \left(\frac{N_j}{V}\right)^{-1} \frac{\Delta n_j}{4\pi r^2 \Delta r},  \qquad (2.19)$$

where $\Delta n_j$ is the number of beads of type $j$ located in a shell of thickness $\Delta r$ at the radial distance $r$ from bead $i$.

The labels $i$ and $j$ could represent beads of the same type or of different types. The ratio of $N_j$ to $V$ corresponds to the particle density of beads of type $j$. Calculation of $g_{ij}(r)$ involves ensemble averaging over a sufficient number of statistically independent configurations, along the simulated trajectory and over statistically independent choices of the reference bead.

For an isotropic system, the RDF, $g(r)$, is related to the potential of mean force between two particles $w^{(2)}(r)$, via (Hansen and McDonald, 2006)

$$g(r) = \exp\left(-\frac{w^{(2)}(r)}{k_B T}\right).  \qquad (2.20)$$

Comparison of $g(r)$ from CGMD and atomistic MD is the means to evaluate the consistency of simulation approaches and refine force-field parameterizations.

Figure 2.13 shows snapshots of equilibrated microstructures of the ionomer–water system at different water content $\lambda$, as indicated in the plot. Figure 2.14 shows corresponding RDFs. These structural correlation functions can be analyzed in view of (i) ionomer aggregation into hydrophobic domains, (ii) formation of ionic domains, (iii) distribution of water and hydronium ions with respect to sidechains and backbones, and (iv) connectivity of water and ionomer networks.

Pair correlation functions of backbone beads, $g_{BB}$, depicted in Figure 2.14a, exhibit several peaks at $r < 1.5$ nm. Figure 2.14a suggests that backbone aggregates are smaller than $\sim$3 nm. This size is consistent with the result of the theory of bundle formation (Melchy and Eikerling, 2014) and experimental data from scattering studies (Gebel, 2000; Rubatat et al., 2004). Pair correlation functions of sidechains, $g_{SS}$,

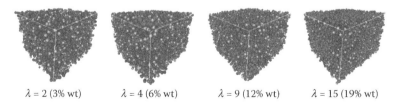

$\lambda = 2$ (3% wt)  $\qquad$ $\lambda = 4$ (6% wt)  $\qquad$ $\lambda = 9$ (12% wt)  $\qquad$ $\lambda = 15$ (19% wt)

**FIGURE 2.13**  Equilibrated microstructures obtained in CGMD simulations of mixtures of Nafion ionomer (equivalent weight of 1100 g of dry polymer per mole of ion exchange sites) with water, obtained at different $\lambda$. Backbone beads are shown in dark and charged sidechain beads in light. Hydrophilic domains containing water and hydronium ions are represented by medium grey beads. (Reprinted from Malek, K. et al. 2008. *J. Chem. Phys.*, **129**, 204702, Figures 1,2,5,6,9,10. Copyright (2008), American Institute of Physics. With permission.)

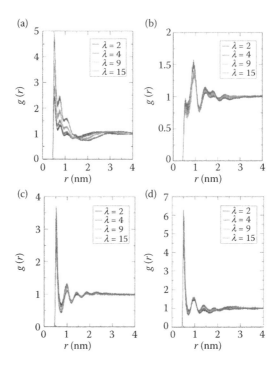

**FIGURE 2.14** Site–site RDFs for equilibrated structures of the ionomer–water system at different $\lambda$. RDFs are depicted for (a) B–B, (b) S–S, (c) S–W and (d) W–W, bead–bead correlations. W: water, S: side chain, H: hydronium and B: Nafion backbone. (Reprinted from Malek, K. et al. 2008. *J. Chem. Phys.*, **129**, 204702, Figures 1,2,5,6,9,10. Copyright (2008), American Institute of Physics. With permission.)

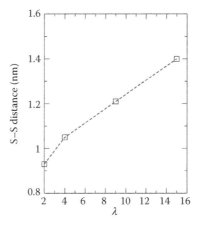

**FIGURE 2.15** Mean separation of sidechains on polymeric aggregates as a function of $\lambda$. (Reprinted from Malek, K. et al. 2008. *J. Chem. Phys.*, **129**, 204702, Figures 1,2,5,6,9,10. Copyright (2008), American Institute of Physics. With permission.)

shown in Figure 2.14b indicate a clustering of sidechains due to the aggregation and folding of polymer backbones (Allahyarov and Taylor, 2007). The pair correlation functions $g_{SW}$ and $g_{WW}$ between sidechain and water beads, shown in Figure 2.14c and d, reveal the existence of three hydration layers at the ionomer–water interface. The degree of ordering of water at the interface decreases with increasing $\lambda$.

Figure 2.15 shows sidechain separations as a function of $\lambda$. As the hydration level increases, sidechains drift apart. Their mean separation is seen to increase from ~1 nm for $\lambda < 4$ to 1.3 nm for $\lambda > 15$. On a single ionomer chain, sidechain separations are in the range from 1.5 to 1.7 nm. Assembly of backbones into bundles or fibrils increases the net density of sidechains at the bundle surface. The observed increase in the interfacial density of sidechains resulting from polymer aggregation indicates that fibrils consist of three to six backbones. Increasing sidechain separations with larger $\lambda$ can be explained by aggregates becoming thinner and more stretched. This trend in sidechain separation could have important implications for proton transport at polymer–water interfaces in water-filled pores, as discussed in the section "Proton Transport."

### Simulation of Transport Properties

Equilibrated structures from CGMD simulations could be used as input for simulations of transport properties in PEMs. The *self-diffusion coefficient of water* is obtained by taking the slope of the *mean-square displacement* (MSD) at long time

$$D = \frac{1}{2dN_\alpha} \lim_{t \to \infty} \left\langle \frac{1}{t} \sum_{i=1}^{N_\alpha} \left| r_i^\alpha(t) - r_i^\alpha(0) \right|^2 \right\rangle, \tag{2.21}$$

where $N_\alpha$ is the number of molecules of component $\alpha$, $d$ is the spatial dimension, $t$ is time, and $r_i^\alpha$ is the center of mass of molecule $i$ of component $\alpha$. The self-diffusion coefficient can, also, be calculated from the time integral of the velocity autocorrelation function

$$D = \frac{1}{dN_\alpha} \int_0^\infty \left\langle \sum_{i=1}^{N_\alpha} v_i^\alpha(t) v_i^\alpha(0) \right\rangle dt, \tag{2.22}$$

where $v_i^\alpha$ is the center of mass velocity of molecule $i$ of component $\alpha$. The first equation is known as the Einstein equation and the second is the Green–Kubo relation (Dubbeldam and Snurr, 2007).

At short timescales ($t < 500$ ps) the MSD exhibits a quadratic time dependence. This is known as the ballistic regime, where particle collisions are infrequent. In nanoporous materials, an intermediate regime starts when particles are colliding but only with a subset of the other particles due to the confinement. When particles are able to escape the local environment and explore the full macroscopic network, the diffusive regime is reached. In the diffusive regime, the MSD becomes linear with time, exhibiting a slope of one on a log-log scale.

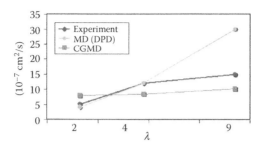

**FIGURE 2.16** The self-diffusion coefficient of water as a function of $\lambda$. Values calculated from simulated MD and CGMD trajectories are compared to PFG-NMR data. (Reprinted from Malek, K. et al. 2008. *J. Chem. Phys.*, **129**, 204702, Figures 1,2,5,6,9,10. Copyright (2008), American Institute of Physics. With permission.)

For the calculation of self-diffusivity in hydrated Nafion, an atomistic structure of the water channel can be generated, based on the mesoscopic structure obtained from DPD or CGMD calculations. This remapped atomistic pore model can be optimized further with atomistic MD calculations. This procedure was employed in Malek et al. (2008) at different $\lambda$.

Figure 2.16 compares diffusion coefficients of water calculated from atomistic MD and CGMD simulations to values from PFG–NMR measurements. There is an agreement between the two simulation methods for $\lambda < 9$. In reference calculations of bulk water, both types of approaches reproduce the experimental value of $2.3 \cdot 10^{-5}$ cm$^2$ s$^{-1}$. In simulations of the hydrated ionomer, discrepancies from experimental values arise at $\lambda > 4$. This deviation can be explained as a finite-size effect of the box used in MD simulations. The discrepancy grows at higher water content, where a longer trajectory length is required for the sampling.

## Morphology Description Based on CGMD

By combining the information from structural analysis and water diffusivity at different $\lambda$, we obtain a model for the morphology of the water network in hydrated Nafion membrane. At low $\lambda$, the membrane interior includes small poorly connected water clusters. At higher $\lambda$, water clusters increase in size due to coalescence of small clusters, forming a highly interconnected random network of hydrophilic domains of water and hydronium ions. Owing to confinement effects, water diffusion coefficients calculated by DPD/MD and CGMD simulations are reduced in comparison to bulk water. The CG model for hydronium ions does not include electrostatic and hydrogen bonding interactions between hydronium ions and water. These simulations overestimate diffusion coefficients of hydronium ions, which are, therefore, not discussed in this section.

Many features of the morphology of hydrated Nafion in CGMD simulations are in agreement with structural characteristics derived from small-angle scattering experiments. At low water content ($\lambda < 4$), semispherical hydrophilic domains are embedded in the hydrophobic domains of backbones. At higher water content ($\lambda > 4$), hydrophobic domains deform into cylindrical shapes due to the emergence and widening of necks. In simulations, the average diameter of hydrophilic domains increases from 1 nm at $\lambda = 2$ to 3–4 nm at $\lambda = 9$. These values lie within the ranges of diameters of water channels found from scattering experiments.

Both CGMD and DPD simulations indicate a *percolation threshold* for hydrophilic domains at $\lambda \sim 4$, which corresponds to a $\sim 10\%$ volume fraction of water. The low value of the conductivity percolation threshold in Nafion PEMs has been concluded as well from proton conductivity studies (Cappadonia et al., 1994, 1995). The high interconnectivity of water channels and the peculiarities of swelling and reorganization of the polymer matrix upon water uptake promote percolation at low water content (Eikerling et al., 1997, 2008).

Other MD simulations predicted markedly larger values of the percolation threshold (Devanathan et al., 2007b; Elliott and Paddison, 2007). Discrepancies in calculated percolation thresholds could be artifacts of overly simplistic representations of ionomer chains in molecular-level simulations. Atomistic models fail in reproducing sizes and shapes of water clusters and polymer aggregates as well as in predicting percolation thresholds, swelling behavior, and related transport properties, if the monomeric sequences that they employ are too short. Notably, for the same reason, many simulations would be inept to reproduce the persistence length of the base ionomer.

## Molecular Modeling of PEMs: What Is Next?

The fundamental objective of molecular modeling of ionomer–water systems is to rationalize dependences of transport properties of water and protons upon changes in the hydration level. A practical goal is to provide predictive models that can be utilized for membrane material selection. Experiments provide empirical insights into the structural evolution upon water uptake and the transport properties. Efforts in materials design demand a systematic understanding of how the chemical architecture affects effective physicochemical properties and electrochemical performance. Any individual simulation technique described in this section falls short in making exact predictions for the morphology and effective properties of PEM materials.

The requirements for self-consistent approaches in molecular modeling and computational PEM research are (i) appropriate structural representation of the primary polymer architecture, (ii) adequate treatment of molecular interactions between components, (iii) sufficient size of the simulated system, allowing effects of nanoscale confinement and random network morphology on water and proton transport to be addressed, and (iv) sufficient statistical sampling of structural configurations or elementary transport processes to obtain reliable estimates of thermodynamic properties and transport parameters. With respect to (i), it is vital that the length of monomeric sequences of the ionomer is larger than the persistence lengths of the polymer backbone, which significantly exceeds 10 nm. Atomistic models often fail in reproducing sizes and shapes of water clusters and polymer aggregates, as well as in predicting percolation properties and swelling behavior of the hydrated membrane because the monomeric sequences they rely on are too short.

The problem of structure formation in PEMs demands a hierachically built modeling framework. Starting with quantum mechanical calculations at atomistic scale, one is able to develop simulation methodologies for proton transport and local electrostatic interactions. Using information on molecular interactions and pair correlation functions obtained from these simulations, models and effective force field parameterizations can be derived for classical atomistic molecular dynamics simulations addressing larger scales. In a further step, built upon kinetic and energetic data, a coarse description can be developed that captures essential parameters in materials synthesis, structural characterization, and transport phenomena of membrane materials for PEFCs.

## WATER SORPTION AND SWELLING OF MEMBRANES

As described in previous sections, primary chemical ionomer structure and thermodynamic conditions control aggregation phenomena in ionomer solution. Molecular modeling and physical theory allow identifying conditions, under which inomer bundles of finite size could form. Moreover, they provide tools to rationalize geometrical, electrostatic, and mechanical properties of such bundles. The properties of ionomer bundles determine microscopic stress–strain relations as well as the density of anionic surface groups at pore walls. As it turns out, the rudimentary steps of bundle formation in ionomer solutions are already complex. Their understanding, thus, remains partially inconclusive. For instance, the role of chemical composition, length, grafting density, and flexibility of sidechains is unresolved. The effect of discrete charge distributions along a hydrophobic backbone is not clear.

The understanding of further steps in the structure formation process, specifically, forming a network of water-filled pores or ionic domains, is bound to much greater uncertainty. A look at debates in the literature and at conferences leaves no doubt about this. This chapter strives to explain how the interplay of elastic and electrostatic properties of bundles determines the total uptake and the distribution of water in the porous network. This interplay holds the key toward unraveling the relation between micro- and macroscopic swelling and toward rationalizing transport properties of PEMs.

### WATER IN PEMs: CLASSIFICATION SCHEMES

The morphology of Nafion has been heavily debated upon over three decades. Likewise, the related problem of *water structure and distribution* has stimulated efforts in experiment and theory. Major *classification schemes of water* distinguish (i) surface and bulk water (Eikerling, 2006; Eikerling et al., 1997, 2007a, 2008), (ii) nonfreezable, freezable-bound, and free water (Nakamura et al., 1983; Seung et al., 2003; Siu et al., 2006), and (iii) water vapor or liquid water (Choi et al., 2005; Elfring and Struchtrup, 2008; Weber and Newman, 2003). *Surface water in pores* strongly interacts with polar surface groups forming a highly oriented network with strong hydrogen bonds at internal polymer–water interfaces (Narasimachary et al., 2008; Roudgar et al., 2008). *Bulk-like water* is identified by the liquid water-like dynamics of protons and water molecules, as discussed in the section "Dynamic Properties of Water and Protons in PEMs."

The existence of surface and bulk-like water can explain the effect of the water content on the microscopic mobility of protons, indicated by the dramatic increase in the activation free energy of proton transport from $\sim 0.1$ eV at $\lambda > 4$ to $\sim 0.36$ eV at $\lambda < 2$, shown in Figure 2.6 (Eikerling et al., 2008; Ioselevich et al., 2004). Moreover, a statistical model of membrane water uptake and proton conductivity, explored in Eikerling et al. (1997, 2001), suggests that conducting elements with restricted surface-like mobility of protons control membrane conductivity at low $\lambda$, while the proton current at large $\lambda$ is carried by percolating clusters of nanoscale elements, which exhibit bulk-water-like proton mobility.

*Pore network models*, based on Gierke's structural model and on cylindrical pore-type models, were developed to account for the transition from surface- to bulk-like conductivity (Eikerling et al., 2001). Relations of conductivity versus water content, calculated with pore network models, are in agreement with experimental data for PFSA-type membranes.

The categorization into nonfreezable, freezable bound, and free water is based on observations of the freezing behavior of water by differential scanning calorimetry (DSC) and NMR (Yoshida and Miura, 1992). DSC has been used to determine the amounts of the different types of water. Nonfreezable water is tightly bound to sulfonic acid head groups. It plasticizes the polymer and lowers its glass transition temperature $T_g$. Freezable water is loosely bound to the polymer, exhibiting a freezing point suppression by up to $\sim 20°C$, which can be explained by the Gibbs–Thomson relation.

A freezing point suppression has also been found by Cappadonia et al. in Arrhenius plots of conductivity data (Cappadonia et al., 1994, 1995). Free water in PEMs possesses the same melting point as bulk water, and it sustains high bulk-like mobility of protons and water. In comparison of the different classification schemes, there seems to be a correlation between surface water and nonfreezable/freezable bound water, but the correspondence is not unique. The distinction of freezable-bound and free water is vague.

The classification into liquid water and vapor inside of the PEM has no physical bearings. It is misguided by empirical efforts in understanding the role of externally controlled conditions on *vapor sorption isotherms*. In employing this distinction, the state of water in the phase outside of the membrane is confused with the state of water in the membrane. Further contributing to this misconception is the frequently cited *Schröder's paradox* (Onishi et al., 2007), which, indeed, refers to a *de facto* difference in membrane water uptake under different thermodynamic conditions. Neither is it realistic to assume the existence of water vapor in the membrane nor is it justified to refer to Schröder's observation as a paradox (Eikerling and Berg, 2011; Freger, 2009; Onishi et al., 2007). Consistent physical models of water sorption equilibria in PEMs eliminate both of these issues.

## PHENOMENOLOGY OF WATER SORPTION

The experimental basis of sorption studies includes isopiestic vapor sorption isotherms (Morris and Sun, 1993; Pushpa et al., 1988; Rivin et al., 2001; Zawodzinski et al., 1993c) and *capillary isotherms*, measured by standard porosimetry (Divisek et al., 1998; Vol'fkovich and Bagotsky, 1994; Vol'fkovich et al., 1980). A number of *thermodynamic models of water uptake* by vapor-equilibrated PEMs have been suggested by various groups (Choi and Datta, 2003; Choi et al., 2005; Elfring and Struchtrup, 2008; Freger, 2002, 2009; Futerko and Hsing, 1999; Meyers and Newman, 2002b; Thampan et al., 2000; Weber and Newman, 2004b). The models account for interfacial energies, elastic energies, and entropic terms in the Gibbs energy functional.

A shortcoming of the majority of *water sorption models* is that they employ a single equilibrium condition, expressed through activities of water in PEMs and adjacent reservoirs of liquid water or vapor. These models concur poorly with morphological membrane models and they often invoke the presence of water vapor in PEMs, which is justified by postulating the existence of hydrophobic pores with contact angle $\theta$ slightly above $90°$, to obtain a balance with external pressure conditions (Choi and Datta, 2003; Choi et al., 2005; Elfring and Struchtrup, 2008; Weber and Newman, 2004b). However, such approaches are thermodynamically inconsistent and they are inadequate to describe the state of water in PEMs. There is no evidence for hydrophobic gas porosity or for the presence of water vapor inside of PEMs. Gas tightness of pertinent membranes and the collapse of their porous structure upon dehydration are arguments against these notions. Moreover, experimental studies employed to estimate wetting angles give wetting angles at external membrane surfaces, which are expected to be predominantly hydrophobic (Hiesgen et al., 2012; Zawodzinski et al., 1993b). These data are, however, irrelevant for microscopic polymer/water interfaces inside of PEMs.

For the membrane interior, independent insights exclude the existence of vapor and hydrophobic pores. Structural data of Gebel's group show no indication of hydrophobic pores but of pore walls that are formed by cylindrical fibrils or ribbons, with densely distributed protogenic surface groups (Gebel, 2000; Gebel and Diat, 2005; Gebel and Moore, 2000; Rubatat et al., 2002). *Gibbs energies of water sorption*, $\Delta G^s(\lambda)$, can be extracted from isopiestic vapor sorption isotherms (Morris and Sun, 1993; Pushpa et al., 1988; Rivin et al., 2001; Zawodzinski et al., 1993a,c). This analysis shows that $\Delta G^s(\lambda) < \Delta G^w$, where $\Delta G^w = -44.7 \text{ kJ mol}^{-1}$ is the Gibbs energy of vapor sorption at a free water surface at ambient conditions. Water absorbed by the membrane is, therefore, bound stronger than water at a free-bulk water surface, affirming the hydrophilic nature of water sorption in PEMs. Lastly, DFT calculations of water binding to a dense interfacial array of protogenic surface groups, representing acid-terminated sidechains in PEMs (Roudgar et al., 2006, 2008), revealed a strong binding of water molecules at the interfacial array under normal conditions, when the packing density of SGs is not too high.

## A MODEL OF WATER SORPTION

To build a model of *water sorption and swelling in PEMs*, three microscopic equilibrium conditions of water must be accounted for in the PEM and the adjacent medium. The global equilibrium state corresponds to the minimum of the appropriate thermodynamic free energy, in this case the Gibbs energy.

For the system of water in PEM pores, the intensive state variables are temperature, pressure, and chemical potential. In equilibrium, each of these variables must be uniform and constant, balanced with the corresponding values in external compartments. The independent equilibrium conditions that correspond to these variables are thermal, chemical, and mechanical equilibrium (Bellac et al., 2004). Any description of the membrane state that employs a smaller number of equilibrium conditions is inconsistent with Gibbs' phase rule.

The water sorption model proposed in Eikerling and Berg (2011) treats the water-filled PEM as a *poroelectroelastic medium*. Starting from a definition of the laws of swelling of such a medium, the model establishes a correlation between microscopic and macroscopic swelling. It correlates vapor sorption data, structural data, polymer elasticity, and volumetric swelling.

While several simplifying assumptions needed to be made so as to derive an analytical model, the model captures all relevant physical processes. Specifically, it employed thermodynamic equilibrium conditions for temperature, pressure, and chemical potential to derive the equation of state for water sorption by a single cylindrical PEM pore. This equation of state yields the pore radius or a volumetric pore swelling parameter as a function of environmental conditions. Constitutive relations for elastic modulus, dielectric constant, and wall charge density must be specified for the considered microscopic domain. In order to treat ensemble effects in equilibrium water sorption, dispersion in the aforementioned materials properties is accounted for.

## THE ROLE OF CAPILLARY CONDENSATION

At sufficient water contents, exceeding the amount of surface water, $\lambda > \lambda_s$, equilibrium water uptake is established by the action of *capillary forces*. To support this hypothesis, isopiestic vapor sorption isotherms for Nafion, in Figure 2.17a, are compared with data on pore size distributions in Figure 2.17b, obtained by standard porosimetry. In Figure 2.17a, a simple fit function,

$$\lambda = 3.0 \cdot \left(\frac{P^v}{P^s}\right)^{0.2} + 11.0 \cdot \left(\frac{P^v}{P^s}\right)^4, \tag{2.23}$$

where $P^s$ is the saturated vapor pressure at the given temperature, gives good agreement with experimental sorption data (Zawodzinski et al., 1993a,c). The first term in Equation 2.23 exhibits a weak dependence on the external vapor pressure. It can be identified with strongly bound water near the charged polymer surface. Equation 2.23 implies that the amount of surface water corresponds to $\lambda_s = 3$. The second term can be identified with bulk-like water, which exhibits a strong dependence on $P^v$.

Figure 2.17b reproduces porosity data with a log-normal pore size distribution, as suggested in Divisek et al. (1998),

$$\lambda = \frac{\lambda_{max}}{\Lambda} \int_0^{R_c} dR \exp\left(-\left(\frac{\log (R/R_m)}{\log s}\right)^2\right)$$

$$\text{with } \Lambda = \int_0^{\infty} dR \exp\left(-\left(\frac{\log (R/R_m)}{\log s}\right)^2\right), \tag{2.24}$$

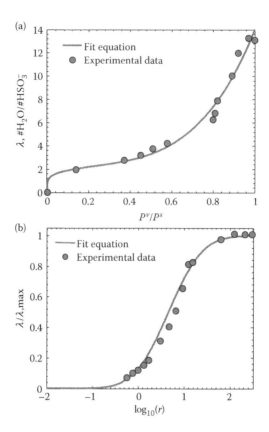

**FIGURE 2.17** Water uptake of Nafion 117. (a) Isopiestic water sorption data (extracted from Springer et al. (1991) and fit by Equation 2.23 and (b) capillary isotherms (extracted from Divisek et al. (1998) and fit by Equation 2.24. (Eikerling, M. and Berg, P. 2011. *Soft Matter*, **7**(13), 5976–5990, Figures 1–7. Reproduced by permission of The Royal Society of Chemistry.)

where $R$ is the pore radius, $R_c$ the capillary radius, $\lambda_{max}$ the maximum water content, $R_m = 0.75$ nm, and $s = 0.15$.

For the sorption isotherm, the Gibbs energy of water sorption $\Delta G^s$ is computed from the thermodynamic relation

$$\Delta G^s = \Delta G^w + R_g T \ln \frac{P^v}{P^s}, \tag{2.25}$$

where $\Delta G^w$ is the Gibbs energy of water sorption at a free water surface. For water uptake by capillary condensation, the Gibbs energy of water in pores is related to the capillary pressure

$$P^c = \frac{2\gamma \cos \theta}{R_c} \tag{2.26}$$

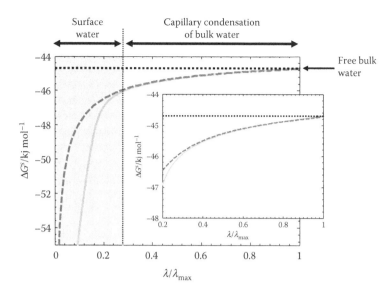

**FIGURE 2.18** Gibbs energy of water sorption by Nafion 117, obtained from isotherms in Figure 2.17. Comparison of energy is obtained from sorption isotherms (solid line) with energies obtained from capillary isotherms (dashed line). (Eikerling, M. and Berg, P. 2011. *Soft Matter*, **7**(13), 5976–5990, Figures 1–7. Reproduced by permission of The Royal Society of Chemistry.)

by

$$\Delta G^c = \Delta G^w - P^c \bar{V}_w, \tag{2.27}$$

where $\gamma$ is the surface tension of water in pores and $\bar{V}_w$ the molar volume of water. Inversion of the experimental relations $\lambda = f(P^v/P^s)$ and $\lambda = g(R_c)$ gives two expressions for the Gibbs free energy of water sorption as a function of $\lambda$, $\Delta G^s(\lambda) = \Delta G^w + R_g T \ln\left(f^{-1}(\lambda)\right)$ and $\Delta G^c(\lambda) = \Delta G^w - 2\gamma \bar{V}_w \cos\theta/g^{-1}(\lambda)$. Figure 2.18 compares these two expressions. At $\lambda/\lambda_{max} > 0.2$, the two graphs are indistinguishable, that is, $\Delta G^s(\lambda) \approx \Delta G^c(\lambda)$. In this range, the modest decrease of $\Delta G^s(\lambda)$ with decreasing $\lambda$ is an effect of the confinement of water in hydrophilic pores. This supports the supposition that *capillary condensation* controls water uptake in this range. The agreement fails for low water content, $\lambda/\lambda_{max} < 0.2$. The steeply increasing strength of water binding, seen in $|\Delta G^s(\lambda)|$ at the lowest water contents, is caused by interfacial effects, which are not accounted for in $|\Delta G^c(\lambda)|$. The large energies of water binding at low $\lambda$, observed in Figure 2.18, are consistent with values found from *ab initio* calculations of water molecules at hydrated arrays of charged surface groups in Roudgar et al. (2006).

## EQUILIBRIUM WATER UPTAKE BY A SINGLE PORE

This section focuses on water uptake and swelling of a single pore. The model employs a continuum description at the nanoscale. It is believed that the accuracy of this approach breaks down as the pore radius decreases below 1 nm. The Stern layer is not modeled explicitly, which leads to slight deviations from the actual proton distribution within the pore. Notwithstanding these simplifications, key phenomena in water sorption are captured.

### Equilibrium Conditions

Water in the pore is in equilibrium with the surrounding phase that is kept at controlled temperature, $T$, vapor pressure, $P^v$, and gas pressure, $P^g$. Corresponding to these thermodynamic variables, equilibrium of water in the pore entails three independent microscopic conditions, the first two of which are obvious (Bellac et al., 2004): (i) thermal equilibrium of the ionomer–water system implying zero heat flux and uniform $T$; (ii) chemical equilibrium implying zero water diffusion and uniform chemical potential of water in the PEM, $\mu_w^{PEM}(\lambda)$, which is in equilibrium with the external water phase; if the external phase is gaseous with a controlled relative humidity, chemical equilibrium corresponds to $\mu_w^{PEM}(\lambda) = R_g T \ln (P^v/P^s)$; if the external water phase is liquid water, the chemical potential is given by the standard value of water; (iii) mechanical equilibrium which can be expressed as a balance of pressures across interfaces in the system; pressures to consider involve $P^g$ and $P^c$, as well as the liquid water pressure in the pore, $P^l$, the osmotic pressure, $P^{osm}$, and the elastic pressure exerted by the polymer matrix, $P^{el}$ (Silberstein, 2008).

The model domain is a unit cell that contains a single pore, as shown schematically in Figure 2.19. The dimensionless microscopic swelling variable of the unit cell model is given by $\eta = v_p/v_0$, where $v_p$ is the volume of the cylindrical pore and $v_0$ is the volume of the unit cell under dry conditions. The polymer volume fraction per unit cell is given by $\varphi_p = (\eta + 1)^{-1}$.

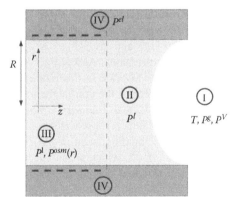

**FIGURE 2.19** Single pore model. Four regions are introduced to describe the processes of water sorption, all of which are characterized by pressures and a (uniform) temperature. Region I: external environment (vapor or liquid water). Region II: pore water void of protons. Region III: pore water including protons. Region IV: elastic membrane backbone. (Eikerling, M. and Berg, P. 2011. *Soft Matter*, **7**(13), 5976–5990, Figures 1–7. Reproduced by permission of The Royal Society of Chemistry.)

The unit cell contains four regions. Region I is the environmental phase with fixed values of $T$, $P^v$, and $P^g$. The temperature determines the saturated vapor pressure of water $P^s$ and other parameters of the polymer–water system, like the surface tension of water, $\gamma$, the dielectric constant of water, $\varepsilon = \varepsilon_0 \varepsilon_r$, and the shear modulus of the polymer phase, $G$. Region II denotes an entrance portion of the nanopore with uncharged walls and zero proton concentration. The liquid pressure $P^l$ determines the state of water in this region. Region III corresponds to the normal pore region, in which dissolved protons balance the charges of anionic surface groups at pore walls. The basic version of the model assumes uniformly charged pore walls. It neglects the impact of the discreteness of the anionic charges. In region III, $P^l$ and $P^{osm}$ add up to the total fluid pressure, $P^{fl} = P^l + P^{osm}$, where $P^{osm}$ is a function of the distance from the pore walls. Finally, region IV denotes the polymeric phase formed by aggregated ionomer backbones, which experiences strain and deformation as the pore swells.

The introduction of region II allows separate equilibrium conditions to be defined at interfaces between regions I and II, as well as between regions II and III, rendering the model amenable for analytical treatment. Two constraints are necessary for the introduction of region II: (i) a fictitious semipermeable mesh, permeable to water molecules and impermeable to protons, placed at the interface between regions II and III to keep region II proton-free, (ii) the elastic constant of polymer walls confining region II, adjusted so as to keep the width of the pore uniform. Owing to these constraints, a capillary at the pore entrance is well defined and the Kelvin–Laplace equation determines the liquid pressure in the pore. This approach highly simplifies effects at the PEM surface. However, region II makes an insignificant contribution to the total water uptake by the PEM; thus, bulk effects in water sorption and swelling are captured.

A more realistic picture of the pore neck would be needed to rationalize pore swelling at the PEM surface. This treatment would include a complex, nonuniform arrangement of sulfonic acid groups, a nonuniform pore radius, and a finite proton concentration near the meniscus. Inclusion of these effects would render an analytical treatment of the problem impossible.

Equilibrium at the gas–liquid interface between regions I and II is expressed by

$$P^l\Big|_{II} = P^g\Big|_{I} - P^c. \tag{2.28}$$

At the liquid–polymer interface between regions III and IV, the balance of elastic pressure of the polymer and fluid pressure is given by

$$P^{el}\Big|_{IV} = P^{fl}\Big|_{III}. \tag{2.29}$$

The decisive trick to link water uptake in the PEM with external thermodynamic conditions is the introduction of region II in Figure 2.19. The semipermeable and fixed mesh at the interface between regions II and III lifts the condition of mechanical equilibrium at this interface. The liquid pressure is uniform across the interface, while the total fluid pressure undergoes a discontinuous transition, that is, $P^l\big|_{II} \rightarrow P^{fl}\big|_{III}$.

### Equation of State of Water in a Pore

The equilibrium conditions discussed so far can be employed to establish the equation of state of water in a PEM pore for arbitrary pore geometry. Even though a realistic description of a membrane should encompass a range of possible pore geometries, the derivation is given for a straight cylindrical pore, as shown in Figure 2.19. The pore radius $R$ corresponds to the region occupied by bulk-like water, excluding surface water in pores.

Using Equation 2.26 for the capillary pressure and employing the Kelvin–Laplace equation

$$P^v = P^s \exp\left(-\frac{2\gamma \bar{V}_w \cos\theta}{R_g T R_c}\right) \tag{2.30}$$

gives the relation for the liquid pressure inside the pore,

$$P^l = P^g + \frac{R_g T}{\bar{V}_w} \ln\left(\frac{P^v}{P^s}\right). \tag{2.31}$$

Henceforth, ideal wetting of internal pore surfaces will be assumed, $\cos\theta = 1$. Moreover, intrinsic properties of the ionomer–water system, like the dielectric permittivity $\varepsilon_r$, the surface charge density $\sigma$, and the shear modulus $G$, are assumed as uniform at the pore level.

The equilibrium condition at the internal pore wall is obtained as

$$P^{el} = P^{fl}(R) = P^l + P^{osm}(R), \tag{2.32}$$

where $P^{osm}(R)$ is the osmotic pressure at the pore wall, for which an expression was derived in Eikerling and Berg (2011).

Effects of *membrane elasticity* on swelling were incorporated in several models of water sorption (Choi and Datta, 2003; Choi et al., 2005; Elfring and Struchtrup, 2008). The approach of Freger treats swelling as a nonaffine inflation of the hydrophobic polymer matrix by small water clusters or "droplets," which are separated by thin polymer films of polymeric aggregates (Freger, 2002). The model is a modification of the classical Flory–Rehner theory of polymer elasticity (Flory and Rehner, 1943). In the swelling process, polymer walls that encapsulate water droplets experience a compression in thickness direction. Assuming volume conservation of the polymer phase, the films must expand in in–plane directions. For isotropic in-plane expansion, the relation between elastic pressure and pore swelling is

$$P^{el} = \frac{2}{3}G\left\{\left(\frac{1}{\eta+1}\right)^{1/3} - \left(\frac{1}{\eta+1}\right)^{7/3}\right\}. \tag{2.33}$$

The procedure for obtaining this expression is a straightforward exercise in statistical thermodynamics of polymer systems (Freger, 2002). For a case of anisotropic in-plane expansion in one direction, along the width of the wall, the relation between

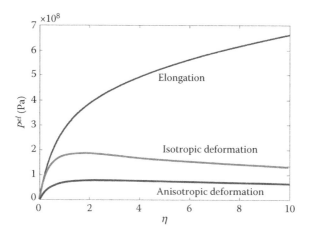

**FIGURE 2.20** Microscopic stress versus strain relation. Elastic pressure $P^{el}$ is depicted as a function of the swelling parameter $\eta$ for the three scenarios of pore wall deformation, discussed in this section. They correspond to isotropic deformation (Equation 2.33), anisotropic deformation (Equation 2.34), and elongation (Equation 2.35).

elastic pressure and pore swelling is

$$P^{el} = \frac{2}{3}G\left\{\left(\frac{1}{\eta+1}\right)^{1/3} - \left(\frac{1}{\eta+1}\right)^{5/3}\right\}. \qquad (2.34)$$

For the case of elongation of bundles or walls of ionomer domains with uniform contraction of the crosssection, the relation becomes

$$P^{el} = \frac{2}{3}G\left\{\left(\frac{1}{\eta+1}\right)^{-1/3} - \left(\frac{1}{\eta+1}\right)^{5/3}\right\}. \qquad (2.35)$$

Relations between $P^{el}$ and $\eta$, discussed in this section, are illustrated in Figure 2.20. All three deformation scenarios predict that $P^{el}$ decreases in the limit of dehydration. At zero swelling, that is, for $\eta = 0$, they approach $P^{el} = 0$. These characteristics of microscopic stress versus strain relations are in qualitative agreement with experimental observations (Freger, 2002; Silberstein, 2008). The shear modulus $G$ of polymer walls is related to Young's modulus $E$ by $E = 2(1 + v)G$, with a value of $v = 0.5$ proposed for Poisson's ratio in Choi et al. (2005).

An important relation is the scaling law of wall charge density, with pore radius. This law controls the balance of osmotic and elastic pressures. The following power law dependence was proposed,

$$\sigma(R) = \sigma_0 \left(\frac{R_0}{R}\right)^{\alpha}, \qquad (2.36)$$

where $\sigma_0$ is the wall charge density given at a reference radius $R_0$, chosen as $R_0 = 1$ nm. The exponent $\alpha$ represents the degree of *surface group reorganization* upon pore swelling. It is conjectured that $\alpha$ could take values in the range $0 < \alpha \leq 1$. As will be demonstrated below, analysis of single pore swelling and water sorption data provide the criteria to define the range of $\alpha$. A value of $\alpha$ close to 1 represents a strong reorganization of anionic surface groups upon swelling, relevant for pores with curved walls, for example, spherical pores. A value of $\alpha$ close to 0 is indicative of weak surface group reorganization, more likely representing lamellar or slab-like pores.

Knowledge of $\alpha$ requires microscopic information on the pore geometry and structural evolution of the hydrated ionomer upon water uptake. Experimentally, this knowledge could be obtained by monitoring the evolution of pore sizes, shapes, and surface area per SG with small-angle scattering studies, as described in the section "Insights from Scattering Experiments." Molecular simulations and theoretical studies specifically focusing on microstructural reorganization during water uptake could help rationalizing the value of $\alpha$.

For the range of considered water contents, excluding the case of completely dry conditions, it can be assumed that the total wall charge is conserved. A strong conservation law requires that the number of sulfonate anions per pore, $n_{SO_3^-}$, should be conserved. A weak requirement demands conservation of the number of anionic charges only at the PEM level, allowing redistribution of charged groups between pores. The invariance condition under the strong requirement is

$$n_{SO_3^-} = 2\pi q^{-1} \sigma(R) RL = 2\pi q^{-1} \sigma_0 R_0 L_0. \tag{2.37}$$

Equations 2.36 and 2.37 lead to the following relation between pore radius and length:

$$\frac{L}{L_0} = \left(\frac{R_0}{R}\right)^{1-\alpha}. \tag{2.38}$$

Using this scaling law, the microscopic swelling function of a pore with radius $R$ is

$$\eta = \frac{\pi R^2 L}{v_0} = \xi \left(\frac{R}{R_0}\right)^{1+\alpha}, \tag{2.39}$$

with $\xi = R_0 \rho_{SO_3^-}/2\bar{\sigma}_0$, where $\rho_{SO_3^-} = -\rho_{H^+}^{max} = -F \cdot IEC \cdot \rho_p^{dry}$ and $\bar{\sigma}_0$ is the average wall charge density defined through a statistical density distribution. Pore radius and wall charge density can be expressed as functions of $\eta$, $R/R_0 = (\eta/\xi)^{\frac{1}{1+\alpha}}$, and $\sigma = \sigma_0 (\xi/\eta)^{\frac{\alpha}{1+\alpha}}$.

Writing the capillary radius $R_c$ in terms of the swelling variable $\eta_c$, the liquid pressure at capillary equilibrium is given by

$$P^l = P^g - \frac{2\gamma}{R_0} \left(\frac{\xi}{\eta_c}\right)^{\frac{1}{1+\alpha}}. \tag{2.40}$$

Finally, Berg and Ladipo (2009) presented an expression for the osmotic pressure, which gives the following relation after a few simple steps:

$$P^{osm} = R_g T c_{H^+}(R) = 2\sigma_0 \left[ \frac{\sigma_0}{4\varepsilon} - \frac{R_g T}{FR_0} \left( \frac{\xi}{\eta} \right)^{\frac{1-\alpha}{1+\alpha}} \right] \left( \frac{\xi}{\eta} \right)^{\frac{2\alpha}{1+\alpha}}, \qquad (2.41)$$

where $c_{H^+}(R)$ is the proton concentration at the pore wall.

Inserting Equations 2.33 (or 2.34, 2.35), 2.40, and 2.41 into Equation 2.32 gives the equation of state for *equilibrium water uptake* by a pore with radius $R_c$:

$$\frac{2}{3} G \left( \left( \frac{1}{\eta_c + 1} \right)^{1/3} - \left( \frac{1}{\eta_c + 1} \right)^{7/3} \right) = P^g - \frac{2\gamma}{R_0} \left( \frac{\xi}{\eta_c} \right)^{\frac{1}{1+\alpha}}$$

$$+ 2\sigma_0 \left[ \frac{\sigma_0}{4\varepsilon} - \frac{R_g T}{FR_0} \left( \frac{\xi}{\eta_c} \right)^{\frac{1-\alpha}{1+\alpha}} \right] \left( \frac{\xi}{\eta_c} \right)^{\frac{2\alpha}{1+\alpha}}, \qquad (2.42)$$

where Equation 2.33 was used for the elastic pressure (isotropic in-plane expansion).

The solution of this equation can be written as an explicit relation between the wall charge density in a pore, in which capillary equilibrium exists, $\sigma_{0,c}$,* and the swelling variable $\eta_c$:

$$\sigma_{0,c} = \frac{2\varepsilon R_g T}{FR_0} \left( \frac{\xi}{\eta_c} \right)^{\frac{1-\alpha}{1+\alpha}} \left\{ 1 - \sqrt{1 + \left( \frac{F}{R_g T} \right)^2 \frac{R_0^2 \Phi}{2\varepsilon} \left( \frac{\xi}{\eta_c} \right)^{\frac{-2}{1+\alpha}}} \right\}, \qquad (2.43)$$

where

$$\Phi = \left[ \frac{2}{3} G \left( \left( \frac{1}{\eta_c + 1} \right)^{1/3} - \left( \frac{1}{\eta_c + 1} \right)^{7/3} \right) - P^g + \frac{2\gamma}{R_0} \left( \frac{\xi}{\eta_c} \right)^{\frac{1}{1+\alpha}} \right]. \qquad (2.44)$$

Equations 2.43 and 2.44 describe the water uptake phenomenon for a single pore in equilibrium with an external gas phase. The value of $\eta_c$ can be related to the relative humidity, $P^v / P^s$, through Equation 2.30. Equations 2.43 and 2.44, thus, establish the relation between environmental conditions $(T, P^v, P^g)$ and $\eta_c$. Parameters that control the water uptake by the pore are the charge density, $\sigma_{0,c}$, the dielectric constant of water, $\varepsilon$, and the shear modulus of polymer walls, $G$.

Using this solution, the modified equilibrium relation can easily be obtained between $\sigma_0$ and the swelling parameter $\eta_l$ for the liquid-equilibrated (LE) case,

---

* Equation 2.43 uses a notation $\sigma_{0,c}$, for the wall charge density, obtained as the solution of Equation 2.42. It is necessary to distinguish $\sigma_{0,c}$, a characteristic of pores, in which capillary equilibrium is established, from $\sigma_0 (<\sigma_{0,c})$ of pores that swell because of further water uptake at a given $\sigma_{0,c}$. The swelling variable $\eta$ is a two variable function of $\sigma_0$ and $\sigma_{0,c}$.

when water in the pore is in equilibrium with an external bulk water phase at liquid pressure $P_{ext}^l$:

$$\sigma_0 = \frac{2\varepsilon R_g T}{FR_0} \left(\frac{\xi}{\eta_l}\right)^{\frac{1-\alpha}{1+\alpha}} \left\{ 1 - \sqrt{1 + \left(\frac{F}{R_g T}\right)^2 \frac{\Phi_0 R_0^2}{2\varepsilon} \left(\frac{\xi}{\eta_l}\right)^{\frac{-2}{1+\alpha}}} \right\}, \qquad (2.45)$$

where

$$\Phi_0 = \left[ \frac{2}{3} G \left( \left(\frac{1}{\eta_l + 1}\right)^{1/3} - \left(\frac{1}{\eta_l + 1}\right)^{7/3} \right) - P_{ext}^l \right]. \qquad (2.46)$$

In Equations 2.45 and 2.46, the external liquid pressure, $P_{ext}^l$, has been inserted in lieu of $P^l$. Moreover, the subscript "c" at $\sigma_0$ has been dropped. Both of theses modifications are made because under LE conditions liquid pressure must be uniform across the system including pore interior and external reservoir.

Replacement of Equations 2.43 and 2.44 by Equations 2.45 and 2.46 corresponds to a transition to zero capillary pressure. Replacement of water vapor by liquid water in the surrounding phase, that is, the VE to LE transition, entails a discontinuous transition in the fluid pressure of water in the pore, giving rise to a step in the swelling variable. This behavior represents a first-order phase transition at the pore level.

To summarize, the mechanism of water uptake by each pore is as follows. At insufficient relative humidity, a pore is in a collapsed state. Once a critical relative humidity is reached for a pore wall charge density $\sigma_{0,c}$, the pore fills with water up to a swelling level $\eta_c$. Replacing saturated vapor by liquid water in the surrounding phase eliminates capillary equilibrium. Pores, therefore, undergo a discontinuous swelling transition to a larger value of $\eta_l$, given by Equations 2.45 and 2.46 and a correspondingly larger equilibrium radius.

## Evaluation of the Single Pore Model

The reference set of parameters was provided in Eikerling and Berg (2011). In the case of material-specific parameters, Nafion is the benchmark system. The pore walls should be considered as hydrophilic. Hence, $\theta = 0$ is a reasonable choice. For the mean value of the wall charge density, a value $\bar{\sigma}_0 = -0.08$ C m$^{-2}$ was taken at the fixed reference radius of $R_0 = 1$ nm. This value depends on the IEC and the structure of ionomer bundles.

Elastic properties of Nafion-type PEMs have been discussed in Silberstein (2008) and Jalani and Datta (2005). Data are available only for macroscopic effective elastic moduli. These properties vary significantly with $T$ and RH, ionomer chemistry, and membrane preconditioning. In the model, information is required for the elastic modulus of microscopic polymer fibrils or walls. These values should be significantly larger than macroscopic values, presumably lying in the range of $G = 200$ to 400 MPa.

In the following study of pore filling and swelling, it is assumed that each pore is characterized by one value of $\sigma_0$. The value of $\sigma_0$ is used in Equations 2.43 and 2.44 under VE or Equations 2.45 and 2.46 under LE conditions to find the swelling

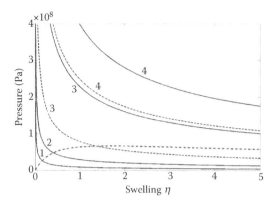

**FIGURE 2.21** Pressure equilibrium between fluid pressure (solid and dashed lines), $P^{fl}$, and elastic pressure (dash-dotted line), $P^{el}$, at the pore wall for $\alpha = 0.5$ in the case of anisotropic in-plane expansion in Figure 2.20. The pore fluid pressure of both vapor-equilibrated (dashed) and liquid-equilibrated (solid) membranes increases with the wall charge density (here, $-0.05$, $-0.1$, $-0.2$, $-0.4\,C\,m^{-2}$). (Eikerling, M. and Berg, P. 2011. *Soft Matter*, **7**(13), 5976–5990, Figures 1–7. Reproduced by permission of The Royal Society of Chemistry.)

parameter $\eta$. The model predicts that pores with lowest $\sigma_0$ possess the lowest $\eta$ and, thus, the smallest equilibrium radius in the swollen state. Considering water uptake under VE conditions this translates into a rule that pores with lowest $\sigma_0$ fill at lowest RH and pores with highest $\sigma_0$ at highest RH. The reason is that swelling requires a sufficiently large osmotic pressure, which increases with $\sigma_0$, forcing pores to attain larger equilibrium radii.

Figure 2.21 shows $P^{el}$ and $P^{fl}$ in a pore for $\alpha = 0.5$ in the case of anisotropic in-plane expansion in Figure 2.20, as a function of $\eta$ at different values of $\sigma_0$. Dotted lines correspond to VE conditions with $P^{fl} = P^g - P^c + P^{osm}$ and solid lines to LE conditions with $P^{fl} = P^g + P^{osm}$. Intersections of the fluid pressure curves with the elastic pressure curve (dashed-dotted line) correspond to solutions of Equations 2.43 and 2.44 or Equations 2.45 and 2.46, respectively.

These solutions provide the relation between $\eta_c$ and $\sigma_{0,c}$ in the VE case and between $\eta_l$ and $\sigma_0$ in the LE case, as shown in Figure 2.22. This figure exposes, at the single-pore level, the difference in water uptake between VE and LE conditions. The origin of the difference is the discontinuity in the fluid pressure.

## Macroscopic Effects in Water Sorption and Swelling

SAXS, SANS, porosimetry, and water sorption studies provide ample evidence for the dispersion in pore size and the evolution of the *pore size distribution in the PEM* upon water uptake. The changes in the pore space morphology upon water uptake translate into variations in transport properties of the PEM, as is well known (Eikerling et al., 1997, 2007a, 2008; Kreuer et al., 2004). There is, however, uncertainty regarding the mechanism of these macroscopic swelling phenomena.

This section will focus on *ensemble effects* and *pore network reorganization* upon water uptake. It will be demonstrated how these structural changes are related to

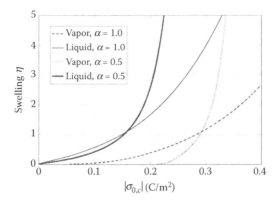

**FIGURE 2.22** Swelling parameter $\eta$ versus wall charge density $\sigma_0$, for two different values of $\alpha$, and for the VE and LE scenarios. (Eikerling, M. and Berg, P. 2011. *Soft Matter*, **7**(13), 5976–5990, Figures 1–7. Reproduced by permission of The Royal Society of Chemistry.)

spatial variations in microscopic materials parameters, focusing mostly on the effect of $\sigma_0$.

## Mechanism of Water Uptake by Pore Ensemble

Different values of $\sigma_0$ correspond to different equilibrium values of $\eta_c$, as illustrated in Figure 2.22. Statistical spatial fluctuations of $\sigma_0$ give rise to the evolution of the *pore radius distribution* (PRD) upon water sorption. The PRD evolution is influenced as well by dispersions in elastic and dielectric properties. Larger values of $G$ will give smaller equilibrium pore radii due to the stronger elastic forces that constrain the swelling. Below, only the effect of fluctuations in $\sigma_0$ is evaluated.

A spatial dispersion in the distribution of wall charges is accounted for by introducing a density distribution function $n(\sigma_0)$, defined as the number of pores per infinitesimal interval of $\sigma_0$. Knowledge of $n(\sigma_0)$ is a prerequisite for predicting the water sorption behavior of a PEM. The following integral expressions relate $n(\sigma_0)$ to the number of unit cells $N_{uc}$ in a sample of total dry volume $V_0$:

$$N_{uc} = \int_0^{\sigma_0^{\max}} d\sigma_0\, n(\sigma_0), \tag{2.47}$$

where $N_{uc} = V_0/v_0$, to the average wall charge density

$$\bar{\sigma}_0 = \frac{1}{N_{uc}} \int_0^{\sigma_0^{\max}} d\sigma_0\, n(\sigma_0)\, \sigma_0, \tag{2.48}$$

and to the volumetric expansion upon swelling of the PEM

$$\frac{\Delta V}{V_0} = \frac{1}{N_{uc}} \int_0^{\sigma_{0,c}} d\sigma_0\, n(\sigma_0)\, \eta\left(\sigma_0, \sigma_{0,c}\right), \tag{2.49}$$

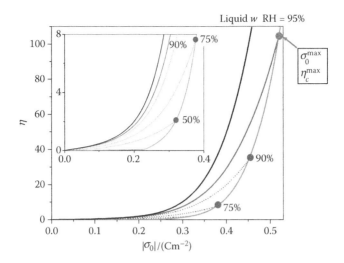

**FIGURE 2.23** Phenomenology of water sorption and swelling in a pore ensemble, calculated with typical parameters. The rightmost curve corresponds to initial pore filling upon attaining capillary equilibrium in a vapor atmosphere (VE conditions). The leftmost curve corresponds to water uptake upon equilibration with a liquid water reservoir (LE conditions). The dotted curves show swelling of pores at different values of the wall charge density, $\sigma_{0,c}$, in the maximally swollen pores, corresponding to relative humidities as indicated at the curves. The inset zooms in to the region of low $\sigma_0$. (Eikerling, M. and Berg, P. 2011. *Soft Matter*, **7**(13), 5976–5990, Figures 1–7. Reproduced by permission of The Royal Society of Chemistry.)

where $0 < \sigma_{0,c} < \sigma_0^{\max}$ and $V_0$ is the dry volume of the PEM. Equation 2.49 entails a convolution of the microscopic swelling parameter $\eta$, which accounts for the pressure balance in a single pore, with the density distribution $n(\sigma_0)$ in the pore ensemble. The notation $\eta(\sigma_0, \sigma_{0,c})$ indicates that the swelling parameter is a two-variable function of the local $\sigma_0$ and of $\sigma_{0,c}$ in the largest pore, in which capillary equilibrium prevails.

Figure 2.23, obtained with the reference parameters taken from Eikerling and Berg (2011), illustrates the phenomenology of water sorption and swelling in a pore ensemble. The corresponding schematic illustration of the involved processes is shown in Figure 2.24 for a highly simplified system of two connected pores with distinct values of $\sigma_0$. Water uptake in PEM involves three processes corresponding to (i) pore filling, (ii) continuous pore swelling, and (iii) a discontinuous swelling transition upon replacing a vapor atmosphere by a liquid water bath.

Pore filling denotes the discontinuous process by which an initially dry pore attains its capillary radius $R_c$ in equilibrium with water vapor, determined by the Kelvin–Laplace equation, Equation 2.30. This process corresponds to the rightmost curve in Figure 2.23. With increasing $P^v/P^s$, capillary equilibrium advances along this curve to pores with larger $R_c$ and $\eta_c$. The liquid pressure increases with $\sigma_{0,c}$ according to Equation 2.40. The values of $R$ and $P^l$ are given in Figure 2.24 at different stages of pore filling and swelling.

Pressure equilibrium demands that the liquid pressure, established in the critical pores at $\sigma_{0,c}$, is homogeneous and isotropic.* Thus, the same liquid pressure applies in all pores with $\sigma_0 < \sigma_{0,c}$ that have been already filled with water. As a result of the increase in $P^l$, these pores will undergo continuous swelling to attain new equilibrium

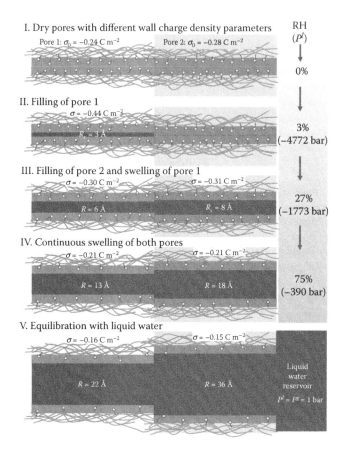

**FIGURE 2.24** Processes involved in water sorption and swelling in a pore ensemble. The sequence of steps in the swelling process is illustrated schematically for a basic system of two equilibrated pores with different values of the reference charge density parameter. Surface water is indicated by a light grey background. Bulk-like water corresponds to dark grey regions. Values of *relative humidity* (RH), liquid pressure $P^l$, wall charge density $\sigma$ and pore radius $R$, have been calculated with the reference parameters presented in Eikerling and Berg (2011). (Eikerling, M. and Berg, P. 2011. *Soft Matter*, **7**(13), 5976–5990, Figures 1–7. Reproduced by permission of The Royal Society of Chemistry.)

---

* Homogeneous and isotropic liquid pressure is warranted for high connectivity of the network of water-filled pores. This requirement, which would be indicated experimentally by the absence of hysteresis effects in water absorption/desorption isotherms, is a prerequisite for the proposed mechanism of pore swelling.

values of $\eta$, determined in implicit form by

$$\sigma_0 = \frac{2\varepsilon R_g T}{F R_0} \left(\frac{\xi}{\eta}\right)^{\frac{1-\alpha}{1+\alpha}} \left\{ 1 - \sqrt{1 + \left(\frac{F}{R_g T}\right)^2 \frac{\Phi_1 R_0^2}{2\varepsilon} \left(\frac{\xi}{\eta}\right)^{\frac{-2}{1+\alpha}}} \right\} \tag{2.50}$$

with

$$\Phi_1 = \left[ \frac{2}{3} G \left( \left(\frac{1}{\eta+1}\right)^{1/3} - \left(\frac{1}{\eta+1}\right)^{7/3} \right) - P^l \right] \tag{2.51}$$

and $P^l = P^g - \frac{2\gamma}{R_0} \left(\frac{\xi}{\eta_c}\right)^{\frac{1}{1+\alpha}}$, corresponding to the liquid pressure in the largest water-filled pore at the capillary radius. Swelling of pores with $\sigma_0 < \sigma_{0,c}$ is indicated in Figure 2.23 at three intermediate values of $\sigma_{0,c}$ by dotted curves. Values of relative humidities $P^v/P^s$, shown at curves, are obtained from Equations 2.31 and 2.40. Each of the dotted curves begins its swelling from the initial pore filling, thereby, starting off on the line that corresponds to $\eta = \eta_c$. In Figure 2.24, continuous swelling under increasing $P^v/P^s$ occurs for pore 1 in step III and for both pores in step IV. Pore sizes can significantly increase between initial filling and maximal swelling.

As can be seen in Figure 2.23, water uptake upon initial pore filling is small, that is, $\eta_c < 1\%$, in pores with wall charge densities of $|\sigma_0| < 0.2$ C m$^{-2}$. However, upon increasing $P^v/P^s$, these pores swell substantially to as much as $\eta \approx 100\%$, as can be seen in the inset of Figure 2.23. This observation underlines the fact that pores with relatively low $\sigma_0$ contribute substantially to total water uptake at high RH, even though they hardly appear in Figure 2.22, where pore ensemble effects are not considered.

Maximal swelling of a PEM in a vapor atmosphere is reached when $\sigma_{0,c} = \sigma_0^{max}$, shown as the solid blue curve in Figure 2.23. *Ab initio* simulations of arrays of charged surface groups suggest that the maximal wall charge density is $\sigma_0^{max} \approx -0.5$ C m$^{-2}$ (Roudgar et al., 2006, 2008). Exceeding $\sigma_0^{max}$ would require that water molecules and hydronium ions be expelled from the space between anionic head groups. This situation is prohibited by unfavorable electrostatic interactions at the interface, which would destabilize the polymer matrix.

Under equilibrium in a saturated vapor atmosphere, $\sigma_0^{max}$ determines the maximal values of radius, swelling parameter and liquid pressure of swollen pores. A typical set of parameters gives $R_c^{max} = 21$ nm, $\eta_c^{max} = 105$ and $P^l \approx -67$ atm. For this case, water sorption from vapor would level off at $P^v/P^s \approx 0.95$. The leveling-off of integral pore volume distributions above pore radii of $\sim 10$ nm, observed in Divisek et al. (1998), is in qualitative agreement with this behavior.

Further swelling can occur if saturated vapor is replaced by a liquid water bath, indicated in step V of Figure 2.24. During this transition, the liquid pressure changes discontinuously from $P^{l,max} = P^g - \frac{2\gamma}{R_0} \left(\frac{\xi}{\eta_c^{max}}\right)^{\frac{1}{1+\alpha}}$ in saturated vapor to $P^l = P^g$ in liquid water. The swelling parameter $\eta$ in Figure 2.23 jumps from the values on the solid blue curve, calculated for $\sigma_0^{max}$, to values on the solid black curve. This discontinuity in liquid pressure and water uptake explains Schröder's paradox.

## Vapor Sorption Isotherms

A necessary consistency check of a water sorption model is the calculation of vapor sorption isotherms. Equation 2.49 gives $\Delta V/V_0$ as a function of $\sigma_{0,c}$. The input for the calculation includes the environmental parameters $(T, P^g, P^v)$, materials parameters $(\varepsilon_r, G, \rho_{SO_3^-})$, and the probability density distribution $n(\sigma_0)$. The value of $\eta_c$ can be converted into relative humidity with Equations 2.31 and 2.40.

In order to simplify the calculation, a modified version of Equation 2.49 is used:

$$\frac{\Delta V}{V_0} = \frac{1}{N_{uc}} \int_0^{\eta_c} d\eta\, n(\sigma_0)\, \eta\, \frac{d\sigma_0}{d\eta} \tag{2.52}$$

where $\sigma_0$ is substituted as a function of $\eta$ and $\eta_c$, using Equation 2.50. Maximal swelling under saturated VE conditions gives

$$\frac{\Delta V_{VE}^{max}}{V_0} = \frac{1}{N_{uc}} \int_0^{\eta_c^{max}} d\eta\, n(\sigma_0)\, \eta\, \frac{d\sigma_0}{d\eta}, \tag{2.53}$$

where the upper integration limit is obtained as the solution of $\sigma_0(\eta_c) = \sigma_0^{max}$, using Equation 2.43. Similarly, swelling under LE conditions gives

$$\frac{\Delta V_{LE}^{max}}{V_0} = \frac{1}{N_{uc}} \int_0^{\eta_l^{max}} d\eta_l\, n(\sigma_0)\, \eta_l\, \frac{d\sigma_0}{d\eta_l}, \tag{2.54}$$

with $\eta_l^{max}$ found as the solution of $\sigma_{0,w}(\eta_l) = \sigma_0^{max}$, using Equation 2.45.

Plots of $\Delta V/V_0$ are shown in Figure 2.25 using a log-normal distribution of $n(\sigma_0)$ as input. The $n(\sigma_0)$ distribution is cutoff at $\sigma_0^{max} = 0.52$ C m$^{-2}$ and normalized according to the finite width of the integration region. Corresponding differential pore size distributions are obtained from the relation

$$\frac{d}{dR}\left(\frac{\Delta V}{V_0}\right) = \frac{(1+\alpha)\,\xi^2}{N_{uc}} \left(\frac{R}{R_0}\right)^{1+2\alpha} \left[n(\sigma_0)\, \frac{d\sigma_0}{d\eta}\right]_{\eta \to f(R)}, \tag{2.55}$$

where the term in square brackets must be considered at the value of $\eta$ that corresponds to the considered value of $R$, using the unique relation between $\sigma_0$ and $\eta$ in Equations 2.50 and 2.51, as well as the definition of $\eta$ as a function of $R$ in Equation 2.39.

Figure 2.25a shows the impact of different deformation scenarios, corresponding to the isotropic in-plane deformation, anisotropic deformation, and uniaxial stretching. Figure 2.25b illustrates the impact of the coefficient $\alpha$, introduced in Equation 2.36. It should be noted that a value of $\alpha < 0.5$ that represents weak reorganization of sidechains upon swelling results in an unphysical behavior in single pore swelling. Increasing the IEC from 0.91 to 1.2 mmol g$^{-1}$ corresponds to a significant increase in water uptake, as seen in Figure 2.25c . Imposing a larger value of the shear modulus of polymer walls suppresses pore swelling, as can be seen in Figure 2.25c for $G = 450$ MPa.

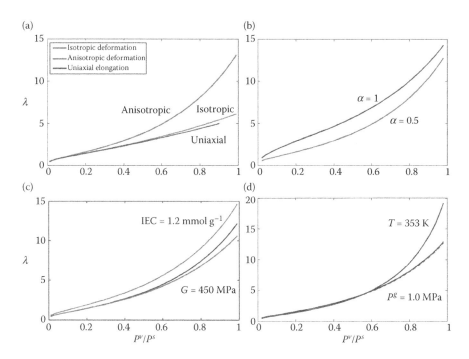

**FIGURE 2.25** Impact of various parameters in the poroelectroelastic theory on water sorption isotherms. Parameters of the reference case, anisotropic deformation in part (a), are $IEC = 0.91$ mmol g$^{-1}$, $G = 380$ MPa, $P^g = 0.1$ MPa, $T = 298$ K, and $\alpha = 0.5$. A log-normal distribution of the surface charge density in the reference state is used as input. Part (a) compares the different elastic deformation scenarios given by Equations 2.33 through 2.35, where the anisotropic deformation scenario will be considered as the reference case in (b) to (d). Part (b) illustrates the impact of the degree of surface group reorganization upon swelling, represented by the parameter $\alpha$ in Equation 2.36. Part (c) shows the impact of an increase in elastic modulus (to $G = 450$ MPa, lower line) or an increase in IEC (to $IEC = 1, 2$ mmol g$^{-1}$, upper line) relative to the reference case (center line). Part (d) shows the effect of an increase in gas pressure (to $P^g = 1.0$ MPa, lower line which is indistinguishable from the reference case) and in temperature (to $T = 353$ K, upper curve) on water sorption isotherms.

Figure 2.25d illustrates the impact of varying thermodynamic conditions on water sorption. Water sorption is rather insensitive to an increase in $P^g$ from 1 atm to 10 atm. Temperature could have a larger impact. The calculation incorporated variations of $\varepsilon_r$, $\gamma$, and $G$ with $T$. The effect of $T$ could be ambiguous resulting from different pressure contributions that depend on it. Regarding the insensitivity of pore filling to changes in $P^g$, it supports experimental evidence. The reason for it is that the normal range of gas pressures is two to three orders of magnitude smaller than $P^{osm}$, $P^c$, and $P^{el}$. This explains the seemingly conflicting finding that, albeit pore filling in PEMs being controlled by a pressure balance, a variation in $P^g$ has only a minuscule impact on water sorption.

The insensitivity of water sorption and swelling to $P^g$ should not be seen as an argument for neglecting pressure driven flux mechanisms in models of PEM operation. Small variations in $P^v/P^s$ cause large variations in internal liquid pressures in pores, establishing substantial driving forces for hydraulic water fluxes. Using the reference case parameters for a simple demonstration, liquid pressures at $P^v/P^s = 0.95$ ($\Delta V/V_0 = 0.34$) and $P^v/P^s = 0.80$ ($\Delta V/V_0 = 0.23$) are $P^l = -67$ bar and $-304$ bar, respectively, establishing a driving force for hydraulic permeation of $\Delta P^l \simeq 250$ bar from the high to the low RH region. This example supports the notion that *hydraulic water flux*, generated by an external RH difference or by electro-osmotic drag, constitutes the major mechanism that controls the water balance in a PEM under PEFC operation (Eikerling et al., 1998, 2007a).

Figure 2.26 compares calculated water sorption isotherms to two independent sets of experimental data for Nafion (Maldonado et al., 2012; Zawodzinski et al., 1991). A fixed amount of surface water is subtracted in experimental isotherms, since the theory describes bulk-like water only. After this modification, the agreement that can be achieved is good.

The model provides relations between macroscopic and microscopic swelling. For the isotherm in Figure 2.26, statistical distributions of pore size and elastic pressure in pores are shown at various values of the relative humidity. Both distributions shift to larger values and their width increases with increasing RH. The trend in the pressure distribution upon increasing RH represents increasing internal stresses and increasing heterogeneity of the stress distribution at the pore level. These trends are important in view of understanding mechanical degradation effects that could limit the lifetime of the PEM and, thus, the PEFC. The increasing elastic pressure exerted at pore walls, upon increasing RH, translates into a higher probability of bundle or wall breakage, as well as fracture formation and propagation.

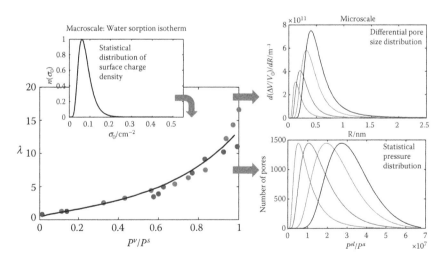

**FIGURE 2.26** Left: Sorption isotherm from theory with the statistical distribution of the surface charge density as input. Comparison with exp. data in (Zawodzinski et al., 1991) (dark grey dots) and (Maldonado et al., 2012) (light grey dots). Right: Microscopic statistical distributions of pore size and pressure (left to right: $P^v/P^s = 0.25$, $0.5$, $0.75$, and $0.90$).

## CAPABILITIES AND LIMITATIONS OF THE WATER SORPTION MODEL

At the single pore level, the balance of liquid, osmotic, and elastic pressures regulate water sorption and swelling. This balance leads to the equation of state that relates the microscopic swelling parameter $\eta$ to thermodynamic conditions ($T$, $P^v$, $P^g$) and microscopic structure-related properties, including the density of anionic groups at pore walls, the dielectric permittivity of water in pores, and the shear modulus of polymer walls. The model includes, moreover, an empirical scaling law for the reorganization of wall charges during swelling, given in Equation 2.36.

Upon increasing the external relative humidity ($P^v/P^s$), capillary equilibrium advances from pores with small $\sigma_0$ to those with large $\sigma_0$. Weakly charged pores fill first and they attain smaller equilibrium radii because they have the smaller osmotic pressure to push pore walls apart against the elastic pressure.

The extent of surface group reorientation during pore swelling, represented by $\alpha$ in Equation 2.36, has a strong impact on the balance of pressures and, thus, on water uptake and swelling in the pore. Weak reorganization of anionic surface groups upon swelling, which corresponds to a small value of $\alpha$, causes large swelling since a large osmotic pressure is maintained during pore growth. Specifically, the model implies that lamellar-like pores with rigid arrays of surface groups ($\alpha = 0$) at opposing flat walls could not attain a finite stable size. In this case, the model predicts unlimited pore growth. Experimentally, block-copolymer PEM with lamellar-like structures take up more water than PFSA-type PEM that form cylindrical superstructures. The reorganization of anionic surface groups or sidechains at pore walls upon swelling, indicated by $\alpha > 0$, is a necessary condition for the stabilization of finite-size water-filled pores. At the other extreme, in pores with strong reorganization of wall charges upon swelling, indicated by $\alpha \simeq 1$, the osmotic pressure exhibits a strong decrease upon pore growth, which limits the pore expansion. This case results in small dispersions in pore sizes, weak swelling, and leveling off of water sorption isotherms at relative humidities below 70%. A qualitative understanding of surface group reorganization upon water uptake and, thus, of the value of $\alpha$, could be gleaned from vapor sorption isotherms.

The model reconciles microscopic swelling at the single pore level with macroscopic swelling effects in an ensemble of pores. The macroscopic volumetric expansion $\Delta V/V_0$ involves a convolution of single pore swelling, represented by $\eta$, with the density distribution $n(\sigma_0)$. Water uptake and swelling of the PEM is a collective phenomenon. In a nutshell, large pores with $\sigma_0 \simeq \sigma_0^{max}$ are rare, but they define the highest liquid pressure in the membrane. For well-connected pores, the liquid pressure will be uniform and isotropic. This liquid pressure allows pores with $\sigma_0 \approx \bar{\sigma}_0$ to swell significantly. These pores make the major contribution to the macroscopic swelling of the PEM.

The transition from equilibrium under saturated vapor conditions to liquid water conditions, implies a discontinuity in the liquid pressure in the pore, which debunks Schröder's paradox as a first-order phase transition at the pore level. At the PEM level, the jump in water uptake is tied to the maximal wall charge density $\sigma_0^{max}$ that is physically possible.

The model predicts huge internal liquid and osmotic pressures, with absolute values in the range of $10^2$ bar. These pressures depend strongly on $P^v/P^s$ and $T$, as well as on the microstructure in pores. The PEM responds to small differences in $P^v/P^s$, applied between PEM surfaces, with huge internal pressure gradients. The impact of external pressure differences of the order of 10 bar is insignificant compared to huge internal pressures. The model, thereby, resolves a seeming contradiction that membrane water uptake is insensitive to external pressures, whereas hydraulic flux is the dominant mechanism of internal water fluxes.

Quantitative predictions of the model are sensitive to the shape of pores, the distribution of fixed-charged groups at pore walls, the reorganization of charged groups at pore walls upon swelling, proton distribution effects in pores, and microscopic elastic properties of the polymer matrix. A consistent model must account for, and properly validate, all of these details. However, there is significant experimental uncertainty related to all of these properties and there are statistical spatial fluctuations in all of them.

Future work should scrutinize these structural aspects and refine corresponding approaches. The effective permittivity of water in pores increases with pore size, resulting in lower osmotic pressures, compared to the constant permittivity case due to a modified proton distribution. Consequently, the water uptake should be less steep at large RH. The inclusion of a Stern layer would further affect the proton distribution and, hence, water sorption. In fact, a modified Poisson–Boltzmann approach might be better suited to describe the electrostatic phenomena in such PEM pores.

The bottom line is that it is difficult, if not impossible, to reproduce quantitative values unambiguously. This is because of the intricate interplay of membrane composition, microscopic structure, reorganization upon swelling, and effective materials properties ($\varepsilon, \sigma_0, G$), along with their statistical fluctuations in a pore ensemble. This random aspect can be captured by an appropriate choice of the probability distribution of the surface group density. One key goal would be to establish probability distributions for each membrane type, which defines a target for structural studies. Reversely, analysis of water sorption isotherms could provide the pore size distribution and, further, the surface charge density distribution of a given PEM type.

Macroscale models of PEM operation that do not include the proper pressure-controlled equilibrium conditions at the single pore level fail in predicting correctly the responses of membrane water sorption, transport properties, and fuel cell operation to changes in external conditions. Single pore models, on the other hand, that do not account for statistical spatial fluctuations in microscopic membrane properties must fail because they cannot predict the dispersion in pore sizes and the evolution of the pore size distribution upon water uptake.

The presented mechanism for PEM water sorption is consistent with thermodynamic principles, while predicting experimentally found trends correctly. Currently, full-scale molecular dynamics simulations would not be able to capture wall charge density effects and elastic effects in an ensemble of pores.

The theory of bundle formation in the section "Aggregation Prenomena in Solutions of Charged Polymers" provides sizes, as well as electrostatic and elastic properties of ionomer bundles. The theory of water sorption and swelling, described in this section, gives a statistical distribution of pore size and local stress in pores. The merging point of both theories is a *theory of fracture formation* in charged polymer membranes. Fracture formation is an important area of statistical physics (Alava et al., 2006; Gardel et al., 2004; Shekhawat et al., 2013; Yoshioka et al., 2010). The formation and propagation of fractures determines the lifetime of materials. Specific challenges related to fracture formation in PEMs, stem from the interplay of elastic and osmotic effects and the presence of a fluctuating internal stress caused by water sorption. The latter point is illustrated in Figure 2.26, which shows the statistical distribution of elastic pressure in pores. Elastic properties of ionomer bundles, including the local stress they experience, are, thus, subject to statistical fluctuations.

## PROTON TRANSPORT

*Proton transport* is of paramount importance for a plethora of materials and processes encompassing acid–base chemistry, biological energy transduction, corrosion processes, and electrochemical energy conversion (Proton transport, 2011). The ubiquitous relevance and the everlasting fascination with the concerted nature of the underlying mechanisms drive research across disciplines from fundamental physics via chemistry and biology to chemical engineering.

Strong coupling of excess protons with molecular degrees of freedom of solvent and host materials can enable *fast proton motion in water*, at biomembranes, at Langmuir monolayers of protogenic surface groups, and along one-dimensional hydrogen-bonded chains, so-called proton wires, in hydrophobic nanochannels (Nagle and Morowitz, 1978; Nagle and Tristam-Nagle, 1983). On the other hand, this coupling could, also, create negative synergies. The solvent coupling renders proton mobility in biomaterials and electrochemical proton-conducting media highly sensitive to the abundance and structure of the proton solvent. In PEMs, in which protons are bound to move in nanosized water channels, membrane dehydration caused by evaporation or by electro-osmotic coupling is responsible for a dramatic decrease in proton conductivity.

The high concentration of bound protons in aqueous electrolytes enables a high proton mobility due to *structural diffusion*. It, however, does not help in increasing the overall proton conductivity, which requires the presence of excess charge carriers or stoichiometric defects. Related to its poor autodissociation, chemically pure water has a proton conductivity of only $10^{-8}$ S cm$^{-1}$ at ambient conditions. The low value is due to the small concentration of free protons ($10^{-7}$ mol L$^{-1}$). The mobility of an excess proton in water at 25°C is $\mu_{H^+}^b = 3.63 \cdot 10^{-3}$ cm$^2$ V$^{-1}$ s$^{-1}$. In the temperature range of aqueous systems, water is the medium with the highest intrinsic proton mobility. The mobility of protons is approximately 7 times higher than that of a sodium ion and 5 times higher than that of a potassium ion, which are objects of similar size as the hydronium ($H_3O^+$) ion (Erdey-Gruz, 1974).

Excess protons in aqueous electrolyte result from the dissociation of acid molecules or molecular groups in solutions of strong acids, hydrated polymer electrolytes, or proteins. In acidic solutions, both protons and counter-anions are mobile. In PEMs, only protons are mobile while anions are immobilized at the macromolecular matrix or skeleton of the pore network.

Boldly stated, the objectives in the design of excellent *water-based proton conductors are* to (1) charge them up with extra protons as much as possible; (2) ensure that the maximal number of protons could acquire bulk water-like mobility. As discussed at the beginning of this chapter, current PEM seem to be close to achieving the second objective, as long as the level of hydration is above a critical value. Improvements in proton density are possible; they could be achieved by increasing the ion exchange capacity of ionomeric resins. However, a high IEC reduces the propensity for ionomer aggregation, which is essential for forming a stable porous matrix. Moreover, high proton and anion densities in pores lead to high osmotic pressures, which cause instabilities because of excessive pore swelling and increased mechanical stress on polymer aggregates or walls. These effects accelerate mechanical degradation.

This section provides a systematic account of proton transport mechanisms in water-based PEMs, presenting studies of proton transport phenomena in systems of increasing complexity. The section on *proton transport in water* will explore the impact of molecular structure and dynamics of aqueous networks on the basic mechanism of proton transport. The section on proton transport at highly acid-functionalized interfaces elucidates the role of chemical structure, packing density, and fluctuational degrees of freedom of hydrated anionic surface groups on concerted mechanisms and dynamics of protons. The section on proton transport in random networks of water-filled nanopores focuses on the impact of pore geometry, the distinct roles of surface and bulk water, as well as percolation effects.

## PROTON TRANSPORT IN WATER

Owing to the importance for processes in biology, chemistry, materials science, and energy technology, experimental and theoretical studies of proton mobility in water have a long and lively history. The question how protons are translocated along hydrogen-bonded networks of water has captivated the imagination of scientists for centuries (de Grotthuss, 1806; Eucken, 1948; Franck et al., 1965; Gierer, 1950; Gierer and Wirtz, 1949; Noyes, 1910; Noyes and Johnston, 1909). Gierer (1950) gives an overview of early studies of *proton conductivity in aqueous systems*, elucidating differences of the anomalous proton mobility from classical hydrodynamic proton motion or Stokes diffusion of "normal" cations in water. Several studies pondered the possibility of a coexistence between unconventional proton mobility and a remnant contribution owing to the classical hydrodynamic motion of $H_3O^+$ ions. However, recent detailed studies based on *ab initio* molecular dynamics refute any classical contribution.

The value of the activation energy of proton transport in PEMs is low ($\sim 0.1$ eV), for water contents above $\lambda \sim 3$, and it is similar to the value in bulk water, as depicted in Figure 2.6. This similarity suggests that the widely studied relay-type mechanism of prototropic mobility in bulk water is relevant for PEMs above a critical water content.

The molecular mechanism of *proton transfer in water* was unraveled in the mid-1990s. The recent history is reviewed in Marx (2006). The invention of the "Car–Parrinello technique" of molecular dynamics (Car and Parrinello, 1985) opened the field of atomistic computer simulations for a myriad of applications in solid-state science, soft matter science, chemistry, biophysics, molecular electronics, and engineering (Marx and Hutter, 2009). About 10 years after its introduction, the first simulations of this type were applied to "distilled water." In that pioneering work, the properties of water were simulated on a cluster model of 32 water molecules (Laasonen et al., 1993).

Soon after the first *ab initio* molecular dynamics study of a pure water cluster, applications of the CPMD methodology started to explore phenomena of ion solvation and transport in water. A series of studies on excess proton migration in water was performed starting from 1994 (Marx et al., 1999; Tuckerman et al., 1994, 1995, 2002). The mechanism proposed is termed "*structural diffusion*" to distinguish it from classical ion motion (or "vehicle mechanism"), and emphasize its distinct nature from the traditional concept that is understood as "*Grotthuss mechanism*." The mechanism found immediate wide acceptance due the coincidental publication of a detailed analysis of spectroscopic data by Agmon (1995), which supported the main findings of theoretical–computational studies by Tuckerman, Laasonen, Sprik, and Parrinello.

More specifically, the detailed spectroscopic analysis by Agmon found that the timescale of water molecule rotation, which takes 1–2 ps at room temperature, is similar to proton hopping times determined from the analysis of the $^{17}O$ resonance in NMR studies by Meiboom (1961). Using the hopping time of $\tau_p = 1.5$ ps and a hopping length of $l_p = 2.5$ Å, an estimate of proton mobility in three-dimensional networks can be obtained from the Einstein relation, $D_{H^+} = l_p^2/6\tau_p = 7.0 \cdot 10^{-5}$ cm$^2$ s$^{-1}$. This estimate is close to the experimental value of $9.3 \cdot 10^{-5}$ cm$^2$ s$^{-1}$. From this closeness, Agmon concluded that proton mobility in water is an incoherently Markovian process.

The next question concerned the nature of the rate-limiting step in proton motion. From the small observed kinetic isotope effect on proton mobility (Erdey-Gruz, 1974), as well as other observations, Agmon concluded that the reaction coordinate of proton transfer does not involve proton motion. Instead, hydrogen bond cleavage was identified as the rate-determining step in proton transfer. He found the NMR proton hopping time to agree with the timescale of a single water molecule rotation, which could trigger hydrogen bond cleavage. The activation energy of the proton transfer of ~0.1 eV is similar to the directional enthalpy that is associated with water reorientation, bolstering the conjecture that the rate-determining hydrogen bond cleavage is caused by water molecule rotation.

In order to devise a molecular-level scheme of proton transport in water, the questions remaining concern the nature of *excess proton structures* and their transformations during proton transitions. Identification of relevant proton complexes and characterization of their lifetimes and interconversions is a particular strength of *ab initio* molecular dynamics. The primary hydrated structure of the excess proton is the hydronium ion, $H_3O^+$. It was long thought that transport of $H_3O^+$ as a rigid unit in the style of hydrodynamic ion transport could make a significant contribution of about 20%, to the total rate of proton transport in water. However, $H_3O^+$ is not a rigid molecular species but merely a metastable proton state. It forms hydrogen bonds with three neighboring water molecules to build a complex known as *Eigen cation* (Eigen, 1964), $H_9O_4^+$, depicted in Figure 2.27. The hydrogen bonds in $H_9O_4^+$ are significantly stronger than in free bulk water and correspondingly shorter, with

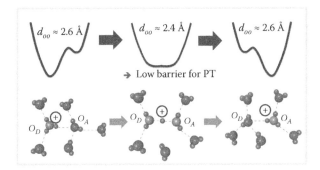

**FIGURE 2.27** A schematic illustration of proton transport via structure diffusion in water. The top panel shows the free energy of the excess proton as a function of the displacement coordinate along the central hydrogen bond. Values of the distance of donor ($O_D$) and acceptor ($O_A$) oxygen nuclei, $d_{OO}$, are indicated as parameters. For the Eigen cation complexes $H_9O_4^+$, on left and right, this distance is $d_{OO} \approx 2.6$ Å. These complexes correspond to localized proton states. Hydrogen bond cleavage in the second hydration shell leads to the formation of a Zundel cation $H_5O_2^+$, illustrated in the middle. The Zundel cation has a shortened hydrogen bond distance $d_{OO} \approx 2.4$ Å, which leads to a shallow (or possibly single well) free energy profile with a delocalized proton in the center. (Reprinted by permission from Macmillan Publishers Ltd., *Nature*, Tuckerman, M. E., Marx, D., and Parrinello, M. 2002. 417, 925–929, Figure 1(a). Copyright (2002).)

the OH• • •O distance of $d_{OO} \approx 2.6$ Å, compared to $d_{OO} \approx 2.8$ Å in bulk water. Agmon conjectured that strong hydrogen bonds in the first solvation shell of $H_3O^+$ must remain intact during the proton transition, since their breaking would involve activation energies that are significantly larger than those found in measurements of proton transport.

However, this conclusion imposes the next question: how could a proton relocate from a stable localized position on a donor Eigen cation D to an acceptor Eigen cation A, without ever leaving its solvation shell and disrupting the hydrogen-bonded network, which should require the breaking of at least one strong hydrogen bond? Here, another limiting structure of an excess proton complex comes into play, the Zundel cation (Zundel and Fritsch, 1986), $H_5O_2^+$, also shown in Figure 2.27. The Zundel cation has a further reduced hydrogen bond length of $d_{OO} \approx 2.4$ Å. The free energy profile of a relocating proton can be represented as a two-variable function. It depends on $d_{OO}$ as well as on the proton displacement coordinate, which can be defined as $\delta_{H^+} = d_{O_DH^+} - d_{H^+O_A}$, where $d_{O_DH^+}$ and $d_{H^+O_A}$ represent separation distances of the transferring proton to donor and acceptor oxygen nuclei. At the short O–O distance of the Zundel complex, the free energy of a proton as a function of $\delta_{H^+}$ reduces to a single well form. It corresponds to a symmetric configuration, in which the central proton is found with the highest probability in a centrosymmetrical position between both water molecules. For the structural diffusion of a proton in water, the Zundel cation represents a transition structure.

Combining spectroscopic insights and CPMD simulations by Tuckerman et al., the following molecular mechanism of proton transport was conceived, as depicted in

Figure 2.27. The top panel in Figure 2.27 shows the modulation of the proton energy as a function $\delta_{H^+}$, corresponding to different values of $d_{OO}$. The different stages of the proton transfer event are

(1) In the initial state, shown on the left-hand side, the excess is proton localized on the Eigen cation, and the excess proton charge is centered at the position of the donor oxygen atom $O_D$. The free energy profile of the proton as a function of $\delta_{H^+}$ exhibits a nonsymmetric double-well character.

(2) Water molecule rotation (the rate-determining step) in the second hydration shell of $H_3O^+$, induces the cleavage of a single hydrogen bond between second and first hydration shells. The energetic cost of this process is ~0.1 eV, mainly incurred by water molecule rotation, which occurs on the timescale of ~1 ps.

(3) The reduction of the number of hydrogen bonds of a water molecule in the first hydration shell of $H_3O^+$ from four to three entails a strengthening and associated shortening of the hydrogen bond between $O_D$ and $O_A$, from $d_{OO} \approx 2.6$ Å to $d_{OO} \approx 2.4$ Å. This ultrafast reorganization that occurs on a femtosecond timescale forms the Zundel cation shown in the middle panel of Figure 2.27.

(4) In the Zundel cation, the most probable proton position lies in the center between $O_D$ and $O_A$, corresponding to the single well energy function illustrated in Figure 2.27. The Zundel cation represents a delocalized proton state.

(5) The formation of a new hydrogen bond between a water molecule and $O_D$ destabilizes the Zundel cation, enforcing the formation of a new Eigen cation at the acceptor proton complex centered at $O_A$. This process completes the elementary proton transfer event. The net proton charge has migrated from the center of Eigen cation D to the center of Eigen cation A, corresponding to a net charge displacement of about 5 Å.

Overall, sequences of fluctuation-induced breaking and making of hydrogen bonds, local reorientation of water molecules, and barrierless proton transfer in $H_5O_2^+$ establish this Eigen–Zundel–Eigen mechanism of the anomalously high protonic mobility in bulk water. In this mechanism, Eigen and Zundel cations exist as short-lived or "limiting" structures (Marx et al., 1999).

## Surface Proton Conduction: Why Bother?

The simplest approximation to describe the operation of a PEM was discussed in the section "PEM Conductivity: Simply a Function of Composition?" It involves the following basic instructions:

(1) Fill the pores of a polymeric host material with bulk-like water and assume an average proton density in pores that is consistent with the IEC of the PEM.

(2) Assume that water retains its dynamic properties, that is, water molecule rotation and hydrogen bond fluctuations occur on similar timescales as in free bulk water.

As discussed, this approach works surprisingly well above a critical water content $\lambda_c$.

Efforts of polymer scientists and fuel cell developers alike are driven by one question: What specific properties of the polymeric host material determine the transport properties of a PEM, especially proton conductivity? The answer depends on the evaluated regime of the water content. At water content above $\lambda_c$, relevant structural properties are related to the porous PEM morphology, described by volumetric composition, pore size distribution and pore network connectivity. As seen in previous sections, effective parameters of interest are IEC, $pK_a$, and the tensile modulus of polymer walls. In this regime, approaches familiar from the theory of porous media or composites (Kirkpatrick, 1973; Stauffer and Aharony, 1994), can be applied to relate the water distribution in membranes to its transport properties. *Random network models* and simpler models of the porous structure were employed in Eikerling et al. (1997, 2001) to study correlations between pore size distributions, pore space connectivity, pore space evolution upon water uptake, and proton conductivity, as will be discussed in the section "Random Network Model of Membrane Conductivity."

At a water content below $\lambda_c$, the specific molecular structure at polymer–water interfaces dictates the transport properties of PEMs. Relevant details of the molecular interfacial structure include chemical composition and length of ionomer sidechains, packing density of sidechains, and structure of the interfacial hydrogen-bonded network that forms between sulfonic acid head groups and interfacial water. At $\lambda < \lambda_c$ and low interfacial density of sidechains, referred to hereafter as SGs, protons will be trapped at interfaces and cannot generate a significant proton conductivity.

However, intriguing phenomena arise if the SGs density at polymer–water interfaces is increased. In the regime of high SG density, proton transport in PEMs become similar to proton transport at acid-functionalized surfaces. *Surface proton conduction* phenomena are of importance to processes in biology. Yet, experimental findings of ultrafast proton transport at densely packed arrays of anionic SG have remained controversial. Theoretically, understanding of the underlying mechanisms is less advanced than for proton transport in bulk water.

## SURFACE PROTON CONDUCTION IN BIOLOGY AND AT MONOLAYERS

Proton transfer in water that surrounds acidic residues in protein channels and at mitochondrial membranes has drawn tremendous attention from the 1960s, starting with Mitchell's chemiosmotic hypothesis (Heberle et al., 1994; Mitchell, 1961). Teissie et al. (1985), using a pH-sensitive fluorescence probe to monitor changes in proton concentration at a lipid monolayer, had obtained an estimate for the surface proton diffusion coefficient that was 20 times higher than the value of bulk water, which is $9.3 \cdot 10^{-5}$ cm$^2$ s$^{-1}$. This result stirred up controversy (Polle and Junge, 1989). Zhang and Unwin, using SECM with a proton feedback method, found a surface proton diffusion coefficient of $6 \cdot 10^{-6}$ cm$^2$ s$^{-1}$ (Zhang and Unwin, 2002). Serowy et al. (2003) using flash photolysis to produce protons and a lipid-bound fluorescent dye to monitor changes in local pH determined a surface proton diffusion coefficient of $5.8 \cdot 10^{-5}$ cm$^2$ s$^{-1}$. Morgan et al. (1991) measured surface pressure and lateral proton conductance at lipid monolayer–water interfaces as a function of surface area per molecule. They observed that during surface compression the surface potential and interfacial conductance of various monolayers exhibited pronounced increases at a critical value of the surface area per SG. They concluded that the increase in lateral conductance during monolayer compression was due to the transport of protons along a two-dimensional hydrogen-bonded network between monolayer headgroups and adjacent water molecules. This conclusion agrees with findings of Sakurai and Kawamura (1987), who measured lateral proton conductance along a phosphatidyl-choline monolayer. Infrared spectroscopy studies of Leberle et al. (1989) support the findings of Morgan et al. and Sakurai et al. Their results show that phosphatidylserine headgroups are hydrogen-bonded with large proton polarizabilities, which supports efficient proton transfer between proton donor–acceptor groups by creating lateral electrochemical gradients.

In spite of diverging predictions of surface proton diffusion coefficients, the studies quoted above provide consistent accounts of the impact of monolayer composition, reduced dimensionality, and interfacial ordering on proton dynamics. Altogether, there is ample evidence for efficient surface proton transport, which is sensitive to the packing density and chemical nature of acid headgroups. Surface pressure, surface electrostatic potential, and lateral proton conductivity increase dramatically upon monolayer compression below a critical area with typical values in the range of 25 to $40 \text{ Å}^2$ per SG. This critical area corresponds to a nearest-neighbor separation distance of SG of 6.5–7 Å (Leite et al., 1998; Mitchell, 1961).

Figure 2.28 shows typical data for the proton conductance, $G$, at a stearic acid monolayer, as a function of the area per surface molecule. This figure reveals a correlation between a structural transition (indicated by surface pressure, $\pi$) and a transition in proton dynamics. The proton conductance exhibits a sharp increase at the critical density of SGs; it levels off to a background value at a smaller density of SGs.

Studies of surface proton transport in charged polymeric systems are sketchy. *Surface proton transport* is believed to prevail at very low $\lambda$. At low relative humidity or temperature $>100°C$ the proton conductivity drops to a small residual value, $\sigma_{H^+} \approx 10^{-5} \text{ S cm}^{-1}$ (Wu et al., 2011), which is insufficient for PEFC operation. The activation energy of proton transport in Nafion, obtained from Arrhenius plots at fixed $\lambda$, increases to a value of $E_a = 0.36$ eV under minimal hydration, as depicted in Figure 2.6. Thus, surface proton conduction is widely considered irrelevant for PEM operation and the usual explanations are that (i) the hydrogen-bonded network of acid head groups and residual water at interfaces is rather stiff, resulting from strong hydrogen bonds, (ii) molecular fluctuation modes that usually trigger proton motion are suppressed at interfaces, (iii) a reduced dielectric constant at the interface leads to strong electrostatic trapping of excess proton charges at anionic sites, as studied for instance in Eikerling and Kornyshev (2001) and Commer et al. (2002). However, these conclusions are misleading, since they ignore insights of proton conductivity studies at monolayer systems, discussed above. It is essential to keep in mind that rates of surface proton transport depend strongly on the packing density of acid-functionalized SGS.

A targeted experimental study on interfacial proton transport with a view to designing PEM with improved water retention and proton conductivity was reported by Matsui et al. (2011). The authors performed surface pressure and conductivity

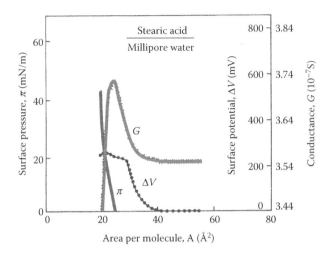

**FIGURE 2.28** Proton conductance, $G$, and surface pressure, $\pi$, at a stearic acid monolayer as a function of the area per surface group. Both properties exhibit a transition at a critical surface area per SG of $\sim$25 Å$^2$. (Adapted from Oliveira, O. N., Leite, A., and Riuland V. B. P.. 2004. *Braz. J. Phys.*, 34, 73–83, Figure 1. Copyright (2004), Sociedade Brasilliera de Fisica. With permission.)

measurements at nanosheet assemblies of monolayers with sulfonic acid head groups. This work is one of the first explicit monolayer studies of surface proton transport with SGs that are terminated by sulfonic acid head groups. Previous experimental studies of surface proton transport, cited above, had utilized carboxylic acid head groups. This preference is not surprising since the main driver for these studies was an interest in biological energy transduction. As demonstrated below, the trigonal structure of the sulfonate anion is essential for the formation of an ordered hydrogen-bonded network that strongly influences interfacial proton transitions and dynamics. Matsui et al. found conductivity values in the order of $10^{-2}$ S cm$^{-1}$ at 70°C, for an ordered multilayer film composed of their nanosheet assemblies. The activation energy of surface proton conductivity in this study was estimated to lie in the range of 0.3–0.35 eV.

## PROTON CONDUCTION AT SIMULATED SURFACES: THEORY AND COMPUTATION

For ordered interfaces or monolayer systems, high proton conductivity has been explained by collective proton dislocations along strong hydrogen bonds, forming solitary traveling waves (Leite et al., 1998; Pnevmatikos, 1988). Soliton mechanisms have been studied in ice (Gordon, 1990; Woafo et al., 1995), one-dimensional hydrogen-bonded chains (Kavitha et al., 2011; Peyrard and Flytzanis, 1987; Zolotaryuk et al., 1991), and two-dimensional Langmuir films (Bazeia et al., 2001). A variety of soliton Hamiltonians has been proposed. Similar mathematical formalisms were applied to collective phenomena in other systems, such as nonlinear dynamics of conformational transitions in DNA (Forinash et al., 1991) and domain wall diffusion in ferromagnetics (Collins et al., 1979).

The application of molecular modeling approaches to studies of structure and proton dynamics at highly charged interfacial systems is complicated. It faces a notorious dilemma: what is the best compromise between (i) the number of degrees of freedom that are included in the structural model and (ii) the level of sophistication of the computational methodology employed? Studies of proton dynamics in condensed matter demand first principles-based computational approaches (Marx and Hutter, 2009). This requirement restricts the structural complexity that can be afforded. In spite of the dramatic growth of infrastructures for high-performance computing and tremendous progress in computational methodology, it is still "but a dream" to perform full *ab initio* calculations of proton and water transport within realistic pores or pore networks relevant for PEMs. The structure of the pore system is too complex and proton transfer events are very rare. Simulations require a highly simplified structural model and computational techniques that permit an efficient sampling of rare events (Bolhuis et al., 2002; Dellago et al., 1998; Mills et al., 1995; Torrie and Valleau, 1974).

An extensive amount of work by Paddison et al. utilized molecular-level simulations at the level of DFT to examine acid solvation and dissociation, water-mediated sidechain correlations, and direct proton exchange between water molecules and sidechain-bound acid groups. A series of elaborate DFT studies focused on the role of neighboring group substitutions in sidechain moieties on proton dissociation and ion separation, studied as a function of the amount of water molecules added (Eikerling et al., 2002; Paddison, 2001). For a detailed account of the literature, see Clark et al. (2012). Further DFT studies by Paddison and Elliott of several sidechains attached to a single polymer backbone are insightful in view of fundamental ionomer–water interactions (Elliott and Paddison, 2007; Paddison and Elliott, 2006). They allude to the importance of an appropriate flexibility of hydrated sidechains as a key factor in proton motion. On the other hand, simulations on single sidechain groups or backbone fragments with a small number of sidechains are inept to account for 2D interfacial correlation effects that arise in self-organized PEM architectures. As will be discussed below, these effects influence hydrogen bond formation, acid dissociation, and flexibility of SGs at hydrated interfaces.

A *trifluoromethane sulfonic acid monohydrate* (TAM) solid was explored by Eikerling et al. (2003). Although this system does not resemble a structure for surface proton conduction in the PEM, it allows studying correlation effects at high density of triflic acid groups, the sidechain head groups in PFSA ionomer membranes, and under conditions of low hydration.

The regular structure of the TAM crystal (Spencer and Lundgren, 1973) is depicted in Figure 2.29a. The Vienna *Ab Initio* Simulation Package (VASP) was used to study the dynamics in the system (Kresse and Furthmüller, 1996a,b; Kresse and Hafner, 1993, 1994a,b). Overall, an MD trajectory of >200 ps was simulated. This trajectory is too short for direct observations of proton transfer events, which occur on timescales >1 ns. Intermittent introduction of a proton–hole defect triggered the transition from the native crystal structure with localized excess proton states, to an activated state with two delocalized protons, illustrated in Figure 2.29b. One of these protons resides within a Zundel cation $H_5O_2^+$, whereas the other one is accommodated between two $SO_3^-$ groups, which approach each other at hydrogen-bond distance. The formation of this sulfonate $O \cdots H \cdots O$ complex, requires rearrangement of the

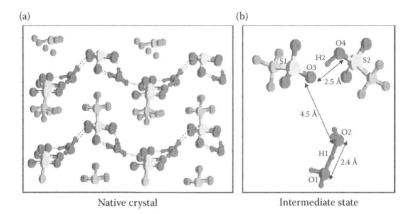

Native crystal                                           Intermediate state

**FIGURE 2.29** *Ab initio* molecular dynamics simulation of a triflic acid monohydrate crystal. (a) shows the structure of the native crystal. (b) shows the intermediate state with two delocalized protons is 0.3 eV higher in energy than the ordered conformation of the native crystal (a). (Reprinted from *Chem. Phys. Lett.*, **368**, Eikerling, M. et al. Defect structure for proton transport in a triflic acid monohydrate solid, 108–114, Figure 1,2,4, Copyright (2003) Elsevier. With permission.)

crystal structure. Simultaneous formation of these two proton complexes stabilizes the intermediate state. The energy of formation for the defect state is approximately 0.3 eV. These calculations suggested that an appropriate flexibility of anionic sidechains is vital for high proton mobility in PEM under conditions of minimal hydration and high density of fixed anions. Furthermore, a drift of the Zundel cation was observed, alluding to its possible role as a relay group for proton shuttling between hydronium ions and/or sulfonate anions.

Higher hydrates of triflic acid were studied by Hayes, Paddison, and Tuckerman using path integral Car–Parrinello MD (Hayes et al., 2009, 2011). The larger amount of water available per acidic proton led to the formation of larger proton complexes and more versatile proton defect structures. Formation of such defects involves local proton transfer events. The observed defects correspond to local structures. They are energetically expensive to form, and it seems unlikely that they could propagate in the crystal.

The conclusions from these studies are sound but they could be misleading if interpreted as evidence for the inability of highly acid functionalized surfaces to sustain long-range proton transport. In the case of triflic acid hydrates, rigid order imposed by the crystalline configuration prohibits long-range proton motion. This system could not explain high rates of surface proton motion that have been observed at biomembranes and Langmuir monolayers. Explanation of these observations demands a more flexible model system. In particular, a viable model of surface proton transport should allow for sidechain fluctuations.

### SIMULATING PROTON TRANSPORT IN A PORE

In order to simulate proton transport in a realistic pore model that more closely resembles the structure of polymer–water interfaces, one has to resort to classical or semiempirical approaches. Continuum dielectric approaches and semiclassical MD simulations have been utilized to explore the effects of interfacial anion distributions on proton mobility in single pore environments of PEMs.

The *empirical valence bond* (EVB) approach, contrived by Warshel and coworkers more than three decades ago (Aqvist and Warshel, 1993; Warshel, 1991; Warshel and Weiss, 1980), provides a powerful methodology for studying solvent effects involved in breaking and making of chemical bonds in solution. The approach utilizes empirical parameterizations of interactions between reactant states, product states, and, where appropriate, a number of intermediate states. The interaction parameters, corresponding to off-diagonal matrix elements of the classical Hamiltonian, are calibrated using *ab initio* studies of potential energy functions of relevant species or complexes in solution. This procedure significantly reduces computational expenses of molecular modeling in comparison to direct *ab initio* calculations. EVB approaches have found a wide range of applications in catalysis, biochemistry, and proton conductivity studies.

Petersen et al. (2005), and Petersen and Voth (2006), and Kornyshev, Spohr, and Walbran (Commer et al., 2002; Spohr, 2004; Walbran and Kornyshev, 2001), adopted EVB-based models to study the effect of confinement in nanometer-sized pores of the PEM and rationalize the role of acid-functionalized polymer walls on proton solvation and transport. Findings of the Voth group alluded to an inhibiting effect of sulfonate anions at pore surfaces on proton motion. The EVB model developed by Kornyshev, Spohr, and Walbran explored the impact charge delocalization within $SO_3^-$ groups and fluctuations of sidechains and sulfonic acid headgroups on proton mobility. This group found that proton mobility increased with increasing delocalization of the negative countercharge on $SO_3^-$. Fluctuational motion of sidechains and sulfonate anions enhanced the mobility of protons. EVB-based studies could, therefore, qualitatively rationalize the increase in proton conductance of a pore with increasing water content.

Continuum dielectric approaches have been utilized to calculate electrostatic effects of charged pore walls on proton distribution and proton mobility in model pores of the PEM (Commer et al., 2002; Eikerling and Kornyshev, 2001; Eikerling et al., 2008; Spohr, 2004). A continuum theory in Eikerling and Kornyshev (2001) evaluated electrostatic proton trapping at fixed surface anions in slab-like pore geometries. The model assumed pores of thickness $L$, filled with continuum water with uniform dielectric constant $\varepsilon_r$. Polymer sidechains and anionic countercharges ($SO_3^-$ groups) were represented by a static square lattice array of point charges on the opposite surfaces of the pore. The density of anions and $L$ were evaluated as parameters. The mean-field Poisson–Boltzmann equation was solved for the distribution of the proton density as a function of position, $\rho^+ (z)$. The surface charge density distribution due to interfacial point charges determines the depths of attractive potential wells for protons at the surface. A low density of anions corresponds to strong localization of surface protons at anionic sites. The proton density steeply decreases from locations near the charged surface toward the center of the pore. Strong electrostatic pinning of protons in potential wells at anionic sites is confined to a layer with thickness of 3–5 Å at pores walls. Upon increasing the density of anionic point charges, the potential modulation becomes shallower.

The mobility in the pore includes molecular mechanisms of proton transport in bulk water and along the *array of charged surface groups*. Coulomb barriers for proton mobility could arise in the vicinity of the anionic surface groups in the case of strong electrostatic pinning of protons.

Electrostatic contributions to the activation Gibbs energy of proton transport, $G_a = (E_r + \Delta G)^2 / (4E_r)$ (Krishtalik, 1986), were evaluated in Eikerling and Kornyshev (2001). The reorganization energy $E_r$ is the difference between the solvation free energy of the proton in its equilibrium solvation shell and the solvation energy of the deprotonated state that still retains the equilibrium solvation shell of the proton. This is excluding the response of the degrees of freedom with characteristic frequencies higher than $k_B T/h$. $\Delta G$ is the Gibbs energy difference between initial and final equilibrium states of the proton. It is zero in the bulk liquid.

Near the surface, $\Delta G$ is dominated by the Coulomb energy profile and, therefore, it is approximately equal to the difference of the electrostatic potentials at the proton positions before and after the transfer. This difference depends strongly on the distance of the proton from the surface. Values of $\Delta G$ were found in the range of 0.5 eV. This value decreases, however, to the activation energy of proton transport in bulk water when the proton–surface separation exceeds ~3 Å (the thickness of one monolayer of water). Moreover, the electrostatic activation energy is a function of the separation between surface charges, which lies in the range of 7 to 15 Å.

A simple *model of pore conductance*, presented in Eikerling et al. (2001) can reproduce a continuous transition from surface-like to bulk-like proton conductance upon increasing the water content in a pore. In calculations of pore conductances, it was taken into account that the average separation between $SO_3^-$ surface groups varies with pore size, and that the dielectric constant is a function of pore size. In nanopores, the reduced orientational flexibility in the stiffer hydrogen bond network offers more resistance to water reorganization resulting in a decrease of the dielectric constant (Booth, 1951; Kornyshev and Leikin, 1997; Paul and Paddison, 2001).

The bulk conductance is mainly affected by the reduced concentration of protons $\rho^+(z)$, which decreases from pore surface to center. On the other hand, surface mobility of protons in the vicinity of the $SO_3^-$ groups could be suppressed due to the existence of Coulomb barriers. A higher density of $SO_3^-$ groups diminishes Coulomb barriers and, thus, facilitates proton motion near the surface. With increasing water content in the pore, the trade-off between proton concentration and mobility shifts in favor of the bulk conductance.

A refined electrostatic approach presented in Commer et al. (2002) addressed several shortcomings. The main modifications were the use of a smeared distribution of the negative excess charge at sulfonate anions, via a form factor; considering conformational fluctuation modes of sidechains and anionic head groups, via a Debye–Waller factor; accounting for the finite size of proton complexes, which form Zundel or Eigen cations. With these changes, the strong dependence of the activation energy of proton mobility on the size of pores was not recovered. MD simulations with a dynamic all-atom model of sulfonate groups supported this conclusion.

EVB-based MD simulations as well as continuum dielectric approaches involve empirical correlations between the structure of acid-functionalized interfaces in PEM and proton distributions and mobilities in aqueous domains. The results remain inconclusive with respect to the role of packing density, fluctuations, and charge delocalization of sidechains and $SO_3^-$ groups on molecular mechanisms and rates of proton conduction. Most importantly, they do not describe proton conduction in PEM under conditions of low hydration with $\lambda < 3$, where interfacial effects prevail.

This topic will be addressed in the following section. The presented double-tracked strategy combines *ab initio* molecular dynamics studies of proton dynamics at highly acid-functionalized interfaces, with the development of *a soliton theory*.

### *AB INITIO* STUDY OF PROTON DYNAMICS AT INTERFACES

Molecular modeling of PT at dense arrays of protogenic surface groups (SGs) demands *ab initio* quantum mechanical calculations. The starting point for the development of a viable model of *surface proton conduction in PEM* is the self-organized PEM morphology at the mesoscopic scale. Figure 2.30a illustrates the random array of hydrated and ionized sidechains that are anchored to the surface of ionomer bundles.

A model for *ab initio* computational studies was created in Roudgar et al. (2006, 2008) on the basis of the following considerations. It was assumed that the interfacial dynamics of sidechains decouples from the dynamics of polymer aggregates. The supporting backbone aggregates are assumed to form an inert basal plane; hydrophobic tails of SGs are fixed by their terminal carbon atoms at positions of a regular hexagonal lattice on this basal plane.

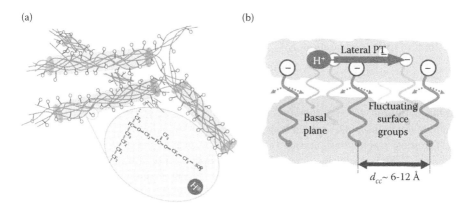

**FIGURE 2.30** Structure and processes at a dense array of protogenic surface groups. Comb-shaped ionomer molecules self-assemble into fibrous bundles that form the matrix of a hydrophilic pore network (a). The dense array of anionic SGs of the type $R–SO_3H$ in (b) emulates the structure of the bundle interface at high SG density. Terminating carbon atoms of *surface groups* (SGs) are fixed at positions of a hexagonal lattice. The basic system considered in Roudgar et al. (2006, 2008) and Vartak et al. (2013) is $R=CF_3$ (trifluoromethane sulfonic acid) with one water molecule added per SG. (Reprinted with permission from Vartak, S. et al. 2013. Collective proton dynamics at highly charged interfaces studied by ab initio metadynamics. *J. Phys. Chem. B*, **117**(2), 583–588. Copyright (2013), American Institute of Physics.)

The resulting model consists, thus, of a regular hexagonal array of acidic SGs with fixed end points, as depicted in Figure 2.30b. In spite of the described simplifications, the model retains essential characteristics for studying structural conformations as well as the dynamics of polymer sidechains, water, and protons. The approach implies that the effect of polymer dynamics on processes inside of pores is primarily caused by variations in chemical architecture, packing density, and vibrational flexibility of SGs.

### Structural Transitions at Dense Arrays of Protogenic Surface Groups

In a first set of studies, quantum mechanical calculations using VASP (Kresse and Furthmüller, 1996a,b; Kresse and Hafner, 1993, 1994a,b) and CPMD (Car and Parrinello, 1985; Marx and Hutter, 2009) were applied to elucidate effects of molecular structure and packing density of SGs on spontaneous ordering, acid dissociation, and water binding at the surface. The lattice constant $d_{CC}$ of the hexagonal lattice of anchoring C atoms of SGs was varied in the range from 5 to 15 Å. The surface area per SG, $A_{SG} = \frac{\sqrt{3}}{2}d_{CC}^2$, varied from 22 to 190 Å$^2$. Initially, the hexagonal arrangement was selected based on symmetry considerations. Thereafter, it was reasoned that, based on the structural simulations performed, self-organization among $H_3O^+$ and $SO_3^-$ ions, mediated by strong interfacial hydrogen bonds, enforces hexagonal ordering at high SG density.

Calculations were performed on a unit cell with three SGs and three water molecules. The main model system studied was of an array of SGs of the type $CF_3SO_3H$ (triflic acid). Figure 2.31 shows the formation energy per unit cell, $E_f^{uc}$. It is defined as the difference between optimized total energy $E_{total}(d_{CC})$ and the total energy of a system of independent surface groups, each with one water molecule, in the limit of infinite separation, $E_f^\infty$,

$$E_f^{uc}(d_{CC}) = E_{total}(d_{CC}) - E_f^\infty. \tag{2.56}$$

In this definition, $E_f^{uc}$ incorporates correlation energies between SGs due to electrostatic interactions and hydrogen bonding in the interfacial layer.

The most stable configuration of the 2D array was found at $d_{CC} \approx 6.2$ Å with $E_f^{uc} = -2.78$ eV. As can be seen in Figure 2.31, in this configuration acidic SGs attain a fully dissociated state. They are oriented in an upright position relative to the basal plane. $H_3O^+$ cations and $SO_3^-$ head groups form a highly ordered hydrogen-bonded network. Each ionic species has saturated its demand for hydrogen bonds with neighboring species. Strong directional hydrogen bonds render this structure rather stiff, resembling a 2D monohydrate crystal. The high electronegativity of hydrophobic sidechain groups repels oxygen in $H_3O^+$ ions from the interface. The average vertical separation between $H_3O^+$ cations (position of O atom) and anionic $SO_3^-$ head groups (position of S atom) is 1.0 Å. The average hydrogen bond length is $d_{OO} = 2.6$ Å and the average OH bond length in $H_3O^+$ ions is 1.02 Å, a value that is slightly larger than OH bond lengths in water (0.98 Å).

Upon increasing $d_{CC}$, the fully dissociated "upright" structure destabilizes. The average hydrogen bond distance increases from $d_{OO} = 2.6$ Å at $d_{CC} \approx 6.2$ Å to

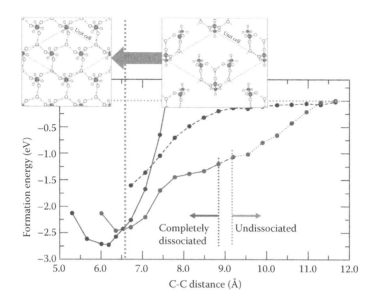

**FIGURE 2.31** Configuration energy of minimally hydrated interfacial array of triflic acid. The configuration energy is shown as a function of the spacing between surface groups in the hexagonal array, $d_{CC}$. Upon densification of SGs, the transition from undissociated to dissociated state occurs at $d_{CC} \approx 9$ Å. The transition to the condensed surface state occurs at $d_{CC} \approx 6.5$ Å. (Reprinted from Roudgar, A., Narasimachary, S. P., and Eikerling, M. 2006. *J. Phys. Chem. B*, **110**(41), 20469–20477, Figures 3 and 8(a). Copyright (2006) American Chemical Society. With permission.)

$d_{OO} = 2.7$ Å at $d_{CC} \approx 7.1$ Å, resulting in weaker hydrogen bonds and, therefore, lower absolute values of the formation energy, $E_f^{uc} = -1.67$ eV. The "upright" conformation becomes unstable at $d_{CC} \approx 6.5$ Å, corresponding to $A_{SG} = 37$ Å$^2$.*

At $d_{CC} = 6.5$ Å, a transition to an energetically more favorable fully dissociated "tilted" structure occurs. In this structure, the three surface groups in the unit cell are inclined toward each other. In comparison to the "upright" structure, SGs are rotated around their C–S axis and one $H_3O^+$ ion is shifted laterally. The tilting angle decreases monotonically from 75° at $d_{CC} \approx 6$ Å to 14° at $d_{CC} > 10$ Å. At $d_{CC} = 7.4$ Å, the number of hydrogen bonds per unit cell decreases from 9 to 7. At this point, the interunit cell hydrogen bonds are broken, leading to the formation of disconnected clusters of anionic SGs and $H_3O^+$ ions.

A transition from fully dissociated to partially dissociated "tilted" configuration occurs at $d_{CC} = 8.7$ Å with two $H_2O$ and one $H_3O^+$ ion remaining per unit cell. At $d_{CC} = 9.2$ Å, another transition from a partially dissociated to a completely nondissociated state occurs. At $d_{CC}$ beyond this point, every acid group retains only one

---

* Initial optimization studies were performed with VASP (Vienna *Ab Initio* Simulation Protocol). Later on, optimization studies were repeated with Car–Parrinello MD (CPMD). With CPMD, the transition from "upright" to "tilted" shifts to a slightly larger value, $d_{CC} \approx 6.7$ Å.

hydrogen bond with the closest water molecule. The intracluster hydrogen bonds are broken and the formation energy approaches $E_f^{uc} = 0$. As is known from hydration studies for clusters representing single hydrated sidechain moieties, adding a single water molecule per sidechain is not sufficient for the dissociation of triflic acid (Clark et al., 2012; Eikerling et al., 2002; Paddison, 2001).

Qualitatively similar sequences of structural transitions were observed for interfacial arrays with different SGs, for which either the chemical composition ($CH_3SO_3H$), or the length of the tail group was varied ($CF_3CF_2SO_3H$, $CF_3CF_2CF_2SO_3H$, and $CF_3OCF_2CF_2SO_3H$) (Narasimachary et al., 2008). Worthy of note, the last surface group listed resembles short sidechains in Dow and Aquivion membranes.

The array of $CH_3SO_3H$ is less stable than the $CF_3 SO_3H$ array and the transition from "upright" to "tilted" conformation shifts to a smaller value of $d_{CC}$ by about 0.3 Å. Moreover, for the array of $CH_3SO_3H$, the average vertical separation between $H_3O^+$ ions (position of O atom) and $SO_3^-$ groups (position of S atom) is only 0.1 Å, which is 0.9 Å smaller than the value found for the $CF_3SO_3H$ system. At the same value of $d_{CC}$, $H_3O^+$ ions are, thus, more deeply embedded into the interfacial layer, owing to the fact that interactions between $H_3O^+$ ions and hydrophobic sidechain tails are significantly less repulsive.

Plots of $E_f^{uc}(d_{CC})$ for arrays of $CF_3CF_2SO_3H$, $CF_3CF_2CF_2SO_3H$, and $CF_3OCF_2CF_2SO_3H$, look similar to the plot for the $CF_3SO_3H$ system. These systems undergo the same series of transitions as the triflic acid system. Most importantly, the transition from "upright" to "tilted" surface structure occurs at an almost identical value of $d_{CC}$. Increasing the length of SGs renders the interfacial array slightly less stable. The range of interfacial correlations increases with the length of SGs.

Upon the addition of extra water, a wetting transition was found from a hydrophilic surface state for $d_{CC} > 7.5$ Å to a superhydrophobic surface state for $d_{CC} < 6$ Å. Figure 2.32 shows the binding energy of an extra water molecule at the minimally hydrated array. For comparison, a vertical line is drawn that represents the Gibbs energy of water sorption at a free water surface. In the condensed superhydrophobic state, the number of interfacial hydrogen bonds is saturated, explaining the very weak interaction with extra water molecules. In the "normal" hydrophilic surface state, the strength of water binding at the minimally hydrated array is markedly enhanced over that of a free water surface (Roudgar et al., 2006).

The formation of the condensed state is related to the perfect match between trigonal structures of $-SO_3^-$ and $H_3O^+$, which enables a saturated number of interacial hydrogen bonds. The value $d_{CC} \approx 6.5$ Å, for the concomitant structural and wetting transition agrees with critical packing densities for surface pressure and surface conductance in experiments at monolayer systems as seen in Figure 2.28 (Leite et al., 1998; Oliveira et al., 2004). These findings indicate that water binding, acid dissociation, and proton dynamics in the condensed state are highly sensitive to fluctuations in SG density. This sensitivity should be an important aspect to evaluate in future experimental studies.

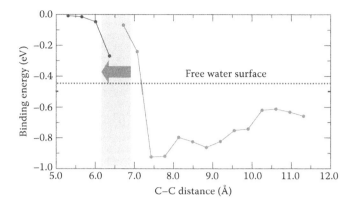

**FIGURE 2.32** The energy of interaction of an extra water molecule with the minimally hydrated interfacial array of triflic acid groups. The Gibbs energy of water sorption at a free water surface is shown for comparison. Upon densification of SGs, the transition from a normal hydrophilic regime to a superhydrophobic regime occurs at $d_{CC} \approx 6.5$ Å. It coincides with the transition to the condensed surface state of the array. (Reprinted from Roudgar, A., Narasimachary, S. P., and Eikerling, M. 2006. *J. Phys. Chem. B*, **110**(41), 20469–20477, Figures 3 and 8(a). Copyright (2006) American Chemical Society. With permission.)

## Mechanisms of Proton Transport at the Interfacial Array

It is of foremost interest to understand, whether the ordered and stable array of densely packed SGs could support long-range proton transport. Figure 2.33 introduces an intuitive graphical representation of possible interfacial $H_3O^+$ ion transitions at the critical value $d_{CC} = 6.7$ Å for the interfacial structural transition, as determined by CPMD (Vartak et al., 2013). At this $d_{CC}$, fluctuations in local SG density trigger proton transitions between proton states with similar energy.

The representation in Figure 2.33 allows for a simple evaluation of the hydrogen bond structure during transitions. Filled green triangles represent mobile $H_3O^+$ ions. Vertices (red dots) correspond to $SO_3^-$ ions. The corners of filled triangles correspond to hydrogen bonds that point to a sulfonate anion. Figure 2.33a shows the perfectly ordered condensed state. The individual translocation of a $H_3O^+$ ion in Figure 2.33b corresponds to a translocation of a filled triangle to an empty neighboring triangle. This defect-type translocation leaves behind a hydrogen bond deficit at the donor SG, and it creates a hydrogen bond excess at the acceptor SG. Any *interfacial proton transfer* on the lattice can be represented as a concerted or disconcerted sequence of these elementary $H_3O^+$ ion moves.

Figure 2.33c and d illustrates final structures after collective single file and multifile $H_3O^+$ ion transfers. These transitions conserve the number of interfacial hydrogen bonds. They involve a concerted orientational motion of donor and acceptor SGs. Collective motion is enforced by the increasing strength of interfacial hydrogen bonds.

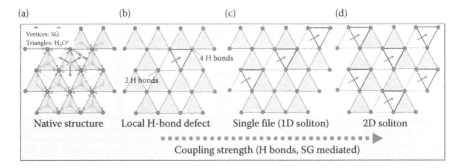

**FIGURE 2.33** Lattice configurations of the 2D two-component lattice of $H_3O^+$ ions and $SO_3^-$ ions. The native 2D crystal structure in (a) is shown at the critical SG density of $d_{CC} \approx$ 6.7 Å. The green triangles represent positions of $H_3O^+$ ions and the red dots represent positions of $SO_3^-$ ions. A single $H_3O^+$ ion move, as shown in (b), creates a hydrogen bond deficit at the donor site and a hydrogen bond excess at the acceptor site. Collective $H_3O^+$ ion translocations that conserve the number of hydrogen bonds, are shown in (c) and (d), denoted as single file and parallel file proton transfer. (Reprinted with permission from Vartak, S. et al. 2013. Collective proton dynamics at highly charged interfaces studied by ab initio metadynamics. *J. Phys. Chem. B*, **117**(2), 583–588. Copyright (2013), American Institute of Physics.)

### Metadynamics Study of Interfacial Proton Transport

The molecular simulation study in Vartak et al. (2013) utilized the *ab initio* Lagrangian metadynamics method of Laio and Parrinello (Ensing et al., 2005) to explore reaction pathways and free energy surfaces of interfacial hydronium ion transitions. Metadynamics is an efficient method of exploring rare events in complex molecular systems and condensed media. It utilizes a coarse-grained dynamics in a space of a few time-dependent collective variables (CVs) to simulate the system trajectory. Adding small Gaussian functions along the molecular dynamics trajectory, creates a history-dependent potential. The height and width of these Gaussian functions must be optimized in view of sampling efficiency and precision of the free energy calculation. Furthermore, a harmonic coupling between real and fictitious CVs in the metadynamics Lagrangian ensures uniform sampling of configurations in the well regions of the free energy landscape. This leads to an even distribution of the history-dependent potential. After the addition of a suitable number of Gaussian functions, the system attains a diffusive state with horizontal energy profile. At this point the simulation is terminated. Gaussian functions are subtracted from the total free energy in the final state, revealing the free energy as a function of CVs.

*Ab initio* metadynamics simulations were performed at the DFT level using the CP2K package, which implements a mixed *Gaussian and plane wave method* (GPW method) (VandeVondele et al., 2005). Using this methodology, activation energies and reaction pathways of local defect-type (cf. Figure 2.33b) and highly collective $H_3O^+$ ion transitions (cf. Figure 2.33d) were determined. In both cases, the initial structure was the perfectly ordered condensed state at $d_{CC} = 6.7$ Å.

In these calculations, valence electrons are represented by double-$\xi$ augmented Gaussian basis sets (DZVP-MOLOPT) (VandeVondele and Hutter, 2007). The pseudopotential of *Goedecker, Teter, and Hutter* (GTH) represents core electrons (Goedecker et al., 1996). The energy cut-off of plane wave expansions was 300 Ry. Exchange and correlation energies were computed within the GGA approximation, using the BLYP functional (Becke, 1988). For every time step, the electronic structure was quenched to an accuracy of $10^{-7}$ Hartree. Starting each metadynamics run from the optimized geometry of the condensed state, the system was thermalized in the NVT ensemble for around 3 ps. The temperature was set to 300 K using a Nóse–Hoover thermostat. The time step was set at 0.3 and 0.5 fs for simulations of collective and local defect-type mechanisms, respectively. The coupling constant of the metadynamics Lagrangian was $k = 0.5$ and 0.4 a.u., and the fictitious particle mass was $M = 50$ and 75 a.u. for collective and local defect-type simulations, respectively. Gaussian potentials with height $h = 0.013$ eV and width $\Delta\delta = 0.02$ Å were used to create the history-dependent potential. A Gaussian function was added to the history-dependent potential every time when the displacement of the CV, relative to the previous state, reached $3\Delta\delta/2$.

$H_3O^+$ ion transitions involve rapid hydrogen bond fluctuations, orientational fluctuations of SGs and translational $H_3O^+$ ion motion. Laborious metadynamics runs were performed to find appropriate CVs from the set of degrees of freedom and to fine-tune metadynamics parameters. As a conclusion from these benchmarking studies, the lateral $H_3O^+$ ion shift was chosen as a sole metadynamics CV, defined by $d_{CV} = d_{12} - d_{23}$ in Figure 2.34. Relocation distances of $H_3O^+$ ions during transitions are 3–4 Å.

Figure 2.35 compares local defect-type and collective proton transitions. Transitions are completed within 140 and 45 ps for local and collective pathways, respectively. Reconstructed Helmholtz energy profiles, as a function of the CV, are shown together with snapshots of configurations during transitions. Activation and reaction Helmholtz energies of the defect-type transition are $\Delta F_a = 0.6$ eV and $\Delta F_r = 0.5$ eV. The final state of this transition is metastable. For the collective transition, activation, and reaction Helmholtz energies are $\Delta F_a = 0.3$ eV and $\Delta F_r = 0$ eV.

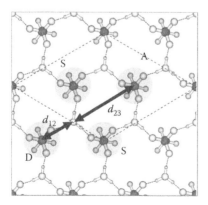

**FIGURE 2.34** Interfacial configuration considered in studies of proton transport mechanisms. Donor (D), acceptor (A), and spectator (S) SGs are indicated. The collective variable $d_{CV} = d_{12} - d_{23}$ is shown as well.

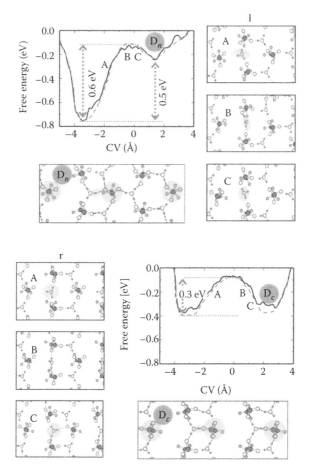

**FIGURE 2.35** Free energy profiles of local defect-type (top panel, l) and collective $H_3O^+$ ion transfers (bottom panel, r), obtained from *ab initio* metadynamics. Snapshots show structures at points A, B, C, and Dn/c, marked at the free energy profiles. (Reprinted from Vartak, S. et al. 2013. *J. Phys. Chem. B*, **117**(2), 583–588, Figures 1,2,4. Copyright (2013) American Chemical Society. With permission.)

The value of $\Delta F_a$ for the collective transition is 2–3 times larger than the activation energy of proton transport in bulk water (0.1 eV), as expected, based on the increased hydrogen bond strength. It is to be seen in refinements of metadynamics simulations whether different choice of CVs and evaluation of longer SGs will reduce $\Delta F_a$ significantly.

Configurations in Figure 2.35 illustrate intermediate structures during transitions, taken at points A–D in Helmholtz energy profiles. Orientational fluctuations of the donor SG trigger hydrogen bond breaking with the transiting $H_3O^+$ ion (A). Hydrogen bond breaking and reformation is responsible for the steep ascending and descending flanks of free energy along the CV. The $H_3O^+$ ion shift occurs in the saddle point region (B). It involves a flip of the $H_3O^+$ ion, while its remaining two HBs with spectator SGs remain intact. Frame C shows the formation of the HB with the acceptor SG. Relaxation of the interfacial network occurs in the well region of the final state ($D_n$ or $D_c$), involving further rotation and tilting of SGs. In the case of the collective transition, each donor SG rotates to accept another HB from the left, simultaneously with HB formation between $H_3O^+$ ion and acceptor SG. The final state ($D_n$) attained at completion of the local transition exhibits HB defects at donor (undersaturated) and acceptor SG (oversaturated), rendering this state highly unstable.

The metadynamics method allows one to use several CVs to reconstruct the free energy surface of a system under investigation. The computational costs of simulations are strongly dependent on the number of CVs used. Compared to results of static calculations (Roudgar et al., 2008), which gave an activation energy of $\sim$0.5 eV for collective proton motion, metadynamics gave a significantly smaller activation energy of $\sim$0.3 eV for the same transition. The inclusion of dynamic effects in simulations of interfacial proton transfer is, thus, of utmost importance. Moreover, results show that the collective nature of the interfacial proton transfer could be essential for achieving high proton conductance.

## Solitons Coming Alive in Surface Proton Conduction

Experimental and molecular modeling studies have provided compelling evidence for fast proton motion at densely packed arrays of protogenic surface groups. The purported collective nature of interfacial proton translocations has opened the field for soliton models (Davydov, 1985). Several soliton approaches were developed to explain collective proton transport in hydrogen-bonded systems. They have been met, however, with limited success, for reasons that will be explained below. The low proton conductivity predicted by soliton models has led to a withering interest in recent years. However, these predictions may be misleading due to oversimplifying assumptions in models from which they follow.

The majority of studies of collective proton motion in ice or at interfaces resorted to model systems represented by infinite quasi-one-dimensional hydrogen-bonded chains (Gordon, 1990; Pnevmatikos, 1988; Tsironis and Pnevmatikos, 1989; Weiner and Askar, 1970). The chain consists of two coupled sublattices of alternating mobile protons and heavy anions. Each mobile proton is located between a pair of heavy anions, referred to, as before, as SGs. SGs are considered to be fixed with their tail group but their anionic head groups are allowed to fluctuate about their equilibrium positions. They act as proton relay groups.

Mobile protons could be transferred in two types of motions: (1) Protons could rattle back and forth between symmetric minima of the effective substrate potential energy. The minima are located at hydrogen bond distance from either of the two neighboring SGs. The double well potential, experienced by the intermittent proton, depends on the equilibrium separation of SGs and on their fluctuations. Similar to what happens in the formation of a Zundel ion in water, the double well potential may transform into a single well potential upon close approach of neighboring SGs. Spontaneous symmetry-breaking, associated with these proton motions, leads to the

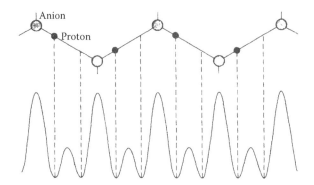

**FIGURE 2.36**  A two-component system of anions and protons in linearly alternating configuration. Anions, representing SGs, occupy fixed lattice positions. Mobile protons are located at
interstitial sites in hydrogen bond distances to one of the neighboring SGs. The bottom graphic
shows the doubly periodic deformable substrate potential, considered in the classical paper
of Pnevmatikos (1988). (Reprinted from Pnevmatikos, S. *Phys. Rev. Lett.*, 60, 1534–1537,
1988, Figure 1. Copyright (1988) by the American Physical Society. With permission.)

formation of two equilibrium configurations of the infinite chain, in which all protons
are located either at hydrogen bond distance from SGs to the left or from SGs to the
right. In the case of strong proton–proton coupling, transitions between these two
equilibrium configurations of the proton ensemble would involve the propagation of
a soliton kink. However, this kink motion of protons alone does not constitute long-
range proton transport. (2) The second type of proton motion involves rotations of
SGs, together with their hydrogen-bonded proton. This motion relocates the proton
to a neighboring cell. It leads to the formation of Bjerrum-type ionic defects, where
the interstitial space between two SGs is proton free (L-Bjerrum defect) or occupied
by two protons (D–Bjerrum defect).

An effective substrate potential for protons that accounts for both types of proton
motions exhibits a doubly periodic form, as illustrated in Figure 2.36 (Pnevmatikos,
1988): low activation barriers separate equivalent positions of a proton in the space
between neighboring SGs, whereas large activation barriers are encountered at equilibrium positions of SGs. The barriers at SG positions are higher since they involve
the creation of hydrogen bond defects. The high potential barriers at the SG sites of
ions stifle long-range proton transport in one-dimensional systems.

The decisive modification that is thought to bring solitons to life in surface proton
conduction is the consideration of a two-dimensional model system in Golovnev and
Eikerling (2013b). This system eliminates high (basically unsurmountable) potential
barriers at the sites of anionic surface groups. The two-dimensional model system
corresponds to Figure 2.34. It consists of an array of SGs with sulfonic acid head
groups. At high SG density and under minimal hydration, with one water molecule per
SG, the array assembles into a perfectly dissociated and hexagonally ordered ground
state, as shown. Close packing of SGs leads to enhanced strength of hydrogen bonds,
rendering the interfacial configuration stable and favoring collective proton motion
over the creation of local proton defects.

A graphical representation of $H_3O^+$ ion translocations at the interface is shown in Figure 2.33. All elementary moves of $H_3O^+$ ions are structurally equivalent and any mechanism of long-range proton transport can be conceived entirely as a sequence of elementary moves. If moves of $H_3O^+$ ions in adjacent cells occur collectively, with almost simultaneous hydrogen bond breaking and reformation at donor and acceptor sites, the number of interfacial hydrogen bonds is conserved. The activation energy for these collective hydronium ion translocations is reduced significantly, to a value of about 0.3 eV, as discussed previously.

The concertedness of $H_3O^+$ ion translocations depends on the strength of interfacial hydrogen bonds. Weak hydrogen bonds favor disconcerted $H_3O^+$ ion transitions, whereas strong hydrogen bonds, corresponding to a strong coupling of SG and hydronium ion sublattices, foster concerted mechanisms. In the case of a concerted motion, collectively moving $H_3O^+$ ions create a traveling wave, which retains a stable shape while it travels, a soliton. A soliton could make the charge transport more efficient, since it relocates a large number of $H_3O^+$ ions simultaneously. Also, a collective motion minimizes the probability of a back transition to the initial state.

## Long-Range Proton Transport at Interfaces: Soliton Theory

A formalism to describe highly correlated lattice operations that could enable long-range proton transport was developed in Golovnev and Eikerling (2013b). Albeit, the lattice being two-dimensional, the formalism evaluated the propagation of one-dimensional solitons that represent a single file proton motion, as depicted in Figure 2.33c.

The Hamiltonian for *single file proton motion* is

$$H = \sum_i \frac{m}{2} \dot{u}_i^2 + \frac{k}{2}(u_{i+1} - u_i)^2 + V(u_i), \tag{2.57}$$

where $i$ is the label of $H_3O^+$ ions along the soliton track, $u_i$ the displacement of the $i$th $H_3O^+$ ion, $m$ the $H_3O^+$ ion mass, $k$ a hydronium–hydronium coupling constant, and $V(u_i)$ the effective substrate potential created by the SG sublattice and experienced by $H_3O^+$ ions. As a result of the periodicity of the system, $V(u_i)$ is a sequence of potential wells that are located at the equilibrium positions of $H_3O^+$ ions. Assuming that the size of the soliton is much larger than the lattice constant $d_{CC}$, the continuum limit can be taken of Equation 2.57:

$$H = \frac{1}{\sqrt{3}a} \int dx \left( \frac{m}{2}(\partial_t u)^2 + \frac{ka^2}{2}(\partial_x u)^2 + V(u) \right), \tag{2.58}$$

where $\sqrt{3}a = d_{CC}$, $x$ is a spatial coordinate along the kink path, and $\partial_t$ denotes a derivative with respect to time $t$. Using the Euler–Lagrange formalism, the equation of motion is found,

$$m\ddot{u} = ka^2 \partial_{xx} - \frac{\partial V(u)}{\partial u}. \tag{2.59}$$

To describe traveling wave solutions, it is customary to introduce a wave coordinate $\xi = x - vt$, with soliton velocity $v$. Integrating the equation of motion once gives

$$\partial_\xi u = \pm\sqrt{2}\Omega\sqrt{V(u) + \varepsilon},$$

(2.60)

where $\varepsilon$ is a potential well depth and $\Omega^2 = \left[m\left(v_0^2 - v^2\right)\right]^{-1}$ with the maximal possible soliton velocity $v_0 = \sqrt{ka^2/m}$.

During the transition, a $H_3O^+$ ion is relocated over a distance $a = d_{CC}/\sqrt{3}$. At $d_{CC} = 6.7$ Å, the transfer distance is $a = 3.75$ Å. Kink motion relocates all $H_3O^+$ ions along the file by the same displacement $a$. In the course of the transition, $H_3O^+$ ions have to overcome the same potential barrier. The substrate potential $V(u)$ accounts for the influence of rotation and tilting motions of SGs on $H_3O^+$ ion motion. It could be assumed that $V(u)$ is a singly periodic function, which is symmetric with respect to the center between the minima at $u \equiv a/2$. The form of $V(u)$ could be determined from *ab initio* studies of surface proton dynamics, discussed in the section "Metadynamics Study of Interfacial Proton Transport." The configuration of SGs changes during the transition from "upright" to "tilted" configuration, and the potential is, therefore, not strictly symmetric. However, it was shown that this assumption is not crucial. For mathematical convenience, the potential function is required to satisfy $V(a/2) = 0$ and $V(0) = V(a) = -\varepsilon$.

Solution of Equation 2.60 gives expressions for the soliton energy,

$$E = \frac{E_0}{\sqrt{1 - \left(\dfrac{v}{v_0}\right)^2}},$$

(2.61)

and the number of $H_3O^+$ ions per kink,

$$N = N_0\sqrt{1 - \left(\frac{v}{v_0}\right)^2},$$

(2.62)

where $E_0 = \sqrt{2\varepsilon ka^2/3}$ and $N_0 = \sqrt{ka^2/2\varepsilon}$ are energy and number of $H_3O^+$ ions of a static soliton. The product $EN = E_0N_0 = ka^2/\sqrt{3}$ is an invariant of soliton motion. It is independent of the soliton velocity.

A standard approach was adopted to find the soliton mobility: in the equation of motion, additional terms representing a viscous friction and an external force $f$, acting on each hydronium ion, were added. The terminal drift velocity was calculated. The ratio of drift velocity to external force, evaluated in the limit $f \to 0$, gives the mobility of the soliton,

$$\mu_v \equiv \lim_{f \to 0}\left|\frac{v_k}{f}\right| = \frac{1}{b}\sqrt{\frac{ka^2}{2\varepsilon}},$$

(2.63)

where $b$ is the viscous friction coefficient.

The expressions obtained relate soliton properties, such as energy and mobility, to microscopic parameters of the interfacial model, that is, $k$, $a$ and $\varepsilon$. These parameters are determined by the packing density of the interfacial array of SGs and by the strength of hydrogen bonds between SGs and hydronium ions. Interdependences exist between the three parameters. Dependence of $\varepsilon$ on $a$, studied in Vartak et al. (2013), is engendered by a change of the hydrogen bond length. Dependence of $k$ on $a$ has not been studied yet, but should exhibit a high sensitivity to the hydrogen bond length as well. Knowledge of these dependencies will leave only one parameter, namely, $a$. Finding the optimal value of $a$ is the key to synthesizing new materials with high interfacial proton mobility.

The constants $k$ and $a$ appear only in the form of a product $ka^2$, representing a convenient parameter for theoretical discussion. Upon increasing $ka^2$, the soliton mobility increases, whereas the probability of soliton creation decreases, since the soliton energy increases. Thus, there is either a large amount of nonmobile excitations (small $ka^2$) or a small amount of highly mobile solitons (large $ka^2$). Therefore, there must be an optimum value of $ka^2$ that maximizes the conductivity. Finding this optimum value requires a formalism to describe soliton creation and annihilation processes, which is a subject for further research.

The steps described above were followed through for various different potential functions $V(u)$, which are amenable to analytical solution and could represent different local forces acting on hydronium ions. Microscopic parameters describing the collective nature of $H_3O^+$ ion motion, such as soliton size and shape, were found to depend strongly on the potential profile. Surprisingly, energy and mobility of solitons are rather independent of the potential profile, being determined merely by the depth of the potential well and the separation and coupling strength between hydronium ions.

## Mechanism of Energy Loss: Soliton Mobility

The key property in studies of proton transport at acid-functionalized monolayers or in PEMs is the *proton conductivity*. It is determined by proton mobility and concentration. In the simplest case, conductivity could be expressed as a product of these properties.

Some of parameters of the surface model needed in Equation 2.63 are already known. The elementary transfer distance of a hydronium ion, $a$, is related to the lattice spacing $d_{CC}$ by $\sqrt{3}a = d_{CC}$; see Figure 2.35. The critical value $d_{CC} \approx 6.7$ Å corresponds to $a = 3.9$ Å. The parameter corresponding to the depth of the potential well $\varepsilon$ was found in Vartak et al. (2013) as $\varepsilon = 0.3$ eV.

A further stride toward relating microscopic surface structure to effective proton mobility was made in Golovnev and Eikerling (2013a). The theoretical formalism developed therein focused on finding the viscous friction coefficient $b$ and the hydronium–hydronium coupling constant $k$.

The essential element of the formalism is a mechanism of energy loss of solitons. The traveling 1D soliton (single file in Figure 2.33c) transfers energy to the SG sublattice. Soliton motion creates a perturbation of the SG sublattice, which emanates along the soliton path and propagates perpendicular to it. Assuming that the perturbation of the SG sublattice is small, it could be treated as a harmonic wave.

The following expression of the viscous friction coefficient was obtained,

$$b = 2\sqrt{kM}\frac{\delta^2}{a^2}, \tag{2.64}$$

where $M$ is the mass of a SG and $\delta$ is the maximal SG displacement in the surface plane, that is, the difference of its coordinates in "tilted" and "upright" structures, projected onto the surface plane. The expression for $b$ is independent of the form of the substrate potential $V(u)$ in Equation 2.58.

Inserting Equation 2.64 into Equation 2.63 gives a new expression for the soliton mobility,

$$\mu = \frac{a^3}{\delta^2\sqrt{8\varepsilon M}}. \tag{2.65}$$

This expression of the mobility depends neither on the coupling constant between hydronium ions, $k$, nor on the hydronium ion mass, $m$, but it depends on the SG mass $M$. For a simple estimate, $M$ can be assumed to represent the mass of an $SO_3^-$ ion with $M = 80m_p$, where $m_p$ is the mass of a proton. Using $\varepsilon = 0.3$ eV and $\delta = 0.09a$ gives $\mu = 1.6 \cdot 10^{14}$ m (Ns)$^{-1}$ as an upper estimate of the soliton mobility. The corresponding proton diffusion coefficient is $D = 8 \cdot 10^{-3}$ cm$^2$ s$^{-1}$. It is significantly larger than the proton diffusion coefficient in bulk water ($9.3 \cdot 10^{-5}$ cm$^2$ s$^{-1}$). However, it was obtained for an idealized model without any defects at the optimal interfacial density of SGs. These conditions will provide the optimal value of proton mobility. Nonidealities in experimental systems, due to variations in interfacial SG density and SG lengths, are expected to diminish the efficiency of interfacial proton transport and lead to smaller values of proton mobility. Moreover, a larger value of $M$ should be used for interfacial arrays with longer SGs.

The second major challenge in finding a value of the surface proton conductivity in the soliton approach is the development of a soliton statistics. Having found an expression for the soliton mobility, the statistics of solitons will yield a soliton density. To this end, it will be necessary to theorize and simulate mechanisms of soliton creation and annihilation. A key aspect in this context could be the relation of soliton creation and annihilation to spatiotemporal fluctuations in SG density and the occurrence of spontaneous lattice defects.

## RANDOM NETWORK MODEL OF MEMBRANE CONDUCTIVITY

The effective PEM conductivity depends on the random heterogeneous morphology, namely, the size distribution and connectivity of the proton-containing aqueous pathways. Random network model of PEMs was developed in Eikerling et al. (1997). It included effects of the swelling of pores and the evolving connectivity of the pore network upon water uptake. The model was applied to study the dependence of membrane conductivity on water content and temperature. It could rationalize trends in swelling properties and conductivity of PEM with distinct ionomer structures, which were found to be in agreement with experiments.

In general, pores swell nonuniformly, as seen in the section "Water Sorption and Swelling of PEMs." As a simplification, the random network was assumed to consist of two types of pores. Nonswollen or "dry" pores (referred to as "red" pores) permit only a small residual conductance resulting from tightly bound surface water. Swollen or "wet" pores (referred to as "blue" pores) contain extra water with high "bulk-like" conductance. Water uptake corresponds to the swelling of "wet" pores and to the increase of their relative fraction. In this model, proton transport in the PEM is mapped as a percolation problem, wherein randomly distributed sites represent pores of variable size and conductance. The distinction of "red" and "blue" pores accounts for variations of proton transport properties due to different water environments at the microscopic scale, as discussed in the section "Water in PEMs: Classification Schemes."

The number fraction of "blue" pores as a function of the water content $\lambda$ is given by

$$x(\lambda) = \frac{N_b}{N}, \tag{2.66}$$

in which both the total number of pores $N$ as well as the number of "blue" pores $N_b$ are functions of $\lambda$. The bond probabilities between two "blue" pores, a "blue" and a "red" pore, and two "red" pores, respectively, are

$$p_{bb}(\lambda) = x(\lambda)^2, \quad p_{br}(\lambda) = 2x(\lambda)(1 - x(\lambda)), \quad p_{rr}(\lambda) = (1 - x(\lambda))^2. \tag{2.67}$$

At the time the model was developed about 15 years ago, it utilized a phenomeno-logical law of pore swelling based on Gierke's experimental data for the structural reorganization of the membrane upon water uptake (Gierke et al., 1981; Hsu and Gierke, 1982, 1983). Empirical relations were extracted from data for the average number of SGs per pore, $n(\lambda) = n_0 (1 + \alpha\lambda)$, and the average volume of water-filled pores, $v(\lambda) = v_0 (1 + \beta\lambda)^3$, where $n_0$ is the average number of SGs per pore in a dry membrane, $v_0$ is the average pore volume in the dry membrane, and $\alpha$ and $\beta$ are fitting parameters. Invoking the conservation of the total number of dissociated $SO_3^-$ groups in the membrane and the proportionality between total water content and the volume increase of "blue" pores, an *empirical swelling law* was derived:

$$x(\lambda) = \frac{\gamma\lambda}{(1 + \beta\lambda)^3 - \alpha\gamma\lambda^2}. \tag{2.68}$$

The parameters $\alpha$, $\beta$, and $\gamma$ can be adjusted in order to reproduce the extent of swelling and reorganization of the polymer matrix upon water uptake. This law accounts for the merging of smaller pores into larger pores upon swelling. It could represent distinct elasticity of the polymeric membrane matrix that leads to distinct water distributions. In a soft polymer matrix, water would be distributed rather hetero-geneously with individual pores swelling to large equilibrium radii and, thus, taking up a lot of water. In a more elastic polymer matrix, pores swell more homogeneously with smaller equilibrium radii.

A rigid microporous morphology with monodisperse pores, which does not reorganize upon water uptake, corresponds to a linear law $x(w) = \gamma w$. In this case, the model resembles the typical percolation problem in random porous media with rigid walls (Stauffer and Aharony, 1994). Swelling causes deviations from this law. The universal percolation exponents are, therefore, not warranted.

Pore-size-dependent conductances are assigned to individual pores and channels. Three possible types of bonds between pores exist. The corresponding bond conductances, specifically, $\sigma_{bb}(\lambda)$, $\sigma_{br}(\lambda)$, and $\sigma_{rr}(\lambda)$, can be established. The model was extended toward the calculation of the complex impedance of the membrane by assigning capacitances in parallel to conductances to pores. The probability distribution of bonds to have conductivity $\sigma_{bb}$, $\sigma_{br}$, or $\sigma_{rr}$ is

$$f(\sigma_b) = p_{bb}(\lambda)\delta(\sigma_b - \sigma_{bb}) + p_{br}(\lambda)\delta(\sigma_b - \sigma_{br}) + p_{rr}(\lambda)\delta(\sigma_b - \sigma_{rr}), \quad (2.69)$$

where $\delta$ is the Dirac delta distribution.

The simplest method of solution of the Kirchhoff equations, corresponding to the random network of conductance elements, is the single-bond effective medium approximation (SB-EMA), wherein a single effective bond between two pores is considered in an effective medium of surrounding bonds. The conductivity $\sigma_b$ of the effective bond is obtained as the self-consistent solution of the equation

$$\int d\bar{\sigma} f(\bar{\sigma}) \frac{\sigma_b - \bar{\sigma}}{\bar{\sigma} + (d-1)\sigma_b} = 0, \quad (2.70)$$

which corresponds to an averaging of voltage fluctuations across the effective bond, where $d = \frac{z}{2}$ and $z$ is the number of channels connected to a single pore. Using Equation 2.69, the effective bond conductivity $\sigma_b$ is obtained from

$$p_{bb}(\lambda)\frac{\sigma_b - \sigma_{bb}(\lambda)}{\sigma_{bb}(\lambda) + q\sigma_b} + p_{br}(\lambda)\frac{\sigma_b - \sigma_{br}(\lambda)}{\sigma_{br}(\lambda) + q\sigma_b} + p_{rr}(\lambda)\frac{\sigma_b - \sigma_{rr}(\lambda)}{\sigma_{rr}(\lambda) + q\sigma_b} = 0, \quad (2.71)$$

where $q = (d-1)$ represents the connectivity of the pore network. When the conductivities of "red" pores and channels vanish, the true percolation behavior is obtained

$$\sigma_b(\lambda) = \frac{1}{q}\left[(1+q)x^2(\lambda) - 1\right]\sigma_{bb}(\lambda), \quad (2.72)$$

with percolation threshold

$$x_c = \sqrt{\frac{1}{1+q}}. \quad (2.73)$$

A value $q = 24$ reproduces the quasi-percolation behavior of Nafion with $IEC = 0.9$ mmol g$^{-1}$.

The random network model explains differences in $\sigma_b(\lambda)$ relations for various sulfonated ionomer membranes (Eikerling et al., 1997). It rationalizes effects of membrane elasticity and swelling behavior on performances under varying degrees of hydration. The EMA solution of the random network model, if implemented correctly, reproduces the percolation behavior observed in Nafion-type membranes and Nafion-composite membranes (Eikerling et al., 1997; Yang et al., 2004). High elasticity of the membrane matrix and high connectivity of pores give small values of the percolation threshold and, thus, beneficial relations $\sigma_b(\lambda)$. In a soft polymer matrix, on the other hand, the fraction of water-filled pores $x(\lambda)$ increases slowly at low to intermediate water contents, indicating that pores swell rather heterogeneously. In this respect, Nafion seems to offer a favorable elasticity of the polymer matrix.

In Eikerling et al. (2001), different pore network models (random network, serial and parallel pore models) were compared to each other. A morphology of equally swelling parallel cylindrical pores gives the most beneficial relations $\sigma_b(\lambda)$, with the steepest increase of proton conductivity at small water contents. Results obtained for this morphology are in good agreement with conductivity data for short sidechain PEMs, for example, the Dow experimental membrane. In order to adapt the model of pore space evolution to different structural membrane models, the major task is to find the applicable law of swelling $x(\lambda)$.

## ELECTRO-OSMOTIC DRAG

Proton flow in water-filled nanopores of PEM induces water transport through the *electro-osmotic coupling*. The phenomenological coupling coefficient, the *electro-osmotic drag coefficient* $n_d$, is a function of $\lambda$. It incorporates contributions of the molecular diffusion of protonated water clusters, that is, $H_3O^+$ ions or larger hydrated proton complexes, and of a hydrodynamic coupling in nanometer-sized water channels (Lehmani et al., 1997; Rice and Whitehead, 1965). Typical values of $n_d$, obtained by volume flux measurement, radiotracer methods, streaming potential measurements, or electrophoretic NMR, are found in the range of $n_d \sim 1-3$ (Kreuer et al., 2004; Pivovar and Pivovar, 2005; Pivovar, 2006).

Electro-osmotic drag phenomena are closely related to the distribution and mobility of protons in pores. The molecular contribution can be obtained by direct molecular dynamics simulations of protons and water in single ionomer pores, as reviewed in the sections "Proton Transport in Water" and "Stimulating Proton Transport in a Pore." The hydrodynamic contribution to $n_d$ can be studied, at least qualitatively, using continuum dielectric approaches. The solution of the *Poisson–Boltzmann equation*

$$\nabla \cdot (\varepsilon_0 \varepsilon_r \nabla \phi\,(r)) = -\rho_0 \exp\left(-\frac{F\phi\,(r)}{R_g T}\right), \tag{2.74}$$

under a boundary condition for the charge density at the pore wall gives the proton charge density distribution

$$\rho\,(r) = \rho_0 \exp\left(-\frac{F\phi\,(r)}{R_g T}\right) \tag{2.75}$$

and the electric potential $\phi(r)$ in pores. Here, it is assumed that pores have cylindrical geometry with radius $R$. The nondimensionalized PB equation in cylindrical pores is

$$\frac{1}{x}\frac{d}{dx}\left(x\frac{d\bar{\psi}(x)}{dx}\right) = -\left(\kappa r_p\right)^2 \exp\left(-\bar{\psi}(x)\right), \tag{2.76}$$

with $x = r/R$, $\bar{\psi} = \frac{F\phi}{R_g T}$, and $\kappa = \left(\frac{F\rho_0}{\varepsilon_0 \varepsilon_r R_g T}\right)^{1/2}$.

The radial component of the electric field is

$$E_r = -\frac{R_g T}{FR}\frac{d\bar{\psi}}{dx}, \tag{2.77}$$

with $E_r(x = 0) = 0$ in the pore center and $E_r(x = 1) = \frac{\sigma_s}{\varepsilon \varepsilon_0}$ at the pore wall, where $\sigma_s$ is the surface charge density. With the substitution $u = \ln x$ and $g = 2u - \bar{\psi}$, Equation 2.76 can be transformed into the one-dimensional PB equation

$$\frac{d^2 g}{du^2} = \left(\kappa r_p\right)^2 \exp\left(g\right). \tag{2.78}$$

The solution was obtained in Eikerling and Kornyshev (2001) and Berg and Ladipo (2009). It is given by

$$\phi(r) = \frac{2k_B T}{e_0} \ln\left(1 - \frac{(\kappa r)^2}{8}\right) \tag{2.79}$$

and

$$\rho(r) = \rho_0 \left(1 - \frac{(\kappa r)^2}{8}\right)^2, \tag{2.80}$$

where $\kappa$ is the inverse Debye length, that is,

$$\lambda_D = \kappa^{-1} = \sqrt{\frac{\varepsilon_0 \varepsilon_r k_B T}{e_0 \rho_0}} \tag{2.81}$$

and

$$\rho_0 = \frac{8\varepsilon_0 \varepsilon_r k_B T \sigma_s}{e_0 \sigma_s R^2 - 4\varepsilon_0 \varepsilon_r k_B TR}. \tag{2.82}$$

It is convenient to introduce a characteristic pore parameter, similar to the Debye length

$$\lambda_p = \sqrt{\frac{4\varepsilon_0 \varepsilon_r k_B TR}{e_0 \sigma_s}}. \tag{2.83}$$

Using this parameter transforms Equation 2.82 into

$$\rho_0 = \frac{2\sigma_s}{R}\frac{\lambda_p^2}{R^2 - \lambda_p^2}. \tag{2.84}$$

The parameter ranges are $\lambda_p \simeq 0.3\text{--}0.5$ nm, $2\sigma_s/R \simeq 0.4 \cdot 10^9$ C m$^{-3}$ to $0.7 \cdot 10^9$ C m$^{-3}$ and $\rho_0 \simeq 0.05 \cdot 10^9$ C m$^{-3}$ for $\sigma_s \simeq 0.5$ C m$^{-2}$, $R \simeq 1.5$ nm, $T \simeq 300$ K, and $\varepsilon_r = 80$.

Pore size and the dielectric constant $\varepsilon_r$ of water in pores exhibit a strong effect on proton distributions, as studied in Eikerling and Kornyshev (1999). Model variants that take into account the effect of strongly reduced $\varepsilon$ near pore walls (Booth, 1951) and the phenomenon of dielectric saturation (Cwirko and Carbonell, 1992a,b) could lead to nonmonotonous profiles in proton concentration with a maximum in the vicinity of the pore wall.

The hydrodynamic equation of motion (Navier–Stokes equation) for the stationary axial velocity $v_z(r)$ of an incompressible fluid in a cylindrical pore under the influence of pressure gradient $dP^l/dz$ and axial electric field $E_z$, is

$$\frac{1}{r}\frac{d}{dr}\left(r\frac{dv_z}{dr}\right) = \frac{1}{\mu}\frac{dP^l}{dz} - \frac{E_z}{\mu}\rho(r), \tag{2.85}$$

where $\mu$ is the dynamic viscosity. For vanishing pressure gradient, water transport is entirely driven by electro-osmotic drag, with a velocity determined by

$$\frac{1}{r}\frac{d}{dr}\left(r\frac{dv_z}{dr}\right) = \varepsilon_0\varepsilon_r\frac{E_z}{\mu}\frac{1}{r}\frac{d}{dr}\left(r\frac{d\phi}{dr}\right). \tag{2.86}$$

Using obvious boundary conditions of zero flux velocity at the pore wall (no slip) $v_z(R) = 0$ and symmetry in the pore center $dv_z/dr|_{r=0} = 0$, the velocity profile is related to the potential profile by

$$v_z(r) = -\varepsilon_0\varepsilon_r\frac{E_z}{\mu}[\phi(R) - \phi(r)]$$

$$= \frac{2\varepsilon_0\varepsilon_r k_B T}{e_0\mu}E_z \ln\left[\frac{8 - (\kappa r)^2}{8 - (\kappa R)^2}\right], \tag{2.87}$$

where on the second line the solution provided in Equation 2.79 was used. Equation 2.87 is valid for cylindrical pores with constant values of $\varepsilon_r$ and $\mu$. The volumetric water flux from electro-osmosis is given by

$$V_{eo} = 2\pi \int_0^R r\,dr\,v_z(r) = -\frac{\varepsilon_0\varepsilon_r\phi(R)}{\mu}\pi R^2 I_g E_z, \tag{2.88}$$

where

$$I_g = \frac{2}{R^2}\int_0^R r\,dr\left[1 - \frac{\phi(r)}{\phi(R)}\right] \tag{2.89}$$

is a geometry factor. Using the solution for $\phi(R)$ gives

$$I_g = -\frac{(\kappa R)^2 + 8\ln\left[1 - (\kappa R)^2/8\right]}{(\kappa R)^2 \ln\left[1 - (\kappa R)^2/8\right]}. \tag{2.90}$$

For $\kappa R < 2.5$, the value of this factor lies in the range $0.5 \leq I_g < 0.6$. For $(\kappa R)^2 \rightarrow 8$, it approaches 1.

Using Ohm's law of proton transport in a pore $j_{\text{pore}} = \sigma_{\text{pore}} E_z$ and the hydrodynamic flux density $j_{hydr} = -\frac{Fc_w \varepsilon_0 \varepsilon_r \phi(R)}{\mu} I_g E_z$, where $c_w$ is the concentration of water, $c_w = 55$ mol L$^{-1}$, results in

$$n_{hydr} = \frac{j_{hydr}}{j_{\text{pore}}} = -\frac{Fc_w \varepsilon \varepsilon_0}{\mu \sigma_{\text{pore}}} I_g \phi(R)$$

$$= -\frac{2c_w \varepsilon \varepsilon_0 R_g T}{\mu \sigma_{\text{pore}}} I_g \ln \left( 1 - \frac{(\kappa R)^2}{8} \right). \tag{2.91}$$

Equation 2.91 reveals dependences of $n_{hydr}$ on dielectric constant, $\varepsilon_r$, viscosity of water, $\mu$, surface charge density of the pore, $\sigma$, pore radius, $R$, and the proton conductivity of the pore, $\sigma_{\text{pore}}$. The hydrodynamic electro-osmotic coefficient for a typical pore with $R = 1$ nm is found in the range $n_{hydr} \simeq 1\text{--}10$.

The total electro-osmotic coefficient $n_d = n_{hydr} + n_{mol}$ includes terms due to hydrodynamic coupling $n_{hydr}$, and a molecular coupling $n_{mol}$, that is related to the structural diffusion of protonic defects. The relative contributions of $n_{hydr}$ and $n_{mol}$ depend on the mechanism of proton transport in pores.

Similar to proton conductivity, the effective membrane parameter $n_d$ is determined by the membrane molecular architecture, in particular, by the concentration of acid protons in water-filled channels. The structural effects translate into characteristic dependencies of $n_d$ on $\lambda$. In general, it is observed that $n_d$ increases with increasing $\lambda$. The prevalence of surface proton transport for low $\lambda$ suggests $n_d \approx n_{mol} \approx 1$. $n_d$ is reduced in membranes with narrow channels and strong polymer–solvent interactions, for example, in S-PEEK as compared to Nafion. These trends can be explained with the decrease of the hydrodynamic component $n_{hydr}$, as discussed above.

## CONCLUDING REMARKS

The grand challenge of PEM modeling is to establish predictive relations between the chemical structure of the ionomer, membrane structure, and physicochemical properties. Addressing these objectives involves a hierarchy of phenomena. Major challenges and approaches along this path were discussed in this chapter. Applied topics in membrane research will be picked up again in Chapter 5.

### SELF-ORGANIZATION OF PHASE-SEGREGATED MEMBRANE MORPHOLOGY

Ionomer molecules aggregate and phase-segregate into hydrophobic polymer domains and hydrophilic water-filled pathways at the nanometer scale. These phenomena are understood on the basis of scattering data (SANS, SAXS, USAXS). However, the interpretation of such data is still controversial. Structural pictures of the membrane morphology, focusing on sizes, distribution, and connectivity of water-filled nanochannels, are still going through revisions. Relatively well established is the formation of ionomer backbones into cylindrical or ribbon-like bundles or fibrils. These fibrils define elastic properties of the polymer phase. They are lined on their surface by a dense array of polymer sidechains, which contain (or are terminated by) sulfonic acid head groups.

Free hydrated protons produced by acid dissociation move inside of porous domains, relative to the interfacial layer of anionic surface groups. Self-consistent theories and coarse-grained molecular simulations can help to rationalize this structural picture. After demonstrating consistency with existing data on structure and transport, these modeling approaches could be employed as predictive tools in polymer design.

Molecular simulations yield unrealistic morphologies (pore sizes, shapes, connectivity) if they employ insufficient representations of ionomer molecules. Results of simulations depend on interaction parameters that are provided as input. Parameters have to be acquired from fundamental modeling studies (DFT-based calculations) and experimental studies (e.g., adsorption studies). CGMD simulations offer a sound trade-off of computational efficiency and adequate structural representation. The coarse-grained treatment implies simplification in interactions, which can be systematically improved with advanced force-matching procedures, but it allows simulations of systems with sufficient size and sufficient statistical sampling. Structural correlations, thermodynamic properties and transport parameters of PEMs can be studied.

Theoretical studies of ionomer aggregation in solution can rationalize and predict stable configurations of ionomer bundles as a function of basic ionomer properties. Theoretical results can be highly insightful to narrow down the configuration space for molecular simulations.

## Water Sorption and Swelling in Response to External Conditions

Membrane structure and external conditions determine water sorption and swelling. The resulting water distribution determines transport properties and operation. Water sorption and swelling are central in rationalizing physical properties and electrochemical performance of the PEM. The key variable that determines the thermodynamic state of the membrane is the water content $\lambda$. The equilibrium water content depends on the balance of capillary, osmotic, and electrostatic forces. Relevant external conditions include the temperature, relative humidity, and pressure in adjacent reservoirs of liquid water or vapor. The theoretical challenge is to establish the equation of state of the PEM that relates these conditions to $\lambda$. A consistent treatment of water sorption phenomena, presented in the section "A Model of Water Sorption," revokes many of the contentious issues in understanding PEM structure and function.

## Structure and Distribution of Water

The minimal distinction that has to be made to rationalize effects in water uptake and transport properties of the PEM is between chemisorbed surface water that strongly interacts with charged surface groups at polymer/water interfaces, and bulk-like water that has weak interactions with the polymer. Capillary condensation is essential for equilibration of bulk-like water in the PEM. Analysis of water sorption and transport of protons and water (e.g., percolation threshold; transition from high to low activation energy of proton transport) suggest that surface water corresponds roughly to a single monolayer of water at the polymer/water interface.

## TRANSPORT MECHANISMS OF PROTONS AND WATER

The distinction of transport mechanisms follows the distinction of surface and bulk-like water. In typical PEMs, proton transport in the layer of surface water exhibits high activation energies and, thus, low mobilities. In theoretical studies based on DFT, it was found that the surface mobility exhibits a strong dependence on the density of charged surface groups. At high surface group densities, proton transport at the surface can become highly efficient, even under conditions of minimal hydration. Understanding the underlying concerted mechanisms of surface proton transport is of great fundamental interest. On the practical side, it could bolster efforts in designing membranes that are conducive to high proton conductivity under almost dry conditions.

The dynamics of protons and water molecules in the bulk of pores are very similar to those in free bulk water. Percolation models and random network simulations can be used to study the impact of pore swelling and changes in pore connectivity upon water uptake on proton conductivity. Existing pore network models provide a good agreement with conductivity data. Updates of these models are needed to account for advances in understanding of the membrane morphology.

Water transport occurs through diffusion of surface water, relative to the polymer matrix, and hydraulic permeation of bulk water. The water content determines the relative contributions of diffusion and hydraulic permeation to the net rate of water transport.

The electro-osmotic coupling of proton and water transport depends on the molecular mechanism of proton transport. It is useful to distinguish a molecular and a hydrodynamic contribution to the electro-osmotic drag coefficient. The latter contribution increases strongly with water uptake and temperature.

# 3 Catalyst Layer Structure and Operation

There is more to catalyst layer operation than electrocatalysis, a lot more! The design of fuel cell electrodes with high performance, long lifetime, and low cost is about embedding the catalyst, usually the most expensive and least stable material in the cell, into a porous composite host medium. It turns out that material selection and structural design of the host medium is as important as that of the catalyst material itself.

The objective of catalyst layer design is twofold: from a *materials scientist's perspective*, the objective is to maximize the electrochemically active surface area (ECSA) per unit volume of the catalytic medium $S_{ECSA}$, by (i) catalyst dispersion in nanoparticle form or as an atomistically thin film and (ii) optimization of access to the catalyst surface for electroactive species consumed in surface reactions. From a *fuel cell developers point of view*, the objective is to optimize pivotal performance metrics like voltage efficiency, energy density, and power density (or specific power) under given cost constraints and lifetime requirements. These performance objectives are achievable by integration of a highly active and sufficiently stable catalyst into a structurally well-designed layer.

This chapter provides a systematic account of the pertinent challenges and approaches in catalyst layer design. The hierarchy of structural effects and physical phenomena discussed includes materials design for high surface area and accessibility, statistical utilization of Pt evaluated on a per-atom basis, transport properties of charged species and neutral reactants in composite media with nano- to mesoporosity, local reaction conditions at internal interfaces in partially electrolyte-filled porous media, and global performance evaluated in terms of response functions for electrochemical performance and water handling.

## POWERHOUSES OF PEM FUEL CELLS

The catalyst layers represent the major competitive ground in view of the multivariate and hierarchical design challenge of PEFCs. All species, structural components, and processes that appear somewhere in the cell show up as well in CLs at the anode and, first and foremost, the cathode.

The main processes are electrochemical reactions at electrified metal–electrolyte interfaces; reactant diffusion through porous networks; proton transport in water and at aggregates of ionomer molecules; electron transport in electronic support materials; water transport by gasous diffusion, hydraulic permeation, and electro-osmotic drag in partially saturated porous media; and vaporization/condensation of water at interfaces between liquid water and gas phase in pores.

Key structural characteristics determining these processes are atomistic surface structure and electronic structure of the catalyst, morphology of the pore network, surface structure and wettability of the support, catalyst nanoparticle shape and size, ionomer structure, mixed wettability of the composite layer, and, last but not least, the electrode thickness, $l_{CL}$.

## PRINCIPLES OF CATALYST LAYER STRUCTURE AND OPERATION

This section gives an overview of the main principles of CL structure and operation.

### Principle 3.1

High-performance electrodes for PEFCs are evaluated by (i) their Pt mass loading per unit electrode surface area, $m_{Pt}$ (in mg cm$^{-2}$), (ii) voltage efficiency, $\epsilon_v = E_{cell}/E_{O_2,H_2}^{eq}$, (iii) (volumetric) energy density, $W_{cell} = FE_{cell}/l_{CL}$ (in J L$^{-1}$), (iv) (volumetric) power density, $P_{cell} = j_0E_{cell}/l_{CL}$ (in W L$^{-1}$), and (v) specific power, $P_{cell}^s = j_0E_{cell}/m_{Pt}$ (in W g$_{Pt}^{-1}$). These parameters are related, as is obvious from their definitions. Yet, it is possible to adjust their values independently to specific requirements of energy-consuming appliances.

### Principle 3.2

Electrochemical reactions in CLs are interfacial processes. High values of $P_{cell}$ and $W_{cell}$ demand a huge value of the metal–electrolyte interfacial area per unit volume, denoted by the parameter $S_{ECSA}$. Therefore, maximization of $S_{ECSA}$ is a key objective in structural design and fabrication of CLs.

### Principle 3.3

The primary property of a catalyst material is the *intrinsic exchange current density* $j_*^0$, normalized to a unit of the active catalyst surface area. It is determined by electronic structure effects, which are sensitive to catalyst bulk composition as well as catalyst particle size and shape for nanoparticle catalysts. In the case of nanostructured thin catalyst films, it depends on film thickness, surface morphology, as well as properties of the support material. The *electrocatalytic activity per unit volume* of the electrode is given by $i^0 = j_*^0 S_{ECSA}$. For nanoparticle-based catalysts, this relation is valid, strictly speaking, only for monodispersed particle size distributions. Owing to the dependence of the intrinsic activity on particle size and shape, a particle average of the activity should be used. The exchange current density per unit surface area of the electrode is given by $j^0 = j_*^0 S_{ECSA} l_{CL}$. A key objective in catalyst layer design is to maximize $j^0$ with a minimal amount of catalyst.

### Principle 3.4

The need to maximize the parameter $S_{ECSA}$, as well as $P_{cell}$ and $W_{cell}$, enforces the use of nanostructured media with small feature size of nanoparticles, nanopores, or nanorods. Therefore, success in developing highly performing electrocatalyst systems hinges on progress in nanomaterials design.

## Principle 3.5

No homogeneous phase could fulfill the conflicting needs for high electrochemical activity and high rates of transport of reactant gases, electrons, protons, and product water simultaneously. In general, CLs demand composite morphologies of several interpenetrating phases with nanoscopic feature sizes. CL designs can be classified into two main types:

1. *Type I electrodes*, the prevailing type, are *three-phase composite media* that consist of a solid phase of Pt and electronic support material, an electrolyte phase of ionomer and water, and the gas phase in the porous medium. Gas diffusion is the most effective mechanism of reactant supply and water removal. Yet, CLs with sufficient gas porosity, usually in the range $X_P \sim 30–60\%$, have to be made with thickness of $l_{CL} \simeq 10\,\mu m$. In this thickness range, proton transport cannot be provided outside of the electrolyte environment. *Porous gas diffusion electrodes* are, therefore, impregnated with proton-conducting ionomer. The concept of a triple-phase boundary, often invoked for such electrodes, is however inadequate. The amount of the electrochemically active interface is usually controlled by two-phase boundary effects at the interface between Pt and water.

2. *Type II electrodes* are *two-phase composite media* that consist of a nanoporous and electronically conductive medium filled with liquid electrolyte or ionic liquid. The electrochemically active interface forms at the boundary of the two phases. The electrolyte phase must provide pathways for diffusion and permeation of protons, water, and reactants. Flooded two-phase CLs could work well when they are made extremely thin, not significantly exceeding a thickness of $l_{CL} \simeq 200\,nm$. Rates of diffusion of reactant molecules and protons in liquid water are then sufficient to provide uniform *reaction rate distributions* over the thickness of the layer.

## Principle 3.6

The main target in CL design is to achieve uniform reaction conditions. Distributions of gaseous reactants, protons, and electrostatic potentials in metal and electrolyte phases are determined by microstructure, composition, and thickness. Uniform reaction conditions demand that the rates of transport of oxygen (on the cathode side) and protons are high compared to the rate at which these species are consumed in interfacial reactions. Key parameters in this context are the ratios of oxygen diffusion coefficient and proton conductivity to thickness i.e., $D_{O_2}/l_{CL}$ and $\sigma_{H^+}/l_{CL}$. The uniformity of reaction rate distributions is evaluated by calculating characteristic *reaction penetration depths* and comparing them with $l_{CL}$.

## Principle 3.7

The design space for CLs is divided by the thickness regime considered. Layers with $l_{CL}$ in the range of 1–10 μm must be fabricated as type I electrodes. Ultrathin layers with $l_{CL}$ in the range of 100 to 300 nm can be designed as type II electrodes. The electrochemical performance of different CL designs can be evaluated by the *effectiveness factor of Pt utilization*, $\Gamma_{CL}$. It is well corroborated that the conventional cathode catalyst layer design severely underutilizes Pt, resulting from partial wetting of Pt, percolation effects, and transport limitations. Past progress in design optimization of conventional CLs does not appear to leave much room for improving the main optimization parameter $j^0$.

## Principle 3.8

*Water management* is a key issue in the optimization of the CCL, involving two-phase flow of liquid water and water vapor as well as the rate of *vaporization exchange* at liquid–vapor interfaces. Water is the only product of the fuel cell reaction. Any high-performance PEFC will face the challenge that water must be removed at the same rate as it is produced. Thus, it goes without saying that water removal is as important for the operation of the cell as any of the electrocatalytic processes. The place at which water is formed is the CCL. The CCL does not only determine the effective rate of electrochemical current conversion and a major proportion of the irreversible voltage losses in the PEFC but it also plays a key role for the water balance of the whole cell, controlling water distribution, as well as vapor and liquid fluxes in all cell components. Conventional CCLs (type I electrodes) possess a benign porous structure with bimodal and bifunctional pore size distribution, and large proportion of pores in the nanometer range. These features predispose the conventional CCL as the favorite water (and heat!) exchanger in a PEFC. A major challenge for MEAs with UTCLs is that the vaporization capability of type II electrodes is insufficient. The interplay of liquid permeation, vapor diffusion, and interfacial vaporization exchange in porous composite CCLs bears similarities to charge fluxes and interfacial charge conversion in porous electrodes. Both these types of mass or charge conversion phenomena can be described with a classical transmission line approach.

## Principle 3.9

Durability and lifetime of catalyst layer materials are key concerns of fuel cell developers. Reaction conditions that favor rapid conversion of reactants in the ORR also accelerate *Pt dissolution* and *support corrosion* processes. These processes continuously transform structure, composition, and water distribution of the catalyst layer. Changes that are usually observed involve "melting" and "shifting" of the Pt particle radius distribution, decrease in both $S_{ECSA}$ and $i^0$, loss of solid Pt mass, and decrease in CL thickness and porosity. Also observable were increases in hydrophilicity and associated liquid saturation. The impact of these changes on local reaction conditions and performance is usually detrimental. These processes are accelerated under highly oxidizing conditions that prevail at high electrode potentials. Particularly harmful is rapid potential cycling through high potential regions, as well as frequent start-up and shut-down of the cell.

## FORMATION OF STRUCTURE AND FUNCTION IN CATALYST LAYERS

As mentioned in the previous subsection, the key classification parameter of distinct CL designs is the thickness $l_{CL}$. The currently prevailing conventional design of CLs (type I electrodes) is illustrated in Figure 3.1. The thickness of these CLs ranges from $l_{CL} \simeq 5$ to 10 μm. Two major improvements in conventional CLs have been, the incorporation of Pt or Pt group metal nanoparticles and the impregnation or colloidal mixing of the high-surface-area particle dispersion of Pt/carbon with ionomer. The former step provides a significantly enhanced electrocatalytically active surface area. The latter step ensures more uniform access of protons to the active Pt surface throughout the layer.

As first demonstrated by Ian D. Raistrick at Los Alamos National Laboratory (LANL), these improvements enabled a dramatic reduction of catalyst loading (Raistrick, 1986, 1989). The *ink-based fabrication approach*, later refined at LANL (Wilson and Gottesfeld, 1992), evolved into the standard method of catalyst

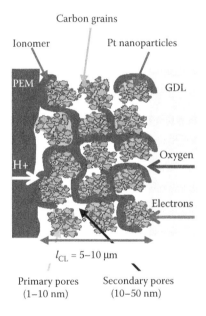

**FIGURE 3.1** The currently prevailing conventional design of CLs. These layers, referred to as type I electrodes, are three-phase composite media. The stable matrix is formed by a porous support material that must be a good electronic conductor, with conductivity $>0.01$ S cm$^{-1}$. Platinum nanoparticles are deposited on the surface of the support, exhibiting formation of stable bonds and good dispersion. The porous structure is impregnated with ionomer electrolyte that confers high intrinsic proton concentration and sufficient proton conductivity. Water exists at interfaces and in the open pore space. Most catalyst layers of this type exhibit aggregation of primary Pt/C particles, resulting in agglomerated mesostructures with bimodal pore size distributions. These involve primary pores with radii of 1–10 nm inside of agglomerates and secondary pores with sizes of 10–50 nm between agglomerates. The challenge in CL modeling is to understand the hierarchy of interrelated phenomena at nanoparticle level, agglomerate level, and macroscale.

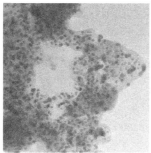
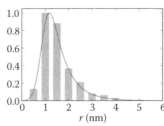

**FIGURE 3.2** A high-resolution TEM, obtained from image analysis of well-dispersed Pt nanoparticles. They are deposited on a high-surface-area carbon support, together with the normalized particle radius distribution of Pt.

layer and MEA fabrication. It has been adopted and continuously improved by laboratories worldwide. Nowadays, ionomer completely replaces PTFE that had formed a major component as a binder and hydrophobizing agent in early fabrication approaches of CLs. Catalyst loadings in CCLs have been reduced from about 10 mg Pt cm$^{-2}$ (in the 1980s) to 0.4 mg Pt cm$^{-2}$ today.

Catalyst nanoparticles, at which electrical current is generated, have typical particle size distributions from 2 to 5 nm, as shown in Figure 3.2. The most important support materials are carbon black or graphitized carbon black, such as Ketjen Black and Vulcan XC-72, which can be produced with a specific surface area of up to 1500 m$^2$ g$^{-1}$ (Kinoshita, 1988).

How can the process of *catalyst layer formation* be depicted? During ink fabrication, self-organization of ionomer, which is usually Nafion, and primary Pt/C particles in colloidal ink solution leads to the formation of phase-segregated agglomerated morphologies. Agglomerates with radii ranging from $R_a \sim 30$ to 100 nm consist of primary Pt/C particles, with sizes in the range from 5 to 10 nm. It was concluded from coarse-grained molecular dynamics studies, reported in (Malek et al., 2007, 2011), that ionomer molecules form a dense skin layer with thickness of 3–10 nm, enveloping the outer surface of agglomerates. Penetration of ionomer into agglomerates depends on the fabrication technique, ionomer concentration, type of carbon, and the kinetics of the agglomeration process. Usually, it is assumed that ionomer molecules do penetrate into intra-agglomerate pores to a significant extent. As explored in Malek et al. (2007, 2011), microstructure formation depends on materials properties of Pt, carbon and ionomer, composition of the catalyst layer ink, dielectric properties of the dispersion medium, and fabrication conditions like temperature and pressure.

Impregnation of the porous network with PFSA-type ionomer imparts high proton concentration and conductivity to the layer. The high proton concentration is beneficial for the ORR activity of the CCL but it is detrimental for catalyst stability. Simultaneous percolation in the interpenetrating networks of the solid Pt/carbon matrix, ionomer electrolyte, and pores ensures efficient transport of electrons, protons, chemical reactants (H$_2$, methanol, O$_2$, etc.), and product water to or from Pt.

*Agglomeration processes in ink mixtures* of carbon/Pt and ionomer result in bimodal pore size distribution (PSD) functions. Primary pores with pore radii from $R_\mu \simeq 1$ to 10 nm exist between primary Pt/C particles inside of agglomerates. Secondary pores with pore radii from $R_M \simeq 10$ to 50 nm form between agglomerates. There is a competition between secondary pore volume and ionomer volume to occupy the voids between agglomerates. Agglomeration and spontaneous formation of the dual porosity network are essential for catalyst layer operation. The notion of agglomerates is consistent not only with findings of coarse-grained molecular dynamics simulations (Malek et al., 2007, 2011) but also with experimental studies (Soboleva et al., 2010, 2011; Suzuki et al., 2011; Uchida et al., 1995a,b).

In recent years, various *support materials* have been tested as replacements for carbon black, including a range of nanostructured carbon allotropes and metal oxides (Malek et al., 2007; Wieckowski et al., 2003; Zhang et al., 2010, 2012). *Carbon nanotubes* (CNTs) have been explored for numerous applications in electrochemical energy storage and conversion (Lota et al., 2011). CNTs with good electrical conductivity and enhanced electrochemical stability, under conditions of the ORR, have been tested as Pt support materials, so far, without resounding success (Soin et al., 2010; Wen et al., 2008; Zhang et al., 2010). Metal-oxide-based support materials that have been tried include various oxides of Nb and Ti (Orilall et al., 2009; Sasaki et al., 2008; Zhang et al., 2010, 2012). Usually, these oxides possess improved electrochemical and thermal stability in comparison to carbon-based materials but they often exhibit insufficient electrical conductivity. For example, the bulk electronic conductivity of undoped $TiO_2$ is in the range of 0.1 S cm$^{-1}$, which is significantly smaller than the conductivity of commercial Vulcan XC-72 carbon that lies in the range of 4 S cm$^{-1}$.

The required electronic conductivity of the support material depends strongly on $l_{CL}$. A simple model of the interplay of local current generation and transport of electrons in the support phase can be used to obtain an estimate of the minimal electronic conductivity required. The underlying model is identical to models that evaluate the interplay of proton transport limitations and reaction in CCLs, discussed for instance in Eikerling et al. (2007a).

The model shows that the electron transport effects do not affect the CL performance if the conductivity of the electron-conducting phase obeys the following requirement:

$$\sigma_{el}^{\min} \approx \frac{j_0 l_{CL}}{2b} \exp\left(-\frac{\Delta\eta_{\max}}{b_{eff}}\right). \tag{3.1}$$

Here, $\Delta\eta_{\max}$ is the voltage loss tolerance due to finite electronic conductivity, $b = R_g T/(\alpha_{eff} F)$ is the Tafel parameter with the effective electronic transfer coefficient $\alpha_{eff}$ of the ORR, and $j_0$ is the operating current density. For instance, at $l_{CL} = 10$ μm, $j_0 = 1$ A cm$^{-2}$, $\alpha_{eff} = 1$, $T = 333$ K, and $\Delta\eta_{\max} = 1$ mV, the electronic conductivity requirement of the CL is $\sigma_{el} > 0.01$ S cm$^{-1}$. In an ultrathin catalyst layer (UTCL) with thickness $L = 100$ nm, this bound on $\sigma_{el}$ is lower, namely, $\sigma_{el} > 10^{-4}$ S cm$^{-1}$. This estimate explains why, in CLs fabricated with the NSTF of the company 3M, a thin film of sputter-deposited Pt provides sufficient electronic conductivity. UTCLs are much less sensitive to the support conductivity.

As these estimates show, conductivity requirements on the substrate are modest and met by currently employed carbon-based materials. Other selection criteria for support materials are the corrosion resistance and pore size distribution, which, together with their thickness, determines available interfacial area and transport properties of CLs.

The anode catalyst layer (ACL) is responsible for the oxidation of the fuel. In the case of hydrogen-fueled PEFCs, the anode is usually of little concern. Overpotential losses incurred by the hydrogen oxidation reaction (HOR) in hydrogen fuel cells are negligible. In direct methanol, ethanol, or formic acid fuel cells, the ACL incurs substantial voltage losses, which diminish the power density and, therefore, restrict severely the range of possible applications of these cells (Chapter 4). In the subsequent treatment, CCLs that carry out the ORR will be focused upon. The majority of theoretical concepts and modeling studies presented could be easily adopted and modified for ACLs. Efforts in fuel cell research over the past decade have revealed, however, that drastic improvements in power density, durability, and cost reduction of PEFCs are impossible without major leaps in CCL design.

## Outline and Objectives of This Chapter

The hierarchy of structural effects in CLs from macroscale down to atomistic scale is illustrated in Figure 1.16. The main driver for efforts in CL research is the platinum dilemma. Of all materials in the PEFC, Pt in the CCL has the biggest impact on performance, cost, durability, and lifetime of the cell. Furthermore, Pt is a scarce commodity residing among the least abundant elements in the Earth's upper continental crust. At the same time, it is very poorly utilized in current CLs, especially the CCL.

This chapter will cover major topics of CL research, focusing on (i) electrocatalysis of the ORR, (ii) porous electrode theory, (iii) structure and properties of nanoporous composite media, and (iv) modern aspects in understanding CL operation. *Porous electrode theory* is a classical subject of applied electrochemistry. It is central to all electrochemical energy conversion and storage technologies, including batteries, fuel cell, supercapacitors, electrolyzers, and photoelectrochemical cells, to name a few examples. Discussions will be on generic concepts of porous electrodes and their percolation properties, hierarchical porous structure and flow phenomena, and rationalization of their impact on reaction penetration depth and effectiveness factor.

A separate section on electrocatalysis focuses on electrochemical processes at the cathode catalyst, including ORR and corrosive dissolution of Pt. Of central importance for both these processes is the formation of intermediate oxygen species at the Pt surface. ORR proceeds in the presence of and through these intermediates with their formation and reduction accelerating Pt dissolution processes. It is thus of primary interest to understand processes of Pt oxide formation and reduction. This understanding is critical in view of the following conundrum: How should catalyst layer design and reaction conditions be modified in order to accelerate the ORR and minimize corrosive dissolution of Pt? A possible solution involving a thorough understanding of reaction conditions in nanoporous media will be discussed in the section "ORR in Water-Filled Nanopores: Electrostatic Effects."

In the section "Structure Formation in Catalyst Layers and Effective Properties" aspects related to the *self-organization phenomena in CL inks* will be discussed. These phenomena determine the effective properties for transport and electrocatalytic activity. Thereafter, catalyst layer performance models that involve parameters related to structure, processes, and operating conditions will be presented.

In this chapter, connections will be established between electrocatalytic surface phenomena and porous media concepts. The underlying logics appear simple, at least at first sight. Externally provided thermodynamic conditions, operating parameters, and transport processes in porous composite electrodes determine spatial distributions of reaction conditions in the medium, specifically, reactant and potential distributions. Local reaction conditions in turn determine the rates of surface processes at the catalyst. This results in an effective *reactant conversion rate* of the catalytic medium for a given electrode potential.

Mathematically, this procedure represents a self-consistency problem. The main variable is the metal phase potential. Dependent variables include metal surface charge density, proton concentration in the electrolyte phase, oxygen concentration as well as variables for the surface oxidation state of the metal. All of these variables are functions of spatial coordinates and time. These interrelated functions determine the interfacial faradaic current, generated in surface reactions. The relations among these functions depend on electrode composition, porous structure, surface structure, electronic structure of catalyst and support materials, and properties of the electrolyte phase. As seen below, the electrolyte phase itself is a composite, consisting of "near-ionomer" regions at the agglomerate surface with high proton concentration, and "bulk-water" regions inside of agglomerates with low proton concentration.

To simplify the problem, a hierarchy of assumptions will be made use of; they are, listed in order of decreasing generality: (i) isothermal operation can be assumed under all relevant conditions; temperature gradients may be significant at MEA and stack levels, but they are small in catalyst layers, that is, $\Delta T < 0.5\,\mathrm{K}$, as determined by Kulikovsky (2006); (ii) for most intents and purposes, models will be limited to consideration of steady-state operation; (iii) owing to the high electronic conductivity, the metal phase in the CL can be considered as equipotential; (iv) it is sufficient to consider the CL as a one-dimensional system; the thickness of the CL is very small compared to feature sizes of flow field channels (length and width) that could give rise to in-plane variations of reactants and reaction rates; (v) most models of ionomer-impregnated CLs assume a constant proton concentration, neglecting the influence of metal surface charging on proton density; a special section, the section "ORR in Water-Filled Nanopores: Electrostatic Effects" is devoted to ionomer-free UTCLs, for which this assumption is invalid.

From the insights presented, conclusions will be drawn about conceivable improvements of catalyst effectiveness, voltage efficiency, power density, water handling capabilities, and stability through optimized operating conditions and advanced structural design.

## THEORY AND MODELING OF POROUS ELECTRODES

Theory and modeling have explored in-depth the effects of composition, porous morphology, and thickness on performance of conventional CLs (Chan et al., 2010; Eikerling, 2006; Eikerling and Kornyshev, 1998; Eikerling and Malek, 2009; Eikerling et al., 2004, 2008; Liu and Eikerling, 2008; Perry et al., 1998; Springer et al., 1993). This section follows the development of these models starting from their early roots in porous electrode theory.

### BRIEF HISTORY OF POROUS ELECTRODE THEORY

A timeline of major developments in porous electrode theory that led to the current approaches in catalyst layer modeling is depicted in Figure 3.3.

The importance of structural effects in gas diffusion electrodes had been realized long before the development of the current generation of CLs for PEFCs started. The basic theory of gas diffusion electrodes, including the interplay of reactant transport through porous networks and electrochemical processes at dispersed electrode–electrolyte interfaces, dates back to the 1940s and the 1950s (Frumkin, 1949). Later work identified the importance of total surface area and utilization of electrocatalyst in porous electrodes (Mund and Sturm, 1975).

A series of seminal contributions by R. De Levie paved the way for the widespread use of electrochemical impedance measurements in the characterization of porous electrodes (Levie, 1963, 1967; Raistrick, 1990). The transmission line approach of De Levie constituted the seed for the study of interfacial phenomena in electrodes with fractal surfaces (Halsey, 1987; Kaplan et al., 1987; Pajkossy and Nyikos, 1990; Sapoval, 1987; Sapoval et al., 1988; Wang, 1988).

Major contributions to the development of the macrokinetic or *macrohomogeneous theory of porous electrodes* were made by Yu.A. Chizmadzhev, Yu.G. Chirkov,

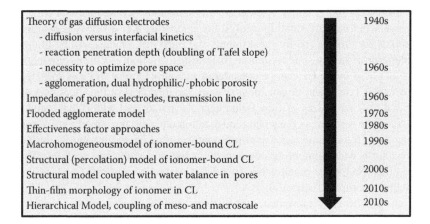

FIGURE 3.3  A timeline of developments in porous electrode theory and catalyst layer modeling.

and their colleagues from the 1970s to the 1980s (Chizmadzhev et al., 1971). These works explored the interplay of oxygen diffusion in porous media and interfacial kinetics at the catalyst surface. For oxygen reduction electrodes, a large electro-catalytically active surface area per unit volume, $S_{ECSA}$, has to compensate for the smallness of the intrinsic activity per unit real-surface area, i.e. $j_*^0$, of any known oxygen reduction electrocatalyst. The characteristic pore dimension $R$ should be small in order to guarantee a large $S_{ECSA}$, which scales as $S_{ECSA} \sim 1/R$.

On the other hand, the rates of electrochemical reaction should be distributed uniformly over the entire thickness of the electrode, requiring high rates of transport of reactants and products via diffusion paths. In the limits of fast reactant diffusion, the internal electrode surface would be utilized uniformly for current conversion, resulting in a simple proportionality of the current density, $j \propto j_*^0 S_{ECSA} l_{CL}$. This proportionality is only valid for electrodes with thickness $l_{CL} \ll \delta_{CL}$, where $\delta_{CL}$ is the *reaction penetration depth*. All in all, the necessity to optimize the pore space in view of gaseous transport and electrochemical reaction was already well comprehended at that time (Chizmadzhev et al., 1971).

The importance of microstructural optimization of hydrophobic gas diffusion electrodes was emphasized in an article by Tantram and Tseung (1969). Tantram and Tseung considered porous electrodes that consisted of mixtures of finely dispersed Pt black particles, bonded by polytetrafluoroethylene (PTFE). Hydrophobic and hydrophilic parts formed interconnected networks of porous PTFE and porous catalyst aggregates. The two authors recognized the importance of agglomeration and dual hydrophobic/hydrophilic porosity.

The functional distinction of hydrophobic mesopores and hydrophilic microporores led to the development of the *flooded agglomerate model* by Giner and Hunter (1969). These authors explored the effects of intrinsic catalyst activity, internal porosity, and real catalyst surface area on electrochemical performance of Teflon-bonded electrodes. Their original model treated the electrode as a group of parallel flooded cylinders that consist of a homogeneous mixture of catalyst and liquid electrolyte. Elongated, sample-spanning void spaces between cylinders provided the pathways for gaseous diffusion of reactants.

Similar agglomerate approaches were adopted by Iczkowski and Cutlip (1980) and by Björnbom (1987). Those works identified the doubling of the apparent Tafel slope as a universal signature of the interplay of mass transport limitations and interfacial electrochemical kinetics. Flooded agglomerate models have been employed ever since to analyze sources of irreversible potential losses, optimum electrode thickness, and *effectiveness factors of catalyst utilization*. Moreover, it was suggested that electrodes with a continuous distribution of catalyst and electrolyte cannot give rise to diffusion-limited current densities, as long as fixed values of exchange current density $j^0$ and transfer coefficient $\alpha_{eff}$ are assumed.

Simple pore models (Srinivasan et al., 1967), thin-film models (Srinivasan and Hurwitz, 1967), macrohomogeneous models, and refined variants of agglomerate models are still being applied and further developed for the present generation of *ionomer-bound composite catalyst layers* in PEFCs (Gloaguen and Durand, 1997; Jaouen et al., 2002; Karan, 2007; Kulikovsky, 2002a, 2010b; Sun et al., 2005). Effectiveness factor approaches have been elaborated as quantitative tools to compare the performance of active layers (Stonehart and Ross, 1976). Macrohomogeneous electrode models have been integrated into simulation approaches for complete PEFCs (Bernardi and Verbrugge, 1992).

In spite of the widely recognized importance of microstructural effects in CLs, many earlier approaches in fuel cell modeling, specifically those employing CFD at single cell and stack levels, used a so-called interface approximation. This approximation treats the CL as an infinitesimally thin interface without internal structure. The only physical property of such a CL that is explicitly considered is the net rate of current conversion, neglecting transport and two-phase effects in the CL. This oversimplified treatment could lead to wrongful or misleading conclusions about the origin of the main detrimental effects in PEFC performance. Nowadays, pertinent CFD models of fuel cells, implemented in FLUENT and CFD Ace+, account for structural details and multiphase flow phenomena in CLs.

In 1986, Raistrick developed the first catalyst layer based on carbon-supported Pt nanoparticles that incorporated Nafion ionomer as a proton conductor and bonding agent (Raistrick, 1986). The first detailed model for composition effects in this new type of *ionomer-bound catalyst layers* of PEFCs was developed by Tom Springer and colleagues at LANL (Springer et al., 1993). Springer's one-dimensional macrohomogeneous model accounts for potential losses due to reactant diffusion, proton migration, and interfacial kinetics at the Pt/ionomer interface. It relates these losses to operating conditions and effective parameters of transport and reaction in PEFCs. A fit between model and experimental data allowed the quantification of potential losses caused by kinetic processes and transport of oxygen and protons.

Since then, other groups have developed similar models (Eikerling and Kornyshev, 1998; Perry et al., 1998). The main features of this approach are: (i) it relates global performance of CCLs to spatial distributions of reactants, electrolyte phase potential, and reaction rates; (ii) it defines a reaction penetration depth, or the active zone; and (iii) it suggests an optimum range of current density and catalyst layer thickness, with minimal performance losses and highest utilization of the catalyst.

Combination of the macrohomogeneous approach for porous electrodes with a statistical description of effective properties of random composite media rests upon concepts of *percolation theory* (Broadbent and Hammersley, 1957; Isichenko, 1992; Stauffer and Aharony, 1994). Involving these concepts significantly enhanced capabilities of CL models in view of a systematic optimization of thickness, composition, and porous structure (Eikerling and Kornyshev, 1998; Eikerling et al., 2004). The resulting structure-based model correlates the performance of the CCL with volumetric amounts of Pt, C, ionomer, and pores. The basis for the percolation approach is that a catalyst particle can take part in reaction only if it is connected simultaneously to percolating clusters of carbon/Pt, electrolyte phase, and pore space. Initially, the electrolyte phase was assumed to consist of ionomer only. However, in order to properly describe local reaction conditions and reaction rate distributions, it is necessary to account for water-filled pores and ionomer-phase domains as media for proton transport.

## Misapprehensions and Controversial Issues

In spite of the long history of porous electrode theory and recent progress in developing structure-based models of ionomer-bound CLs in PEFCs, misapprehensions and controversial issues persist. To begin with, it has remained obscure for a long time, as to which type of interface, namely, Pt–gas, Pt–liquid water, and Pt–ionomer, is most abundant and makes a major contribution to electrochemical current generation in CLs. Pt–vapor interfaces are disconnected from the proton-supplying network and must be deemed inactive. Water in CCLs is produced in liquid form. Nevertheless, many modeling approaches in the past included a misleading assumption that water is produced in vapor form, and that the issue of water management is to prevent condensation.

The relative contribution to electrocatalytic conversion arising from Pt–ionomer and Pt–water interfaces depends on phase segregation, porous morphology, and wetting properties of pores. A larger fraction of hydrophilic micropores will increase the contribution of Pt–water interfaces, relative to that of Pt–ionomer interfaces (Eikerling, 2006; Liu and Eikerling, 2008).

The nature of electrocatalytically active catalyst surfaces is related to the extent of aggregation of Pt/C and spontaneous phase segregation between these Pt/C aggregates and dense ionomer phase domains. Interest in understanding these mesoscale self-organization phenomena in CL inks motivated recent molecular dynamics studies. Experimentally, agglomerates are difficult to identify and characterize. The majority of data supporting the existence of agglomerates are based on TEM micrographs and porosimetry (Uchida et al., 1995a; Xie et al., 2004, 2005).

Lasting controversy evolves around size and composition of agglomerates. The structure and distribution of ionomer and interactions between ionomer and carbon/Pt are of key importance in this context. Commonly encountered views assume (i) agglomerates as uniform mixtures of ionomer and Pt/C particles, (ii) a thin uniform ionomer film (skin layer) surrounding agglomerates of C/Pt, or (iii) uncorrelated percolating phases of C/Pt and ionomer. In fact, many approaches in agglomerate modeling adhere to the first option, describing agglomerate properties by the so-called Thiele modulus (Perry et al., 1998; Thiele, 1939). It demands uniform penetration of agglomerate pores by an intrinsically proton-conductive medium. However, on the basis of size considerations, it seems improbably that ionomer fibrils could penetrate into nanometer-sized pores of Pt/C aggregates and form an embedded proton-conducting phase. Relevance of the other two options, that is, assuming a thin ionomer film coating agglomerates or independent percolating phases, depends on type and strength of adhesive forces between carbon and ionomer.

Another important aspect, which has been slowly moving into the focus of interest of modeling efforts, is the role of bimodal porous morphology, heterogeneous wettability, and water accumulation in pores. A large fraction of primary pores inside of Pt/C aggregates, with radii in the range of 4–10 nm, is advantageous in that it provides a large surface area for Pt deposition. These pores should be flooded with water to maximize the wetted surface fraction of Pt. For this purpose, it is beneficial if they are hydrophilic. Secondary pores between agglomerates with radii of 10–50 nm are vital for the gaseous transport of reactants. These pores should be hydrophobic. As emphasized in Eikerling (2006) and Liu and Eikerling (2008), understanding the interplay between composition, porous structure, wetting properties of internal surfaces, and operating conditions is crucial for determining distribution, fluxes, and phase change rates of water in CCLs and MEAs.

*Wetting phenomena* in secondary pores of a CCL determine the transition from a partially saturated to a fully saturated state upon increasing current density. This transition is accompanied by a significant drop in fuel cell voltage (Eikerling, 2006; Liu and Eikerling, 2008). In this context, limiting current behavior frequently observed in polarization curves of PEFCs, has been often ascribed to flooding of the CCL. However, a fully saturated CCL does not give rise to limiting current behavior. The oxygen diffusion coefficient of a fully-flooded CCL, $D_{fl}^o$, is reduced by about two orders of magnitude compared to a partially saturated CCL. The reaction penetration depth, introduced in Chapter 1, $(\delta_{CL} \propto D_{fl}^o)$ and the effective exchange current density $(j_{eff}^0 \propto \sqrt{D_{fl}^o})$, will be reduced correspondingly. The apparent effect will be a downshift of the polarization curve by $\sim$100–200 mV, depending on the value of the effective Tafel slope. Flooding of the CCL could cause a drastic decrease in fuel cell voltage, only if it occurs in a CCL with very poor proton conductivity ($\ll 0.01\,\mathrm{S\,cm^{-1}}$) due to high ohmic losses incurred in the inactive region of the CCL. The transition region from a partially saturated to a fully saturated state of CCL operation in steady-state polarization curves, could exhibit bistable behavior. Bistability means that at a given current density two values of the potential response are obtained, corresponding to the coexistence of partially saturated and fully saturated states. These nonlinear phenomena play an important role for the selection of operating conditions, and they should be accounted for in systematic efforts to improve the structure and performance of CCLs.

## HOW TO EVALUATE THE STRUCTURAL DESIGN OF CCLs?

*Evaluation of CL performance* requires a number of parameters that define the ideal electrocatalyst performance, allowing deviations from ideal behavior to be rationalized and quantified. Ideal electrocatalyst performance is achieved when the total Pt surface area per unit volume, $S_{tot}$, is utilized and when reaction conditions at the reaction plane (or Helmholtz layer) near the catalyst surface are uniform throughout the layer. These conditions would render each portion of the catalyst surface equally active. Deviations from ideal behavior arise due to statistical underutilization of catalyst atoms, as well as nonuniform distributions of reactants and reaction rates at the reaction plane that are caused by transport effects. This section introduces the *effectiveness factor of Pt utilization* and addresses the hierarchy of structural effects from atomistic to macroscopic scales that determine its value.

### STATISTICAL MOMENTS OF THE PARTICLE RADIUS DISTRIBUTION

The total amount of catalyst in the CL is specified as the *Pt loading*, given in mass of Pt per unit apparent (external) surface area of electrode or MEA, $m_{Pt}$, with units of mg cm$^{-2}$. The main structural characteristic of dispersed catalyst nanoparticles is the *particle radius distribution* (PRD) function, $f(r_{Pt})$. It is defined as the number of catalyst particles per unit radius interval and per unit volume of the CL, assuming spherically shaped particles. Experimentally, $f(r_{Pt})$ can be obtained from high-resolution TEM studies or from x-ray diffraction data. The function $f(r_{Pt})$ is the main observable to monitor during CL degradation. Within spherical particle assumption, $f(r_{Pt})$ determines mean particle radius $\bar{r}_{Pt}$, total catalyst surface area $S_{tot}$, and Pt loading $m_{Pt}$, via

$$\bar{r}_{Pt} = \frac{\displaystyle\int_0^\infty r_{Pt} f(r_{Pt}) \, dr_{Pt}}{\displaystyle\int_0^\infty f(r_{Pt}) \, dr_{Pt}}, \tag{3.2}$$

$$S_{tot} = 4\pi \int_0^\infty r_{Pt}^2 f(r_{Pt}) \, dr_{Pt}, \tag{3.3}$$

$$m_{Pt} = l_{CL} \frac{4\pi}{3} \rho_{Pt} \int_0^\infty r_{Pt}^3 f(r_{Pt}) \, dr_{Pt}. \tag{3.4}$$

Under degradation conditions, the time-dependent distribution function $f(r_{Pt}, t)$ will determine the temporal evolution of the moments $\bar{r}_{Pt}$, $S_{tot}$, and $m_{Pt}$.

## EXPERIMENTAL ASSESSMENT OF PT UTILIZATION

A survey of studies that focus on Pt utilization and effectiveness factors can be found in Eikerling et al. (2008) and Xia et al. (2008). In the past, ambiguous definitions of catalyst utilization have been circulating with inconsistent values having been reported. This could lead to a wrongful assessment as to how much the performance of fuel cells could be improved by advanced structural design of catalyst layers. Without any doubt, the accurate distinction and determination of catalyst utilization and effectiveness factors have a major impact on defining priorities in catalyst layer research.

*Pt utilization* is a statistical property of the CCL. It can be obtained from *ex situ* electrochemical studies. It is defined as the ratio of the electrocatalytically active surface area, accessible to electrons and solvated protons, to the total surface area of Pt:

$$\Gamma_{stat} = \frac{S_{ECSA}}{S_{tot}}. \tag{3.5}$$

The *statistical surface area utilization factor* $\Gamma_{stat}$ has been considered under different conditions, specifically in catalyst powders and in MEAs of operational PEFCs. The electrocatalytically active surface area in the catalyst powder can be obtained from the charge under the H-adsorption or CO-stripping waves measured by thin-film rotating disk electrode (TF-RDE) voltammetry in $H_2SO_4$ solution (Easton and Pickup, 2005; Schmidt et al., 1998; Shan and Pickup, 2000). The total surface area $S_{tot}$ can be calculated from the mean particle size determined by XRD experiments (Easton and Pickup, 2005; Shan and Pickup, 2000) or from HR-TEM images (Schmidt et al., 1998), based on the assumption that particles are spheres with surface area $4\pi \bar{r}_{Pt}^2$. For catalyst powders of carbon-supported Pt nanoparticles, values of $\Gamma_{stat}$ were reported to be 109% by Easton and Pickup (2005), 125% by Shan and Pickup (2000), and 100% by Schmidt et al. (1998). The fact that estimated values of $\Gamma_{stat}$ could be greater than 100% is attributed to contributions from background currents (e.g., including those for $H_2$ evolution) to the total measured charge.

The purpose of studying Pt utilization in the catalyst powder is to evaluate the electronic connectivity of carbon/Pt and maximum ionic accessibility of Pt nanoparticles. Pt utilization of 100% means that all Pt nanoparticles are connected to the electronic conduction network, with the entire Pt surface accessible for protons. Close to 100% Pt utilization in the catalyst powder of carbon-supported Pt, immersed in liquid electrolyte, is possible, since only a negligible fraction of the Pt surface is covered by carbon particles. For Pt nanoparticles, supported by a polymer material such as poly(3,4-ethylenedioxythiphene)/poly(styrene-4-sulfonate), Pt utilization was reported to be 43–62%.

The value of $\Gamma_{stat}$ in the CCL of an operational MEA can be estimated from the charge under the H-adsorption waves in *cyclic voltammetry* (CV) using driven-cell mode (Schmidt et al., 1998). For example, to determine the accessible Pt surface area of a cathode, the cathode compartment of a single cell is purged with humidified nitrogen, serving as the working electrode. At the same time, the anode compartment is purged with humidified hydrogen, serving as both counter and reference electrode. In some studies, $S_{tot}$ was calculated from the mean particle size, and Pt utilization in the cathode was reported to be 34% by Dhathathreyan et al. (1999), 45% by Cheng et al. (1999), 52% by Sasikumar et al. (2004), and 55–76% by Li and Pickup (2003). As mentioned above, $S_{tot}$ could also be obtained by the TF-RDE method, and Pt utilization in the cathode was reported to be 86–87% by Schmidt et al. (1998) and 90% by Gasteiger and Yan (2004). Compared with Pt utilization in the catalyst powder, the reduced Pt utilization in the CL of a fully functional MEA originates from two effects: (i) some Pt nanoparticles are encapsulated by ionomer and, therefore, loose electronic contact during fabrication, and (ii) a fraction of Pt particles is neither covered by ionomer electrolyte nor wetted by liquid water and, therefore, disconnected from the proton-supplying network.

The foregoing discussion reveals large uncertainties in determining $\Gamma_{stat}$ in experiment. The total charge transferred upon hydrogen adsorption may involve nonwetted catalyst surface area, which will be inactive under PEFC operation, or include the catalytically inert substrate surface from spillover effects (Zhdanov and Kasemo, 1997). Generally, the surface area obtained from hydrogen adsorption and CO-stripping methods could be different from the active area in an operating fuel cell. Moreover, the values of $\Gamma_{stat}$ do not account for nonuniform reaction rate distributions, which arise at finite rates of reactant transport and consumption under stationary fuel cell operation.

## CATALYST ACTIVITY

Various parameter definitions are used to specify the *intrinsic activity* of a catalytic material. Many practical studies, focusing on comparing catalyst materials under conditions that are typical for PEFC operation, provide values of current densities measured at a certain cathode potential, often chosen as 0.9 V vs. SHE.

The main basic parameter of catalyst evaluation is the *specific exchange current density* $j_*^0$, which, by definition, is normalized to the unit surface area of the electrocatalyst. This property is the target of many fundamental studies in electrocatalysis, too numerous to be listed (Adzic et al., 2007; Debe, 2013; Gasteiger and Markovic, 2009; Kinoshita, 1992; Paulus et al., 2002; Stamenkovic et al., 2007a,b; Tarasevich et al., 1983; Zhang et al., 2005, 2008).

A myriad of experimental studies have explored the impact of size and shape of catalyst nanoparticles as well as of substrate properties on $j_*^0$ (Wieckowski et al., 2003). Predictive relations between particle size and activity are, however, difficult to be established, since the size of particles affects electronic and geometric properties at their surface. The advantage of using $j_*^0$ for catalyst evaluation, is that it can be obtained from *ex situ* Tafel analysis conducted under reproducible conditions. As a result of changes in the reaction pathway with potential, $j_*^0$ must be found for the potential region of interest.

The effective exchange current density of the porous composite CL, given per unit of apparent (external) electrode surface area, $j^0 = j_*^0 S_{ECSA} l_{CL} = j_*^0 S_{tot} l_{CL} \Gamma_{stat}$, is the key physical property of a CL. This parameterization reveals two main avenues for enhancing the value of $j^0$: via improving the intrinsic electrocatalytic activity $j_*^0$ or via optimizing the structural CL design, embodied in $S_{ECSA}$ and $l_{CL}$.

For monodisperse PRD with mean radius $\bar{r}_{Pt}$, the relation $m_{Pt} = \frac{1}{3} S_{tot} l_{CL} \rho_{Pt} \bar{r}_{Pt}$ can be established, where $\rho_{Pt}$ is the mass density of Pt. Thus, one can write

$$j^0 = j_*^0 \frac{3 m_{Pt}}{\rho_{Pt} \bar{r}_{Pt}} \Gamma_{stat}. \tag{3.6}$$

An alternative expression can be derived

$$j^0 = j_*^0 \frac{m_{Pt} N_A}{M_{Pt} \nu_{Pt}} \Gamma_{np} \Gamma_{stat}, \tag{3.7}$$

where $M_{Pt}$ is the atomic mass of Pt and $\nu_{Pt}$ is the number of Pt atoms per unit surface area of the catalyst. The factor $\Gamma_{np}$ is the surface-to-volume atom ratio of catalyst nanoparticles. It is an effective geometrical parameter, averaged over the PRD, that represents catalyst utilization at the nanoparticle level. The fraction on the right-hand side of Equation 3.7 is the ratio of ideal catalyst surface area to real-geometric surface area of the electrode.

The ideal catalyst surface area is obtained by spreading out catalyst atoms as a perfect monolayer. Assuming the densest packing of surface atoms with Pt(111) surface structure and lattice constant $a_{Pt} = 3.92$ Å, it is found that

$$j^0 \approx j_*^0 2060 \, [m_{Pt}] \, \Gamma_{np} \Gamma_{stat}, \tag{3.8}$$

where the Pt loading must be provided as a dimensionless number normalized to $1$ mg cm$^{-2}$.

The parameters $j_*^0$, $\nu_{Pt}$, and $\Gamma_{np}$ depend on size, shape, and atomic surface structure of catalyst nanoparticles. The proposed factorization in Equation 3.8 is an approximation. It provides a useful estimate of catalyst utilization only if the PRD is reasonably monodisperse, that is, if particles of a certain size and shape prevail.

## ATOM-BASED UTILIZATION FACTOR AT THE LEVEL OF PT NANOPARTICLES

Different *facetted-type Pt nanoparticles* with crystalline configuration and sizes $<2$ nm are shown in Figure 3.4a (Burda et al., 2005; Frenkel et al., 2001; Narayanan and El-Sayed, 2008; Narayanan et al., 2008; Rioux et al., 2006; Wang et al., 2009). For large, roughly spherical particles, the *surface atom ratio* $\Gamma_{np} = N_S/N$, introduced in Equation 3.7, is expected to be proportional to $N^{-1/3}$, where $N_S$ is the number of surface atoms and $N$ is the total number of atoms. For particles with atom numbers $N < 200$ with the different shapes depicted in Figure 3.4a, $\Gamma_{np}$ can be represented well by

$$\Gamma_{np} = A + aN^{-1/3}, \tag{3.9}$$

as illustrated in Figure 3.4b. The parameters $a$ and $A$ assume different values for different particle shapes. The tetrahedral structure, enclosed by four Pt(111) facets, corresponds to the largest $\Gamma_{np}$. At a given number of catalyst atoms per particle, the difference in $\Gamma_{np}$ is about 15–20% between tetrahedral and cubo-octahedral particles, corresponding to the highest and lowest values, respectively. In general, more nonspherical shapes of nanoparticles correspond to larger values of $\Gamma_{np}$.

The variation of $\Gamma_{np}$, seen in Figure 3.4b, reveals that shape control could be a significant factor for the mass activity of Pt nanoparticles (Wang et al., 2009). The surfaces of small nonspherical nanoparticles that give higher values of $\Gamma_{np}$ possess, however, larger fractions of undercoordinated corner and edge atoms. They are thermodynamically less stable and more chemically active, implying a higher propensity for Pt dissolution. Moreover, the ORR activity is affected by particle shape. Undercoordinated surface atoms at edge and corner sites possess the highest *oxygen adsorption energies*, an important descriptor of the ORR activity. In the case of Pt, an enhanced oxygen adsorption energy renders the catalyst less active for the ORR.

Comparison of Equations 3.6 and 3.7 provides a simple approximate relation for $\Gamma_{np}$

$$\Gamma_{np} \approx \frac{3M_{Pt}\nu_{Pt}}{\rho_{Pt}N_A}\frac{1}{\bar{r}_{Pt}} \approx \frac{\sqrt{3}\,\bar{a}_{Pt}}{\bar{r}_{Pt}}, \tag{3.10}$$

where $\rho_{Pt} \approx 4m_{Pt\,atom}/\left(a_{Pt}^b\right)^3$ and $\nu_{Pt} \approx 4/\sqrt{3}\left(a_{Pt}^s\right)^2$ have been used with effective lattice constants $a_{Pt}^b$ and $a_{Pt}^s$ in the bulk and at the particle surface, respectively, and $\bar{a}_{Pt} = \left(a_{Pt}^b\right)^3/\left(a_{Pt}^s\right)^2$. Particles with diameters $<3$ nm give $\Gamma_{np} > 50\%$.

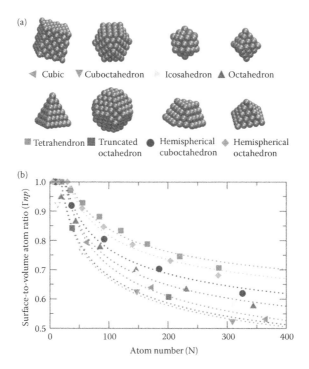

**FIGURE 3.4** (a) Facetted-type Pt nanoparticles with regular crystalline structure and sizes <2 nm. (b) Surface-to-volume atom ratio, $\Gamma_{np}$, as a function of atom number per particle for the distinct particle shapes shown in (a). (Reprinted from Wang, L., Roudgar, A. and Eikerling, M. *J. Phys. Chem. C*, **113**(42), 17989–17996, 2009, Figures 1,2,3,4,5,9. Copyright (2009) by the American Physical Society. With permission.)

## STATISTICAL UTILIZATION FACTOR

What is the physical meaning of the factor $\Gamma_{stat}$ in Equation 3.6? The statistical utilization factor $\Gamma_{stat}$ stands for the fraction of the catalyst surface that is wetted by electrolyte and accessible for protons needed in the ORR. It corresponds to the utilization factor introduced in the section "Experimental Assessment of Pt Utilization."

In the original structure-based CCL model of Eikerling and Kornyshev (Eikerling and Kornyshev, 1998; Eikerling et al., 2004, 2007a), an expression for $\Gamma_{stat}$ was derived based on the *theory of active bonds* in *random three-phase composite media*.

In this approach, discussed in the section "Effective Catalyst Layer Properties from Percolation Theory," $\Gamma_{stat}$ corresponds to the statistical fraction of Pt particles at or near the *triple-phase boundary* of solid carbon/Pt phase (volume fraction $X_{PtC}$), ionomer phase ($X_{el}$), and pore space ($X_P = 1 - X_{PtC} - X_{el}$).

In realistic CCL structures, application of *percolation laws* to parameterize the composition dependence of $\Gamma_{stat}$ is disputable due to two crucial assumptions, which start to seem insufficient in the light of more detailed knowledge of the CL structure. The first assumption is that protons needed for the ORR are available only inside the ionomer phase. The second assumption is that oxygen diffusion is significant only in the gas phase, whereas condensed phases of water, ionomer, and Pt/C are impermeable for oxygen.

As for the first assumption, the electrolyte phase must be treated as a mixed phase. It consists of a thin-film structure of ionomer at the surface of Pt/C agglomerates and of water in ionomer-free intra-agglomerate pores. The proton density is highest at the ionomer film (pH $\sim$ 1 or smaller), and it is much smaller in water-filled pores (pH > 3). However, the proton density distribution is not incorporated in the statistical utilization $\Gamma_{stat}$, but in an *agglomerate effectiveness factor*, defined in the section "Hierarchical Model of CCL Operation."

As concerns the second assumption, direct access to the gas pore network is not a stringent requirement for keeping the Pt surface active. The condensed phases of water and ionomer possess finite oxygen permeabilities, which will render oxygen concentrations finite at electrolyte-covered Pt particles. Again, this is an issue of nonuniform reaction rate distributions that will be dealt with in the section "Hierarchical Model of CCL Operation." In that section, a two-scale performance model will be presented, which couples transport and distribution of protons and oxygen at the agglomerate level and at the macroscopic scale.

In summary, the statistical factor $\Gamma_{stat}$ accounts for the catalyst surface fraction covered by electrolyte and liquid water.

## Nonuniform Reaction Rate Distributions: Effectiveness Factor

At this point, it is important to realize that the ultimate optimization target of electrode design is neither Pt utilization nor is it intrinsic activity. The two optimization functions of catalyst layer design are the net activity (exchange current density) per unit electrode volume and the effectiveness factor of Pt utilization. The effectiveness factor accounts for statistical effects and nonuniform reaction rate distribution. The latter arise from *mass transport phenomena* at finite current densities of fuel cell operation. In simple one-dimensional electrode theory, the interplay of $j^0$ with a slow transport process, involving the sluggish diffusion of oxygen or poor transport of protons, determines a *reaction penetration depth* $\delta_{CL}$. The criterion for uniform reaction rate distributions is that the reaction penetration depth for any of the transport phenomena considered is larger than the thickness of the CL, $\delta_{CL} > l_{CL}$. This requirement implies that the thickness of the layer must be adapted well to the transport properties of the medium (Chan et al., 2010; Eikerling, 2006; Eikerling and Kornyshev, 1998; Eikerling and Malek, 2009; Eikerling et al., 2004, 2008).

A 1D model of the effects of structure and processes on steady-state performance of CLs requires a minimum of two phenomenological parameters: the exchange current density $j^0$ and the reaction penetration depth $\delta_{CL}$. Each of them is a function of structure and operating conditions. Neglecting further complicating traits, for instance, issues related to the liquid water balance, these two parameters uniquely define the potential loss in the CCL. Thereby, voltage efficiency, energy density, power density, and effectiveness of catalyst utilization at given current density and catalyst loading can be calculated.

The exchange current density $j^0$ is a static materials property defined in the section "Catalyst Activity." The reaction penetration depth $\delta_{CL}$ is a steady-state property determined by the interplay of transport properties of the layer and local electrocatalytic activity, embodied in $j^0$. Together, both parameters $j^0$ and $\delta_{CL}$ determine the overall effectiveness of catalyst utilization. For illustration purposes, a simple scenario of this interplay for the case of severely limited oxygen diffusion will be considered below.

The overall *effectiveness factor of Pt utilization* of the CL, including transport effects, can be defined by

$$\Gamma_{CL} = \Gamma_{np}\Gamma_{stat}\frac{j_0}{j_0^{id}}, \tag{3.11}$$

with $\Gamma_{np}$ and $\Gamma_{stat}$ given above. The quotient $j_0/j_0^{id}$ follows from the solution of the performance problem. The actual current density, $j_0$, incorporates the impact of transport phenomena. The ideal current density, $j_0^{id}$, would be attained if all transport effects were negligible; it represents virtually infinite rates of transport, and reaction rates that are distributed perfectly uniformly.

In the section "Hierarchical Model of CCL Operation," a hierarchical modeling framework will be presented; it couples multiple effects of proton and oxygen transport at the mesoscopic scale of agglomerates and at the macroscopic scale of the layer. In that general case, the CL effectiveness factor can be defined by

$$\Gamma_{CL} = \Gamma_{np}\Gamma_{stat}\bar{\Gamma}_{agg}, \tag{3.12}$$

where the *average effectiveness factor of agglomerates*, $\bar{\Gamma}_{agg}$, is obtained by averaging the local effectiveness factors, $\Gamma_{agg}(z)$, over the thickness of the CCL:

$$\bar{\Gamma}_{agg} = \frac{1}{l_{CL}} \int_0^{l_{CL}} \Gamma_{agg}(z)\, dz. \tag{3.13}$$

The consideration of effects included in $j_*^0$ and $\Gamma_{CL}$ suggest that improvements in structure and function of catalyst layers should be pursued in three areas: (1) nanoparticle electrocatalysis (interplay of $j_*^0$ and $\Gamma_{np}$), (2) statistical utilization of the catalyst ($\Gamma_{stat}$), and (3) mixed transport in composite media ($\bar{\Gamma}_{agg}$). These areas encompass a hierarchy of kinetic and transport processes that span many scales.

## Effectiveness Factor in Oxygen Depletion Regime: A Simple Case

For the case of severely limited oxygen transport and ideal proton transport, one may write

$$\Gamma_{CL} = \Gamma_{np}\Gamma_{stat}\Gamma_\delta. \tag{3.14}$$

A detailed discussion of the *diffusion-limited case* will be presented in the section "Ideal Proton Transport" of Chapter 4. In the *oxygen-depleted regime*, an explicit expression for $\Gamma_\delta$ can be derived. It is assumed here that proton transport limitations are negligible and that oxygen is consumed in a thin sublayer with thickness $\delta_{CL} \ll l_{CL}$. Under these conditions, the set of transport equations that governs the distribution of oxygen partial pressure, $p(x)$, and proton current density, $j(x)$, is given by

$$\frac{dj}{dx} = -j^0 \frac{p(x)}{p_L} \exp\left(-\frac{\eta_0}{b_c}\right) \qquad (3.15)$$

$$\frac{dp}{dx} = \frac{j_0 - j_p(x)}{4fD^o}, \qquad (3.16)$$

with $f = \frac{F}{R_g T}$, $p_L$ being the oxygen partial pressure at $x = l_{CL}$, $\eta_0$ being the CCL overpotential, $D^o$ being the oxygen diffusion coefficient, and $b_c = \frac{R_g T}{\alpha_c F}$ being the effective Tafel slope of the ORR (*cf.* the section "Deciphering the ORR"). These equations represent a subset of the general set of equations that defines the macrohomogeneous model, discussed in the section "Macrohomogeneous Model with Constant Properties." Specific conditions assumed are: fixed composition and transport parameters; excellent proton transport, corresponding to constant electrolyte potential; and severely restricted oxygen transport.

Taking the first derivative of Equation 3.16 and inserting Equation 3.15 into it gives a second-order ordinary differential equation

$$\frac{d^2 p}{dx^2} = \frac{1}{\delta_{CL}^2} p(x), \qquad (3.17)$$

with a *reaction penetration depth* defined for this case as

$$\delta_{CL} = \sqrt{\frac{4fD^o p_L}{j^0}} \exp\left(\frac{\eta_0}{2b_c}\right). \qquad (3.18)$$

The solution of Equation 3.17 with the boundary conditions $p(x = l_{CL}) = p_L$ and $\left.\frac{dp}{dx}\right|_{x=0} = 0$ is

$$p(x) = p_L \exp\left[-\frac{l_{CL}}{\delta_{CL}}\left(1 - \frac{x}{l_{CL}}\right)\right] \qquad (3.19)$$

and

$$j(x) = I\frac{l_{CL}}{\delta_{CL}}\left\{1 - \exp\left[-\frac{l_{CL}}{\delta_{CL}}\left(1 - \frac{x}{l_{CL}}\right)\right]\right\}, \qquad (3.20)$$

with

$$I = \frac{4fD^o p_L}{l_{CL}}. \qquad (3.21)$$

The relation between fuel cell current density and overpotential is given by

$$j_0 = \sqrt{Ij^0} \exp\left(-\frac{\eta_0}{2b_c}\right) \left\{1 - \exp\left[-\sqrt{\frac{j^0}{I}} \exp\left(-\frac{\eta_0}{2b_c}\right)\right]\right\}. \qquad (3.22)$$

The expression for $\Gamma_\delta$ is

$$\Gamma_\delta = \frac{\delta_{CL}}{l_{CL}} \left\{1 - \exp\left(-\frac{l_{CL}}{\delta_{CL}}\right)\right\}. \qquad (3.23)$$

For this case, $\Gamma_\delta$ can be expressed as a simple analytical function of $\eta_0$,

$$\Gamma_\delta = \sqrt{\frac{I}{j^0}} \exp\left(\frac{\eta_0}{2b_c}\right) \left\{1 - \exp\left[-\sqrt{\frac{j^0}{I}} \exp\left(-\frac{\eta_0}{2b_c}\right)\right]\right\}. \qquad (3.24)$$

Another useful relation is

$$\delta_{CL} = l_{CL}\frac{I}{j_0} = \frac{4fD^o p_L}{j_0}, \qquad (3.25)$$

immediately showing the impact of ineffective oxygen diffusion on the reaction penetration depth. The overall effectiveness factor is

$$\Gamma_{CL} = \Gamma_{np}\Gamma_{stat}\frac{\delta_{CL}}{l_{CL}} \left\{1 - \exp\left(-\frac{l_{CL}}{\delta_{CL}}\right)\right\}. \qquad (3.26)$$

For a rough illustration, considering a poorly designed CCL with $D^o \sim 10^{-5}$ cm$^2$ s$^{-1}$, $\Gamma_\delta \sim 0.1$ is found at $j_0 = 1$ A cm$^{-2}$. Equation 3.26 highlights the importance of adjusting thickness and effective properties of oxygen transport in CLs in such a way that $\delta_{CL} \geq l_{CL}$. This analysis could be applied to analyze a CL design, in which protons and electrostatic potential are distributed uniformly.

## STATE-OF-THE-ART IN THEORY AND MODELING: MULTIPLE SCALES

The objective of catalyst layer modeling is to establish relations between fabrication procedures and conditions; microstructure; effective properties of transport and reaction; and performance (Eikerling et al., 2007a). The foregoing sections illustrated that structure and function of catalyst layers evolve over a wide range of scales. The modeling of structure and operation of CLs is, therefore, a multiscale problem.

The challenges for theory and modeling of CL operation can be rationalized much more clearly and physical-mathematical challenges can be simplified accordingly, if the main structural effects occur at scales that are separated by at least one or two orders of magnitude. The main scales correspond to catalyst nanoparticles

($r_{Pt} \sim 2$ nm), agglomerates composed of Pt and carbon ($R_a \sim 100$ nm), and the macroscopic device level ($l_{CL} \sim 10$ µm), at which CLs can be considered as effective homogeneous media.

At the nanoparticle level, the specific exchange current density $j_*^0$ is sensitive to size, surface structure, and surface electronic structure of Pt nanoparticles as well as to the electronic properties of the substrate (Boudart, 1969; Eikerling et al., 2003; Housmans et al., 2006; Maillard et al., 2004). A better understanding of the relationships between particle size and activity, including those effects, is critical in view of the design of highly performing catalysts (Cherstiouk et al., 2003; Maillard et al., 2007, 2005). Obviously, a reduction in particle size improves the utilization factor $\Gamma_{np}$. The relation between particle size and activity is non-monotonic, since the size of the particles affects electronic and geometric properties at their surface (Hansen et al., 1990; Wang et al., 2009). A complex interplay of elementary surface processes, including molecular adsorption, surface diffusion, charge transfer, and desorption, determines the net rates of surface reactions.

Studies of stable conformations of supported nanoparticles, as well as of the elementary processes on their surface demand DFT calculations and kinetic modeling based on Monte Carlo simulations or mean field approaches (Andreaus and Eikerling, 2007; Andreaus et al., 2006). Because of the complex nature of structural effects in electrocatalysis, only few theoretical studies have been successful in rationalizing systematic trends in structure versus reactivity relations. The most notable contribution is the *d-band model* of Hammer and Norskov (Hammer and Norskov, 1995; Hammer and Nørskov, 2000). It relates trends in chemisorption energies for adsorbates on transition metal surfaces to the position of the d-band center, the first moment of the density of states from the Fermi level. Systematic DFT calculations and experiments on series of polycrystalline alloy films of the type $Pt_3M$ (M=Ni, Co, Fe, and Ti) have confirmed predicted correlations between the position of the d-band center, oxygen chemisorption energies, and electrode activities for the ORR (Stamenkovic et al., 2006). This success of the d-band model has fostered efforts in devising DFT-based combinatorial screening schemes for identifying highly active electrocatalyst materials (Greeley et al., 2006).

Usually, insights gained from DFT calculations are not straightforwardly applicable to materials design. In electrocatalysis, direct DFT calculations have been applied for studying elementary surface processes on catalyst systems with single-crystalline surface structure. Practical catalyst systems that employ supported nanoparticles represent special challenges (Gross, 2006; Kolb et al., 2000). Supported catalyst nanoparticles are relatively large systems, usually consisting of several hundreds of atoms. They exhibit irregular surface structures with different surface facets and significant fractions of lowly coordinated surface sites, viz. edge, corner, or defect atoms. Moreover, support effects have to be accounted for. Recent efforts employing DFT calculations focused explicitly on morphologies and electrocatalytic properties of small metal nanoclusters (Song et al., 2005; Xiao and Wang, 2004). *Ab initio* calculations in Roudgar and Groß (2004) have explored the strongly modified chemical properties of Pd nanoclusters supported on Au(111). Such studies help in evaluating the electronic structure effects exerted by the substrate and the consequences of the low coordination of surface atoms on energetics and kinetics of binding atomic or molecular species, for example, $OH_{ad}$, $CO_{ad}$, and $H_{ad}$.

At the mesoscopic scale, interactions between molecular components control the self-organization of molecular components that leads to random phase segregation during fabrication of CLs (Malek et al., 2007, 2011). Mesoscale simulations can describe the morphology of heterogeneous materials and rationalize their effective properties, beyond length- and time-scale limitations of atomistic simulations. A recently introduced coarse-grained computational approach allows the key factors during fabrication of CLs to be evaluated. These simulations rationalize structural properties such as pore sizes, internal porosity, and wetting angles of internal/external surfaces of agglomerates. They help to elucidate whether or not ionomer is able to penetrate into primary pores inside of agglomerates (Fernandez et al., 2005). Moreover, dispersion media with distinct dielectric properties can be evaluated in view of capabilities for controlling sizes of Pt/carbon agglomerates, ionomer domains, and the resulting pore network topology. Relative contributions to the total electrocatalytically active surface area of wetted Pt at the walls of water-filled pores inside agglomerates, at the surfaces of agglomerates, or at interfaces with wetted ionomer can be rationalized.

In a porous catalyst layer, treated as a random heterogeneous and nanoporous medium, macroscopic properties such as diffusion and reaction rates are defined as averages of the corresponding microscopic quantities (Sahimi et al., 1990). These averages must be taken over *representative elementary volume* elements (REVs), which are large compared to microscopic structural elements (pores, particles). At the same time, they are small compared to the system size. The morphology of the composite layer can be related to effective properties that characterize transport and reaction, using concepts from the theory of random heterogeneous media (Torquato, 2002).

Finally, conditions for stationary operation at the macroscopic device level can be defined. Balance equations for involved species, that is, electrons, protons, reactant gases, and water, can be established on the basis of fundamental conservation laws. The general ingredients of a macrohomogeneous model of CL operation are the source term for electrochemical current conversion, using, for example, Butler–Volmer equation or, at a more fundamental level, transition state theory, the source/sink term for the phase change of water at interfaces (vaporization, condensation), as well as terms that account for the transport of species, that is, the migration of electrons/protons in conduction media, diffusion of dissolved oxygen and protons in water-filled pores, and diffusion of oxygen and vapor in gas-filled pores.

The two-step strategy in the *physical modeling of catalyst layer operation* is depicted in Figure 3.5. The first step relates structure to the physical properties of the layer, considered as an effective medium. The second step relates these effective properties to electrochemical performance. Relations between structure and performance are complicated by the formation of liquid water, affecting effective properties and performance. Solutions for such a model provide relations between structure, properties, and performance. These relations allow predictions of architectures of materials and operating conditions that optimize catalyst layer and fuel cell operation to be made.

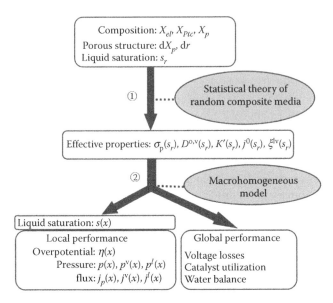

**FIGURE 3.5** A general framework for structure-based catalyst layer modeling. The first step relates primary parameters of structure and composition to physicochemical properties, using the statistical theory of random composite media. The second step links these properties to performance. The performance model provides local functions as well as global performance metrics.

## NANOSCALE PHENOMENA IN FUEL CELL ELECTROCATALYSIS

### PARTICLE SIZE EFFECTS

The electrical, magnetic, optical, and catalytic properties of transition metal nanoparticles differ decisively from those of the bulk phase. The peculiar effects of nanoparticle size, shape, and electronic structure are of interest both from the perspective of fundamental research and applications in energy research (Roduner, 2006; Yacaman et al., 2001). Specific phenomena at particle sizes below ∼2 nm arise from the confinement of (quasi)-free electrons and the increasingly discrete nature of the electronic structure (Halperin, 1986). In electrocatalysis, the primary interests are to understand (i) the classical effects of atom arrangement and (ii) the heterogeneous electronic structure at the nanoparticle surface, controlling interfacial adsorption and charge transfer phenomena.

A key to achieving improvements in electrocatalytic activity in fuel cell electrodes lies in understanding how the structure of the catalyst determines the rates of relevant surface processes (Bayati et al., 2008; Eikerling et al., 2007a; Maillard et al., 2004; Mulla et al., 2006; Vielstich et al., 2003). On nanoparticles, the proportions of the different surface sites, located at facets, edges, and corners, and the surface electronic structure, are closely related to the particle size (Bradley, 2007; Mayrhofer et al., 2005; Mukerjee, 1990; Mukerjee and McBreen, 1998; Somorjai, 1994). These particle size effects exhibit varying trends for different reactions (Ahmadi et al., 1996;

Kinoshita, 1990; Lee et al., 2008; Lisiecki, 2005; Meulenkamp, 1998; Susut et al., 2008; Xiong et al., 2007). Oxygen reduction on Pt shows a maximum in mass-specific activity for a particle diameter of ~3 nm (Ross, 2003; Sattler and Ross, 1986); the mass activity for hydrogen reduction on Pd particles increases monotonically with decreasing Pd cluster size (Eikerling et al., 2003); the $CO_{ad}$ electro-oxidation rate on Pt nanoparticles decreases when the particle size decreases to <5 nm due a reduction in the surface diffusivity of $CO_{ad}$ (Andreaus and Eikerling, 2007; Andreaus et al., 2006; Maillard et al., 2004, 2007, 2005).

As a result of the heterogeneous surface morphology of nanoparticles and peculiar size effects on the electronic structure, the applicability of insights from studies at well-defined extended surfaces to small Pt nanoparticles or clusters is limited. In systems of supported catalyst nanoparticles, quantum confinement effects, irregular surface structure, and substrate effects have to be accounted for (Jiang et al., 2008; Lin et al., 2008; Lopez et al., 2004; Meier et al., 2002; Rupprechter, 2007). Claus and Hofmeister (1999) pointed out that the electrocatalytic activity of nanoparticles depends on the local environment of surface sites, namely, their local coordination geometry and electronic structure.

The nanocrystallite particles in Figure 3.4 are enclosed by (111) and (100) facets. Surface atoms on distinct facets, corner atoms, and edge atoms, possess varying coordination numbers. The proportion of undercoordinated corner and edge atoms increases strongly with decreasing particle sizes below 2–3 nm. This renders the nanoparticles physically unstable, causing the strong size dependence of particle dissolution rates.

Owing to the geometric and electronic heterogeneity of the particle surface, the activities of surface reaction processes exhibit significant spatial fluctuations at the atomistic scale. Another important influence on the electronic structure and surface processes is lattice contraction, arising from the undercoordination of surface atoms. It is a foremost challenge for theory and modeling to understand the relations between nanoparticle size and shape, on the one hand, and their stability and electrocatalytic activity, on the other.

## COHESIVE ENERGY OF PT NANOPARTICLES

Wang et al. (2009) studied the stability of Pt nanoparticles using *ab initio* calculations at the density functional theory level (DFT–GGA), with the Vienna *Ab Initio* Software Package (VASP) (Kresse and Furthmüller, 1996a,b; Kresse and Hafner, 1993, 1994a,b). The ionic cores were represented by *projected augmented waves* (PAW). The Kohn–Sham one-electron wave functions were expanded in a plane wave basis set with kinetic energy cut-off of 250 eV. Dividing the total energy for the optimized particle conformation, obtained through DFT calculations, by the atom number $N$, gives the *cohesive energy of Pt nanoparticles*:

$$E_{coh} = E_{tot}/N - E_{tot}^{gas}, \tag{3.27}$$

where $E_{tot}^{gas}$ is the total energy of one Pt atom in the gas phase.

For ideal spherical particles, the *Gibbs–Thompson equation* relates the Gibbs energy to the radius. The chemical potential per metal atom of a particle with radius $r_{Pt}$ is given by

$$\mu_{Pt}(r_{Pt}) = \mu_{Pt}^{b} + \frac{2\gamma_{Pt}}{n_{Pt}r_{Pt}}, \qquad (3.28)$$

with $\mu_{Pt}(r_{Pt}), \mu_{Pt}^{b} < 0$, where $\mu_{Pt}^{b}$ is the chemical potential, $\gamma_{Pt}$ the surface tension, and $n_{Pt}$ the atomic number density of the bulk metal (Kuntova et al., 2005; Wynblatt and Gjostein, 1976). Assuming for simplicity that the atomic number density and the surface tension are independent of particle size, Equation 3.28 can be transformed into

$$|\mu_{Pt}(N)| = \left|\mu_{Pt}^{b}\right| - 2\left(\frac{4\pi}{3}\right)^{1/3} \frac{\gamma}{n_{Pt}^{2/3}} N^{-1/3}. \qquad (3.29)$$

Entropy contributions to the Gibbs energy caused by vibrations of atoms in the nanoparticle are small and vary little with particle size. The relation $|E_{coh}(N)| \approx |\mu_{Pt}(N)|$ is expected to be valid in the range of particle sizes, in which the *spherical particle approximation holds*.

In a plot of $|E_{coh}|$ versus $N^{-1/3}$, the intersection with the ordinate (i.e., in the limit $N \to \infty$) corresponds to the bulk chemical potential of the metal $\left|\mu_{Pt}^{b}\right|$, and the slope gives $\gamma_{Pt}/n_{Pt}^{2/3}$. Lin et al. (2001) tested the relation between $|E_{coh}|$ and $N^{-1/3}$ for Pt nanoparticles with fewer than 25 atoms. Although their data could be represented reasonably well by a linear relation, these particle sizes are too small for a meaningful extrapolation to the bulk limit.

Figure 3.6a depicts results for $|E_{coh}|$ versus $N^{-1/3}$ from the study of Wang et al., for particles with up to 92 atoms and with shapes shown in Figure 3.4. A line according to the linear relation proposed in Equation 3.29 represents the data well. Extrapolation of this linear relation to $N \to \infty$ provides an estimate of the *bulk cohesive energy* $|E_{coh}(N \to \infty)| = 5.83$ eV, which is in excellent agreement with the experimental value for the cohesive energy of bulk Pt, 5.85 eV (Lide, 1990). The purported scaling of cohesive energy with the specific surface area extends to clusters with as few as nine atoms, corresponding to cluster sizes of $\sim$7 Å. Using the slope of the graph and the bulk density of Pt, $n_{Pt} = 6.62 \cdot 10^{22}$ cm$^{-3}$, an estimate of the *surface tension* was obtained, $\gamma_{Pt} = 3.8$ J m$^{-2}$, that agrees reasonably well with the experimental value for the surface tension of Pt *in vacuo*, $\gamma_{Pt} \approx 3.2$ J m$^{-2}$ (Linford, 1973). These findings suggest that metallic bonding, mediated by delocalized electrons, persists down to the sizes of these small metal clusters.

Figure 3.6a does not reveal any systematic trend of particle shape on $|E_{coh}|$. At a given $N$, variations in $|E_{coh}|$ for different shapes are <0.07 eV. This relative shape invariance indicates that for large particle ensembles, one should expect to find a statistical distribution of shapes; furthermore, it is to be expected that thermal effects could incur shape fluctuations in any given narrow interval of $N$ (Iijima and Ichihashi, 1986).

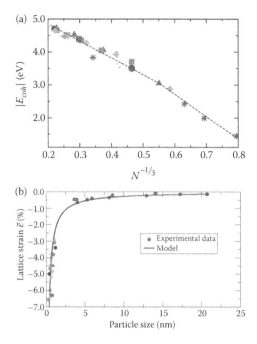

**FIGURE 3.6** (a) The absolute value of the cohesive energy $|E_{coh}|$ obtained from DFT-based optimization for Pt nanoparticles with up to 92 atoms, as a function of $N^{-1/3}$, where $N$ is the number of atoms per particle (Wang et al., 2009). Different symbols represent particle shapes in Figure 3.4a. As can be seen in this representation, the cohesive energy follows a linear relationship that is largely insensitive to the particle shape. The stars indicate results from Lin et al. (2001). (b) The average lattice contraction in nanoparticles (same symbols as in (a)), as a function of particle size, in comparison with experimental data (Wasserman and Vermaak, 1972) and a thermodynamic model of particle formation (Jiang et al., 2001). (Reprinted from Wang, L., Roudgar, A. and Eikerling, M. *J. Phys. Chem. C*, **113**(42), 17989–17996, 2009, Figures 1,2,3,4,5,9. Copyright (2009) by the American Physical Society. With permission.)

The variation of $|E_{coh}|$ with $N$ causes the drift in the statistical particle size distribution of Pt toward larger sizes, through particle dissolution and Pt ion redeposition, as explored in Rinaldo et al. (2010, 2012). The net effect is particle growth, known as Ostwald ripening. It is a kinetic phenomenon that is driven by the difference in surface energies between small and large particles. The dependence of the particle dissolution rate on particle radius is given by

$$k_{diss}(r, t) \propto \exp\left( \frac{\beta_{Ost} \gamma_{Pt} \bar{V}_{Pt}}{R_g T} \frac{1}{r_{Pt}} \right), \tag{3.30}$$

where $\beta_{Ost}$ is a phenomenological Ostwald coefficient, similar to a Brønsted coefficient. According to Equation 3.30, smaller particles are thermodynamically less stable and so dissolve at a higher rate. A corresponding expression with a negative argument of the exponential function could be written for the rate of Pt-ion redeposition, describing the growth of particles. Overall, larger particles will grow at the expense of small particles. The change in the slope of $|E_{coh}|$ versus $N^{-1/3}$, seen in Figure 3.6a at $N \approx 9$, suggests that the smallest clusters dissolve at a significantly enhanced rate.

The cohesive energy of small metal nanoclusters and nanoparticles is strongly dependent on the atom coordination number and interatomic distance (van Santen and Neurock, 2006). The deviation in interatomic distance from bulk values is quantified by the lattice contraction $\varepsilon = (a_{np} - a_{cryst})/a_{cryst}$, where $a_{np}$ and $a_{cryst}$ are the lattice constants of the nanoparticle and the extended crystal configuration, respectively (Mavrikakis et al., 1998; Rigsby et al., 2008). Both the *coordination number* and the *lattice contraction* in a nanoparticle exhibit fluctuations at the atomistic scale. In order to identify general trends of variations in particle size and shape, it is insightful to evaluate these effects using particle-averaged coordination number $\bar{z}$ and lattice contraction $\bar{\varepsilon}$.

Figure 3.6b shows experimental data of $\bar{\varepsilon}$ as a function of particle size in the range of 3–30 nm (Wasserman and Vermaak, 1972). The solid line is the lattice contraction obtained from a simple thermodynamic model that assumes the particles as spherical (Jiang et al., 2001). Predictions of the thermodynamic model are in very good agreement with experimental data from Wasserman and Vermaak (1972). In the range of particle sizes <3 nm, the lattice contraction strongly increases with decreasing sizes. Lattice strains obtained from DFT calculations of Wang et al. (2009) for particles with sizes <1.5 nm (same symbols as in Figure 3.4a) are in good agreement with the thermodynamic model.

The increasing undercoordination of surface atoms with decreasing particle size reduces the average binding strength of surface atoms. This effect competes with the impact of the lattice contraction, which enhances the average binding strength. Figure 3.7 shows $|E_{coh}|$ as a function of $\bar{z}$ and $\bar{\varepsilon}$:

$$|E_{coh}| = g\left(\bar{z}, \bar{\varepsilon}\right) \tag{3.31}$$

$|E_{coh}|$ increases with increasing $\bar{z}$ and decreasing $\bar{\varepsilon}$. The effect of undercoordination supersedes the effect of lattice contraction. The net effect of reducing particle size is a destabilization of the particle. For $-7\% < \bar{\varepsilon} \leq -6\%$, corresponding to a size range from 0.5 to 1 nm, a large irregular variation of $\bar{z}$, $5 < \bar{z} < 8$ is found. In this range, $|E_{coh}|$ is strongly dependent on the specific particle morphology. For $\bar{\varepsilon} > -6\%$, $|E_{coh}|$ decreases monotonically with decreasing $\bar{z}$.

Results discussed in this section reveal important trends in the stability of Pt nanoparticles. They identify the *surface tension* as a valid descriptor of nanoparticle stability. The surface tension must play an important role in the kinetic modeling of nanoparticle dissolution (Rinaldo et al., 2010, 2012). However, the main kinetic mechanisms that contribute to Pt nanoparticle dissolution proceed via formation and reduction of surface oxide intermediates at Pt. This well-founded observation suggests that stability studies, reported here for bare Pt nanoparticles evaluated *in vacuo*, should be expanded to Pt nanoparticles of varying surface oxidation state as well as conditions that mimic electrochemical conditions that the fuel cell catalyst is exposed to.

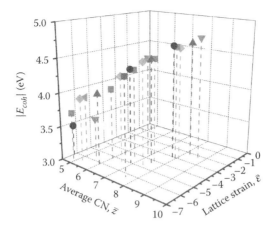

**FIGURE 3.7** Absolute value of the cohesive energy $|E_{coh}|$ obtained from DFT-based optimization for Pt nanoparticles with up to 92 atoms. It is function of the average coordination number $\bar{z}$ and average lattice strain $\bar{\varepsilon}$, for Pt nanoparticles in the size range of 1–2 nm (Wang et al., 2009). Different symbols represent particle shapes in Figure 3.4a. (Reprinted from Wang, L., Roudgar, A. and Eikerling, M. *J. Phys. Chem. C*, **113**(42), 17989–17996, 2009, Figures 1,2,3,4,5,9. Copyright (2009) by the American Physical Society. With permission.)

## ACTIVE AND INACTIVE SITES IN $CO_{ad}$ ELECTRO-OXIDATION

A complex interplay of elementary surface processes includes (i) molecular adsorption, (ii) surface diffusion, (iii) charge transfer, (iv) recombination of adsorbed species, and (v) desorption of reaction products. These processes determine observable rates of reactions in PEFCs, that is, reduction of oxygen and oxidation of hydrogen, methanol, or carbon dioxide.

All relevant electrochemical reactions in PEFCs exhibit peculiar sensitivities to the surface structure of the catalyst (Boudart, 1969). The abundances of the different surface sites, for example, edge sites, corner sites, or sites on crystalline facets, are related to the size of nanoparticles (Kinoshita, 1990). Support–particle interactions may alter the electronic structure of catalyst surface atoms at the rims with the support (Mukerjee, 2003). Moreover, the support may serve as a source or sink of reactants via the so-called spillover effect (Eikerling et al., 2003; Liu et al., 1999; Wang et al., 2010; Zhdanov and Kasemo, 2000).

In the following part, approaches in kinetic modeling that explore the links between particle size, surface heterogeneity, and electrocatalytic activity of supported Pt nanoparticles will be discussed. At atomistic resolution, the rates of electrocatalytic processes may differ significantly among different surface sites. A strong differentiation in the activity of catalyst surface atoms implies that only a fraction of sites constitute active sites. The remaining surface atoms should be deemed "inactive" (Solla-Gullón et al., 2006). This effect is most obvious for alloyed catalyst particles, when a second catalyst material is added to act as active sites, through the bifunctional mechanism (Watanabe and Motoo, 1975). For $CO_{ad}$ electro-oxidation on PtRu alloy catalysts, Ru atoms on the surface constitute the active sites. They promote the rate-determining step of $OH_{ad}$ formation through water splitting (Gasteiger et al., 1993b; Marković and Ross, 2002), whereas Pt surface atoms represent the inactive sites, acting as a reservoir for adsorbed $CO_{ad}$. The distinct components of the alloy, thus, promote different reactions (Liu and Nørskov, 2001; Marković and Ross, 2002).

*$CO_{ad}$ monolayer (ML) electrooxidation* on Pt nanoparticles has a long history as a prototype electrochemical reaction. Furthermore, irreversibly adsorbed $CO_{ad}$ is a catalyst poison in PEFCs. Numerous studies have revealed strong effects of particle size and morphology on the electrocatalytic activity of $CO_{ad}$ electro-oxidation (Arenz et al., 2005; Cherstiouk et al., 2003; Friedrich et al., 2000; Maillard et al., 2004; Mayrhofer et al., 2005; Solla-Gullón et al., 2006).

Controversy has evolved, in particular, around the question as to whether *$CO_{ad}$ surface mobility* could be a limiting process for the overall activity (Kobayashi et al., 2005; Koper et al., 2002; Lebedeva et al., 2002). Widely different values of $CO_{ad}$ surface mobilities have been determined, with relatively high values found on flat surfaces (Feibelman et al., 2001) and considerably smaller values found on surfaces of small nanoparticles (Ansermet, 1985; Becerra et al., 1993). Maillard et al. (2004) concluded that significant restrictions in $CO_{ad}$ surface diffusivity arise on the smallest Pt nanoparticles.

A *heterogeneous surface model* for $CO_{ad}$ electro-oxidation on Pt nanoparticles should incorporate the heterogeneous surface morphology and the limited mobility of $CO_{ad}$. The minimalist's modeling approach is to use a simple two-state model with a fraction $0 < \xi_{tot} \leq 1$ of electrocatalytically active sites as exclusive sites, at which $OH_{ad}$ can be formed by water splitting. A fraction $(1 - \xi_{tot})$ of inactive sites merely serves as a reservoir of $CO_{ad}$. Finite surface mobility of adsorbed $CO_{ad}$ is included, defining the time for $CO_{ad}$ on inactive sites to reach the active sites (Andreaus et al., 2006; Maillard et al., 2004).

To create a tractable kinetic model of $CO_{ad}$ electro-oxidation, the heterogeneous 3D nanoparticle surface is mapped onto a regular hexagonal 2D array with the two surface states, as shown in Figure 3.8. The total number of surface sites $N_s$ is fixed to match the total number of surface sites on the nanoparticle of a certain shape. For a cubo-octahedral particle, a particle diameter of 3 nm corresponds to $N_s = 397$. Further structure-defining factors of the model are the fraction of active sites $\xi_{tot}$, and the numbers of nearest neighbors (NN) of an active site with other active sites, $z_{aa}$, or inactive sites, $z_{an}$. These NN numbers allow the degree of active site clustering on the surface to be accounted for.

The considered reaction scheme of $CO_{ad}$ electro-oxidation follows the *Langmuir–Hinshelwood mechanism*. It distinguishes the reaction steps of CO adsorption, $OH_{ad}$ formation on active sites due to water splitting, surface mobility of $CO_{ad}$, and $COOH_{ad}$ formation and removal. The inclusion of finite surface mobility of adsorbed reactants to active sites leads to an interplay of reactant surface mobility and on-site reactivity. The state variables of the model are the surface coverage of $CO_{ad}$, $\theta_{CO}$, on inactive sites, the $CO_{ad}$-free fraction of active sites $\theta_\xi$, and the fraction of active sites covered by $OH_{ad}$, $\theta_{OH}$. In the mean-field approach, these coverages represent local averages, normalized to the disjoint surface fractions of active and inactive sites. The ranges of variation are $0 \leq \theta_{CO} \leq 1$, normalized to $(1 - \xi_{tot})$, as well as $0 \leq \theta_{OH} \leq 1$ and $0 \leq \theta_\xi \leq 1$, each normalized to $\xi_{tot}$. The model, moreover, involves a law of

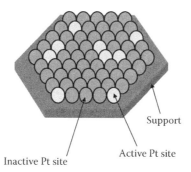

Support

Active Pt site

Inactive Pt site

**FIGURE 3.8** Mapping of a 3D nanoparticle model onto a 2D surface model with active and inactive sites. The active site concept is the integral part of the heterogeneous surface model of nanoparticle activity, explored here for $CO_{ad}$ electro-oxidation. Immobile $OH_{ad}$ adsorbs, preferentially, on specific sites called the active sites. The mobile $CO_{ad}$ is found on the remaining inactive catalyst sites. (Reprinted with permission from Andreaus, B.et al. 2006. Kinetic modeling of COad monolayer oxidation on carbon-supported platinum nanoparticles. *J. Phys. Chem. B*, **110**, 21028–21040, Figure 2,7,8,9, American Institute of Physics.)

$CO_{ad}$ removal from active sites (nucleation process), and it accounts for finite surface diffusion of $CO_{ad}$.

The balance equations for $\theta_\xi$, $\theta_{OH}$, and $\theta_{CO}$ were formulated and solved with two approaches: a mean-field model with nucleation processes on active sites and kinetic Monte Carlo simulations, as illustrated in Figure 3.9.

The calculated transient current is

$$j = e_0 \gamma_s \xi_{tot} \left( v_{ox}^{a-a} + v_{ox}^{a-n} + 2v_N + v_f - v_b \right), \tag{3.32}$$

where $v_{ox}^{a-a}$, $v_{ox}^{a-n}$ are oxidation rates (between $OH_{ad}$ on active and $CO_{ad}$ on active sites, or $OH_{ad}$ on active and $CO_{ad}$ on inactive sites), $v_N = k_N \left( 1 - \theta_\xi \right)$ is the nucleation rate, and $v_f$ and $v_b$ are forward and backreaction rates of $OH_{ad}$ formation. Figure 3.10 shows the simulated experiment.

It should be noted that the active site model reduces to the well-known homogeneous mean-field (MF) model for the limit of fast $CO_{ad}$ mobility and $\xi_{tot} = 1$, that is, with all surface sites being equally active (Bergelin et al., 1999; Koper et al., 1998; Petukhov, 1997). At the other end of the mobility scale, for vanishing $CO_{ad}$ mobility on a homogeneous surface (that is, for $\xi_{tot} = 1$), the active site model is equivalent to nucleation and growth (NG) models (Bewick et al., 1962; McCallum and Pletcher, 1976). Accordingly, the active site model represents a generalization of homogeneous surface models toward structured surfaces.

The general solution of the model can be obtained using kinetic Monte Carlo (kMC) simulations. This stochastic method has been successfully applied in the field of heterogeneous catalysis on nanosized catalyst particles (Zhdanov and Kasemo, 2000, 2003). It describes the temporal evolution of the system as a Markovian random walk through configuration space. This approach reflects the probabilistic nature of many-particle effects on the catalyst surface. Since these simulations permit atomistic

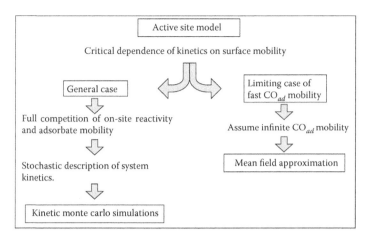

**FIGURE 3.9** Approaches used to find the solution of the active site model of surface activity. The main distinction is made based on surface mobility. For the general case, the full interplay between on-site reactivity and extremely low $CO_{ad}$ surface diffusivity unfolds. All processes, including nucleation of active sites (rate constant $k_N$), forward and reverse rates of $OH_{ad}$ formation ($k_f$, $k_b$), surface diffusion of $CO_{ad}$ ($k_{diff}$), and oxidative removal of $CO_{ad}$ ($k_{ox}$), are important for the overall kinetics. The solution for the general case, demands kinetic Monte Carlo simulations, where evolution of the system is described stochastically and positions of adsorbed $CO_{ad}$ and $OH_{ad}$ are relevant. Modeling is substantially simplified in the limit of fast $CO_{ad}$ mobility. If one assumes $CO_{ad}$ mobility to be infinitely fast, a mean field approximation can be implemented. In such a mean-field approach, positions of $CO_{ad}$ and $OH_{ad}$ are irrelevant. (Reprinted with permission from Andreaus, B.et al. 2006. Kinetic modeling of COad monolayer oxidation on carbon-supported platinum nanoparticles. *J. Phys. Chem. B*, **110**, 21028–21040, Figures 2,7,8,9, American Institute of Physics.)

resolution, any level of structural detail may easily be incorporated. Moreover, kMC simulations proceed in real time. The simulation of current transients or cyclic voltammograms is, thus, straightforward. In order to simulate chronoamperometric current transients, a variable time-step algorithm was implemented in Andreaus et al. (2006), as proposed by Gillespie (1976).

As a limiting case of the general active site model, an MF approximation with active sites in the limit of infinite $CO_{ad}$ surface diffusion was considered by Andreaus et al. (2006) and Andreaus and Eikerling (2007). The model evaluated in this limit accounts for the heterogeneous surface structure of the catalyst, but assumes uniform coverages of adsorbates on disjunct surface fractions of active and inactive sites. This simplification permits a straightforward deterministic formulation of the kinetic equations. It provides full analytical solutions in the case, when oxidative $OH_{ad}$ formation is fast in comparison with $CO_{ad}$ removal. Systematic fitting can be used with this model, which can provide starting values for more elaborate modeling of experimental data with the kMC approach. As discussed in Andreaus et al. (2006), correlation effects cause a mismatch between MF approximation and kMC solution, in the case of extensive active site clustering.

The active site model was solved for $CO_{ad}$ monolayer oxidation, and results agree with chronoamperometric current transients that had been measured at various

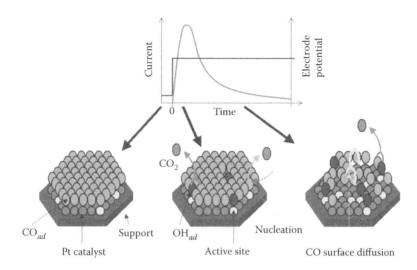

**FIGURE 3.10** Illustration of a typical transient experiment of $CO_{ad}$ electro-oxidation at nanoparticle catalysts. The current (grey) is measured in response to an applied voltage step at a catalyst surface initially fully covered with a monolayer of $CO_{ad}$. Snapshots at the bottom illustrate the evolution of the surface state in the 2D model of the heterogeneous particle surface. (Reprinted with permission from Andreaus, B.et al. 2006. Kinetic modeling of COad monolayer oxidation on carbon-supported platinum nanoparticles. *J. Phys. Chem. B*, **110**, 21028–21040, Figures 2,7,8,9, American Institute of Physics.)

particle sizes and electrode potentials (Andreaus et al., 2006). In each considered case, the active site model was first solved with the heterogeneous MF model that corresponds to the limit of infinite $CO_{ad}$ surface diffusion. This approach is sufficient for particles with sizes >5 nm. Moreover, the MF model gives reasonable fits for ~3.3 nm particles at small potentials <0.8 $V_{SHE}$.

However, the MF approach is insufficient for particles with ~3.3 nm size, if the potential exceeds 0.8 $V_{SHE}$, that is, in the potential range with rapid kinetics of surface reactions. The MF approach fails for particles with sizes in the range of 1.8 nm. In these cases, it is necessary to account for the finite surface diffusivity of $CO_{ad}$ and, thus, solve the active site model with the kinetic Monte Carlo simulation approach. Figure 3.11a shows typical results of current transients for particles with mean size of 3.3 nm that are matched closely with the model. Analysis of the data with the kinetic model allows important structural and dynamic parameters of the catalytic system to be extracted and analyzed.

In comparison with alternative surface models (homogeneous MF and NG approach), the active site model clearly shows a superior agreement with the experiment, as illustrated in Figure 3.12.

For small particles (1.8–3.3 nm), the fraction of active sites is ~10%. It increases only for larger particles with nanograined structure. These findings suggest that active sites are likely to be related to defect sites, rather than low-coordination sites of an idealized crystallite structure. All electrochemical steps, that is, oxidative $OH_{ad}$ formation, nucleation of active sites, and recombination of $OH_{ad}$ and $CO_{ad}$, can be

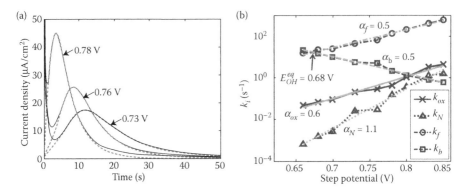

**FIGURE 3.11** (a) Fits (dashed grey lines) of experimental (thin black lines) current transients for CO electro-oxidation at Pt nanoparticles, with a size distribution centered around 3.3 nm at the indicated values of the electrode potential (versus SHE). (b) Tafel plots for kinetic rates obtained from the analysis of chronoamperometric transients at a range of electrode potentials. Extracted values of the equilibrium potential of oxidative $OH_{ad}$ formation from water $E_{OH}^{eq}$, as well as electron-transfer coefficients $\alpha_i$ for the considered electrochemical surface processes, are shown in the graphs. (Reprinted with permission from Andreaus, B.et al. 2006. Kinetic modeling of COad monolayer oxidation on carbon-supported platinum nanoparticles. *J. Phys. Chem. B*, **110**, 21028–21040, Figures 2,7,8,9, American Institute of Physics.)

described by Tafel laws, as shown in Figure 3.11b. Thereby, the equilibrium potential of oxidative $OH_{ad}$ formation from water, $E_{OH}^{eq}$, electron-transfer coefficients for the various steps, and generic rate constants have been determined. In general, it was found that the nucleation process is the limiting step, whereas the formation of $OH_{ad}$

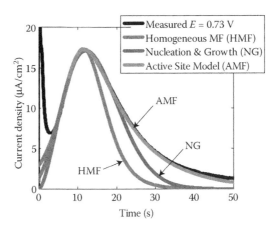

**FIGURE 3.12** A comparison of best-fits to experimental data (black solid line) at $E = 0.78$ V for 3.3 nm particles, using homogeneous MF, NG, and active site models. (Reprinted with permission from Andreaus, B.et al. 2006. Kinetic modeling of COad monolayer oxidation on carbon-supported platinum nanoparticles. *J. Phys. Chem. B*, **110**, 2102–21040, Figures 2,7,8,9, American Institute of Physics.)

is the fastest process. Most importantly, $CO_{ad}$ diffusivity was found to decrease by at least two decades with decreasing particle size, from $>10^{-14}$ cm$^2$ s$^{-1}$ for particles with diameter $>5$ nm to $\sim 10^{-16}$ cm$^2$ s$^{-1}$ for particles with diameter 1.8 nm.

The active site approach to the modeling of the electrocatalytic activity on heterogeneous surfaces of nanoparticles exhibits consistency with experimental data. It establishes vital links between particle size, surface morphology, and kinetic processes. Applied to systematic experimental data, the model provides useful diagnostic capabilities for analyzing the impact of surface structure on activity and extracting kinetic parameters of surface processes. Extension of the model to alloy PtRu nanoparticles is straightforward. In this case, active sites can be identified with Ru surface atoms. Therefore, the structural analysis of fitting results reveals information on the fraction and clustering of Ru surface atoms.

Future studies have to clarify the role of active sites and active site clustering as well as the mechanism of active site nucleation. In this regard, the model would benefit from electronic structure calculations for systems that mimic heterogeneous nanoparticle surfaces. These calculations could form the basis for model refinements in terms of surface structure representation, reaction pathways, surface mobility, and kinetics of charge transfer steps. Moreover, the role of adsorbate interactions and anion adsorption effects should be evaluated.

### Surface Heterogeneity for Oxide Formation at Pt Nanoparticles

Structural effects in nanoparticle electrocatalysis are complex. For extended surface catalysts surface the activity toward electrochemical reactions can be rationalized using a small set of intuitive activity descriptors (Greeley et al., 2009; Nørskov et al., 2004; Suntivich et al., 2011). In the case of the ORR at metal-based catalysts, the adsorption energy of atomic oxygen, $\Delta E_O$, has been singled out as the primary *activity descriptor*. For d-band metals, the elegant d-band model of Hammer and Nørskov established a fundamental linear relation between adsorption energies of reaction intermediates and the energy of the d-band center of the metal, a property readily obtained from electronic structure calculations (Hammer and Norskov, 1995; Hammer and Nørskov, 2000). The validity of this descriptor has been demonstrated in catalyst screening studies that have predicted most active material combinations and compositions of Pt–based metal alloys (Greeley et al., 2009; Stamenkovic et al., 2007a; Stephens et al., 2011).

The surface heterogeneity of nanoparticles calls for detailed spatial maps of the energetics and kinetics of elementary surface processes. Han et al. (2008) studied O and OH adsorption energies on cubo-octahedral Pt nanoparticles, with sizes of 1 and 2 nm using DFT. For both adsorbents, they found that adsorption energies change dramatically from the surface of Pt(111) to the surface of nanoparticles. The highest adsorption energies were found at undercoordinated edge atoms. However, this study did not provide a separate discussion of Pt–Pt bond distortion and particle relaxation effects.

Wang et al. performed a similar study, in which they calculated the detailed map of oxygen adsorption energies at hemispherical cubo-octahedral nanoparticles with

(111) basal planes (Wang et al., 2009). These particles provide the highest variability of coordination structures of surface sites, compared to other particle shapes in Figure 3.4. Two sizes with 37 ($\sim$1 nm) and 92 atoms ($\sim$1.5 nm) were considered. The potential energy associated with dissociative chemisorption of atomic oxygen was used to probe the heterogeneity of the surface electronic structure. Two covalent bonds can be formed between the two unpaired valence electrons in an oxygen atom and metal d-electrons. Variations of the nearest-neighbor structure of distinct adsorption sites on the metal create the well-known selectivity of O adsorption among fcc, hcp, bridge, and top sites on Pt(111) and among on-top, bridge, and fourfold hollow sites on Pt(100) (Eichler et al., 2000; Feibelman, 1997; Jacob et al., 2004).

The topological heterogeneity of nanoparticle surfaces amplifies this site-selectivity. In order to distinguish between the effects of electronic structure reorganization and nanoparticle relaxation on the adsorption energy, calculations of oxygen adsorption were performed in two steps. In the first step, positions of Pt atoms were fixed at the optimum configuration of the relaxed bare particle. The oxygen atom was positioned above a well-defined adsorption site on the particle. The lateral position was fixed, while the optimum position of the oxygen atom in the direction perpendicular to the surface was found by DFT-based optimization. This procedure was repeated for all relevant adsorption sites on each of the distinct surface facets: on-top, bridge, and fourfold hollow sites on the Pt(100) side facet; and on-top, bridge, hpc, and fcc sites on Pt(111) side and top facets. From these optimizations studied, contour plots of the potential energy of oxygen adsorption for the unrelaxed particles were generated.

In the second step, full optimizations were performed with the relaxation of all atoms in the system, where the oxygen atom was again positioned initially at all high-symmetry surface sites. Thereby, the contribution of particle relaxation to the adsorption energy could be quantified. The potential energy of adsorption after either of these steps was calculated with respect to the energy of the free $O_2$ molecule:

$$\Delta E_O = E_{PtO} - E_{Pt} - \frac{1}{2}E_{O_2}. \tag{3.33}$$

The total energy of $O_2$ was determined as $-9.975$ eV, using thermochemical data (Rossmeisl et al., 2005).

The 3D map of $\Delta E_O$ for a Pt nanoparticle with 92 atoms at fixed atomic configurations is shown in Figure 3.13. The complex topological distribution of adsorption energies reflects the underlying periodic arrangement of surface atoms as well as electronic effects resulting from nanoscale confinement. The highest absolute values of adsorption energies occur at the edges. Oxygen adsorption at Pt37 (not shown) is stronger compared to Pt92, as expected.

From studies at extended surfaces, it is well known that undercoordination of surface atoms induces a narrowing of the d-band, causing an increase in the adsorption energy. A lattice contraction, on the other hand, induces a broadening of the d-band, which causes a decrease in the adsorption energy. Both of these effects compete on nanoparticle surfaces. As stated previously, in the case of nanoparticles, the effect of surface-atom undercoordination supersedes the lattice strain effect.

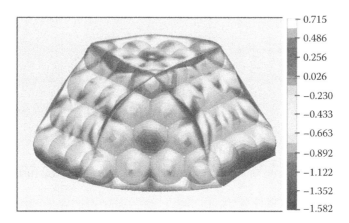

0.715
0.486
0.256
0.026
−0.230
−0.433
−0.663
−0.892
−1.122
−1.352
−1.582

**FIGURE 3.13** A 3D map of the adsorption energy of atomic oxygen (in eV) at the surface of a hemispherical cubo-octahedral Pt nanoparticles with 92 atoms.

In order to compare oxygen adsorption energies at unrelaxed nanoparticles with the values for an unrelaxed, extended Pt(111) surface, DFT-GGA-based calculations were performed for oxygen atomic adsorption on a $3 \times 3$ unit cell of a Pt(111) slab with four layers of Pt at the oxygen coverage of 0.1 ML. The surface Brillouin zone was sampled by a Monkhorst–Pack **k**-point set of $3 \times 3 \times 1$. Figure 3.14 shows the normalized site density distribution of geometrically equivalent adsorption sites versus the potential energies of oxygen adsorption for Pt92(111). The dashed lines correspond to adsorption at the same sites on an unrelaxed extended Pt(111) surface.

*Nanoscale confinement* causes a pronounced shifting and broadening of the site density distribution of adsorption energies. The preferred adsorption sites change from fcc on extended Pt(111) to bridge and hcp on Pt92(111). The shift in adsorption energies between nanoparticle and extended surface is particularly striking for the fcc site. This becomes a less favorable site for adsorption on the nanoparticle. However, for a number of bridge, hcp, and on-top sites, nanoscale effects lead to significantly stronger adsorption, that is, enhanced $|\Delta E_O|$, in comparison to the corresponding site on the extended surface. The maximal nanoscale enhancements of $|\Delta E_O|$ are 0.62, 0.33, and 1.11 eV for bridge, hcp, and on-top sites. Bridge sites become the most favorable sites for adsorption. The density distribution is expected to become narrower with increasing particle size, approaching the limiting values for the analogous sites on an extended surface.

To evaluate whether trends in chemisorption energies on Pt nanoparticles are consistent with the d-band model, d-band densities of states were projected out for different adsorption sites to determine the corresponding d-band centers relative to the Fermi level. No correlation was observed between the adsorption energies and site-specific d-band centers. Even though metal nanoparticles possess a continuous electronic band structure and, thus, metal-like electronic properties, their catalytic surface properties are not controlled by band structure effects but by the local electronic structure of the adsorption sites. An important conclusion from this study is that

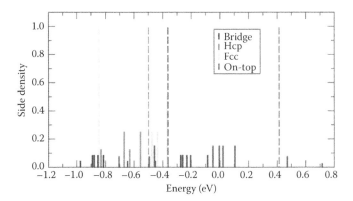

**FIGURE 3.14** Normalized density distribution of surface sites with distinct geometry overpotential energies of adsorption for Pt92(111) facets. The dotted lines correspond to adsorption on the corresponding sites of an extended Pt(111) surface. (Reprinted from Wang, L., Roudgar, A. and Eikerling, M. *J. Phys. Chem. C*, **113**(42), 17989–17996, 2009, Figures 1,2,3,4,5,9. Copyright (2009) by the American Chemical Society. With permission.)

catalytic properties of metal nanoparticles are not predictable from the properties of the corresponding bulk material.

Full relaxation of Pt particles and extended surfaces in the presence of the adsorbed oxygen atom, allowed the maximum oxygen adsorption energy for every facet and the contribution of particle relaxation to be determined. For Pt37, the most favorable adsorption site is at the top corner of the Pt(100) facet. For Pt92, it is at the bottom corner of the Pt(111) side facet. The net effect of particle relaxation is smaller for Pt37 than Pt92 because of the larger number of atoms that are involved in the geometry relaxation of Pt92. Owing to particle relaxation, $|\Delta E_O|$ on Pt(100) facet, side Pt(111) facet, and top Pt(111) facet of the Pt37 particle increased by 0.62, 0.52, and 0.14 eV, respectively. For the Pt92 particle, these increases are 0.28, 0.74, and 0.41 eV, respectively.

## ELECTROCATALYSIS OF THE OXYGEN REDUCTION REACTION AT PLATINUM

Oxygen reduction and evolution reactions are crucial for energy conversion in biological and electrochemical systems (Koper and Heering, 2010). The ability of Pt to catalyze them is unsurpassed, albeit, only incompletely understood. Accordingly, *Pt electrochemistry and electrocatalysis* is a topic of foremost importance in fundamental electrochemistry and applied electrochemical materials science. Platinum remains the benchmark material for any new catalyst material. The key property that decides over the success or failure of catalyst layer and fuel cell design is the mass-specific electrocatalytic activity, given in units of A $mg_{Pt}^{-1}$.

An overarching requirement for Pt-based catalysts is that they must strike the optimal balance between performance and durability. Deconvoluting the mechanism of *platinum oxide formation and reduction* is the key challenge in understanding basic properties of Pt that determine both its oxygen-reducing capability along with its propensity for dissolution. The dependence on rapid oxide formation and reduction seems to lead any catalyst material with high ORR activity to invariably exhibit high Pt dissolution activity. Hence, it becomes a precarious task to identify materials, electrode designs, and operating conditions that enhance the ORR activity and fulfill fuel cell requirements in terms of durability and lifetime. From this point of view, it is obvious why oxide formation and reduction at Pt have attracted the attention of generations of researchers since the beginnings of fuel cell electrochemistry (Alsabet et al., 2006; Angerstein-Kozlowska et al., 1973; Birss et al., 1993; Clavilier et al., 1991; Conway and Jerkiewicz, 1992; Conway, 1995; Conway and Gottesfeld, 1973; Conway et al., 1990; Damjanovic and Yeh, 1979; Damjanovic et al., 1980; Farebrother et al., 1991; Feldberg et al., 1963; Gilroy, 1976; Gilroy and Conway, 1968; Harrington, 1997; Harris and Damjanovic, 1975; Heyd and Harrington, 1992; Jerkiewicz et al., 2004; Katsounaros et al., 2012; Markovic et al., 1997, 1999; Sun et al., 153; van der Geest et al., 1997; Vetter and Schultze, 1972a,b; Wakisaka et al., 2010; Wang et al., 2006; Ward et al., 1976; Yamamoto et al., 1979; Yeager et al., 1978; Zeitler et al., 1997; Zolfaghari and Jerkiewicz, 1999; Zolfaghari et al., 1997).

The ORR is notoriously known as a complex reaction. An oxygen molecule needs to be associated with four electrons and four protons:

$$O_2 + 4H^+ + 4e^- \rightleftarrows 2H_2O. \tag{3.34}$$

In the process, electrons lose potential energy by occupying energy levels in the newly formed water molecules. The total Gibbs energy change for all four electrons is $-4.92$ eV, which corresponds to $1.23$ eV per electron.

The ORR proceeds at the catalyst surface, where electrons are readily available, at a concentration that is determined by the electronic density of states of the metal. Protons are supplied from the electrolyte, with a concentration determined by the composition of the electrolyte and by the distribution of the electrolyte potential. The ORR involves, at least, three surface-adsorbed intermediate species. They are, surface oxide, $O_{ad}$, hydroxide, $OH_{ad}$, and superoxide, $OOH_{ad}$. Surface processes that transform these species into one another and ultimately into water involve kinetic barriers that determine the net rate of the overall reaction. In the electrode potential range of interest, from 0.6–1.0 V versus SHE at the cathode, the formation of surface oxides as intermediates of the ORR interferes with the formation of surface oxides from the splitting of water.

Thanks to a plethora of efforts in theory and experiment, the ORR has lost some of its enigmatic appearance. Especially, electronic structure calculations at the level of DFT have brought about tremendous advances in understanding of surface electrochemical processes at metallic catalysts. In this section, the following questions will be explored: Does a cohesive picture of the ORR pathway and mechanisms exist? Is the significance of electronic structure effects, formation of surface-adsorbed reaction intermediates, and kinetic limitations for the overall process understood?

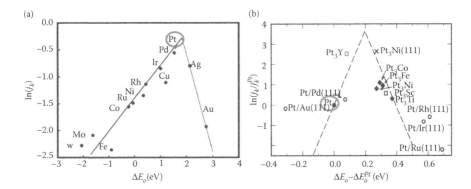

**FIGURE 3.15** Trends in ORR activity plotted as a function of the oxygen adsorption energy, $\Delta E_O$ for (a) a range of elemental transition metal catalysts (Nørskov et al., 2004) and (b) transition metal alloys. (a) is reprinted with permission from Nørskov, J. K. et al. *J. Phys. Chem. B.*, **108**(46), 17886–17892, 2004. Copyright (2004) American Physical Society. Part (b) is reprinted from Greeley, J. et al. *Nat. Chem.*, **1**(7), 552–556, 2009, Figure 1, by permission from Macmillan Publishers Ltd., Copyright (2009).)

### SABATIER-VOLCANO PRINCIPLE

Since the ORR is a surface reaction that involves adsorbed intermediates, it is subdued to the Sabatier-volcano principle of heterogeneous catalysis (Balandin, 1969; Sabatier, 1920). This principle states that in order for a catalyzed surface reaction to proceed, some bonding of adsorbed intermediate(s) is necessary, whereas a bonding that is too strong will block the surface and slow down the reaction. Several examples of reactions in heterogeneous catalysis and electrochemistry that obey these principles are discussed by Parsons (2011).

As illustrated in Figure 3.15, a plot of the logarithm of exchange current density versus standard free energy of adsorption of the intermediate results in a volcano-shaped curve. It exhibits a linearly decreasing flank toward more negative adsorption energy, corresponding to strong adsorbate bonding and slow rate of desorption, and a linearly decreasing flank toward more positive adsorption energy, corresponding to weak adsorbate bonding and small rate of adsorbate formation. The implicit message is that for any catalytic system, an optimal value of the adsorption energy exists, maximizing the rate of the reaction. It is a challenge for materials design to reach the peak of the volcano curve. The implication for the theoretical studies of catalytic surface processes is that the standard free energy of adsorption is a valid descriptor of the surface activity.

In recent years, the Sabatier-volcano principle and the theoretical methodology based thereon have been developed into a practical approach of catalyst screening. Adsorption energies of reaction intermediates can be readily calculated with DFT methods. These calculations allow predictions on the reactivity to be made, which can be compared with experimental activity studies. Several studies have demonstrated an agreement between predicted volcano curves and measured surface reactivities (Greeley et al., 2009; Nørskov et al., 2004).

For transition metals, Nørskov and coworkers have established the d-band center of the electronic band structure of the catalyst (relative to the position of the Fermi level) as a descriptor of catalytic activity. Based on their d-band model, they demonstrated that a linear relation exists between oxygen adsorption energy and the d-band center, a lumped parameter representing electronic structure effects on adsorbate binding (Hammer and Nørskov, 2000). A convincing indication of success of the d-band model is the extensive exploration of a range of Pt-based transition metal alloys, illustrated in Figure 3.15. The volcano plot in Figure 3.15a reveals why pure Pt is the best choice for the ORR among elemental catalysts: it lies close to the peak of the volcano curve; however, being slightly left of the peak, it binds oxygen species too strongly by about 100 meV (Nørskov et al., 2004). For Pt, Pd, or even more so, Ni or Co, formation and desorption of the reduction product will be rate-determining reaction steps. Au and Ag, on the other hand, are found on the other side of the volcano peak because of the high activation energy for the initial step of formation of oxygen intermediates.

Structural modification or alloying of Pt with other transition metals could be used to tune the strength of oxide binding at metal surfaces. The main underlying physical phenomena are (i) variation in local atomic configuration (and electronic structure) of the adsorption site, a geometric effect, and (ii) modulation of the electronic structure of Pt related to lattice strain. As for the geometric effect, an increase in the fraction of undercoordinated surface atoms, located at edges and corners of Pt nanoparticles, incurs a net increase in oxygen adsorption energy with decreasing particle size, as seen in Wang et al. (2009). It was shown for studies at extended surfaces that an undercoordinated surface atom experiences an upshift of the center of the calculated d-projected density states. This shift could increase oxygen chemisorption energies by values of 1 eV.

Lattice strain effects have been studied systematically for model systems of several atomic layers of Pt deposited pseudomorphically on another metal or metal oxide. The lattice strain exerted by the substrate modifies the lattice constant in the Pt layer. An increase of the lattice constant in a strained Pt layer, decreases the width of the d-band. This band-narrowing causes an upshift of the d-band center, and as the d-band center is shifted toward the Fermi-level, the strength of oxygen binding increases. The same reasoning applies for hydrogen binding. An additional ligand effect exists as a result of hybridization of d-states of surface Pt atoms with atoms of the second metal in subsurface layers. In fact, in Pt monolayer catalysts (Zhang et al., 2005) and bulk Pt alloy catalyst, such as $Pt_3Ni$ (Stamenkovic et al., 2007a,b), lattice strain and ligand effects are inseparable (Stephens et al., 2011).

To evaluate and exploit predictions of the d-band model in catalyst design, oxygen adsorption energies were calculated by DFT for thin overlayer structures ("skins") of Pt on a series of transition metals to form alloys of composition $Pt_3X$. Figure 3.15b shows the corresponding volcano plot. For $Pt_3Y$ or $Pt_3Ni$, oxygen binding energies are weaker than that for elemental Pt. The effect of Pt deposition on transition metals to the left of Pt in the periodic table leads to a downshift of the d-band center of Pt. This modification causes weaker oxygen binding, thereby, explaining the increase in ORR activity for these materials (Greeley et al., 2009; Nørskov et al., 2004).

However, the correlation between $\ln j_k$ and $\Delta E_O$, in Figure 3.15b, is not as clear-cut as that in Figure 3.15a. Furthermore, the slope on the adsorption branch (right) is different between the two plots, and could not be explained based on any existing theory. In general, other effects determining activity beyond the correlation with d-band center, or $\Delta E_O$, must be accounted for in studies of thin film structures of Pt on another metal or metal oxide. The mismatch in electronic work functions at the contact between these materials leads to a redistribution of electronic charge, affecting charging properties at the Pt-solution interface. These effects are usually not accounted for in DFT-based screening studies. It was, however, evaluated in a combined experimental and DFT-based modeling study that explored activity trends in thin layers of Pt deposited on niobium oxides, $Nb_xO_y$, with $y/x = 0, 1, 2, 2.5$ (Zhang et al., 2010). In this case, the experimental ORR activity does not follow the trend predicted by the d-band model, indicating that interfacial charging effects could play a significant role.

An obvious question can be asked. What is the reason for the success of a catalyst screening strategy for the ORR that relies on a single activity descriptor, specifically, the oxygen chemisorption energy (or the d-band center)? As explained above, the ORR involves three adsorbed oxygen intermediate species. Initially, this suggests that computational screening studies should explore a three-dimensional space, spanned by the three corresponding adsorption energies, to find the material with the highest ORR activity. A higher-dimensional search space should enhance the chances of discovering more active catalyst materials. However, as it turns out for transition metal-based catalyst materials, the number of phenomenological descriptors of ORR activity can be reduced to a single one, the adsorption energy for $O_{ad}$. This point will be returned to after a discourse on experimental studies of ORR mechanism and intermediates.

### EXPERIMENTAL OBSERVATIONS

To obtain a detailed microscopic understanding of structure and processes at interfaces, one needs a "microscope" with sufficient resolution. *Cyclic voltammetry* is the metaphorical microscope of surface electrochemistry (Gileadi, 2011). The voltage scan rate, viz. the linear rate of potential change (in $V\ s^{-1}$), is its *resolution* and the considered potential window is the *focus*. A cyclic voltammogram (CV) of a Pt(111) surface, in contact with a nonadsorbing aqueous electrolyte, is shown in Figure 3.16.

Features in Figure 3.16 that correspond to *Pt oxide formation and reduction* are labeled and explained in the pictorial legend. It should be noted for the illustrated CV that the surface processes do not involve any supplied reactants. They correspond to the transformation of interfacial water molecules into surface species and vice versa. A CV shows the amount of electronic charge withdrawn from the metal surface in anodic scan direction (oxidative current, $j > 0$) or transported to it in cathodic scan direction (reductive current, $j < 0$). The electrical charge flux generated or consumed at the interface is the result of double layer charging and faradaic processes. An interesting aspect of cyclic voltammetry is that the surface is never in a steady state. The potential is continuously ramped up and down, at a constant scan rate, $v_s$, and between precisely controlled upper and lower potential bounds.

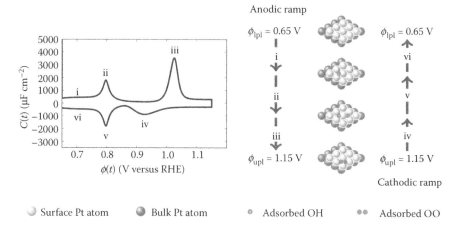

**FIGURE 3.16** A schematic cyclic voltammogram of bulk Pt catalyst with Pt(111) surface structure in nonadsorbing acidic electrolyte for low scan rates electrolyte. It shows the interfacial current density, divided by the scan rate, on the ordinate ($C(t) = j(t)/v_s$), as a function of the electrode potential on the abscissa. The pictorial legend on the right illustrates various oxidation states that the Pt surface passes through upon potential cycling. (Reprinted from *Electrocatalysis*, Mechanistic principles of platinum oxide formation and reduction, 2014, 1–11, Rinaldo et al. Copyright (2014) Springer. With permission.)

Peaks in Figure 3.16 represent various oxidation states that the Pt surface passes through upon potential cycling. As the electrode potential increases in the relevant range for the ORR, which lies between 0.6–1.0 V versus SHE, oxidation or discharge of surface water molecules $H_2O_{ad}$ occurs, leading to the formation of surface oxide species. The onset potential for surface oxide formation lies in the range of 0.6–0.8 V versus SHE. It shifts to smaller values at nanoparticle surfaces because of the stronger binding of oxygen intermediate species in comparison to extended surfaces. Mirrored symmetric peaks i and vi as well as ii and v give the so-called butterfly feature commonly observed in the potential range of 0.7–0.8 V. The diagram illustrates that these peaks arise as a result of sequential steps in water oxidation, involving the formation of $OH_{ad}$ and further transformation into $O_{ad}$ at specific surface sites. Nonsymmetric peaks iii and iv are indicative of an irreversible (meaning kinetically hindered) process. Figure 3.16 suggests that the formation of OO is the source of this irreversibility.

At the upper range of potentials displayed in Figure 3.16, that is, in the range of 1.1 V versus SHE, strong electric fields at the Pt surface transform the surface oxide layer into a quasi-3D lattice by way of so-called place exchange mechanism (Jerkiewicz et al., 2004). The place-exchanged PtO layer comprises a staggered configuration of $Pt^{2+}$ and $O^{2-}$. A peculiarity of Pt is the thickness of the place-exchanged layer: it grows to a thickness of two atomic layers of Pt.

The interpretation of CV data is ambiguous. However, if complemented with data from surface microscopy, x-ray absorption, Auger or photoelectron spectroscopy, and mass-sensitive techniques using electrochemical quartz crystal nanobalance (EQCN), definitive conclusions about abundances of surface adsorbates and processes can be drawn. Moreover, insights from DFT studies and kinetic modeling can be used to furnish reaction pathways and mechanisms involved in oxide formation and reduction.

In order to identify and quantify surface species, Wakisaka et al. (2010) used x-ray photoelectron spectroscopy (XPS) in combination with an electrochemical cell to apply a potential. The electrochemical XPS (EC-XPS) approach provides surface sensitivity and energy resolution of O 1s spectra. Recorded spectra were deconvoluted into four distinct surface species which could be fingerprinted by the value of their binding energy: Pt-$O_{ad}$ (529.6 eV), Pt-$OH_{ad}$ (530.5 eV), Pt-$H_2O_{ad,1}$ (531.1 eV), and Pt-$H_2O_{ad,1}$ (532.6 eV). XPS signal analysis was performed for varying potential and different catalyst materials and structures. Results of this analysis for Pt(111) and polycrystalline Pt are shown in Figure 3.19 in support of a kinetic model of Pt oxide formation and reduction.

## MODELING PT OXIDE FORMATION AND REDUCTION

In aqueous electrolytes, Pt surface oxides could be formed by oxidation of interfacial water or by reduction of oxygen molecules. Formation and reduction of adsorbed surface oxygen intermediates depend on the surface structure of the catalyst, the structure of interfacial water, electrolyte ion concentrations, pH, and oxygen concentration. Interfacial water at an ordered Pt(111) surface is believed to exhibit the structure of hexagonal ice (Ih). It can be characterized by the strength of hydrogen bonds and by the preferential orientation of water dipoles in the near-surface region. Toward more positive electrode potential, the preferential orientation of water dipoles should be pointing away from the interface ($O^{-2\delta}$ closer to surface) and vice versa. Oriented water molecules at the interface could make a significant contribution to the surface potential and, thus, the Helmholtz capacitance at the interface (Schmickler and Santos, 2010).

Positions, heights, and widths of the peaks in Figure 3.16 are functions of the catalyst surface structure. Depending on the degree and type of crystalline ordering, the surface offers a number of distinct adsorption sites, distinguished by the adsorption energies. Single-crystal surfaces offer a small number of well-defined surface sites. Disordered surfaces or surfaces of nanoparticles possess larger fractions of undercoordinated surface sites forming stronger bonds with adsorbed oxide species.

The challenge involved in finding the correct oxidation state of a metal as a function of potential, is illustrated in Figure 3.17. The metal phase potential, $\phi_M$, determines the *surface oxidation state* and the charge distribution at the catalyst surface, including surface charge density, $\sigma_M$, surface dipole moment, and higher moments. These moments determine the solution phase potential, $\phi_{el}$, on the electrolyte side of the interfacial region. This potential, in turn, determines the concentrations of ions in the electrolyte.

Obtaining the solution for these coupled functions represents a self-consistency problem, the subject of research in an emerging scientific field of *first-principles electrochemistry*. The main objective is to employ *ab initio* methods for the modeling of

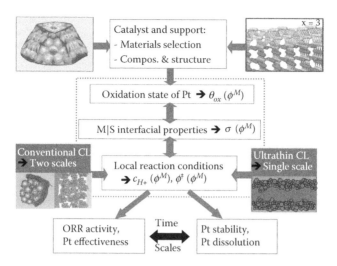

**FIGURE 3.17** The self-consistency problem in Pt electrocatalysis. The metal phase potential determines oxidation state and charging properties at the catalyst surface. These properties in turn determine the local reaction conditions at the Helmholtz or reaction plane. At this point, structural design and transport properties of the catalyst layer come into play (as illustrated for conventional and ultrathin catalyst layers). Newly developed methods in the emerging field of first-principles electrochemistry attempt to find self-consistent solutions for this coupled problem.

electrochemical systems. However, so far, none of the developed approaches is able to capture the full coupling between $\phi_M$ and the functions depending on it (Janik et al., 2008; Jinnouchi and Anderson, 2008; Taylor et al., 2006). Existing approaches do not properly account for interfacial water and surface oxidation state. They fail to reproduce the pH dependence or, in the simplest case, control $\sigma_M$ instead of $\phi_M$. This is easier to do in computations but leads to uncontrollable results. A recently developed methodology that employs a generalized computational hydrogen electrode strives to overcome these difficulties (Rossmeisl et al., 2013).

A potentiodynamic kinetic model of Pt oxide formation that is consistent with CV data over a wide range of scan rates and for different catalyst surface structures is an important step in understanding Pt oxide formation and reduction. Evaluation of the model against a range of electrochemical and spectroscopic data and comparison with theoretical calculations of reaction pathways and energetics could help furnishing details of reaction mechanisms.

Upon increasing the electrode potential in the anodic direction, the first step of the reaction sequence considered in Rinaldo et al. (2014) involves adsorption of hydroxide from interfacial water on two distinct adsorption sites $A$ and $B$:

$$Pt_A + HOH_{aq} \rightleftharpoons Pt_A OH + H^+ + e^- \quad \text{(step 1a)}, \tag{3.35}$$

and

$$Pt_B + HOH_{aq} \rightleftharpoons Pt_B OH + H^+ + e^- \quad \text{(step 1b)}. \tag{3.36}$$

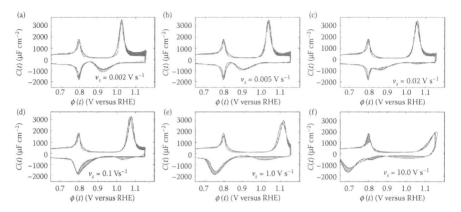

**FIGURE 3.18** A comparison of the proposed model of *surface oxide formation and reduction at Pt(111)* in nonadsorbing acidic electrolyte with CV data of Gomez-Marin et al. (2013). The scan rate varies by four orders of magnitude from (a) to (f), as indicated in the plots. (Reprinted from *Electrocatalysis*, Mechanistic principles of platinum oxide formation and reduction, 2014, 1–11, Rinaldo et al. Copyright (2014) Springer. With permission.)

Here, $Pt_A$ and $Pt_B$ denote free A and B sites. $Pt_AOH$ and $Pt_BOH$ originate from partial oxidation of hexagonal (Ih) water (Clay et al., 2004; Iwasita and Xia, 1996; Nie et al., 2010; Ogasawara et al., 2002; Su et al., 1998).

At higher potentials, in the range from approximately 0.9 to 1.1 V, adsorbed hydroxides $Pt_AOH$ and $Pt_BOH$ are oxidized further via

$$Pt_AOH \rightleftharpoons Pt_AO + H^+ + e^- \quad \text{(step 2a)},\qquad(3.37)$$

and

$$Pt_BOH \rightleftharpoons Pt_BO + H^+ + e^- \quad \text{(step 2b)}.\qquad(3.38)$$

Association of $Pt_AO$ and $Pt_BO$ leads to the formation of $Pt_COO$ on a third site C,

$$Pt_AO + Pt_BO \rightarrow Pt_COO \quad \text{(step 1a)}.\qquad(3.39)$$

This is the last step in the oxidation sequence considered. The set of kinetic equations for this reaction sequence (including reduction processes) was formulated, parameterized, and solved numerically. By comparison to CV data over a wide range of scan rates, a consistent set of kinetic parameters was obtained. Results of the comparison with experimental CVs are shown in Figure 3.18.

The solution space of the kinetic model includes the electrical current as a function of the transient potential, normalized to the scan rate, in addition to coverages of different surface oxide species considered. Using parameters obtained from the fits of CV scans, steady-state coverages as a function of potential in the limit of $v_s \rightarrow 0$ can be evaluated. Figure 3.19 compares these coverages with coverage values obtained by analysis of EC-XPS data (Wakisaka et al., 2010). Among parameters obtained

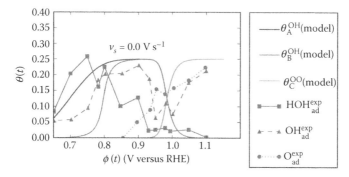

**FIGURE 3.19** Species coverage in the oxide formation and reduction model in the limit of zero scan rate (potentiostatic) compared with EC-XPS data from Wakisaka et al. (2010). (Reprinted from *Electrocatalysis*, Mechanistic principles of platinum oxide formation and reduction, 2014, 1–11, Rinaldo et al. Copyright (2014) Springer. With permission.)

from fits to CV data are values of adsorption energies of different oxide species at distinct adsorption sites. Adsorption energy differences, obtained from the CV fitting, agree well with adsorption energy differences that were calculated in a DFT study of oxygen adsorption at Pt(111) (Wang et al., 2009). According to this comparison, the sites for initial Pt-OH formation correspond to FCC and HCP sites, respectively, with the FCC site having the more negative adsorption energy. A similar analysis of Pt oxide formation was performed for polycrystalline Pt, where a consistent set of model parameters was also obtained.

## ASSOCIATIVE MECHANISM OF THE ORR

The ORR proceeds in four or five steps. Four of these elementary steps involve proton and electron transfer. The main *ORR mechanisms* are known as dissociative and associative mechanism. In the *dissociative mechanism*, the oxygen molecule first adsorbs onto the metal and then dissociates by the breaking of the O–O bond. Dissociation of $O_2$ is followed by the transfer of two electrons and two protons to form two adsorbed $OH_{ad}$. Another coupled proton–electron transfer process transforms each of the $OH_{ad}$ into water. However, DFT studies have shown that direct dissociation of $O_2$ has an activation barrier of >0.5 eV, rendering this process an unlikely reaction step (Hyman and Medlin, 2005).

The subsequent treatment focuses on the *associative mechanism of the ORR*. In this mechanism, $O_2$ first adsorbs on the metal surface, followed immediately by the first proton–electron transfer to form an *adsorbed superoxide intermediate*,

$$O_2 + H^+ + e^- \rightleftarrows OOH_{ad} \quad \text{(step 1)}. \tag{3.40}$$

$OOH_{ad}$ is reduced further via two alternative pathways. In the first pathway, two subsequent steps transform $OOH_{ad}$ into $O_{ad}$ and $OH_{ad}$ through a chemical step,

$$OOH_{ad} \rightleftarrows O_{ad} + OH_{ad} \quad \text{(step 2)}, \tag{3.41}$$

followed by a proton–electron transfer step,

$$O_{ad} + H^+ + e^- \rightleftarrows OH_{ad} \quad \text{(step 3)}. \tag{3.42}$$

The second pathway involves a step

$$OOH_{ad} + H^+ + e^- \rightleftarrows 2OH_{ad} \quad \text{(step 2}') \tag{3.43}$$

that will lead to the formation of two Pt-$OH_{ad}$. An alternative to the latter step is the formation of hydrogen peroxide

$$OOH_{ad} + H^+ + e^- \rightleftarrows 2H_2O_2 \quad \text{(step 2}''), \tag{3.44}$$

an undesired side product formed via a two-electron pathway. Hydrogen peroxide plays a key role as a precursor of radical formation and attack in the PEM.

Continuing with the first pathway through step 2 and step 3, two identical reactions involving the last two concerted proton–electron transfers transform adsorbed $OH_{ad}$ into water

$$2\left(OH_{ad} + H^+ + e^-\right) \rightleftarrows 2H_2O \quad \text{(steps 4 and 5)}, \tag{3.45}$$

thereby completing the four-electron pathway.

Each of the elementary reactions has a corresponding standard equilibrium potential, the value of which could be obtained from first-principles calculations. The free energy balance at equilibrium, written for each elementary reaction step, leads to a simple relation between these standard equilibrium potentials. For the four electron sequence that follows the first pathway, involving Equation 3.43, this relation is

$$\frac{E^0_{O_2,OOH} + E^0_{OOH,OH} + 2E^0_{OH,H_2O}}{4} = E^0_{O_2,H_2} = 1.23V. \tag{3.46}$$

The overall reaction proceeds through three intermediates: $OOH_{ad}$, $OH_{ad}$, and $O_{ad}$. This suggests that a Sabatier-volcano analysis of the activation energy of the kinetically most hindered step should evaluate the highest activation Gibbs energy of any step as a function of adsorption energies of all three intermediates. However, a detailed DFT study by Nørskov and coworkers revealed scaling relationships between adsorption energies of these intermediates (Rossmeisl et al., 2005). These linear relations are

$$\Delta G\left(OH_{ad}\right) \approx 0.5\Delta G\left(O_{ad}\right) + C_1 \tag{3.47}$$

and

$$\Delta G\left(OOH_{ad}\right) \approx 0.5\Delta G\left(O_{ad}\right) + C_2 \tag{3.48}$$

with constants $C_1 = 0.04$ eV and $C_2 = 3.18$ eV for Pt(111). The factor 0.5 on the right hand side of Equations 3.50 and 3.51 occurs because $O_{ad}$ forms a double bond with the surface, whereas $OH_{ad}$ and $OOH_{ad}$ form a single bond. The scaling relations reduce the number of descriptors of the ORR activity of metallic catalysts from three to one.

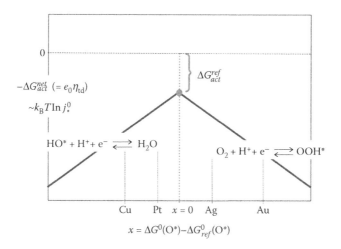

**FIGURE 3.20** A schematic Sabatier-volcano curve of the ORR. It shows the negative intrinsic activation barrier of the thermodynamically least favorable reaction step $-\Delta G_{act}^{net}$ as a function of the reaction Gibbs energy of $O_{ad}$ adsorption $\Delta G\left(O_{ad}\right)$. For the decreasing linear branch on the right-hand side, formation of the first oxygen intermediate species is the thermodynamically limiting step (adsorption is too weak). For the increasing linear branch on the left-hand side, desorption of an oxygen intermediate species is the thermodynamically limiting step (adsorption is too strong).

A schematic Sabatier-volcano curve for the ORR is shown in Figure 3.20. It illustrates the negative net activation potential of the ORR, $-\Delta G_{act}^{net}$, as a function of the Gibbs energy of $O_{ad}$ adsorption, $\Delta G\left(O_{ad}\right)$. $\Delta G_{act}^{net}$ is a composite parameter. It represents the sum of reaction Gibbs energies of quasi-equilibrium steps and of the activation Gibbs energy of the reaction step with the largest barrier. The meaning of $\Delta G_{act}^{net}$ will become clear in the discussion of the *free energy diagram* in the section "Free Energy Profile of the ORR" and from Equations 3.56 and 3.57. $\Delta G_{act}^{net}$ is related to the intrinsic exchange current density through $k_B T \ln j_*^0 \sim \left(-\Delta G_{act}^{net}\right)$. For the linearly decreasing branch at $\Delta G\left(O_{ad}\right) > \Delta G^{ref}\left(O_{ad}\right)$, the formation of $OOH_{ad}$ through (step 1), Equation 3.40, has the highest intrinsic activation barrier. The activation energy on this branch is

$$\Delta G_{act}^{net} = \Delta G_{act}^{ref} + \beta_r \left(\Delta G\left(O_{ad}\right) - \Delta G^{ref}\left(O_{ad}\right)\right),  \tag{3.49}$$

where $\beta_r$ is a Brønsted coefficient of $OH_{ad}$ formation, with a value between 0 and 1, and $\Delta G_{act}^{ref}$ is the activation energy that corresponds to the peak of the volcano curve.

For $\Delta G\left(O_{ad}\right) < \Delta G^{ref}\left(O_{ad}\right)$, desorption of the reaction product (steps 4 and 5), Equation 3.45, is the step with the highest activation energy. The activation energy on this branch is

$$\Delta G_{act}^{net} = \Delta G_{act}^{ref} - \beta_l \left(\Delta G\left(O_{ad}\right) - \Delta G^{ref}\left(O_{ad}\right)\right),  \tag{3.50}$$

where $\beta_l$ is a Brønsted coefficient with value between 0 and 1. The volcano plot in Figure 3.20 suggests that too strong an adsorbate adsorption is encountered, for instance, at Cu. To a lesser degree, it is encountered at Pt, whereas Ag and, even more so, Au exhibit poor thermodynamic conditions for oxide formation at their surface.

What is the value of $\Delta G_{act}^{ref}$? For the case of the HOR at transition metals, the purported value is $\Delta G_{act}^{ref} \approx 0$ eV. For the ORR at transition metals, a value of $\Delta G_{act}^{ref} \approx 0.4$ eV has been proposed by Nørskov and his coworkers, as discussed in Koper (2011). These values imply that exchange current densities of HOR and ORR should differ by about six orders of magnitude, which is in agreement with the experimentally found activity ratio.

What is the effect of an overpotential at the electrode on volcano curve and associated activity values? An overpotential will shift the activation energy on either branch of the volcano curve, according to $\Delta G_{act}^{net}(\eta) = \Delta G_{act}^{net}(\eta = 0) + \alpha e_0 \eta$. If the electronic transfer coefficient is the same for both branches of the volcano curve, the ORR mechanism remains unaltered, and reaction conditions at the electrode are invariant with respect to variation of $\eta$, then the volcano curve will be shifted parallelly along the vertical axis. Under these conditions, a comparison of catalyst activity at finite value of $\eta$ will give the same results as a comparison at $\eta = 0$. However, under violation of any of these conditions, a simple extrapolation of the volcano curve from $\eta = 0$ to $\eta > 0$ is not feasible.

## FREE ENERGY PROFILE OF THE ORR

Specific mechanisms of the ORR have been proposed and studied theoretically by various groups (Jacob, 2006; Rossmeisl et al., 2005; Roudgar et al., 2010). Calculated free energy profiles for a mechanism explored in Rossmeisl et al. (2005) are shown in Figure 3.21. The mechanism is slightly different from the one formulated in the previous section, but essential aspects are the same. The elaborate procedure for calculating the reaction Gibbs energy for each step involves thermochemical modeling to filter out properties that can be calculated using DFT.

A shortcoming of the *Gibbs energy profile* in Figure 3.21 is that it shows reaction Gibbs energies of elementary steps only. It does not show activation Gibbs energies for the individual reaction steps. Therefore, it is not possible to identify the step with the highest activation barrier. This would be necessary in order to determine the dominant reaction pathway and to accurately describe the kinetics of the ORR. In a similar approach, a reaction mechanism of the ORR at a hydrated $Pt_4$ cluster was studied in Roudgar et al. (2010). In that work, six reaction steps were considered and activation energies due to concerted proton–electron transfer processes were calculated explicitly, using the nudged elastic band method.

The reaction Gibbs energy of each elementary step consists of an internal energy $\Delta U$ and an energy shift $\Delta G_E$ attributable to an applied electrode potential, i.e., $\Delta G = \Delta U + \Delta G_E$. Contributions from entropy change and zero point energy are negligible. Equilibrium corresponds to zero reaction Gibbs energy of the overall reaction $O_2 + 4H^+ + 4e^- \rightleftarrows H_2O$, with a corresponding overpotential $\eta = E - E_{O_2,H^+}^{eq} = 0$. Application of a negative overpotential, $\eta < 0$, shifts the reaction Gibbs energy of

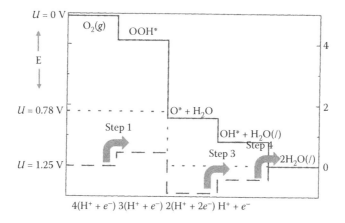

**FIGURE 3.21** A calculated Gibbs energy profile for the ORR shown at different values of the electrode potential relative to the SHE. Close to equilibrium (dashes), the profile contains two uphill sequences, corresponding to step 1 and steps 3 and 4. Below an electrode potential of 0.78 V all steps in the sequence are downhill, as shown by the profile in the middle (short dashes). (Adapted from *Chem. Phys.*, **319**(1), Rossmeisl, J., Logadottir, A., and Nørskov, J. K. Electrolysis of water on (oxidized) metal surfaces, 178–184, Figure 2, Copyright (2005) Elsevier. With permission.)

each reaction step involving coupled proton–electron transfer down by an equal amount $\Delta G_E = e_0 \eta$.

The constructed Gibbs energy diagram along the reaction path connects the initial state of $O_2$ in aqueous solution with the final state with two newly formed water molecules. The free energy profile at equilibrium $E = E^{eq}_{O_2,H^+} = 1.23$ V versus SHE exhibits two uphill sequences: the first one being the formation of $OOH_{ad}$ from molecular oxygen (step 1) and the second one corresponding to the transformation of $O_{ad}$ into water (steps 3 and 4). The intermittent step 2 (not shown in Figure 3.21) transforms the unstable intermediate $OOH_{ad}$ into a more stable species on the surface. As stated in Rossmeisl et al. (2009), an ideal catalyst for the ORR would exhibit a perfectly flat Gibbs energy profile at equilibrium. However, this has not been achieved for any ORR catalyst. The potential, at which all steps become downhill for the evaluated mechanism of the ORR at Pt (111), is $E = 0.78$ V versus SHE.

## DECIPHERING THE ORR

So far presented is a basic mechanistic picture of the ORR. It comprises discussions of potential-controlled formation of surface-adsorbed oxygen intermediates, alternative reaction pathways, and free energy profiles of typical stepwise reaction sequences. The overarching questions are: How can these ingredients be cast into a consistent picture of the *ORR at Pt-based catalysts*? What controls the net rate of the reaction? How are phenomenological parameters, which are employed in the Butler–Volmer equation, related to fundamental mechanisms and parameters?

Here, a *kinetic model of the ORR* under steady-state conditions is discussed and analyzed. The focus is on the associative mechanism, involving the steps in Equations 3.40, 3.41, 3.42, and 3.45. The rate equations for these processes are

$$v_1 = k_1[O_2][H^+](1 - \theta_O - \theta_{OH} - \theta_{OOH}), \tag{3.51}$$

$$v_2 = k_2\theta_{OOH}(1 - \theta_O - \theta_{OH} - \theta_{OOH}), \tag{3.52}$$

$$v_3 = k_3[H^+]\theta_O - k_{-3}\theta_{OH}(1 - \theta_O - -\theta_{OOH}), \tag{3.53}$$

$$v_4 = k_4[H^+]\theta_{OH} - k_{-4}(1 - \theta_O - \theta_{OH} - \theta_{OOH}). \tag{3.54}$$

In agreement with spectroscopic evidence (Wakisaka et al., 2010; Wang et al., 2007), negligible $OOH_{ad}$ coverage is assumed. Therefore, the back reaction in Equation 3.51, could be neglected. The back reaction for the second step, Equation 3.52, was neglected because $OOH_{ad}$ is unstable and spontaneously reacts to form $O_{ad}$ and $OH_{ad}$ (Rossmeisl et al., 2005, 2009).

At steady state, reaction rates must fulfill the following balance: $v_{ORR} = v_1 = v_2 = v_3 = \frac{1}{2}v_4$, representing three independent equations for the three coverage variables, assuming that $k_1[O_2][H^+] \ll k_2$ corresponds to $\theta_{OOH} \approx 0$. This assumption eliminates the process represented by $v_2$, which instantaneously follows the much slower first reaction step, in which $OOH_{ad}$ is formed. The following solution for the *overall rate of the ORR* can, thus, be found,

$v_{ORR}$

$$= \frac{k_1 k_3 k_4 [O_2][H^+]^3}{k_1[O_2][H^+](2k_{-3} + k_4[H^+] + 2k_3[H^+]) + k_3 k_4[H^+]^2 + k_3 k_{-4}[H^+] + k_{-3}k_{-4}}. \tag{3.55}$$

This cumbersome expression can be transformed into a more appealing form, allowing for a simple statistical interpretation and being convenient for the analysis of possible kinetic regimes. The following expression for the inverse ORR rate is equivalent to Equation 3.55,

$$\frac{1}{v_{ORR}} = \frac{1}{k_1^*} + \frac{1}{k_3^*} + \frac{2}{k_4^*} + \frac{2}{K_3 k_4^*} + \frac{1}{K_4 k_1^*} + \frac{1}{K_3 K_4 k_1^*}, \tag{3.56}$$

where $k_1^* = k_1[O_2][H^+]$, $k_3^* = k_3[H^+]$, and $k_4^* = k_4[H^+]$ are rate constants that depend on oxygen and proton concentration, and $K_3 = k_3^*/k_{-3}$ and $K_4 = k_4^*/k_{-4}$.

Formally, Equation 3.56 for the inverse overall rate of the ORR represents the net resistance of a serial resistor network. All of the terms in Equation 3.56 contribute, but the term with the smallest rate, therefore largest resistance, dominates in $v_{ORR}$. Each term in the denominator has the form $K_3^\nu K_4^\mu k_i^*$ with exponents $\nu, \mu = 0, 1$, and $i = 1, 3, 4$. Capital $K$s are quasiequilibrium rate constants for uphill reaction steps that are not kinetically rate-determining. They represent thermodynamic barriers determined by reaction Gibbs energies of the corresponding step. The small $k_i^*$ is the kinetic rate constant of the uphill step with the largest activation barrier $\Delta G_i^{\ddagger}$.

Each net rate in Equation 3.56 has the form of an effective resistance: (i) the abundance of "charge carriers" available to pass the highest energy barrier is represented by the product of $K$ values, and (ii) the mobility of these carriers is represented by the kinetic rate $k^*$. In the considered mechanism, three possible uphill steps exist, which all involve concerted proton–electron transfer processes. They correspond to adsorptive formation of $OOH_{ad}$ via Equation 3.40 with rate constant $k_1^*$, the intermediate step in Equation 3.42 with rate constant $k_3^*$, and desorptive formation of water in Equation 3.45 with rate constant $k_4^*$.

As a general rule, contributions with a larger number of factors will exhibit a smaller rate. Varying the electrode potential $E$ or the oxygen concentration $[O_2]$ will change all rates and induce transitions in the dominant term in Equation 3.56. These transitions correspond to phenomenological transitions in kinetic parameters of the Butler–Volmer equation, namely, $\alpha_c, j_*^0$ and reaction orders. In principle, the net rate for each resistance term in Equation 3.56 could be determined from the Gibbs energy profile, as studied by DFT. However, DFT studies to date do not account for effects of abundance of adsorbed oxygen intermediates on the Gibbs energy, let alone interaction effects of surface oxygen intermediates. Strictly speaking, they could handle only Langmuirian adsorption in a consistent manner.

The main classifications of terms in Equation 3.56 are related to the effects of oxygen concentration and electrode potential. With regard to $[O_2]$, Equation 3.56 contains a mix of terms that are of 0th and first-order in $[O_2]$. Thus, the effective order should lie between 0 and 1. At sufficiently small $[O_2]$, first-order terms will prevail. In this case, the rate of adsorptive $OOH_{ad}$ formation is the kinetic process with the highest activation barrier, that is, $k_1^* \ll k_3^*, k_4^*$. Experimental findings consistently show that the $[O_2]$ dependence of the ORR is of first order in the normal potential range for the ORR (Ihonen et al., 2002). If this adsorption step represents the highest barrier at equilibrium (or small overpotential), it must remain the step with the highest activation barrier at any reduction overpotential.

How can Equation 3.56 be reconciled with the volcano plot in Figure 3.20, suggesting that strong bonding of $O_{ad}$ or $OH_{ad}$ is the thermodynamically limiting factor? The Gibbs energy profile in Figure 3.21 could give the answer: the initial $OOH_{ad}$ formation (step 1) exhibits the highest activation barrier of any single step and, therefore, represents the most severe kinetic limitation; however, the two last uphill steps corresponding to $K_3$ and $K_4$, if taken together, constitute a larger Gibbs energy difference representing a thermodynamic barrier.

Further considerations will be restricted to the case that only terms with first-order dependence on $[O_2]$ need to be accounted for in Equation 3.56. Under this condition, the overall rate of the ORR will undergo the following sequence of transitions upon increasing the absolute value of the cathode overpotential $|\eta_{ORR}| = E_{O_2,H^+}^{eq} - E$,

$$\left(v_{ORR} \sim K_3 K_4 k_1^*\right) \rightarrow \left(v_{ORR} \sim K_4 k_1^*\right) \rightarrow \left(v_{ORR} \sim k_1^*\right). \tag{3.57}$$

Normalized steady-state coverages of oxide species* are given by

$$\theta_O = \frac{1}{1 + K_3 + K_3 K_4}, \qquad \theta_{OH} = \frac{K_3}{1 + K_3 + K_3 K_4}, \qquad \theta_{OOH} \approx 0. \qquad (3.58)$$

For comparison with experimental data, it is convenient to define a total coverage, given by

$$\theta_t = \theta_O + \theta_{OH} = \frac{1 + K_3}{1 + K_3 + K_3 K_4} \approx 1 - K_3 K_4. \qquad (3.59)$$

In principle, reaction and activation free energies required in Equation 3.59 could be obtained from *ab initio* studies of reaction pathways and mechanisms, such as those performed in Rossmeisl et al. (2005), Jacob (2006), and Roudgar et al. (2010). Values of these energy parameters for the appropriate sequence of steps, evaluated at equilibrium, will determine the effective activation potential and the exchange current density of the ORR. These relations are complicated by adsorbate interaction effects and site-specific dispersion of adsorption energies of oxygen intermediates. Currently, these effects are not accounted for in *ab initio* studies.

The analysis is straightforward in cases where adsorbate interaction effects do not play a role. This is the case either under Langmuirian adsorption conditions or in the saturation limit, $\theta_O \to 1$. The former is an unrealistic assumption, since it is known that *Temkin conditions* prevail, illustrated in the seminal work of Sepa et al. (1981). Under the latter assumption, one can proceed to determine values of the effective transfer coefficient in the Butler–Volmer equation. Each quasi-equilibrium step, represented by a factor $K$ in $v_{ORR}$, contributes an amount of $\frac{1}{4}(E - E^{eq}_{O_2,H^+})$ to the potential-dependent part of the net activation energy. The step with the highest barrier is a concerted proton–electron transfer reaction, for which a transfer coefficient $\alpha_{k_1} = 1/2$ can be assumed. Thus, the proportionality can be written as

$$v_{ORR} \propto \exp\left(-\frac{\alpha_c F \left(E - E^{eq}_{O_2,H^+}\right)}{R_g T}\right), \qquad (3.60)$$

with an *effective transfer coefficient*

$$\alpha_c = \frac{v + \mu + 2}{4}. \qquad (3.61)$$

At room temperature, upon increasing the overpotential, the sequence of transitions in Equation 3.57 will give transfer coefficients $\alpha_c$ and effective Tafel slopes $b = R_g T/(\alpha_c F)$ of (i) $\alpha_c = 1$ and $b = 60$ mV at high electrode potential, (ii) $\alpha_c = 0.75$ and $b = 80$ mV at intermediate electrode potential, and (iii) $\alpha_c = 0.5$ and $b = 120$ mV at low electrode potential. These values agree with those found by Sepa and Damjanovic.

---

* Note that a normalization to the saturated oxide coverages is used here, in contrast to the section "Modeling Pt Oxide Formation and Reduction," where normalization was done to the number of available adsorption sites.

Overall, the kinetic model and the statistical analysis of Equation 3.59 represent a way to reconcile kinetic modeling of surface reactions with *ab initio* studies of reaction pathways and free energy profiles, experimental studies of the abundance of reaction intermediates, and macroscopic effective parameters used in surface reactivity models based on the phenomenological Butler–Volmer equation. A specific sequence of elementary reaction steps has been focused on, corresponding to the widely accepted associative mechanism. The same formalism could be used for different reaction sequences.

As for effective kinetic parameters of the ORR that should be used in macroscopic models, it seems reasonable to assume that the *reaction order for oxygen concentration* will be $\gamma_{O_2} = 1$ for conditions of interest. The effective transfer coefficient of the ORR, $\alpha_c$, will transition through a sequence of discrete values between 1 and 0.5, as a function of electrode potential. The *reaction order for proton concentration*, $\gamma_{H^+}$, depends strongly on the adsorption regime and, therefore, a prediction of the value is not trivial. The difference $\alpha_c - \gamma_{H^+}$ is a key determinant of electrostatic effects in water-filled nanopores inside of catalyst layers, as discussed in the section "ORR in Water-Filled Nanopores: Electrostatic Effects."

## CRITICAL REMARKS

The presented procedure for "deciphering" the ORR does not represent a general theory but, merely, a recipe for analyzing reaction mechanisms and pathways. It shows how to incorporate insights from kinetic modeling and thermodynamic parameters of elementary steps from thermochemical modeling and *ab initio* simulations. It was demonstrated how an expression for the effective electron transfer coefficient of the ORR, $\alpha_c$, could be constructed. Similarly, the net thermodynamic activation potential of the ORR, $\Delta G_{act}^{net}$, could be obtained through further analysis.

There are two challenges associated with the Gibbs energy profile shown in Figure 3.21 and the correlation between $\Delta G_{act}^{net}$ and $\Delta G(O_{ad})$ shown in Figure 3.20. First, as mentioned above, the depicted Gibbs energy profile of the ORR only deals with reaction Gibbs energies of elementary steps. It does not show activation barriers. Thus, from this profile, it would not be possible to identify the step in the reaction sequence with the highest activation energy, which would be necessary in order to evaluate the separate terms in Equation 3.56. This shortcoming is most obvious for the Gibbs energy profile at, or below, the electrode voltage of $0.78\,V_{SHE}$. Figure 3.21 implies that at (or below) this potential, one should have $\alpha_c = 0$ and infinite Tafel slope. However, if an activation barrier remains finite for the kinetically inhibited step 1, it could be expected that $\alpha_c = 0.5$ persists at smaller potential. This assertion is consistent with a recent study by Zalitis et al. (2013), which reported a finite Tafel slope of the ORR over a wide potential range.

Second, an issue may be raised with respect to the consistency of the volcano plot in Figure 3.20. First-order reaction kinetics in oxygen concentration implies that the first step of $OOH_{ad}$ formation in Figure 3.21 is the step with the highest activation barrier. If that is the case, why does Pt show up on the left-hand side of the volcano peak, suggesting that strong adsorption of surface oxygen intermediates is the main inhibiting phenomenon? The answer is that $\Delta G_{act}^{net}$ is a composite parameter, and it accounts for the sluggishness of steps 3 and 4, as well as step 1. Step 1 should have the highest activation energy, rendering it the kinetically most hindered step, but steps 3 and 4 combined make the major contribution to $\Delta G_{act}^{net}$.

## ORR IN WATER-FILLED NANOPORES: ELECTROSTATIC EFFECTS

Reaction conditions in CLs are dictated by composition and microstructure. The normal approach in CL modeling is to treat the porous composite electrode layer as a macrohomogeneous medium. Its effective physical properties at the macroscale can be obtained by averaging over spatially varying microstructural properties, using percolation theory, effective medium approximations, random network simulations (Sahimi, 2003; Stauffer and Aharony, 1994; Torquato, 2002) or mathematically rigorous homogenization approaches (Schmuck and Bazant, 2012). The major postulate underlying any of the classical approaches in porous electrode theory is electroneutrality.

This section emphasizes the importance of *local reaction conditions in nanopores* of CLs. In this case, spatially varying charge distributions ions exert a major impact on electrochemical processes at internal pore surfaces. Such charge distributions, occurring in the region of the electrochemical double layer, invalidate the assumption of electroneutrality. In fact, the double layer concept itself becomes meaningless when the nominal thickness of the double layer, that is, the Debye length, is of the same order as the pore radius.

This section elucidates the impact of variations in proton density (or pH) and associated variations in solution phase potential in a nanoporous catalyst layer medium. These variations are significant at the mesoscopic (or agglomerate) scale in conventional CLs (type I electrodes), cf. Figure 3.1. A significant contribution to the total ORR current is generated at Pt nanoparticles inside of primary pores in Pt/C agglomerates. With its large molecular size, ionomer may not be able to penetrate into these pores, but can form a thin adhesive skin layer on the agglomerate surface (Malek et al., 2007). As a consequence, most of the Pt particles in agglomerates will not be surrounded by ionomer but by water.

The importance of proton distribution and transport in *water-filled nanopores* with charged metal walls is most pronounced in *ionomer-free UTCLs* (type II electrodes), cf. the main case considered in this section. In either type of CLs, proton and potential distribution at the nanoscale are governed by *electrostatic phenomena*.

A model of the ORR in a single water-filled pore with charged walls of Pt will be presented. It affords the definition of an effectiveness factor, by which the performance of any nanoporous CL material could be evaluated. Furthermore, a remarkable conclusion is drawn in view of the coupling of ORR kinetics and metal corrosive dissolution.

### IONOMER-FREE ULTRATHIN CATALYST LAYERS

In recent years, advanced designs of UTCLs have shown great promise in view of achieving a tremendous Pt loading reduction in CCLs, from 0.4 mg cm$^{-2}$ in conventional CCLs, down to 0.1 mg cm$^{-2}$ in UTCLs. Enabled by a combination of changes in Pt surface structure, supports effects, and unique reaction conditions in nanopores, these layers could bring about dramatic improvements in mass-specific activity of Pt over ionomer-impregnated Pt/C layers, as demonstrated for UTCLs fabricated with 3M's nanostructured thin film technology (3M NSTF) (Debe, 2013; Debe et al., 2006). MEAs with 3M NSTF layers on the cathode side are comparable to standard Pt/C-based CCLs in terms of power density and exhibit excellent durability and longevity.

The main attributes that distinguish UTCLs from conventional CLs are that they are ionomer-free and at least an order of magnitude thinner, with $l_{CL}$ ranging from 20 to 500 nm. The distinct composition and thickness regime leads to different distributions of proton density and solution phase potential, as well as different transport properties of electroactive species, comprising of protons, electrons, and dissolved oxygen. As a result of reduced Pt loading and thickness, the ECSA is significantly smaller than that of conventional CLs. Typical surface area enhancement factors (ratio of ECSA to apparent electrode surface area) for UTCLs, determined from cyclic voltammetry, are in the range of 10–40. This is much lower than the values for conventional ionomer-impregnated CCLs. However, the small ECSA in conventional CCLs could be compensated by better catalyst utilization, improved local reaction conditions, and higher area-specific activity (A cm$^{-2}$).

Interest in the development of catalyst layers with low Pt-loading has spurred development of a range of support-free and supported UTCLs. *Support-free UTCLs* are fabricated by sputtering or ion-beam-assisted deposition of Pt directly on PEM or diffusion media (Gruber et al., 2005; O'Hayre et al., 2002; Saha et al., 2006). In the category of *supported UTCLs*, electronically conductive as well as insulating support materials have been explored. Both options are illustrated in Figure 3.22. In the former case, nanoislands or nanoparticles of Pt are deposited on the conductive support to maximize Pt utilization at the nanoscale. This approach enables a more radical reduction of Pt loading. Support materials that have been tested include carbon nanotubes (Ramesh et al., 2008; Tang et al., 2007), nanoporous foils of Au (Zeis et al., 2007), or ordered nanoporous electrodes of metals and metal oxides (Kinkead et al., 2013). Pt can be deposited using reduction of Pt salts, electroless deposition, or sputtering. Carbon-free support layers offer an additional benefit resulting from the mitigation of carbon corrosion.

The case of insulating support materials is exemplified by the 3M NSTF design (Debe, 2013; Debe et al., 2006). These layers utilize a crystalline organic pigment, perylene red, as the support. Sputter deposition of a continuous layer of Pt transforms the organic substrate into an electronically conductive and electrocatalytically active medium. The minimal Pt loading required to form a continuous Pt layer is 20 μg cm$^{-2}$. The resulting layer consists of densely packed crystalline whiskers, with cross sections of about 50 nm, aspect ratio from 20 to 50, and packing density of 3–5 billion whiskers per cm$^2$.

The break-in process of 3M NSTF films involves voltage cycling to create a smooth polycrystalline Pt surface on the whiskers. The surface area enhancement factors of resulting structures are from 10 to 25, determined from cyclic voltammograms in the potential region of hydrogen underpotential deposition. The surface area enhancement factor (or roughness factor or real-to-apparent surface area ratio) of a

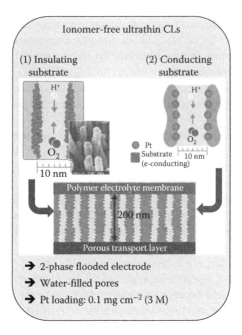

**FIGURE 3.22** An illustration of design and key properties of ionomer-free ultrathin catalyst layers with insulating or electronically conductive support materials. The typical thickness is in the range of 200 nm.

typical UTCL is thus a factor of 10–40 smaller than it is for conventional CCLs. This downside is compensated for by an improved catalyst utilization and an enhanced area-specific activity for the ORR, which is 5–10 times higher for polycrystalline Pt than for Pt nanoparticles with ~3 nm diameter.

Figure 3.23 shows SEM images of a platinized NSTF layer before (a) and after (b) transfer to the PEM surface. Excellent power densities have been obtained with Pt loadings as low as $0.1$ mg cm$^{-2}$. A further asset of 3M NSTF layers is their dramatically improved resistance to ECSA loss. Superior durability and longevity of MEAs fabricated with 3M NSTF layers have been clearly demonstrated (Debe et al., 2006).

A question of general interest is whether tremendous gains in Pt loading reduction, durability, and longevity for 3M NSTF could be realized or exceeded with further improved designs, for example, using nanoparticles deposited on conductive supports, possibly acting as catalytic enhancers. To address this question, a theory that rationalizes fundamental processes in ionomer-free UTCLs is needed, especially to understand mechanisms of proton transport and reaction rate distributions in nanoporous media with charged walls. This theory will be presented in subsequent sections.

It is also widely known that MEAs utilizing 3M NSTF technology face severe water management challenges. They show poorer performance than conventional CCLs at low RH, presumably caused by poor proton transport in insufficiently hydrated layers. Moreover, NSTF MEAs exhibit an increased propensity for flooding of porous diffusion media, shutting down pathways for gaseous supply of reactants. These issues are believed to be common for other realizations of ionomer-free UTCLs as well and they have been studied recently by Chan and Eikerling (2014).

An estimate of the minimal electronic conductivity required in UTCLs can be obtained using a simple model of the interplay between the distributed rate of current generation and transport of electrons with conductivity $\sigma_{el}$ in the support phase. The underlying model is identical to models that evaluate the interplay of proton transport limitations and reaction in conventional CCL, discussed, for instance, in Eikerling et al. (2007a). Using Equation 3.1 with $l_{CL} = 100$ nm yields $\sigma_{el} > 10^{-4}$ S cm$^{-1}$. Other relevant properties for the selection of support materials are (i) the characteristic feature size of pores or solid domains, which determines the ECSA, (ii) the pore network morphology, which determines the transport properties, (iii) the thickness, determining the interplay of transport and reaction, (iv) intrinsic support stability, and (v) catalytic properties in view of ORR activity and catalyst dissolution.

Since UTCLs contain no added electrolyte, the mode of proton transport in such layers remains a debated question. It was postulated in Chan and Eikerling (2011) that protons in water-filled UTCL pores undergo bulk-water-like transport, similar to ion transport in charged nanofluidic channels (Daiguji, 2010; Stein et al., 2004) and gold nanoporous membranes (Nishizawa et al., 1995). The proton conductivity of the pore is then determined by the electrostatic interaction of protons with the surface charge of pore walls.

In order to investigate the impact of *metal surface charge density* on the performance of UTCLs, Chan and Eikerling (2011) developed a continuum *single pore model for oxygen reduction* in water-filled, cylindrical nanopores with platinum walls. At one end, pores are in contact with the PEM, acting as a reservoir of protons. Electrostatic interaction of protons with the metal surface charge density at pore walls, $\sigma_M$, drives proton migration into the pore. A *Stern model of the metal-solution interface* was used to relate $\sigma_M$ to the metal phase potential $\phi^M$.

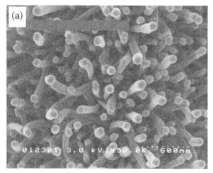
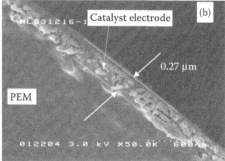

**FIGURE 3.23** SEM images of platinized ultrathin catalyst layers fabricated with the NSTF technology of 3M (a) before transfer to the PEM surface and (b) after transfer to PEM surface. The magnification in both images is 50,000. (Reprinted from *J. Power Sources*, **161**, Debe et al., High voltage stability of nanostructured thin film catalysts for PEM fuel cells, 1002–1011, Figures 1 and 2, Copyright (2006) Elsevier. With permission.)

Proton and potential distributions within the pore are governed by *Poisson–Nernst–Planck theory* and the oxygen distribution by Fick's law. The metal surface charge density and the corresponding proton conductivity of the pore are tuned by the departure of the electrode potential from the *potential of zero charge of the metal*, $\phi^{pzc}$. This is the key determinant of the current conversion effectiveness of the pore. Other determinants of pore performance are the Helmholtz capacitance, $C_H$, kinetic parameters of the ORR, and pore size and length. A simple upscaling of the pore model allows comparison with experimental polarization data for MEAs with UTCLs to be made. Implications of the model for materials selection and nanostructural design of UTCLs will be discussed.

## MODEL OF A WATER-FILLED NANOPORE WITH CHARGED METAL WALLS

As shown in Figure 3.24, a water-filled nanopore in an ionomer-free UTCL is modeled as a straight, water-filled cylinder of radius $R_p$ and length $L_p$, with smooth walls of Pt. The model neglects effects that result from geometric surface roughness and atomic-scale heterogeneities at pore walls. The pore is bounded at one opening, $z = 0$, by the PEM, supplying protons, and at the other end, $z = L_p$, by MPL or GDL, supplying oxygen. Charge transfer occurs at the internal Pt–solution interface.

Without an embedded ionomer phase in the UTCL, the surface charge at the pore walls is the driving force for proton migration into the pore. Since the focus is on the bulk UTCL response, effects of the electric double layer at the PEM/UTCL interface are ignored. These effects could become significant only if the bulk of the UTCL is essentially inactive.

It is worth mentioning that proton transport phenomena in UTCL pores resemble those in water-filled pores of PEM, discussed in the section "Proton Transport" in

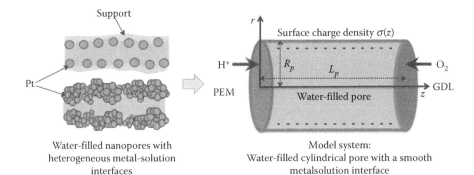

**FIGURE 3.24** Model representation of a *UTCL nanopore*. The pore is assumed to be straight and cylindrical with charged Pt walls. The pore is bounded by the PEM at one end and the porous transport layer (MPL or GDL) at the other end. In real pores in UTCLs, Pt could be deposited in the form of nanoparticles at a conductive support or as a continuous layer on an insulating support, as indicated on the left. (Reprinted from Chan, K. and Eikerling, M. 2011. *J. Electrochem. Soc.*, **158**(1), B18–B28, Figures 1,2,3,4,5,6. Copyright (2011), the Electrochemical Society. With permission.)

Chapter 2. In both cases, liquid water acts as the proton shuttle with the proton concentration controlled by the charge density at pore walls. The essential difference is that, in the PEM pore, the interfacial charge density is a materials property implemented at the fabrication stage. It corresponds to the packing density of fixed anionic surface groups. In the UTCL pore, the surface charge density is a function of the applied metal potential and of the metal structure.

Determination of the electrochemical performance of the pore requires the relation between the faradaic current density at the interface, $j_F$, and $\phi^M$. In order to establish this relation, one needs the following: (i) an electric double layer model to relate $\phi^M$ to $\sigma_M$, (ii) transport equations to relate $\sigma_M$ to reactant concentrations, and (iii) charge transfer kinetic equations to relate local reactant concentrations and potentials to $j_F$. The continuum model developed by Chan and Eikerling (2011) consists of coupled relations for reactant transport, metal surface charge, and charge transfer kinetics.

The continuum approach is expected to be applicable when the size of the pore is large compared to the size of a hydrated proton. In studies of biological ion channels, results of continuum models and molecular dynamics simulations agree well in channels of $> 15\,\text{Å}$ diameter (Noskov et al., 2004). In smaller channels, corrections were required (Corry et al., 2003; Graf et al., 2004; Nadler et al., 2003). As typical pore sizes in UTCL should be larger than 5 nm, the breakdown of the continuum model is not a relevant concern.

## GOVERNING EQUATIONS AND BOUNDARY CONDITIONS

The distribution of proton concentration $c_{H^+}$ and potential $\Phi$ in solution is governed by the Poisson–Nernst–Planck (PNP) model, widely used in the theory of ion transport in biological membranes (Coalson and Kurnikova, 2007; Keener and Sneyd, 1998). Oxygen diffusion is determined by Fick's law. Inside the pore, the continuity and transport equations for protons and oxygen are

$$\frac{\partial c_{H^+}}{\partial t} = -\nabla \cdot \mathbf{N}_{H^+}, \quad \mathbf{N}_{H^+} = -D_{H^+}\left(\nabla c_{H^+} + \frac{F}{R_g T} c_{H^+} \nabla \Phi\right), \tag{3.62}$$

$$\Delta \Phi = -\frac{F}{\varepsilon_0 \varepsilon_r} c_{H^+} \tag{3.63}$$

and

$$\frac{\partial c_{O_2}}{\partial t} = -\nabla \cdot \mathbf{N}_{O_2}, \quad \mathbf{N}_{O_2} = -D_{O_2} \nabla c_{O_2}, \tag{3.64}$$

where $c_i$ and $\mathbf{N}_i$ denote concentrations and fluxes of species $i$. The governing equations are time-dependent transport equations in two spatial dimensions, namely, axial direction $z$ and radial direction $r$ (Chan and Eikerling, 2011, 2012).

In the remainder of this section, the steady-state solution will be the focus. A model for the linear impedance response of UTCLs has been developed by Chan and Eikerling (2012).

*Boundary conditions.* At the PEM–pore interface, $z = 0$, $c_{H^+}$ is assumed to be the volume-averaged proton concentration of the bulk PEM. This assumption neglects diffuse layer effects within the PEM. The potential $\Phi$ at the interface is assumed uniform and the flux of oxygen is assumed to be zero,

$$c_{H^+}(r, 0) = c_{H^+}^0, \quad \Phi(r, 0) = \Phi^0, \quad \left.\frac{\partial c_{O_2}}{\partial z}\right|_{z=0} = 0. \tag{3.65}$$

At the interface of the UTCL pore with the porous diffusion medium, $z = L_p$, the concentration of dissolved oxygen is assumed to be fixed. It can be determined from the oxygen partial pressure in the reactant gas using Henry's law, $c_{O_2}^0 = H_{O_2}p_{O_2}^0$. Moreover, the proton flux must be zero,

$$c_{O_2}(r, L_p) = c_{O_2}^0, \quad \left.\frac{\partial c_{H^+}}{\partial z}\right|_{z=L_p} = 0, \quad \left.\frac{\partial \Phi}{\partial z}\right|_{z=L_p} = 0. \tag{3.66}$$

The problem has axial symmetry, so that at the pore center, $r = 0$,

$$\left.\frac{\partial \Phi}{\partial r}\right|_{r=0} = \left.\frac{\partial c_{H^+}}{\partial r}\right|_{r=0} = \left.\frac{\partial c_{O_2}}{\partial r}\right|_{r=0} = 0. \tag{3.67}$$

As for the boundary conditions at the internal pore wall, it is assumed that $r = R_p$ coincides with the position of the reaction or Helmholtz plane, that is, the plane of the closest approach of hydrated protons to the interface. Quantities are indicated at $r = R_p$ by a superscript "s." Components of proton and oxygen fluxes normal to the reaction plane are related to the local faradaic current density by

$$\mathbf{N}_{H^+}(R_p, z) \cdot \mathbf{n}_r = \frac{j_F(z)}{F}, \quad \mathbf{N}_{O_2}(R_p, z) \cdot \mathbf{n}_r = \frac{j_F(z)}{4F}, \tag{3.68}$$

where the faradaic current density is assumed as the cathodic branch of the Butler–Volmer equation,

$$j_F(z) = j_*^0 \left(\frac{c_{O_2}^s}{c_{O_2}^0}\right)^{\gamma_{O_2}} \left(\frac{c_{H^+}^s}{c_{H^+}^0}\right)^{\gamma_{H^+}} \exp\left(-\frac{\alpha_c F \eta^s(z)}{R_g T}\right), \tag{3.69}$$

with $j_*^0$ as the exchange current density at the reference concentrations $c_{O_2}^0$ and $c_{H^+}^0$, and $\gamma_{O_2}$ and $\gamma_{H^+}$ as the reaction orders.

The local overpotential at the reaction plane is

$$\eta^s(z) = \left(\phi^M - \Phi^s(z)\right) - \left(\phi_{eq}^M - \Phi_{eq}^0\right), \tag{3.70}$$

where subscripts "eq" indicate values at equilibrium. Using the total overpotential of the CCL

$$\eta_c = \left(\phi^M - \Phi^0\right) - \left(\phi_{eq}^M - \Phi_{eq}^0\right) \tag{3.71}$$

transforms Equation 3.70 into

$$\eta^s(z) = \eta_c - \left(\Phi^s(z) - \Phi^0\right), \tag{3.72}$$

with the term in brackets corresponding to a Frumkin-type diffuse layer correction (Bard and Faulkner, 2000).

The boundary condition for potential at the pore wall requires special consideration. Basically, this boundary condition defines the electrostatic reaction conditions in the pore. It captures the interaction between charged metal walls and protons in solution. The boundary condition should relate $\phi^M$ to $\Phi^s(z)$ and further to $\sigma_M(z)$. Deriving this relation requires a *model of the metal–solution interface*. This problem is complicated by the formation of adsorbed oxygen species, particularly difficult in the potential region for ORR, where various Pt oxides are formed at the surface. Oxide formation modulates the metal surface charge. It leads to pseudocapacitance effects, which make an accurate determination of $\sigma_M(z)$ impossible with the current level of understanding.

A simplified approach, based on the *Stern–Grahame double layer model*, gives the following relation for the *metal surface charge density*,

$$\sigma_M(z) = \int_{\phi^{pzc} - \Phi^0}^{\phi^M - \Phi^s(z)} C_H(\varphi)\, d\varphi. \tag{3.73}$$

Assuming constant Helmholtz capacitance, $C_H$, leads to the following Robin boundary condition for potential,

$$\sigma_M(z) = \varepsilon_0 \varepsilon_r \left. \frac{\partial \Phi}{\partial r} \right|_{r=R_p} = C_H \left[ \left( \phi^M - \phi^{pzc} \right) - \left( \Phi^s(z) - \Phi^0 \right) \right]. \tag{3.74}$$

In this form, the value of the solution phase potential at the reaction plane, $\Phi^s(z)$, at given $\phi^M$, is determined by $C_H$ and $\phi^{pzc}$.

A similar approach to the boundary condition for the potential at the metal–solution interface has been applied by Biesheuvel et al., in consideration of diffuse charge effects in galvanic cells, desalination by porous electrodes, and transient response of electrochemical cells (Biesheuvel and Bazant, 2010; Biesheuvel et al., 2009; van Soestbergen et al., 2010). However, their treatment neglected the explicit effect of $\phi^{pzc}$. In principle, the PNP model could be modified to incorporate size-dependent and spatially varying dielectric constants in nanopores, as well as ion saturation effects at the interface. However, in a heuristic fashion, such variations could be accounted for in the Helmholtz capacitance of the Stern double layer model.

Equations 3.62 through 3.74 form a closed set of equations that can be solved for the functions $\Phi(r, z), c_{H^+}(r, z), c_{O_2}(r, z)$. Using these functions, the faradaic current density $j_F(z)$ can be obtained. This function can be employed to calculate the *effectiveness factor of Pt utilization* of the pore, defined as the total current produced by the pore, normalized by an "ideal" current that would be obtained, if reactant and potential distributions were completely uniform, with $\Phi^s(z) = \Phi^0$, $c_{H^+}^s = c_{H^+}^0$, and $c_{O_2}^s = c_{O_2}^0$, that is

$$\Gamma_{pore} = \frac{1}{L_p j_{id}} \int_0^{L_p} j_F(z)\, dz \quad \text{with } j_{id} = j_*^0 \exp\left( -\frac{\alpha_c F \eta_c}{R_g T} \right). \tag{3.75}$$

## MODEL SOLUTION IN STEADY STATE

Numerical solutions for $\Phi(r, z)$, $c_{H^+}(r, z)$, and $c_{O_2}(r, z)$ can be readily obtained using standard mathematical software packages like MATLAB® or Simulink®. These solutions could be utilized in further analysis to calculate the effectiveness factor of the pore and study the response function between catalyst layer overpotential $\eta_c$ and fuel cell current density $j_0$. Thereby, the dependence of these effective performance quantifiers on local reaction conditions in pores could be rationalized.

Usually, the ORR is the culprit in the eyes of fuel cell researchers and technology developers. However, for the solution of the model presented above, the sluggishness of the ORR brings a blessing. It allows the governing equations to be decoupled into an electrostatic problem and a standard oxygen diffusion equation.

Under conditions relevant for fuel cell operation, the reaction current density of the ORR is small compared to separate flux contributions caused by proton diffusion and migration in Equation 3.62. Therefore, the electrochemical flux term $\mathbf{N}_{H^+}$ on the left-hand side of the Nernst–Planck equation in Equation 3.62 can be set to zero. In this limit, the PNP equations reduce to the Poisson–Boltzmann equation (PB equation). This approach allows solving for the potential distribution independently and isolating the electrostatic effects from the effects of oxygen transport.

In dimensionless form, the PB equation for the cylindrical geometry of the channel is

$$\frac{1}{\gamma} \frac{\partial}{\partial \gamma} \left( \gamma \frac{\partial \varphi}{\partial \gamma} \right) + \frac{R_p^2}{L_p^2} \frac{\partial^2 \varphi}{\partial \xi^2} = -\frac{R_p^2}{\lambda_D^2} \exp\left(-[\varphi - \varphi_0]\right) \tag{3.76}$$

where $\gamma = r/R_p$, $\xi = z/L_p$, $\varphi = F\Phi/(R_g T)$, $\lambda_D = \sqrt{\varepsilon_0 \varepsilon_r R_g T/(F^2 c_{H^+}^0)}$, and the dimensionless proton concentration is related to potential via

$$C_{H^+} = \frac{c_{H^+}}{c_{H^+}^0} = \exp\left(-[\Phi - \Phi_0]\right). \tag{3.77}$$

This transforms the effectiveness factor into

$$\Gamma_{pore} = \int_0^1 \left(C_{O_2}^s(\xi)\right)^{\gamma_{O_2}} \exp\left[(\alpha_c - \gamma_{H^+})\left(\varphi^s(\xi) - \varphi_0\right)\right] d\xi, \tag{3.78}$$

where the dimensionless oxygen concentration $C_{O_2}^s = c_{O_2}^s/c_{O_2}^0$ was introduced, and $\varphi^s$ is the dimensionless potential at the reaction plane.

A Donnan potential difference exists at the PEM–UTCL interface. The extent of this region in $z$ direction is on the order of the Debye length, $\lambda_D = 4$ Å. Beyond this potential decay region, the proton concentration and solvent potential vary only in the radial direction. This small decay region does not make a significant contribution to the current generation in the pore. If it is neglected, Equation 3.76 reduces further to an electrostatic problem in radial direction only. With this simplification, $\Gamma_{pore}$ can be separated into an *electrostatic factor* $\Gamma_{elec}$ and a factor due to nonuniform oxygen distribution $\Gamma_{O_2}$, with

$$\Gamma_{elec} = \exp\left[(\alpha_c - \gamma_{H^+})\left(\varphi^s(\xi) - \varphi_0\right)\right] \quad \text{and} \quad \Gamma_{O_2} = \int_0^1 \left(C_{O_2}^s(\xi)\right)^{\gamma_{O_2}} d\xi. \quad (3.79)$$

The expression for $\Gamma_{elec}$ embodies two competing trends of the solution phase potential at the reaction plane. An increase in solution phase potential results in a larger driving force for electron transfer in cathodic direction. This effect is proportional to the cathodic transfer coefficient $\alpha_c$. At the same time, a more positive value of $(\varphi^s(\xi) - \varphi_0)$ corresponds to lower proton concentration at the reaction plane, following a Boltzmann distribution (Equation 3.77). The magnitude of this effect is determined by the reaction order $\gamma_{H^+}$. It is, therefore, of primary interest to know the difference of kinetic parameters, $\alpha_c - \gamma_{H^+}$.

Pathways, mechanisms, and corresponding kinetic parameters of the ORR have been discussed in the section "Electrocatalysis of the Oxygen Reduction Reaction at Platinum." In a highly simplified picture, derived originally on the basis of a series of experimental studies by Damjanovic and coworkers (Damjanovic, 1992; Gatrell and MacDougall, 2003; Sepa et al., 1981, 1987), it was proposed that the rate-determining reaction step is the initial adsorption,

$$O_2 \rightleftharpoons O_{2,ads}, \quad (3.80)$$

followed immediately by the first electrochemical step

$$O_{2,ads} + H^+ + e^- \rightarrow OOH_{ads}. \quad (3.81)$$

A doubling of Tafel slopes from 60 to 120 mV/decade at room temperature results from a transition from Temkin to Langmuirian kinetics upon decreasing the electrode potential. This causes a shift in the effective transfer coefficient and reaction order from $\alpha_c = 1$ and $\gamma_{H^+} = 3/2$ at low current density to $\alpha_c = 1/2$ and $\gamma_{H^+} = 1$ at high current density. The transition between these two regions occurs near the onset of oxide species formation around $\eta_c \approx -0.4$ V. Moreover, different values of the exchange current density must be used for Temkin and Langmuirian regions. The values are adopted from Parthasarathy's microelectrode study for a Pt|Nafion interface (Parthasarathy et al., 1992). The value of $j_*^0$ depends on particle composition, size, shape, and surface structure and it varies with the type of support material (Mayrhofer et al., 2008).

Using the values of kinetic parameters proposed in the seminal paper by Sepa et al. (1981), gives $\alpha_c - \gamma_{H^+} = -1/2$, under both Temkin- and Langmuir-type adsorption conditions. This means that the net effect of increasing solution phase potential and decreasing proton concentration on the rate of the ORR and on $\Gamma_{elec}$ is negative. In other words, an increase in proton concentration $C_{H^+}^s$ at the pore wall along with a decrease in $\varphi^s$ has a net positive impact on electrochemical performance. It is, thus, of primordial importance to understand the factors that influence the *proton affinity of the channel*.

The 1D radial PB equation has an analytical solution that is given by

$$\varphi(\gamma) - \varphi_0 = 2\ln\left(1 - b\gamma^2\right) + \varphi_c \tag{3.82}$$

with the potential at the pore center

$$\varphi_c = \ln\left(\frac{R_p\left(R_p + R_c\right)}{8\lambda_D^2}\right). \tag{3.83}$$

The *electrostatic effectiveness factor*, as a function of $\sigma_M$, is

$$\Gamma_{elec} = \left[\frac{\sigma_M^2}{2c_{H^+}^0\varepsilon_0\varepsilon_r R_g T}\left(1 + \frac{R_c}{R_p}\right)\right]^{-(\alpha_c - \gamma_{H^+})}. \tag{3.84}$$

The functional dependence of $\sigma_M$ on $\phi^M$ is a specific relation of a catalyst material. Note that $\sigma_M$ is restricted to negative values since it must balance the positive charge of protons in the pore. The characteristic radius $R_c = -4\varepsilon_0\varepsilon_r R_g T/(\sigma_M F)$, introduced in Equations 3.83 and 3.84, represents the radius, below which finite size effects start to strongly affect the electrostatic properties of the nanopore. For relevant values of $\sigma_M$, in the range of $-0.01$ C m$^{-2}$ to $-0.1$ C m$^{-2}$, the value of $R_c$ lies between 7 nm and 1 nm.

Figure 3.25a shows the impact of $\sigma_M$ and $R_p$ on $\Gamma_{elec}$. As the surface charge attains more negative values, the *proton affinity of the channel* increases, improving $\Gamma_{elec}$. A smaller pore radius $R_p$ increases the proton affinity caused by the confining pore geometry, as illustrated in Figure 3.25b. This enhancement of the current conversion efficiency is significant for $R_p \sim R_c$.

In order to establish the relation between $\Gamma_{elec}$ and $\phi^M$, the relation $\sigma_M = f\left(\phi^M\right)$ needs to be inserted into the solution of the PNP equations. This dependence is of fundamental importance in nanoscale electrochemistry. In principle, the function $\sigma_M = f\left(\phi^M\right)$ could be obtained from experimental studies or it could be derived from basic theoretical considerations and *ab initio* simulations for the considered metal-solution interface.

## INTERFACIAL CHARGING BEHAVIOR

Equation 3.74 is a parameterization of $\sigma_M = f\left(\phi^M\right)$ based on the Stern–Grahame double layer model. This relation requires as input the potential of zero charge and the Helmholtz capacitance. As mentioned, finding $\sigma_M = f\left(\phi^M\right)$ is complicated by the formation of various surface and bulk oxide species, occurring simultaneously with double layer charging.

It is questionable whether a *potential of zero charge* could be defined uniquely and measured unambiguously. Indeed, in spite of a tremendous interest in the literature, this subject has remained inconclusive. The presence of adsorption processes led to the definition of a potential of zero total charge $\phi^{pztc}$ and a potential of zero free

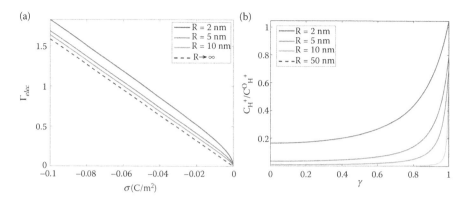

**FIGURE 3.25** Solution of the single pore model in the 1D Poisson–Boltzmann limit: (a) the electrostatic effectiveness factor, $\Gamma_{elec}$, as a function of the metal surface charge density, $\sigma_M$, for various values of $R_p$; (b) radial variation of the normalized proton concentration in the pore for various values of $R_p$ at $\sigma_M = -0.05$ C m$^{-2}$. (Reprinted from Chan, K. and Eikerling, M. 2011. *J. Electrochem. Soc.*, **158**(1), B18–B28, Figures 1,2,3,4,5,6. Copyright (2011), the Electrochemical Society. With permission.)

charge $\phi^{pzfc}$ (Climent et al., 2006). The former refers to the potential, at which the sum of the excess free charge and the charge that has crossed the interface arising from the adsorption is zero. The latter refers to the potential, at which the excess free charge is zero. For Equation 3.74, the value of $\phi^{pzfc}$ is required.

Different measurement methods have yielded a range of values for $\phi^{pzfc}$. Measurements using the CO displacement method obtained $\phi^{pztc} = 0.33$ V versus RHE (Climent et al., 2006). The value of $\phi^{pzfc}$ was estimated to be close to this value (Weaver, 1998). Using an *ex situ* immersion method at a clean Pt(111) surface and in 0.1 M HClO$_4$, Hamm et al. found $\phi^{pztc} = 0.84$ V versus SHE and estimated an even higher $\phi^{pzfc} = 1.1$ V versus SHE (Hamm et al., 1996). Friedrich's second harmonic generation study of the Pt(111) electrode in 0.1 and 0.001 M HClO$_4$ solutions found a negatively charged surface up to a potential of 0.6 V versus RHE. Frumkin and Petrii proposed a second "inversed" $\phi^{pzfc}$ in the oxide adsorption region, the point at which the surface charge shifts from positive to negative upon increasing potential, to explain findings of ion adsorption studies at Pt (Frumkin and Petrii, 1975; Petrii, 1996). Their results are illustrated in Figure 3.26. A more complex charging behavior in the Pt oxide region was also proposed in a recent theoretical study of Pt(111), where $\phi^{pzc}$ was reported to increase with the degree of surface oxidation (Tian et al., 2009). Overall, these results suggest that in the case of oxide-covered Pt, the relation $\sigma_M = f\left(\phi^M\right)$ is more complicated than that implied by Equation 3.74. With a second $\phi^{pzc}$ in the oxide region, the relation would no longer be monotonic.

The approach of Chan and Eikerling (2011) emphasizes the importance of charging phenomena at the metal–solution interface, but it does not account for the intricate and largely unsettled effects that could arise from the progressive oxidation of Pt at high $\phi^M$. It treats the potential of zero charge as a variable parameter that could attain values in the range from $\phi^{pzc} = 0.3$ to 1.1 V$_{SHE}$.

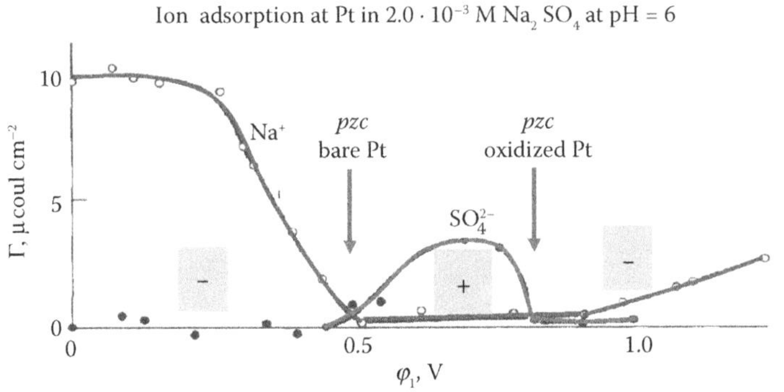
Ion adsorption at Pt in 2.0 · 10⁻³ M Na₂ SO₄ at pH = 6
Γ, μcoul cm⁻²
10
5
0
0
0.5
1.0
φ₁, V
Na⁺
pzc
bare Pt
pzc
oxidized Pt
SO₄²⁻
−
+
−

**FIGURE 3.26** The surface excess of sodium cations and sulfate anions as a function of metal phase potential, determined from radiotracer measurements. In the normal region on the left, the cation concentration decreases with increasing potential from the resulting decrease in negative surface charge on the metal. The normal potential of zero charge is found at 0.5 $V_{SHE}$. Above this potential, oxide formation commences. An inverted region is seen in the potential range >0.9 $V_{SHE}$. The accumulation of sodium ions suggests that the fully oxidized surface exhibits a net negative excess free charge. (Adapted from *Electrochim. Acta*, **20**(5), Frumkin, A. N. and Petrii, O. A. Potentials of zero total and zero free charge of platinum group metals, 347–359, Figure 5, Copyright (1975) Elsevier. With permission.)

## Electrostatic Effectiveness as a Function of Potential

Using the metal charging boundary condition in the previously discussed formalism, an implicit relationship for $\Gamma_{elec}$ can be derived,

$$\frac{\ln \Gamma_{elec}}{\alpha_c - \gamma_{H^+}} + \frac{\varepsilon_0 \varepsilon_r}{R_p C_H}\left(2 - \sqrt{4 + \frac{2R_p^2}{\lambda_D^2 \left(\Gamma_{elec}\right)^{1/(\alpha_c - \gamma_{H^+})}}}\right) = \frac{F}{R_g T}\left\{\phi^M - \phi^{pzc}\right\}. \tag{3.85}$$

Figure 3.27 shows the *electrostatic effectiveness* $\Gamma_{elec}$ obtained as the solution of Equation 3.85 for $R_p = 2, 5$, and 10 nm and in the limit of infinite radius, as a function of $\phi^M - \phi^{pzc}$. The corresponding variation of $\sigma^M$ is shown in the inset.

In the potential range well below the potential of zero charge, in which $\Gamma_{elec} \sim 1$, the following approximate relation can be obtained,

$$\Gamma_{elec} \approx A\left[\frac{2\varepsilon_0 \varepsilon_r}{R_p C_H} - \frac{F}{R_g T}\left(\phi^M - \phi^{pzc}\right) + 2\right], \tag{3.86}$$

where $A \approx \left[2 + \frac{F}{C_H}\sqrt{\frac{2\varepsilon_0 \varepsilon_r c^0_{H^+}}{R_g T}}\right]$. The first term in Equation 3.86 accounts for the enhancement effect in proton concentration due to double layer overlap in small pores. The second term in Equation 3.86 describes the increase in proton affinity as the metal phase potential is lowered relative to the potential of zero charge. The constant $A$ contains the dependence of $\Gamma_{elec}$ on $C_H$. An increase in $C_H$ increases $A$,

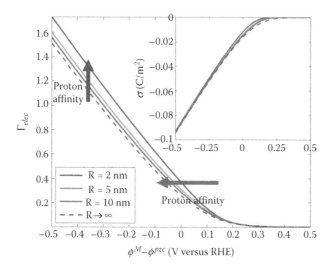

**FIGURE 3.27** The electrostatic effectiveness $\Gamma_{elec}$, obtained as the solution of Equation 3.85 for $R_p = 2, 5$, and 10 nm and in the limit of infinite radius, as a function of $\phi^M - \phi^{pzc} = E - E$. The inset shows the surface charge density as a function of metal phase potential. (Reprinted from Chan, K., and Eikerling, M. 2011. *J. Electrochem. Soc.*, **158**(1), B18–B28, Figures 1,2,3,4,5,6. Copyright (2011), the Electrochemical Society. With permission.)

meaning that $\Gamma_{elec}$ increases faster with decreasing $\phi^M$. This is consistent with the physical interpretation of more extensive pore wall charging with a given shift in potential, when $C_H$ is higher.

For $\phi^M$ close to or above $\phi^{pzc}$, $\Gamma_{elec} \sim 0$ and the following approximate relation can be found:

$$\Gamma_{elec} \approx \exp\left[-\frac{F}{2R_gT}\left(\phi^M - \phi^{pzc}\right)\right].$$  (3.87)

It shows that $\Gamma_{elec}$ and $c_{H^+}$ both asymptotically approach zero as $\phi^M$ increases above the nominal value of the pzc.

The expressions discussed above emphasize the importance of the metal charging behavior for the performance of the pore. In the presented approach, $\phi^{pzc}$, roughly proportional to the metal work function, is a measure of the propensity of a given metal–solution interface to retain electrons at its surface. For a given operating electrode potential $\phi^M$, higher $\phi^{pzc}$ corresponds to more negative $\sigma_M$ and higher $\Gamma_{elec}$. In the linear region, given by Equation 3.86, a shift in $\phi^{pzc}$ by 0.3 V would shift $\Gamma_{elec}$ by $\sim 0.8$. This implies that even an electrocatalytically inactive catalyst support may exert a considerable influence on $\Gamma_{elec}$, through its surface charge. Since $\Gamma_{elec}$ has been shown to exhibit particle size (Mayrhofer et al., 2005, 2008) and roughness dependencies (Climent et al., 2006), this aspect has expected implications for the choice of Pt particle size and shape as well.

Equations 3.86 suggest that $\Gamma_{elec}$ increases indefinitely as $\phi^M$ decreases. However, at high-enough current densities, transport limitations will affect the overall pore effectiveness factor $\Gamma_{pore}$. Therefore, oxygen transport has to be considered. The solution of the diffusion problem is straightforward. It leads to the definition of an effectiveness factor due to oxygen transport, given by

$$\Gamma_{O_2} = \int_0^1 C_{O_2}^s(\xi)\,d\xi = \frac{2R_p}{L_p} \sum_{k=1}^\infty \frac{R_p\Lambda}{\lambda_k\left[\lambda_k^2 + \left(R_p\Lambda\right)^2\right]} \tanh\left(\frac{\lambda_k L_p}{R_p}\right), \qquad (3.88)$$

where $\Lambda = j_{id}\Gamma_{elec}/(4c_{O_2}^0 D_{O_2}F)$. The values of $\lambda_k$ are obtained as the roots of

$$R_p\Lambda J_0\left(\lambda_k\right) = \lambda_k J_1\left(\lambda_k\right), \qquad (3.89)$$

with $J_i$ as Bessel functions of the first kind. The impact of oxygen transport effects on $\Gamma_{pore}$ is more pronounced with increasing pore length $L_p$ and larger catalyst layer overpotential $\eta_c$.

Using COMSOL multiphysics, it was verified that the approximate analytical solution, presented in this section, is in excellent agreement with the full numerical solution of coupled PNP and transport equations in 2D. Generally, the approximation $\Gamma_{pore} \approx \Gamma_{elec}\Gamma_{O_2}$ with expressions for effectiveness factors given in Equations 3.84 and 3.88, is sufficiently accurate for relevant structures and conditions in UTCLs.

The effects of $L_p$ and $\phi^{pzc}$ on the effectiveness factor $\Gamma_{pore}$ are illustrated in Figure 3.28. As can be seen in Figure 3.28a, the effect of oxygen transport limitations becomes significant at $|\eta_c| \geq 0.4$ V. It depends on the value of $L_p$ and leads to the occurrence of a maximum in $\Gamma_{pore}$. For $L_p \sim 100$ nm or smaller, the impact of oxygen diffusion is small. For $L_p \sim 1\,\mu$m, oxygen transport exerts a significant impact on $\Gamma_{pore}$, resulting from a severe depletion of oxygen along the pore, seen in the inset.

As discussed above, $\phi^{pzc}$ represents the metal charging properties and exerts the most significant effect on $\Gamma_{pore}$, as can be deduced from Figure 3.28b. Generally, higher $\phi^{pzc}$ leads to a higher proton affinity of the nanopore, higher electrostatic effectiveness $\Gamma_{elec}$, and higher overall $\Gamma_{pore}$. Acceptable values of $\Gamma_{pore}$ demand $\phi^{pzc} > 0.7$ V$_{SHE}$, whereas $\phi^{pzc} < 0.3$ V$_{SHE}$ gives $\Gamma_{pore} \approx 0$, corresponding to insignificant utilization of Pt.

Based on the results presented in this section, the design criteria for UTCLs can be refined: (i) the thickness should be 200 nm (or less), (ii) charging properties of catalyst and support materials should be such that $\phi^{pzc} > 0.7$ V$_{SHE}$, (iii) the pore radius is not a significant parameter, as long as it remains within a range of 2 nm $< R_p <$ 20 nm, and (iv) too small a radius may affect the intrinsic transport properties of the pore (not accounted for in the present model), whereas too large a radius could infringe on requirements in terms of thickness and ECSA.

## EVALUATION OF NANOPORE MODEL

An obvious approach in evaluating the nanopore model would involve a measurement of the proton conductivity of a nanoporous layer with charged metallic walls as a function of the applied voltage. Proton concentration in a conductive nanoporous membrane is a function of $\phi^{pzc}$. Assuming bulk-like proton transport, the variation of

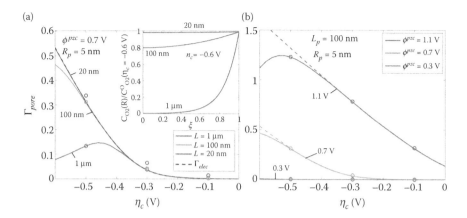

**FIGURE 3.28** The impact of pore length and metal charging properties on the pore effectiveness factor $\Gamma_{pore}$. For a reference set parameter with $\phi^{pzc} = 0.7$ V versus SHE, $R_p = 5$ nm, and $L_p = 100$ nm, the graphs show the effects of (a) pore length and (b) potential of zero charge, with values of these parameters shown in the legends. (Reprinted from Chan, K., and Eikerling, M. 2011. *J. Electrochem. Soc.*, **158**(1), B18–B28, Figures 1,2,3,4,5,6. Copyright (2011), the Electrochemical Society. With permission.)

proton concentration is expected to be proportional to proton conductivity. In principle, the tunability of the ion conductivity of porous gold matrices with $\phi^{pzc}$ has already been demonstrated in Nishizawa et al. (1995) and Kang and Martin (2001). Proton conductivity could be probed by EIS, a standard tool for the characterization of porous electrodes (Barsoukov and Macdonald, 2005). The theoretical basis of EIS applied to UTCLs has been developed by Chan and Eikerling (2012). However, application to UTCLs is not straightforward.

An initial evaluation of the nanopore model can be conducted by comparison with electrochemical performance data for MEAs that utilize UTCLs on the cathode side. The two types of experimental materials considered are nanoporous gold leafs plated with varying amounts of Pt (Pt-NPGL) (Zeis et al., 2007) and Pt NSTF layers of 3M (Pt-NSTF) (Debe et al., 2006).

A simple scale-up procedure gives the relation between UTCL overpotential $\eta_c$ and fuel cell current density $j_0$,

$$j_0(\eta_c) = j_{id}(\eta_c)\,\Gamma_{pore}S_{ECSA}, \tag{3.90}$$

with $\Gamma_{pore}$ as the solution of the nanopore model. The ECSA of the catalyst, $S_{ECSA}$, can be deduced from information on pore space morphology and catalyst dispersion. Using $\eta_c = \phi^M - \phi^M_{eq}$ and the definition of $\Gamma_{pore}$, an expression is obtained for the fuel cell potential:

$$E_{fc}(j_0) = \phi^M_{eq} + \eta_c - j_0(\eta_c)\,R_{PEM} - \eta_{other}. \tag{3.91}$$

where $E^{eq} \equiv \phi_{eq}^{M}$ and $R_{PEM}$ is the PEM resistance. The term $\eta_{other}$ includes anode overpotential, mass transport losses due to reactant transport in diffusion media, as well as contact resistances. In the following analysis, $\eta_{other}$ is neglected.

For simplification, UTCLs are assumed to be composed of cylindrical water-filled pores with monodisperse $R_p$ and $L_p$. The potential of zero charge, $\phi^{pzc}$, is considered as an adjustable parameter. Further input parameters needed in the model include $R_{PEM}$ and kinetic parameters of the ORR, namely, $j_*^0$, $\gamma_{H^+}$, and $\alpha_c$ (Bonakdarpour et al., 2007; Sarapuu et al., 2008).

Figure 3.29 compares model and experimental polarization data for Pt-NPGL layers (Zeis et al., 2007) with Pt loadings of $m_{Pt} = 20, 25, 51$ μg cm$^{-2}$. The main effect seen in this figure is the downward shift in polarization curves with decreasing Pt loading. Normally such a shift in $E_{fc}$ would be attributed to a variation in the exchange current density, a kinetic effect. However, for the model results in Figure 3.29 the observed downward shift of $E_{fc}$ has been reproduced by decreasing the value of $\phi^{pzc}$ from 0.9 V$_{SHE}$ at $m_{Pt} = 51$ μg cm$^{-2}$ to 0.45 V$_{SHE}$ at $m_{Pt} = 20$ μg cm$^{-2}$. The decrease of $\phi^{pzc}$ corresponds to a decrease of the proton concentration in pores, producing the same effect as a decrease of $j_*^0$. In this scenario, it is not the electrode kinetics but local electrostatic reaction conditions in pores that cause the performance change.

Figure 3.30 shows a comparison of calculated polarization curves with experimental data for Pt-NSTF UTCL (Debe et al., 2006). The pore radius $R_p$ was assumed to be 20 nm. A value of $\phi^{pzc} = 0.7$ V$_{SHE}$ gives good agreement with experimental data. The deviation of the model from experimental data at high current densities is attributed to excessive water accumulation in porous diffusion media on the cathode side, strongly inhibiting the oxygen supply. The inset shows variations of the effectiveness factor

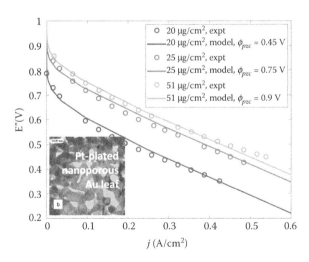

**FIGURE 3.29**  A comparison of nanopore model and experimental polarization data for Pt-NPGL layers. (Reprinted from Chan, K., and Eikerling, M. 2011. *J. Electrochem. Soc.*, **158**(1), B18–B28, Figures 1,2,3,4,5,6. Copyright (2011), the Electrochemical Society. With permission.)

as a function of $E_{fc}$. As can be seen, the total effectiveness $\Gamma_{pore}$ exhibits a maximum of $\Gamma_{pore} \approx 0.15$ at $E_{fc} \approx 0.75$ V. This value suggests that the *effectiveness factor of Pt utilization* in Pt-NSTF catalyst layers is still rather low, leaving significant room for improvement. This value can be compared with those obtained for conventional CCLs in the section "Hierarchical Model of CCLs Operation."

The Pt effectiveness in Pt-NSTF-type CL remains quite low because of a combination of the relatively low value of $\phi^{pzc} = 0.7$ V$_{SHE}$, diminishing the electrostatic effectiveness, and the relatively high thickness of $\sim$250 nm. This incurs considerable oxygen depletion effects at $E_{fc} < 0.85$ V.

Figure 3.31 is a key result of the presented nanopore model for UTCLs. It compares the calculated proton concentration in Pt-NSTF layers, displayed as pH versus $E_{fc}$, with that assumed for ionomer-impregnated CLs. In comparison to proton concentrations in ionomer-impregnated CLs, proton concentrations in ionomer-free UTCLs are significantly lower and exhibit a strong dependence on the fuel cell potential. Of course, the metal charging behavior will have an impact on electrostatic reaction conditions in ionomer impregnated CLs as well, albeit weaker than in UTCLs. The essence of the UTCL model is that the proton concentration in nanopores improves with decreasing value of $\phi^M$ and $E_{fc}$.

Arising from the coupling of proton concentration and solution phase potential, implied by Equation 3.77, the overall reaction order of the ORR with respect to proton concentration is $\gamma_{H^+}^{eff} = \gamma_{H^+} - \alpha_c$, as seen in Equation 3.69. A value $\gamma_{H^+}^{eff} = 1/2$ represents experimental kinetic data well. Therefore, the dependence of ORR activity

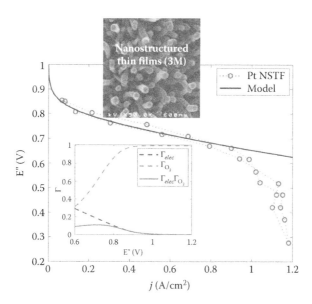

**FIGURE 3.30** A comparison of nanopore model and experimental polarization data for Pt-NSTF layers. The inset shows the effectiveness of Pt utilization. (Reprinted from Chan, K. and Eikerling, M. 2011. *J. Electrochem. Soc.*, **158**(1), B18–B28, Figures 1,2,3,4,5,6. Copyright (2011), the Electrochemical Society. With permission.)

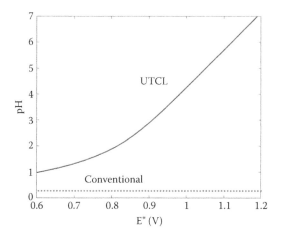

**FIGURE 3.31** A comparison of the calculated proton concentration in Pt-NSTF layers, displayed as pH versus $E_{fc}$, with that encountered in ionomer impregnated CLs. (Reprinted from Chan, K. and Eikerling, M. 2011. *J. Electrochem. Soc.*, **158**(1), B18–B28, Figures 1,2,3,4,5,6. Copyright (2011), the Electrochemical Society. With permission.)

on proton concentration is relatively weak. In the normal potential region of PEFC operation, the reduction in proton concentration in UTCLs has a modest impact on performance that is compensated well by the higher intrinsic activity of NSTF catalysts. An ideal electrocatalyst would have $\gamma_{H^+}^{eff} = 0$. It is an open question to be answered by detailed studies of ORR reaction mechanisms, as to whether such a catalyst could be found.

In the considered model, electrostatic interaction between protons and metal surface charge determines the distributions of protons and electrostatic potential in the pore. These phenomena distinguish the present pore model from the gas- and electrolyte-filled single pore models pioneered by Srinivasan et al. (1967), Srinivasan and Hurwitz (1967) and De Levie (Levie, 1967). With the explicit consideration of the pore wall surface charge, the potential of zero charge $\phi^{pzc}$ of the catalyst material emerges as the crucial parameter. It represents the propensity of the metal to retain electronic charge. The lower the applied potential relative to $\phi^{pzc}$, the more negative the surface charge, and the higher the proton concentration and current density produced at the pore walls. Thus, at a given $\phi^M$, a higher $\phi^{pzc}$ is desirable since it increases the net activity of the nanoporous medium for reduction processes. UTCLs can outperform conventional catalyst layers provided that $\phi^{pzc}$ is not too small.

Pt dissolution mainly occurs in the high potential region where proton concentrations in UTCLs are significantly smaller than those in ionomer-impregnated CLs, as can be seen in Figure 3.31. Since Pt dissolution exhibits a strong dependence on proton concentration $k_{diss} \propto (c_{H^+})^2$ (Rinaldo et al., 2010), Pt dissolution rates in water-filled nanochannels of UTCLs should be vanishingly small for $\phi^M > \phi^{pzc}$. This model-based prediction is consistent with experimental data for 3M NSTF layers (Debe et al., 2006).

As discussed, variation of $\phi^{pzc}$ by materials selection and design could be a viable route to fine-tune UTCLs, in view of both high ORR activity and low Pt dissolution rate. Where $\phi^{pzc}$ is high, that is, $\phi^M - \phi^{pzc} < 0$ and $|\phi^M - \phi^{pzc}| \gg R_g T/F$, a high Helmholtz capacitance $C_H$ would further improve the Pt effectiveness. For these conclusions to be valid, kinetic parameters of the ORR must be such that $\gamma_{H^+} - \alpha_c > 0$ so that an increasing proton concentration results in higher reaction rate of the ORR. Kinetic ORR data suggest that $\gamma_{H^+} - \alpha_c = 1/2$, as discussed.

## NANOPROTONIC FUEL CELLS: A NEW DESIGN PARADIGM?

Basically, the nanoporous water-filled medium with chargeable metal walls works like a tunable proton conductor. It could be thought of as a nanoprotonic transistor. In such a device, a nanoporous metal foam is sandwiched between two PEM slabs, acting as proton source (emitter) or sink (collector). The bias potential applied to the metal phase $\phi^M$ controls proton concentration and proton transmissive properties of the nanoporous medium. The value of $\phi^M$ needed to create a certain proton flux depends on surface charging properties and porous structure of the medium. Moreover, coating pore walls with an electroactive material, for example, Pt, would transform it from a tunable proton conductor into a catalytic layer with proton sinks at the interface. Owing to the intrinsically small reaction rate of the ORR, it would not significantly affect the proton transport properties.

For simple metals, the metal charging behavior can be described by the potential of zero charge $\phi^{pzc}$. If $\phi^M - \phi^{pzc} > 0$ for such a medium, proton transport will be suppressed. For $\phi^M - \phi^{pzc} < 0$, it should transmit protons with a protonic resistance that decreases upon decreasing $\phi^M$. This tunability of proton conductivity could be applied as a method for determining $\phi^{pzc}$ of porous metallic materials, for example, by using the linear relation of Equation 3.74. This setup would allow for systematic studies of effects of materials composition, surface roughness, and surface heterogeneity on $\phi^{pzc}$.

In conclusion, the main challenge in the design of nanoporous materials for UTCLs is to fine-tune the proton concentration in order to optimize the interplay of ORR and Pt dissolution. This fundamental principle has not been widely realized, as most efforts in fuel cell electrocatalysis compare candidate catalyst materials at identical and normally high proton concentration. Proton concentration is usually not considered a parameter to tinker with although it is the key card in the game.

As a further device modification, a thin porous metal foam with tunable proton concentration and vanishing electrocatalytic activity could be inserted between PEM and UTCLs. This layer could reduce proton concentration in the UTCL to a level that optimizes the interplay of ORR rate and Pt dissolution rate.

## STRUCTURE FORMATION IN CATALYST LAYERS AND EFFECTIVE PROPERTIES

Discussed in this section are computational approaches to study structural correlations and dynamical behavior of distinct materials and phases in CLs. Figure 3.32

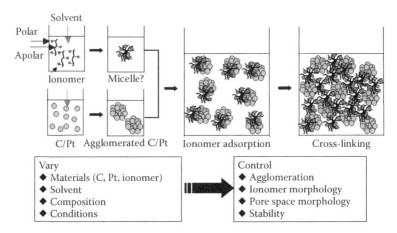

**FIGURE 3.32**  A schematic illustration of the catalyst layer formation process by mixing of Pt/C particles and ionomer in a solvent.

is a schematic illustration of the *catalyst layer formation process*, by way of self-assembly in colloidal solution. Primary carbon particles, with Pt nanoparticles deposited onto their surface, aggregate to form a high-surface-area porous substrate. Ionomer molecules assemble into a network structure that is embedded into the porous medium (Uchida et al., 1995b, 1996).

Various modeling approaches have been developed to describe the present generation of ionomer-bound CLs. Newer CL models account for agglomerate formation and bimodal porous architecture (Eikerling, 2006; Sadeghi et al., 2013b). In spite of progress in modeling and characterization, important structural details of CLs are still unresolved (Eikerling et al., 2008). Debates evolve around the composition and size of Pt/C agglomerates; structure, distribution, and function of ionomer; and heterogeneous wetting properties of internal pore surfaces.

It is generally understood that the ionomer phase in conventional CLs plays a versatile role in determining the density distribution and transport properties of protons, the water distribution in the porous medium and the utilized fraction of the total Pt surface. It is speculated, albeit not reliably known, that the ionomer structure in CLs deviates decisively from the self-aggregated ionomer structure in PEMs. Models of proton conductivity should be based on the specific ionomer structure in CLs. It is often believed that only Pt at Pt–ionomer interfaces is active. However, such a condition is overly restrictive and misleading. The primary medium for proton transport in pores of PEMs or CLs is water. A sufficient condition for rendering the catalyst surface electrocatalytically active is the wetting by liquid water, as this ensures continuous supply of protons to reaction sites. Thus, the foremost objective of CL design can be subsumed as the requirement to provide a large water-wetted fraction of Pt surface area that is continuously connected to the proton supplying ionomer network.

Improvements in structure versus dynamics relationships of CLs demand a high level of control over parameters and processes involved in the fabrication process. For good references on ink formation and *CL fabrication*, the reader is referred to

Wang et al. (2004) and Zhang (2008). Systematic modifications at the fabrication stage include (i) the selection of materials (e.g., carbon or metal-oxide-based support materials, Pt or Pt-alloy catalyst materials, perfluorinated or alternative ionomer materials), (ii) size dispersion of catalyst and support particles, (iii) gravimetric composition of the ink (amounts of carbon, ionomer, and Pt) and solvent properties, (iv) thickness of the CL, and (v) fabrication conditions (temperature, solvent evaporation rate, pressure, and tempering procedures).

These parameters and conditions determine complex interactions between Pt nanoparticles, carbon support, ionomer molecules, and solvent, which control the catalyst layer formation process. Self-organization of ionomer and carbon/Pt in the colloidal ink leads to the formation of phase-segregated and agglomerated morphologies. The choice of a dispersion medium determines whether ionomer exists in solubilized, colloidal, or precipitated form. This influences the microstructure and pore size distribution of the CL (Uchida et al., 1996). It is believed that mixing of ionomer with dispersed Pt/C catalysts in the ink suspension, prior to deposition to form a CL, enhances the interfacial area of Pt with water in pores and with Nafion ionomer.

This section presents a review of atomistic simulations and of a recently introduced mesoscale computational method to evaluate key factors affecting the morphology of CLs. The bulk of molecular dynamics studies in PEFC research has concentrated on proton and water transport in hydrated PEMs (Cui et al., 2007; Devanathan et al., 2007a,b,c; Elliott and Paddison, 2007; Jang et al., 2004; Spohr et al., 2002; Vishnyakov and Neimark, 2000, 2001). There has been much less effort in using MD techniques for elucidating structure and transport properties of CLs, particularly in three-phase systems of Pt/carbon, ionomer, and gas phase.

*Coarse-grained molecular dynamics* (CGMD) simulations have become a viable tool to unravel self-organization phenomena in complex materials and to analyze their impact on physicochemical properties (Malek et al., 2007; Marrink et al., 2007; Peter and Kremer, 2009). Various MD simulations to study microstructure formation in catalyst layers will be discussed. The impact of structures obtained on pore surface wettability, water distribution, proton density distribution, and Pt effectiveness will be evaluated.

## MOLECULAR DYNAMICS SIMULATIONS

In classical molecular dynamics (MD) simulations, the system of $N$ atoms is treated as a set of interacting particles (Allen and Tildesley, 1989). Atoms are represented by spherical nuclei that attract or repel each other. Their electronic structure is not considered explicitly. After assigning point charges to each particle, as appropriate, the forces acting on the particles are derived from a combination of bonding, nonbonding, and electrostatic potentials. The motions of particles (atoms and ions) are calculated using the laws of classical mechanics.

Before starting a simulation, a model system is built that consists of all chemical components present at the desired composition and density in a simulation box. Just like any real experiment, this system needs to be carefully prepared. It should be a realistic representation of the system that is to be studied. The result of a molecular dynamics simulation is a time trajectory of positions and velocities of all $N$ particles in the system. If simulated with an appropriate time step and for a sufficiently long time, thermodynamic properties, spatial and temporal correlation functions, and transport properties can be determined.

The meaningful time of the simulated trajectory depends on the characteristic length scales in the system under study and the timescale needed for calculation of physical parameters. Nowadays, computationally feasible time trajectories in atomistic MD simulations extend from a few nanoseconds to microseconds.

The time trajectory of an MD system is obtained from solving a system of second-order differential equations that follow from Newton's second law $m_i \frac{d^2 \mathbf{r}_i}{dt^2} = \sum_j \mathbf{F}_{ij} + \sum \mathbf{F}_k, i = 1 \ldots N$. Here, $i$ is the particle label, $m_i$ the particle mass, and $\mathbf{r}_i$ the position vector of this particle. The forces $\mathbf{F}_{ij}$ represent two-body interactions between atoms $i$ and $j$, while forces $\mathbf{F}_k$ are a result of the action of external fields, for example, electric fields. Determining these forces as functions of atomic and molecular degrees of freedom for all atoms defines the force field of the atomistic model. Given all force terms, the equations of motion can be integrated.

The choice of the force field is the key to accurate results that reproduce or predict physical properties and phenomena in the system under study. The forces acting on the nuclei are derived from the gradients of the potential energy function

$$\mathbf{F}_i = -\nabla_{r_i} U. \tag{3.92}$$

The force field involves contributions resulting from nonbonded interactions between all nuclei and bonded interactions between nuclei that are part of the same molecule. The nonbonded interactions consist of electrostatic interactions, van der Waals interactions, and polarization effects. Polarization effects are the result of spatially varying electron densities. They cannot be described explicitly using force field methods, which invariably ignore electron dynamics. It is common practice to include them implicitly in the van der Waals interactions. This leaves two terms for the nonbonded interactions. The first term corresponds to Coulomb interactions

$$U_{el}\left(r_{ij}\right) = \sum_{ij} \frac{q_i q_j}{4\pi \varepsilon_0 r_{ij}} \tag{3.93}$$

between two charged spheres at a distance $r_{ij}$ from each other. In this equation, $q_i$ represents the charge of particle $i$ and $\varepsilon_0$ is the dielectric permittivity of vacuum. Particles can be assigned partial charges or integer values in the case of ions.

The second type of nonbonded interactions are commonly described by a Lennard–Jones potential with the standard form

$$U_{LJ}\left(r_{ij}\right) = 4\varepsilon_{ij} \left[ \left(\frac{\sigma_{ij}}{r_{ij}}\right)^{12} - \left(\frac{\sigma_{ij}}{r_{ij}}\right)^{6} \right]. \tag{3.94}$$

The first term in brackets in Equation 3.94 represents a strong repulsion of atoms at short range, attributable to their overlapping electron densities. It follows from the Pauli exclusion principle of quantum mechanics. The functional form used for this term ($\propto r^{-12}$) is empirical. The attractive long-range term corresponds to dispersion or van der Waals forces. It has the functional form $\propto r^{-6}$. In the equation above, $\varepsilon_{ij}$ represents the depth of the potential at the minimum $r_{\min} = 2^{1/6}\sigma$ and $\sigma_{ij}$ is the point at which $U_{LJ} = 0$.

For atoms that are part of molecules, bonded interactions have to be calculated, conferring flexibility to the molecular structure. These interactions account for effects of bond stretching (two-body interaction), bond angle variations (three-body interaction), and dihedral angle variations (four-body interaction). The first two interactions can be described using a harmonic potential. Dihedral interactions cannot be described using a harmonic potential but rather by a periodic function to account for the rotational symmetry. They play an important role in simulations of hydrated Nafion ionomer, as discussed in more detail in the section "Molecular Modeling of Self-Organization in PEMs."

## ATOMISTIC MD SIMULATIONS OF CLs

The Pt/C composite of a typical CL is fabricated by a "colloidal crystal templating technique" (Moriguchi et al., 2004). Besides carbon black (Vulcan XC-72 by Cabot Corp. or Ketjen black by Tanaka), synthetic porous carbon particles have recently been used to improve surface area and electrochemical activity. In a standard fabrication process, Pt particles are deposited on carbon surfaces by reducing $H_2PtCl_6$ on the carbon surface (Yamada et al., 2007). Carbon particles are dispersed in a tetrahydrofuran solution of $H_2PtCl_6$ by an ultrasonic treatment, while formic acid is added as the reducing agent. Filtering the solution collects the dispersed Pt/carbon particles.

In MD simulations, the molecular adsorption concept is used to describe Pt–C interactions during the fabrication processes. The Pt complexes are mostly attached to the hydrophilic sites on carbon particles, namely, carbonyl or hydroxyl groups (Hao et al., 2003), involving both physical and chemical adsorption. Carbon particle preparation, impregnation, and reduction are three main steps of catalyst preparation. The *point of zero charge* (PZC, not to be confused with the potential of zero charge, pzc) specifies the pH range, at which the impregnation step should be carried out. The PZC is an important parameter in catalyst preparation.

Several MD simulations have focused on Pt nanoparticles, adsorbed on carbon, in the presence or absence of ionomer (Balbuena et al., 2005; Chen and Chan, 2005; Lamas and Balbuena, 2003, 2006). Lamas and Balbuena (2003) performed classical MD simulations for a simple model of the interface between graphite-supported Pt nanoparticles and hydrated Nafion. In MD studies of CLs, the equilibrium shape and structure of Pt clusters have been simulated using the *embedded atom method* (EAM). Semiempirical potentials such as the many-body *Sutton–Chen potential* (SC) (Sutton and Chen, 1990) are useful choices for simulations of close-packed metal clusters or nanoparticles. These potentials include the effect of the local electron density that is generated by many-body terms. The SC potential for Pt–Pt and Pt–C interactions provides a reasonable description of the properties of small Pt clusters. The potential energy in the SC potential is

$$U_{pp}\left(r_{ij}\right) = \varepsilon_{pp}^{SC} \sum_{1}^{N} \left[ \frac{1}{2} \sum_{j \neq i}^{N} \left( \frac{\sigma_{pp}^{SC}}{r_{ij}} \right)^{n} - c\sqrt{\rho_i} \right], \text{ with } \rho_i = \sum_{j \neq i}^{N} \left( \frac{\sigma_{pp}^{SC}}{r_{ij}} \right)^{m}, \quad (3.95)$$

where for Pt–Pt interaction $\varepsilon_{Pt-Pt}^{SC} = 0.019833$ eV, $\sigma_{Pt-Pt}^{SC} = 3.92$ Å, $n = 10$, $m = 8$, and $c = 34.408$. The first term represents a pairwise repulsive potential, while the second represents the metallic bonding energy associated with the local electron density. The weak interaction between a Pt cluster and a graphitic substrate could be accounted for using 12-6 LJ interactions with $\varepsilon_{Pt-C} = 0.022$ eV and $\sigma_{Pt-C} = 2.905$ Å in Equation 3.95. In order to obtain reasonable results with this empirical description, the size of Pt clusters should be sufficiently large. Fitting the LJ equation to the SC potential provides LJ parameters for Pt–Pt interactions, given by $\varepsilon_{Pt-Pt} = 2336$ K and $\sigma_{Pt-Pt} = 2.41$ Å. These simulations, based on LJ interactions adapted from the SC potential, allow one to account for interactions between Pt nanoclusters as well.

Nafion is typically represented by oligomer models. Interactions in ionomer systems can be described using the Dreiding force field (Goddard et al., 2006; Jang et al., 2004). This force field parameterizes interactions due to bond stretching, bond angle variations, and dihedral angle variations, including nonbonded (LJ and electrostatic) interactions of components involved. A two-site model and single-point-charge (SPC) model are usually employed for oxygen and water, respectively (Wu et al., 2006).

## MESOSCALE MODEL OF SELF-ORGANIZATION IN CATALYST LAYER INKS

Because of computational limitations, atomistic models are not able to probe the morphology of catalyst layers. Despite undeniable progress in the development of multiscale modeling approaches (Morrow and Striolo, 2007), enormous challenges remain in bridging atomistic simulations and continuum models of materials for PEFC applications. CGMD methods have become a vital link to bridge this gap. These methods have proved to be highly insightful for the understanding of self-organization phenomena and adhesion properties that have to be considered in fabrication and operation of CLs.

The remainder of this section describes a *CGMD methodology* used to unravel *self-organization phenomena in the CL* and to analyze their impact on physicochemical properties (Malek et al., 2007; Marrink et al., 2007). In particular, the focus will be on *structure and distribution of ionomer*. Moreover, it will explore the implications of ionomer morphology and porous structure on water distribution (wettability), Pt utilization, and proton transport properties. Validation of the emerging structural picture by experimental data on adsorption and transport properties will be discussed briefly.

The methodology for performing CGMD studies of self-organization in catalyst layer mixtures has been introduced by Malek et al., (2007; 2011).

The *coarse-grained model* is developed in two main steps. First, all atomistic and molecular species, that is, Nafion ionomer chains, solvent molecules, water, hydronium ions, carbon, and Pt particles, are replaced by spherical beads with predefined subnanoscopic length scale. Each bead represents a small group of atoms or molecules. Normally, four main types of spherical beads can be distinguished: metallic, polar, nonpolar, and charged beads (Marrink et al., 2007). Second, renormalized interaction energies between beads must be determined, leading to the definition of a force field for the CGMD model.

Clusters of four water molecules are represented by polar beads. Clusters of three water molecules plus a hydronium ion form charged beads with charge $e_0$ (elementary charge). Sidechains of ionomer, terminated by sulfonic acid head groups, are coarse-grained as charged spherical beads with a charge $-e_0$ (Eikerling and Malek, 2009; Malek et al., 2007; Wescott et al., 2006). The hydrophobic Nafion backbone is replaced by a coarse-grained chain of 40 nonpolar beads; each bead represents a four monomeric unit ($-[-CF2-CF2-CF2-CF2-CF2-CF2-CF2-CF2-]-$) of the backbone. The corresponding stretched length of the backbone chain is $\sim$30 nm (Malek et al., 2007). For computational convenience, all beads in the simulation are assumed as spherical with radius 0.47 nm and volume 0.43 nm$^3$.

The work of Malek et al. (2011) employed a new multiscale coarse-graining strategy to implement explicit Pt particles into the CG model. It considered Pt nanoparticles of cubo-octahedral shape, enclosed by (111) and (100) facets, as shown in Figure 3.4. This shape is believed to be among the most stable conformations of nanoparticles with fcc structure (Antolini, 2003; Antolini et al., 2002; Ferreira et al., 2005; Lee et al., 1998). Coarse-grained cubo-octahedral Pt nanoparticles of size $\sim$2 nm were modeled using an approximate 5:1 mapping of Pt atoms in particles, replacing 201 Pt atoms with 38 Pt beads, as shown in Figure 3.33. It should be noted that this mapping preserves the fcc lattice geometry and the shape of the original Pt nanoparticle.

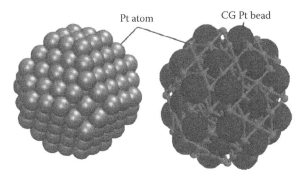

Pt atom CG Pt bead

**FIGURE 3.33** Atomistic and coarse-grained configurations of cubo-octahedral Pt nanoparticles. The coarse-grained representation is obtained by an approximate 5:1 mapping of atoms on CG sites, preserving the particle shape. (Reprinted from *Electrocatalysis*, Microstructure of catalyst layers in PEM fuel cells redefined: A computational approach, **2**(2), 2011, 141–157, Figures 1,2,4,6,7,8,9,10,13, Malek, K., Mashio, T., and Eikerling, M. Copyright (2011) Spinger. With permission.)

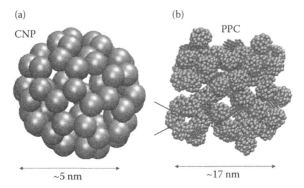

**FIGURE 3.34** A coarse-grained model representation of Pt/C particles. (a) shows a coarse-grained carbon nanoparticle (CNP) after energy minimization. The initial particle configuration was a C540 atomistic fullerene system that was mapped onto the CG representation with a mapping ration of 9:1. (b) is an aggregated primary Pt/C particle (PPC), formed from 52 CNPs and 8 Pt nanoparticles (represented by smaller beads), after energy minimization. (Reprinted from *Electrocatalysis*, Microstructure of catalyst layers in PEM fuel cells redefined: A computational approach, **2**(2), 2011, 141–157, Figures 1,2,4,6,7,8,9,10,13, Malek, K., Mashio, T., and Eikerling, M. Copyright (2011) Spinger. With permission.)

Carbon particles can be coarse-grained in various ways, building upon a technique called multiscale coarse-graining (MS-CG) (Izvekov and Violi, 2006; Izvekov and Voth, 2005; Izvekov et al., 2005). In this method, the CG potential parameters are obtained from atomistic-level interactions by a systematic procedure. Using this technique, a model for approximately spherical carbon particles can be generated. The simulation in Malek et al. (2011) started by defining CG sites in the C540 atomistic fullerene system, using a 9:1 mapping (atom/bead ratio) of carbon atoms. Each of the C nanoparticles in the coarse-grained model consisted of 60 beads of radius 0.47 nm (Malek et al., 2011), preserving the hexagonal arrangement of carbon atoms. The CG-C60 configuration has a bead–bead distance of ∼0.94 nm. The diameter of the initial carbon nanoparticles was ∼5 nm.

Energy minimization of initially spherical CG carbon particles forms primary particles with more irregular shapes, as indicated in Figure 3.34(a). These primary carbon nanoparticles are denoted as CNP. Subsequently, a simulation batch of 52 CNPs and 8 CG Pt particles equilibrates to form primary Pt/C particles, denoted as PPC, illustrated in Figure 3.34(b).

The simulation protocol for the formation of Pt-decorated primary C particles (PPCs) mimics catalyst dispersions obtained by pertinent fabrication techniques. In practice, two methods are used to obtain PPCs: (1) impregnation of carbon nanoparticles with Pt precursor or (2) adsorption of Pt oxide or Pt metal colloids onto the carbon surface (Antolini, 2003; Antolini et al., 2002). In the case of impregnation with Pt precursor, diffusion into the pores of each individual support particle can occur. For the second mechanism, colloidal Pt or Pt oxide particles adsorb on the external surface of the support particles; as a result of size exclusion, the accessibility of the inner pores is limited and, therefore, Pt particles are mostly formed on the surface of CNPs.

The structural model of PPCs was fine-tuned to resemble the properties of *Vulcan carbon XC-72* (hydrophobic carbon particles with low internal porosity) (Kinoshita, 1988). These PPC are included in the initial configuration of the simulation box. Although PPCs were allowed to relax during simulations, they retained their initial structure.

## Parameterization of the Coarse-Grained Force Field

The parameterization of CG force fields is not trivial. For the CG study of complex systems such as CLs, the loss of atomistic detail is a real drawback. In fact, this difficulty is not specific to the system under study in this work, but it applies to classical atomistic and coarse-grained molecular modeling. Employing these methods relies on averaging out microscopic degrees of freedom. Electronic structure effects are not explicitly treated. These limitations are well apprehended (Markvoort, 2010). An adequate approach should incorporate essential physicochemical properties of interest, while ignoring certain less important details to keep the calculation computationally viable.

In the approach developed in Malek et al. (2007, 2011), CG interaction parameters were derived mainly from atomistic MD simulations, based on phenomenological and MS–CG methods. In the phenomenological method, the parameters from the Martini force field are used as initial guesses. The initial interactions are chosen to distinguish between polar, charged, hydrophilic, and hydrophobic particles. Although simple, this method predicts well the geometry and structural behavior of composite systems (Everaers and Ejtehadi, 2003). Interaction parameters were improved using an iterative scheme, based on Boltzmann inversion (Reith et al., 2003) for ionomer, water, and hydronium ions, and MS–CG for Pt and carbon. Reviews on approaches used to derive FF parameters can be found in Voth (2008), Peter and Kremer (2009) and Murtola et al. (2009).

In the MS–CG scheme described above (Izvekov and Violi, 2006; Izvekov and Voth, 2005; Izvekov et al., 2005), the positions $\mathbf{R}_{i,m}^{CG}$ of CG beads and forces $\mathbf{F}_{i,m}^{CG}$ acting on CG centers are obtained from reference atomistic MD forces $\mathbf{F}_{i,m}^{ref}$. The reference MD force is calculated by independent, short MD simulations for subsystems of Nafion–water, Pt–carbon, Pt–carbon-water, Pt–Nafion, and carbon–Nafion. The applied *force matching procedure* involves M selected configurations from atomistic simulations (running index m) and N coarse-grained sites (running index i). The forces of the coarse-grained model are

$$\mathbf{F}_{i,m}^{CG} = \sum_{j \neq i} f_{ij}^{CG}\left(\left|\mathbf{R}_{ij,m}^{CG}\right|\right) \frac{\mathbf{R}_{ij,m}^{CG}}{\left|\mathbf{R}_{ij,m}^{CG}\right|}, \tag{3.96}$$

where the parameters $f_{ij}^{CG}(r)$ are chosen in order to minimize the following expression:

$$\varepsilon = \frac{1}{3MN} \sum_{i=1}^{N} \sum_{m=1}^{M} \left|\mathbf{F}_{i,m}^{ref} - \mathbf{F}_{i,m}^{CG}\right|^2 .$$

## TABLE 3.1

**Phenomenological CG interaction parameters between Water (W), Hydronium ion (H), Backbone (B), Sidechain (S), and Carbon (C) in Carbon Particles**

|     | W ($C_6$, $C_{12}$) | H ($C_6$, $C_{12}$) | B ($C_6$, $C_{12}$) | S ($C_6$, $C_{12}$) | C ($C_6$, $C_{12}$) | Pt ($C_6$, $C_{12}$) |
|-----|------------|------------|------------|------------|------------|------------|
| W   | 0.22, 0.0023, | 0.22, 0.0023 | 0.077, 0.00084 | 0.22, 0.0023 | 7.2, 7.2 | 0.22, 0.0023 |
| H   |            | 0.78, 0.00084 | 0.078, 0.00084 | 0.18, 0.0020 | 7.2, 7.2 | 0.22, 0.0023 |
| B   |            |            | 0.15, 0.0016 | 0.77, 0.00084 | 13.6, 13.6 | 0.15, 0.0016 |
| S   |            |            |            | 7.2, 7.2 | 0.22, 0.0023 |
| C   |            |            |            |            | 0.24, 0.0026 |
| Pt  |            |            |            |            |            | 0.086, 0.00093 |

*Note:* This includes Pt beads, initially used in CGMD simulations, with $C_6 = 4\varepsilon\sigma^6$ and $C_{12} = 4\varepsilon\sigma^{12}$.

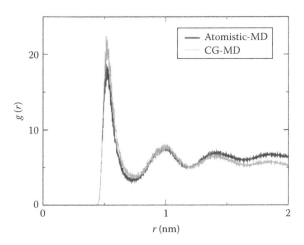

**FIGURE 3.35** A water–water (W–W) radial distribution function obtained from CGMD simulation in comparison with that of an atomistic MD simulation. (Reprinted from *Electrocatalysis*, Microstructure of catalyst layers in PEM fuel cells redefined: A computational approach, **2**(2), 2011, 141–157, Figures 1,2,4,6,7,8,9,10,13, Malek, K., Mashio, T., and Eikerling, M. Copyright (2011) Spinger. With permission.)

Table 3.1 summarizes initial CG interaction parameters used in the parameterization, based on the MS–CG scheme, with $C_6 = 4\varepsilon\sigma^6$ and $C_{12} = 4\varepsilon\sigma^{12}$. During the parameterization, the largest deviation from these initial values was observed for interactions between ionomer backbone and carbon beads. Figure 3.35 compares the water–water RDF obtained from a CGMD simulation with that of the atomistic MD simulation.

### Computational Details

The initial simulation box evaluated in Malek et al. (2011) had a size of $500 \times 500 \times 500 \text{ nm}^3$. It contained eight PPCs, including in total 416 CNPs and varying numbers of coarse-grained Pt, Nafion oligomers and CG hydronium ions, to obtain the desired ionomer and Pt content. Before energy minimization, Pt particles were removed randomly from the equilibrated configuration of PPCs in order to generate simulation boxes with Pt:C mass ratios of 1:1, 1:3, and 1:9. The number of water beads was fixed at a value that corresponded to approximately nine water molecules per ionomer sidechain (i.e., $\lambda = 9$). The ionomer-to-carbon mass ratio, in short I:C ratio, was varied from 0.4 to 1.5 to reproduce the lowest (0.4), typical (0.9), and highest (1.5) ratios in CL fabrication.

Simulations were performed using a modified version of the GROMOS96 force field by adapting the nonbonding and bonding parameters of CG beads (Berendsen et al., 1995; Lindahl et al., 2001; Peter and Kremer, 2009). In the case of nonbonded interactions involving charged sidechains, a negative charge $(-1e_0)$ and parameters for electrostatic repulsion and attraction are assigned to each sidechain. Bonded interactions consist of bond, angular, and dihedral terms. The force constant of the harmonic bond potential is 1250 kJ mol$^{-1}$ nm$^{-2}$. A harmonic potential of cosine type is used for the angular terms, with a force constant of 25 kJ mol$^{-1}$ rad$^{-2}$. The dihedral potentials represent the energy involved in changing the out-of-plane angles.

LJ potentials for nonbonded interactions were allowed to have five levels of interaction strength corresponding to (i) attractive ($\varepsilon_{ij} = 5$ kJ mol$^{-1}$), (ii) semiattractive ($\varepsilon_{ij} = 4.2$ kJ mol$^{-1}$), (iii) intermediate ($\varepsilon_{ij} = 3.4$ kJ mol$^{-1}$), (iv) semirepulsive ($\varepsilon_{ij} = 2.6$ kJ mol$^{-1}$), and (v) repulsive ($\varepsilon_{ij} = 1.8$ kJ mol$^{-1}$) interaction energies. The minimal bead separation was fixed at the value of the bead diameter, that is, $\sigma_{ij} = 0.94$ nm. A cut-off of 1.0 nm was used for van der Waals interactions.

A typical simulation protocol started with an optimization of the initial structure for 100 ps at $T = 0$, using a 20 fs time step in the course of a steep integration procedure. This short energy minimization slightly displaced the overlapped beads of Pt, CNP, ionomer, water molecules, and hydronium ions from their initial positions. Equilibration of the structure was performed using an annealing procedure, during which the system was first expanded over a period of 50 ps by gradually increasing the temperature from 295 to 395 K. Thereafter, a short MD simulation was performed for another 50 ps in an NVT ensemble (thermalization), followed by a cooling down to 295 K.

Following this procedure, MD simulations were conducted for an NPT ensemble. In these simulations, the total energy decreased steeply during an initial equilibration period of $\sim 0.08$ μs, after which it converged and became stable, with a total energy variation $\Delta E <500$ kJ mol$^{-1}$. This variation is small compared to the total energy of the system, typically in the order of $10^6$ kJ mol$^{-1}$. After reaching a stable mass density, production runs were performed in the NPT ensemble for up to 5 μs with an integration time step of 0.04 ps. Statistically independent configurations were extracted from the simulated trajectory for structural analysis.

The temperature was controlled by the Berendsen algorithm, simulating a weak coupling to an external heat bath with given temperature $T_0$. The weak coupling algorithm was applied separately for each component (polymer, carbon particles, water, hydronium ions) with a time constant of 0.1 ps and a temperature of 295 K. During the production run, structures were saved every 500 steps (25 ps) and used for the analysis. All simulations were carried out using a modified GROMACS package (http://www.gromacs.org) (Babadi et al., 2006; Markvoort, 2010).

## Microstructural Analysis

The microstructure has been characterized using various identifiers, including (i) pore size distribution, (ii) size distributions of ionomer and carbon domains, (iii) RDFs, and (iv) two-dimensional density maps. RDFs, $g_{ij}$, between distinct components of the CL provide information on the phase segregation. The definition of RDF was provided in the section "Analysis of the Coarse-Grained Membrane Structure" in Chapter 2. Monte Carlo techniques were employed to determine pore size distribution (Kisljuk et al., 1994) and domain size distributions of ionomer and carbon particles.

Furthermore, a distance analysis method was used to evaluate water and ionomer coverage on the outer surface of C and Pt beads in the CL blend. In this method, the water coverage is determined based on the minimum distance between water and carbon beads. Water beads, which are closer than 0.5 nm to carbon beads, are considered to be attached to the surface of the carbon. Similarly, the coverage of ionomer beads on carbon is determined. The water and ionomer coverages $\theta_W$ and $\theta_I$ are defined by the number of water or ionomer beads in contact with carbon divided by the total number of carbon beads on the surface, that is

$$\theta_W = \frac{\text{\# of water beads in contact with C beads}}{\text{total \# of C beads on surface}}$$

and

$$\theta_I = \frac{\text{\# of ionomer beads in contact with C beads}}{\text{total \# of C beads on surface}}.$$

The total surface area of carbon is evaluated by an independent simulation, in which the carbon phase is position-restrained (fixed), with ionomer and Pt phases removed from the system. After a relatively fast simulation run, water propagates quickly around the carbon phase. In this case, the total number of carbon beads covered by water is proportional to the total carbon surface area.

## Microstructure Formation in CLs

Figure 3.36 shows snapshot of catalyst layer blends obtained after CGMD equilibration in cases where (a) Pt was not included or (b) Pt was included in the simulation blend. Interaction parameters of C particles were selected to mimic properties of VULCAN-type C/Pt particles. They are hydrophobic, with a repulsive interaction with water and Nafion sidechains, and semiattractive interactions with other carbon particles and Nafion backbones.

Final microstructures obtained reveal a high sensitivity of carbon agglomeration and ionomer structure formation to the wetting properties of carbon particles and the strength of ionomer–carbon interactions. While ionomer sidechains are confined in hydrophilic domains, with a weak contact to carbon domains, ionomer backbones are preferentially attached to the surface of carbon agglomerates for the given hydrophobic type of C. As expected, the correlation between hydrophilic species

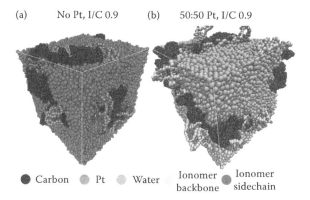

(a)    No Pt, I/C 0.9        (b)    50:50 Pt, I/C 0.9

● Carbon    ● Pt    ● Water    Ionomer    ● Ionomer
                                backbone      sidechain

**FIGURE 3.36** A snapshot of final microstructures of catalyst layer ink mixtures from the CGMD simulations obtained (a) without and (b) with Pt nanoparticles included in the simulation box. The I:C mass ratio in both simulations was 0.9 and the Pt:C mass ratio in (b) was 1. Colors use to represent beads: *Green* solvent beads, *black* carbon beads *blue* ionomer backbone, *red* ionomer sidechain, *gold* Pt. In order to see this figure in color, please, go to the online version of Malek et al. (2011), http://link.springer.com/article/10.1007%2Fs12678-011-0047-0. (Reprinted from *Electrocatalysis*, Microstructure of catalyst layers in PEM fuel cells redefined: A computational approach, **2**(2), 2011, 141–157, Figures 1,2,4,6,7,8,9,10,13, Malek, K., Mashio, T., and Eikerling, M. Copyright (2011) Springer. With permission.)

(water, hydronium ions) and ionomer is significantly stronger than that between those species and C particles. The compactness of carbon agglomerates depends strongly on the dielectric properties of the solvent.

The key observation in the analysis of CGMD structures is the formation of agglomerates with a core–shell structure. The core region consists of nanoporous aggregates of carbon and Pt. Ionomer backbones assemble into a *thin adhesive* film that forms the shell. The thickness of the *ionomer shell* ranges from ~3 to 4 nm. This *ionomer skin layer* exhibits well-packed morphology consisting of several layers of backbone chains. As a result of their hydrophilic nature, sidechains are expelled from the vicinity of carbon. Water and hydronium ions tend to maximize their separation from carbon, while trying to stay in the vicinity of sidechains.

Simulations in Malek et al. (2007) were analyzed to examine the effect of the solvent dielectric constant on structural correlations. Polar solvents ($\varepsilon_r = 20, 80$) behave similarly, while the effect of the apolar solvent ($\varepsilon_r = 2$) is markedly different. Low $\varepsilon_r$ implies stronger correlations between C and hydrophobic polymer backbones, exhibiting stronger phase separation into hydrophobic and hydrophilic domains. The magnitude of short-range interactions and the carbon agglomerate size steadily decrease upon increasing $\varepsilon_r$. Therefore, the carbon agglomerates become more separated, as the pore size between agglomerates increases. The peak positions for long-distance correlations are shifted to larger distances for the apolar solvent. In the presence of apolar solvents, the structural organization into separate hydrophilic and hydrophobic domains spans to higher distances compared to polar solvents.

In an extended CGMD simulation study, Pt nanoparticles in various realistic amounts were included, with varying ionomer content (Malek et al., 2011).

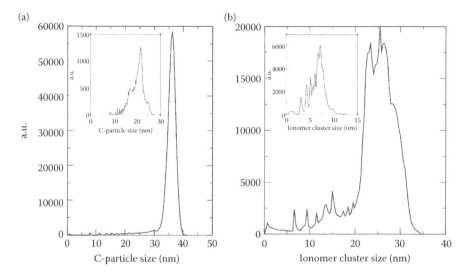

**FIGURE 3.37**   Size distributions of (a) Pt/C aggregates and (b) ionomer clusters in equilibrated CGMD structures that incorporate Pt nanoparticles. The corresponding distributions obtained in Pt-free CL blends are shown in the insets for comparison. (Reprinted from *Electrocatalysis*, Microstructure of catalyst layers in PEM fuel cells redefined: A computational approach, **2**(2), 2011, 141–157, Figures 1,2,4,6,7,8,9,10,13, Malek, K., Mashio, T., and Eikerling, M. Copyright (2011) Spinger. With permission.)

Interaction parameters were refined to describe varying compositions and a larger simulation box was considered.

Figure 3.37 depicts the size distributions of ionomer and carbon domains in the presence of Pt particles. Similar to the pronounced effect of dielectric solvent properties (Malek et al., 2007), Pt particles affect the size and connectivity of ionomer and carbon phase domains. Sizes of Pt/C aggregates range from 30 to 40 nm. Owing to the uniform and small PPCs used in simulations (∼17 nm) and the finite size of the simulation box, computed sizes of Pt/C aggregates are smaller than those determined from TEM analyses (Carmo et al., 2007; Soboleva et al., 2011). Pore size distributions (PSD) were examined using a Monte Carlo procedure (Kisljuk et al., 1994). Results are shown in Figure 3.38. Increasing the Pt content shifts the PSD to larger sizes, with a relatively narrow peak occurring at ∼75 nm.

Pt particles influence the structure of the ionomer film, as shown in Figure 3.39. In the presence of hydrophobic carbon (e.g., Vulcan XC-72), ionomer mostly covers the external surface of Pt/C aggregates. The core region of agglomerates consists of 10–15 aggregated PPCs, resulting in sizes ∼50 nm. As for the case without Pt, a thin ionomer film forms at the agglomerate surface; ionomer domains are connected by fibrous ionomer aggregates. In the absence of Pt nanoparticles, the external surface of the ionomer film is mostly hydrophilic. Here, charged sidechains form a highly ordered array with sidechains pointing away from the carbon surface, that is, preferentially oriented toward the pore interior space. In the presence of Pt nanoparticles,

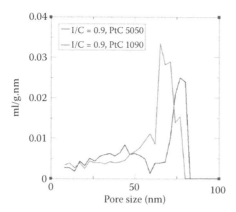

**FIGURE 3.38** Pore size distributions in CL blends obtained by CGMD simulations at different Pt:C mass ratios of 5:5 (PtC 5050) and 1:9 (PtC 1090). (Reprinted from *Electrocatalysis*, Microstructure of catalyst layers in PEM fuel cells redefined: A computational approach, **2**(2), 2011, 141–157, Figures 1,2,4,6,7,8,9,10,13, Malek, K., Mashio, T., and Eikerling, M. Copyright (2011) Spinger. With permission.)

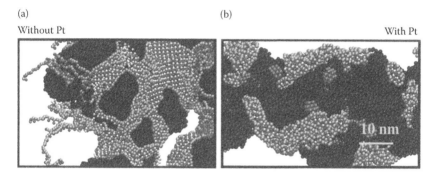

**FIGURE 3.39** The effect of Pt on the ionomer structure in CL blends. (a) shows a Pt-free structure with a hydrophobic C substrate. Ionomer forms a highly ordered layer with hydrophobic backbones attached to the substrate and sidechains oriented toward the interior pore space. The addition of Pt, illustrated in (b), creates a mixed hydrophobic–hydrophilic surface leading to a less uniform spreading of the ionomer film and a greater disorder in the orientation of sidechains. (Reprinted from *Electrocatalysis*, Microstructure of catalyst layers in PEM fuel cells redefined: A computational approach, **2**(2), 2011, 141–157, Figures 1,2,4,6,7,8,9,10,13, Malek, K., Mashio, T., and Eikerling, M. Copyright (2011) Spinger. With permission.)

(Figure 3.39b) the ionomer phase is more clustered and less connected with more anionic sidechains pointing toward the Pt/C surface.

Various RDFs are shown in Figure 3.40 at Pt:C mass ratios from 0 to 1. Figure 3.40 shows that Pt affects the interaction between water and sidechain beads (S–W) and sidechain–sidechain interactions (S–S) significantly. A trend can be discerned that peak heights at low radial separations decrease, with increasing amount of Pt. This affirms the conjecture made previously that the addition of Pt decreases the uniform dispersion and sidechain ordering of the thin adhesive ionomer film.

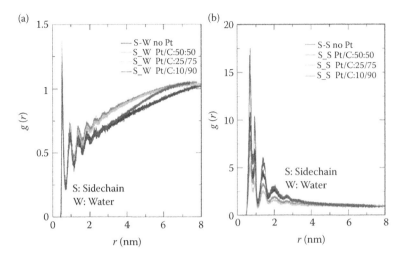

**FIGURE 3.40** Radial distribution functions (RDFs) for (a) sidechain–water and (b) sidechain–sidechain interactions, as a function of Pt content at I:C = 0.9. (Reprinted from *Electrocatalysis*, Microstructure of catalyst layers in PEM fuel cells redefined: A computational approach, **2**(2), 2011, 141–157, Figures 1,2,4,6,7,8,9,10,13, Malek, K., Mashio, T., and Eikerling, M. Copyright (2011) Spinger. With permission.)

## Ionomer Structure in Catalyst Layers Redefined

The structural analysis provided above reveals a refined structural picture of ionomer in CL. Ionomer forms an adhesive thin film (or skin layer) with a thickness of 3–4 nm on the surface of small aggregates of primary Pt/C particles. Ionomer covers up to 40% of carbon at the aggregate. For a highly hydrophobic carbon surface, such as Vulcan XC-72, ionomer attaches to it with the backbone groups. Increasing hydrophilicity of Pt/C aggregates, for example, due to increased amount of Pt, induces the reorientation of ionomer groups. Observed was a regular, dense array of sidechains on the surface of this film, with few water molecules inside of the ionomer region. Pt/C, water, and ionomer form a layered interfacial structure.

Figure 3.40 implies that a regular arrangement of densely packed sidechains is established. The nearest-neighbor separation distance between sidechains is ~0.7 nm, as inferred by the sidechain–sidechain correlation function $g_{ss}$. This ordering is relatively insensitive to the amount of Pt. Notice that the sidechain arrangement is denser than that assumed for well-hydrated pores of a Nafion membrane, where separation distances of anionic sidechains lie in the range of 0.8–1.0 nm (Gebel and Diat, 2005). The square lattice arrangement of sidechains found in CGMD simulations is an artifact, caused by the neglect of hydrogen bonding between anionic heads of sidechains and hydronium beads. At high interfacial sidechain density, strong hydrogen bonds between sulfonate ions and hydronium ions as well as the matching trigonal symmetries of these groups will enforce the formation of a hexagonal arrangement of sidechains (Roudgar et al., 2006, 2008; Vartak et al., 2013).

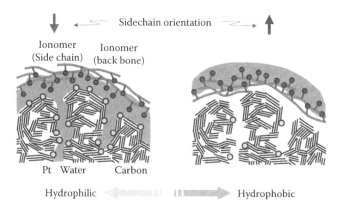

**FIGURE 3.41** Sketch of the inversion of ionomer orientation for diametrical wetting scenarios of Pt/C. (Reprinted from *Electrocatalysis*, Microstructure of catalyst layers in PEM fuel cells redefined: A computational approach, **2**(2), 2011, 141–157, Figures 1,2,4,6,7,8,9,10,13, Malek, K., Mashio, T., and Eikerling, M. Copyright (2011) Spinger. With permission.)

The sidechain density has a vital impact on hydrogen bonding and ordering at the interface. It determines water binding and mechanisms of proton transport at the ionomer. A separation of 0.7 nm and hexagonal ordering optimize hydrogen bond interactions between sidechains (Roudgar et al., 2006, 2008). The high sidechain density could be an indicative of a surface-dominated mechanism of proton transport, as discussed in the section "Ab Initio Study of Proton Dynamics at Interfaces" of Chapter 2.

Higher Pt loading causes a transition of the Pt/C surface from predominantly hydrophobic to predominantly hydrophilic. The change in surface wettability of Pt/C triggers a structural inversion in the ionomer film. This inversion of the orientation of the ionomer film has implications for catalyst utilization, density distribution and conductivity of protons, electrocatalytic activity, and water balance in catalyst layers.

Two extreme scenarios for the orientation of the ionomer film are illustrated in Figure 3.41. The cartoon depicted on the left-hand side corresponds to a highly beneficial situation. In this case, the Pt/C surface is assumed to be predominantly hydrophilic. Consequently, a water film forms between Pt/C and ionomer skin layer at the agglomerate surface. The hydrophilic surface groups at the ionomer film will be oriented predominantly toward the Pt/C surface, donating protons to this intermediate water film. Formation of the continuous water film indicates a high wetted fraction of Pt and high proton density in the intermediate water layer. As a result, for this ionomer film orientation, both the electrocatalytic activity of the pH-sensitive ORR and the proton conductivity will exhibit high values at the agglomerate level.

On the other hand, for a predominantly hydrophobic Pt/C surface, depicted on the right-hand side of Figure 3.41, hydrophobic ionomer backbones will be preferentially oriented toward the Pt/C surface. This unfavorable scenario exerts a negative impact on (i) proton conductivity resulting from the relatively low density of sulfonic acid head groups at the ionomer surface, (ii) electrocatalytic activity due to the low concentration of protons in the dehydrated region between Pt surface and ionomer film, and (iii) oxygen supply through the porous medium due to increased wettability and water accumulation in secondary pores, which are less hydrophobic.

The structural inversion of the ionomer skin inferred by results of CGMD simulations is supported by recent water sorption studies of Vol'fkovich et al. (2010). Using their well-established method of standard porosimetry, these authors studied a range of ionomer–carbon composites with varying wettability of the carbon material. Self-organization in a mixture of Nafion ionomer with a highly hydrophobic carbon was observed to result in the formation of a hydrophilic porous medium. This finding appears counterintuitive at first but it agrees with the situation illustrated in Figure 3.41. A hydrophilic substrate surface, enforcing the preferential sidechain orientation toward this surface, renders inter-agglomerate pores hydrophobic, as seen in Vol'fkovich et al. (2010). Thus, the latter situation not only enhances the electrocatalytic activity of the layer, as discussed previously, but it also improves water-handling capabilities by keeping water out of secondary pores.

A viable approach to optimize the dispersion and structure of ionomer in catalyst layers is, thus, to make the surface of PPC agglomerates sufficiently hydrophilic, as to achieve the favorable orientation of the ionomer film, depicted on the left-hand side of Figure 3.41. Essentially, in this configuration, water exists where it is needed, at the interface with Pt. The ionomer structure is optimally utilized to provide high proton concentration in the interfacial water film. At the same time, the secondary pore space remains hydrophobic and thus water-free.

Another important consequence of this structural picture is that due to small thickness and incomplete coverage, the ionomer phase is unlikely to constitute a significant diffusion barrier for oxygen; limiting current behavior caused by a high resistance due to oxygen diffusion through ionomer in CL seems unrealistic. The so-called "flooded agglomerate model," inappropriately implies the filling of agglomerate pores with liquid electrolyte. In general, agglomerates in PEFC CLs are filled with either water or ionomer, depending on how one defines agglomerates. CGMD results suggest that agglomerates, the building blocks of the CL, exhibit pore sizes of less than 10 nm. Small intra-agglomerate pores should exclude ionomer, at least in its proton-conductive form.

The structural picture of ionomer in catalyst layers, unraveled in this section, suggests that extrapolation of bulk membrane properties in terms of water uptake, water binding, and proton transport skews specific properties of ionomer in CLs and is not generally feasible for the purpose of CL modeling. One needs to adapt mechanisms of water and proton transport to the thin-film ionomer morphology, where (i) proton transport is dominated by surface properties of ionomer and (ii) electrocatalytic properties are determined by the interfacial thin-film structure formed by Pt/C surface, ionomer film, and a thin intermediate water layer.

Since ionomer is unlikely to penetrate into micropores, the major portion of Pt particles will not be in contact with ionomer, but with water in pores. Therefore, connectivity of ionomer phases should not be a critical requirement for proton transport. It is merely the continuity of water channels that is required. Proton diffusion through water in pores, between disconnected ionomer domains, may contribute significantly to the proton conductivity.

Finally, to understand properties of various types of ionomers in CLs (Astill, 2008; Astill et al., 2009; Holdcroft, 2014), one has to study, using techniques proposed in this section, their self-organized morphologies in the presence of Pt/C, their tendency to form a thin, adhesive, and well-connected film at aggregated Pt/C particles, and the arrangement of sidechains in terms of their density and orientation relative to the Pt/C surface.

## SELF-ORGANIZATION IN CATALYST LAYERS: CONCLUDING REMARKS

CGMD simulations have become a viable tool in studying self-organization processes in catalyst layers of PEFCs. Structural parameters of interest for such studies involve composition and size distributions of Pt/C agglomerates, pore space morphology, surface wettability, as well as the structure and distribution of ionomer. The latter aspect has important implications for electrochemically active area, proton transport properties, and net electrocatalytic activity of the CL.

The morphology of ionomer aggregates in CL is clearly different from that in the PEM, even if the base ionomer is the same. A simple extrapolation of ionomer properties in CLs, from properties of the bulk PEM, using percolation theory is generally insufficient. Ionomer forms a thin adhesive skin layer at agglomerate surfaces with undetectable internal porosity. The Pt loading could play a significant role in the transition of pore surface properties from hydrophobic to hydrophilic.

Evolution of the structural attributes of CLs through self-organization is particularly important for the further analysis of the transport characteristics of protons, electrons, reactant molecules ($O_2$), and water. This encompasses the distribution of electrocatalytic activity at Pt–water interfaces. In principle, mesoscale simulations can establish relations between these characteristics and the intrinsic properties of various solvents, carbon materials, and ionomer materials. They can also help clarify the dependency of these characteristics on the gravimetric composition and the level of hydration. There is still a lack of comprehensive experimental data, with which simulation results could be compared. Versatile experimental techniques have to be employed to study particle–particle interactions, structural characteristics of phases and interfaces, and phase correlations of carbon, ionomer, and water.

## STRUCTURAL MODEL AND EFFECTIVE PROPERTIES OF CONVENTIONAL CCL

This part focuses on CCLs with conventional design, illustrated in Figure 3.1. The composition is specified by volume fractions of solid Pt/C, $X_{PtC}$, ionomer, $X_{el}$, and porosity, $X_p = 1 - X_{PtC} - X_{el}$. In the pore space, one can furthermore distinguish primary and secondary pores with volume fractions $X_\mu$ and $X_M$, respectively, that fulfill the relation $X_P = X_\mu + X_M$.

The volumetric filling factor of the pore space by water is defined as the liquid water saturation, $S_r$. It depends on the PSD and wettability of pores. Moreover, it varies with environmental conditions and the current density of fuel cell operation. The amount of liquid water arriving in the CCL is roughly proportional to $j_0$, due to the combined effect of water production in the ORR and water flux to the cathode via electro-osmotic drag. The general trend is therefore an increase of $S_r$ with $j_0$. Moreover, $S_r$ is a spatially varying function at $j_0 > 0$. Local values of $S_r$ are determined by distributions of gas and liquid pressures in the porous medium. The dependence of transport properties on $S_r$, which itself is a function of spatially varying conditions, leads to a highly nonlinear coupling in the system of transport equations that must be solved.

For conventional CCLs with thickness of $l_{CL} = 5$–$10$ μm, adequate gas porosity is indispensable. They must be operated as gas diffusion electrodes (GDEs). In completely flooded CCLs with $S_r \approx 1$, reaction penetration depths would be in the range of 0.5 μm or smaller, which would render the major part ($>90\%$) inactive for the ORR.

Random three-phase composites, as depicted in Figure 3.1, concur best with competing requirements of large catalyst surface area, $S_{ECSA}$, achieved by high dispersion of Pt nanoparticles on the carbon support, good electron transport through the carbon matrix, high proton conductivity through embedded ionomer, and good gaseous transport in water-free pore space.

Experimental data (Soboleva et al., 2010; Uchida et al., 1995a,b) as well as coarse-grained MD simulations, discussed in the section "Mesoscale Model of Self-Organization in Catalyst Layer Ink," imply that catalyst layer inks self-organize into an agglomerated structure with *bimodal pore size distribution* in the mesoporous region. Carbon particles (5–20 nm) aggregate and form agglomerates. Primary pores (2–10 nm radius) exist inside agglomerates; larger, secondary pores (10–50 nm) form the pore space between agglomerates.

The structure of a conventional CCL stipulates that the competition between volume fractions of ionomer for proton conduction and gas pore space for gaseous diffusion mainly unfolds in the secondary pores. Even if some ionomer penetration into micropores within agglomerates might be possible, that ionomer would merely act as a binder, not as a proton conductor. Modeling studies based on the structural pictures shown in Figure 3.1 have explored how thickness and composition of CCLs should be adjusted in order to obtain best performance and highest catalyst utilization (Eikerling, 2006; Eikerling and Kornyshev, 1998; Eikerling et al., 2004, 2007a; Liu and Eikerling, 2008; Xia et al., 2008; Xie et al., 2005).

Prior to 2006, modeling of CCL neglected the impact of liquid water formation in pores. Results of these studies are valid in a regime of sufficiently low current density. In this regime, the amount of water accumulation in secondary pores is too small to exert a significant impact on performance, whereas water in hydrophilic primary pores is strongly bound by capillary forces. Therefore, the assumption of constant and uniform composition and effective properties is justified in this regime.

More recent work explicitly incorporated the effect of the water accumulation in secondary pores, determined by porous structure, wettability, current density, and environmental conditions. In a regime of high current density, increasing $j_0$ leads to higher overall liquid water saturation and more nonuniform water distribution in porous electrode layers (Eikerling, 2006; Liu and Eikerling, 2008).

A common characteristic of all performance models is that they treat the heterogeneous layer as a continuous effective medium. All processes in it are averaged over micro- and mesoscopic domains, referred to as REVs. Effective properties of REVs are obtained from composition and pore network structure using the theory of random composite media. The size of REVs should be considerably larger than the typical scales of structural heterogeneities, that is, $>100$ nm. At the same time, it should be much smaller than the macroscopic dimensions of the CCL, at which continuous distributions of physical properties and processes can, therefore, be studied.

## EXPERIMENTAL STUDIES OF CATALYST LAYER STRUCTURE

The composition of CLs can be precisely controlled at the fabrication stage. It is usually specified in terms of Pt mass loading, Pt-to-carbon mass ratio, and ionomer-to-carbon mass ratio. If in addition to these parameters the thickness is also known, composition in terms of volumetric amounts can be obtained.

Porous composite morphologies of catalyst layers in PEFCs have been evaluated with mercury *porosimetry*, gas adsorption, standard porosimetry, and *water sorption studies* (Holdcroft, 2014; Rouquerol et al., 2011; Soboleva et al., 2010; Uchida et al., 1995a,b; Vol'fkovich et al., 2010). Both mercury porosimetry as well as nitrogen adsorption can be used to explore wide ranges of pore sizes. Owing to large pressures required for mercury injection into nanometer pores, the lower bound for this method is about 3 nm (Xie et al., 2004). Generally, the method is destructive and the inaccuracy due to irreversible sample deformation is difficult to assess. Nitrogen adsorption provides a suitable method for measuring pore size distributions from $\sim 1$ nm to about 300 nm. Measurements probe the quantity of gas adsorbed onto, or desorbed from a solid surface at a selected equilibrium vapor pressure by means of the static volumetric method.

More recently, electron microscopy and tomography studies have been applied as well for *visualization of microstructures* formed by agglomerated carbon and Pt particles (Thiele et al., 2013). The resolution of imaging and tomographical techniques has seen tremendous progress. FIB–SEM can be employed to obtain cross-sectional images, which can be used to reconstruct the 3D structure of CLs.

Sorption isotherms of carbon powders and catalyst layers can be analyzed in view of BET surface area and pore size distributions. Using these analysis tools, effects of carbon content, Pt content, and ionomer content on aggregation characteristics, and volume fractions of different types of pores can be rationalized. Earlier observations by Uchida et al. (1995a) suggested that ionomer impregnation mainly affects the secondary pore volume between agglomerates, leaving the primary pores in agglomerates largely unaffected. $N_2$ adsorption studies by Soboleva et al. found a more uniform decrease in measured pore volume across the whole size range with increasing ionomer loading. The impact of ionomer content on the volume fraction of primary pores was explained either as a pore blocking effect or as being due to ionomer penetration into these small pores.

Figure 3.42 shows PSDs measured by Suzuki et al. (2011) and Soboleva et al. (2011). The distributions are qualitatively similar and they both concur with the picture of an agglomerated microstructure. Mathematically, distributions of this type can

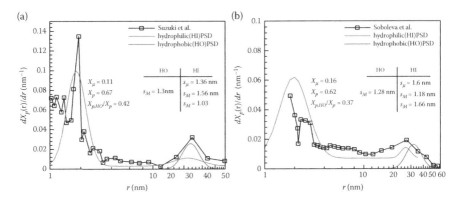

**FIGURE 3.42** PSDs measured by Suzuki et al. (2011) (a) and Soboleva et al. (2011) (b) using nitrogen physisorption studies and BET analysis. The proposed parameterizations of the experimental PSDs distinguish hydrophilic (bimodal distribution function) and hydrophobic (monomodal distribution function with peak at ∼35 nm) porosities.

be represented by a *bimodal log-normal PSD*, given by

$$\frac{dX_P(r)}{dr} = \frac{1 - X_{PtC} - X_{el}}{\sqrt{\pi}\left[\ln s_\mu + \chi_M \ln s_M\right]}\frac{1}{r}$$

$$\times \left\{ \exp\left[-\left(\frac{\ln\left(r/r_\mu\right)}{\ln s_\mu}\right)^2\right] + \chi_M \exp\left[-\left(\frac{\ln\left(r/r_M\right)}{\ln s_M}\right)^2\right] \right\}. \quad (3.97)$$

Here, $r_\mu$ and $r_M$ determine the positions of the two peaks and $s_\mu$ and $s_M$ determine their widths. The parameter $\chi_M$ controls the relative contributions of primary and secondary pores. The distribution in Equation 3.97 is normalized to 1. Porosities due to primary and secondary pores are given by

$$X_\mu = \int_0^{r_{cut}} \frac{dX_p(r)}{dr}\,dr \quad \text{and} \quad X_M = \int_{r_{cut}}^{\infty} \frac{dX_p(r)}{dr}\,dr \quad (3.98)$$

where $r_{cut}$ ($r_\mu < r_{cut} < r_M$) separates the two peak regions.

A refinement of the bimodal PSD incorporates a distinction of hydrophilic (HI) and hydrophobic (HO) pores

$$\frac{dX_{P,HI}(r)}{dr} = \frac{1 - X_{PtC} - X_{el}}{\sqrt{\pi}\left[\ln s_\mu + \chi_M \ln s_M\right]}\frac{1}{r}$$

$$\times \left\{ \exp\left[-\left(\frac{\ln\left(r/r_\mu\right)}{\ln s_\mu}\right)^2\right] + \chi_{M,HI} \exp\left[-\left(\frac{\ln\left(r/r_M\right)}{\ln s_M}\right)^2\right] \right\}$$

$$(3.99)$$

and

$$\frac{dX_{P,HO}(r)}{dr} = \frac{1 - X_{PtC} - X_{el}}{\sqrt{\pi}\left[\ln s_\mu + \chi_M \ln s_M\right]} \frac{\chi_{M,HO}}{r} \exp\left[-\left(\frac{\ln(r/r_M)}{\ln s_M}\right)^2\right] \quad (3.100)$$

Using these hydrophobic and hydrophilic pore size distributions, the (local) liquid water saturation can be calculated as

$$S_r = \frac{1}{X_P}\left[\int_0^{r_{c,HI}} \frac{dX_{P,HI}(r)}{dr} dr + \int_{r_{c,HO}}^\infty \frac{dX_{P,HO}(r)}{dr} dr\right], \quad (3.101)$$

where $r_{c,HI}$ and $r_{c,HO}$ are capillary radii of hydrophilic and hydrophobic pores.

It can be assumed that all primary pores in the CCL are hydrophilic. Under normal conditions, these pores are filled with water. However, for secondary pores between agglomerates, it is assumed that a fraction of these pores are hydrophobic. This fraction is estimated to be 50% based on experimental results of Kusoglu et al. (2012), which show that 20–60% of all CCL pores could be hydrophobic.

Primary hydrophilic pores have a strong tendency to retain water due to their small size. For secondary hydrophobic pores in CCL with wetting angle of 100°, a liquid water pressure of $P^l \sim 5$ bar is required to start filling them; having larger wetting angle the required liquid water pressure for pore filling increases. Such high liquid pressures do not occur under realistic conditions in CLs.* Therefore, it is reasonable to assume that hydrophobic pores in catalyst layers remain water-free at relevant current densities. Progressive liquid water formation could occur only in secondary hydrophilic pores.

## KEY CONCEPTS OF PERCOLATION THEORY

Percolation theory represents the most advanced and most widely used statistical framework to describe structural correlations and effective transport properties of random heterogeneous media (Sahimi, 2003; Torquato, 2002). Here, briefly described are the basic concepts of this theory (Sahimi, 2003; Stauffer and Aharony, 1994) and its application to catalyst layers in PEFCs.

Let us consider a container with a large number, $N \gg 1$, of randomly dispersed and noninteracting guest objects, embedded in a host medium. The characteristic size of individual objects must be very small, that is, microscopic, compared to the macroscopic size of the container. Objects could represent phase domains of a solid or liquid substance or they could represent pores in a porous medium. The roles of guest and host media are interchangeable and components of both media may be represented by spherical particles on a lattice.

Individual objects possess a characteristic physical property like electronic conductivity, ion conductivity, magnetic susceptibility, gas diffusivity, or liquid permeability. Mixing of at least two distinct types of objects (of guest and host material) that

---

\* Excess liquid pressures needed to fill hydrophobic pores in gas diffusion layers are significantly smaller due to the larger pore sizes.

differ markedly, that is, by several orders of magnitude, in the value of the characteristic physical property of interest creates a percolation system. Percolating clusters form through connected objects of one type that are nearest neighbors on a lattice or that form continuous domains in a continuum percolation model. A variation in composition of this system leads to variations in sizes, connectivities, and tortuosities of these clusters.

At a critical value of the fraction of objects of one type, these objects would form an extended cluster that connects the opposite external faces of the sample. At this so-called percolation threshold, the corresponding physical property represented by the connected objects would start to increase above zero. Thereby percolation theory establishes constitutive relations between composition and structure of heterogeneous media and their physical properties of interest. For porous electrodes or catalyst layers in PEFC, these properties are electrical conductivities of electrons and protons, diffusivities of gaseous reactants and water vapor, and liquid water permeability.

Typical examples of percolation systems are (i) random mixtures of electronically conducting metallic spheres in an insulating host medium, for example, a polymer electrolyte, (ii) a network of gas pores providing high diffusivity in a porous matrix of a gas-tight material, or (iii) a porous electrode partially saturated with a liquid electrolyte.

Percolation theory represents a random composite material as a network or lattice structure of two or more distinct types of microscopic objects or phase domains. These objects will be referred to as "black" and "white," representing mutually exclusive physical properties of some sort. The network onto which "black" and "white" elements of the composite medium are distributed could be continuous (continuum percolation) or discrete (discrete or lattice percolation); it could be a disordered or regular network. With a probability $p$, a randomly chosen percolation site will be occupied by a "white" element. With the complementary probability $(1 - p)$, the element occupying the site will be "black."

Percolation theory rationalizes sizes and distribution of connected "black" and "white" domains and the effects of cluster formation on macroscopic properties, for example, electric conductivity of a random composite or diffusion coefficient of a porous rock. A percolation cluster is defined by a set of connected sites of one color (e.g., "white") surrounded by percolation sites of the complementary color (i.e., "black"). If $p$ is sufficiently small, the size of any connected cluster is likely to be small compared to the size of the sample. There will be no continuously connected path between the opposite faces of the sample. On the other hand, the network should be entirely connected if $p$ is close to 1. Therefore, at some well-defined intermediate value of $p$, the percolation threshold, $p_c$, a transition occurs in the topological structure of the percolation network that transforms it from a system of disconnected "white" clusters to a macroscopically connected system. In an infinite lattice, the *site* percolation threshold is the smallest occupation probability $p$ of sites, at which an infinite cluster of "white" sites emerges.

In the metal/insulator composite, $p_c$ is the threshold, above which the first conducting path between two opposing sites of the macroscopic sample forms, representing the transition from insulating to conducting sample. At the microscopic scale, the conductivity of the sample will depend on the contact resistance at the

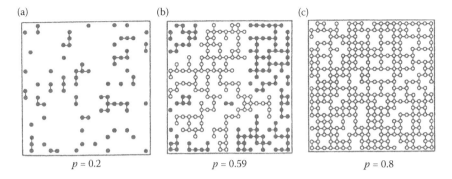

(a)                              (b)                              (c)

p = 0.2                         p = 0.59                         p = 0.8

**FIGURE 3.43**  A finite size sample with occupied sites (dots) and bounds between them on a square lattice. The samples is shown at different site occupation probability, $p$, that is (a) below the percolation threshold, $p = 0.2$, (b) at the percolation threshold, $p_c = 0.59$, and (c) above the percolation threshold, $p = 0.8$. Open symbols in (b) and (c) represent sites that belong to the infinite percolation cluster.

interface of two conducting elements. As $p$ increases further, more connections between nearest-neighbor sites will be established, leading to a monotonic increase in conductivity.

For lattice percolation, two types of problems are distinguished. For the hitherto described site percolation problems, clusters are formed by "white" or "black" sites of the lattice. For bond percolation, the same statistical concept is applied to the connections or bonds between lattice sites: with probability $p$, a randomly chosen bond will be occupied by a "white" element and otherwise, that is, with the complementary probability $(1 - p)$, the bond will remain "black." Such a system is depicted in Figure 3.43a.

In the vicinity of the percolation threshold, at $p \geq p_c$, the effective conductivity of the metal/insulator composite is determined by a critical law,

$$\sigma_{dc}(p) \propto (p - p_c)^{\mu} \, \Theta \, (p - p_c) \tag{3.102}$$

where $\Theta \, (x)$ is the Heaviside step function. The same percolation law would be obeyed if the liquid permeability of a liquid-saturated porous medium were evaluated as a function of the porosity.

The relation between diffusion coefficient and dc conductivity is given (approximately) by the Nernst–Einstein equation,

$$\sigma_{dc} = \frac{e_0^2}{k_B T} nD, \tag{3.103}$$

where $n$ represents the density of diffusing particles (charge carriers). For a percolating system, the approximate relation $n \sim P_\infty \sim (p - p_c)^{\beta}$ is valid, where $P_\infty$ is the probability that a site belongs to the infinite cluster of conducting sites that forms at the percolation threshold; $P_\infty$ is also referred to as the density of the infinite cluster.

**TABLE 3.2**

**Critical Fraction of Lattice Points (Percolation Threshold) That Must Be Filled to Create a Continuously Connected Path of Nearest Neighbors from One Side to Another**

| dim. | lattice type | site percolation | bond percolation |
|---|---|---|---|
| 2 | square | ≈0.59 | 0.5 (exact) |
| 2 | triangular lattice | 0.5 (exact) | $2\sin(\pi/8)$ (exact) |
| 2 | honeycomb lattice | ≈0.7 | $1 - 2\sin(\pi/18)$ (exact) |
| 3 | simple cubic | ≈0.31 | ≈0.25 |
| 3 | bcc | ≈0.25 | ≈0.18 |
| 3 | fcc | ≈0.20 | ≈0.12 |
| 3 | diamond | ≈0.43 | ≈0.39 |

*Source:* Values are taken from Stauffer and Aharony (1994).

Combination of these relations yields an expression for the effective self-diffusivity,

$$D_e(p) \propto \frac{\sigma_{dc}(p)}{P(p)} \propto (p - p_c)^{\mu - \beta} \Theta (p - p_c). \qquad (3.104)$$

A diffusing species can move on all macroscopically connected clusters of the conductive phase. Above $p_c$, however, only the largest cluster, the so-called sample-spanning cluster significantly contributes to transport.

Other physical properties like correlation lengths and percolation probabilities follow as well power laws in the vicinity of $p_c$, however with different critical exponents. Values of the critical exponent $\mu$ in Equation 3.102 are known in 2D and 3D from computer simulations (Isichenko, 1992). For lattice percolation in 2D, it is $\mu \approx 1.3$ and in 3D it is $\mu \approx 2.0$. For the exponent $\beta$, a value of $\beta \approx 0.4$ was proposed. Critical exponents are universal for lattice percolation models, that is, they are independent of the topology of the network and only depend on the dimensionality of the sample. This universality of critical exponents does not extend to continuum percolation problems. For continuum percolation in 3D (Swiss cheese model), it was suggested that $\mu \approx 2.38$ (Bunde and Kantelhardt, 1998).

In contrast to critical exponents, values of percolation thresholds depend on the lattice topology and the type of the percolation problem considered. Percolation thresholds for a few known lattice types are listed in Table 3.2. In 1D, it is trivially $p_c = 1$. In 2D, values for $p_c$ are known exactly for specific lattice types. In 3D, values of $p_c$ can only be found with the help of computer simulations.

For continuum percolation, the percolation probability is replaced by a percolation volume fraction, given by $X_c = p_c f$, where $f$ is the filling factor (Hunt, 2005). For a polydisperse medium, $X_c$ is a monotonically decreasing function of the polydispersity.

Percolation theory has been successfully applied to such diverse phenomena as hopping conductivity in semiconductors (Shklovskii and Efros, 1982), gelation in polymer melts (de Gennes, 1979), permeability of porous rocks (Sahimi, 1993;

Torquato, 2002), spreading of epidemics (Grassberger, 1983), and spreading of wildfires (MacKay and Jan, 1984). In PEFC research, percolation theory has been employed for establishing relations between water uptake of polymer electrolyte membranes and their proton conductivities (Eikerling et al., 1997). Moreover, percolation concepts play an important role in theoretical efforts to unravel structure versus function relations of catalyst layers, as will be described next (Eikerling, 2006; Eikerling and Kornyshev, 1998; Eikerling et al., 2004).

## EFFECTIVE CATALYST LAYER PROPERTIES FROM PERCOLATION THEORY

CCL operation entails transport of gases, water, electrons, and protons, as well as interfacial transformation of species due to electrochemical reaction and evaporation. Effective parameters that steer the interplay of these processes are proton and electron conductivity; diffusion coefficients of oxygen, water vapor, and residual gaseous components; liquid water permeability; as well as exchange current density and vaporization rate per unit volume. These parameters incorporate information about composition, pore size distribution, pore surface wettability, and liquid water saturation. This section introduces functional relationships between effective properties and structure.

Percolation theory can be applied to parameterize the effective properties of CLs. Specific parameters employed could be determined from structural diagnostics, including porosimetry measurements, water sorption studies, as well as rapidly evolving tomographical approaches using scanning electron microscopy and transmission electron microscopy (Thiele et al., 2013).

Percolation relations for effective transport properties use the lowest order of structural information for the layer, namely, its composition. In principle, more detailed structural models could be devised to incorporate higher-order structural information, that is, pore and particle shapes, and correlations in distributions of distinct components. Random network simulations, variational principles, and effective medium theory could be involved to study these relations (Milton, 2002; Torquato, 2002).

There is a certain ambiguity in parameters used to define percolation properties of CCLs. Parameterizations of effective properties presented reproduce major qualitative trends in experimental observations of composition effects. Moreover, they have proven useful in systematic optimization studies. However, progress in structural characterization using advanced experimental analysis methods and computational approaches will certainly lead to refinements of the proposed relations.

### Effective Proton Conductivity

The proton conductivity of a conventional CL is determined by the amount and network topology of the ionomer phase. It can be expressed as

$$\sigma_{el} = \sigma_0 \left( \frac{X_{el} - X_c}{1 - X_c} \right)^{\mu} \Theta(X_{el} - X_c) \qquad (3.105)$$

where $\sigma_0$ represents a bulk electrolyte conductivity and $X_c$ is the volume fraction at the percolation threshold. Previous parameterization gave reasonable results with

$X_c \approx 0.12$ (Eikerling, 2006; Eikerling and Kornyshev, 1998; Eikerling et al., 2004, 2007a). It should be noted that this parameterization does not take into account the specific thin-film morphology of ionomer in CLs discussed in the section "Ionomer Structure in Catalyst Layers Redefined."

## Effective Diffusivity

The cathodic gas mixture consists of interdiffusing oxygen (superscript o), vapor (v), and residual (r) components. For all practical purposes, it is reasonable to assume that gas diffusion through liquid water-filled pores or ionomer makes a small contribution to species transport at the macroscale. Diffusion in liquid water has a significantly smaller diffusion coefficient compared to diffusion in the gas phase and it requires solution and exsolution of the diffusing species. At length scales of the order of $\sim$1 μm, gas effectively diffuses only through gas-filled pores. Furthermore, contributions of primary pores to effective diffusivity are relatively small due to small sizes of these pores and them being filled with water under normal conditions. Transport of gaseous species is given by

$$D^{o,v,r}(S_r) = D_0^{o,v,r} \frac{(X_P - X_\mu - X_c)^{2.4}}{(1 - X_c)^2 (X_P - X_c)^{0.4}}$$

$$\times \left\{ \left[ \frac{(1 - S_r) X_P - X_c}{X_P - X_\mu - X_c} \right]^{2.4} \Theta \left( S_r - \frac{X_\mu}{X_P} \right) + \Theta \left( \frac{X_\mu}{X_P} - S_r \right) \right\} + D^{res},$$

$$(3.106)$$

similar to expressions suggested in Hunt and Ewing (2003) and Moldrup et al. (2001) for percolation in the open pore space of a partially saturated porous medium. The percolation threshold is assumed to be the same as in Equation 3.105. The prefactors $D_0^{o,v,r}$ for the distinct gaseous species in Equation 3.106 are given by an expression from kinetic gas theory:

$$D_0^{o,v} = \sqrt{\frac{2RT}{\pi M^{o,v}}} \frac{4}{3} r_{crit}, \qquad (3.107)$$

where $r_{crit}$ is a critical pore radius, for example, obtained from critical path analysis (Ambegaokar et al., 1971) and $M^{o,v}$ is the molar mass of the diffusing gas molecules. Equation 3.107 dwells on Knudsen-type diffusion as the main mechanism of gas transport. In the Knudsen regime, molecule–wall collisions predominate over molecule–molecule collisions. The condition for Knudsen diffusion is that the mean-free path of diffusing gas molecules, $\lambda_m = R_g T / (\sqrt{2}\pi d_m^2 N_A P^g)$, is large compared to the pore diameter, that is, $\lambda_m \geq 2r_{crit}$, or, using the definition of the Knudsen number, $Kn = \lambda_m / 2r_{crit} \geq 1$. Here, $d_m$ is the molecule diameter, and $N_A$ is the Avogadro number. For oxygen at a pressure of 1 atm, one finds $\lambda_m = 70$ nm (Kast and Hohenthanner, 2000; Mezedur et al., 2002). Knudsen diffusion, thus, prevails if $r_{crit} < 35$ nm, a condition that is likely to be fulfilled in CCLs.

Above a critical liquid water saturation $S^{crit} = 1 - X_c/X_p$, only a residual diffusion due to transport of dissolved species remains finite, accounted for by $D^{res}$.

## Effective Liquid Permeability

Water-filled pores in the ionomer (radius $r_{el}$, corresponding water volume fraction $\varepsilon_{el}$), and primary and secondary pores contribute to the liquid permeability, which is written in the following form

$$
K^l(S_r) = \frac{\delta}{24\tau^2} \left\{ r_{el}^2 \varepsilon_{el} X_{el} + r_\mu^2 \left[ S_r X_p \Theta \left( \frac{X_\mu}{X_p} - S_r \right) + X_\mu \Theta \left( S_r - \frac{X_\mu}{X_p} \right) \right] \right.
$$
$$
\left. + \tau^2 r_M^2 \frac{[S_r X_p - X_\mu - X_c]^2}{(1 - X_c)^2} \Theta \left( S_r - \frac{X_\mu}{X_p} \right) \Theta \left( S_r X_p - X_\mu - X_c \right) \right\}
$$

$$(3.108)$$

In this expression, a percolation dependence is assumed in the mesopore space; $\delta$ is a constrictivity factor and $\tau$ is a tortuosity factor (Dullien, 1979).

## Interfacial Vaporization Exchange Area

The interfacial vaporization exchange area is an important property, which exerts a marked impact on CCL performance. Few experimental and theoretical studies have explored this property. It depends on fine topological details of the pore network. The menisci separating liquid and gas phases in pores contribute to the liquid–vapor interfacial area. The dependence on porosity and pore radii is written in the form

$$
\xi^{lv}(S_r) = \Upsilon l_{CL} \int_0^\infty \frac{dX_p(r)}{dr} \frac{1}{r} h(r_c, r) \, dr \tag{3.109}
$$

Convolution with the function $h(r_c, r)$ takes into account that the interface does not advance completely to pores with radius equal to the capillary radius, $r_c$, due to hysteresis effects. Capillary equilibrium shifts gradually from smaller pores to larger pores with increasing $S_r$ for a hydrophilic medium. As local capillary equilibrium advances to pores with size $r_c$, a portion of liquid–vapor interfaces persists in smaller pores with $r < r_c$. This effect depends on PSD and connectivity. It can be incorporated in a simple phenomenological way by using $h(r_c, r) = \Theta(r_c - r)\Theta(r - \zeta r_c)$, where $\zeta$ is a factor that depends on the pore space topology. A poorly connected, heterogeneous pore space results in larger $\zeta$ and, thus, larger $\xi^{lv}(S_r)$. $\Upsilon$ is a factor of order 1, which depends on the geometry and wettability of pores.

## EXCHANGE CURRENT DENSITY

A basic variant for the parameterization of the exchange current density is written as

$$
j^0 = j_*^o \frac{m_{Pt} N_A}{M_{Pt} v_{Pt}} \Gamma_{np} \Gamma_{stat}, \tag{3.110}
$$

as discussed in the section "Catalyst Activity." In earlier CCL modeling, it was proposed that the statistical utilization could be written as

$$
\Gamma_{stat} = g(S_r) \frac{f(X_{PtC}, X_{el})}{X_{PtC}}, \tag{3.111}
$$

where $g(S_r)$ is the wetted fraction of pore surface area and $f(X_{PtC}, X_{el})/X_{PtC}$ represents the statistical fraction of Pt particles at or near the triple-phase boundary.

The factorization into two functions $f$ and $g$ was rationalized by considering a separation of length scales. At the macroscopic scale, utilization of catalyst sites requires triple-phase accessibility to interpenetrating phases of Pt/C, ionomer, and pore network. These requirements are expressed as a function of the statistical geometry of the composite (Eikerling, 2006; Eikerling and Kornyshev, 1998; Eikerling et al., 2004, 2007a):

$$f(X_{PtC}, X_{el}) = P(X_{PtC}) P(X_{el})$$

$$\times \left\{ (1 - \chi_{ec}) \left( 1 - [1 - P(X_p)]^M \right) + \chi_{ec} [1 - P(X_p)]^M \right\} \quad (3.112)$$

with the density of the percolating cluster

$$P(X) = \frac{X}{(1 + \exp[-a(X - X_c)])^b} \quad (3.113)$$

where $a = 53.7$, $b = 3.2$, and $M = 4$, as considered in Ioselevich and Kornyshev (2001, 2002). The parameter $\chi_{ec}$ allows for the effect of residual activity at nonoptimal reaction spots. The largest possible value of $f(X_{PtC}, X_{el})$, corresponding to the optimal active area, is $f(X_{PtC}, X_{el}) \approx 0.1$, obtained with $X_{PtC} = 0.38$ and $X_{el} = 0.38$. The optimum value of $f(X_{PtC}, X_{el})/X_{PtC}$ is obtained with $X_{PtC} = 0.18$ and $X_{el} = 0.53$.

At the mesoscopic scale, corresponding to the size of an agglomerate ($\sim$100 nm), the reaction front expands from the true triple-phase boundary toward catalyst sites inside agglomerates. At this scale, oxygen diffusion incurs no mass-transport penalty in water-filled agglomerates. Electrostatic effects control the distribution of protons and reaction rates in agglomerates, as explored in the section "ORR in Water-Filled Nanopores: Electrostatic Effects." The wetting of pores inside agglomerates is critical, however, in view of the accessibility of catalyst sites to protons. These effects lead to the factor that accounts for the fraction of water-filled pores

$$g(S_r) = \frac{1}{\Pi} \int_0^{r_c} \frac{1}{r} \frac{dX_p(r)}{dr} dr, \quad (3.114)$$

where $\Pi$ is a normalization factor that ensures $g \to 1$ in the limit of complete wetting, $S_r \to 1$. The function $g(S_r)$ varies in the range $0 \le g(S_r) \le 1$. Catalyst utilization is thus determined by the internal wetted pore fraction.

Applicability of the statistical approach for $\Gamma_{stat}$ in Equation 3.110 depends primarily on the ionomer structure. A thin-film morphology of ionomer in CLs, suggested by results in the section "Ionomer Structure in Catalyst Layers Redefined," invalidates the statistical law derived from percolation theory. Instead, the statistical utilization factor is determined by the coverage of the ionomer skin layer on agglomerates. A random statistical approach to describe the catalyst layer morphology misrepresents this organized structure and it underestimates $\Gamma_{stat}$. However, as long as $\Gamma_{stat}$ remains reasonably constant (upon variation of $j_0$), predictions derived based on the statistical law for $\Gamma_{stat}$ will remain essentially correct.

The structure versus property relations, presented in this chapter, show how the random composite morphology can steer effective transport properties of CLs on the basis of percolation theory. Notice that the bimodality of the pore network (primary and secondary pores) plays a key role in balancing the different functions of a CL. A large porosity due to primary pores is beneficial for the exchange current density, since it guarantees that a large fraction of catalyst sites inside agglomerates could be reached by protons. Moreover, increasing the fraction of primary pores results in a larger liquid/vapor interfacial area and, thus, higher rates of evaporation. Secondary pores on the other hand regulate the gas diffusivity.

It will remain a pivotal yet scientifically highly challenging task to make exact predictions for the effective properties of random heterogeneous media. In research on fuel cell materials, both continuum as well as discrete network models have been utilized for this purpose. Continuum models represent the classical approach to describing transport properties in materials of complex and irregular morphology. They provide the effective properties of the medium as volume averages of the corresponding microscopic quantities. The shortcoming of discrete models such as random network models, Bethe lattice models, and so on compared to continuum modes is the demand for large computational effort to represent a realistic physical model of the material and simulate its effective properties. Finally, progress in the development of advanced structural models for heterogeneous fuel cell materials hinges on the availability of experimental microstructure information.

## CONCLUDING REMARKS

This chapter has provided a detailed treatment of the structure, properties, and function of catalyst layers in PEFCs. The practical objective behind it is to develop architectures of catalyst layers with high catalytic activity, low catalyst loading, and high stability.

Any expedient development in electrocatalyst materials and structural design of catalyst layers relies on understanding self-organization phenomena that take place in catalyst layer inks, electrokinetics of ORR and HER at highly dispersed catalyst surfaces, transport phenomena in porous composite media, integration between CLs and other components in the fuel cell in particular with respect to water management, and degradation phenomena in view of the CL stability. A fundamental guiding principle in this endeavor is the scale paradigm that was presented in Figure 1.16. Major structural effects in CLs occur at well-separated scales. This observation is vital for the development of theory and diagnostic approaches to understand CL properties and function. The main scales correspond to catalyst nanoparticles (a few nanometers), water-filled nanopores ($\sim 10$ nm), agglomerates of carbon/Pt ($\sim 100$ nm) encapsulated within an ionomer skin layer, and the macroscopic device level, at which CLs can be considered as effective media.

At the atomistic or molecular scale (Å to nm), *ab initio* simulations are being employed to identify electrocatalyst and support materials that offer adequate activity and stability. Current methods, based on density functional theory, can be applied to screen catalyst materials in view of electronic band structure, electron affinity (Fermi energy or work function), and density of electronic states at the Fermi level. Inclusion of an electrolyte environment allows electrostatic charging properties (surface charge density, potential of zero charge) and adsorption properties (structure and stability of adsorbates, adsorption energies, adsorbate interactions) of the electrified interface to be studied, at the expense of markedly expanded computational needs. The combination of DFT-based simulations with thermochemical and kinetic modeling makes it possible to decipher elementary steps and pathways of complex reaction mechanisms and determine the governing parameters. Approaches in first-principles electrochemical studies that incorporate these steps are rapidly evolving. Progress in this field will be crucial to furnish trends between catalyst structure and surface reactions in fuel cells.

Simulations of physical properties of realistic Pt/support nanoparticle systems can provide interaction parameters that are used by molecular-level simulations of self-organization in CL inks. Coarse-grained MD studies presented in the section "Mesoscale Model of Self-Organization in Catalyst Layer Inks" provide vital insights on structure formation. Information on agglomerate formation, pore space morphology, ionomer structure and distribution, and wettability of pores serves as input for parameterizations of structure-dependent physical properties, discussed in the section "Effective Catalyst Layer Properties From Percolation Theory." CGMD studies can be applied to study the impact of modifications in chemical properties of materials and ink composition on physical properties and stability of CLs.

Understanding of the mesoscopic structure is needed in order to rationalize local reaction conditions in nanopores. The conclusive step is to relate the hierarchy of structural effects to the performance of a catalyst layer. Owing to the importance of catalyst layer performance modeling, Chapter 4 will be devoted to this topic.

The most promising approach to the optimization of catalyst layers would be a concerted experimental–theoretical strategy. Since theory and modeling inevitably have to invoke simplifying assumptions, offering a pure theoretically driven optimization would be irresponsible. *Ex situ* diagnostics is needed to characterize structural details and explore their relations to effective properties. The availability of such experimental data defines the level of detail of structure–property relationships that a theory should be permitted to employ.

*In situ* experimental studies, exploring the performance and comparing it with the theoretical predictions provide the essential benchmarks for the optimization. The theory, corroborated by these systematic experimental procedures, could be used to (i) identify salient features of good or bad catalyst layer performance, (ii) explain causes of catalyst layer failure, (iii) identify the needs in view of new material designs, and (iv) suggest catalyst particle sizes, porous structure, wetting properties, thickness, composition, operating conditions, and so on for attaining highest fuel cell efficiencies and power densities. Examples of how to employ instructive theoretical tools in this process, leading from understanding to new design, will be discussed in Chapters 4 and 5.

# 4 Modeling of Catalyst Layer Performance

Chapter 3 treated fundamental properties of electrocatalyst materials. Dwelling upon fundamental electrocatalysis, elementary surface processes and kinetic models of Pt oxide formation and reduction as well as of the ORR were discussed. These topics could lead to more detailed treatises of catalyst design, fabrication, and testing. The section "ORR in Water-Filled Nanopores: Electrostatic Effects" in Chapter 3 dealt with the ORR in nanopores of the catalytic medium. This topic brings into play the surface charging properties of the porous matrix. Coarse-grained MD studies, presented in the section "Mesoscale Model of Self-Organization in Catalyst Layer Links" in Chapter 3, provide vital insights on self-organization in CLs that serve as input for parameterizations of structure-dependent physical properties, discussed in the section "Effective Catalyst Layer Properties from Percolation Theory" in Chapter 3. A specific finding of these MD studies is the formation of an ionomer skin layer at the agglomerate surface.

The transport properties of water-filled nanopores inside of agglomerates and the properties of the ionomer film at the agglomerate surface define local reaction conditions at the mesoscopic scale. These local conditions, which involve distributions of electrolyte phase potential, proton density (or pH), and oxygen concentration, determine the kinetic regime, under which interfacial electrocatalytic processes must be considered. Combining this information, a local reaction current can be found, which represents the source term to be used in performance modeling of the cathode catalyst layer.

This chapter is devoted entirely to performance models of conventional catalyst layers (type I electrodes), which rely on reactant supply by gas diffusion. It introduces the general modeling framework and employs it to discuss the basic principles of catalyst layer operation. Structure-based models of CCL rationalize distinct regimes of performance, which are discernible in polarization curves. If provided with basic input data on structure and properties, catalyst layer models reproduce PEFC polarization curves. Consistency between model predictions and experimental data will be evaluated. Beyond polarization curves, performance models provide detailed maps or shapes of reaction rate distributions. In this way, the model-based analysis allows vital conclusions about an optimal design of catalyst layers with maximal catalyst utilization and minimal transport losses to be drawn.

For the purpose of physical modeling of CL operation, it is expedient to treat the layer as a two-scale system as illustrated in Figure 4.1. Referring to the hierarchical scheme in Figure 1.16, the emphasis will be on the two rightmost scales. At the mesoscopic scale, operation is determined by the properties of agglomerates composed

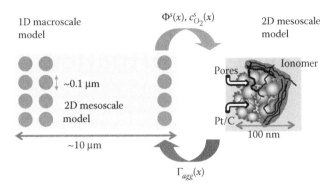

**FIGURE 4.1** Schematic illustration of the two-scale model of CCLs showing, at the macroscale level (left), the assembly of spherical agglomerates, and, at the mesoscale level, the internal porous structure of a single agglomerate. (Reprinted from Sadeghi, E., Putz, A., and Eikerling, M. 2013b. *J. Electrochem. Soc.*, **160**, F1159–F1169, Figures 1,2,5,6,8,12. Copyright (2013), the Electrochemical Society. With permission.)

of carbon and Pt particles. Agglomerates ($\oslash$ 50–100 nm) are coated partially or completely by a skin layer of ionomer (3–4 nm thick), as indicated in Figure 4.1. Primary pores in agglomerates ($\oslash$ 2–20 nm) provide a large Pt surface area. At the macroscopic scale, gas diffusion of oxygen is vital in order to provide a large reaction penetration depth, which ensures uniform utilization of catalyst throughout the complete layer. This requires the volume fraction of gas-filled secondary pores ($\oslash$ 20–100 nm) to be above the percolation threshold. In an ideal layer, these pores would be hydrophobic. Agglomeration, ionomer-film formation, bimodal pore size distribution, and mixed wettability are essential attributes of a CCL that achieves the optimal interplay of transport and reaction.

## FRAMEWORK OF CATALYST LAYER PERFORMANCE MODELING

Basically, a fuel cell electrode is a highly dispersed and heterogeneous interface between Pt nanoparticles and an electrolyte phase composed of acid-loaded ionomer and water. The structure is designed to maximize the interfacial area between catalyst and proton-supplying electrolyte.

Random statistical composition, phase-segregated morphology, pore size distribution, and wetting properties of pores determine spatial distributions of potentials, concentrations of reactants, and liquid water saturation. A subtle electrode theory has to link these distributions with the overall CCL performance.

The strategy in developing physical models of electrochemical performance and water balance in CCLs is illustrated schematically in Figure 3.5. At the materials level, it requires constitutive relations between the variables that represent composition and structure, and the variables that represent effective physicochemical properties of the layer. The most important properties, scrutinized below, involve proton conductivity, gas diffusivities, liquid permeabilities, electrochemical source term, and vaporization source term. The set of relationships between structure and physicochemical properties has been discussed in the section "Effective Catalyst Layer Properties from Percolation Theory" in Chapter 3.

Specific challenges in CCL modeling arise because structure and composition of the layer are not fixed. They undergo changes on long timescales due to degradation as well as rapid changes due to variations in water content. Proton transport, gas diffusion, and electrocatalytic activity respond to these changes. Since the liquid water saturation $s$ is a spatially varying function at $j_0 > 0$, physicochemical properties become spatially varying functions in an operating cell. This creates a nonlinear coupling of properties and performance that demands self-consistent solution schemes. Mathematically, the problem is formulated as a set of continuity equations to satisfy conservation of mass and charge of all species involved. Transport equations of these species and source or sink terms complete the modeling framework.

This chapter focuses on steady-state phenomena during electrode operation. The structure that is formed during self-organization in catalyst layer inks and during MEA fabrication is considered to be constant. Unless stated otherwise, one-dimensional modeling approaches will be explored. This means that species transport occurs in through-plane direction only, perpendicular to the electrode plane. Moreover, in most part of this chapter, conditions are assumed to be isothermal.

The self-consistent solution can be obtained in analytical form under simplifying assumptions. In the general case, the solution requires numerical software tools. The solution provides spatial distributions (maps) of electrolyte phase potential, pressures (concentrations), and fluxes of species, as indicated in Figure 3.5. These distributions can be related further to global performance, rated in terms of voltage efficiency, power density, effectiveness factor of Pt utilization, and water handling capabilities. The ability to relate local performance variables to global performance metrics is a unique signature of physical modeling. These relations are not amenable to direct experimental investigation and they are of tremendous value for systematic improvements in CCL structure and operation.

## APPROACHES TO CATALYST LAYER PERFORMANCE MODELING

The simplest way of modeling the CCL performance is to consider the structure depicted in Figure 3.1 as a homogeneous medium with effective parameters for electron, proton, and oxygen transport (macrohomogeneous model, MHM). Particularly, the membrane phase potential $\Phi$ and carbon-phase potential $\phi$ are assumed to be continuous functions of coordinates.* Physically, this means that each representative volume of the CCL contains many carbon and Pt particles, Nafion phase domains, and voids, so that one may speak about volume-averaged concentrations, potentials, transport coefficients, currents, reaction rates, and so on.

Several approaches considered single agglomerate as an elementary conversion unit within the MHM (Baschuk and Li, 2000; Dobson et al., 2012; Jaouen et al., 2002; Pisani et al., 2003; Schwarz and Djilali, 2007; Sun et al., 2005; Tabe et al., 2011).

---

* Note that the electronic conductivity of the CCL is usually large, so that the potential loss from electron transport can be safely ignored.

These flooded agglomerate models (FAM) assumed that agglomerates are spherical and filled uniformly with a mixture of catalyst species and ionomer electrolyte. Solutions to an auxiliary problem of oxygen transport and consumption in ionomer-filled agglomerates yield equations for the potential-dependent conversion function (the ORR rate), which includes the agglomerate radius and composition as well as the thickness of the Nafion film covering the agglomerate (Harvey et al., 2008; Karan, 2007). This conversion function could then be used in the through-plane, MHM-like model as a source/sink function. A shortcoming of this approach is that it makes no distinction between ionomer and water phases. In any well-functioning catalyst layer for PEFC operation at $T < 100°C$, water is the active medium for proton supply while ionomer acts as the proton donor.

A newer variant of a water-filled FAM, discussed in the sections "Mesoscale Model of Self-Organization in Catalyst Layer Inks" in Chapter 3 and "Hierarchical Model of CCL Operation," explicitly distinguishes between the roles of ionomer and water (Sadeghi et al., 2013a,b). It is based on mesoscale modeling of the CCL structure (Malek et al., 2011). The results suggest that the polymer electrolyte (Nafion) does not penetrate an agglomerate; therefore, inside the agglomerate, the ORR runs at the Pt/water interface. The key differences between the two FAM variants are (i) in the nature of the interface, at which the OR proceeds, and (ii) in the transport media for oxygen and protons inside the agglomerate, via Nafion or water.

Is the FAM (incorporated as a submodule in MHM) more accurate than the MHM alone? The MHM alone could give a sufficient representation of polarization curves. However, the MHM is inept to account for water fluxes and distribution effects in the CCL, which depend on agglomeration and pore space morphology. Moreover, as demonstrated in the section "Hierarchical Model of CCL Operation," the MHM overestimates the effectiveness factor of Pt utilization. Processes of oxygen diffusion and proton transport at the agglomerate level (or the level of water-filled nanopores in agglomerates) must be accounted for to obtain an accurate estimate of this structure-sensitive function, particularly at high current densities (Sadeghi et al., 2013b). Nonetheless, in most of the cases, the MHM is a good starting point for CCL characterization and understanding its basic function.

## WATER IN CATALYST LAYERS: PRELIMINARY CONSIDERATIONS

Accounting for the role of water in physical models of CCLs requires knowledge of the following two aspects in theory and experiment.

**Local equilibrium of water:** By which mechanism does water attain local equilibrium? The approach in Eikerling (2006) and Liu and Eikerling (2008) presumes that capillary forces at the liquid–gas interfaces in pores equilibrate the local water content. This approach neglects surface film formation or droplet formation in pores of CLs. *Ex situ* diagnostics, probing porous structures, and water sorption characteristics under equilibrium condition, are needed to establish relations between porous structure, operating conditions, and local water distribution.

**Water transport and transformation:** What mechanisms of water transport are involved and what are the values of the transport parameters? The relevant mechanisms include diffusion, convection, electro-osmotic drag, and vaporization exchange. The corresponding parameters are amenable to evaluation by *ex situ* diagnostics. The statistical theory of random composite media (Kirkpatrick, 1973; Milton, 2002; Torquato, 2002) and especially percolation theory (Broadbent and Hammersley, 1957; Isichenko, 1992; Stauffer and Aharony, 1994) provide instructive tools for studying effective parameters of transport and interfacial processes, as discussed in the section "Effective Catalyst Layer Properties from Percolation Theory" in Chapter 3.

On the cathode side of a PEFC, electro-osmotic influx of water from the PEM and water production in the ORR create an excess of liquid water under normal conditions, even if the reactant at the cathode inlet is dry. Under normal conditions, it is reasonable to assume that primary hydrophilic pores and ionomer in the CCL are well hydrated. The proton conductivity can be assumed to be relatively constant. At high rate of water formation and insufficient water removal, excessive accumulation of water occurs in diffusion media and flow fields, which blocks critical pathways for gas diffusion of reactants.

The main mechanism of equilibration of water with the porous medium is by capillary condensation. Pore filling is therefore determined by the Young–Laplace equation that relates the capillary pressure, $p^c$, or the capillary radius, $r^c$, to the local gas pressure, $p^g$, and liquid pressure, $p^l$, at static menisci in pores,

$$p^c = \frac{2\gamma_w \cos(\theta)}{r^c} = p^g - p^l, \tag{4.1}$$

where $\gamma_w$ is the surface tension of water and $\theta$ the wetting angle. It is assumed that small primary pores inside agglomerates of CLs are hydrophilic (Vol'fkovich et al., 2010). Under any realistic operating conditions, these pores are fully saturated with liquid water.

With increasing current density, the liquid water accumulation in secondary pores is bound to increase. This process depends on their pore size and wetting angle. The wettability of secondary pores is, therefore, vital for controlling the water formation in CCLs. If all of the secondary pores were considered hydrophilic, the liquid water saturation would be given by an integral expression containing the differential PSD, $dX_p(r)/dr$,

$$s = \frac{1}{X_p} \int_0^{r^c} dr \frac{dX_p(r)}{dr}. \tag{4.2}$$

Pores with radius $r \le r^c$ are filled with liquid water, while pores with $r > r^c$ are filled with gas. In an operating fuel cell, $s$ depends on the pore size distribution, wettability distribution, and the distributions of pressures, i.e., $p^g$ and $p^l$. The pressure distributions are coupled to stationary fluxes of species as well as to rates of current generation and evaporation via the set of flux and conservation equations that will be presented in the section "Macrohomogeneous Model with Constant Properties."

When it becomes necessary to distinguish hydrophilic and hydrophobic pores in CCLs (Kusoglu et al., 2012), the liquid saturation is given by Equation 3.101.

However, as discussed in the paragraph below that equation, liquid water formation in hydrophobic CLs pores is virtually impossible, requiring huge liquid excess pressures. Therefore, during operation, the saturation $s$ could increase only in secondary pores if they are hydrophilic. The volume fraction of hydrophobic pores will determine whether the CL performance depends significantly on the liquid water saturation.

## MODEL OF TRANSPORT AND REACTION IN CATHODE CATALYST LAYERS

From the modeling perspective, all catalyst layers perform the same job: with the aid of neutral molecules, they convert a flux of ions or protons into a flux of electrons, or vice versa. Based on these principles, the CL model can be formulated in rather general terms, which would be suitable for catalyst layers of different types. Specific features or requirements of the CL of interest can be taken into account by model parameters, or by adding new terms, if necessary.

The operation of generic catalyst layer is illustrated schematically in Figure 4.2. It shows the *shapes* of local variables and fluxes in a layer, where the axis $x$ is directed from the interface with the electrolyte medium, viz., the PEM, toward the interface with a porous gas diffusion medium, viz., a porous transport layer (PTL) or gas diffusion layer (GDL). The ionic current arrives or leaves the CL at the electrolyte interface, while the feed molecules are supplied from the porous transport medium (Figure 4.2).

Most of this chapter concentrates on the operation of the CCL in a PEFC. The fluxes of oxygen molecules, protons, electrons, and water molecules are depicted in Figure 4.3. The overall operation can be thought of as the conversion of an electron current into a proton current, sustained by the supply of oxygen and kept in balance by the removal of water. In a steady state, the fluxes are related by stoichiometric

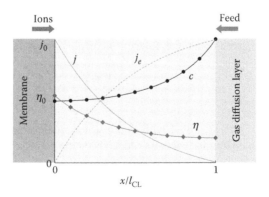

**FIGURE 4.2** A schematic of a generic catalyst layer. The through-plane shapes of the ionic current density $j$, the electron current density $j_e$, the feed molecules concentration $c$, and the local overpotential $\eta$.

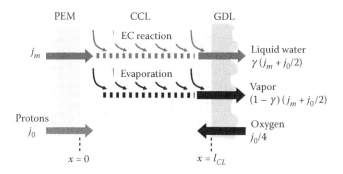

**FIGURE 4.3** Illustration of the coupled fluxes of species in operational CCLs. (Reprinted from Eikerling, M. 2006. *J. Electrochem. Soc.*, 153(3), E58–E70. Copyright (2006), the Electrochemical Society. With permission.)

coefficients, which are given most conveniently in units of an equivalent current density. The conversion occurs in the CCL volume along the thickness direction $x$.

The proton current density $j$ decreases from $j_0$ at the membrane/CL interface, to zero at the CL/GDL interface. The electron current density $j_e$ grows in this direction, from zero at $x = 0$ to $j_0$ at $x = l_{CL}$. The electrochemical conversion is driven by the local overpotential $\eta$, which increases toward the membrane (Figure 4.2).* Feed molecules (oxygen) are consumed in the reaction, and their molar concentration $c$ decreases toward the membrane. Their flux is given by a transport equation, usually a diffusion equation.

According to Ohm's law $j = -\sigma_p \partial \eta / \partial x$, $\eta$ must grow monotonically toward the membrane to provide a positive proton current in the CL (Figure 4.2). Similarly, Ohm's law should be considered for the electron flux in the electronic conductor phase. However, the conductivity of this phase is usually high so that the potential shape of this phase will be uniform. The shape of the overpotential follows the shape of the electrolyte phase potential. The function $\eta(x)$ peaks at the membrane interface, and obviously $\eta_0$ is the total potential loss in the catalyst layer.

Of foremost interest is the dependence of $\eta_0$ on $j_0$, the CL *polarization curve*. This dependence shows how much of the cell open-circuit potential should be spent for the electrochemical conversion of $j_0$ from ionic into electronic form, or vice versa. Huge efforts of CL designers and modelers have been directed toward lowering $\eta_0$.

However, minimization of $\eta_0(j_0)$ is hardly possible without the detailed knowledge of the local shapes of currents and potentials through the CL. Unfortunately, state-of-the-art experimental techniques cannot resolve these shapes. This is why, modeling plays a key role in understanding the laws of CL function. This problem is the main subject of Chapter 4.

---

* In Chapters 4 and 5, a positively defined ORR overpotential is assumed. Thus the local overpotential $\eta(x)$ corresponds to the difference of electrolyte and metal phase potentials. This convention reverses the definition used in Chapters 1 and 3. Owing to the constancy of the metal phase potential, which is the default case, changes in $\eta(x)$ exactly match changes in the electrolyte phase potential, that is, $\Delta \Phi(x) = \Delta \eta(x)$. This makes the redefinition of the overpotential a convenient choice in performance modeling.

## STANDARD MODEL OF CCL OPERATION

Processes inside the layer, shown in Figure 4.3, are related by mass and charge conservation laws for involved species, expressed in the form of continuity equations. The general form of the continuity equation is

$$\frac{\partial \rho}{\partial t} + \nabla \cdot \mathbf{j} = R \tag{4.3}$$

where $\rho(\mathbf{r}, t)$ is the density field of the considered species, $R(\mathbf{r}, t)$ is a volumetric source/sink rate, and $\mathbf{j}(\mathbf{r}, t)$ is a flux density. Steady-state operation corresponds to $\partial \rho / \partial t = 0$.

Continuity equations including all species and processes have been presented in Eikerling (2006). Several simplifications led to the set of steady-state equations, which will be presented in two blocks (A and B). Block A comprises the governing equations for electrochemical processes, involving proton transport,

$$\frac{d\eta}{dx} = -\frac{j(x)}{\sigma_p(s(x))}, \tag{4.4}$$

interfacial charge transfer due to faradaic reaction,

$$\frac{dj}{dx} = -R_{reac}(x), \tag{4.5}$$

and oxygen diffusion,

$$\frac{dp}{dx} = \frac{j_0 - j(x)}{4fD(s(x))}, \tag{4.6}$$

where $f = F/(R_g T)$ and $D$ is the oxygen diffusion coefficient. Fluxes of oxygen and protons are related by $j_{ox}(x) = (j_0 - j(x))/4$; all fluxes are given in units of A cm$^{-2}$.

Below in this chapter, an equivalent system of equations, written in terms of the oxygen concentration $c$, will also be used:

$$\frac{\partial j}{\partial x} = -R_{reac}(x), \tag{4.7}$$

$$j = -\sigma_p \frac{\partial \eta}{\partial x}, \tag{4.8}$$

$$D\frac{\partial c}{\partial x} = \frac{j_0 - j}{4F}. \tag{4.9}$$

Block B comprises governing equations for coupled water fluxes in the CCL, including water formation and vaporization exchange:

$$\frac{dj^l}{dx} = \frac{1}{2}R_{reac}(x) - R_{lv}(x), \tag{4.10}$$

liquid water transport

$$\frac{dp^l}{dx} = \frac{1}{B_0 f K^l (s(x))} \left[ \left( n_d + \frac{1}{2} \right) (j_p(x) - j_0) + j^v(x) + n_d j_0 - j_m \right], \quad (4.11)$$

vaporization exchange

$$\frac{dj^v}{dx} = R_{lv}(x), \quad (4.12)$$

and vapor diffusion

$$\frac{dq}{dx} = -\frac{j^v(x)}{f D^v (s(x))}, \quad (4.13)$$

where $B_0 = R_g T / (\bar{V}_w \mu^l)$ with the dynamic viscosity of water, $\mu^l$. The solution of Equations 4.4 through 4.13 gives the local partial pressure of oxygen, $p(x)$, and water vapor, $q(x)$, the liquid water pressure, $p^l(x)$, the local electrode overpotential, $\eta(x)$, as well as the flux (or current) densities of protons, $j(x)$, liquid water, $j^l(x)$, and water vapor, $j^v(x)$. Equation 4.11 includes a term for water transport due to electro-osmotic drag with a drag coefficient $n_d$.

Block A contains one free parameter, $j_0$, the total current density of the PEFC. Fixing this parameter defines the working point of the PEFC. Block B has two free parameters, the total water flux and the liquid water fraction of the total water flux at $x = l_{CL}$. A self-consistent solution to find these two parameters demands a full MEA model.

The oxygen partial pressure is related to the oxygen concentration in the gas phase via the ideal gas law, $p(x) = R_g T c(x)$. To obtain the oxygen concentration in solution, Henry's law constant (solubility) of oxygen must be used.*

In the remainder of this book, the cathodic overpotential is defined as the positive difference

$$\eta = E^{eq}_{O_2,H^+} - E_{O_2,H^+} = \Phi(x) - \phi(x) - \left( \Phi^{eq} - \phi^{eq} \right), \quad (4.14)$$

where $E^{eq}_{O_2,H^+}$ is the equilibrium cathode potential, defined via the Nernst equation.

Employing the usual assumption of high electronic conductivity of the Pt/C phase ($>10$ S cm$^{-1}$), this phase can be considered equipotential, $\phi(x) = \text{const}$. Using this condition, the total overpotential incurred by the cathode is given by the local value of the overpotential at $x = 0$:

$$\eta_0 = \eta(0) \quad (4.15)$$

---

* In gas diffusion electrodes, it is usually assumed that the mass transport resistance caused by diffusion of dissolved oxygen is negligible, since the diffusion lengths in solution is small compared to the diffusion lengths in the gas phase. Under this assumption, Henry's constant can be absorbed as a constant factor in the exchange current density. This is a common practice in PEFC modeling.

Moreover, since $\phi(x) = $ const. and, therefore, $d\Phi/dx = d\eta/dx$, the local electrolyte potential $\Phi(x)$ that should be used in Ohm's law of proton transport in Equation 4.4 can be replaced by $\eta(x)$.

In this section, the practically relevant case of high cathode overpotentials, $\eta \geq 3R_gT/F$ will be focussed on. The electrochemical source term or faradaic reaction term, $R_{reac}(x)$, can thus be represented by the cathodic branch of the Butler–Volmer equation:

$$R_{reac}(x) = \frac{j^0}{l_{CL}} \left(\frac{p(x)}{p^{FF}}\right)^{\gamma_{O_2}} \exp\left(\alpha_c f \eta(x)\right), \tag{4.16}$$

where $\alpha_c$ is the effective transfer coefficient of the ORR, defined in the section "Deciphering the ORR" in Chapter 3, $j^0$ is the effective exchange current density, and $p^{FF}$ is the oxygen partial pressure at the flow field. The reaction order with respect to oxygen partial pressure (or concentration) is $\gamma_{O_2} = 1$, as discussed in the section "Deciphering the ORR" in Chapter 3. For ionomer-impregnated catalyst layers (type I electrodes), a reasonable assumption can be made that the proton concentration is constant; its effect is therefore not explicitly accounted for in Equation 4.16. A refinement to this parameterization will be discussed in the section "Hierarchical Model of CCL Operation."

The source term of water vapor is

$$R_{lv}(x) = \frac{Fk_v}{l_{CL}} \xi^{lv}(s) \left\{q_r^s(T) - q(x)\right\}, \tag{4.17}$$

where $k_v$ denotes the intrinsic rate constant of evaporation and $q_r^s(x)$ is the saturated vapor pressure in pores at the capillary radius $r^c(x)$, given by the Kelvin equation,

$$q_r^s(x) = q^{s,\infty} \exp\left(-\frac{2\gamma_w \cos(\theta)\bar{V}_w}{RTr^c(x)}\right), \quad q^{s,\infty} = q^0 \exp\left(-\frac{E_a}{k_BT}\right) \tag{4.18}$$

with saturated pressure $q^{s,\infty}(T)$ for a planar vapor–liquid interface. The activation energy of vaporization is approximately $E_a \approx 0.44$ eV (corresponding roughly to the strength of two hydrogen bonds in water).

The water balance problem requires a closure relation that expresses $s(x)$ as a function of the pressure distribution. The Young–Laplace equation,

$$p^c(x) = \frac{2\gamma_w \cos(\theta)}{r^c(x)} = p^g(x) - p^l(x) = p(x) + q(x) + p^r - p^l(x), \tag{4.19}$$

relates the local distributions of liquid and gas pressures to the local capillary radius, $r^c$, which determines the local liquid water saturation, $s(x)$, via Equation 4.2. Thereby, all transport coefficients can be obtained, rendering a closed system of equations.

The two blocks of equations presented above represent coupled problems of mixed transport and conversion of charged species and water. The electrochemical problem

(Block A), Equations 4.4 through 4.6, is the standard MHM of catalyst layer operation, developed in different variants by Springer et al. (1993), Perry et al. (1998), and Eikerling and Kornyshev (1998). Block B, Equations 4.10 through 4.13, describing the water balance problem, has a formal structure similar to the electrochemical problem.

Both blocks, when considered separately, are analogous to transmission line models of porous electrodes that were introduced by De Levie in the 1960s (de Levie, 1963, 1967). Two types of coupling exist between them. An explicit form of coupling is due to the electrochemical source term, $R_{reac}$, that appears in both sets. An implicit coupling is due to the dependence of the solution on the spatially varying liquid saturation, $s(x)$. The relations $p^c \rightarrow r^c \rightarrow s$ and $D(s(x))$, $D^v(s(x))$, $K^l(s(x))$, $j^0(s(x))$, $\xi^{lv}(s(x))$ make the system of equations highly nonlinear. Obtaining self-consistent solutions warrants, in general, numerical procedures.

The MHM and the water balance model decouple when the liquid saturation is constant; with this assumption, effective parameters of transport and reaction will be constant as well. This is the situation normally evaluated in CCL modeling. It will be considered next. Specific effects due to the complex coupling between porous morphology, liquid water formation, oxygen transport, and reaction rate distributions will be discussed in the section "Water in Catalyst Layers: The Watershed."

Approaches to solving this problem can be distinguished as to how they incorporate the distinct scales and how they treat the water balance in the CCL. This chapter presents three model variants:

1. The basic MHM assumes a CL with fixed structure and composition and thus constant properties of transport and reaction processes; these properties are represented by effective parameters, which are averaged over micro- and mesoscopic heterogeneities in the layer; for this scenario, the main regimes and limiting cases of operation will be discussed and results of optimization studies will be presented.
2. A modification of the basic model is to evaluate the impact of progressive water-filling of secondary pores with increasing current density; in this scenario, the composition of the layer and associated transport properties are spatially varying functions; they depend on the water distribution, which varies with the current density; the nonlinear coupling of external conditions, liquid water distribution, local properties, and voltage response of the CCL triggers a transition from partially saturated to fully saturated state of operation; the transition could involve a bistability phenomenon.
3. In the hierarchical model, macroscale transport processes will be coupled to transport and reaction at the mesoscale, as illustrated in Figure 4.1; an explicit treatment of agglomerate effects is needed to properly assess the effectiveness factor of Pt utilization, which transpires as the key parameter in the structural optimization of CCLs.

Further treatment in this section will evaluate general principles and capabilities of the MHM. Moreover, the MHM will be applied for an analysis of the impact of CCL structure and composition on performance.

## MACROHOMOGENEOUS MODEL WITH CONSTANT PROPERTIES

Focusing on Equations 4.4 through 4.6, Neumann boundary conditions are imposed on potential and a Dirichlet boundary condition on oxygen partial pressure:

$$\sigma_p \left.\frac{d\eta}{dx}\right|_{x=0} = -j_0, \quad \sigma_p \left.\frac{d\eta}{dx}\right|_{x=l_{CL}} = 0, \quad p|_{x=l_{CL}} = p_L = p^{FF} - \frac{j_0 L_{PTL}}{4fD^{PTL}} \quad (4.20)$$

As for the last condition, excellent diffusion properties of porous transport media will be implied for the remainder of section, that is, $p(l_{CL}) = p^{FF}$.

It is convenient to rewrite the MHM set of equations (Block A) in dimensionless form:

$$\frac{dP}{d\zeta} = -(1 - \iota) \tag{4.21}$$

$$\frac{d\Gamma}{d\zeta} = -g\Gamma\iota \tag{4.22}$$

$$\frac{d\iota}{d\zeta} = P\Gamma \tag{4.23}$$

with dimensionless and normalized variables

$$\zeta = \frac{j_0}{I}\left(1 - \frac{x}{l_{CL}}\right), \quad P = \frac{p}{p_L}, \quad \iota = \frac{j}{j_0}, \quad \Gamma = \frac{j^0 I}{j_0^2}\exp\left(\frac{\eta}{b}\right),$$

where $I = \frac{4fDp_L}{l_{CL}}$ is an effective oxygen transport parameter with units of A cm$^{-2}$ and $b = (\alpha_c f)^{-1}$ is a Tafel parameter.

The solution of Equations 4.21 through 4.23 is subject to the boundary conditions

$$P(\zeta = 0) = 1, \quad \iota(\zeta = 0) = 0, \quad \text{and } \iota\left(\zeta = \frac{j_0}{I}\right) = 1. \tag{4.24}$$

Now it is evident that the shape of the solution is determined by a single parameter

$$g = \frac{4fDp_L}{\sigma_p b}, \tag{4.25}$$

which allows classifying diffusion-limited ($g \ll 1$), proton transport-limited ($g \gg 1$) and mixed ($g \sim 1$) cases. Parameter estimates in Eikerling and Kornyshev (1998) and Eikerling et al. (2007a) suggest that the mixed case is the most likely scenario to be encountered in actual CCL.

Solution of the problem at a given $j_0$ gives a value of the dimensionless variable $\Gamma_0 = \Gamma\left(\zeta = \frac{j_0}{I}\right)$, which can be rearranged to obtain an expression for the CCL overpotential

$$\eta_0 = b \ln \Gamma_0 + 2b \ln \frac{j_0}{I} - b \ln \frac{j^0}{I}. \tag{4.26}$$

The last term on the right-hand side is a constant. In general, the first term on the right-hand side will depend on $j_0$. However, as shown in numerical solutions, this term will approach a constant value (for $g \ll 1$) or a constant value plus a linear ohmic term (for $g \gg 1$) at high current densities, $j_0 \gg I$, when severe transport limitations leave most of the CCL electrocatalytically inactive. Equation 4.26 implies that the occurrence of a term with double Tafel slope characteristic (the second term) is a generic signature of CCL under severely impeded oxygen diffusion or proton conduction. The emergence of this behavior will be demonstrated by rigorously derived analytical expressions for limiting cases of $g$ that will be discussed in later sections of this chapter.

For the case of poor oxygen diffusion, which arises at $j_0 \gg I$, only a thin sublayer of a thickness

$$l_D = \frac{I}{j_0} l_{CL} = \frac{4fDp_L}{j_0} \tag{4.27}$$

adjacent to the PTL (at $x \approx l_{CL}$) will be active, see Equations 3.18 and 4.91. In this oxygen depletion regime, the performance is given by

$$\eta_0 \approx b \ln \Gamma_c + 2b \ln \frac{j_0}{I} + \frac{l_{CL}}{\sigma_p} j_0 - b \ln \frac{j^0}{I} \tag{4.28}$$

with a constant $\Gamma_c$. The part of the CCL adjacent the PEM side with thickness $(l_{CL} - l_D) \approx l_{CL}$, is catalytically inactive due to the oxygen starvation. This highly nonuniform reaction rate distribution implies a drastic underutilization of the catalyst. The current density versus voltage response function is dominated by a term with double Tafel slope and by an ohmic term on the right-hand side.

The general solution of the MHM equations can be obtained only numerically. The prevailing transport limitation depends on the structure of the CCL. A CCL with low gas porosity or liquid water saturation close to the flooding limit, but with high ionomer content, will have low oxygen diffusivity and high proton conductivity; this case corresponds to the limit $g \ll 1$. On the other hand, a CCL with high open gas porosity and low amount of ionomer, which forms a poorly percolating network, will exhibit high oxygen diffusivity and low proton conductivity, corresponding to the limit $g \gg 1$. Approximate analytical solutions have been obtained in these limits (Eikerling and Kornyshev, 1998; Eikerling et al., 2007a; Springer et al., 1993). These solutions will be discussed in the section "MHM with Constant Coefficients: Analytical Solutions."

Based on this discussion, three current density regimes can be distinguished in electrode polarization curves:

1. A kinetic regime at small current densities, $j_0 \ll I$, with simple Tafel dependence, $\eta_0 \sim b \ln \left( j_0 / j^0 \right)$; in this regime, transport limitations of any kind are

negligible; it represents the simple limiting case of uniform utilization of the active catalyst surface.

2. An intermediate regime for $j_0 \gtrsim I$ (if $g < 1$) or $j_0 \gtrsim \sigma_p b / l_{CL}$ (if $g > 1$) with prevailing double Tafel slope characteristic, $\eta_0 \sim 2b \ln (j_0/I)$, given by the second term on the right-hand side in Equation 4.28.

3. An oxygen-depletion regime for $j_0 \gg I$, in which oxygen is consumed in a sublayer of thickness $l_D \ll l_{CL}$; if the proton conductivity is poor, $\sigma_p \ll 0.1 \ S \ cm^{-1}$, the voltage loss will be dominated by the linear ohmic term in Equation 4.28.

The three regimes are indicated in Figure 4.4a, which shows $\eta_0$, as a function of $j_0/I$ for $g = 1$.

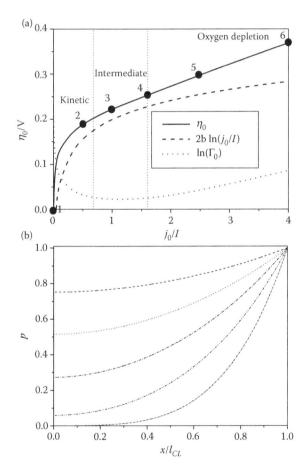

**FIGURE 4.4** (a) CCL overpotential, $\eta_0$, as a function of $j_0/I$ for the mixed case of $g = 1$. The three performance regimes, corresponding to kinetic, intermediate, and oxygen depletion regime are indicated. In addition to total overpotential, the separate contributions due to the first term and the "double Tafel slope" term are also plotted separately. (b) Oxygen concentration profiles at points 1 through 6 labeled in (a).

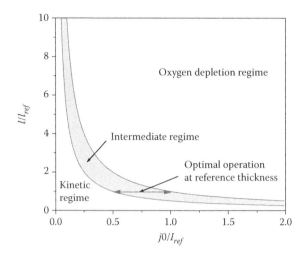

**FIGURE 4.5** Configuration diagram of a cathode catalyst layer, showing the different regimes of operation assumed in dependence of catalyst layer thickness (ordinate) and fuel cell current density (abscissa). Thickness and current density are normalized to reference parameters with typical values $l_{ref} \approx 10$ μm and $I_{ref} \approx 1$ A cm$^{-2}$.

Operation in the intermediate regime is most desirable. It represents the best trade-off between requirements of a large catalyst surface area and a sufficient rate of supply of protons and oxygen. For given target current density $j_0$, composition and thickness should be adjusted in order to operate the CCL in the intermediate regime. Although reaction rate distributions exhibit a pronounced nonuniformity in this regime, all parts of the layer are used for reactions. There are, thus, no inactive parts. As long as the CCL is operated in the intermediate regime, overpotential losses are almost independent of the thickness. In Eikerling et al. (2007a), these findings were displayed in the form of an operational map that is reproduced in Figure 4.5. The existence of a maximum thickness beyond which the performance degrades is due to the concerted impact of oxygen and proton transport limitations. Considered separately, each of these limitations would only serve to define a minimum thickness, below which performance worsens due to an insufficient electroactive surface.

### INTERMEDIATE REGIME: TWO LIMITING CASES

The voltage response of the CCL in the intermediate regime is dictated by the "double Tafel slope" term, $\eta_0 \sim 2b \ln j_0$. This regime leads to similar expressions for reaction penetration depth and differential CCL resistance in the limiting cases of (i) rapid proton conduction and poor oxygen diffusion ($g \ll 1$) and (ii) rapid oxygen diffusion and poor proton conduction ($g \gg 1$).

In the first case ($g \ll 1$), the reaction penetration depth and the differential resistance of the layer are given by

$$l_D = l_{CL}\sqrt{\frac{I}{j^0}}\exp\left(-\frac{\eta_0}{2b}\right) \tag{4.29}$$

and

$$R_D = b\sqrt{\frac{1}{Ij^0}} \exp\left(-\frac{\eta_0}{2b}\right), \tag{4.30}$$

respectively. In the second case ($g \gg 1$), reaction penetration depth and differential resistance are given by

$$l_\sigma = \sqrt{\frac{l_{CL}\sigma_p b}{j^0}} \exp\left(-\frac{\eta_0}{2b}\right) \tag{4.31}$$

(see also Equation 4.154) and

$$R_\sigma = b\sqrt{\frac{l_{CL}}{\sigma_p b j^0}} \exp\left(-\frac{\eta_0}{2b}\right). \tag{4.32}$$

It should be emphasized that the characteristic lengths $l_D$ and $l_\sigma$ are independent of $l_{CL}$, as they should be. This becomes obvious when the scaling of the characteristic current densities is considered: $I \propto l_{CL}^{-1}$, and $j^0 \propto l_{CL}$. Intrinsic lengths and differential resistances are shown together because they reveal practically useful relationships:

$$l_D = \frac{4fDp_L}{b}R_D, \quad l_\sigma = \sigma_p R_\sigma. \tag{4.33}$$

The reaction penetration depths, $l_D$ or $l_\sigma$, are highly insightful parameters to evaluate catalyst layer designs in view of transport limitations, uniformity of reaction rate distributions, and the corresponding effectiveness factor of Pt utilization, as discussed in the sections "Catalyst Layer Designs" in Chapter 1 and "Nonuniform Reaction Rate Distributions: Effectiveness Factor" in Chapter 3. Albeit, these parameters are not measurable. The differential resistances, $R_D$ or $R_\sigma$, can be determined experimentally either as the slope of the polarization curve or from electrochemical impedance spectra (Nyquist plots) as the low-frequency intercept of the CCL semicircle with the real axis. The expressions in Equation 4.33 thus relate the reaction penetration depths to parameters that can be measured.

In the intermediate regime, the CCL overpotential is completely determined by the relevant values of the reaction penetration depth and the differential resistance (except for an additive constant). The dominance of one transport limitation, be it impeded proton or oxygen transport, will result in a logarithmic current–voltage plot with double Tafel slope, $2b = 2/(\alpha_c f)$, at large $j_0$. The effective exchange current densities, characteristic of the intermediate regime, are $j_\sigma^0 = 2\sqrt{\sigma b j^0}$ in the limit of fast oxygen diffusion and $j_D^0 = 2\sqrt{Ij^0}$ in the limit of high proton conductivity. Both of these characteristic current densities are independent of $l_{CL}$, demonstrating that in the regime with double Tafel behavior the CCL performance is thickness-invariant.

## Structural Optimization Using MHM

The MHM was exploited in optimization studies of CCL composition and thickness, using the relations specified in the section "Effective Catalyst Layer Properties from Percolation Theory" in Chapter 3. From a statistical point of view, the effective properties of random composite media depend on the volume fractions of the distinct components, as well as further parameters that represent structural correlations. In experimental investigations, the composition is usually specified in terms of weight fractions, which are readily controlled during fabrication. It is, thus, useful to link the performance directly to the weight fractions of the catalyst layer components. The volume fractions of distinct components in CCLs can be expressed via mass densities of Pt, C, and ionomer ($\rho_{Pt}$, $\rho_C$, $\rho_{el}$), weight fractions ($Y_{Pt}$, $Y_{el}$), Pt loading per geometric surface area ($m_{Pt}$), and $l_{CL}$ using the following relations:

$$X_{Pt} = \frac{m_{Pt}}{l_{CL}} \frac{1}{\rho_{Pt}}, \tag{4.34}$$

$$X_C = \frac{m_{Pt}}{l_{CL}} \frac{1 - Y_{Pt}}{Y_{Pt}\rho_C}, \tag{4.35}$$

$$X_{el} = \frac{m_{Pt}}{l_{CL}} \frac{Y_{el}}{(1 - Y_{el}) Y_{Pt}\rho_{el}}. \tag{4.36}$$

The theory of composition-dependent CCL performance reproduces experimental trends (Lee et al., 1998; Passalacqua et al., 2001; Uchida et al., 1995b; Xie et al., 2005; Wang et al., 2004). The dependence of the fuel cell voltage, $E_{cell}$, on $Y_{el}$ is shown at various values of $j_0$ in Figure 4.6a; a CCL with uniform composition was considered. The value of $Y_{el}$ that gives the highest $E_{cell}$ depends on $j_0$. At intermediate current densities, $0.5$ A cm$^{-2} \leq j_0 \leq 1.2$ A cm$^{-2}$, the best performance is obtained with $Y_{el} \simeq 35$ wt%. Increasing the ionomer content to $Y_{el} > 35$ wt% reduces $E_{cell}$ ($j_0$) because of its negative impact on oxygen diffusion. At low current densities of $j_0 \approx 0.2$ A cm$^{-2}$, an increase in $Y_{el}$ enhances $E_{cell}$ ($j_0$), owed to the improved ECSA and the higher proton conductivity, whereas the oxygen diffusivity is not affected

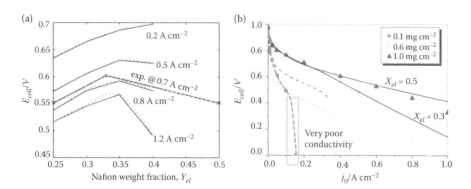

**FIGURE 4.6** (a) Effect of the Nafion weight fraction, $Y_{el}$, in CCL with uniform composition on the fuel cell voltage, $E_{cell}$, evaluated at different values of the current density, $j_0$. Experimental data taken from Passalacqua et al. (2001) (crosses) are shown for comparison. (b) Comparison of polarization curves, calculated in the model of composition-dependent performance, with experimental data of Uchida et al. (1995a,b).

significantly. The observed trends agree with experimental data by Passalacqua et al. (2001).

Figure 4.6b compares calculated plots of $E_{cell}$ versus $j_0$ with experimental data of Uchida et al. (1995a,b) for CCL with different ionomer content, as specified in the legend. Composition-dependent properties were parameterized using the functions proposed in the section "Effective Catalyst Layer Properties from Percolation Theory" in Chapter 3. The fuel cell voltage was assumed to be of the form

$$E_{cell} = 1.23 \text{ V} - \eta_0 (j_0) - \left(0.1 \text{ cm}^2 \text{ S}^{-1}\right) j_0,$$

where the third term on the right-hand side accounts for ohmic voltage loss in the membrane and all other voltage loss contributions are assumed negligible. The ionomer loading $m_{el}$ is converted into ionomer volume fraction using $X_{el} \approx \left(0.5 \text{ cm}^2 \text{ mg}^{-1} m_{el}\right)$. The percolation approach in modeling of composition effects reproduces the main experimental trends. It fails at very low ionomer loading, where the ionomer content might fall below the percolation threshold, and at high current densities, where flooding of porous diffusion media blocks the gaseous supply of oxygen.

The structure-based MHM was also used to explore novel design modifications. It was predicted and confirmed in experiment that functionally graded layers result in improved performance compared to standard CCLs with uniform composition (Xie et al., 2005; Wang et al., 2004). In this design, the catalyst layer is fabricated as a sublayer structure with gradually varying composition. Compared with the layer of uniform composition with $Y_{el} = 35$ wt%, corresponding to the optimum identified Figure 4.6, a simple three-sublayer structure with 30 wt% Nafion content in the sublayer at the GDL side, 35 wt% Nafion content in the middle sublayer, and 40% wt Nafion content at the PEM side was seen to improve $E_{cell} (j_0)$ by about 5%. Another advantage of this design is the reduced ohmic resistance at the PEM|CCL interface due to the improved contact area between electrolyte phases in PEM and CLL at higher ionomer loading in the CCL. In the interface between GDL and CCL, lower Nafion loading will decrease the probability of blockage of pores by Nafion and, thereby, facilitates water removal via the GDL.

As discussed in the section "Ionomer Structure in Catalyst Layers Redefined" in Chapter 3, a theory of composition-dependent effective properties that incorporates recent insights into structure formation in CCLs is yet to be developed. At present, the relations presented in the section "Effective Catalyst Layer Properties from Percolation Theory" in Chapter 3 do not account for agglomerate formation and skin-type morphology of the ionomer film at the agglomerate surface. Qualitative trends predicted by the simple structure-based catalyst layer theory should be correct, as confirmed by the results discussed in this section.

## WATER IN CATALYST LAYERS: THE WATERSHED

Until about 10 years ago, CCL models had neglected effects related to formation and transport of water. Up to that point, large drops in fuel cell voltage as the current density was ramped up to $\sim 1$ A cm$^{-2}$ or above had been frequently observed in experiments (Uchida et al., 1995a,b; Vielstich et al., 2003). Since this current density range is essential for attaining high power density, interest in understanding the so-called limiting current density phenomena has grown quickly. These water management challenges are aggravated in MEAs that employ UTCLs on the cathode side, as discussed in Chan and Eikerling (2014).

Based on empirical observations, water management challenges arising in PEFC operation at high current density, high relative humidity, or low temperature have been loosely associated with excessive flooding of the CCL. However, water balance models of MEAs with conventional or ultrathin CCLs show that this hypothesis is uncorroborated and indeed fundamentally wrong as explained below (Chan and Eikerling, 2014; Eikerling, 2006; Liu and Eikerling, 2008). This section deals with water management issues in conventional CCL. The model is an extension to the MHM with constant properties.

The effects of porous structure and liquid water accumulation on steady-state performance of conventional CCLs were explored in Eikerling (2006) and Liu and Eikerling (2008). In these modeling works, uniform wetting angle was assumed in secondary pores, with a value $\theta < 90°$. The full set of equations presented in the section "Macrohomogeneous Model with Constant Properties" are solved with the following boundary conditions:

1. At the CCL–PTL boundary $(x = l_{CL})$, the electrode potential is fixed, $\eta(l_{CL}) = \eta^L$; this defines the working point of the PEFC, that is, the proton flux at the PEM–CCL boundary $(x = 0)$, $j_0 = j(0)$ as well as the oxygen flux at the CCL–PTL boundary.

2. In Eikerling (2006), the oxygen partial pressure at the CCL–PTL interface, $p_L$, was assumed to be fixed. Instead, in Liu and Eikerling (2008), the oxygen partial pressure $p^{FF}$ at the flow field (FF) was assumed to be controlled, and diffusive gas transport PTL was accounted for as given by Equation 4.20. This modification allowed distinct signatures in $E_{cell}(j_0)$ due to limited oxygen transport in either CCL or PTL to be compared.

3. The proton flux is converted completely into electron flux in the CCL and, therefore, $j(l_{CL}) = 0$.

4. Assumed gas tightness of the PEM implies $j^v(0) = 0$.

5. A value of the liquid pressure at the PEM–CCL boundary is assumed, that is, $p^l(0) = p^{l0}$.

6. The liquid flux at the PEM–CCL boundary is $j^l(0) = j_m$.

7. At the CCL–PTL boundary, the liquid flux is $j^l(l_{CL}) = \gamma(j_m + j_0/2)$. The parameter $\gamma$, thus, represents the proportion of the water flux that leaves the CCL toward the GDL side in liquid form. The remaining proportion, that is $j^v(l_{CL}) = (1 - \gamma)(j_m + j_0/2)$, is transported out over the cathode side in vapor form. Complete liquid-to-vapor conversion in the CCL corresponds to $\gamma = 0$.

A full MEA model would be needed to solve for the parameters $p_0^l$, $j_m$, and $\gamma$. Since the model presented here focuses exclusively on the CCL, these parameters cannot be determined self-consistently. In order to reduce the number of undetermined parameters, a condition is imposed on $\gamma$: for any set of operating conditions, maximal conversion of liquid water into water vapor is assumed, that is, $\gamma$ should be minimal. This condition defines a critical current density of fuel cell operation, below which the CCL could completely vaporize liquid water. This critical current density could be used to assess the liquid-to-vapor conversion capability of a CCL. The condition of minimal $\gamma$ implies optimum vapor removal out of the MEA through PTL and FF. Such conditions are realistic for operation of the PEFC with dry reactant on the cathode side. Moreover, the PTL should possess high gas diffusivity.

As a further simplification, the continuous PSD is replaced by a bimodal $\delta$-distribution:

$$\frac{dX_p(r)}{dr} = X_\mu \delta(r - r_\mu) + X_M \delta(r - r_M). \tag{4.37}$$

This assumption permits a full analytical solution and readily reveals major principles of water handling and performance in CCLs (Eikerling, 2006). At the same time, it still captures physical processes, critical phenomena, operating conditions, and structural features such as distinct pore sizes ($r_\mu$, $r_M$) and porosity contributions of primary and secondary pores ($X_\mu$, $X_M$).

In terms of liquid water saturation and water management in the CCL, the bimodal $\delta$-distribution leads to a three-state model. The three states that any REV could attain are the dry or water-free state ($s \approx 0$); the ideally wetted state ($s = X_\mu/X_p$), in which primary pores are water-filled while secondary pores are water-free; and the fully saturated state ($s = 1$); these states are illustrated in Figure 4.7. In the ideally wetted state, catalyst utilization and exchange current density are high and the interfacial area for vaporization exchange, $\xi^{lv}$, is large. Moreover, diffusion coefficients will still be high, since secondary pores remain water-free. In the fully saturated state, major parts of the CCL are deactivated due to the blocking of oxygen diffusion pathways by liquid water.

The characteristic current density, below which complete liquid-to-vapor conversion is possible, is

$$j_{crit}^{lv} = \frac{2q_2^s f D_2^v}{\lambda_v} \tanh\left(\frac{l_{CL}}{\lambda_v}\right) - 2j_m, \tag{4.38}$$

"Dry" state          Ideally wetted state          Fully saturated state

**FIGURE 4.7** The three states in which the CCL could operate if a bimodal, $\delta$ function-like pore size distribution with uniform hydrophilic contact angle is assumed. The layer will poorly perform in the fully saturated or dry states. The optimum performance is attained in the ideally wetted state, in which primary pores are flooded and secondary pores provide the space for gaseous diffusion of oxygen and water vapor. (Reprinted from Eikerling, M. 2006. *J. Electrochem. Soc.*, **153**(3), E58–E70, Figure 3 and 5. Copyright (2006), the Electrochemical Society. With permission.)

with an effective evaporation penetration depth

$$l_v = \sqrt{\frac{D_2^v l_{CL}}{k_v \xi_2^{lv}}}, \qquad (4.39)$$

where $q_2^s$, $D_2^v$, and $\xi_2^{lv}$ represent the saturated vapor pressure, the vapor diffusion coefficient, and the liquid–vapor interfacial area in the ideally wetted state. For a well-designed conventional CCL with sufficient gas porosity, $l_v$ will be in the range of 300–700 nm, indicative of very good vaporization capabilities of such layers.

The model reveals sensitive dependencies of CCL operation on porous structure, thickness, wetting angle, gas pressure at the cathode side, and net liquid water flux from the membrane. With rather favorable parameters (10 μm thickness, 5 atm cathodic gas pressure, 89° wetting angle), the critical current density of CCL flooding is found in the range 2–3 A cm$^{-2}$. For increased thickness, significantly smaller and thus more realistic cathode gas pressure, or slightly reduced wetting angles of secondary pores, CCLs could be fully flooded at current densities below 1 A cm$^{-2}$. The contact angle is an important parameter in this context, highly influential but difficult to control in fabrication (Alcañiz-Monge et al., 2001; Studebaker and Snow, 1955; Vol'fkovich et al., 2010). The assumption of a uniform contact angle, valid for all pores, is unrealistic; recent experimental studies suggest that a significant fraction of hydrophobic pores (above the percolation threshold) exist (Kusoglu et al., 2012). In this case, flooding of the CCL would be virtually impossible; the excess liquid pressure required to fill hydrophobic pores with sizes in the nanometer range would be exceedingly large.

An extension of the structure-based model of CCL operation for general continuous pore size distributions was presented in Liu and Eikerling (2008). Continuous PSDs allow relating global performance effects (limiting currents, bistability) to spatial distributions of water; fluxes and concentrations of reactants; and reaction rates. It was found that a CCL alone cannot give rise to a limiting current behavior in polarization curves. The explanation is simple and intuitive: in the fully saturated state, the CCL retains a residual oxygen diffusivity through liquid–water filled pores, $D \sim 10^{-6}$ cm$^2$ s$^{-1}$; the effect of flooding will thus be a drastic decrease in the reaction penetration depth $l_D$ to values in the range of 100 nm $\ll l_{CL}$; the overpotential of the CCL will increase so that the correspondingly reduced fraction of active Pt particles in the thin active sublayer could sustain the total current; the increase in overpotential is related logarithmically to the factor by which the reaction penetration depth decreases, following the Tafel law dependence.

Figure 4.8 illustrates the different signatures of flooding occurring in either CCL or GDL. The red curve corresponds to insignificant diffusion resistance in either medium. Full flooding of the GDL leads to a knee in the polarization curve and to the occurrence of a limiting current. Full flooding of the CCL causes a voltage drop of a finite value and it could be accompanied by bistability in the transition region; however, it does not lead to a limiting current behavior.

Upon increasing the fuel cell current density, a transition between two principal states of operation occurs, as illustrated in Figure 4.9. The ideally wetted state at low

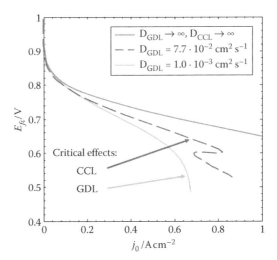

**FIGURE 4.8** Different signatures of flooding and severely reduced oxygen diffusivity in either CCL or GDL. Values of oxygen diffusion coefficients used are shown in the legend. The top curve is a reference curve for the case of ideal oxygen transport in CCL and GDL. Flooding of the GDL causes limiting current behavior. Flooding of the CCL causes a drop in fuel cell voltage accompanied by a bistability in the transition region. (Reprinted from *Electrochim. Acta*, **53**(13), Liu, J., and Eikerling, M. Model of cathode catalyst layers for polymer electrolyte fuel cells: The role of porous structure and water accumulation. 4435–4446. Copyright (2008), Elsevier. With permission.)

current densities exhibits levels of liquid water saturation below the critical value for pore blocking. In this state, reaction rate distributions are relatively uniform and the effectiveness factor of Pt utilization is high. In the fully saturated state, liquid water saturation exceeds the critical value. The CCL could sustain only low residual gas diffusivity. Correspondingly, the reaction rate distribution will be highly nonuniform, rendering most of the CCL inactive. The transition between the two states of operation can occur monotonously or it could involve a bistable transition region. Bistability means that two stable-steady state solutions of the continuity equations coexist. It occurs as a result of the nonlinear coupling between liquid water accumulation, gaseous oxygen diffusion, and electrochemical conversion rate.

The critical current density of the transition from ideally wetted state to transition region or fully saturated state is the major optimization target of CCLs in view of their water handling capabilities. A larger critical current density gives higher voltage efficiency and power density. Critical current densities depend on structural parameters and operating conditions. Stability diagrams have been introduced for assessing effects of these parameters on CCL performance. The stability diagram in Figure 4.10 displays the effect of the pore volume fraction due to secondary pores (assuming a constant, slightly hydrophilic wetting angle in these pores) on fuel cell operation. The diagram distinguishes between the ideally wetted state, the bistability region, and the fully saturated state. The task of water management in CCLs is to push back capillary equilibrium to small enough pores so that liquid water formation cannot

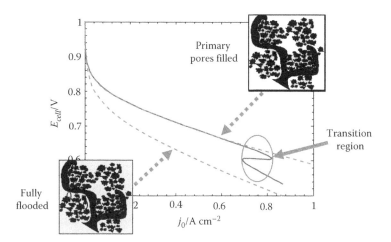

**FIGURE 4.9** Polarization curve calculated with a CCL model that accounts for the full coupling between porous structure, liquid water accumulation, and performance. The dashed lines represent limiting scenarios corresponding to the ideally wetted state and the fully saturated state. Bistability occurs in the transition region. (Reprinted from *Electrochim. Acta*, **53**(13), Liu, J., and Eikerling, M. Model of cathode catalyst layers for polymer electrolyte fuel cells: The role of porous structure and water accumulation. 4435–4446. Copyright (2008), Elsevier. With permission.)

block gaseous transport in secondary pores. Beneficial conditions in view of this objective are high total porosity, large volume fraction of secondary pores, a wetting angle that closely approaches or exceeds 90°, a high total gas pressure, and high operating temperature.

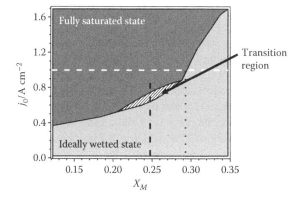

**FIGURE 4.10** Stability diagram showing the stable state of operation of the CCL as a function of the pore volume fraction of secondary pores, $X_M$, and $j_0$. (Reprinted from *Electrochim. Acta*, **53**(13), Liu, J., and Eikerling, M. Model of cathode catalyst layers for polymer electrolyte fuel cells: The role of porous structure and water accumulation. 4435–4446. Copyright (2008), Elsevier. With permission.)

It can be seen in Figure 4.9 that deviations of performance from the reference curve for the ideally wetted state are negligible at current densities below the transition region. An important conclusion from this observation is that the assumption of constant and uniform effective properties is justified in this range of current densities.

## HIERARCHICAL MODEL OF CCL OPERATION

The last statement in the previous section implies that to a good approximation, constant and uniform effective properties could be used to describe transport and reaction in CCLs under relevant operating conditions. However, the MHM could be refined by (i) using a description of the porous agglomerated morphology that is based on findings from CGMD studies, discussed in the section "Mesoscale Model of Self-Organization in Catalyst Layer Inks" in Chapter 3; and (ii) by incorporating electrostatic interactions between electroactive protons and charged metal walls of pores that define local reaction conditions, as discussed in the section "ORR in Water-Filled Nanopores: Electrostatic Effects" in Chapter 3.

These modifications lead to a two-scale model. It consists of a two-dimensional mesoscale model linked to a one-dimensional macroscale model, as illustrated schematically in Figure 4.1. Solution of the macroscale model gives spatially dependent functions for electrolyte phase potential, $\Phi(x)$, and oxygen concentration, $c_{O_2}(x)$, along the thickness direction $x$. These two functions are applied as boundary conditions on the agglomerate surface. Solution of the mesoscale model for an agglomerate at position $x$ gives the local agglomerate effectiveness factor $\Gamma_{agg}(x)$. This factor is the electrochemical source term to be used again in the macroscale model. A self-consistent solution is found, when a convergence criterion for $\Gamma_{agg}(x)$ is fulfilled. This solution provides the distributions of potential and oxygen at meso- and macroscale. The corresponding reaction rate distribution can be determined, providing a spatial map of active and inactive zones in the CCLs in agglomerates and at the macroscale. This information can be condensed into the calculation of an overall effectiveness factor of the CCL.

Experimental findings (Soboleva et al., 2010) as well as CGMD simulations (Malek et al., 2011) suggest that each agglomerate consists of packed primary Pt/C particles with diameter of 20–30 nm. Considering a spherical agglomerate with a diameter of 75 nm and spherical carbon particles with a diameter of 25 nm, the number of primary carbon particles that can be packed in the agglomerate is ∼13. This packing results in an agglomerate porosity of 52%. Experimental data of Suzuki et al. (2011) show a smaller porosity of 31%, which could be caused by the polydispersity of sizes of primary Pt/C particles not accounted for in the model.

An example of a representation of the pore space in agglomerates by conical pores is shown in Figure 4.11a. By varying size and opening angle of the conical pores, intra-agglomerate porosity and pore surface area can be adjusted to match experimental data. In a basic structural model, it was assumed that intra-agglomerate pores are hydrophilic and filled with water; ionomer was considered to form a continuous film of uniform thickness on the agglomerate surface.

Owing to the assumed spherical symmetry of agglomerates, the agglomerate structure can be represented by a unit cell that is repeated throughout the agglomerate. This

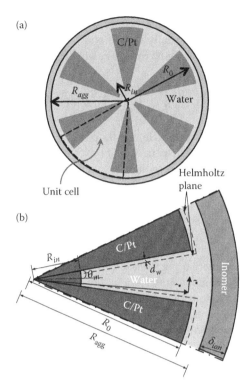

**FIGURE 4.11** (a) Schematic 2D cross-sectional view of an agglomerate of Pt/C that is surrounded by a bilayer film of water and ionomer; (b) unit cell at the agglomerate level with a single conically shaped pore that is considered to represent the porous structure of agglomerates; main variables and position of the Helmholtz (reaction) plane are indicated. (Reprinted from Sadeghi, E., Putz, A., and Eikerling, M. 2013b. *J. Electrochem. Soc.*, **160**, F1159–F1169, Figures 1,2,5,6,8,12. Copyright (2013), the Electrochemical Society. With permission.)

unit cell includes a single conical pore and the outer surface region at the ionomer interface, as shown in Figure 4.11b.

The macroscale model is almost identical to the MHM discussed in the section "Macrohomogeneous Model with Constant Properties." In the electrochemical source term of the MHM Equation 4.5, a spatial variation in the agglomerate effectiveness factor must be accounted for

$$R_{reac}(x) = \frac{j^0 \Gamma_{agg}(x)}{l_{CL}} \exp\left(\alpha_c f \eta(x)\right). \tag{4.40}$$

The model of water-filled nanopores, presented in the section "ORR in Water-Filled Nanopores Electrostatic Effects" in Chapter 3, was adopted to calculate the agglomerate effectiveness factor. As a reminder, this model establishes the relation between metal-phase potential and faradaic current density at pore walls using Poisson–Nernst–Planck theory, Fick's law of diffusion, and Butler–Volmer equation for the electrochemical kinetics. Polar coordinates can be used to solve the corresponding set of equations in conically shaped pores. The boundary condition at the Helmholtz (or reaction) plane is found from the cathodic branch of the Butler–Volmer equation:

$$j_i^* (\hat{r}) = j_{uc}^0 \left( \frac{c_{O_2}}{c_{O_2}^s} \right)^{\gamma_{O_2}} \left( \frac{c_{H^+}}{c_{H^+}^s} \right)^{\gamma_{H^+}} \exp \left( \alpha_c f \eta^* (\hat{r}) \right), \tag{4.41}$$

where $*$ indicates quantities at the Helmholtz plane, and $\hat{r}$ is a normalized radial coordinate. Assuming uniform dispersion of Pt over the total available support surface of the agglomerate and uniform conditions on the outer agglomerate surface, the exchange current density in an agglomerate unit cell (cf. Figure 4.11b) is given by

$$j_{uc}^0 = j_*^0 \frac{m_{Pt} N_A}{M_{Pt} \nu_{Pt}} \Gamma_{np} \frac{A_{CL}}{N_{agg} N_p (A_{in} + A_{out})}, \tag{4.42}$$

where $A_{in}$ and $A_{out}$ are internal pore surface area and external agglomerate surface area in a unit cell of an agglomerate, $A_{CL}$ is the cross-sectional area of the CCL, $N_{agg}$ is the number of agglomerates in the CCL, and $N_p$ is the number of pores per agglomerate. The agglomerate effectiveness factor at each location along the CCL is defined as the total current produced by agglomerates at the location $x$, normalized to the ideal current that would be obtained, if reactant and potential distributions were uniform in the CCL

$$\Gamma_{agg} (x) = \frac{\left( j_{out}^* A_{out} \int_{A_{in}} j_i^* (\hat{r}) \, dA \right)}{j_{uc}^0 (A_{in} + A_{out}) \exp (\alpha_c f \eta_0)} \frac{p(x)}{p(l_{CL})}, \tag{4.43}$$

with

$$j_{out}^* = j_{uc}^0 \exp (\alpha_c f \eta(x)). \tag{4.44}$$

The total effectiveness factor of the CCL is thus

$$\Gamma_{CCL} = \Gamma_{np} \Gamma_{stat} \frac{1}{l_{CL}} \int_0^{l_{CL}} \Gamma_{agg}(x) \, dx. \tag{4.45}$$

In addition to the detailed treatment of CCL structure and processes, the model developed in Sadeghi et al. (2013b) employed a simple empirical description of pore blocking in GDL by liquid water.

The coupled hierarchical model was evaluated by comparison with experimental data of Suzuki et al. (2011) and Soboleva et al. (2011). Both of these studies provided experimental data on CL structure as well as electrochemical performance, which were used to parameterize the model. The pore size distributions of the catalyst layers are depicted in Figure 3.42. Figure 4.12a shows polarization curves from both experimental studies compared to the curves obtained from the hierarchical model. Experimental trends are reproduced within the model. It is evident that flooding of the GDL is responsible for the knee in fuel cell voltage at high current density.

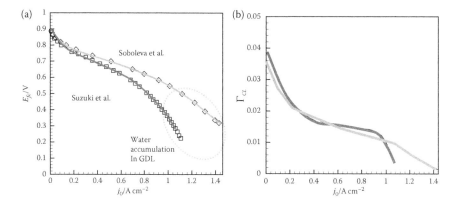

**FIGURE 4.12** (a) Polarization curves obtained from experimental studies of Suzuki et al. (2011) and Soboleva et al. (2011) in comparison with the hierarchical model. (b) Corresponding total effectiveness factors of the CCL as a function of current density. (Reprinted from Sadeghi, E., Putz, A., and Eikerling, M. 2013b. *J. Electrochem. Soc.*, **160**, F1159–F1169, Figures 1,2,5,6,8,12. Copyright (2013), the Electrochemical Society. With permission.)

Values of the agglomerate effectiveness factor, $\Gamma_{agg}(x)$, are found in the range from 0.35 to 0.1, as seen in Figure 4.13. They exhibit their highest values close to the interface with the PEM at moderate current density.

Figure 4.12b shows the variation of the total effectiveness factor of the CCL with current density for the polarization curves in Figure 4.12a. $\Gamma_{stat}$ and $\Gamma_{np}$ are functions of composition and microstructure of the CCL; neglecting degradation effects, these parameters should remain constant. However, the agglomerate effectiveness factor decreases with current density. The dependence of the effectiveness factor on current density is stronger at low and high current densities, $j_0 < 0.4$ A cm$^{-2}$ and $j_0 > 1$ A cm$^{-2}$. Effectiveness factor values are similar over a wide range of $j_0$ for the studies of Suzuki et al. (2011) and Soboleva et al. (2011). However, the higher propensity for flooding of the GDL results in a sharper drop of the effectiveness factor at $j_0 > 1$ A cm$^{-2}$ in Suzuki et al. (2011).

The catalyst layers evaluated in this model-based analysis are not intended to represent the best-in-class in terms of performance. Instead, the experimental studies were picked out from the literature because they provided porosimetry as well as performance data. Nevertheless, the low value of the CL effectiveness factor is a striking result of this analysis. The value of $\Gamma_{CL}$ decreases from ~4 % at $j_0 < 0.4$ A cm$^{-2}$ to ~1 % at $j_0 \simeq 1$ A cm$^{-2}$. This parameter incorporates statistical effects and transport phenomena across all scales in the CCL. The values found are consistent with an experimental evaluation of effectiveness factors by Lee et al. (2010); if the values found in that study are corrected with the atom utilization factor $\Gamma_{np}$, the agreement is very good. The low value of $\Gamma_{CL}$ suggests that tremendous improvements in fuel cell performance and Pt loading reduction could be achievable through advanced structural design of catalyst layers.

This section has demonstrated that refined structural models, together with kinetic modeling of catalyst surface reactions and fluid dynamics modeling of transport

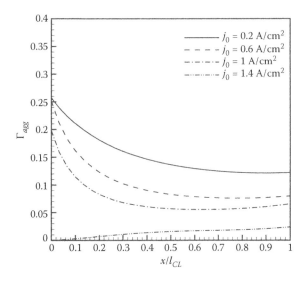

**FIGURE 4.13** Variation of agglomerate effectiveness factor, $\Gamma_{agg}$ $(x)$, as a function of the CL coordinate $x$ at different values of $j_0$, obtained for the analysis of structural and performance data in Soboleva et al. (2011). (Reprinted from Sadeghi, E., Putz, A., and Eikerling, M. 2013b. *J. Electrochem. Soc.*, **160**, F1159–F1169, Figures 1,2,5,6,8,12. Copyright (2013), the Electrochemical Society. With permission.)

processes in nanoporous media, could provide detailed maps of active zones in catalyst layers. Hierarchical models of catalyst layer structure provide images of reaction rate distributions at the agglomerate scale and in nanopores inside of agglomerates. These images can be correlated with polarization curves and effectiveness factors of Pt utilization. Such correlations allow conclusions with respect to performance-relevant effects of pore network morphology; wettability; ionomer structure and distribution; Pt distribution; and water distribution to be drawn. This model-based analysis is the next best thing to direct experimental imaging (*in operando*) at the relevant scales, which will not be available within the forseeable future.

## MHM WITH CONSTANT COEFFICIENTS: ANALYTICAL SOLUTIONS

In the following, a fixed composition of the CCL will be assumed. This section will present an analytical solution of the MHM in limiting cases (Kulikovsky, 2010b). It will provide expressions for the shapes depicted in Figure 4.2 that could serve as a fingerprint of good or bad CL performance.

As a starting point, it is useful to reload the familiar expression for the reaction or faradiac current per unit CCL volume, given in Equation 4.40. The slightly modified version of this equation is

$$R_{reac} = i_* \left[ \frac{c}{c_h^0} \exp\left( \frac{\alpha_c F\eta}{R_g T} \right) - \exp\left( -\frac{\beta_c F\eta}{R_g T} \right) \right]. \tag{4.46}$$

where

$$i_* = \frac{j^0}{l_{CL}} \tag{4.47}$$

is the volumetric exchange current density (per unit of electrode volume, A cm$^{-3}$), and $c$ and $c_h^0$ are local and reference oxygen concentrations that replace partial pressures considered previously. The reaction order of the ORR with respect to the oxygen concentration is assumed to be $\gamma_{O_2} = 1$, as follows from fundamental electrochemical studies, cf. the section "Deciphering the ORR" in Chapter 3. The proton concentration is determined by the presence of ionomer and it is considered a constant parameter, which does not need to be considered explicitly. Similarly, the activity of water is equal to the standard value for liquid water, which means that it can be omitted as a prefactor of the second term in brackets in Equation 4.46. Thus the oxygen concentration remains as the only relevant variable to describe the chemical composition in the CCL.

The first, cathodic, term on the righ-thand side of Equation 4.46 describes the electrochemical transformation of oxygen into water, accomplished by the oxygen reduction reaction. The second, anodic, term represents the reverse transformation of water into oxygen via water electrolysis,

$$H_2O \longrightarrow \frac{1}{2}O_2 + 2H^+ + 2e^-. \tag{4.48}$$

Under supply of electrical work, that is, operating at $\eta < 0$, protons and electrons are transported to the opposite electrode, where the formation of molecular hydrogen completes the process of water electrolysis.

As seen in the section "Deciphering the ORR" in Chapter 3, the effective parameters $\alpha_c$ and $j^0$ of the ORR depend on potential. Usually, they can be considered as piecewise constant, which is a sufficient condition for modeling purposes. Upon lowering of the cathode potential relative to the equilibrium potential, $\alpha_c$ had been found to shift from 1 to 0.75 and further to 0.5. Similarly, $j^0$ assumes different values in different potential regions. The transfer coefficients in Equation 4.46 ought to be treated as effective parameters. In general, it should be assumed that $\alpha_c + \beta_c \neq 1$, deviating from the simple textbook case for one-electron outer-sphere electron transfer. Moreover, it should be noted that the precise values of electrode kinetic parameters in the limit of $R_{reac} \to 0$ are not known. Progressive oxidation of the Pt surface blurs the picture in the equilibrium limit. On the other hand, the approach to equilibrium is of interest, as close to the equilibrium, the contribution of transport processes to the polarization potential is negligible, which allows to estimate kinetic parameters of the ORR from polarization curve fitting.

Nonetheless, because of the smallness of the intrinsic exchange current density of the ORR, $j_*^0 \sim 10^{-9}$ A cm$^{-2}$, CCLs of low-temperature fuel cells usually operate far from equilibrium. In the range of normal ORR overpotentials, $\eta \sim 400$ mV, the ratio of anodic-to-cathodic reaction terms in Equation 4.46 is $\lesssim \exp(-16)$. Thus, no harm is done if one replaces $\beta_c = \alpha_c$. In formal mathematical terms, this replacement represents a tremendous simplification, as the following expression could now be considered:

$$R_{reac} = i_* \frac{c}{c_h^0} \left[ \exp \left( \frac{\eta}{b} \right) - \exp \left( -\frac{\eta}{b} \right) \right] \approx 2 i_* \left( \frac{c}{c_h^0} \right) \sinh \left( \frac{\eta}{b} \right), \tag{4.49}$$

with

$$b = \frac{R_g T}{\alpha_c F}. \tag{4.50}$$

For the sole purpose of algebraic simplification, the concentration terms were dragged in front of the square brackets. Equation 4.49 will be extensively used in Chapters 4 and 5.

At $\eta \gtrsim R_g T / (\alpha_c F)$, the second exponent in Equation 4.49 is negligible, while at $\eta \to 0$, Equation 4.49 reduces to a linear polarization curve. In both these cases, $\alpha_c$ in the equations below may have an arbitrary value.

Subsequently, Equation 4.49 will be used in Equations 4.7 through 4.9. It is convenient to introduce dimensionless variables

$$\tilde{x} = \frac{x}{l_{CL}}, \quad \tilde{\eta} = \frac{\eta}{b}, \quad \tilde{c} = \frac{c}{c_h^0}, \quad \tilde{j} = \frac{j}{j_{ref}}, \tag{4.51}$$

where

$$j_{ref} = \frac{\sigma_p b}{l_{CL}} = \sigma b \tag{4.52}$$

is the characteristic (reference) current density.

With these variables, Equations 4.7 through 4.9 take the form

$$\varepsilon^2 \frac{\partial \tilde{j}}{\partial \tilde{x}} = -\tilde{c} \sinh \tilde{\eta}, \tag{4.53}$$

$$\tilde{j} = -\frac{\partial \tilde{\eta}}{\partial \tilde{x}}, \tag{4.54}$$

$$\tilde{D} \frac{\partial \tilde{c}}{\partial \tilde{x}} = \tilde{j}_0 - \tilde{j}. \tag{4.55}$$

Here

$$\varepsilon = \sqrt{\frac{\sigma_p b}{2 j^0 l_{CL}}} = \sqrt{\frac{\sigma_p b}{2 i_* l_{CL}^2}} = \frac{l_N}{l_{CL}}, \tag{4.56}$$

is a dimensionless characteristic thickness and

$$\tilde{D} = \frac{n F D c_h^0}{\sigma_p b} \tag{4.57}$$

is the dimensionless diffusion coefficient. Note that the subscripts 0 and 1 mark the values at $\tilde{x} = 0$ and $\tilde{x} = 1$, respectively (Figure 4.2).

One of the variables $\tilde{\eta}$ or $\tilde{j}$ can be eliminated from the system of Equations 4.53 through 4.55. Usually, it is convenient to eliminate overpotential $\tilde{\eta}$ (the section "Ideal Oxygen Transport"). The resulting system of equations for $\tilde{j}$ and $\tilde{c}$ should be supplemented by the following boundary conditions:

$$\tilde{j}(0) = \tilde{j}_0, \quad \tilde{j}(1) = 0, \tag{4.58}$$

$$\tilde{c}(1) = \tilde{c}_1, \tag{4.59}$$

where $\tilde{c}_1$ is the dimensionless oxygen concentration at the CCL/GDL interface. If the proton current is eliminated, Equation 4.58 should be replaced by alternative boundary conditions

$$\tilde{\eta}(0) = \tilde{\eta}_0, \quad \left.\frac{\partial \tilde{\eta}}{\partial \tilde{x}}\right|_{\tilde{x}=1} = 0. \tag{4.60}$$

The performance equations involve four parameters: $\varepsilon$, $\tilde{D}$, $\tilde{j}_0$ (or $\tilde{\eta}_0$), and $\tilde{c}_1$. Parameters $\tilde{j}_0$ and $\tilde{\eta}_0$ are not independent; they are related by the polarization curve.

Equations 4.53 through 4.55 describe the distribution of local the ionic current, overpotential, oxygen concentration, and ORR rate, through the CCL. Of largest interest is the polarization curve $\tilde{\eta}_0(\tilde{j}_0)$ and its parametric dependence on $\varepsilon$ and $\tilde{D}$. However, in many situations, equally interesting is the shape of the ORR rate $\tilde{R}_{reac}(\tilde{x})$. For example, this shape allows to optimize the distribution of catalyst loading through the CL thickness, as explored in the section "Optimal Catalyst Layer."

## FIRST INTEGRAL

Equations 4.54 and 4.55 can be integrated. Using Equation 4.54 in Equation 4.55 gives

$$\frac{\partial}{\partial \tilde{x}}\left(\tilde{D}\tilde{c} - \tilde{\eta}\right) = \tilde{j}_0, \tag{4.61}$$

where the right-hand side is independent of $\tilde{x}$. Integration leads to

$$\tilde{D}\tilde{c} - \tilde{\eta} = \tilde{j}_0\tilde{x} + \left(\tilde{D}\tilde{c}_0 - \tilde{\eta}_0\right) \tag{4.62}$$

In setting $\tilde{x} = 1$, $\tilde{D}\tilde{c}_1 - \tilde{\eta}_1 = \tilde{j}_0 + \left(\tilde{D}\tilde{c}_0 - \tilde{\eta}_0\right)$ is obtained. By expressing $\tilde{D}\tilde{c}_0 - \tilde{\eta}_0$ from this equation and using the result in Equation 4.62, one arrives at

$$\tilde{D}(\tilde{c}_1 - \tilde{c}) + (\tilde{\eta} - \tilde{\eta}_1) = \tilde{j}_0(1 - \tilde{x}). \tag{4.63}$$

This is the general relation of local $\tilde{c}$ and $\tilde{\eta}$ to the boundary values $\tilde{\eta}_1$, $\tilde{c}_1$, and $\tilde{j}_0$. As before, the subscript 1 marks the values at the CL/GDL interface.

By setting in Equation 4.63 $\tilde{x} = 0$, a useful general relation is acquired for the overpotential drop in the CL:

$$\delta\tilde{\eta} \equiv \tilde{\eta}_0 - \tilde{\eta}_1 = \tilde{j}_0 - \tilde{D}(\tilde{c}_1 - \tilde{c}_0). \tag{4.64}$$

Care should be taken when using this relation in the case of ideal feed transport. In this case, $\tilde{D} \to \infty$, while $\tilde{c}_0 \to \tilde{c}_1$, and the product $\tilde{D}(\tilde{c}_1 - \tilde{c}_0)$ remains finite. In the section "Another Explicit Form of the Polarization Curve" it is shown that at small cell currents, $\lim_{\tilde{D} \to \infty} \tilde{D}(\tilde{c}_1 - \tilde{c}_0) = \tilde{j}_0/2$. With this, Equation 4.64 transforms to

$$\delta\tilde{\eta} = \frac{\tilde{j}_0}{2}, \quad \tilde{j}_0 \ll 1. \tag{4.65}$$

Thus, at small currents, the overpotential drop increases linearly with the cell current density.

From Equation 4.64, it follows that $\delta\tilde{\eta}$ falls into the range $\tilde{j}_0 - \tilde{D}\tilde{c}_1 \le \delta\tilde{\eta} \le \tilde{j}_0$. In dimensional variables, this equation is equivalent to

$$0 \le j_0 - \frac{\sigma_p \delta\eta}{l_{CL}} \le \frac{nFDc_1}{l_{CL}}, \tag{4.66}$$

where the right-hand side will be referred to as the *oxygen-transport current density*. The value $j_{min} = \sigma_p \delta\eta / l_{CL}$ is the proton current density, corresponding to the linearly varying $\eta(x)$. Physically, this is the minimal cell current that can be produced with the overpotential drop $\delta\eta$. Equation 4.66 says that the difference of the cell current and $j_{min}$ cannot exceed the oxygen-transport current density in the CCL.

Rewriting the right inequality of Equation 4.66 in the form

$$j_0 \le \frac{nFDc_1}{l_{CL}} + \frac{\sigma_p \delta\eta}{l_{CL}} \tag{4.67}$$

shows that for any $j_0$, $\delta\eta$ can be increased to fulfill this relation. As the ORR rate increases with $\eta$, it can be concluded that the CCL operation does not limit the cell current density: at any nonzero oxygen concentration, the CCL is able to provide a given cell current at a cost of increased overpotential drop (unless this drop does not exceed the cell open-circuit potential).

If the oxygen-transport current density $nFDc_1/l_{CL}$ is much less than $j_0$, the central term $j_0 - \sigma_p \delta\eta / l_{CL}$ in Equation 4.66 is a small difference of two large values. In this case, the potential drop is linear in the cell current:

$$\delta\eta = \frac{l_{CL} j_0}{\sigma_p}, \quad j_0 \gg \frac{nFDc_1}{l_{CL}}. \tag{4.68}$$

This is the regime with a strong oxygen depletion in the CCL. In the section "Large-Current CCL Resistivity," it will be shown that this regime is realized in the DMFC cathode.

Note that the equations of the section "First Integral" are derived not using the current balance equation Equation 4.53. Above, the oxygen mass balance Equation 4.55 and Ohm's law (4.54) do not contain the ORR rate. Thus, the results of this section are valid regardless of the assumptions on the ORR kinetics.

## IDEAL PROTON TRANSPORT

A general analytical solution to the system of Equations 4.53 through 4.55 has not been found yet. However, in the limiting cases of ideal transport of proton or oxygen, the explicit analytical solutions can be derived. In this section, the case of ideal proton transport in the CCL (infinite $\sigma_p$) is considered. The CCL performance is, hence, determined by the ORR kinetics and by oxygen transport through the CCL.

### REDUCED SYSTEM OF EQUATIONS AND SOLUTIONS

Ideal ionic transport in the CCL means that the overpotential gradient is small. Setting in the system (4.53)–(4.55) $\tilde{\eta} \simeq \tilde{\eta}_0$ and omitting Equation 4.54 yields

$$\varepsilon^2 \frac{\partial \tilde{j}}{\partial \tilde{x}} = -\tilde{c} \sinh \tilde{\eta}_0, \tag{4.69}$$

$$\tilde{D} \frac{\partial \tilde{c}}{\partial \tilde{x}} = \tilde{j}_0 - \tilde{j}. \tag{4.70}$$

Differentiating Equation 4.69 with respect to $\tilde{x}$ and eliminating $\partial \tilde{c}/\partial \tilde{x}$ with the help of Equation 4.70, one arrives at

$$\frac{\partial^2 \tilde{j}}{\partial \tilde{x}^2} = -\zeta_0^2 \left( \tilde{j}_0 - \tilde{j} \right), \quad \tilde{j}(0) = \tilde{j}_0, \quad \tilde{j}(1) = 0, \tag{4.71}$$

where

$$\zeta_0 = \sqrt{\frac{\sinh \tilde{\eta}_0}{\varepsilon^2 \tilde{D}}}. \tag{4.72}$$

The solution to Equation 4.71 is

$$\tilde{j} = \tilde{j}_0 \left( 1 - \frac{\sinh(\zeta_0 \tilde{x})}{\sinh(\zeta_0)} \right). \tag{4.73}$$

Solving Equation 4.69 for $\tilde{c}$, gives

$$\tilde{c} = -\left( \frac{\varepsilon^2}{\sinh \tilde{\eta}_0} \right) \frac{\partial \tilde{j}}{\partial \tilde{x}} = -\left( \frac{1}{\tilde{D} \zeta_0^2} \right) \frac{\partial \tilde{j}}{\partial \tilde{x}}$$

Using Equation 4.73 here yields the $\tilde{x}$ shape of the oxygen concentration

$$\tilde{c} = \frac{\tilde{j}_0 \cosh(\zeta_0 \tilde{x})}{\tilde{D} \zeta_0 \sinh \zeta_0}. \tag{4.74}$$

By definition, the dimensionless rate of the electrochemical conversion (the ORR rate) is

$$\tilde{R}_{reac} = -\frac{\partial \tilde{j}}{\partial \tilde{x}} \tag{4.75}$$

and, for this shape, one obtains

$$\tilde{R}_{reac} = \frac{\tilde{j}_0 \zeta_0 \cosh(\zeta_0 \tilde{x})}{\sinh \zeta_0} = \tilde{D}\zeta_0^2 \tilde{c}(\tilde{x}). \tag{4.76}$$

Thus, $\tilde{R}_{reac}(\tilde{x})$ follows the shape of $\tilde{c}(\tilde{x})$.

The local ionic current, feed concentration, and reaction rate in the CL are shown in Figure 4.14 for the case of a small oxygen diffusivity (large $\zeta_0$, the section "High Cell Current ($\zeta_0 \gg 1$)"). The conversion occurs in a small domain close to the CCL/GDL interface (Figure 4.14). The width of this *conversion domain* (or, alternatively, the reaction penetration depth) $\tilde{l}_D$ is the characteristic length of $\tilde{R}_{reac}$ decay. Equations 4.74 and 4.76 show that the relation for this length is

$$\tilde{l}_D = \frac{1}{\zeta_0} = \sqrt{\frac{\varepsilon^2 \tilde{D}}{\sinh \tilde{\eta}_0}}. \tag{4.77}$$

Thus, the higher $\tilde{\eta}_0$, the smaller $\tilde{l}_D$, that is, the conversion domain shrinks with the increase in the cell current.

To derive an explicit dependence of $\tilde{l}_D$ on $\tilde{j}_0$, the CCL polarization curve is needed. This curve is obtained using the following arguments. Setting $\tilde{x} = 1$ and $\tilde{c} = \tilde{c}_1$ in Equation 4.74 gives

$$\tilde{j}_0 = \tilde{D}\tilde{c}_1 \zeta_0 \tanh \zeta_0. \tag{4.78}$$

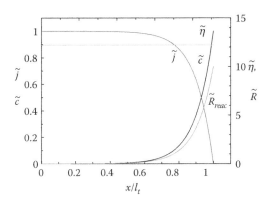

**FIGURE 4.14** Shapes of the local current density $\tilde{j}$, oxygen molar concentration $\tilde{c}$, overpotential $\tilde{\eta}$, and the rate of the electrochemical conversion $\tilde{R}_{reac}$ in the catalyst layer for the oxygen transport-limiting case (large $\zeta_0$). Parameters: $\tilde{\eta}_0 = 12.21, \tilde{D} = 0.1, \varepsilon = 100, \tilde{j}_0 = 1, \tilde{c}_1 = 1$.

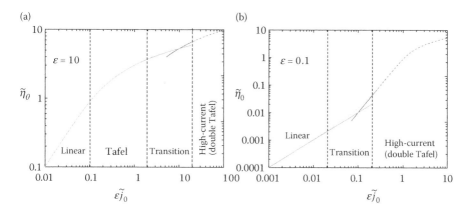

**FIGURE 4.15** Polarization curves of the catalyst layer with ideal ionic transport for $\tilde{D} = \tilde{c}_1 = 1$ and (a) $\varepsilon = 10$, as well as (b) $\varepsilon = 0.1$. The dashed line: the exact curve (4.78). The solid curves: the low current (lower, Equation 4.80) and the high current (upper, Equation 4.86), respectively.

With Equation 4.72, this results in

$$\tilde{j}_0 = \tilde{c}_1 \tilde{D} \sqrt{\frac{\sinh \tilde{\eta}_0}{\varepsilon^2 \tilde{D}}} \tanh \sqrt{\frac{\sinh \tilde{\eta}_0}{\varepsilon^2 \tilde{D}}}. \tag{4.79}$$

This equation is the polarization curve for the CCL with ideal proton transport (Figure 4.15, dashed curve). Of particular interest are the limiting cases of small and large $\zeta_0$, corresponding to low and high cell current, respectively. In these cases, relations (4.77) and (4.78) can be simplified to express $\tilde{\eta}_0$ and $\tilde{l}_D$ through the cell current $\tilde{j}_0$.

## Low Cell Current ($\zeta_0 \ll 1$)

In this case, $\tanh \zeta_0 \simeq \zeta_0$ and Equation 4.78 reduces to $\tilde{j}_0 = \tilde{D}\tilde{c}_1 \zeta_0^2$. With Equation 4.72, this gives $\varepsilon^2 \tilde{j}_0 = \tilde{c}_1 \sinh \tilde{\eta}_0$, yielding

$$\tilde{\eta}_0 = \text{arcsinh}\left(\frac{\varepsilon^2 \tilde{j}_0}{\tilde{c}_1}\right). \tag{4.80}$$

It is seen that $\tilde{D}$ disappears and Equation 4.80 coincides with Equation 4.129, derived below, for the case of ideal oxygen transport at the small current. Small $\zeta_0$ means small $\tilde{\eta}_0$ or large diffusion coefficient. In both these cases, the oxygen transport does not contribute to the potential loss. The respective polarization plots cover linear and Tafel regions (Figure 4.15, lower solid curves).

Equation 4.80 is a generalized Tafel equation. If the argument of the arcsinh function exceeds 2, this function can be replaced by the logarithm of twice the argument and Equation 4.80 simplifies to a standard Tafel equation

$$\tilde{\eta}_0 = \ln\left(\frac{2\varepsilon^2 \tilde{j}_0}{\tilde{c}_1}\right), \tag{4.81}$$

With Equation 4.80, the characteristic thickness of the conversion domain (the reaction penetration depth) is

$$\tilde{l}_D = \sqrt{\frac{\tilde{D}\tilde{c}_1}{\tilde{j}_0}} \tag{4.82}$$

or, in dimensional form

$$l_D = \sqrt{\frac{nFDc_1 l_{CL}}{j_0}}. \tag{4.83}$$

Note that the polarization curve given by Equation 4.81 is independent of the transport coefficient $D$, while the RPD is proportional to $\sqrt{D}$.

Using Equations 4.72 and 4.80, the inequality $\zeta_0 \ll 1$ simplifies to

$$\sqrt{\frac{\tilde{D}\tilde{c}_1}{\tilde{j}_0}} \gg 1, \tag{4.84}$$

equivalent to $\tilde{j}_0 \ll \tilde{D}\tilde{c}_1$ or, in dimensional form

$$j_0 \ll \frac{nFDc_1}{l_{CL}}. \tag{4.85}$$

This shows the validity limits of the low-current approximation. A comparison of Equations 4.84 and 4.82 shows that the dimensionless thickness of the conversion domain must be large: $\tilde{l}_D \gg 1$. In other words, $l_D$ must greatly exceed the CL thickness. This situation is realized if the cell current is much less than the characteristic oxygen-transport current density in the CCL on the right handside of Equation 4.85. The typical shapes of the local parameters are shown in Figure 4.16a. As can be seen, $\tilde{\eta}_0$ and $\tilde{R}_{reac}$ are constant, while $\tilde{j}$ decreases linearly with $\tilde{x}$.

## HIGH CELL CURRENT ($\zeta_0 \gg 1$)

In this case, $\tanh \zeta_0 \simeq 1$ and Equation 4.78 reduces to $\tilde{j}_0 = \tilde{D}\tilde{c}_1\zeta_0$. Using Equation 4.72 gives the CL polarization curve

$$\tilde{\eta}_0 = \text{arcsinh}\left(\frac{\varepsilon^2 \tilde{j}_0^2}{\tilde{D}\tilde{c}_1^2}\right) \tag{4.86}$$

(Figure 4.15, upper solid curve).

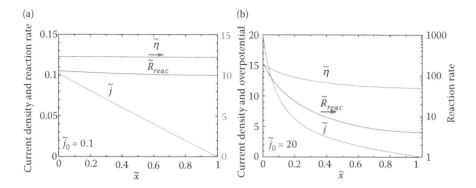

**FIGURE 4.16** The shapes of the dimensionless ionic current density $\tilde{j}$, overpotential $\tilde{\eta}$, and the rate of the electrochemical conversion $\tilde{R}_{reac}$ in the CL with ideal feed molecules transport and large parameter $\varepsilon_*^2 \tilde{j}_0^2$. (a) Small current density ($\tilde{j}_0 = 0.1$). (b) Large current density ($\tilde{j}_0 = 20$). Parameter $\varepsilon = 100$; the membrane is at $\tilde{x} = 0$.

For $\varepsilon^2/(\tilde{D}\tilde{c}_1^2) \gtrsim 1$, the argument of arcsinh is large (see below). Thus, $\mathrm{arcsinh}(y) \simeq \ln(2y)$ and

$$\tilde{\eta}_0 \simeq 2\ln\left(\frac{\varepsilon \tilde{j}_0}{\tilde{c}_1}\sqrt{\frac{2}{\tilde{D}}}\right). \tag{4.87}$$

In dimensional form, Equation 4.87 explicitly exhibits the Tafel slope doubling:

$$\eta_0 = 2b\ln\left(\frac{j_0}{j_{D*}}\right), \tag{4.88}$$

where the characteristic current density $j_{D*}$ is

$$j_{D*} = \sqrt{\frac{i_* nFDc_1^2}{c_h^0}}. \tag{4.89}$$

With Equation 4.86, the characteristic thickness of the conversion domain is inversely proportional to the cell current density:

$$\tilde{l}_D = \frac{\tilde{D}\tilde{c}_1}{\tilde{j}_0} \tag{4.90}$$

or, in dimensional form,

$$l_D = \frac{nFDc_1}{j_0}. \tag{4.91}$$

This parameter is a reaction penetration depth for the case of poor oxygen transport.

According to Equation 4.72, $\zeta_0 \gg 1$ is equivalent to $\sqrt{\sinh \tilde{\eta}_0} \gg \sqrt{\varepsilon^2 \tilde{D}}$. With Equation 4.86, this inequality simplifies to

$$\tilde{j}_0 \gg \tilde{D}\tilde{c}_1, \tag{4.92}$$

which in dimensional form is

$$j_0 \gg \frac{nFDc_1}{l_{CL}}. \tag{4.93}$$

Comparing Equation 4.92 to 4.90, it can be seen that the high-current approximation corresponds to $\tilde{l}_D \ll 1$, that is, in this limit, the conversion occurs in a thin domain at the CCL/GDL interface (Figure 4.14). Equation 4.92 justifies the replacement $\mathrm{arcsinh}(y) \simeq \ln(2y)$ made above to derive Equation 4.87.

In low-temperature cells, liquid water can dramatically reduce oxygen transport through the CCL and GDL. In this situation, the cathode turns into the high-current, oxygen-limiting regime, considered in this section. Note that the high-current relations, obtained above, give a qualitative picture only, as they have been derived assuming infinite proton conductivity of the CL. At high currents, poor proton transport may also limit the CCL performance.

To conclude this section, note that the low- and high-current polarization curves can be derived directly by dividing Equation 4.69 by Equation 4.70. This yields

$$\varepsilon^2 \frac{\partial \tilde{j}}{\partial \tilde{c}} = -\frac{\tilde{D}\tilde{c} \sinh \tilde{\eta}_0}{\tilde{j}_0 - \tilde{j}}.$$

Separating variables and integrating the resulting equation, one arrives at the following polarization curve:

$$\tilde{\eta}_0 = \mathrm{arcsinh} \left( \frac{\varepsilon^2 \tilde{j}_0^2}{\tilde{D} \left( \tilde{c}_1^2 - \tilde{c}_0^2 \right)} \right). \tag{4.94}$$

At small cell current, $\tilde{D}(\tilde{c}_1 - \tilde{c}_0) \simeq \tilde{j}_0/2$ and $\tilde{c}_1 + \tilde{c}_0 \simeq 2\tilde{c}_1$, which yields Equation 4.80. At large current, $\tilde{c}_0^2$ can simply be neglected, giving Equation 4.86.

## TRANSITION REGION

The inequalities describing the limits of validity of the low- and high-current approximations (4.84) and (4.92), respectively, are unduly restrictive. Inspection of Equation 4.78 plotted together with the low- and high-current curves (4.80) and (4.87) shows that these inequalities can be relaxed to some extent. To a satisfactory level of accuracy, the transition region between the low- and high-current domains is given by

$$0.2\tilde{D}\tilde{c}_1 \leq \tilde{j}_0 \leq 2\tilde{D}\tilde{c}_1 \tag{4.95}$$

(Figure 4.15, the transition region). Equation 4.95 stems from the fact that $\tanh(y) \simeq y$ for $y \lesssim 0.2$ and $\tanh(y) \simeq 1$ for $y \gtrsim 2$.

In dimensional form, Equation 4.95 is

$$0.2 j_D \leq j_0 \leq 2 j_D,$$ (4.96)

where $j_D$ is the oxygen-transport current density in the CCL:

$$j_D = \frac{nFDc_1}{l_{CL}}.$$ (4.97)

Note that if $j_0 \gg j_D$, the oxygen concentration drops to vanishingly small values at the membrane interface. In this case, the CCL still converts the current, though at a cost of a doubled Tafel slope.

## IDEAL OXYGEN TRANSPORT

### REDUCED SYSTEM OF EQUATIONS AND INTEGRAL OF MOTION

This section deals with the case of the ideal feed molecules transport in the CL. This situation is realized in modern cathode and anode catalyst layers of PEM fuel cells (Eikerling et al., 2007b) and in well-designed DMFC anodes.

In this case, in the system of Equations 4.53 through 4.55, one may set $\tilde{c} = \tilde{c}_1$ and omit Equation 4.55. This system simplifies to

$$\varepsilon_*^2 \frac{\partial \tilde{j}}{\partial \tilde{x}} = -\sinh \tilde{\eta},$$ (4.98)

$$\tilde{j} = -\frac{\partial \tilde{\eta}}{\partial \tilde{x}},$$ (4.99)

where

$$\varepsilon_* = \frac{\varepsilon}{\sqrt{\tilde{c}_1}}.$$ (4.100)

The boundary conditions to this system are given by Equation 4.58.

Multiplying Equations 4.98 and 4.99 together results in

$$\varepsilon_*^2 \tilde{j} \frac{\partial \tilde{j}}{\partial \tilde{x}} = \sinh \tilde{\eta} \frac{\partial \tilde{\eta}}{\partial \tilde{x}}$$

or

$$\varepsilon_*^2 \frac{\partial (\tilde{j}^2)}{\partial \tilde{x}} = 2 \frac{\partial (\cosh \tilde{\eta})}{\partial \tilde{x}}.$$ (4.101)

Integrating Equation 4.101 gives

$$2 \cosh \tilde{\eta} - \varepsilon_*^2 \tilde{j}^2 = 2 \cosh \tilde{\eta}_0 - \varepsilon_*^2 \tilde{j}_0^2 = 2 \cosh \tilde{\eta}_1$$ (4.102)

that is, the sum $(2 \cosh \tilde{\eta} - \varepsilon_*^2 \tilde{j}^2)$ is constant along $\tilde{x}$.

To analyze Equations 4.98 and 4.99, it is convenient to transform this system into a single equation for the ionic current (Kulikovsky, 2009a). Differentiating

Equation 4.98 over $\tilde{x}$, using Equation 4.99 and the identity $\cosh^2 \tilde{\eta} - \sinh^2 \tilde{\eta} = 1$, yields

$$\varepsilon_*^2 \frac{\partial^2 \tilde{j}}{\partial \tilde{x}^2} = \tilde{j} \sqrt{1 + \varepsilon_*^4 \left(\frac{\partial \tilde{j}}{\partial \tilde{x}}\right)^2}. \tag{4.103}$$

Below, it will be shown that the upper estimate for the second term under the root sign is $\varepsilon_*^2 \tilde{j}_0^2$. In limiting cases of small and large product $\varepsilon_*^2 \tilde{j}_0^2$, Equation 4.103 can be solved.

## CASE OF $\varepsilon_* \ll 1$ AND $\varepsilon_*^2 \tilde{j}_0^2 \ll 1$

In dimensional form, the inequality $\varepsilon_*^2 \tilde{j}_0^2 \ll 1$ reads

$$j_0^2 \ll j_\sigma^2, \tag{4.104}$$

where

$$j_\sigma = \sqrt{2i_* \sigma_p b(c_1/c_h^0)}. \tag{4.105}$$

### x shapes

For small $\varepsilon$ and $\varepsilon_*^2 \tilde{j}_0^2$, in Equation 4.103, the second term under the root may be neglected. The resulting linear equation

$$\varepsilon_*^2 \frac{\partial^2 \tilde{j}}{\partial \tilde{x}^2} = \tilde{j}, \quad \tilde{j}(0) = \tilde{j}_0, \quad \tilde{j}(1) = 0, \tag{4.106}$$

can easily be integrated to yield

$$\tilde{j}(\tilde{x}) = \frac{\tilde{j}_0 \sinh ((1 - \tilde{x})/\varepsilon_*)}{\sinh (1/\varepsilon_*)}. \tag{4.107}$$

The shape of $\tilde{\eta}$ now follows from Equation 4.98:

$$\tilde{\eta} = \operatorname{arcsinh} \left(\frac{\varepsilon_* \tilde{j}_0 \cosh ((1 - \tilde{x})/\varepsilon_*)}{\sinh (1/\varepsilon_*)}\right). \tag{4.108}$$

For the rate of conversion $\tilde{R}_{reac} = -\partial \tilde{j}/\partial \tilde{x}$, one finds

$$\tilde{R}_{reac}(\tilde{x}) = \frac{\tilde{j}_0 \cosh ((1 - \tilde{x})/\varepsilon_*)}{\varepsilon_* \sinh (1/\varepsilon_*)}. \tag{4.109}$$

This equation allows establishing the limits-of-validity of the results in this section. Noting that the peak of $\tilde{R}_{reac}$ is located at $\tilde{x} = 0$, the following estimate is obtained from Equation 4.109, $\varepsilon_*^4(\partial\tilde{j}/\partial\tilde{x})_0^2 = \varepsilon_*^4\tilde{R}_{reac}^2(0) = \varepsilon_*^2\tilde{j}_0^2\coth(1/\varepsilon_*)$. Thus, the results in this section are valid if

$$\varepsilon_*^2\tilde{j}_0^2\coth(1/\varepsilon_*) \ll 1. \qquad (4.110)$$

Further, $\varepsilon_* \ll 1$ gives $\coth(1/\varepsilon_*) \simeq 1$ and, hence, Equation 4.106 is valid if

$$\varepsilon_*^2\tilde{j}_0^2 \ll 1 \text{ and } \varepsilon_* \ll 1. \qquad (4.111)$$

Note that if $\varepsilon_* \gg 1$, then $\coth(1/\varepsilon_*) \simeq \varepsilon_*$, and Equation 4.110 reduces to $\varepsilon_*^3\tilde{j}_0^2 \ll 1$. However, with large $\varepsilon_*$, this inequality holds for vanishingly small currents only, and these are of no interest.

It is advisable to calculate the characteristic values of $\varepsilon_*$ and the typical dimensionless current density $\tilde{j}_0$ for different electrodes (Table 5.7). As can be seen, the conditions $\varepsilon_* \ll 1$ and $\varepsilon_*\tilde{j}_0 \ll 1$ are simultaneously fulfilled only on the hydrogen side of a PEFC operating at a current density below 300 mA cm$^{-2}$. Thus, the results of this section are of interest for PEFC anodes only. In other electrodes, for typical working currents, the opposite relation $\varepsilon_*^2\tilde{j}_0^2 \gg 1$ holds (see the next section).

The functions in Equations 4.107 through 4.109 are depicted in Figure 4.17. As can be seen, the characteristic length of decay of all the shapes is $\varepsilon_*$ (Figure 4.17). This gives the physical meaning of $\varepsilon_*$: this parameter determines the reaction penetration depth into the electrode (the characteristic thickness of the conversion domain) at small currents. Note that both $\varepsilon$ and $\varepsilon_*$ are independent of the cell current $\tilde{j}_0$. Below, it is shown that at high currents, the thickness of the conversion domain is determined by the RPD $\tilde{l}_\sigma$, inversely proportional to $\tilde{j}_0$ (the section "Reaction Penetration Depth").

Small $\varepsilon_*$ typically means either a high exchange current density $i_*$ or a large CL thickness (Equation 4.56). Physically, large $i_*$ makes it possible to convert ionic current close to the membrane, with only a minor potential loss (note a small value of $\tilde{\eta}_0$ in Figure 4.17). This leads to rapidly decaying shapes of $\tilde{j}$, $\tilde{R}_{reac}$, and $\tilde{\eta}$ (Figure 4.17). As discussed above, this picture is characteristic of a hydrogen electrode. In this

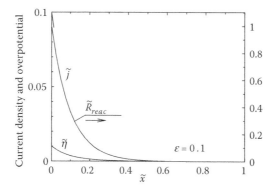

**FIGURE 4.17** The shapes of the dimensionless ionic current density $\tilde{j}$, overpotential $\tilde{\eta}$, and the rate of the electrochemical conversion $\tilde{R}_{reac}$, in the catalyst layer, with ideal feed transport for $\varepsilon_* \ll 1$ and $\varepsilon_*\tilde{j}_0 \ll 1$ (PEFC anode). Parameters are $\varepsilon_* = 0.1$ and $\tilde{j}_0 = 0.1$. The membrane is at $\tilde{x} = 0$.

regime, the reaction "feels" the small, but nonnegligible proton transport limitation and the reaction runs close to the membrane, where protons are "cheaper." The case of large $l_{CL}$ is typical for the solid oxide fuel cell anode (see Kulikovsky, 2010b for further details).

### Polarization Curves

Setting $\tilde{x} = 0$ in Equation 4.108 results in the polarization curve of the CL

$$\tilde{\eta}_0 = \text{arcsinh}\left(\varepsilon_*\tilde{j}_0 \coth\left(1/\varepsilon_*\right)\right). \tag{4.112}$$

As $\varepsilon_* \ll 1$ gives $\coth(1/\varepsilon_*) \simeq 1$, and Equation 4.112 reduces to

$$\tilde{\eta}_0 = \text{arcsinh}\left(\varepsilon_*\tilde{j}_0\right), \quad \varepsilon_* \ll 1.$$

Further, as in this section $\varepsilon_*\tilde{j}_0 \ll 1$, one obtains

$$\tilde{\eta}_0 = \varepsilon_*\tilde{j}_0, \quad \varepsilon_* \ll 1, \quad \tilde{j}_0 < 1. \tag{4.113}$$

Thus, under the conditions of the section "The Case of $\varepsilon_* \ll 1$ and $\varepsilon_*^2\tilde{j}_0^2 \ll 1$," the CL polarization curve is linear. Equation 4.113 in the dimensional form reads

$$\eta_0 = R_{act}j_0, \tag{4.114}$$

where $R_{act}$ is the *CL activation resistivity*

$$R_{act} = \sqrt{\frac{b}{2\sigma_p i_*\left(c_1/c_h^0\right)}}. \tag{4.115}$$

Comparing this to the ideal-transport CCL resistivity given in Equation 1.79, shows that Equation 4.115 contains proton conductivity and does not depend on the CL thickness. The latter feature is due to the presence of the internal space scale $\varepsilon_*$ in the problem.

## LARGE $\varepsilon_*^2\tilde{j}_0^2$

It should be emphasized that $\varepsilon_*^2\tilde{j}_0^2 \gg 1$ does not mean that the dimensionless cell current is large. Depending upon $\varepsilon_*$, $\tilde{j}_0$ in this inequality could be either small or large. In the dimensional form, the inequality discussed reads

$$j_0^2 \gg j_\sigma^2, \tag{4.116}$$

where $j_\sigma$ is given by Equation 4.105. Equation 4.116 holds for small $j_0$ if, for example, the oxygen concentration $c_1$ or the CCL proton conductivity $\sigma_p$ is small. Thus, taking Equation 4.116 as granted, the cases of small and large $j_0$ have to be considered separately (the case of small $j_0$ corresponds to a small $j_\sigma$).

## x shapes

In the limit of $\varepsilon_*^2 \tilde{j}_0^2 \gg 1$, the unity under the square root in Equation 4.103 may be neglected and this equation reduces to

$$\frac{\partial^2 \tilde{j}}{\partial \tilde{x}^2} = -\tilde{j}\frac{\partial \tilde{j}}{\partial \tilde{x}}$$

or

$$2\frac{\partial^2 \tilde{j}}{\partial \tilde{x}^2} + \frac{\partial (\tilde{j}^2)}{\partial \tilde{x}} = 0. \qquad (4.117)$$

Note that $\partial \tilde{j}/\partial \tilde{x} < 0$ and, hence, a negative value is taken for the square root in Equation 4.103.

Integrating Equation 4.117 results in

$$2\frac{\partial \tilde{j}}{\partial \tilde{x}} + \tilde{j}^2 = 2\left.\frac{\partial \tilde{j}}{\partial \tilde{x}}\right|_0 + \tilde{j}_0^2 = 2\left.\frac{\partial \tilde{j}}{\partial \tilde{x}}\right|_1 \equiv -\gamma^2, \quad \tilde{j}^{(l)} = 0, \qquad (4.118)$$

where $\gamma$ is independent of $\tilde{x}$ (see below). The solution to Equation 4.118 is

$$\tilde{j} = \gamma \tan\left(\frac{\gamma}{2}(1-\tilde{x})\right). \qquad (4.119)$$

With Equation 4.119, for the rate of conversion $\tilde{R}_{reac} = -\partial \tilde{j}/\partial \tilde{x}$, one finds

$$\tilde{R}_{reac} = \frac{\gamma^2}{2}\left(1 + \tan^2\left(\frac{\gamma}{2}(1-\tilde{x})\right)\right) = \frac{\gamma^2 + \tilde{j}^2}{2}. \qquad (4.120)$$

Using Equation 4.119 in Equation 4.98 yields $\tilde{\eta}$:

$$\tilde{\eta} = \text{arcsinh}\left(\frac{\gamma^2 \varepsilon_*^2}{2}\left(1 + \tan^2\left(\frac{\gamma}{2}(1-\tilde{x})\right)\right)\right)$$

$$= \text{arcsinh}\left(\frac{\varepsilon_*^2}{2}\left(\gamma^2 + \tilde{j}^2\right)\right). \qquad (4.121)$$

Parameter $\gamma$ can be expressed in terms of $\tilde{j}_0$. Setting in Equation 4.119 $\tilde{x} = 0$ yields

$$\tilde{j}_0 = \gamma \tan\left(\frac{\gamma}{2}\right). \qquad (4.122)$$

Small $\tilde{j}_0$ gives $\tan(\gamma/2) \simeq \gamma/2$ and from Equation 4.122, we obtain

$$\gamma \simeq \sqrt{2\tilde{j}_0}. \qquad (4.123)$$

Expanding the tan function in Equation 4.119 and using Equation 4.123, it can be observed that at small currents, the shape of $\tilde{j}(\tilde{x})$ is linear:

$$\tilde{j} = \tilde{j}_0(1 - \tilde{x}). \tag{4.124}$$

At small $\tilde{j}_0$, the overpotential and the reaction rate are nearly constant along $\tilde{x}$ (Figure 4.16a).

Large $\tilde{j}_0$ gives $\gamma \lesssim \pi$. For this $\gamma$, the asymptotic expansion of tan function is $\tan(\gamma/2) \simeq 2/(\pi - \gamma)$. Using this in Equation 4.122 and solving for $\gamma$, one arrives at

$$\gamma \simeq \frac{\pi \tilde{j}_0}{2 + \tilde{j}_0}. \tag{4.125}$$

To summarize, the limiting solutions for $\gamma$ are

$$\gamma = \begin{cases} \sqrt{2\tilde{j}_0}, & \tilde{j}_0 \ll 1 \\ \dfrac{\pi \tilde{j}_0}{2 + \tilde{j}_0}, & \tilde{j}_0 \gg 1 \end{cases}. \tag{4.126}$$

Equation 4.127 well approximates the exact numerical solution to Equation 4.122 in the whole range of current densities:

$$\gamma = \frac{\sqrt{2\tilde{j}_0}}{1 + \sqrt{1.12\tilde{j}_0} \exp\left(\sqrt{2\tilde{j}_0}\right)} + \frac{\pi \tilde{j}_0}{2 + \tilde{j}_0}. \tag{4.127}$$

As can be seen, at small and large $\tilde{j}_0$, Equation 4.127 reduces to the correct asymptotic solutions (4.126).

The shapes $\tilde{j}(\tilde{x})$, $\tilde{\eta}(\tilde{x})$, and $\tilde{R}_{reac}(\tilde{x})$ for $\tilde{j}_0 = 20$ are shown in Figure 4.16b. Note that at large $\tilde{j}_0$, the shapes $\tilde{j}(\tilde{x})$ (4.119) and $\tilde{R}_{reac}(\tilde{x})$ (4.120) do not depend on the RPD $\varepsilon_*$. In this regime, a new internal scale of the problem arises, as discussed in the section "Reaction Penetration Depth."

However, the overpotential $\tilde{\eta}(\tilde{x})$ "feels" $\varepsilon_*$: the variation of $\varepsilon_*$ merely shifts $\tilde{\eta}(\tilde{x})$ as a whole along the potential axis. This can be shown explicitly when the argument of arcsinh in Equation 4.121 exceeds 2. For these arguments, $\mathrm{arcsinh}(y) \simeq \ln(2y)$ and the factor $\gamma^2 \varepsilon_*^2/2$ shifts the overpotential by $2\ln(\gamma\varepsilon_*)$. Below, it will be shown that this value is the overpotential at the CCL/GDL interface $\tilde{\eta}_1$ (Equation 4.135). Physically, variation of $\varepsilon_*$ means a variation of the exchange current density. The shift in $\tilde{\eta}$ compensates for the change in $i_*$.

## Polarization Curve

Setting $\tilde{x} = 0$ in Equation 4.121, yields

$$\tilde{\eta}_0 = \text{arcsinh}\left(\frac{\varepsilon^2\left(\gamma^2 + \tilde{j}_0^2\right)}{2\tilde{c}_1}\right).$$ (4.128)

This is the general polarization curve of the CCL with ideal oxygen transport for the case of large $\varepsilon_*^2\tilde{j}_0^2$. Equation 4.128 describes normal Tafel kinetics at small currents, double Tafel kinetics at large currents and the transition region between these two regimes. Moreover, this equation reduces to a correct limit at $\tilde{j}_0 \to 0$.

To show this, note that for small currents, $\gamma^2 = 2\tilde{j}_0$ (Equation 4.126). Using this in Equation 4.128 and neglecting the term $\tilde{j}_0^2$, one finds

$$\tilde{\eta}_0 = \text{arcsinh}\left(\frac{\varepsilon^2\tilde{j}_0}{\tilde{c}_1}\right),$$ (4.129)

which in the limit of $\tilde{j}_0 \to 0$ reduces to $\tilde{\eta}_0 = \varepsilon^2\tilde{j}_0/\tilde{c}_1$.

The case of $\tilde{j}_0 \ll 1$, $\varepsilon_*^2\tilde{j}_0^2 \gg 1$ implies that $\varepsilon_* \gg 1$. Thus, the arcsinh function in Equation 4.129 can be replaced by the logarithm of twice the argument, which gives

$$\tilde{\eta}_0 = \ln\left(\frac{2\varepsilon^2\tilde{j}_0}{\tilde{c}_1}\right).$$ (4.130)

In dimensional form, this equation reads

$$\eta_0 = b \ln\left(\frac{j_0}{l_{CL}i_*(c_1/c_h^0)}\right), \quad j_\sigma \ll j_0 \ll j_{ref},$$ (4.131)

which is a standard Tafel equation. Note that this equation is valid if the cell current is sufficiently small. In this regime, the CCL works as an ideal ORR electrode, with the uniform distribution of the reaction rate (Figure 4.16a).

At large current, $\gamma \to \pi, \tilde{j}_0^2 \gg \gamma^2$ and Equation 4.128 simplifies to

$$\tilde{\eta}_0 = \text{arcsinh}\left(\frac{\varepsilon^2\tilde{j}_0^2}{2\tilde{c}_1}\right).$$ (4.132)

In Equation 4.132, the argument of the arcsinh function is large and this function can be replaced by the logarithm of twice the argument:

$$\tilde{\eta}_0 = 2\ln\left(\frac{\varepsilon\tilde{j}_0}{\sqrt{\tilde{c}_1}}\right).$$ (4.133)

The factor 2 exhibits the Tafel slope doubling at high currents resulting from insufficient rate of proton transport in the CCL. In dimensional form, Equation 4.133 reads

$$\eta_0 = 2b \ln\left(\frac{j_0}{j_\sigma}\right), \quad j_0 \gg \max\{j_{ref}, j_\sigma\}, \tag{4.134}$$

where $j_\sigma$ is given by Equation 4.105.

Figure 4.16b shows the physical origin of the effect of Tafel slope doubling. Low $\sigma_p$ forces the electrochemical conversion to occur close to the membrane, where the expenditure for the ionic transport is lower. The nonuniform electrochemical conversion appears to be expensive in terms of $\tilde{\eta}_0$.

The exact numerical and the approximate (Equation 4.128) polarization curves of the CCL with $\gamma$ from Equation 4.127 are depicted in Figure 4.18. As can be seen, Equation 4.128 with $\gamma$ from Equation 4.127 provides excellent accuracy in the whole range of cell currents.

Setting in Equation 4.121 $\tilde{x} = 1$, yields a useful relation for the overpotential $\tilde{\eta}_1$ at the CCL/GDL interface:

$$\tilde{\eta}_1 = \text{arcsinh}\left(\frac{\gamma^2 \varepsilon_*^2}{2}\right) \simeq 2\ln(\gamma \varepsilon_*). \tag{4.135}$$

Large $\tilde{j}_0$ gives $\gamma \to \pi$ and $\tilde{\eta}_1$ tends to the limiting value

$$\tilde{\eta}_1^{\lim} = 2\ln(\pi \varepsilon_*), \quad \tilde{j}_0 \gg 1. \tag{4.136}$$

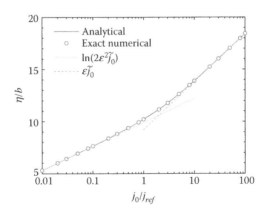

**FIGURE 4.18** The exact numerical (open circles) and the analytical (solid line, Equation 4.128) polarization curves of the CCL with the ideal oxygen transport. Dashed lines: the low- and high-current curves, Equations 4.129 and 4.133, respectively. The exact numerical points are calculated with $\gamma$ being the exact solution to Equation 4.122, while the analytical curve (4.128) is calculated with the approximate $\gamma$ given by Equation 4.127. The current density is normalized to $j_{ref} = \sigma_p b/l_{CL}$. Parameter $\tilde{c}_1 = 1$ and $\varepsilon = 100$ (PEFC cathode, Table 5.7). Note the transition from normal to double Tafel slope at the current densities around $\tilde{j}_0 = 2$.

With the data from Table 5.7, for the PEFC cathode, $\tilde{\eta}_1^{\lim}$ is in the range from $11.5b$ to $16b$. With $b = 50$ mV, it can be seen that $\eta_1^{\lim}$ varies in the range from 0.55 to 0.8 V.

### Transition Region

Equations 4.129 and 4.132 allow calculating the position and length of the transition region between the Tafel and double Tafel domains of the polarization curve (Figure 4.18). Equating Equations 4.129 and 4.132 gives the point of intersection of the Tafel and double Tafel polarization curves (dashed curves in Figure 4.18):

$$\tilde{j}_0 = 2. \tag{4.137}$$

Equation 4.137 and Figure 4.18 give the following ranges for the cell current density in the Tafel and double Tafel regimes:

$$j_0 < \frac{\sigma_p b}{l_{CL}}, \quad \text{Tafel regime,} \tag{4.138}$$

$$j_0 > \frac{4\sigma_p b}{l_{CL}}, \quad \text{double Tafel regime.} \tag{4.139}$$

Therefore, the transition from the Tafel to double Tafel regime occurs in the region

$$\frac{\sigma_p b}{l_{CL}} \le j_0 \le \frac{4\sigma_p b}{l_{CL}}, \quad \varepsilon \gg 1.$$

### ANOTHER EXPLICIT FORM OF THE POLARIZATION CURVE

The first integral in Equation 4.64 of the general system of Equations 4.53 through 4.55 and the integral in Equation 4.102 of the reduced system of Equations 4.98 and 4.99 lead to another form of the polarization curve.

Suppose for a moment that Equation 4.65 holds for all the currents. This relation allows $\tilde{\eta}_1$ to be eliminated from Equation 4.102, yielding

$$\varepsilon_*^2 \tilde{j}_0^2 = 2 \left( \cosh \tilde{\eta}_0 - \cosh \left( \tilde{\eta}_0 - \tilde{j}_0/2 \right) \right).$$

In the limit of $\varepsilon_*^2 \tilde{j}_0^2 \gg 1$, cosh may be replaced by half of the leading exponent, and from the equation above, one finds

$$\tilde{\eta}_0 = \ln \left( \frac{\varepsilon^2 \tilde{j}_0^2}{\tilde{c}_1 \left( 1 - \exp(-\tilde{j}_0/2) \right)} \right). \tag{4.140}$$

A "minor defect" of this equation (no limit at $\tilde{j} \to 0$) can be corrected by replacing the ln function by the arcsinh function of half of the argument:

$$\tilde{\eta}_0 = \text{arcsinh} \left( \frac{\varepsilon^2 \tilde{j}_0^2}{2\tilde{c}_1 \left(1 - \exp(-\tilde{j}_0/2)\right)} \right). \qquad (4.141)$$

The great advantage of the explicit form of Equation 4.141 is that the parameter $\gamma$ has been eliminated from it. Equation 4.141 is the simplest explicit form of the CCL polarization curve in the limit of ideal oxygen transport. The plot of this function is indistinguishable with the exact numerical solution depicted in Figure 4.18. It is easy to show that in the limits of small and large $\tilde{j}_0$, Equation 4.141 reduces to the correct limiting expressions given by Equations 4.130 and 4.133, respectively. Indeed, in the limit of small $\tilde{j}_0$, the exponent in Equation 4.141 can be expanded and this equation reduces to Equation 4.130. Further, Equation 4.141 provides a correct linear CCL resistivity at $\tilde{j} \to 0$. At high $\tilde{j}_0$, the exponent in Equation 4.141 can be neglected, which yields Equation 4.133.

It should be emphasized that Equation 4.65 does not hold at high currents. This is seen by naked eye in Figure 4.16b. Using Equation 4.119 in Equation 4.55, it can be shown that at large $\tilde{j}_0$, the following relation holds:

$$\tilde{\eta}_0 - \tilde{\eta}_1 = 2\ln \left( \frac{2 + \tilde{j}_0}{\pi} \right), \quad \tilde{D} \to \infty, \quad \tilde{j} \gg 1.$$

However, at large $\tilde{j}_0$, $\exp \tilde{\eta}_1$ is much less than $\exp \tilde{\eta}_0$, and the exact dependence of $\tilde{\eta}_1$ on $\tilde{j}_0$ does not play any role in the polarization curve. Thus, Equation 4.141 is approximate in the transition region between the small and large currents only. For small and large currents, this equation is asymptotically exact. Numerical calculations confirm that the accuracy of Equation 4.141 is high in the whole range of cell currents, as long as diffusion effects can be neglected.

In dimensional form, Equation 4.141 reads

$$\eta_0 = b \, \text{arcsinh} \left( \frac{j_0^2/(2j_\sigma^2)}{1 - \exp(-j_0/(2j_{ref}))} \right), \qquad (4.142)$$

where $j_\sigma$ and $j_{ref}$ are given by Equations 4.105 and 4.52, respectively.

Table 5.7 shows that the PEFC cathode typically works in the transition regime with $\tilde{j}_0 \simeq 1$. A typical polarization curve of the PEFC spans the region from 0 to 2 A cm$^{-2}$, corresponding to the range of the dimensionless current density of $0 \leq \tilde{j}_0 \leq 2$. In this range, however, Equation 4.142 has to be used with precaution, as on the right-hand side of this range, oxygen transport effects in the CL must be accounted for (the section "Polarization Curve Fitting" in Chapter 5).

## THE PARAMETRIC FORM OF THE POLARIZATION CURVE

Using Ohm's law in Equation 4.102 to exclude $\tilde{j}$, one obtains the following equation for $\tilde{\eta}$:

$$\varepsilon_*^2 \left( \frac{\partial \tilde{\eta}}{\partial \tilde{x}} \right)^2 = 2 \cosh \tilde{\eta} - 2 \cosh \tilde{\eta}_1$$

Taking the square root of this equation and noting that $\partial \tilde{\eta} / \partial \tilde{x} < 0$, one arrives at

$$\varepsilon_* \frac{\partial \tilde{\eta}}{\partial \tilde{x}} = -\sqrt{2 \cosh \tilde{\eta} - 2 \cosh \tilde{\eta}_1}, \quad \tilde{\eta}(0) = \tilde{\eta}_0. \tag{4.143}$$

From this equation, it follows that $\tilde{\eta}$ is a function of the stretched coordinate $\tilde{x}/\varepsilon_*$. This is explicitly seen in the low-current solution (Equation 4.108). The high-current solution will be discussed below.

A general solution to Equation 4.143 is given by an implicit equation

$$\frac{\tilde{x}}{\varepsilon_*} = \int_{\tilde{\eta}}^{\tilde{\eta}_0} \frac{d\phi}{\sqrt{2 \cosh \phi - 2 \cosh \tilde{\eta}_1}}. \tag{4.144}$$

Setting $\tilde{x} = 1$ here results in an implicit relation of $\tilde{\eta}_0$ and $\tilde{\eta}_1$

$$\frac{1}{\varepsilon_*} = \int_{\tilde{\eta}_1}^{\tilde{\eta}_0} \frac{d\phi}{\sqrt{2 \cosh \phi - 2 \cosh \tilde{\eta}_1}}. \tag{4.145}$$

Writing Equation 4.102 as

$$\varepsilon_*^2 \tilde{j}_0^2 = 2 \cosh \tilde{\eta}_0 - 2 \cosh \tilde{\eta}_1, \tag{4.146}$$

gives a pair of equations (Equations 4.145 and 4.146), which is a parametric form of the polarization curve $\tilde{\eta}_0(\tilde{j}_0)$.

Unfortunately, an attempt to calculate an integral in Equation 4.144, leads to elliptic integrals that are difficult to analyze. A simple explicit solution to Equation 4.143 can be obtained when the CCL works far from equilibrium. In that case, cosh functions can be replaced by half of the leading exponents and Equation 4.143 transforms to

$$\varepsilon_* \frac{\partial \tilde{\eta}}{\partial \tilde{x}} = -\sqrt{\exp \tilde{\eta} - \exp \tilde{\eta}_1}, \quad \tilde{\eta}(0) = \tilde{\eta}_0. \tag{4.147}$$

Below, it will be shown that the following relation holds*:

$$\tilde{\eta}_0 = \tilde{\eta}_1 + \ln \left( 1 + \tan^2 \left( \frac{1}{2\varepsilon_*} \exp \left( \frac{\tilde{\eta}_1}{2} \right) \right) \right) \tag{4.148}$$

---

* Equation 4.148 is obtained if we solve Equation 4.147 "as is" with the boundary condition $\tilde{\eta}(0) = \tilde{\eta}_0$, and then solve $\partial \tilde{\eta}/\partial \tilde{x}|_1 = 0$ for $\tilde{\eta}_0$.

With this, the solution to Equation 4.147 is

$$\tilde{\eta}(\tilde{x}) = \tilde{\eta}_1 + \ln\left(1 + \tan^2\left(\frac{(1-\tilde{x})}{2\varepsilon_*}\exp\left(\frac{\tilde{\eta}_1}{2}\right)\right)\right). \tag{4.149}$$

Note that Equation 4.148 follows from this equation by setting $\tilde{x} = 0$. Differentiating Equation 4.149 over $\tilde{x}$ yields the $x$ shape of the ionic current density

$$\tilde{j}(\tilde{x}) = \frac{1}{\varepsilon_*}\exp\left(\frac{\tilde{\eta}_1}{2}\right)\tan\left(\frac{(1-\tilde{x})}{2\varepsilon_*}\exp\left(\frac{\tilde{\eta}_1}{2}\right)\right). \tag{4.150}$$

Setting $\tilde{x} = 0$ here, one obtains

$$\tilde{j}_0 = \frac{1}{\varepsilon_*}\exp\left(\frac{\tilde{\eta}_1}{2}\right)\tan\left(\frac{1}{2\varepsilon_*}\exp\left(\frac{\tilde{\eta}_1}{2}\right)\right). \tag{4.151}$$

Equations 4.148 and 4.151 represent a parametric polarization curve of the CL when far from equilibrium (Eikerling and Kornyshev, 1998).

It is advisable to compare Equation 4.149 and Equation 4.121. As the argument of the arcsinh function in Equation 4.121 is large, this function can be replaced by a logarithm of twice the argument. Comparing the resulting expression with Equation 4.149 yields the following relation of $\gamma$ and $\tilde{\eta}_1$:

$$\tilde{\eta}_1 = 2\ln(\gamma\varepsilon_*),$$

which coincides with Equation 4.135.

The Tafel approximation is valid provided that $\tilde{\eta}_1 \geq 2$. In that case, $\sinh\tilde{\eta}_1 \simeq \cosh\tilde{\eta}_1 \simeq (\exp\tilde{\eta}_1)/2$. On the other hand, the argument of the tan function in Equations 4.148 and 4.151 must be less than $\pi/2$. Setting $\tilde{\eta}_1 = 2$ in this argument gives a minimal $\varepsilon_*$ for which the parametric form of the polarization curve (4.148) and (4.151) can be used:

$$\varepsilon_*^{\min} \simeq \frac{\exp(1)}{\pi} \simeq 0.87. \tag{4.152}$$

At $\varepsilon_* < \varepsilon_*^{\min}$, part of the CL at the GDL interface works in the Butler–Volmer regime, with the nonnegligible contribution of the reverse reaction. The CL in this regime should be described by the polarization curve (4.145) and (4.146).

## REACTION PENETRATION DEPTH

Equation 4.120 for the rate of ORR does not contain $\varepsilon$. In other words, the characteristic thickness of the conversion domain $l_\sigma$ is independent of RPD $\varepsilon$ (the section "$x$ shapes"). Moreover, in the high-current case, this thickness decreases with the growth of the cell current $\tilde{j}_0$.

By analogy, $l_\sigma$ can be referred to as the *reaction penetration depth*. To calculate $l_\sigma$, we can approximate the tan function Equation 4.120 by the exponent $\tilde{R}_{\exp}$ of the form

$$\tilde{R}_{\exp} = \frac{\gamma^2}{2}\left(1 + \tan^2\left(\frac{\gamma}{2}\right)\right)\exp\left(-\frac{\tilde{x}}{\tilde{l}_\sigma}\right). \tag{4.153}$$

Equation 4.153 provides the exact value of $\tilde{R}_{reac}(0)$, at the electrolyte interface (see Equation 4.120). Note that the approximation (4.153) works well only in the conversion domain, close to the membrane.

Expanding Equations 4.120 and 4.153 in series at $\tilde{x} = 0$, retaining the first-order terms and equating the resulting expressions, we arrive at $\tilde{l}_\sigma = 1/(\gamma\tan(\gamma/2))$. Taking into account Equation 4.122, gives

$$\tilde{l}_\sigma \simeq \frac{1}{\tilde{j}_0}, \quad \text{or} \quad l_\sigma \simeq \frac{\sigma_p b}{j_0}. \tag{4.154}$$

By definition, the RPD $l_N \equiv \varepsilon l_{CL}$ is inversely proportional to the square root of the exchange current density (Equation 4.56). In contrast, the RPD $l_\sigma$ is independent of $i_*$. Physically, this means that the rate-determining process is the impeded proton transport through the CL. In other words, regardless of the available catalyst active surface, the conversion occurs close to the membrane, where the expenditure for proton transport is lower.

The typical parameters and the resulting RPD $l_\sigma$, Equation 4.154 for various electrodes are listed in the last row of Table 5.7. As can be seen, only in a PEFC cathode is $l_\sigma$ in the order of the CL thickness. In DMFC anode and cathode and in a PEFC anode, at typical cell currents, $l_\sigma$ covers only a small fraction of the CL thickness. In other words, at the working cell current, only a small sublayer (conversion domain) at the membrane interface contributes to the current conversion. Furthermore, the higher the cell current, the thinner this sublayer, as revealed by Equation 4.154.

## WEAK OXYGEN TRANSPORT LIMITATION

### THROUGH-PLANE SHAPES

This section derives a more general analytical solution to the system of Equations 4.53 through 4.55 (Kulikovsky, 2012a). This solution corresponds to the mixed case of arbitrary proton transport and weak oxygen transport limitations, and is of particular interest, since the CCL in PEFCs usually works in this regime.

It is convenient to eliminate the overpotential from Equations 4.53 through 4.55. Differentiating Equation 4.53 over $\tilde{x}$ and using the identity $\cosh^2 - \sinh^2 = 1$, one arrives at

$$\varepsilon^2\frac{\partial^2\tilde{j}}{\partial\tilde{x}^2} - \varepsilon^2\left(\frac{\partial\ln\tilde{c}}{\partial\tilde{x}}\right)\frac{\partial\tilde{j}}{\partial\tilde{x}} - \tilde{c}\tilde{j}\sqrt{1 + \left(\frac{\varepsilon^2}{\tilde{c}}\right)^2\left(\frac{\partial\tilde{j}}{\partial\tilde{x}}\right)^2} = 0. \tag{4.155}$$

In this equation, a reasonable estimate for the derivative under the square root is $\partial\tilde{j}/\partial\tilde{x} \simeq \tilde{j}_0$. With $\tilde{c} \leq 1$, the second term under the square root in Equation 4.155 is much greater than unity if $\tilde{j}_0 > 1/\varepsilon^2$. In a PEFC cathode, $\varepsilon \simeq 10^2$–$10^3$ (Table 5.7);

with the characteristic current density $j_{ref} \simeq 1$ A cm$^{-2}$, one finds that this approximation works well for current densities above 0.1 mA cm$^{-2}$.

Neglecting unity under the square root in Equation 4.155 leads to

$$\frac{\partial^2 \tilde{j}}{\partial \tilde{x}^2} + \tilde{j}\frac{\partial \tilde{j}}{\partial \tilde{x}} - \left(\frac{\partial \ln \tilde{c}}{\partial \tilde{x}}\right)\frac{\partial \tilde{j}}{\partial \tilde{x}} = 0. \tag{4.156}$$

Note that in Equation 4.156, a negative value of the square root has been taken, as the positive value results in an unphysical solution.

Consider the case of a small gradient of the logarithm of the oxygen concentration in the CCL, that is, assume that

$$\frac{\partial \ln \tilde{c}}{\partial \tilde{x}} \ll 1. \tag{4.157}$$

Under this condition, the last term in Equation 4.156 can be considered a small perturbation.

Zero flux of oxygen through the membrane means that $\partial \ln \tilde{c}/\partial \tilde{x}|_0 = 0$, while the maximal value of $\partial \ln \tilde{c}/\partial \tilde{x}$ is achieved at $\tilde{x} = 1$ (CCL/GDL interface). Thus, Equation 4.156 can be replaced by an approximate equation

$$\frac{\partial^2 \tilde{j}}{\partial \tilde{x}^2} + \tilde{j}\frac{\partial \tilde{j}}{\partial \tilde{x}} - \epsilon\frac{\partial \tilde{j}}{\partial \tilde{x}} = 0, \quad \tilde{j}(0) = \tilde{j}_0, \quad \tilde{j}(1) = 0 \tag{4.158}$$

where

$$\epsilon \equiv \frac{\partial \ln \tilde{c}}{\partial \tilde{x}}\bigg|_1 \ll 1. \tag{4.159}$$

Integrating Equation 4.158 once gives

$$\frac{\partial \tilde{j}}{\partial \tilde{x}} + \frac{\tilde{j}^2}{2} - \epsilon\tilde{j} = \frac{\partial \tilde{j}}{\partial \tilde{x}}\bigg|_1 \equiv -\frac{\gamma^2}{2}, \tag{4.160}$$

where $\gamma$ is independent of $\tilde{x}$. Equation 4.160 can be integrated further. This gives

$$\tilde{j} = \epsilon + \sqrt{\gamma^2 - \epsilon^2}\tan\left(\frac{\sqrt{\gamma^2 - \epsilon^2}}{2}(1 - \tilde{x}) - \arctan\left(\frac{\epsilon}{\sqrt{\gamma^2 - \epsilon^2}}\right)\right). \tag{4.161}$$

Expanding Equation 4.161 in series over $\epsilon$ and retaining terms that are linear in $\epsilon$, one finds

$$\tilde{j} \simeq \gamma\tan\left(\frac{\gamma}{2}(1 - \tilde{x})\right) - \epsilon\tan^2\left(\frac{\gamma}{2}(1 - \tilde{x})\right). \tag{4.162}$$

Setting $\tilde{x} = 0$ here gives an equation for the parameter $\gamma$:

$$\tilde{j}_0 = \gamma\tan\left(\frac{\gamma}{2}\right) - \epsilon\tan^2\left(\frac{\gamma}{2}\right). \tag{4.163}$$

The oxygen concentration and parameter $\epsilon$ can now be calculated using Equation 4.162 in equation for $\tilde{c}$ (4.55):

$$\tilde{D}\frac{\partial \tilde{c}}{\partial \tilde{x}} = \tilde{j}_0 - \gamma \tan\left(\frac{\gamma}{2}(1-\tilde{x})\right) + \epsilon \tan^2\left(\frac{\gamma}{2}(1-\tilde{x})\right), \quad \tilde{c}(1) = \tilde{c}_1. \tag{4.164}$$

Solving this equation gives

$$\tilde{c} = \tilde{c}_1 - \frac{\tilde{j}_0}{\tilde{D}}(1-\tilde{x}) + \frac{1}{\tilde{D}}\ln\left(1 + \tan^2\left(\frac{\gamma}{2}(1-\tilde{x})\right)\right)$$
$$+ \frac{\epsilon}{\tilde{D}}\left(1 - \frac{2}{\gamma}\tan\left(\frac{\gamma}{2}(1-\tilde{x})\right)\right). \tag{4.165}$$

Calculating $\partial \ln \tilde{c}/\partial \tilde{x}$ and setting $\tilde{x} = 1$ in the resulting expression yields

$$\epsilon = \frac{\tilde{j}_0}{\tilde{D}\tilde{c}_1}. \tag{4.166}$$

Thus, the analysis of this section is valid provided that

$$\frac{\tilde{j}_0}{\tilde{D}\tilde{c}_1} \ll 1. \tag{4.167}$$

The shape of the local overpotential follows from Equation 4.53:

$$\tilde{\eta} = \text{arcsinh}\left(-\frac{\varepsilon^2}{\tilde{c}}\frac{\partial \tilde{j}}{\partial \tilde{x}}\right). \tag{4.168}$$

Using here Equations 4.162 and 4.165 gives

$$\tilde{\eta} = \text{arcsinh}\left(\frac{\varepsilon^2\gamma\,(\gamma/2 - \epsilon y)\,(1+y^2)}{\tilde{c}_1 + \frac{1}{\tilde{D}}\left(-\tilde{j}_0(1-\tilde{x}) + \ln(1+y^2) + \epsilon\,(1-\tilde{x} - (2y/\gamma))\right)}\right), \tag{4.169}$$

where

$$y \equiv \tan\left(\frac{\gamma}{2}(1-\tilde{x})\right)$$

and the parameter $\gamma = \gamma(\tilde{j}_0)$ is a solution to Equation 4.163.

In Figure 4.19, the shapes of $\tilde{j}$, given in Equation 4.162, $\tilde{c}$, given in Equation 4.165, and $\tilde{\eta}$, given in Equation 4.169, are compared with the respective shapes resulting from the numerical solution of the exact problem in Equations 4.156 and 4.55. The curves are plotted for typical PEFC parameters and operating conditions, corresponding to $\epsilon = 0.257$. As can be seen, the agreement of analytical and numerical shapes is good. For smaller values of $\epsilon$, the agreement is even better.

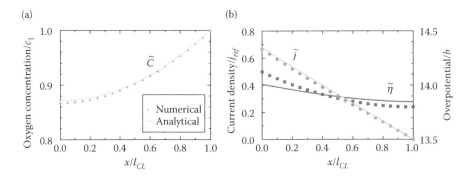

**FIGURE 4.19** The shapes of (a) the oxygen concentration $\tilde{c}$ and (b) the proton current density $\tilde{j}$ and the overpotential $\tilde{\eta}$ through the catalyst layer thickness. The points: the exact numerical solution to Equations 4.156 and 4.55. The solid lines: the analytical results. Parameters used are $\tilde{D} = 2.59$, $\tilde{j}_0 = 0.6667$, $\varepsilon = 866$, and $\tilde{c}_1 = 1$. This set corresponds to $\epsilon = 0.257$, $\gamma = 1.1735$.

## POLARIZATION CURVE

Setting $\tilde{x} = 0$ in Equation 4.169 results in the CCL polarization curve

$$\tilde{\eta}_0 = \text{arcsinh}\left( \frac{f_c \varepsilon^2 \gamma \left( \gamma/2 - \epsilon y_0 \right) \left( 1 + y_0^2 \right)}{\tilde{c}_1 + \dfrac{1}{\tilde{D}} \left( -\tilde{j}_0 + \ln(1 + y_0^2) + \epsilon \left( 1 - (2y_0/\gamma) \right) \right)} \right), \qquad (4.170)$$

where $y_0 = y(0) \equiv \tan(\gamma/2)$, and $f_c$ is a correction factor (see below). Equation 4.170 is the most general polarization curve of the catalyst layer under arbitrary proton transport and weak oxygen transport limitations. It is easy to verify that in the limit of ideal oxygen transport ($\tilde{D} \to \infty$), Equation 4.170 reduces to Equation 4.128.

Figure 4.20 compares the exact numerical and analytical polarization curves (the latter is calculated with Equations 4.170 and 4.127 and $f_c = 1$). The numerical curve is a solution of the system of Equations 4.156 and 4.55, valid at arbitrary oxygen transport limitations in the CCL. As can be seen, Equation 4.170 "as is" describes the exact polarization curve up to $\epsilon \simeq 0.2$ well. However, a simple correction factor of the form

$$f_c = 1 + \frac{\tilde{j}_0}{\tilde{D}\tilde{c}_1} \qquad (4.171)$$

improves the agreement of analytical and numerical curves up to $\epsilon \simeq 0.5$ (Figure 4.20).

## SOLUTION TO EQUATION FOR $\gamma$

Equation 4.163 can be solved using numerical methods. However, a good approximate solution of this equation can be derived as follows. First, note that at $\epsilon = 0$,

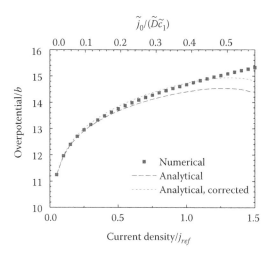

**FIGURE 4.20** Polarization curves of the CCL for the dimensionless diffusion coefficients of 2.59. The points: the exact numerical solution. The dashed line: the analytical Equation 4.170 with $f_c = 1$ and $\gamma$ calculated from Equation 4.127. The dotted line: Equation 4.170 with $f_c$ given by Equation 4.171. The parameter $\varepsilon = 866$.

Equation 4.163 reduces to Equation 4.122, which has the solution given by Equation 4.127. Let this solution be $\gamma_0$. Since $\epsilon$ is small, the solution to Equation 4.163 can be written as

$$\gamma = \gamma_0 + \epsilon\gamma_1. \tag{4.172}$$

Substituting this expansion into Equation 4.163, expanding the left-hand side of the resulting equation over $\epsilon$, and retaining only the term that is linear in $\epsilon$, yields

$$\gamma_0 \tan(\gamma_0/2)$$
$$+ \left( \frac{\gamma_0\gamma_1}{2} + \frac{\gamma_0\gamma_1}{2} \tan^2(\gamma_0/2) - \tan^2(\gamma_0/2) + \gamma_1 \tan(\gamma_0/2) \right) \epsilon = \tilde{j}_0.$$

Taking into account that at the leading order, $\gamma_0 \tan(\gamma_0/2) \simeq \tilde{j}_0$ from the equation above, gives

$$\gamma_1 = \frac{2\tilde{j}_0^2}{\gamma_0 \left( 2\tilde{j}_0 + \tilde{j}_0^2 + \gamma_0^2 \right)}. \tag{4.173}$$

Equation 4.172 with $\gamma_1$ from Equation 4.173 and $\gamma_0$ from Equation 4.127 solve the problem. With this, Equation 4.170 is an explicit analytical polarization curve.

## WHEN CAN THE OXYGEN TRANSPORT LOSS BE IGNORED?

Equation 4.167 in the dimensional form reveals that the effect of oxygen transport on the polarization curve is minimal, if the cell current density is much less than the oxygen transport current density

$$j_0 \ll j_D = \frac{4FDc_1}{l_{CL}}. \tag{4.174}$$

The same condition was obtained in the section "Low Cell Current ($\zeta_0 \ll 1$)," Equation 4.85. This result is confirmed by direct numerical solution of the system (4.53) through (4.55) (Kulikovsky, 2009b).

In the range of $\epsilon \simeq 0.1$–$0.5$, Equation 4.170 is the best approximation for the CCL polarization curve. For $\epsilon \lesssim 0.1$, the effect of oxygen diffusion can be completely ignored and either of Equation 4.128 or 4.141 can be used.

It is insightful to estimate the parameter $\epsilon$ for the CCL of a PEFC and for the ACL of a DMFC. For this estimate, the oxygen transport loss in the GDL is ignored.* With $c_1 \simeq 7.4 \cdot 10^{-6}$ mol cm$^{-3}$ (oxygen concentration in air) and the characteristic for PEFC value of $D$ from Table 5.9, one finds $j_D \simeq 3$ A cm$^{-2}$. For DMFC, with the CCL being 10 times thicker, one obtains $j_D \simeq 0.3$ A cm$^{-2}$.

Typical working currents of PEFC and DMFC are 1 A cm$^{-2}$ and 0.1 A cm$^{-2}$, respectively. Hence, under normal working conditions, in both of these types of cell, one has $\epsilon \simeq 0.3$. In the range of current densities up to the typical working currents, the accurate polarization curve is, thus, given by Equation 4.170. However, for larger currents (in the so-called mass transport-limited region), the effect of oxygen transport *in the CCL* cannot be ignored, and the polarization curve is determined by the full system of Equations 4.53 through 4.55. Note that GDL flooding makes the situation much worse. In that case, $c_1$ in Equation 4.174 can be by a factor of 10 lower. This shifts the onset of the oxygen transport-limited regime to much lower currents.

## POLARIZATION CURVES FOR SMALL TO MEDIUM OXYGEN TRANSPORT LOSS

At large cell currents, when the oxygen transport loss is significant, Equation 4.170 leads to unphysical results (the polarization curve exhibits a negative slope, or even drops to zero, Figure 4.20). This effect limits the usefulness of this equation for polarization curve fitting. Another approach discussed in this section leads to physically meaningful results in the whole range of cell currents. This approach is based on the following idea (Kulikovsky, 2014).

The oxygen-transport-free polarization curve in Equation 4.141 can be considered as the zero-order term in the expansion of the solution to the full system of equations (4.53) through (4.55) over the parameter $1/(\varepsilon^2 \tilde{D})$. The goal of this section will be to find the first-order correction in this series.

---

* Accounting for the oxygen transport in the GDL leads to the relation between $\tilde{c}_1$ and the oxygen concentration in the channel $\tilde{c}_h$ (Equation 5.41). Inserting $\tilde{c}_1$ into an equation $\tilde{j}_D = \tilde{D}\tilde{c}_1$ gives

$$\frac{1}{\tilde{j}_D} = \frac{1}{\tilde{j}_{\lim}^{c0}} + \frac{1}{\tilde{D}},$$

where $\tilde{j}_{\lim}^{c0}$ is given by Equation 5.42. Estimates with the equation above lead to nearly twice the lower values of $j_D$, leaving the conclusions unchanged in this section.

Consider a system of equations, following from Equations 4.53 through 4.55:

$$2\varepsilon^2 \frac{\partial^2 \tilde{\eta}}{\partial \tilde{x}^2} = \tilde{c} \exp \tilde{\eta}, \tag{4.175}$$

$$2\varepsilon^2 \tilde{D} \frac{\partial^2 \tilde{c}}{\partial \tilde{x}^2} = \tilde{c} \exp \tilde{\eta}. \tag{4.176}$$

Equation 4.175 is obtained by substituting Equation 4.54 into Equation 4.53. Equation 4.176 is derived by differentiating Equation 4.55 over $\tilde{x}$ and expressing $\partial \tilde{j}/\partial \tilde{x}$ in the resulting equation through Equation 4.53. Note that on the right-hand sides of Equations 4.175 and 4.176, the cathodic exponential term replaces sinh, as the effects of poor oxygen transport are significant only at large cell currents corresponding to overpotentials $\tilde{\eta} > 2$.

Consider the following expansion of the solution to Equations 4.175 and 4.176:

$$\tilde{\eta} = \tilde{\eta}^0(\tilde{x}) + \xi \tilde{\eta}^1(\tilde{x}), \tag{4.177}$$

$$\tilde{c} = \tilde{c}_1 + \xi \tilde{c}^1(\tilde{x}), \tag{4.178}$$

where

$$\xi = \frac{1}{\varepsilon^2 \tilde{D}}. \tag{4.179}$$

Note that the zero-order solution corresponds to $\varepsilon^2 \tilde{D} \to \infty$, equivalent to $\xi = 0$. The zero-order overpotential $\tilde{\eta}^0$ is given by Equation 4.121, while the zero-order concentration is simply constant $\tilde{c}^0 \equiv \tilde{c}_1$. For further references, the zero-order $\tilde{\eta}^0$ is written in the form

$$\tilde{\eta}^0 = \ln \left( \frac{\gamma^2 \varepsilon^2}{\tilde{c}_1} \left( 1 + \tan^2 \left( \frac{\gamma}{2} (1 - \tilde{x}) \right) \right) \right). \tag{4.180}$$

(cf. Equation 4.121). The functions $\tilde{\eta}^1$ and $\tilde{c}^1$ are the first-order corrections, which take into account a finite value of the diffusivity $\tilde{D}$.

The parameter $\xi$ is small. Indeed, with $\varepsilon \sim 10^2$–$10^3$ and $\tilde{D} \sim 1$, we obtain $\xi \sim 10^{-4}$–$10^{-6}$. Equations 4.177 and 4.178 can, thus, be considered as the two terms in expansion of the solution to the full problem of Equations 4.175 and 4.176 over the parameter $\xi$.

Substituting Equations 4.177 and 4.178 into Equations 4.175 and 4.176 gives

$$2\varepsilon^2 \frac{\partial^2 \tilde{\eta}^0}{\partial \tilde{x}^2} + 2\varepsilon^2 \xi \frac{\partial^2 \tilde{\eta}^1}{\partial \tilde{x}^2} = (\tilde{c}_1 + \xi \tilde{c}^1) \exp(\tilde{\eta}^0)(1 + \xi \tilde{\eta}^1), \tag{4.181}$$

$$2\frac{\partial^2 \tilde{c}_1}{\partial \tilde{x}^2} + 2\xi \frac{\partial^2 \tilde{c}^1}{\partial \tilde{x}^2} = \xi(\tilde{c}_1 + \xi \tilde{c}^1) \exp(\tilde{\eta}^0)(1 + \xi \tilde{\eta}^1). \tag{4.182}$$

Taking into account that $\tilde{c}_1$ is constant, subtracting from Equation 4.181 the zero-order equation $2\varepsilon^2 \partial^2 \tilde{\eta}^0 / \partial \tilde{x}^2 = \tilde{c}_1 \exp \tilde{\eta}^0$, collecting the terms with the first power of

$\xi$ and neglecting the high-order terms, yields a system of equations for the first-order corrections $\tilde{\eta}^1$ and $\tilde{c}^1$:

$$2\varepsilon^2 \frac{\partial^2 \tilde{\eta}^1}{\partial \tilde{x}^2} = (\tilde{c}^1 + \tilde{c}_1 \tilde{\eta}^1) \exp \tilde{\eta}^0, \tag{4.183}$$

$$2\frac{\partial^2 \tilde{c}^1}{\partial \tilde{x}^2} = \tilde{c}_1 \exp \tilde{\eta}^0. \tag{4.184}$$

As can be seen, Equation 4.184 decouples from the system. Substituting 4.180 for $\tilde{\eta}^0$ into Equation 4.184 one finds

$$2\frac{\partial^2 \tilde{c}^1}{\partial \tilde{x}^2} = \varepsilon^2 \gamma^2 \left(1 + \tan^2\left(\frac{\gamma}{2}(1 - \tilde{x})\right)\right). \tag{4.185}$$

The solution to this equation is subjected to the boundary conditions $\tilde{c}^1(1) = 0$, $\partial \tilde{c}^1/\partial \tilde{x}|_{\tilde{x}=0} = 0$. The first condition means that the oxygen concentration at the CCL/GDL interface remains unchanged when the oxygen diffusion is "switched on." The second condition expresses zero oxygen flux in the membrane. Under these conditions, the solution to Equation 4.185 is

$$\tilde{c}^1 = -\varepsilon^2 \tilde{j}_0(1 - \tilde{x}) + \varepsilon^2 \ln\left(1 + \tan^2\left(\frac{\gamma}{2}(1 - \tilde{x})\right)\right), \tag{4.186}$$

where Equation 4.122 has been used.

With this $\tilde{c}^1$, Equation 4.183 can be transformed to

$$\frac{4}{\gamma^2} \cos^2\left(\frac{\gamma}{2}(1 - \tilde{x})\right) \frac{\partial^2 \tilde{\eta}^1}{\partial \tilde{x}^2}$$
$$= -\varepsilon^2 \tilde{j}_0(1 - \tilde{x}) + \varepsilon^2 \ln\left(1 + \tan^2\left(\frac{\gamma}{2}(1 - \tilde{x})\right)\right) + \tilde{c}_1 \tilde{\eta}^1. \tag{4.187}$$

The boundary conditions to this equation are $\partial \tilde{\eta}^1/\partial \tilde{x}|_{\tilde{x}=0} = \partial \tilde{\eta}^1/\partial \tilde{x}|_{\tilde{x}=1} = 0$. The first condition means that the current $\tilde{j}_0 = -\partial \tilde{\eta}^0/\partial \tilde{x}|_0$ is fixed. The second condition follows from the zero proton current at the CCL/GDL interface.

Unfortunately, the exact solution to Equation 4.187 leads to rather cumbersome expressions containing quadratures (Kulikovsky, 2014). A much simpler approximate solution to this equation at $\tilde{x} = 0$, necessary for the polarization curve, can be derived using the following arguments. From Equation 4.122 it follows that at large cell currents $\gamma \to \pi$ and hence at $\tilde{x} = 0$, the cos function on the left-hand side of Equation 4.187 is close to zero. This means that at $\tilde{x} = 0$, the right-hand side of Equation 4.187 is, also, close to zero. Equating the right-hand side to zero, solving for $\tilde{\eta}^1$, setting $\tilde{x} = 0$ in the resulting expression, dividing it by $\varepsilon^2 \tilde{D}$, and taking into account Equation 4.122, one arrives at

$$\tilde{\eta}_D = \frac{1}{\tilde{D}\tilde{c}_1}\left(\tilde{j}_0 - \ln\left(1 + \frac{\tilde{j}_0^2}{\gamma^2}\right)\right), \tag{4.188}$$

where $\tilde{\eta}_D = \tilde{\eta}^1(0)/(\varepsilon^2 \tilde{D})$ is the overpotential due to the finite rate of oxygen transport in the CCL.

Adding this potential loss to Equation 4.141, the total polarization overpotential of the CCL is finally obtained:

$$\tilde{\eta}_0 = \text{arcsinh}\left(\frac{\varepsilon^2 \tilde{j}_0^2}{2\tilde{c}_1\left(1 - \exp(-\tilde{j}_0/2)\right)}\right) + \frac{1}{\tilde{D}\tilde{c}_1}\left(\tilde{j}_0 - \ln\left(1 + \frac{\tilde{j}_0^2}{\gamma^2}\right)\right). \quad (4.189)$$

This equation takes into account all the potential losses in the CCL: the first term accounts for the ORR activation overpotential and the proton transport loss, while the second term represents the oxygen transport loss.

The analytical CCL polarization curve Equation 4.189 is compared to the exact numerical solution of the system (4.53) and (4.54) in Figure 4.21. A "reference" value of $D_{ref} = 1.37 \cdot 10^{-3}$ cm$^2$ s$^{-1}$ is taken from measurements (Shen et al., 2011). The curves in Figure 4.21 correspond to the indicated ratios $D/D_{ref}$. Clearly, as this ratio tends to infinity, the analytical and numerical results tend to the diffusion-free polarization curve Equation 4.141. Note that as $\tilde{D}$ decreases, the overpotential due to oxygen transport increases, and the accuracy of the model drops. Nonetheless, in the region of currents $\tilde{j}_0 \lesssim 1$, the model works well for $D/D_{ref}$ as small as 0.1 (Figure 4.21).

What are the limits-of-validity of Equation 4.189? Clearly, a necessary (though not sufficient) condition is the smallness of the second term as compared to the first term in this equation. Taking the leading-order approximation for the oxygen transport losses $\tilde{j}_0/(\tilde{c}_1\tilde{D})$ and the high-current approximation for the first term, results in the

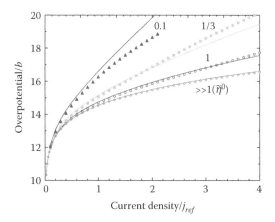

**FIGURE 4.21** Exact numerical (points) and analytical (Equation 4.189) (solid lines) polarization curves of the cathode catalyst layer with the finite rate of oxygen transport. The indicated parameter for the curves is the ratio $D/D_{ref}$, where $D_{ref} = 1.37 \cdot 10^{-3}$ cm$^2$ s$^{-1}$ is the CCL oxygen diffusivity measured in Shen et al. (2011). The bottom solid line is the curve for infinitely fast oxygen transport in the CCL (Equation 4.141).

following estimate:

$$\frac{\tilde{j}_0}{\tilde{D}\tilde{c}_1} \ll \ln\left(\frac{\varepsilon^2 \tilde{j}_0^2}{\tilde{c}_1}\right). \tag{4.190}$$

In contrast to the model of the section "Polarization Curve," this equation does not require the left-hand side to be small as compared to unity. Instead, for $\tilde{D} \simeq 1, \tilde{c}_1 \simeq 1$, and $\varepsilon \simeq 10^3$, Equation 4.190 holds up to $\tilde{j}_0/(\tilde{D}\tilde{c}_1) \simeq 2$, that is, up to current densities $\tilde{j}_0 \simeq 2$, covering the typical range of working cell currents for the PEM fuel cell. This is because the expansion (4.177) and (4.178) does not contain $\tilde{j}_0$ in the expression for the small parameter.

Equation 4.189 can be used for the fitting at experimental polarization curves, provided that (i) the oxygen stoichiometry is large and (ii) the potential loss caused by oxygen transport in the GDL is minimal. The validity of the second condition is usually not known *a priori*. However, it can be easily relaxed by incorporating the respective transport loss into the polarization equation, as discussed in the section "Oxygen Transport Loss in the Gas-Diffusion Layer" in Chapter 5.

## REMARKS TO THE SECTIONS 4.4–4.7

If the cell current is high, the polarization curves of CLs are similar, when either feed or ion transport is insufficient. In both the cases, the curve exhibits doubling of the Tafel slope (Equations 4.88 and 4.134). This makes it difficult to distinguish the physical origin of this doubling, by measuring the polarization curve only.

However, the characteristic current densities under the logarithm in Equations 4.88 and 4.134 are proportional to the square root of the "limiting" transport parameter ($D$ in Equation 4.89 and $\sigma_p$ in Equation 4.105). Thus, the position of the Tafel plot intersection with the current density axis (the point $\tilde{j}_*$, where $\tilde{\eta}_0 = 0$, (Figure 4.22)) depends on the limiting transport coefficient.

Further, in the CCL with poor oxygen transport, the thickness of the conversion domain $l_D$ is proportional to $D$, Equation 4.91, and this domain is located at the

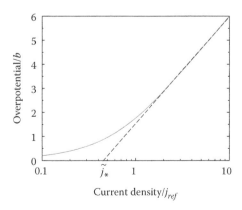

**FIGURE 4.22**  A Tafel plot for the electrode and the characteristic current density $\tilde{j}_*$.

CCL/GDL interface. In the CCL with poor ionic transport, $l_\sigma$ is proportional to the ionic conductivity $\sigma_p$, Equation 4.154, and the conversion domain resides at the membrane interface. It is expected that the long-term degradation processes run faster in the conversion domain. Thus, post mortem pictures of the CCL may give a hint on the prevailing regime of CCL operation.

At high cell current, in both the oxygen- and proton-limiting regimes, the thickness of the conversion domain is inversely proportional to the cell current $j_0$ (Equations 4.91 and 4.154). The higher $j_0$, the smaller the respective RPD. This means, that the domain of heat generation in the CL gets thinner with the growth of the cell current. This may cause problems related to heat transport in the MEA.

Equations 4.91 and 4.154 do not contain the exchange current density. Thus, regardless of the nature of the high-current regime, increasing the catalyst loading cannot change the characteristic thickness of the conversion domain. However, larger catalyst loading lowers $\varepsilon$ and eventually it can switch the regime of CL operation to the low-current mode.

# DIRECT METHANOL FUEL CELL ELECTRODES

## CATHODE CATALYST LAYER IN A DMFC

### Introductory Remarks

In contrast to hydrogen, at standard atmospheric conditions, methanol is liquid. The energy density of liquid methanol exceeds the energy density of hydrogen pressurized to 300 bar. This makes DMFCs an attractive alternative to Li-ion batteries in mobile devices.

In the DMFC anode, the methanol molecule is oxidized, giving six protons and six electrons:

$$CH_3OH + H_2O \longrightarrow CO_2 + 6H^+ + 6e^-. \tag{4.191}$$

In the cathode, the proton and electron currents are transformed into the flux of water in the ORR (Equation 1.4).

A great potential of methanol as a high-energy-density feed for fuel cells has been recognized, seemingly in the early 1960s. One of the first studies of the methanol oxidation reaction (MOR) on a Pt electrode has been performed by Frumkin's colleagues in 1964 (Bagotzky and Vasilyev, 1964). A year later, Frumkin's group reported superior performances of Pt–Ru alloys for methanol oxidation (Petry et al., 1965). Since that time, DMFC has been considered as one of the most promising candidates for powering mobile devices.

One of the key problems in DMFC technology is methanol crossover through the membrane (Dohle et al., 2002; Jiang and Chu, 2004; Qi and Kaufman, 2002; Ravikumar and Shukla, 1996; Ren et al., 2000b; Thomas et al., 2002). Methanol readily permeates from the anode to the cathode, where it is oxidized. The parasitic MOR strongly shifts the regime of CCL operation to higher currents. In addition, the cathodic MOR induces an extra flux of oxygen on the cathode side and poisons the CCL by adsorbed intermediates. Both processes translate into a higher potential loss.

The mechanism of MOR in the DMFC cathode is controversial. Vielstich et al. (2001) reported it to be a purely chemical catalytic methanol–oxygen combustion, while Jusys and Behm (2004) and Du et al. (2007b) provided arguments in favor of an electrochemical pathway. There are indications that the chemical and the electrochemical oxidation may run in parallel (Du et al., 2007b).

In DMFC modeling, the CCL is typically not resolved and the effect of crossover on the cell potential is taken into account in an effective manner, by taking the sum of the useful and equivalent crossover currents as a total current that needs to be converted in the ORR (Garcia et al., 2004; Murgia et al., 2003; Wang and Wang, 2003; Yan and Jen, 2008; Yang and Zhao, 2007, 2008). For the review of DMFC models see Yao et al. (2004). This approach implicitly assumes that the permeated methanol is rapidly converted to proton current close to the membrane interface. The peak proton current in the CCL is then equal to the sum of the useful and crossover currents.

The model below shows that this approach is correct unless the cell current is not large (Kulikovsky, 2012c). At small currents, the MOR runs close to the membrane, while the ORR is shifted toward the GDL. In this regime, the DMFC cathode represents a complete short-circuited fuel cell. However, at large currents, the MOR and ORR share the same domain of the CCL, leading to a rapidly decreasing, resistive-like polarization curve.

## Basic Equations

The schematic of the CCL in a DMFC is depicted in Figure 4.23. The main model assumptions are as follows:

1. The transport of liquid water is ignored. Liquid saturation $s$ is constant though the CCL and it simply lowers the effective oxygen diffusivity.
2. The MOR and ORR reaction rates follow the Butler–Volmer kinetics.

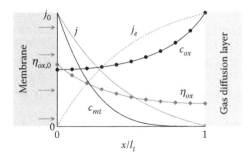

**FIGURE 4.23** Schematic of the cathode catalyst layer in a DMFC, and the expected shapes of the proton $j$ and electron $j_e$ current densities, the local ORR overpotential $\eta_{ox}$, and the oxygen and methanol concentration $c_{ox}$, $c_{mt}$, respectively. Note that the proton current density in the membrane $j_0$, is the cell current density and the total potential loss in the system is $\eta_0$.

The system of equations for the CCL with two simultaneously running electrochemical reactions (ORR and MOR) is

$$\frac{\partial j}{\partial x} = -2i_{ox}\left(\frac{c_{ox}}{c_{ox}^{h0}}\right)\sinh\left(\frac{\eta_{ox}}{b_{ox}}\right) + 2i_{mt}\left(\frac{c_{mt}}{c_{mt}^{h0}}\right)\sinh\left(\frac{\eta_{mt}}{b_{mt}}\right), \tag{4.192}$$

$$-D_{ox}\frac{\partial^2 c_{ox}}{\partial x^2} = -\frac{2i_{ox}}{4F}\left(\frac{c_{ox}}{c_{ox}^{h0}}\right)\sinh\left(\frac{\eta_{ox}}{b_{ox}}\right), \tag{4.193}$$

$$-D_{mt}\frac{\partial^2 c_{mt}}{\partial x^2} = -\frac{2i_{mt}}{6F}\left(\frac{c_{mt}}{c_{mt}^{h0}}\right)\sinh\left(\frac{\eta_{mt}}{b_{mt}}\right), \tag{4.194}$$

$$j = -\sigma_p \frac{\partial \eta_{ox}}{\partial x}. \tag{4.195}$$

In this section, the variables related to oxygen and methanol are equipped with the subscripts $ox$ and $mt$, respectively. Here, $i_{ox}$, $i_{mt}$ are the ORR and MOR volumetric exchange current densities, $c_{ox}$, $c_{ox}^{h0}$ are the local and reference (inlet) oxygen concentrations, $c_{mt}$, $c_{mt}^{h0}$ are the local and reference methanol concentrations, $\eta_{ox}$, $\eta_{mt}$ are the ORR and MOR polarization potentials (overpotentials), $b_{ox}$, $b_{mt}$ are the ORR and MOR Tafel slopes, and $D_{ox}$, $D_{mt}$ are the oxygen and methanol diffusion coefficients, respectively.

Equation 4.192 describes current conservation: the proton current is consumed in the ORR (the first term on the right-hand side) and produced in the MOR (the second term). Equations 4.193 and 4.194 express the mass balance of oxygen and methanol, respectively. Equation 4.195 is Ohm's law relating the proton current density to the ORR overpotential gradient. Below, it will be shown that $\nabla \eta_{ox} = -\nabla \eta_{mt}$, so that any of these gradients with the proper sign can be used in Equation 4.195.

To simplify calculations, dimensionless variables are introduced:

$$\tilde{x} = \frac{x}{l_{CL}}, \quad \tilde{c}_{ox} = \frac{c_{ox}}{c_{ox}^{h0}}, \quad \tilde{c}_{mt} = \frac{c_{mt}}{c_{mt}^{h0}}, \quad \tilde{j} = \frac{jl_{CL}}{\sigma_p b_{ox}}, \quad \tilde{\eta} = \frac{\eta}{b_{ox}}. \tag{4.196}$$

Substituting Equation 4.195 into Equation 4.192 and using these dimensionless variables, Equations 4.192 through 4.194 take the form

$$-\varepsilon^2 \frac{\partial^2 \tilde{\eta}_{ox}}{\partial \tilde{x}^2} = -\tilde{c}_{ox}\sinh(\tilde{\eta}_{ox}) + r_* \tilde{c}_{mt}\sinh(\tilde{\eta}_{mt}/p), \tag{4.197}$$

$$-\varepsilon^2 \tilde{D}_{ox}\frac{\partial^2 \tilde{c}_{ox}}{\partial \tilde{x}^2} = -\tilde{c}_{ox}\sinh(\tilde{\eta}_{ox}), \tag{4.198}$$

$$-\varepsilon^2 \tilde{D}_{mt}\frac{\partial^2 \tilde{c}_{mt}}{\partial \tilde{x}^2} = -r_* \tilde{c}_{mt}\sinh(\tilde{\eta}_{mt}/p). \tag{4.199}$$

Here,

$$\varepsilon = \sqrt{\frac{\sigma_p b_{ox}}{2 i_{ox} l_{CL}^2}}, \quad r_* = \frac{i_{mt}}{i_{ox}}, \quad p = \frac{b_{mt}}{b_{ox}},$$

$$\tilde{D}_{ox} = \frac{4 F D_{ox} c_{ox}^{h0}}{\sigma_p b_{ox}}, \quad \tilde{D}_{mt} = \frac{6 F D_{mt} c_{mt}^{h0}}{\sigma_p b_{ox}}, \qquad (4.200)$$

are dimensionless parameters and diffusion coefficients.

By definition, the ORR and MOR overpotentials are given by

$$\eta'_{ox} = \phi - \Phi - E_{ox}^{eq},$$

$$\eta_{mt} = \phi - \Phi - E_{mt}^{eq},$$

where $\phi$ is the carbon phase (electrode) potential, $\Phi$ is the membrane phase potential, and $E_{ox}^{eq}$ and $E_{mt}^{eq}$ are the equilibrium ORR and MOR potentials (Table 4.1).

## TABLE 4.1
## DMFC Physical and Operation Parameters Used in the Calculations

| | |
|---|---|
| Reference oxygen molar concentration $c_{ox}^{h0}$ (mol cm$^{-3}$) | $7.36 \cdot 10^{-6}$ |
| Reference methanol molar concentration $c_{mt}^{h0}$ (mol cm$^{-3}$) | $10^{-3}$ (1M) |
| Methanol diffusion coefficient in the ABL and CCL $D_{mt}$ (cm$^2$ s$^{-1}$) | $10^{-5}$ |
| Oxygen diffusion coefficient in the CCL $D_{ox}$ (cm$^2$ s$^{-1}$) | $10^{-3}$ |
| Oxygen diffusion coefficient in the GDL $D_b^c$ (cm$^2$ s$^{-1}$) | $10^{-2}$ |
| Limiting current density due to oxygen transport in the GDL $j_{lim}^{c0}$ (A cm$^{-2}$) | 1.42 |
| Limiting current density due to methanol transport in the ABL $j_{lim}^a$ (A cm$^{-2}$) | 0.579 |
| Crossover parameter $\beta_*$ | 0.0–0.5 |
| Catalyst layer thickness $l_{CL}$ (cm) | 0.01 |
| Backing layer thickness $l_b$ (cm) | 0.02 |
| Catalyst layer proton conductivity, $\sigma_p$ ($\Omega^{-1}$cm$^{-1}$) | 0.001 |
| ORR Tafel slope $b_{ox}$ (V) | 0.05 |
| MOR Tafel slope $b_{mt}$ (V) | 0.05 |
| ORR exchange current density $i_{ox}$ (A cm$^{-3}$) | 0.1 |
| MRR exchange current density $i_{mt}$ (A cm$^{-3}$) | 0.1 |
| ORR equilibrium potential $E_{ox}^{eq}$ (V) | 1.23 |
| MOR equilibrium potential $E_{mt}^{eq}$ (V) | 0.028 |
| Reference current density $j_{ref} = \sigma_p b / l_{CL}$ (A cm$^{-2}$) | 0.005 |
| Parameter $\varepsilon$ | 1.58 |
| Dimensionless oxygen diffusivity in the CCL $\tilde{D}_{ox}$ | 56.82 |
| Dimensionless methanol diffusivity in the ABL and CCL $\tilde{D}_{mt}$ | 115.8 |

Subtracting these equations and introducing positive

$$\eta_{ox} = -\eta'_{ox} \geq 0$$

for the dimensionless values yields

$$\tilde{\eta}_{mt} = \tilde{E}^{eq}_{ox} - \tilde{E}^{eq}_{mt} - \tilde{\eta}_{ox}. \qquad (4.201)$$

Note that Equations 4.197 through 4.199 are written for $\tilde{\eta}_{ox} > 0$ and $\tilde{\eta}_{mt} > 0$. With this convention, positive $\tilde{\eta}_{ox,0}$ is the total potential loss in the system. As before, the subscripts 0 and 1 mark the values at the membrane/CCL and CCL/GDL interfaces, respectively, while the superscript 0 indicates the values at the channel inlet.

### Boundary Conditions

The boundary conditions to Equation 4.197 fix the potential loss $\tilde{\eta}_{ox,0}$ at the membrane interface and zero proton current at the CCL/GDL interface:

$$\tilde{\eta}_{ox}(0) = \tilde{\eta}_{ox,0}, \quad \left.\frac{\partial \tilde{\eta}_{ox}}{\partial \tilde{x}}\right|_1 = 0. \qquad (4.202)$$

The boundary conditions to Equation 4.198 express zero oxygen flux in the membrane and fix the oxygen concentration $\tilde{c}_{ox,1}$ at the CCL/GDL interface:

$$\left.\frac{\partial \tilde{c}_{ox}}{\partial \tilde{x}}\right|_0 = 0, \quad \tilde{c}_{ox}(1) = \tilde{c}_{ox,1}. \qquad (4.203)$$

Note that $\tilde{c}_{ox,1}$ is determined by the oxygen transport in the GDL (see below).
    The boundary conditions to Equation 4.199 are

$$-\tilde{D}_{mt}\left.\frac{\partial \tilde{c}_{mt}}{\partial \tilde{x}}\right|_0 = \tilde{j}_{cross,0}, \quad \tilde{c}_{mt}(1) = 0. \qquad (4.204)$$

The first condition means that the incoming diffusion flux of methanol equals the equivalent crossover current density $\tilde{j}_{cross,0}$. The second condition of Equation 4.204 expresses the assumption that all methanol is converted in the CCL.
    For a typical one-molar methanol concentration, the electro-osmotic flux of methanol in the membrane is small as compared to the methanol diffusion. The expression for the crossover current follows from the methanol mass balance in the cell (Kulikovsky, 2002b)

$$\tilde{j}_{cross,0} = \beta_* \left(\tilde{j}^a_{lim} - \tilde{j}_0\right), \qquad (4.205)$$

where

$$\tilde{j}^a_{lim} = \frac{6FD^a_b c_{mt}}{l^a_b} \qquad (4.206)$$

is the limiting current density due to methanol transport in the anode backing layer (ABL) of a thickness $l^a_b$. Here, $D^a_b$ is the methanol diffusion coefficient in the backing layer. Note that $\tilde{j}_{cross,0}$ (4.205) linearly decreases with the growth of the useful cell current $\tilde{j}_0$, a feature, confirmed in many experiments (see, e.g., Ren et al. (2000b)).

The crossover parameter $0 \le \beta_* \le 1$ is given by (Kulikovsky, 2002b)

$$\beta_* = \frac{\beta_{cross}}{1 + \beta_{cross}}, \tag{4.207}$$

where

$$\beta_{cross} = \frac{D_{mt}^m l_b^a}{D_b^a l_m} \tag{4.208}$$

is the ratio of methanol mass transfer coefficients in the membrane and in the ABL. Here, $D_{mt}^m$ is the methanol diffusion coefficient in the membrane of thickness $l_m$. Note that $\beta_*$ should not be mixed up with the methanol drag coefficient in the membrane.

A simple linear diffusion equation for oxygen transport in the GDL leads to the following relation between the oxygen concentration at the CCL/GDL interface $\tilde{c}_{ox,1}$ and the oxygen concentration in the channel $\tilde{c}_{ox}^h$:

$$\tilde{c}_{ox,1} = \tilde{c}_{ox}^h - \frac{\tilde{j}_0 + \tilde{j}_{cross,0}}{\tilde{j}_{\lim}^{c0}}, \tag{4.209}$$

where

$$\tilde{j}_{\lim}^{c0} = \frac{4FD_b^c c_{ox}^{h0}}{l_b^c} \tag{4.210}$$

is the limiting current density due to the oxygen transport in the GDL at the oxygen channel inlet. A typical DMFC operates at a high oxygen flow rate and in Equation 4.209, $\tilde{c}_{ox}^h = 1$ may be set. The resulting equation is used in Equation 4.203.

## Conservation Law

The system (4.197) through (4.199) has a first integral. Replacing the terms on the right-hand side of Equation 4.197 by the left-hand sides of Equations 4.198 and 4.199, and integrating the resulting equation once gives

$$\tilde{j} + \tilde{D}_{ox}\frac{\partial \tilde{c}_{ox}}{\partial \tilde{x}} - \tilde{D}_{mt}\frac{\partial \tilde{c}_{mt}}{\partial \tilde{x}} = \tilde{j}_0 + \tilde{j}_{cross,0}. \tag{4.211}$$

This equation expresses the balance of fluxes in the CCL. Note that the right-hand side of Equation 4.211 is independent of $\tilde{x}$.

Setting in Equation 4.211 $\tilde{x} = 1$, $\tilde{j}(1) = 0$, and assuming that the methanol flux to the cathode GDL is zero, $\tilde{D}_{mt}\partial \tilde{c}_{mt}/\partial \tilde{x}|_1 = 0$ (complete methanol oxidation in the CCL), one arrives at

$$\tilde{j}_0 + \tilde{j}_{cross,0} = \tilde{D}_{ox}\frac{\partial \tilde{c}_{ox}}{\partial \tilde{x}}\bigg|_1. \tag{4.212}$$

This equation simply expresses the stoichiometric requirement that the total oxygen flux coming from the GDL must be equal to the sum of the useful and crossover current densities.

Equation 4.211 can further be integrated. Replacing $\tilde{j}$ by $-\partial \tilde{\eta}_{ox}/\partial \tilde{x}$, one obtains

$$\tilde{D}_{ox} \frac{\partial \tilde{c}_{ox}}{\partial \tilde{x}} - \tilde{D}_{mt} \frac{\partial \tilde{c}_{mt}}{\partial \tilde{x}} - \frac{\partial \tilde{\eta}_{ox}}{\partial \tilde{x}} = \tilde{j}_0 + \tilde{j}_{cross,0}.$$

Integration of this equation from $\tilde{x}$ to 1 yields

$$\tilde{D}_{ox} \left( \tilde{c}_{ox,1} - \tilde{c}_{ox} \right) + \tilde{D}_{mt} \tilde{c}_{mt} + \left( \tilde{\eta}_{ox} - \tilde{\eta}_{ox,1} \right) = \left( \tilde{j}_0 + \tilde{j}_{cross,0} \right) (1 - \tilde{x}). \qquad (4.213)$$

This equation is a conservation law, relating the local concentrations $\tilde{c}_{ox}$ and $\tilde{c}_{mt}$ to the local overpotential $\tilde{\eta}_{ox}$. Setting in Equation 4.213 $\tilde{x} = 0$ gives

$$\tilde{D}_{ox} \left( \tilde{c}_{ox,1} - \tilde{c}_{ox,0} \right) + \tilde{D}_{mt} \tilde{c}_{mt,0} + \tilde{\eta}_{ox,0} - \tilde{\eta}_{ox,1} = \tilde{j}_0 + \tilde{j}_{cross,0}. \qquad (4.214)$$

Importantly, this equation follows from the mass balance in the CCL only, and it does not depend on the assumptions on the ORR and MOR mechanisms.

Below, it will be shown that at sufficiently large currents, the oxygen concentration at the membrane interface tends to zero. Setting in Equation 4.214 $\tilde{c}_{ox,0} = 0$ and noting that $\tilde{D}_{ox}\tilde{c}_{ox,1} \gg \tilde{D}_{mt}\tilde{c}_{mt,0}$ results in

$$\tilde{D}_{ox} \tilde{c}_{ox,1} + \tilde{\eta}_{ox,0} - \tilde{\eta}_{ox,1} = \tilde{j}_0 + \tilde{j}_{cross,0}. \qquad (4.215)$$

This equation determines the CCL polarization curve at large currents, as discussed below.

### Polarization Curves and $x$ Shapes

The system (4.197) through (4.199) with the boundary conditions (4.202), (4.203), (4.204), and (4.209) has been solved numerically using Maple® software. The physical and operational parameters are listed in Table 4.1. Note that the boundary conditions (4.202) and (4.204) contain both the cell current density $\tilde{j}_0$ and the potential loss $\tilde{\eta}_{ox,0}$. However, these values are related by the polarization curve, which *a priori* is unknown. Thus, iterations are required to find $\tilde{\eta}_{ox,0}$, providing the prescribed cell current $\tilde{j}_0$.

Figure 4.24 shows the CCL polarization curves for the range of parameter $\beta_*$ between zero (no crossover) and 0.5 (large crossover). Note that Figure 4.24 shows the electrode potential calculated according to $E_{cath} = E_{ox}^{eq} - \eta_{ox,0}$. The crossover dramatically (by 300–600 mV) lowers the electrode potential at the OCP (Figure 4.24). Note that for $\beta_*$ between 0 and 0.2, there is a range of currents (50 to 150 mA cm$^{-2}$), where the polarization curve is nearly flat (Figure 4.24). This is a feature of DMFC: an increase in the useful current $\tilde{j}_0$ lowers the crossover current $\tilde{j}_{cross,0}$ so that the sum $\tilde{j}_0 + \tilde{j}_{cross,0}$ and hence the cell potential do not change significantly.

Further, the curves for $\beta_*$ from 0 to 0.4 exhibit a drastic change in the slope at higher currents. Consider, for example, the curve for $\beta_* = 0.3$ (Figure 4.24). The electrode potential decreases quite slowly up to the critical current density of $j_0 \simeq 130$ mA cm$^{-2}$, and it drops much faster above this point. Note that in this "supercritical" region, the curves decay linearly with the current.

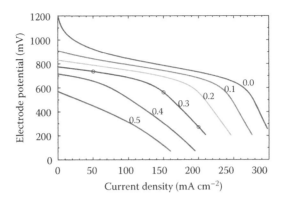

**FIGURE 4.24** Polarization curves of the DMFC cathode ($E_{ox}^{eq} - \eta_{ox,0}$) for the indicated crossover parameter $\beta_*$. Parameters for calculations are listed in Table 4.1. The shapes of local parameters in Figures 4.25 and 4.26 correspond to the points (open circles) on the curve for $\beta_* = 0.3$.

To understand these features, it is advisable to plot the shapes of local parameters through the CCL at several current densities. These shapes for $\beta_* = 0.3$ and $j_0 = 50$, 150, and 200 mA cm$^{-2}$ are shown in Figure 4.25. In addition, the ORR and MOR rates for the same currents are shown in Figure 4.26. To understand these shapes, consider Equation 4.201 relating the ORR and MOR overpotentials. This equation can be rewritten in the form

$$\eta_{mt} = \delta E^{eq} - \eta_{ox} \simeq 1.18 - \eta_{ox}, \tag{4.216}$$

where $\delta E^{eq} = E_{ox}^{eq} - E_{mt}^{eq} \simeq 1.18$ V is constant.

At a small current (50 mA cm$^{-2}$), close to the membrane, the ORR overpotential $\eta_{ox}$ is relatively small (Figure 4.25b) and according to Equation 4.216, the MOR overpotential $\eta_{mt}$ is large. In a small domain close to the membrane ($\tilde{x} < 0.05$), the rate of the MOR greatly exceeds the rate of the ORR (Figure 4.26a), and the crossover flux of methanol is rapidly converted to the proton current. This leads to the growing proton current in this domain (Figure 4.25a). The total proton current then enters the ORR-dominated domain ($\tilde{x} > 0.05$), working as a normal cathode (Figure 4.26a). In other words, the small domain at the membrane interface works as a virtual anode, *within the cathode catalyst layer*, and the CCL represents a complete short-circuited fuel cell.

At large currents (150 and 200 mA cm$^{-2}$), near the membrane, the ORR overpotential increases (Figure 4.25b) and the MOR overpotential decreases. The distinct peaks of the ORR and MOR rate form and, with the growth of the cell current, these peaks shift from the membrane toward the GDL (Figure 4.26b,c). The peaks overlap, so that the virtual anode disappears. In other words, each proton produced in the MOR is converted "in-place" back to the neutral water molecule in the ORR (Figure 4.25a).

As the cell current increases, the crossover current decreases (Equation 4.205). Unexpectedly, because of a lower rate of methanol conversion in the CCL, the

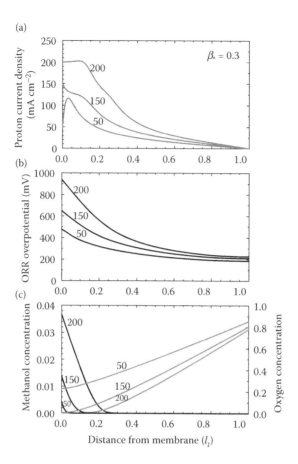

**FIGURE 4.25** The shapes of local parameters through the CCL thickness for the indicated cell current densities (mA cm$^{-2}$). The oxygen and methanol concentrations are normalized to their inlet values (Equation 4.196). The crossover parameter $\beta_* = 0.3$ and the other parameters are listed in Table 4.1. Shown are the through-plane shape of (a) proton current density, (b) ORR overpotential, and (c) methanol concentration.

concentration of methanol at the membrane interface increases. This is a rather paradoxical situation: with the growth of $j_0$, the flux of methanol decreases, while the methanol concentration in the CCL rises and methanol more deeply penetrates the layer (Figure 4.25c).

At $j_0 \gtrsim 150$ mA cm$^{-2}$, the oxygen concentration at the membrane interface becomes very small (Figure 4.25c). From this critical current density on, the electrode potential exhibits very fast linear decay with $j_0$ (Figure 4.24).

## Large-Current CCL Resistivity

The model allows deriving the approximate polarization curve of the CCL in the supercritical (high-current) regime. In this regime, $\tilde{c}_{ox,0} \simeq 0$ and hence

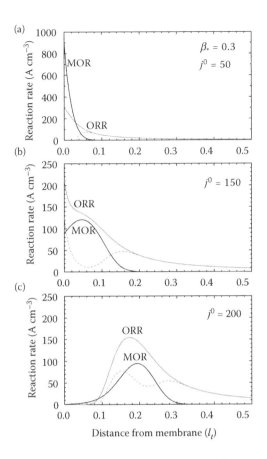

**FIGURE 4.26** The solid lines: the ORR and MOR rates (A cm$^{-3}$) for the indicated cell current densities (mA cm$^{-2}$). The dotted line: the difference of the ORR and MOR rates (A cm$^{-3}$). Each frame shows the ORR and MOR reaction rates and their difference (dashed line) for the current densities of (a) 50 mA cm$^{-2}$, (b) 150 mA cm$^{-2}$, and (c) 200 mA cm$^{-2}$.

Equation 4.215 holds; it is advisable to rearrange this equation as

$$\tilde{\eta}_{ox,0} = \tilde{\eta}_{ox,1} + \tilde{j}_0 + \tilde{j}_{cross,0} - \tilde{D}_{ox}\tilde{c}_{ox,1}. \tag{4.217}$$

The variations of $\tilde{\eta}_{ox,1}$ and $\tilde{c}_{ox,1}$ with $\tilde{j}_0$ are small. Differentiating Equation 4.217 over $\tilde{j}_0$, and chalking out the terms $\partial\tilde{\eta}_{ox,1}/\partial\tilde{j}_0$ and $\tilde{D}_{ox}\partial\tilde{c}_{ox,1}/\partial\tilde{j}_0$, gives the large-current CCL resistivity (the slope of the large-current straight lines in Figure 4.24):

$$\tilde{R}_{ccl} = \frac{\partial\tilde{\eta}_{ox,0}}{\partial\tilde{j}_0} \simeq 1 - \beta_*. \tag{4.218}$$

In dimensional variables, this equation reads

$$R_{ccl} \simeq \frac{l_{CL}(1 - \beta_*)}{\sigma_p}. \tag{4.219}$$

The resistivities fitted to the curves in Figure 4.24 and calculated with Equation 4.219 are listed in Table 4.2. As can be seen, the agreement is good.

The resistivity $R_{ccl}$, Equation 4.219 is proportional to the CCL proton resistivity $l_{CL}/(3\sigma_p)$. The reason is seen in Figure 4.26: At high current, the peak of the ORR rate shifts away from the membrane, which means that protons need to be transported more deeply into the CCL (Figure 4.26c). In addition, the overpotential $\eta_{ox}$ is very high so that the rate of ORR itself does not limit the cathode performance. In this situation, proton transport is the "rate-determining" process, that is, the rates of electrochemical conversion appear to be much higher than the rate of proton transport. In addition, methanol crossover reduces the resistivity (4.219), since the MOR provides protons "in-place," without the need of transporting them from the membrane.

DMFC polarization curves exhibiting linear behavior at large currents have been measured by Argyropoulos et al. (2002). Figure 4.27a shows their experimental curves (points) for several cell temperatures and a linear fit of the large-current regions. Table 4.3 shows the cell resistivity obtained as a slope of the straight lines in Figure 4.27a. As can be seen, the three curves exhibit very similar high-current resistivities of 36–38 $\Omega$ cm$^2$. The resistivity of the curve corresponding to 348.15 K is larger (about 51 $\Omega$ cm$^2$), probably a result of the current limitation by the methanol transport on the anode side. The curve for 343.15 K was not fitted, as it does not exhibit straight lines at high currents (perhaps, for the same reason).

Argyropoulos et al. (2002) reported a thickness of the CCL ($l_{CL} \simeq 0.03$ cm). The numbers from Table 4.3 and Equation 4.219 allow us to estimate the CCL ionic conductivity in their experiments. Taking for the estimate $\beta_* = 0.3$ and the cell resistivity of 37 $\Omega$ cm$^2$ (Table 4.3), one arrives at $\sigma_p \simeq 5.7 \cdot 10^{-4}$ $\Omega^{-1}$ cm$^{-1}$. This value is not far from $\sigma_p \simeq 1 \cdot 10^{-3}$ $\Omega^{-1}$ cm$^{-1}$, measured by Havranek and Wippermann (2004) for the similar catalyst layer with the 10% Nafion content. Note that the temperature variation of Nafion conductivity in the range from 348 to 363 K is about 15% (Silva

**TABLE 4.2**

**Large-Current CCL Resistivity ($\Omega$ cm$^2$) Fitted to the Curves in Figure 4.24 and Calculated with Equation 4.219**

| $\beta_*$ | Fitted to Figure 4.24 | Equation 4.219 |
|-----------|----------------------|----------------|
| 0.0 | 12.38 | 10.0 |
| 0.1 | 10.25 | 9.0 |
| 0.2 | 8.12 | 8.0 |
| 0.3 | 6.27 | 7.0 |
| 0.4 | 5.03 | 6.0 |

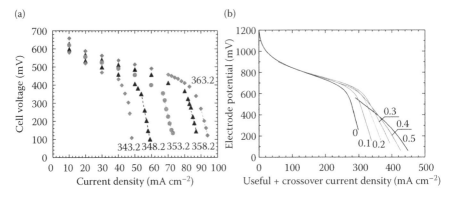

**FIGURE 4.27** (a) The points: the DMFC polarization curves for the indicated working temperatures (data from Argyropoulos et al. (2002)). The dotted lines: the linear fit according to Equation 4.215. The slopes of the straight line (the CCL resistivities) are collected in Table 4.3. (b) The calculated electrode polarization curve versus the sum of the useful and crossover current densities $j_0 + j_{cross,0}$ for the indicated parameter $\beta_*$.

**TABLE 4.3**
**Large-Current Cell Resistivity ($\Omega$ cm$^2$) Fitted to the Curves in Figure 4.27a**

| Temperature (K) | Large-Current Cell Resistivity/$\Omega$ cm$^2$ Fitted to Figure 4.27a |
|---|---|
| 348.15 | 51.44 |
| 353.15 | 37.83 |
| 358.15 | 37.12 |
| 363.15 | 36.31 |

et al., 2004), and the effect of this variation on the slope of the curves in Figure 4.27 is marginal.

To understand the conditions for the onset of the linear regime, it is advisable to plot the model electrode potential for various $\beta_*$ versus the sum of the useful and crossover current densities (Figure 4.27b). As can be seen, in the low-current region, the curves for different $\beta_*$ coincide. In this region, the standard model of DMFC cathode, based on the use of the sum $j_0 + j_{cross,0}$ in the Tafel equation, works.*

In the coordinates of Figure 4.27b, the onset of the linear regime occurs regardless of $\beta_*$ at almost the same value of $j_0 + j_{cross,0} \simeq 320$ mA cm$^{-2}$. This value is close to the oxygen-transport current density $j_D$ through the CCL. Indeed, the data from

---

* The ORR Tafel slope in the current balance equation should be doubled, as due to the high rate of crossover, the cathode always works in the regime with the Tafel slope doubling.

Table 4.1 give $j_D = 4FD_{ox}c^h_{ox}/l_{CL} \simeq 300$ mA cm$^{-2}$, where, for the estimate $c_{ox,1} \simeq c^h_{ox} = 7.6 \cdot 10^{-6}$ mol cm$^{-3}$. Thus, the critical current density $j^0_{crit}$ for the linear regime onset is determined by the equation

$$j^0_{crit} + \beta_* \left( j^a_{lim} - j^0_{crit} \right) = \frac{4FD_{ox}c_{ox,1}}{l_{CL}}. \qquad (4.220)$$

In other words, the onset of the high-current linear regime is determined by the characteristic current of oxygen transport through the CCL, while the polarization curve in this regime is determined by the transport of methanol and protons.

Generally, the cell current density can also be limited by the transport of either methanol or oxygen through the backing layer. The respective limiting current densities are given by Equations 4.206 and 4.210. In order for the cell polarization curve to exhibit the linear regime, the other mechanisms of current limitation must appear at higher currents. If, for example, the limiting current density due to methanol transport in the backing layer is close to $j^0_{crit}$, the linear regime would not appear and the cell polarization curve would decay in a nonlinear manner (presumably, this happens with the lowest temperature curve in Figure 4.27a).

The resistive decay of the cell potential discussed above limits the cell current. The "resistive" limiting current density $j^R_{lim}$ can be calculated from the relation

$$E_{cell}(j^0_{crit}) - R_{ccl}(j^R_{lim} - j^0_{crit}) = 0, \qquad (4.221)$$

where $j^0_{crit}$ is given by Equation 4.220 and $R_{ccl}$ by Equation 4.219. The nature of the cell current limitation is, thus, determined by the smallest of $j^a_{lim}$, $j^c_{lim}$, and $j^R_{lim}$.

## DMFC Triode

The model above has immediate practical implications. It has shown that significant fuel crossover radically changes the regime of the CCL operation: at small cell currents, a virtual anode (VA) within the CCL forms. As a result of a very high rate of MOR, the VA domain is subject to a strong poisoning by the MOR products. Though this poisoning has not been taken into account in the model, qualitatively, this effect can be understood. As a CCL in a DMFC is usually based on a pure Pt catalyst, the poisoning of the virtual anode by MOR products may strongly lower the rate of the MOR in this domain. This means that the poisoning would increase the VA thickness, as more catalyst is needed for a complete conversion of the permeated methanol. Therefore, an interesting option would be to design a two-layer cathode, in which the VA domain is loaded with the poisoning-resistant Pt/Ru catalyst.

A more radical option is to completely separate the VA from the cathode by cutting the CCL along the VA interface and inserting a membrane layer along this interface (Kulikovsky, 2012e). A schematic of such a DMFC triode is depicted in Figure 4.28. As compared to the conventional DMFC, the cell contains an additional electrode $E_*$. The composition of this electrode is that of the anode: it contains Nafion

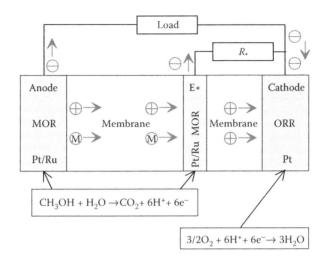

**FIGURE 4.28** Schematic of a three-electrode DMFC. Small circles with plus, minus, and M inside show protons, electrons, and methanol molecules, respectively. Note that the carbon phase of the auxiliary electrode $E_*$ is connected to the cathode through the small resistor $R_*$. This provides a high rate of the methanol oxidation in $E_*$ (Kulikovsky, 2012e). Electrochim Acta, **79**, 52–56.

electrolyte and a carbon-supported Pt/Ru catalyst. Importantly, the carbon (electron conducting) phase of $E_*$ is connected to the cathode through a small resistor $R_*$.

The electrode $E_*$ serves as a second anode, which converts the flux of methanol in the membrane into useful current. Indeed, if $R_*$ is small, the carbon phase potential of $E_*$ is almost equal to the carbon phase potential of the cathode, while the electrolyte phase potential in $E_*$ is not far from that potential in the anode. Thus, the MOR overpotential in $E_*$ would be large, that is, the auxiliary electrode would efficiently convert the flux of methanol in the membrane into the useful ionic and electron currents. Moreover, thanks to $E_*$, no methanol would arrive at the cathode, thereby no poisoning of the ORR electrode by the MOR products would occur.

Ideally, the potential drop between the auxiliary electrode and the cathode should provide complete oxidation of the methanol flux in the membrane. This potential drop is easy to control by varying the load resistor $R_*$. Thus, for any given current in the main load, the value of $R_*$ should be selected to provide the highest cell potential. Note that the resistor $R_*$ could also be a useful load, so that the current in $R_*$ is not wasted.

The cost of the third electrode in the system is an increased resistivity for the proton transport from anode to cathode. However, because of the high overpotential for the MOR, the thickness of $E_*$ could be much smaller than the anode thickness. Further, the thickness of the membrane between $E_*$ and the cathode could also be small, since no methanol crossover is expected through this membrane. Overall, complete blocking of methanol crossover and conversion of the crossover current into the useful current due to the presence of $E_*$-electrode may outweigh the increase in the MEA resistive loss.

## Anode Catalyst Layer in DMFC

### Preliminary Remarks

An experimental study of methanol oxidation reaction kinetics has been a subject of numerous works, most of which studied MOR on well-defined crystalline catalyst surfaces in contact with liquid electrolyte (Bagotzky and Vasilyev, 1967; Gasteiger et al., 1993a; Lamy and Leger, 1991; Tarasevich et al., 1983).

In electrochemical studies, every effort is usually made to eliminate mass transport effects in order to avoid their influence on reaction kinetics. However, owing to sluggish kinetics of MOR, a real DMFC anode requires high catalyst loading, and, hence, the anode is usually 50–100 μm thick. At typical operating currents of 100–300 mA cm$^{-2}$, such a system exhibits quite a substantial resistivity for proton transport, strongly affecting the anode performance. A finite anode thickness together with a relatively poor proton conductivity, make the distribution of the MOR overpotential and rate strongly nonuniform through the anode depth. This nonuniformity is of large interest, as it changes the anode performance and leads to nonuniform electrode degradation.

The range of cell currents, in which the proton transport loss in the anode can be neglected is estimated as $j_0 < \sigma_p b_{mt}/l_{CL}$, where $\sigma_p$ is the proton conductivity, $b_{mt}$ is the Tafel slope, and $l_{CL}$ is the thickness (page 309). With $\sigma_p$ and $b_{mt}$ from Table 4.4 and $l_{CL} = 0.01$ cm (100 μm ACL), we obtain $j_0 < 18$ mA cm$^{-2}$. Thus, for the cell

## TABLE 4.4
### Fitting Parameters for the Curves in Figure 4.31

|  | 70°C | 50°C |
|---|---|---|
| $b_{mt}$ (V) | 0.06 | |
| $l_b^a$ (cm) | 0.02 | |
| $l_{CL}$ (cm) | 0.0014 | |
| $\sigma_p$ ($\Omega^{-1}$ cm$^{-1}$) | 0.003 | |
| $D_b^a$ (cm$^2$ s$^{-1}$) | $8 \cdot 10^{-3}$ | |
| $\eta_*$ (V) | 0.11 | |
| $j_{ref}$ (A cm$^{-2}$) | 0.129 | |
| $i_{mt}$ (A cm$^{-3}$) | 1.0 | 0.2 |
| $\varepsilon$ | 11.98 | 15.15 |
| $\tilde{i}_{ads} \equiv i_{ads}/i_{mt}$ | | |
| (0.1 M) | 500 | 1120 |
| (0.5 M) | 800 | 1500 |
| (1.0 M) | 1100 | 1900 |
| (2.0 M) | 1600 | 2400 |

*Note:* Note that due to the small ACL thickness, in both the cases, $\varepsilon \simeq 10$. For the typical 100 μm-thick ACL, this parameter is in the order of unity. The layer proton conductivity $\sigma_p$ is taken from Havranek, A., and Wippermann, K. 2004. *J. Electroanal. Chem.*, **567**, 305–315.

operating at 100 mA cm$^{-2}$, proton transport in the ACL strongly affects the anode performance.

Numerical modeling of a stepwise MOR mechanism in a porous DMFC anode has been reported in a number of works, assuming uniformity of MOR overpotential in the anode (see e.g., Krewer et al., 2006; Meyers and Newman, 2002a and the references therein). Evidently, the time- and space-dependent numerical codes for the distributed MOR kinetics in the electrode would be time-consuming. In addition, such codes would include a number of poorly known rate constants.

Also of large interest is the nature of the limiting current density $j_{lim}$ in a cell. Typically, DMFC is run at a high oxygen stoichiometry and relatively low methanol concentration (below 2M), so that the cell limiting current density is determined by the anode side (Baldaus and Preidel, 2001; Scott et al., 1999; Xu et al., 2006). Scott et al. (1999) reported proportionality of the limiting current density to the methanol concentration. Xu et al. (2006) published a detailed study of the effect of methanol stoichiometry on the limiting current density and analyzed this effect in terms of the mass transfer resistance of the anode side. In their experiments, the limiting current density was also proportional to the methanol concentration and this proportionality was attributed to the methanol transport in the ABL. Below, it will be shown that current limitation can result from the MOR kinetics as well.

In the modeling of cells, stacks, and systems and in the design of automatic control devices, a physically realistic analytical equation for the anode polarization curve is highly desirable. Many works utilize a Tafel law for the MOR rate in the electrode (Argyropoulos et al., 2002; Baxter et al., 1999; Casalegno and Marchesi, 2008; Cho et al., 2009; Dohle et al., 2000; Lam et al., 2011; Miao et al., 2008; Murgia et al., 2003; Yang and Zhao, 2007; Zhao et al., 2009). This extensive list reflects a great demand for a simple analytical expression for the anode overpotential. However, applicability of the Tafel law to the DMFC anode is questionable. This law neither describes the potential-independent methanol adsorption (see below) nor does it take into account the finite rate of proton transport in the electrode.

Jiang and Kucernak (2005) experimentally studied MOR on a Pt/Nafion interface in a porous electrode. According to their work, methanol electro-adsorption on this interface most probably proceeds through a slow step of chemisorption on a Pt surface followed by a fast electrochemical step:

$$CH_3OH \xrightarrow{slow} (CH_3OH)_{ads}, \tag{4.222}$$

$$(CH_3OH)_{ads} \xrightarrow{fast} (CH_2OH)_{ads} + H^+ + e^-. \tag{4.223}$$

Thus, the overall rate of methanol electroadsorption is determined by the potential-independent chemisorption (4.222). It is, therefore, reasonable to approximate the MOR kinetics in a DMFC anode by a two-step reaction mechanism:

$$CH_3OH \longrightarrow (CH_3OH)_{ads}, \tag{4.224}$$

$$(CH_3OH)_{ads} + H_2O \longrightarrow CO_2 + 6H^+ + 6e^-. \tag{4.225}$$

where Equation 4.225 accumulates all the electrochemical steps.

In this section, a one-dimensional, through-plane model of a DMFC anode is constructed (Kulikovsky, 2013a). The model takes into account the finite rate of proton transport in the ACL, the two-step kinetics of the MOR (4.224), (4.225) and the potential loss resulting from the methanol transport in the ABL.

## Basic Equations

Let the axis $x$ be directed from the membrane to the ABL (Figure 4.29). The main model assumption is that methanol transport in the catalyst layer is fast, that is, the main contribution to the methanol transport loss gives the ABL. Physically, methanol transport in the ACL is fast because of a concerted action of the diffusion and electro-osmotic effect (Jeng and Chen, 2002).

The system of equations governing anode performance is

$$\frac{\partial j}{\partial x} = -R_{MOR},\qquad(4.226)$$

$$R_{MOR} = \frac{i_{mt}\exp(\eta/b_{mt})}{1 + i_{mt}\exp(\eta/b_{mt})/(i_{ads}c_{mt}/c_{mt}^{ref})},\qquad(4.227)$$

$$j = -\sigma_p\frac{\partial\eta}{\partial x},\qquad(4.228)$$

where $\eta$ is the MOR polarization voltage (overpotential) and $i_{ads}$ is the volumetric characteristic rate of methanol adsorption on the catalyst surface (A cm$^{-3}$). The other notations have been introduced in the section "Basic Equations."

Equation 4.226 is a proton current conservation equation. The right side of Equation 4.226 is a rate of proton production (A cm$^{-3}$) in the ACL given by

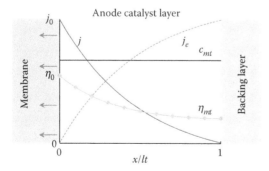

**FIGURE 4.29** A schematic of the anode catalyst layer. The proton current density $j$ and the MOR overpotential $\eta$ increase toward the membrane, while the electron current density $j_e$ increases toward the backing layer. The methanol concentration $c_{mt}$ is assumed to be constant along $x$ (though $c_{mt}$ depends of the cell current).

Equation 4.227. The latter equation results from the reaction scheme (4.224) and (4.225).*

Equation 4.227 takes into account a zero-order dependence on methanol concentration at low overpotentials and the adsorption step (4.224). Indeed, as $i_{mt}/i_{ads} \ll 1$ (see below), at small $\eta$, the second term in the denominator of Equation 4.227 is much less than unity. This equation reduces to $R_{MOR} = i_{mt} \exp(\eta/b_{mt})$, which is a Tafel law with zero-order concentration dependence. In the opposite limit of large $\eta$, unity in the denominator of Equation 4.227 can be neglected and this equation reduces to

$$R_{MOR}^{\lim} = i_{ads} c_{mt}/c_{mt}^{ref}, \tag{4.229}$$

describing the adsorption-limiting rate of MOR. In this regime, the electrochemical conversion steps are fast, and the rate-determining step is the chemisorption process (4.224). In other words, at high overpotential, MOR behaves like a "chemical" rather than an electrochemical reaction.

To simplify calculations, it is convenient to introduce dimensionless variables:

$$\tilde{x} = \frac{x}{l_{CL}}, \quad \tilde{j} = \frac{j}{j_{ref}}, \quad \tilde{\eta} = \frac{\eta}{b_{mt}}, \quad \tilde{c}_{mt} = \frac{c_{mt}}{c_{mt}^{ref}}, \tag{4.230}$$

where

$$j_{ref} = \frac{\sigma_p b_{mt}}{l_{CL}} \tag{4.231}$$

is a characteristic current density. Substituting Equation 4.227 into Equation 4.226 and using the variables (4.230) results in a system of two equations:

$$2\varepsilon^2 \frac{\partial \tilde{j}}{\partial \tilde{x}} = -\frac{\exp \tilde{\eta}}{1 + \xi \exp \tilde{\eta}}, \tag{4.232}$$

$$\tilde{j} = -\frac{\partial \tilde{\eta}}{\partial \tilde{x}}, \tag{4.233}$$

where

$$\varepsilon = \sqrt{\frac{\sigma_p b_{mt}}{2 i_{mt} l_{CL}^2}} \tag{4.234}$$

is the dimensionless reaction penetration depth, and

$$\xi = \frac{1}{\tilde{i}_{ads} \tilde{c}_{mt}}, \quad \tilde{i}_{ads} = \frac{i_{ads}}{i_{mt}} \tag{4.235}$$

are dimensionless parameters. Note that $\xi$ is the inverse characteristic rate of methanol adsorption.

---

* The prototype of Equation 4.227 had been derived by Meyers and Newman (2002a). A similar equation, with a more complicated concentration dependence, was studied by Nordlund and Lindbergh (2004). A simple derivation of Equation 4.227 from the scheme (4.224) and (4.225) is given in Kulikovsky (2005).

### First Integral and the Polarization Curve

Multiplying Equations 4.233 and 4.232, together yields

$$\varepsilon^2 \frac{\partial (\tilde{j}^2)}{\partial \tilde{x}} = \frac{\exp \tilde{\eta}}{1 + \xi \exp \tilde{\eta}} \left( \frac{\partial \tilde{\eta}}{\partial \tilde{x}} \right) = \frac{1}{\xi} \frac{\partial (\ln (1 + \xi \exp \tilde{\eta}))}{\partial \tilde{x}}. \tag{4.236}$$

Integrating this equation once, one finds the first integral of the system

$$\varepsilon^2 \xi \tilde{j}^2 - \ln (1 + \xi \exp \tilde{\eta}) = \varepsilon^2 \xi \tilde{j}_0^2 - \ln (1 + \xi \exp \tilde{\eta}_0)$$
$$= -\ln (1 + \xi \exp \tilde{\eta}_1), \tag{4.237}$$

where the subscripts 0 and 1 mark the values at $\tilde{x} = 0$ and $\tilde{x} = 1$, respectively. Thus, the sum $\varepsilon^2 \xi \tilde{j}^2 - \ln (1 + \xi \exp \tilde{\eta})$ is constant along $\tilde{x}$.

The last equation of (4.237) can be solved for $\tilde{\eta}_0$:

$$\tilde{\eta}_0 = \ln \left( \frac{(1 + \xi \exp \tilde{\eta}_1) \exp(\varepsilon^2 \xi \tilde{j}_0^2) - 1}{\xi} \right). \tag{4.238}$$

Equation 4.238 is the exact polarization curve of the ACL; however, it contains an unknown parameter $\tilde{\eta}_1$.

Replacing $\tilde{j}^2$ by $(\partial \tilde{\eta}/\partial \tilde{x})^2$ in Equation 4.237, and taking the square root of the resulting equation, one arrives at

$$\varepsilon \sqrt{\xi} \frac{\partial \tilde{\eta}}{\partial \tilde{x}} = -\sqrt{\ln \left( \frac{1 + \xi \exp \tilde{\eta}}{1 + \xi \exp \tilde{\eta}_1} \right)}. \tag{4.239}$$

This equation has an implicit solution

$$\tilde{x} = \varepsilon \sqrt{\xi} \int_{\tilde{\eta}}^{\tilde{\eta}_0} \left( \ln \left( \frac{1 + \xi \exp \phi}{1 + \xi \exp \tilde{\eta}_1} \right) \right)^{-1/2} d\phi. \tag{4.240}$$

In setting $\tilde{x} = 1$, a relation of the boundary values $\tilde{\eta}_1$ and $\tilde{\eta}_0$ is acquired

$$1 = \varepsilon \sqrt{\xi} \int_{\tilde{\eta}_1}^{\tilde{\eta}_0} \left( \ln \left( \frac{1 + \xi \exp \phi}{1 + \xi \exp \tilde{\eta}_1} \right) \right)^{-1/2} d\phi. \tag{4.241}$$

This relation can be used to eliminate $\tilde{\eta}_1$ from Equation 4.238. However, a many more convenient explicit approximate expressions for the shapes of parameters and for the polarization curve can be obtained, as discussed below.

## Through-Plane Shapes

Differentiating Equation 4.232 over $\tilde{x}$ gives

$$2\varepsilon^2 \frac{\partial^2 \tilde{j}}{\partial \tilde{x}^2} = -\left( \frac{\exp \tilde{\eta}}{1 + \xi \exp \tilde{\eta}} - \xi \left( \frac{\exp \tilde{\eta}}{1 + \xi \exp \tilde{\eta}} \right)^2 \right) \frac{\partial \tilde{\eta}}{\partial \tilde{x}}.$$

Taking into account Equations 4.232 and 4.233, the equation above transforms to

$$\frac{\partial^2 \tilde{j}}{\partial \tilde{x}^2} + \tilde{j} \frac{\partial \tilde{j}}{\partial \tilde{x}} + 2\varepsilon^2 \xi \tilde{j} \left( \frac{\partial \tilde{j}}{\partial \tilde{x}} \right)^2 = 0, \quad \tilde{j}(0) = \tilde{j}_0, \quad \tilde{j}(1) = 0, \tag{4.242}$$

containing the proton current density only. The boundary conditions to Equation 4.242 are obvious.

It is advisable to rewrite Equation 4.242 in the form

$$\frac{\partial^2 \tilde{j}}{\partial \tilde{x}^2} + \tilde{j} \frac{\partial \tilde{j}}{\partial \tilde{x}} \left( 1 + 2\varepsilon^2 \xi \frac{\partial \tilde{j}}{\partial \tilde{x}} \right) = 0. \tag{4.243}$$

In a DMFC anode, the parameter $\varepsilon \simeq 1\text{--}10$; however, the parameter $\xi$ is typically small: $\xi \sim O(10^{-3})$ (this parameter has been estimated using the data of Nordlund and Lindbergh (2004)). Thus, the product $2\varepsilon^2 \xi \simeq 10^{-1}\text{--}10^{-3}$ and hence at leading order, the term with $\varepsilon^2 \xi$ in Equation 4.243 can be omitted (validity of this approximation is discussed in the section "Model Validation").

This leads to an equation

$$\frac{\partial^2 \tilde{j}}{\partial \tilde{x}^2} + \tilde{j} \frac{\partial \tilde{j}}{\partial \tilde{x}} = 0, \tag{4.244}$$

which can be easily integrated to yield

$$\tilde{j} = \gamma \tan \left( \frac{\gamma}{2} (1 - \tilde{x}) \right), \tag{4.245}$$

where $\gamma$ is given by Equation 4.127. Note that $\gamma$ tends to zero as $\tilde{j}_0 \to 0$.

Equation 4.245, with $\gamma(\tilde{j}_0)$ given by Equation 4.127, provides the leading-order shape of the proton current density through the ACL thickness. Substituting Equation 4.245 into Equation 4.232, calculating the derivative and solving the resulting equation for $\eta$, gives the shape of overpotential through the ACL depth:

$$\tilde{\eta}(\tilde{x}) = \ln \left( \frac{\varepsilon^2 \left( \gamma^2 + \tilde{j}^2 \right)}{1 - \varepsilon^2 \xi \left( \gamma^2 + \tilde{j}^2 \right)} \right), \tag{4.246}$$

where $\tilde{j}(\tilde{x})$ is given by Equation 4.245.

## ACL Polarization Curve

Setting in Equation 4.246 $\tilde{x} = 0$, one obtains the ACL polarization curve

$$\tilde{\eta}_0 = \ln\left(\varepsilon^2\left(\gamma^2 + \tilde{j}_0^2\right)\right) - \ln\left(1 - \varepsilon^2 \xi\left(\gamma^2 + \tilde{j}_0^2\right)\right). \tag{4.247}$$

Obviously, Equation 4.247 fails to describe the overpotential at zero current, when the argument of the first logarithm tends to zero.

In an analogy to the Butler–Volmer equation, this defect can be corrected replacing the first logarithm in Equation 4.247 by arcsinh function of the halved argument. This gives

$$\tilde{\eta}_0 = \operatorname{arcsinh}\left(\frac{\varepsilon^2}{2}\left(\gamma^2 + \tilde{j}_0^2\right)\right) - \ln\left(1 - \frac{\varepsilon^2\left(\gamma^2 + \tilde{j}_0^2\right)}{\tilde{i}_{ads}\tilde{c}_{mt}}\right). \tag{4.248}$$

Here, $\gamma(\tilde{j}_0)$ is given by Equation 4.127 and was taken into account Equation 4.235. Equation 4.248 is valid down to $\tilde{j}_0 = 0$.

The first term in Equation 4.248 is independent of the inverse adsorption rate of methanol $\xi$, meaning that this term describes the activation overpotential for the electrochemical steps of MOR. Note that at large $\tilde{j}_0$, this term exhibits doubling of the apparent Tafel slope, a well-known effect of slow proton transport in the ACL. Indeed, $\gamma \to \pi$ as $\tilde{j}_0 \gg 1$; thus, $\gamma$ can be neglected and the activation term reduces to

$$\tilde{\eta}_0^{act} = 2\ln\left(\varepsilon\tilde{j}_0\right), \tag{4.249}$$

directly showing the Tafel slope doubling. At small overpotentials $\gamma \simeq \sqrt{2\tilde{j}_0}$, the term $\tilde{j}_0^2$ is small and the activation polarization reduces to a concentration-independent Tafel law

$$\tilde{\eta}_0^{act} = \operatorname{arcsinh}\left(\varepsilon^2\tilde{j}_0\right), \tag{4.250}$$

The second term in Equation 4.248 has the form of a transport logarithm representing the overpotential required to bring methanol molecule to the catalyst surface. Though methanol adsorption is independent of overpotential, the electrochemical steps of MOR require methanol to be adsorbed on the catalyst surface. The second term in Equation 4.248 describes the respective quasi-transport voltage loss.

## Half-Cell Polarization Curve and the Limiting Current Density

The methanol concentration $\tilde{c}_{mt}$ in the anode depends on the cell current density. A balance of methanol flux through the DMFC leads to the following relation between $\tilde{c}_{mt}$ and the methanol concentration in the feed channel $\tilde{c}_h$ (Kulikovsky, 2002b):

$$\tilde{c}_{mt} = (1 - \beta_*)\left(\tilde{c}_h - \frac{\tilde{j}_0}{\tilde{j}_{lim}^{ref}}\right), \tag{4.251}$$

where $\beta_*$ is the crossover parameter given by Equation 4.207, and

$$j_{lim}^{ref} = \frac{6FD_b^a c_{mt}^{ref}}{l_b^a} \qquad (4.252)$$

is the limiting current density due to the transport of the reference methanol concentration in the ABL. Here, $D_b^a$ is the methanol diffusion coefficient in the ABL of a thickness $l_b^a$ and $D_m$ is the methanol diffusion coefficient in the membrane of a thickness $l_m$.

Substitution of Equation 4.251 into Equation 4.248 gives

$$\tilde{\eta}_0 = \operatorname{arcsinh}\left(\frac{\varepsilon^2}{2}\left(\gamma^2 + \tilde{j}_0^2\right)\right) - \ln\left(1 - \frac{\varepsilon^2\left(\gamma^2 + \tilde{j}_0^2\right)}{\tilde{i}_{ads}(1 - \beta_*)\left(\tilde{c}_h - \tilde{j}_0/\tilde{j}_{lim}^{ref}\right)}\right). \qquad (4.253)$$

This is the general form of the half-cell polarization potential.

The effective anode limiting current density $\tilde{j}_{lim}$ is obtained if the expression under the logarithm in Equation 4.253 is zero:

$$\frac{\varepsilon^2\left(\gamma^2 + \tilde{j}_0^2\right)}{\tilde{i}_{ads}(1 - \beta_*)\left(\tilde{c}_h - \tilde{j}_0/\tilde{j}_{lim}^{ref}\right)} = 1. \qquad (4.254)$$

As $\gamma$ is a rather complicated function of $\tilde{j}_0$ given by Equation 4.127, in the general case, Equation 4.254 has to be solved numerically. However, in cases of small and large limiting currents, this equation can be simplified.

## Low Limiting Current: $\tilde{j}_{lim} \ll 1$

If $\tilde{j}_0 \ll 1$, then $\gamma = \sqrt{2\tilde{j}_0}$ and the term $\tilde{j}_0^2$ in Equation 4.254 can be omitted. Solving the resulting equation, one finds

$$\frac{1}{\tilde{j}_{lim}} = \frac{1}{\tilde{c}_h \tilde{j}_{lim}^{ref}} + \frac{1}{\tilde{i}_{ads}(1 - \beta_*)\tilde{c}_h/(2\varepsilon^2)}. \qquad (4.255)$$

In the dimensional variables, this equation reads

$$\frac{1}{j_{lim}} = \frac{1}{j_{lim}^{ABL}} + \frac{1}{j_{lim}^{ads,low}}, \qquad (4.256)$$

where

$$j_{\lim}^{ads,low} = \frac{l_{CL} i_{ads}(1 - \beta_*) c_h}{c_{mt}^{ref}}, \tag{4.257}$$

$$j_{\lim}^{ABL} = \frac{6 F D_b^a c_h}{l_b^a}. \tag{4.258}$$

Here, $j_{\lim}^{ads,low}$ is the low-current limiting current density due to the methanol adsorption on the catalyst surface, and $j_{\lim}^{ABL}$ is the limiting current due to the methanol transport in the ABL. Note that $j_{\lim}^{ads,low}$ is proportional to the product of the ACL thickness by the volumetric rate of methanol adsorption. In the low-current regime, the rate of MOR is uniformly distributed through the ACL, leading to this simple proportionality.

## High Limiting Current: $\tilde{j}_{\lim}^2 \gg \pi^2$

In that case, $\gamma^2$ can be omitted in Equation 4.254. Solving the resulting equation for $\tilde{j}_0$, one obtains

$$\tilde{j}_{\lim} = \sqrt{\left(\frac{\tilde{i}_{ads}(1 - \beta_*)}{2\varepsilon^2 \tilde{j}_{\lim}^{ref}}\right)^2 + \frac{\tilde{i}_{ads}(1 - \beta_*)\tilde{c}_h}{\varepsilon^2} - \frac{\tilde{i}_{ads}(1 - \beta_*)}{2\varepsilon^2 \tilde{j}_{\lim}^{ref}}}. \tag{4.259}$$

Equation 4.259 can further be simplified to

$$\tilde{j}_{\lim} = \begin{cases} \sqrt{\dfrac{\tilde{i}_{ads}(1 - \beta_*)\tilde{c}_h}{\varepsilon^2}}, & 2\varepsilon^2 \tilde{j}_{\lim}^{ref}/\tilde{i}_{ads} \gg 1 \\ \tilde{j}_{\lim}^{ref} \tilde{c}_h, & 2\varepsilon^2 \tilde{j}_{\lim}^{ref}/\tilde{i}_{ads} \ll 1 \end{cases}. \tag{4.260}$$

In dimensional variables, Equation 4.260 reads

$$j_{\lim} = \begin{cases} j_{\lim}^{ads,high}, & j_{\lim}^{ref} \gg l_{CL} i_{ads} \\ j_{\lim}^{ABL}, & j_{\lim}^{ref} \ll l_{CL} i_{ads} \end{cases}, \tag{4.261}$$

where

$$j_{\lim}^{ads,high} = \sqrt{2\sigma_p b\, i_{ads}(1 - \beta_*) c_h/c_{mt}^{ref}} \tag{4.262}$$

is the high-current limiting current density resulting from the methanol adsorption on the catalyst surface. The limiting current from methanol transport in the ABL $j_{\lim}^{ABL}$ is still given by Equation 4.258.

Note that $j_{\lim}^{ads,high}$ is independent of the ACL thickness $l_{CL}$ but it depends on the ACL proton conductivity $\sigma_p$ and on the MOR Tafel slope $b_{mt}$. In the case of high cell current, the electrochemical conversion runs near the membrane in a thin conversion domain of a thickness in the order of $\sigma_p b_{mt}/j_0$ (the section "Reaction Penetration Depth"). This introduces an internal space scale to the problem, making it independent of the ACL thickness.

## Remarks

Equations of the section "Half-Cell Polarization Curve and the Limiting Current Density" show that there are three regimes of the anode operation. In the first regime, the anode performance is limited by the rate of methanol adsorption; this regime is realized if $j_{\lim}^{ref} \gg l_{CL}i_{ads}$. In the opposite limit of $j_{\lim}^{ref} \ll l_{CL}i_{ads}$, the anode current is limited by the rate of methanol diffusive transport in the ABL. The third regime is the mixed one: when $j_{\lim}^{ref} \simeq l_{CL}i_{ads}$, both mechanisms of current limitation affect the anode polarization curve with the effective limiting current density given by Equation 4.255 in the low-current regime and by Equation 4.259 in the high-current one.

A comparison of Equations 4.262 and 4.258 suggests a means for distinguishing the mechanism of current limitation on the anode in the high-current mode. Indeed, $j_{\lim}^{ABL}$ is independent of $b_{mt}$ and $\sigma_p$. However, it depends on $l_b^a$. On the other hand, $j_{\lim}^{ads,high}$ is independent of $l_b^a$, but it depends on $b_{mt}$ and $\sigma_p$. Changing $b_{mt}$ or $\sigma_p$ is difficult, yet changing $l_b^a$ is relatively simple by taking the anode backing media of different thickness. Thus, preparing MEAs with different ABL thicknesses could help to understand the origin of current limitation.

## Model Validation

The quality of Equation 4.248 can be checked by comparing this equation to the exact numerical polarization curve, following from the direct solution of the system (4.232) and (4.233). For the sake of comparison, zero transport loss is assumed in the ABL and zero crossover, therefore in Equation 4.248, $\tilde{c}_{mt} = 1$ is set. Figure 4.30 shows that the analytical curve works well up to the limiting current density. Further numerical tests show that Equation 4.248 only fails in a small vicinity of the adsorption-limiting current (Figure 4.30).

Nordlund and Lindbergh (2004) published a set of measured polarization curves of DMFC anode. Equation 4.248 has been fitted to their data and the results are shown in Figure 4.31. In the paper of Nordlund and Lindbergh, the thickness and porosity of the ABL have not been specified and hence a reasonable assumption would be that the voltage loss due to methanol transport in the backing layer is small. For simplicity, the crossover can be neglected by setting $\beta_* = 0$. Thus, in Equation 4.248, $\tilde{c}_{mt} = \tilde{c}_h$ is set, where $c_h$ is the feeding methanol concentration.

All the experimental curves exhibit a constant shift along the $\eta$-axis. This shift has been taken into account by adding a constant $\eta_* = 0.11$ V to the dimensional form of Equation 4.248. The curves have first been fitted using the *NonlinearFit* procedure of Maple, and finally fine tuning of the fitting parameters has been performed in order to maximize a number of common parameters between the curves.

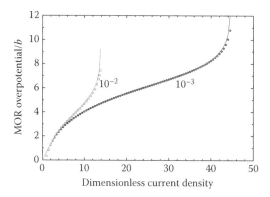

**FIGURE 4.30** Exact numerical (points) and analytical (lines, Equation 4.248) polarization curves of the anode catalyst layer for the indicated values of parameter $\xi$. In both the cases, parameter $\varepsilon = 1$.

Figure 4.31 shows the experimental and model curves for the two cell temperatures (50 and 70°C) and the indicated methanol molar concentrations. The fitting parameters are listed in Table 4.4. As can be seen, all the parameters except $i_{mt}$ and $\tilde{i}_{ads} = i_{ads}/i_{mt}$ were kept fixed, while $i_{mt}$ was taken to be constant for each temperature and $\tilde{i}_{ads}$ was varied to get a best fit for every methanol concentration. The plots of the resulting $\tilde{i}_{ads}$, as a function of the normalized methanol concentration $c_h/c_{mt}^{ref}$, are shown in Figure 4.32. These plots are linear, suggesting that the adsorption rate constant $i_{ads}$ is itself linear in the methanol concentration (Figure 4.32), that is, the rate of adsorption is quadratic in $\tilde{c}_{mt}$. The reason is unclear.

Overall, the quality of fitting is not bad, taking into account that at a fixed cell temperature, parameters for all the curves in Figure 4.32 are the same. However, for different types of catalysts, these parameters may differ, so that the predictive capability of Equation 4.248 should not be overestimated. Nonetheless, this equation can

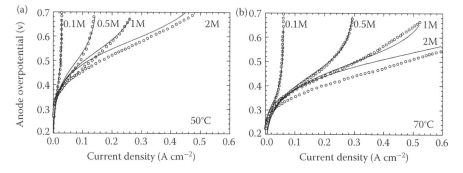

**FIGURE 4.31** Analytical (lines) and experimental polarization curves of DMFC anode measured by Nordlund and Lindbergh (2004) for the indicated methanol concentrations and cell temperature of (a) 50°C and (b) 70°C. The fitting parameters are listed in Table 4.4.

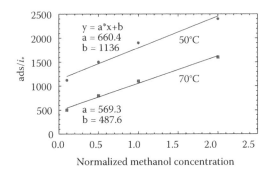

**FIGURE 4.32** Dependence of parameter $\tilde{i}_{ads}$ on the normalized methanol concentration for the two cell temperatures of 50 and 70°C. Linear fits are also displayed.

be useful for impedance calculations, or for analysis of the time variation of the anode parameters in DMFC aging studies.

Nordlund and Lindbergh (2004) suggested a simple analytical polarization curve. In the notations of this section, a kinetic part of their polarization equation (not accounting for the methanol transport in the ABL) has the form

$$j_0 = \frac{j_* \exp(\eta_0/b_{mt})}{1 + \xi \exp(\eta_0/b_{mt})}, \tag{4.263}$$

where $\eta_0 = E_* - E_c$ and $E_c$ is the electrode potential, and $E_*, j_*$, and $\xi$ are the fitting parameters. Solving Equation 4.263 for $\eta_0$, arrives at the polarization curve

$$\eta_0 = \ln\left(\frac{j_0}{j_*}\right) - \ln\left(1 - \frac{\xi j_0}{j_*}\right). \tag{4.264}$$

This equation differs from Equation 4.248 by the absence of a quadratic in $j_0$ terms. Physically, this means that Equation 4.264 ignores the proton transport loss in the ACL.

In other words, Equation 4.264 is justified, provided that the ACL proton conductivity is high. In that case, the rate of MOR is nearly uniform across the ACL and the total current produced by MOR is simply a product of the reaction rate (4.227) by the ACL thickness, eventually leading to Equations 4.263 and 4.264.

According to the model above, in experiments of Nordlund and Lindbergh, the approximation of ideal proton transport is justified for currents below 100 mA cm$^{-2}$. To show this explicitly, it is advisable to plot the model shape of MOR rate along $x$. By definition, $\tilde{R}_{MOR} = -\partial\tilde{j}/\partial\tilde{x}$. Using Equation 4.232 here, one obtains

$$\tilde{R}_{MOR} = \frac{\gamma^2 + \tilde{j}^2}{2}, \tag{4.265}$$

where $\tilde{j}(\tilde{x})$ is given by Equation 4.232 and $\gamma$ by Equation 4.127.

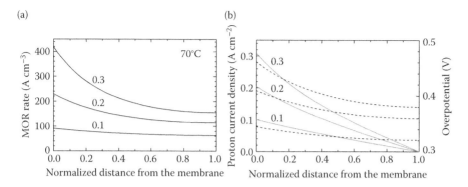

**FIGURE 4.33** The model shapes of (a) the MOR rate and (b) the proton current density (solid curves) and local MOR overpotential (dashed curves) through the catalyst layer thickness, for the indicated mean cell current density. Cell temperature is 70°C. Parameters for the calculations are taken from the curve fitting in Figure 4.31.

Figure 4.33a shows the MOR rate through the ACL thickness for the 1 M methanol concentration and the current densities of 100, 200, and 300 mA cm$^{-2}$. Figure 4.33b depicts the respective shapes of the proton current density and of the local overpotential $\eta$. Parameters for these plots result from fitting of the polarization curves corresponding to 70°C (Table 4.4).

As can be seen, even in a 14-$\mu$m ACL used by Nordlund and Lindbergh, for currents above 200 mA cm$^{-2}$, $R_{MOR}$ and overpotential are far from being uniform. Note that for the thicker layers, quite a substantial nonuniformity is achieved at lower currents. Physically, poor proton transport in the ACL forces the reaction to run faster at the membrane interface, where protons are "cheaper" (see Kulikovsky, 2010b for discussion).

## OPTIMAL CATALYST LAYER

### PRELIMINARY REMARKS

A typical catalyst layer is a composite of four ingredients: (i) voids for neutral molecule transport, (ii) ionic (proton) conductor, and (iii) electron conductor with (iv) catalyst particles hosting the electrochemical conversion. In most cases, such a composite consists of three interpenetrating random clusters of transport avenues for molecules, ions, and electrons, so that each representative volume of the CL contains the same fractions of all the three components.

So far, most of the fuel cell tests reported in the literature have been performed with macroscopically uniform electrodes. However, over the past decade, there has been a growing interest in so-called "gradient" electrodes with a nonuniform distribution of components. One of the first attempts to optimize the distribution of catalyst particles through the CL thickness has been made by the Los Alamos group. Ticianelli et al. (1988) reported that sputtering of nonsupported platinum on the electrode/membrane interface of the standard Pt/C electrode improves the cell performance. The gain in performance was higher at larger cell currents.

Modeling of two-layer (Wang and Feng, 2009) and multilayer (Feng and Wang, 2010) CCLs indicates that the CCLs with a catalyst or Nafion (electrolyte) loading, that grows toward the membrane perform better. A detailed model utilizing relations of transport parameters to the electrode structure taken from Eikerling and Kornyshev (1998), has been studied by Wang et al. (2004). They have tested several profiles of Nafion loading through the electrode thickness and found that those growing linearly toward the membrane shape provide the best cell performance. A similar result was obtained by Mukherjee and Wang (2007), who performed direct numerical simulations of a two-layer ORR electrode with varying Nafion content in the sublayers.

All these works tested trial shapes of Nafion content. However, to the best of our knowledge, none of the existing models were used for *optimization* of these shapes. Optimization of the catalyst loading profile, based on an MHM model has been carried out in Kulikovsky (2009c). The results have shown that a 10-fold increase of catalyst loading near the membrane, with a respective 10-fold decrease of this loading at the GDL interface, dramatically elevates the cell current. This optimization has been performed assuming a uniform profile of Nafion content in the CCL. In this section, optimization of both the proton conductivity and catalyst loading shapes is discussed (Kulikovsky, 2012f).

## Model

In this section, the fast transport of feed molecules (oxygen) in the CL is assumed. The respective condition has been discussed in the section "When can the Oxygen Transport Loss Be Ignored." Another essential model assumption is that the liquid water saturation $s$ is constant through the CL. In that case, $s$ simply rescales the CL transport parameters and the exchange current density. These changes can be easily accounted for in the resulting equations. Below, it will be shown that the optimization is effective mainly at large cell currents. Thus, the reverse exponent in the Butler–Volmer equation can be neglected and the Tafel law for the ORR rate can be used.

The system of equations for the optimization of catalyst and ionomor shapes through the CL depth follows from Equations 4.53 and 4.54:

$$2\varepsilon^2 \frac{\partial \tilde{j}}{\partial \tilde{x}} = -g(\tilde{x})\exp(\tilde{\eta}), \qquad (4.266)$$

$$\tilde{j} = -p^2(\tilde{x})\frac{\partial \tilde{\eta}}{\partial \tilde{x}}. \qquad (4.267)$$

Here, $\varepsilon$ is given by Equation 4.56, where the exchange current density $i_*$, and the proton conductivity $\sigma_p$ corresponds to the reference uniform catalyst and Nafion loadings. Further, $g(\tilde{x})$ is the profile function describing the relative variation of the exchange current density along $\tilde{x}$, and $p(\tilde{x})$ as the profile function for ionomer content. In this section, the variable $p(\tilde{x})$ represents the ionomer content in the catalyst layer, for which the parameter $X_{el}$ had been used previously.

Following Eikerling and Kornyshev (1998) and Wang et al. (2004), the CL proton conductivity is taken to be proportional to the square of the ionomer content; therefore, Equation 4.267 contains the factor $p^2$ on the right-hand side. It can be assumed that the exchange current density is proportional to the catalyst surface, so that $g(\tilde{x})$ represents the normalized shape of the catalyst loading through the electrode depth. Thus, the functions $g$ and $p$ must obey the normalization condition

$$\int_0^1 g\,d\tilde{x} = 1, \quad \int_0^1 p\,d\tilde{x} = 1. \tag{4.268}$$

These equations mean that for any shapes of catalyst and ionomer loadings, the total amount of the catalyst and proton conductor remains the same. It is also assumed that in all cases, the ionomer content is above the percolation threshold.

Multiplying Equations 4.266 and 4.267 together gives

$$2\varepsilon^2 \tilde{j}\frac{\partial \tilde{j}}{\partial \tilde{x}} = p^2 g \exp(\tilde{\eta})\frac{\partial \tilde{\eta}}{\partial \tilde{x}}. \tag{4.269}$$

It is convenient to consider the product $p^2 g$ as a function of $\tilde{\eta}$:

$$p^2 g \equiv \phi(\tilde{\eta}). \tag{4.270}$$

With this, Equation 4.269 can be written in the form

$$\varepsilon^2 \frac{\partial (\tilde{j}^2)}{\partial \tilde{x}} = \phi(\tilde{\eta}) \exp(\tilde{\eta})\frac{\partial \tilde{\eta}}{\partial \tilde{x}}. \tag{4.271}$$

Integrating this equation over $\tilde{x}$ from 0 to 1, and taking into account that $\tilde{j}(1) = 0$, yields

$$\varepsilon^2 \tilde{j}_0^2 = -\int_0^1 \phi(\tilde{\eta}) \exp(\tilde{\eta})\frac{\partial \tilde{\eta}}{\partial \tilde{x}}\,d\tilde{x} = \int_{\tilde{\eta}_1}^{\tilde{\eta}_0} \phi(\tilde{\eta}) \exp(\tilde{\eta})\,d\tilde{\eta}. \tag{4.272}$$

Thus, the problem of maximizing $\tilde{j}_0$ is reduced to finding a function $\phi(\tilde{\eta})$ that maximizes the integral on the right-hand side of Equation 4.272. Direct application of the Euler–Lagrange method to this integral leads to the equation $\exp(\tilde{\eta}) = 0$, which has no solution for $\tilde{\eta} > 0$. This means that the maximum of the integral in Equation 4.272 does not exist. This is a trivial result as the integral contains a growing function $\exp \tilde{\eta}$. However, the function $\phi$ is subject to constrains (4.268), which are not accounted for by the Euler–Lagrange equation.

The problem can be solved using the following arguments. Assuming that $\phi$ has derivatives of all orders, repeated integration by parts gives (Kulikovsky, 2009c)

$$\int \phi(\tilde{\eta}) \exp(\tilde{\eta})\,d\tilde{\eta} = \Phi(\tilde{\eta}) \exp \tilde{\eta}, \tag{4.273}$$

where

$$\Phi \equiv \phi - \frac{\partial \phi}{\partial \tilde{\eta}} + \frac{\partial^2 \phi}{\partial \tilde{\eta}^2} - \dots \tag{4.274}$$

Using Equation 4.273 in Equation 4.272 yields

$$\varepsilon^2 \tilde{j}_0^2 = \Phi(\tilde{\eta}_0) \exp \tilde{\eta}_0 - \Phi(\tilde{\eta}_1) \exp \tilde{\eta}_1. \tag{4.275}$$

From the structure of Equation 4.275, it follows that any nondecreasing function $\Phi(\tilde{\eta})$ improves the CL performance.* However, growing functions $\Phi(\eta)$ do not describe the limiting case of uniform catalyst and electrolyte loading ($g = p = 1$). Indeed, $g = p = 1$ gives $\phi = 1$, and from Equation 4.274 it follows that for all $\tilde{\eta}$, one must have $\Phi = 1$.

Thus, the only remaining option for optimal $\phi$ is

$$\Phi = \alpha, \tag{4.276}$$

where $\alpha$ is a constant determined by one of Equations 4.268. With $\alpha = 1$, this equation provides a correct limiting case of uniform loadings. Further, with $\Phi(\tilde{\eta}) = \alpha > 1$, Equation 4.275 transforms to

$$\varepsilon^2 \tilde{j}_0^2 = \alpha \left( \exp \tilde{\eta}_0 - \exp \tilde{\eta}_1 \right).$$

Neglecting for the estimate the second exponent, one arrives at

$$\varepsilon^2 \tilde{j}_0^2 = \exp(\tilde{\eta}_0 + \ln \alpha).$$

That is, for a fixed $\tilde{j}_0$, optimal $g$ and/or $p$ lower the total potential loss by $\ln \alpha$.

Neglecting the third- and higher-order derivatives in Equation 4.274 gives[†]

$$\phi - \frac{\partial \phi}{\partial \tilde{\eta}} + \frac{\partial^2 \phi}{\partial \tilde{\eta}^2} = \alpha, \quad \phi(\tilde{\eta}_1) = \phi_1, \quad \left. \frac{\partial \phi}{\partial \tilde{\eta}} \right|_{\tilde{\eta}_1} = 0. \tag{4.277}$$

---

* To illustrate this, consider $\Phi(\tilde{\eta}) = \alpha \exp \tilde{\eta}$, where $\alpha$ is constant. Substituting this into Equation 4.275, yields

$$\varepsilon^2 \tilde{j}_0^2 = \alpha(\exp(2\tilde{\eta}_0) - \exp(2\tilde{\eta}_1)).$$

At large currents, the second exponent can be neglected leading to

$$\varepsilon^2 \tilde{j}_0^2 = \exp(2\tilde{\eta}_0 + \ln \alpha).$$

Therefore, this choice of $\Phi(\tilde{\eta})$ halves the polarization potential and, in addition, lowers it by $\ln \alpha$.

[†] With Equation 4.277, the third- and higher-order derivatives in Equation 4.274 vanish identically in groups of three. Indeed, denoting the left-hand side of Equation 4.277 by $\psi$, Equation 4.274 can be written in terms of $\psi$:

$$\Phi = \psi - \frac{\partial^3 \psi}{\partial \tilde{\eta}^3} + \frac{\partial^6 \psi}{\partial \tilde{\eta}^6} - \dots$$

Obviously, with $\psi = \alpha$, all the derivatives in this equation vanish, so that $\Phi = \psi = \alpha$. The same is true if an arbitrary number of terms is retained on the left-hand side of Equation 4.277. However, numerical calculations show that taking into account higher-order terms in Equation 4.277 only leads to a marginal change in the results.

The first of the boundary conditions in Equation 4.277 means that the catalyst and electrolyte loadings at $\tilde{x} = 1$ are fixed. Zero derivative of $\phi$ at $\tilde{\eta}_1$ is required in accordance with the zero derivative of $\tilde{\eta}$ (zero proton current) at $\tilde{x} = 1$.

The solution to Equation 4.277 gives optimal $\phi$ as a function of $\tilde{\eta}$:

$$\phi(\tilde{\eta}) = \alpha \left[ 1 + \left( 1 - \frac{\phi_1}{\alpha} \right) \exp \left( \frac{\tilde{\eta} - \tilde{\eta}_1}{2} \right) \left( \frac{1}{\sqrt{3}} \sin \left( \frac{\sqrt{3}}{2}(\tilde{\eta} - \tilde{\eta}_1) \right) \right. \right.$$

$$\left. \left. - \cos \left( \frac{\sqrt{3}}{2}(\tilde{\eta} - \tilde{\eta}_1) \right) \right) \right].$$
(4.278)

The constant $\alpha$ is obtained from one of Equations 4.268. However, to convert $\phi(\tilde{\eta})$ into $\phi(\tilde{x})$, the shape of $\tilde{\eta}(\tilde{x})$ is needed. This shape obeys an equation that follows from substitution of Equation 4.267 into 4.266[*]

$$2\varepsilon^2 p \frac{\partial^2 \tilde{\eta}}{\partial \tilde{x}^2} + 4\varepsilon^2 p \frac{\partial p}{\partial \tilde{\eta}} \left( \frac{\partial \tilde{\eta}}{\partial \tilde{x}} \right)^2 = g \exp(\tilde{\eta}), \quad \tilde{\eta}(0) = \tilde{\eta}_0, \quad \left. \frac{\partial \tilde{\eta}}{\partial \tilde{x}} \right|_1 = 0. \quad (4.279)$$

Note that in Equation 4.279, the $\tilde{x}$ shape of either $p$ or $g$ must be prescribed. The other variable can be expressed through the optimal $\phi$ by Equation 4.270. It is advisable to consider the cases of either uniform $g$ or $p$ first.

Let $g(\tilde{x}) = 1$. To find optimal $p(\tilde{x})$, the following procedure has to be executed:

- Fix the potential loss $\tilde{\eta}_0$ and guess the initial value of $\alpha > 1$.
- Solve Equation 4.279 numerically with $p = \phi$, where $\phi$ is given by Equation 4.278. Calculate the second integral Equation 4.268.
- Change $\alpha$ and repeat the previous step until the second of Equation 4.268 is satisfied.

If optimal $g$ is searched for, one has to set $p = 1$, $g = \phi$ and repeat the procedure above until the first integral in 4.268 equals 1.

## OPTIMAL LOADINGS

The physical parameters and the resulting dimensionless parameters required for calculations are listed in Table 4.5. In all cases, optimization is performed with the boundary conditions $g(1) = 0.1$ for catalyst loading and $p(1) = 0.5$ for Nafion content. This means that at the CL/GDL interface (Figure 4.2), the optimal $g$ is assumed to be small, while the optimal $p$ is taken to be half of the reference uniform value. This assumptions seems reasonable, as at high currents, the peak of electrochemical conversion resides at the membrane/CL interface. An optimal performance is, therefore, expected under the catalyst loading, and/or proton conductivity growing toward

---

[*] Alternatively, the system (4.266) and (4.267) can be solved as a Cauchy problem, with initial conditions at $\tilde{x} = 1$: $\tilde{j}(1) = 0$ and $\tilde{\eta}(1) = \tilde{\eta}_1$. This requires iterations to find $\tilde{\eta}_1$, providing the given total potential loss $\tilde{\eta}_0$.

**TABLE 4.5**

**Characteristic Physical and Operational Parameters for the PEFC Cathode and DMFC Anode**

| Parameter | PEFC Cathode | DMFC Anode |
|---|---|---|
| Tafel slope $b$ (V) | 0.05 | 0.05 |
| Proton conductivity $\sigma_p^0$ ($\Omega$ cm$^{-1}$) | 0.03 | 0.003 |
| Exchange current density $i_*^0$ (A cm$^{-3}$) | $10^{-3}$ | 1.0 |
| CL thickness $l_{CL}$ (cm) | $10^{-3}$ | 0.01 |
| Working current density $j_0$ (A cm$^{-2}$) | 1.0 | 0.1 |
| $\varepsilon$ | 866 | 0.866 |
| $\tilde{j}_0$ | 0.667 | 6.67 |

this interface. Note that the minimal Nafion content is limited by the percolation threshold. Below, the relative content of 0.5 is assumed to be above the threshold.

Calculations show that solution to the optimization problem strongly depends on the *dimensionless* working current density $\tilde{j}_0$. This parameter varies in a wide range between different types of cells. A discussion of solutions relevant to the CCL in a PEFC and to the ACL in a DMFC is given below. The respective typical $\tilde{j}_0$'s differ by an order of magnitude (0.6667 for PEFC and 6.67 for DMFC, Table 4.5). This leads to dramatic differences in the optimization results. The results for optimal and uniform shapes are compared at the same value of the total potential loss $\tilde{\eta}_0$, that is, optimization raises the cell current density $\tilde{j}_0$.

### PEFC Cathode

Optimization of proton conductivity at uniform catalyst loading prescribes doubling of Nafion content at the membrane interface (Figure 4.34a). However, the gain in the cell current appears to be minimal (Figure 4.34b).

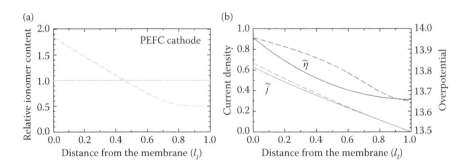

**FIGURE 4.34** (a) The uniform (solid line) and optimal (dashed line) relative ionomer content in the cathode catalyst layer (CCL) of a PEFC. (b) The respective distributions of overpotential (two upper curves) and proton current density (lower curves) through the CCL.

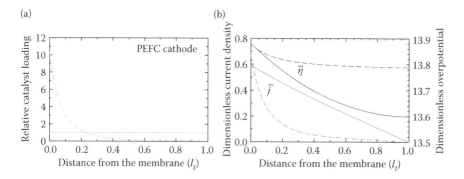

**FIGURE 4.35** (a) The uniform (solid line) and optimal (dashed line) catalyst loading in the CCL of a PEFC. (b) The respective distributions of overpotential (two upper curves) and proton current density (lower curves) through the CCL.

Optimal catalyst loading, at uniform Nafion content, is depicted in Figure 4.35. As can be seen, optimization requests to increase catalyst loading at the membrane by a factor of 10 as compared to the reference uniform value (Figure 4.35a). However, the effect of this strongly nonuniform catalyst distribution is not large, the cell current density increases by 10% only (Figure 4.35b).

## DMFC Anode

Optimization of the ACL in DMFC gives much more impressive results. The optimal shape of the Nafion content (Figure 4.36a) provides a 30% growth of the cell current (Figure 4.36b). Even better results give the optimal distribution of the catalyst loading. The optimal shape $g(\tilde{x})$ (Figure 4.37a) is similar to that of the PEFC cathode (cf. Figures 4.35a and 4.37a). However, in the DMFC anode, optimal $g(\tilde{x})$ nearly doubles the cell current (Figure 4.37b).

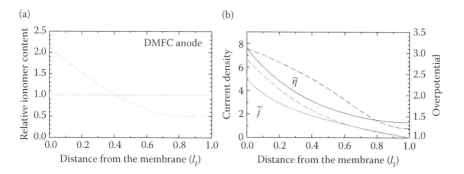

**FIGURE 4.36** The same diagrams, as in Figure 4.34, but for the DMFC anode. (a) The uniform (solid line) and optimal (dashed line) catalyst loading in the ACL of a DMFC. (b) The respective distributions of overpotential (two upper curves) and proton current density (lower curves) through the ACL.

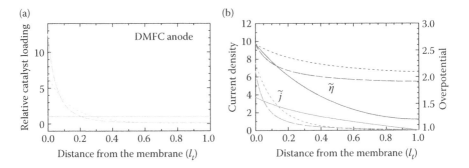

**FIGURE 4.37** The same diagrams as in Figure 4.35 but for DMFC anode. In addition, the dotted line in (a) shows the optimal shape of catalyst loading in the CCL, with the electrolyte loading in Figure 4.36a. Dotted lines in (b) show the respective shapes of overpotential and local proton current.

A comparison of the optimal shapes of electrolyte content for PEFC and DMFC (Figures 4.34a and 4.36a) shows that these shapes are very similar. Thus, it is advisable to optimize catalyst loading by taking the optimal shape of electrolyte loading in Figure 4.36a. This procedure can be considered as a first iteration step in the process of simultaneous optimization of electrolyte and catalyst loadings.

The resulting optimal distribution of the catalyst loading and the shapes of the local current and overpotential are shown in Figure 4.37 by dotted lines. As can be seen, this optimization gives an additional 10% gain in the cell current (Figure 4.37b). Moreover, the optimal shape of Nafion content strongly lowers the gradient of optimal catalyst loading, changing it from the exponential-like to the parabolic-like (Figure 4.37a). At a uniform electrolyte content, the peak value of the normalized catalyst loading at the membrane interface is about 13, while with the optimal shape of the electrolyte content, this peak value is about 7 (Figure 4.37a). This suggests the necessity of simultaneous optimization of electrolyte and catalyst loadings.

Calculations show that the effect of CL performance optimization by the gradient catalyst loading strongly depends on $\tilde{j}_0$. To characterize the effect, the optimization factor $k_{opt}$ is introduced, defined as the ratio of cell currents at optimal and uniform loadings

$$
k_{opt} = \left. \frac{\tilde{j}_0^{opt}}{\tilde{j}_0} \right|_{\tilde{\eta}_0 = \text{const}} .
\tag{4.280}
$$

The function $k_{opt}(\tilde{j}_0)$ for an optimal catalyst loading at the uniform Nafion content is shown in Figure 4.38. As can be seen, the effect of catalyst redistribution is marginal at small $\tilde{j}_0$, and it grows rapidly with the current density. Thus, optimization of the catalyst loading is ultimately effective in cells operated at a large *dimensionless* current. Equation 4.51 shows that $\tilde{j}_0$ is large ($\tilde{j}_0 \gg 1$) if the catalyst layer thickness is large, or the CL proton conductivity is low.

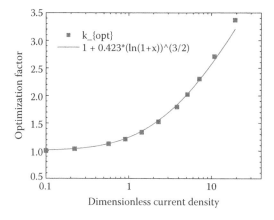

**FIGURE 4.38** The points: the optimization factor (4.280) as a function of the dimensionless cell current density $\tilde{j}_0$. The solid line: the fitting function, indicated in the legend.

Physically, this result is quite clear: if the cell current is low, the reaction rate is distributed uniformly over the CL thickness, and the spatial optimization of catalyst/Nafion loading is not necessary. If, however, the dimensionless current is substantial, the reaction runs mainly in a small conversion domain close to the membrane (the section "Ideal Proton Transport"). Placing more catalyst and Nafion into this domain, greatly improves the performance.

The shapes in Figure 4.37a may suggest replacing the CL by a much thinner one, with 10 times higher catalyst loading properties. Leaving aside practical aspects of such a design, it should be noted that a thin CL would lead to a higher potential loss. The problem is that the proton current at the CL/GDL interface must be zero. In a thin CL with high catalyst loading, this boundary condition would raise $\tilde{\eta}_0$, as compared to the optimal layer shown in Figure 4.37a.

The model above ignores the relation of Nafion and catalyst loadings to the catalyst layer structure. In particular, it does not take into account that in standard CLs, higher Nafion loading usually means lower catalyst content and vice versa. An account

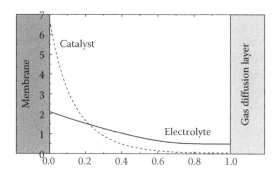

**FIGURE 4.39** Optimal catalyst and electrolyte loadings for DMFC anode.

of these effects requires incorporation of percolation-type relations into the model, similar to that discussed in Wang et al. (2004) and Mukherjee and Wang (2007).

Nonetheless, the simple model above clearly indicates the trends: it prescribes that both Nafion and catalyst loadings should grow toward the membrane surface. The optimal Nafion content is almost linear, while the optimal catalyst loading is parabolic-like (Figure 4.39). Calculations show that even non-optimal shapes of similar types improve the CL performance, that is, real shapes could always be chosen in accordance with structural limitations.

## HEAT FLUX FROM THE CATALYST LAYER

### INTRODUCTORY REMARKS

Most of the transport and kinetic processes in a fuel cell strongly depend on temperature. Thus, calculation of the temperature field in cells is of great importance. In PEFCs, the largest amount of heat is produced in the cathode catalyst layer. This section will consider the heat problem in the CCL.

The vast majority of CCL models have been isothermal and only few works have been devoted to the heat transport in CCLs. The reason is that temperature variation in the CCL is usually small (see below). However, the CCL is thin and the heat flux from the CCL (temperature gradient) is not small. This flux is obtained from the solution of the heat transport problem, coupled to the problem for proton current $j$ and reaction rate $R_{reac}$ distributions across the layer (performance problem).

In this section, the assumption is made that the rate of oxygen transport through the CCL is large and, in addition, the square of the cell current density is much larger than $j_\sigma^2$ (Equation 4.116). In that case, the electrochemical conversion typically occurs in one of the two regimes discussed in the section "Ideal Oxygen Transport."

In the *low-current* (LC) regime $\tilde{j}_0 \ll 1$, the rate of conversion is distributed uniformly along $x$ as the proton current linearly decays with distance. In the *high-current* (HC) regime $\tilde{j}_0 \gg 1$, the conversion runs predominantly in a small conversion domain close to the membrane interface. This is where the proton current and the rate of conversion rapidly decay (Figure 4.16). This situation is a characteristic of the CCL with insufficient proton conductivity. The reaction rate peaks at the membrane interface due to the lower expenditure of proton transport there. Importantly, in the HC regime, the polarization potential is much larger than in the LC regime (Figure 4.18).

An energy conservation equation is a necessary part of numerous CFD models of cells and stacks (Wang, 2004; Weber and Newman, 2004a). Most of CFD models consider the catalyst layer as a thin interface generating heat and current. The usual expression for the heat flux $q$ emitted by the CCL is

$$q = \left( \frac{T|\Delta S_{ORR}|}{nF} + \eta_0 \right) j_0 + \frac{j_0^2 l_{CL}}{\sigma_p}. \qquad (4.281)$$

Here, $T$ is the temperature and $|\Delta S_{ORR}|$ is the entropy change in the half-cell reaction.

The first term, on the right-hand side of Equation 4.281, describes the thermodynamic (the term with $|\Delta S_{ORR}|$) and irreversible (the term with $\eta_0$) heat generated in the ORR. The last term in Equation 4.281 represents the Joule heat produced by the proton current in the CL.* Equation 4.281 implicitly assumes that the proton current is uniform across the CL. This is a rather crude approximation: even in the LC regime, $j$ linearly decays along $x$, and the Joule term in Equation 4.281 is overestimated by a factor of 3. The goal of this section is to derive the correct equation for the heat flux from the catalyst layer working in the LC, HC, or intermediate regime (Kulikovsky and McIntyre, 2011).

## Basic Equation

A general equation for the heat transport in the solid (carbon) phase of the CCL is

$$-\lambda_T \frac{\partial^2 T}{\partial x^2} = \left( \frac{T|\Delta S_{ORR}|}{nF} + \eta \right) R_{reac} + \frac{j^2}{\sigma_p}. \tag{4.282}$$

Here, $\lambda_T$ is the CCL thermal conductivity and $R_{reac}$ is the volumetric rate of electrochemical conversion ($A\ cm^{-3}$). Equation 4.282 says that the variation of conductive heat flux (the left-hand side) equals the sum of the heating rates from the reaction and the Joule dissipation of electric energy. On the other hand, $R_{reac}$ determines the rate of proton current decay along $x$:

$$\frac{\partial j}{\partial x} = -R_{reac}. \tag{4.283}$$

The right-hand side of Equation 4.282 includes the dominant sources of heat in catalyst layers of low-temperature fuel cells. Note that part of the heat flux from the CCL is transported with liquid water produced in the ORR. Being not represented explicitly, this flux is taken into account in the equations of this section (see below).

Generally, evaporation of liquid water is an important mechanism of fuel cell cooling and the evaporation term should be added to the right-hand side of Equation 4.282. For simplicity, in this section, evaporation will be ignored, which is equivalent to the assumption that the water vapor pressure in the CCL is equal to the saturated pressure. Thus, solution to Equation 4.282 gives a maximal heat flux from the CCL, which is of large interest for cell and stack modeling.

The exact solution to Equation 4.282 is rather cumbersome (Kulikovsky, 2007); however, this equation can be simplified. The temperature variation across the CCL is not large; thus, $T(x)$ on the right-hand side may safely be replaced by the temperature at the CCL/GDL interface $T_1 \equiv T(l_{CL})$.

In the LC regime, the variation of $\eta$ with $x$ is small and $\eta(x) \simeq \eta_0$ may be set, where $\eta_0$ is the overpotential at the membrane interface. In the HC regime, reaction runs close to the membrane and thus only a small domain close to the membrane contributes to heat production. In this domain, $\eta \simeq \eta_0$. Thus, in all the cases, $\eta(x)$ may be replaced by $\eta_0$ in Equation 4.282. Taking into account these remarks and

---

* The CCL electron conductivity is much larger than the proton conductivity and, hence, the electron Joule heat can be neglected.

Equation 4.283, one arrives at

$$-\lambda_T \frac{\partial^2 T}{\partial x^2} = -\left(\frac{T_1|\Delta S_{ORR}|}{nF} + \eta_0\right)\frac{\partial j}{\partial x} + \frac{j^2}{\sigma_p}. \tag{4.284}$$

The boundary conditions for Equation 4.284 are

$$\left.\frac{\partial T}{\partial x}\right|_{x=0} = 0, \quad T(l_{CL}) = T_1. \tag{4.285}$$

The first condition means that heat flux in the membrane is zero. The second condition fixes the temperature at the CCL/GDL interface. It should be emphasized that although Equations 4.285 determine the temperature shape in the CL, they do not affect the equation for the *total* heat flux $q_{tot}$, leaving the CL. Physically, $q_{tot}$ is determined only by the rate of heat energy production inside the CCL, and it does not depend on boundary conditions. The adiabatic boundary condition at $x = 0$ simply means that $q_{tot}$ leaves the CCL at the CCL/GDL interface.

The general solutions to the problem (4.284), depend on the shape of proton current density $j(x)$. In the LC and HC regimes, the explicit expressions for $j(x)$ are Equations 4.124 and 4.119, respectively. These shapes allow solving the heat transport problem (4.284) and (4.285). Since the temperature variation across the CCL is small, this variation does not affect $R_{reac}$ and the thermal and performance problems can be decoupled.

## Low-Current Regime

In the LC regime, the proton current linearly decays with $x$. In the dimensional form, Equation 4.124 reads

$$j(x) = j_0\left(1 - \frac{x}{l_{CL}}\right). \tag{4.286}$$

With this relation, Equation 4.284 transforms to

$$-\lambda_T \frac{\partial^2 T}{\partial x^2} = \left(\frac{T_1|\Delta S_{ORR}|}{nF} + \eta_0\right)\frac{j_0}{l_{CL}} + \frac{j_0^2}{\sigma_p}\left(1 - \frac{x}{l_{CL}}\right)^2. \tag{4.287}$$

Integrating this equation from 0 to $x$ and taking into account the first of Equation 4.285, we obtain

$$-\lambda_T \frac{\partial T}{\partial x} = \left(\frac{T_1|\Delta S_{ORR}|}{nF} + \eta_0\right)\frac{j_0 x}{l_{CL}} + \frac{j_0^2 l_{CL}}{3\sigma_p}\left[1 - \left(1 - \frac{x}{l_{CL}}\right)^3\right]. \tag{4.288}$$

Setting $x = l_{CL}$ here gives the total heat flux leaving the CL:

$$q_{tot}^{low} = \left(\frac{T_1|\Delta S_{ORR}|}{nF} + \eta_0\right)j_0 + \frac{j_0^2 l_{CL}}{3\sigma_p}. \tag{4.289}$$

It is advisable to rewrite this equation in the form

$$q_{tot}^{low} = \left( \frac{T_1 |\Delta S_{ORR}|}{nF} + \eta_0 + \eta_{ion} \right) j_0, \qquad (4.290)$$

where

$$\eta_{ion} = j_0 R_{ion}$$

is *voltage loss due to the proton transport*, and $R_{ion} = l_{CL}/(3\sigma_p)$ is the CCL proton resistivity. Physically, overpotentials $\eta_0$ and $\eta_{ion}$ can be treated as potential drops on activation (charge transfer) and ionic resistivities connected in series. In the section "Physical Modeling of Catalyst Layer Impedance" in Chapter 5, solution to the impedance spectroscopy problem confirms this interpretation.

## HIGH-CURRENT REGIME

In the HC regime, the electrochemical conversion occurs in a narrow conversion domain at the membrane interface (Figure 4.16b). The characteristic thickness of this domain is given by $l_\sigma = \sigma_p b/j_0$ (Equation 4.154). In this domain, the square of proton current density can be approximated by the exponent

$$j^2 \simeq j_0^2 \exp \left( -\frac{x}{l_\sigma} \right). \qquad (4.291)$$

Note that Equation 4.291 approximates $j^2$ only close to the membrane. However, this is exactly what is needed for the heat transport problem: outside the conversion domain, the proton current and the reaction rate are small. This region, therefore, is of no interest as it does not contribute to heat production.

Using Equation 4.291 in Equation 4.284 yields an equation for the heat balance:

$$-\lambda_T \frac{\partial^2 T}{\partial x^2} = -\left( \frac{T_1 |\Delta S_{ORR}|}{nF} + \eta_0 \right) \frac{\partial j}{\partial x} + \frac{j_0^2}{\sigma_p} \exp \left( -\frac{x}{l_\sigma} \right). \qquad (4.292)$$

Integrating this equation from 0 to $x$ and taking into account Equation 4.285 gives

$$-\lambda_T \frac{\partial T}{\partial x} = \left( \frac{T_1 |\Delta S_{ORR}|}{nF} + \eta_0 \right) (j_0 - j) + \frac{j_0^2 l_\sigma}{\sigma_p} \left( 1 - \exp \left( -\frac{x}{l_\sigma} \right) \right). \qquad (4.293)$$

Setting $x = l_{CL}$ and taking into account that $j(l_{CL}) = 0$ results in the heat flux

$$q_{tot}^{high} = \left( \frac{T_1 |\Delta S_{ORR}|}{nF} + \eta_0 \right) j_0 + \frac{j_0^2 l_\sigma}{\sigma_p} \left( 1 - \exp \left( -\frac{l_{CL}}{l_\sigma} \right) \right). \qquad (4.294)$$

In the HC regime, $l_{CL}/l_\sigma \gg 1$ and the exponent $\exp(-l_{CL}/l_\sigma)$ is vanishingly small. Neglecting this exponent, one finds heat flux leaving the CCL

$$q_{tot}^{high} \simeq \left( \frac{T_1|\Delta S_{ORR}|}{nF} + \eta_0 \right) j_0 + \frac{j_0^2 l_\sigma}{\sigma_p}. \tag{4.295}$$

Rearranging terms and taking into account Equation 4.154, we finally arrive at

$$q_{tot}^{high} = \left( \frac{T_1|\Delta S_{ORR}|}{nF} + \eta_0 + b \right) j_0. \tag{4.296}$$

In the HC regime, Joule heating manifests itself as an additive constant $b$ in the sum of overpotentials in Equation 4.296. In low-temperature fuel cells, $b$ is in the order of 50–100 mV (Neyerlin et al., 2006), whereas $\eta_0$ is typically in the order of 300–500 mV. The Peltier potential $T_1|\Delta S_{ORR}|/(nF)$ is in the order of 300 mV and, hence, the Joule heating contributions in the HC regime, are in the range of 10–20% of the overall heat flux.

### GENERAL EQUATION FOR THE HEAT FLUX

Substituting Equation 4.154 into Equation 4.294 and rearranging the terms yields

$$q_{tot}^{high} = \left( \frac{T_1|\Delta S_{ORR}|}{nF} + \eta_0 + b \left( 1 - \exp\left( -\frac{j_0 l_{CL}}{\sigma_p b} \right) \right) \right) j_0. \tag{4.297}$$

Equation 4.297 is the exact expression for the HC heat flux from the CL. As discussed above, this expression is simplified to Equation 4.296 since in the HC regime, the parameter $j_0 l_{CL}/(\sigma_p b)$ is large, therefore, the exponent in Equation 4.297 can be neglected.

Equation 4.297 suggests the general expression for the heat flux from the CL, which is valid in the whole range of cell currents:

$$q_{tot} = \left( \frac{T_1|\Delta S_{ORR}|}{nF} + \eta_0 + b \left( 1 - \exp\left( -\frac{j_0 l_{CL}}{3\sigma_p b} \right) \right) \right) j_0. \tag{4.298}$$

Equation 4.298 differs from Equation 4.297 by the factor 3 in the power of exponent.

In the LC regime, the exponent in Equation 4.298 can be expanded in a series. Retaining the linear term, gives the LC equation (4.288). In the HC limit, the exponent in Equation 4.298 can be neglected, which leads to Equation 4.296. Thus, in the LC and HC limits, Equation 4.298 reduces to the respective exact expression and hence this equation interpolates between the limiting solutions.

With the general relation for the activation overpotential in the oxygen-transport-free regime (4.142), an explicit dependence of the heat flux on the cell current density

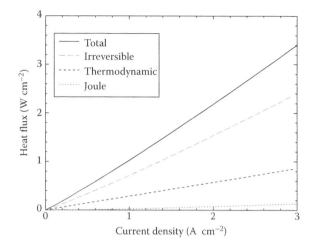

**FIGURE 4.40** Components and the total heat flux from the CCL of a PEFC. $|\Delta S_{ORR}| = 326.36$ J K$^{-1}$ (Lampinen and Fomino, 1993) and the other parameters are listed in Table 5.7.

is obtained:

$$q_{tot} = \left[ \frac{T_1 |\Delta S_{ORR}|}{nF} + b \ln \left( \frac{(j_0/j_\sigma)^2}{1 - \exp(-j_0/(2j_{ref}))} \right) \right.$$
$$\left. + b \left( 1 - \exp \left( -\frac{j_0}{3j_{ref}} \right) \right) \right] j_0. \tag{4.299}$$

Figure 4.40 shows the total heat flux from the CCL of a PEFC and its components. As illustrated, for a typical current density of 1 A cm$^{-2}$, the heat flux is about 1 W cm$^{-2}$, which is close to the useful electric power density generated by the fuel cell at this current density. Further, the dominating contributions give the thermodynamic (the first term in the square brackets in Equation 4.299) and the irreversible (the second term) heating. The Joule term contribution is marginal.

It is worth noting, however, that the curves in Figure 4.40 correspond to the well-humidified CCL. In the case of membrane phase drying, the proton conductivity $\sigma_p$ dramatically drops, and the contribution of the Joule term appears to be much higher.

### REMARKS

In PEFCs, part of the heat is transported with the liquid water produced in the ORR. Physically, ORR generates water molecules in a liquid state and part of the reaction heat is spent for warming up these molecules to working temperature $T_1$.

Although Equation 4.282 does not contain the convective term, the right-hand side of this equation contains all the sources of heat and, hence, "liquid" heat flux is included in the conductive heat flux, on the left-hand side of Equation 4.298. The heat flux transported, with the liquid water produced in the ORR, is

$q_w = M_w c_{pw} T_1 j_0 / (2F)$, where $M_w$ is the molecular weight and $c_{pw}$ is the heat capacity of liquid water. The data from Table 1.3 yield $q_w \simeq 0.134$ W cm$^{-2}$. Comparison of this flux to the total heat flux of $\simeq 1$ W cm$^{-2}$ from the CCL operated at 1 A cm$^{-2}$ shows that the liquid water transports less than 20% of the total heat produced in the CL.*

---

* To be precise, the convective term for the transport of liquid water should be added to the left-hand side of Equation 4.282. This would lower the conductive heat flux by the value $q_w$, and the same value would appear as the convective heat flux. Note that the total heat flux of Equation 4.298 would not change, since this flux is determined by the rate of heat production in the CL only.

# 5 Applications

## INTRODUCTION TO APPLICATIONS CHAPTER

The previous chapters have elucidated detailed relations between structure and properties of essential fuel cell materials. Approaches have been presented that integrate the relevant scales in PEM and CLs. Detailed structural models lead to physical models of properties and function.

Water is the active medium of PEFCs. From a chemical point of view, water is the main product of the fuel cell reaction. It is the only product in hydrogen fuel cells. Cells supplied with direct methanol or ethanol as the fuel produce water and carbon dioxide in stoichiometric amounts. The low operating temperature implies that water is present in liquid form. It mediates direct electrostatic as well as "colloidal" interactions in solutions or ink mixtures containing ionomeric, electronic, and electrocatalytic materials. These interactions control phase separation and structural relaxation phenomena that lead to the formation of PEMs and CLs. Variations in water content and distribution, thus, lead to transformations in stable structures of these media, which incur modifications in their physicochemical properties. It is evident that many of the issues of understanding the structure and function of fuel cell components under operation are intimately linked to water fluxes and distribution.

Structure and water sorption characteristics of fuel cell media determine their transport properties. The dynamic properties of water determine microscopic transport mechanisms and diffusion rates of protons in PEM and CLs. Protons must be transported at sufficiently high rates, away from or toward the active Pt catalyst in anode and cathode catalyst layers, respectively. Effective rates of proton transport in nanoporous PEM and CLs result from a convolution of microscopic transport rates of protons with random network properties of aqueous pathways. Accounting for the geometry of these materials, namely, their external surface area and thickness, gives their resistances.

The main function of porous diffusion media is to homogenize feed gas supply over the cell surface at high rates of gas diffusion. Liquid water, if it accumulates in these media, acts as an asphyxiant. Fuel cell water management should strive to keep these diffusion media water-free. Catalyst layers play an ambivalent role in this regard: they must fulfill simultaneous requirements of high rates of proton transport and rapid diffusion of reactant gases. Ionomer-free ultrathin catalyst layers solve this dilemma by having a low thickness, in the range of 200 nm, which effectively eliminates the oxygen transport resistance. Conventional catalyst layers, however, function as gas diffusion electrodes. Their composition and structural design must optimize competing requirements for proton transport and oxygen transport via gaseous diffusion. Aspects of this complex interplay were treated separately in Chapter 3.

The first part of this chapter returns to the topic of transport processes in a PEM. Proton transport and electro-osmotic water drag in PEMs have been discussed extensively in Chapter 2. To understand the operation of the PEM in a fuel cell, other transport phenomena, including liquid water diffusion, hydraulic permeation, and interfacial vaporization exchange of water, must be addressed as well. For this purpose, the corresponding transport parameters must be found and their impact on performance rationalized.

## POLYMER ELECTROLYTE MEMBRANE IN FUEL CELL MODELING

The water content is the state variable of PEMs. Water uptake from a vapor or liquid water reservoir results in a characteristic vapor sorption isotherm. This isotherm can be described theoretically under a premise that the mechanism of water uptake is sufficiently understood. The main assumption is a distinction between surface water and bulk water. The former is chemisorbed at pore walls and it strongly interacts with sulfonate anions. Weakly bound bulk-like water equilibrates with the nanoporous PEM through the interplay of capillary, osmotic, and elastic forces, as discussed in the section "Water Sorption and Swelling of PEMs" in Chapter 2. Given the amounts and random distribution of water, effective transport properties of the PEM can be calculated. Applicable approaches in theory and simulation are rooted in the theory of random heterogeneous media. They involve, for instance, effective medium theory, percolation theory, or random network simulations.

From a practical perspective, it is important to understand the response of water distribution and fluxes in PEMs to variations in external conditions. Such variations can be applied in a controlled fashion in *ex situ* studies to isolate and extract effective transport parameters, or they could occur spontaneously in fuel cell operation under varying load requirements.

In view of their application in PEFCs, PEMs are assessed by the following criteria:

- The ability of a PEM to deliver high fuel cell performance, evaluated in terms of voltage efficiency and power density. This ability is determined mainly by the proton conductivity, which should be high, and PEM thickness, which should be low, in order to minimize resistive voltage losses.
- The ability to establish the optimal water distribution in the cell and facilitate water removal from the MEA. This ability is determined by the electro-osmotic drag, which should be small, and by the liquid water permeability, which should be high.
- The ability of a PEM to operate at high temperature ($>100°C$) and low relative humidity. These conditions are desirable from a system layout and design perspective. They require the design of PEMs that could provide sufficient proton conductivity with a minimal amount of water tightly bound to the polymeric host.
- The ability of the PEM to adapt to the ranges of PEFC operating conditions for the targeted application, for example, enabling the start-up of fuel cell vehicles at subfreezing temperatures, normal operation at temperatures close to or above $100°C$ and operation at low RH.

- High durability and lifetime.
- Ability to be fabricated by robust high-volume and low-cost processes.

## DYNAMIC WATER SORPTION AND FLUX IN PEMs

Under steady-state operation, local mechanical equilibrium prevails at all microscopic and macroscopic interfaces in the membrane. It fixes the stationary distribution of absorbed water. The condition of chemical equilibrium is, however, lifted to allow for the flux of water. Continuity of the net water flux in the PEM and across its interfaces with adjacent media adjusts the gradients in water activity or pressure in the system. Water fluxes occur by diffusion, hydraulic permeation, and electro-osmotic drag. At external interfaces, vaporization and condensation proceed at rates that match the net water flux. These mechanisms apply to PEM operation in a working cell, as well as to *ex situ* water flux measurements that are conducted in order to investigate the transport properties of PEMs.

### WATER TRANSPORT IN MEAs

The molar flux of water generated in the ORR at the cathode at a typical current density of 1 A cm$^{-2}$ is ~0.05 mol m$^{-2}$ s$^{-1}$. At the same current density, the water flux due to electro-osmosis is ~0.05-0.1 mol m$^{-2}$ s$^{-1}$. These processes, whose rates are linked to the current density, induce nonuniform distributions of water within the MEA, which can have a negative impact on performance. Extreme situations of PEM dehydration on the anode side or excessive accumulation of liquid water in electrode layers on the cathode side could cause fuel cell failure.

A highly water-permeable PEM would facilitate water removal via liquid transport toward the anode, alleviating the problem of cathode flooding and anode dehydration. From a system perspective, it is deemed beneficial to make use of internal humidification of CLs and PEM by water that is produced at the cathode. This mode of internal water management obviates the need for external humidifiers. It demands, however, precise control of water permeation rates through the PEM and of vaporization rates in partially saturated porous electrodes. Therefore, it is crucial to know how relevant parameters of water transport (diffusion, hydraulic permeation, electro-osmotic drag, vaporization, and condensation) depend on PEM morphology and thermodynamic conditions.

During the past decade, a number of experimental approaches have been developed to study water fluxes in PEM, under *ex situ* and *in situ* conditions. The objective is to measure water fluxes for materials with varying structure and composition, as a function of applied gradients, in relative humidity and liquid pressure (Adachi et al., 2009; Duan et al., 2012; Romero and Merida, 2009; Satterfield and Benziger, 2008).

Ideally, water flux data should be complemented by independent data on PEM structure and water sorption. An analysis of the data demands water flux models that provide a means to deconvolute contributions of different flux mechanisms and extract the relevant parameters. The problem of isolation and parameter extraction for water transport mechanisms from water flux measurements is analogous to the electrical problem of analyzing charge transport and electrochemical reactions at interfaces and in porous electrodes. In the latter realm, a typical experiment consists of measuring a time-dependent electrical current as a function of an applied voltage signal. The deconvolution of timescales either in direct time-dependent experiments or in the frequency domain for EIS provides insights into characteristic timescales. Using physical models, timescales may be deconvoluted and kinetic parameters of basic processes could be determined. In porous electrode theory, EIS is a sophisticated and hugely successful analysis method. The development of similar methods and tools for analogous water flux problems in porous media is much less advanced.

## EXPERIMENTAL STUDIES OF WATER PERMEATION IN PEMs

### Electro-Osmotic Drag Coefficient

Numerous measurements of electro-osmotic drag coefficients have been reported for Nafion PEM (Aotani et al., 2008; Fuller and Newman, 1993; Ge et al., 2006; Ise et al., 1999; Xie and Okada, 1995; Ye and Wang, 2007; Zawodzinski et al., 1993). As shown in Table 5.1, electro-osmotic drag coefficients depend on the hydration state of the membrane. In most cases, electro-osmotic drag coefficients were found to increase with the hydration level. However, Aotani et al. reported the reverse trend, that is, an increase in electro-osmotic drag coefficients with a decrease in membrane hydration level (Aotani et al., 2008). Since the mechanisms of electro-osmotic drag of water in Nafion membranes are still under debate, the precise relationship between the electro-osmotic drag coefficient and the hydration state remains unclear.

Table 5.1 (Adachi, et al., 2009), displays a significant variation and, thus, uncertainty in experimental values of the electro-osmotic drag coefficient. This scatter of results is related to the difficulty of separating the electro-osmotic drag effect from

### TABLE 5.1
### Comparison of Reported Electro-Osmotic Drag Coefficients for Nafion PEM

|  | T (°C) | Hydration State | $n_d$ ($H_2O/H^+$) | PEM |
|---|---|---|---|---|
| Zawodzinski et al. (1993) | 30 | 22 ($H_2O/SO_3H$) | ~2.5 | Nafion, 117 |
| Zawodzinski et al. (1993) | 30 | 1–14 ($H_2O/SO_3H$) | ~0.9 | Nafion, 117 |
| Fuller and Newman (1993) | 25 | 1–14 ($H_2O/SO_3H$) | 0.2–1.4 | Nafion, 117 |
| Ise et al. (1999) | 27 | 11–20 ($H_2O/SO_3H$) | 1.5–3.4 | Nafion, 117 |
| Xie and Okada (1995) | Ambient | 22 ($H_2O/SO_3H$) | ~2.6 | Nafion, 117 |
| Ge et al. (2006) | 30–80 | 0.2–0.95 (activity) | 0.3–1.0 | Nafion, 117 |
| Ge et al. (2006) | 30–80 | Contact with liquid water | 1.8–2.6 | Nafion, 117 |
| Aotani et al. (2008) | 70 | 2–6 ($H_2O/SO_3H$) | 2.0–1.1 | Nafion, 115 |
| Ye and Wang (2007) | 80 | 3–13 ($H_2O/SO_3H$) | ~1.0 | Layered Nafion, 115 |

*Source:* Adapted from Adachi, M. et al. 2009. *J. Electrochem. Soc.*, **156**, B782–B790.

water fluxes by diffusion and convection that are driven, respectively, by gradients in chemical potential and hydraulic pressure. If transport parameters of hydraulic flux and chemical diffusion were known as a function of membrane structure and composition, as well as external thermodynamic parameters, these values could be used to analyze experimental data and extract accurate values of electro-osmotic drag coefficients.

The success of such a program of water flux measurements hinges on a number of conditions: (i) the availability of theoretical models that account for the coupling of various water transport mechanisms, (ii) the development of model-based diagnostic tools to separate and isolate electro-osmotic flux from other fluxes, and (iii) experiments that can be conducted under controlled conditions.

### Water Permeation Measurements (*Ex Situ*)

While the body of work on measurements of net water transport through operating PEFCs is quite large, relatively few studies have attempted deconvoluting the net water flux into electro-osmotic drag and water fluxes via chemical diffusion or hydraulic permeation. As a consequence, conditions that promote net water transport to the anode side to offset the effects of anode dehydration and cathode flooding are insufficiently understood (Cai et al., 1006; Janssen and Overvelde, 2001; Liu et al., 2007; Yan et al., 2006). For this reason, studies on water permeation through PEMs have drawn significant interest as part of general strategies for mitigating issues associated with water management and improving the performance of PEM fuel cells.

Net rates of water uptake by a PEM can be extracted from the observation of transient mass changes of the PEM-water system during swelling and deswelling, when the membrane is exposed to water vapor with equal relative humidity on both sides (Burnett et al., 2006; Krtil and Samec, 2001; Morris and Sun, 1993; Pushpa et al., 1988; Rivin et al., 2001). The water permeability of a PEM can be determined by applying a gradient in vapor activity (concentration) or hydraulic pressure, and measuring the responding net water flux (Adachi et al., 2009; Majsztrik et al., 2007; Monroe et al., 2008; Motupally et al., 2000; Romero and Merida, 2009). The challenge of these experiments is to isolate and extract bulk and interfacial transport parameters. This demands well-devised experimental protocols and modeling tools.

### *In Situ* Characterization of Water Fluxes and Distributions

Data on net water fluxes through fuel cell systems are usually provided and discussed in terms of the ratio of the net water flux to proton flux $\beta$, defined by Springer et al. (1991), Zawodzinski et al. (1993), and Ren et al. (2000a). Positive $\beta$ corresponds to a net water flux toward the cathode. Zawodzinski et al. reported $\beta = 0.2$ at 0.5 A cm$^{-2}$ for MEAs with Nafion 117 under operation with fully humidified gases. Choi et al. (2000) reported $\beta = 0.55 - 0.31$ for current densities, varying from 0 to 0.4 A cm$^{-2}$ for an MEA with a fully humidified Nafion 117 PEM. They found a large increase in $\beta$ under dry operating conditions, indicating a change in coupled mechanisms of proton transport and electro-osmosis. Janssen and Overvelde (2001) evaluated $\beta$ using Nafion 105 under combinations of wet, dry, and differential pressure. Dry conditions at the anode resulted in negative $\beta$, whereas other operating conditions gave positive

$\beta$ values. Ren and Gottefeld (2001) operated a Nafion 117-based MEA with over-saturated hydrogen and dry oxygen at 80°C and reported positive values of $\beta$ that changed from 3.0 to 0.6 for current densities varying from 0 to 0.7 A cm$^{-2}$. Yan et al. (2006) observed that $\beta$ increased when the cathode RH decreased, while maintaining the anode gas saturated. Negative $\beta$ were recorded when the cathode gases were saturated and the flow rate of the relatively drier hydrogen gas (20% RH) at the anode was increased. They had also explored effects of applying differences in RH of gas streams and gas pressures to distinguish contributions of concentration-driven diffusion, including pressure-driven permeation, and to distinguish these fluxes from the electro-osmotic flux. Murahashi et al. (2006) followed a similar approach of investigating $\beta$ as a function of the difference in RH between electrodes. General trends of decreasing $\beta$ with an increase in cathode RH and a positive shift in $\beta$ with increased cell temperature were reported.

Cai et al. (1006) conducted a water balance study of Nafion 112-based MEAs, under dry hydrogen and moderately humidified air, and reported that $\beta$ values were negative, decreasing from $-0.06$ to $-0.18$ as the current density was increased from 0.1 to 0.6 A cm$^{-2}$. Liu et al. (2007) monitored the variance of $\beta$ along the flow channel, using a unique setup that incorporated a gas chromatograph. They operated a rectangular cell with 30-μm Gore-select PEM in combination with moderately humidified and dry gases and observed a significant change in $\beta$ along the gas flow channel. Ye and Wang (2007) found that $\beta$ varied from 0.5 to 1.1 for RH $= 0.95 - 0.35$ at current densities up to 1.2 A cm$^{-2}$, for Gore-PRIMEA18-based MEAs with an 18 μm-thick PEM.

More sophisticated techniques reveal detailed information on in-plane and through-plane distributions of water in an operating fuel cell. The 1D water distribution was observed by employing magnetic resonance imaging (MRI) (Teranishi et al., 2006; Zhang et al., 2008). The various degrees of hydration of operational MEA component materials were determined by EIS (Andreaus and Scherer, 2004; Schneider et al., 2005; Springer et al., 1996). EIS was also used to report on water distributions across the MEA (Buechi and Scherer, 2001; Takaichi et al., 2007); neutron imaging has been used to visualize the in-plane and through-plane water distribution in an operating PEFC (Hickner et al., 2006; Mench et al., 2003).

The work of Adachi et al. (2009) represented a first attempt to correlate and validate *ex situ* and *in situ* water permeation phenomena in PEMs. Water permeabilities of Nafion PEMs and water transport in operating PEFCs were investigated under comparable *ex situ* and *in situ* values of temperature and RH. The examined parameters included the type of driving forces (RH, pressure), the phases of water at PEM interfaces, PEM thickness, and the effect of catalyst layers at the membrane interfaces. Several experimental setups and schemes were designed and explored. Water permeability at 70°C was determined for Nafion membranes exposed to either liquid or vapor phases of water. Chemical potential gradients of water across the membrane are controlled through the use of differences in RH (38–100%), in the case of contact with water vapor, and hydraulic pressure (0–1.2 atm), in the case of contact with liquid water. Three types of water permeation experiments were performed, labeled as *vapor-vapor permeation* (VVP), *liquid-vapor permeation* (LVP), and *liquid-liquid permeation* (LLP). *Ex situ* measurements revealed that the flux of water is largest when the membrane is exposed to liquid on the one side and vapor on the other, that is, under LVP conditions. These conditions generate the largest gradient in liquid water content and, thus, internal hydraulic pressure in the PEM.

## MODELING OF *Ex Situ* WATER FLUXES IN PEM

The resistance to water flux in the membrane interior and vaporization-condensation kinetics at its interfaces are important factors for the operation of PEFCs (Eikerling et al., 1998, 2007a; Weber and Newman, 2004b). Specifically, developed models can be applied to water flux measurements (Adachi et al., 2009; Majsztrik et al., 2007; Monroe et al., 2008; Romero and Merida, 2009) to distinguish interfacial sorption-desorption kinetics from the mechanisms of water transport inside of the PEM.

When modeling dynamic water sorption phenomena, information about evaporation and condensation is contained in the boundary conditions that account for water exchange across membrane-gas interfaces. The rate of interfacial water exchange is determined by values of the instantaneous water content on the PEM side of the interface, $\lambda_{int}$, and by the vapor pressure, $P_{int}^v$, of the adjacent gas. The deviation of these local variables from their chemical equilibrium establishes the driving force of interfacial vaporization exchange.

In the past, it has been a common albeit dubious practice to adopt an equilibrium sorption isotherm for the relation between $\lambda_{int}$ and $P_{int}^v$. This approach demands an infinite rate constant of vaporization exchange. It is problematic for two reasons. First, the relative importance of interfacial water exchange grows with decreasing membrane thickness. Below a critical thickness, interfacial kinetics, rather than bulk transport, will limit the net water flux, implying an out-of-equilibrium condition. Second, if gases adjacent to the membrane are moving, water may be convected away from its surfaces. It is inherently contradictory to assume equilibrium in the presence of any kinetic or convective process.

Refined physical approaches to dynamic water exchange between PEM and adjacent gases incorporate a flux boundary condition to quantify water transport across the interfaces, as proposed by Benziger and coworkers (Majsztrik et al., 2007; Satterfield and Benziger, 2008). Although these models describe ionomer water sorption with some success, they involve a number of restrictive assumptions. These include a neglect of the dependence of condensation kinetics on vapor pressure and an assumption that water accumulation in the gas can be ignored. It remains possible, however, to treat both condensation and evaporation explicitly and to include a water balance within the gas phase. A refined physical model that allows determining of interfacial water exchange rates and water permeabilities from measurements involving PEM in contact with flowing gases was proposed by Monroe et al. (2008). It was applied to the experimental system of Romero and Merida (2009), described in the next section.

### Experimental Setup

Figure 5.1 shows the experimental setup devised by Romero and Merida (2009) for which the model, discussed below, has been developed. A PEM of thickness $l_{PEM}$ is sandwiched between two ring seals. The accessible surface of the PEM has a circular

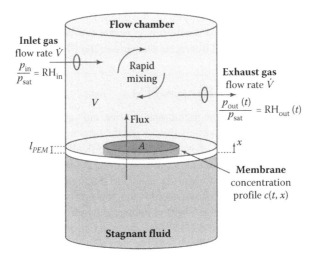

**FIGURE 5.1** Experimental setup for water flux measurements on PEMs devised by Romero and Merida (2009). A PEM is mounted as a separator between a chamber with a stagnant fluid (liquid water to saturated vapor) and a flow chamber. Gas flows through the flow chamber at a steady volumetric flow rate, $\dot{V}$. The partial pressure of water at the inlet, $p_{in}$, is held constant and the exhaust partial pressure, $p_{out}$, is recorded over time. (Reprinted from *J. Membr. Sci.*, **324**, Monroe, C. et al. A vaporization-exchange model for water sorption and flux in Nafion, 1–6, Figures 1,2,3, Copyright (2008) Elsevier. With permission.)

shape with a surface area $A$. The thickness $l_{PEM}$ is assumed to be small compared to the radius of the disk. This means that water fluxes through the PEM can be approximated as one-dimensional. Transport occurs along the coordinate $x$ that is normal to the PEM surface.

During experiments, the PEM surface at $x = 0$ is in contact with a chamber filled with a stagnant fluid. The opposite surface at $x = l_{PEM}$ is exposed to a flow chamber with a steady throughput of gas. Experiments where the stagnant fluid is liquid water are referred to, hereafter, as *liquid-equilibrated* (LE). Those where the stagnant fluid comprises of saturated vapor are labeled as *vapor-equilibrated* (VE). To account for possible differences in water uptake under VE and LE conditions, two different maximal water contents are considered, namely, $c_{max}^{vap}$ and $c_{max}^{liq}$ with a natural expectation that $c_{max}^{liq} > c_{max}^{vap}$.

Gas flows with a regulated volumetric flow rate $\dot{V}$ through an inlet into the flow chamber with total volume $V$. Gas leaves the flow chamber at the same rate. In practice, this is achieved by throttling the gas into the ambient air through an exhaust tube. The nozzles of the inlet and exhaust are small, compared to $A$, and are oriented well away from each other to ensure good mixing within the flow chamber before the gas leaves. Also, both vents are oriented away from the interface to ensure that random gas currents prevail near the membrane surface, preventing a steady convective flow.

Although the areas of the vents are relatively small, the volumetric flow rate of the throughput gas is usually small enough that the Reynolds number is not in the turbulent regime.

Capacitive sensors are employed to measure the dew point temperature of gas in the flow chamber. Accurate empirical correlations exist between dew point temperatures and water vapor pressures. If the characteristic diffusion time for water in the gas is short compared to both the turnover time of gas in the flow chamber and the characteristic timescale for water transport through the bulk membrane, then the water vapor pressure of the outlet stream, $P_{out}^v$, equals the average vapor pressure in the flow chamber, $P^v$.

## Mathematical Model

The water balance in the flow chamber can be approximated by

$$\frac{V}{RT}\frac{dP_{out}^v}{dt} = \left[P_{in}^v - P_{out}^v(t)\right]\frac{\dot{V}}{RT} + AJ(t), \tag{5.1}$$

where $J(t)$ is the molar flow of water in the $x$ direction per surface area $A$. In Equation 5.1, $P_{in}^v$ is the water vapor pressure at the flow chamber inlet, and $P_{out}^v(t)$ is the water vapor pressure in the outlet gas stream. Water transport within the membrane is described by

$$\frac{\partial c}{\partial t} = D_w^{eff}\frac{\partial^2 c}{\partial x^2}, \tag{5.2}$$

where the effective diffusion coefficient $D_w^{eff}$ could incorporate contributions from hydraulic permeation as well as diffusion (Eikerling et al., 1998, 2007a; Weber and Newman, 2004b). The local effective water concentration $c$ in the PEM was introduced in Monroe et al. (2008). It is related to previously defined variables used to specify the amount of water in PEM:

$$c = \frac{X_w}{\bar{V}_w} = \frac{\lambda - \lambda_s}{(\lambda - \lambda_s)\,\bar{V}_w + \bar{V}_p} \tag{5.3}$$

where $\bar{V}_p$ and $\bar{V}_w$ are molar volumes of ionomer (per repeat unit on the backbone that includes one side chain) and water.

The flux boundary condition at the PEM|gas interface at $x = l_{PEM}$ is

$$J(t) = -D_w^{eff}\left.\frac{\partial c(t)}{\partial x}\right|_{x=l_{PEM}}. \tag{5.4}$$

Continuity of water fluxes at the interface implies balanced rates of water permeation to the interface and vaporization exchange at the interface. This continuity condition takes the form

$$-D_w^{eff}\left.\frac{\partial c(t)}{\partial x}\right|_{x=l_{PEM}} = \frac{k_v}{RT}\left\{P^{v,eq}\left(c\left(t, l_{PEM}\right)\right) - P^v(t)\right\}. \tag{5.5}$$

The term on the right-hand side describes vaporization and condensation kinetics at the membrane-gas interface, where $k_v$ is the vaporization rate constant, $P^v$ the actual vapor pressure in the adjacent gas phase, and $P^{v,eq}$ the equilibrium vapor pressure that corresponds to the local water concentration $c$ at $x = l_{PEM}$. The function $P^{v,eq}(c(t, l_{PEM}))$ represents the vapor sorption isotherm, discussed in the section "Water Sorption and Swelling of PEMs" in Chapter 2.

The values of $c(t, l_{PEM})$ and $P^v(t)$ must be obtained self-consistently as the solution of the system of equations that describe coupled water fluxes in PEM and adjacent compartments.

Vaporization is an activated process controlled by the Gibbs energy of water sorption, which determines $P^{v,eq}(c(t, l_{PEM}))$. Condensation is a kinetic process with a rate proportional to the vapor pressure $P^v$ at the interface between PEM and flow chamber. The vaporization exchange rate constant $k_v$ has a physical meaning similar to the exchange current density in electrochemical kinetics, and $P^{v,eq}(c(t, l_{PEM})) - P^v$ acts like an overpotential of the vaporization process.

Equivalently to Equation 5.5, we could write

$$-D_w^{\text{eff}} \left.\frac{\partial c(t)}{\partial x}\right|_{x=l_{PEM}} = \frac{k_v P^s}{RT} \left\{ \exp\left(\frac{\Delta G^s(c(t, l_{PEM}))}{RT}\right) - \frac{P^v}{P^s} \right\}, \qquad (5.6)$$

where the only empirical input is the Gibbs energy of water sorption that could be determined from isopiestic sorption data.

In the LE case, the boundary condition at the interface with the stagnant fluid at $x = 0$ is

$$c(t, 0) = c_{\text{max}}^{liq}, \qquad (5.7)$$

corresponding to the saturated water uptake of the PEM in contact with liquid water.

In the VE case, the boundary condition at $x = 0$ is

$$-D_w^{\text{eff}} \left.\frac{\partial c(t)}{\partial x}\right|_{x=0} = \frac{k_v}{RT} \left\{ P^s - P^{v,eq}(c(t, 0)) \right\}, \qquad (5.8)$$

where $P^{v,eq}(c(t, 0))$ is the equivalent equilibrium vapor pressure that corresponds to the local water concentration at the interface. Again, the term $P^s - P^{v,eq}(c(t, 0))$ can be interpreted as an overpotential that establishes the driving force for the interfacial water transfer.

It is convenient to define a set of dimensionless variables

$$\xi = \frac{x}{l_{PEM}}, \quad \tau = \frac{D_w^{\text{eff}} t}{l_{PEM}^2}, \quad \theta_m(\tau, \xi) = \frac{c - c_0}{c_{\text{max}}^{liq/vap} - c_0}, \quad \theta_v(\tau) = \frac{P_{out}^v - P_{in}^v}{P^s - P_{in}^v}, \quad (5.9)$$

where $\theta_m$ represents the water concentration in the PEM and $\theta_v$ the vapor pressure in the flow chamber. The parameters $c(0, x) = c_0$ and $P_{out}^v(0) = P_{in}^v$ are the initial values of water concentration in PEM and vapor pressure at the outlet, respectively. For initially dry conditions, $c_0 \approx 0$ and $P_{in}^v \approx 0$ is expected.

Essentially the model describes the interplay of bulk water permeation in the PEM and interfacial vaporization exchange. The dimensionless parameter that represents this interplay is

$$\kappa_{PEM} = \frac{l_{PEM} k_v \left(P^s - P^v_{in}\right)}{D^{eff}_w \left(c^{liq/vap}_{max} - c_0\right) RT}. \tag{5.10}$$

The case $\kappa_{PEM} \ll 1$ corresponds to the situation when the overall water transport is limited by interfacial vaporization exchange. For $\kappa_{PEM} \gg 1$, bulk water permeation in the PEM is the limiting process. For a given type of membrane, a critical thickness $l_{PEM}$ can be defined, which marks the transition between these two regimes, as will be shown below. As noted in Monroe et al. (2008), the condition $\kappa_{PEM} \gg 1$ does not imply that absorbed water at the surface of the PEM attains equilibrium with water vapor in the gas.

In the LE case, the governing equations in dimensionless form are

$$\frac{\partial \theta_v}{\partial \tau} = -\varphi \theta_v - \frac{\gamma}{\kappa_{PEM}} \frac{\partial \theta_m}{\partial \xi}\bigg|_{\tau,1}$$

$$\frac{\partial \theta_m}{\partial \tau} = \frac{\partial^2 \theta_m}{\partial \xi^2}$$

with initial conditions $\theta_v(0) = 0$ and $\theta_m(0, \xi) = 0$ and boundary conditions

$$\theta_m(\tau, 0) = 1 \tag{5.11}$$

and

$$\frac{\partial \theta_m}{\partial \xi}\bigg|_{\tau,1} = \kappa_{PEM} \left[\theta_v(\tau) - \theta_m(\tau, 1) - \psi(1 - \theta_m(\tau, 1))\right], \tag{5.12}$$

introducing dimensionless parameters

$$\psi = \frac{c_0 P^s - c^{liq}_{max} P^v_{in}}{c^{liq}_{max} \left(P^s - P^v_{in}\right)}, \quad \gamma = \frac{k_v A l^2_{PEM}}{D^{eff}_w V}, \quad \varphi = \frac{l^2_{PEM} \dot{V}}{D^{eff}_w V}.$$

The second boundary condition in Equation 5.12 involves an approximation of the vapor sorption isotherm in the form of Henry's law:

$$P^{v,eq} \approx \frac{P^s}{c^{liq}_{max}} c(t, l_{PEM}). \tag{5.13}$$

Using this linear approximation might be considered oversimplistic, but it renders the mathematical problem amenable to analytical solution by Laplace transformation. Moreover, it does not affect the agreement with experimental data. Of course, one may always resort to a numerical solution of the complete problem with a more accurate representation of the vapor sorption isotherm.

Using the same linear approximation for the vapor sorption isotherm in Equation 5.8, similar expressions can be obtained as well for the VE case. In Monroe et al.

(2008), steady-state solutions were analyzed for LE and VE cases. For the LE case, the steady-state solutions for vapor pressure and water concentration in the PEM are

$$\theta_v^{ss} = \frac{\gamma}{\gamma + \varphi \left[1 + \kappa_{PEM} \left(1 - \psi\right)\right]} \tag{5.14}$$

and

$$\theta_m^{ss}\left(\xi\right) = 1 - \frac{\kappa_{PEM}\varphi}{\gamma + \varphi \left[1 + \kappa_{PEM} \left(1 - \psi\right)\right]}\xi. \tag{5.15}$$

The main results of the model analysis are linear response functions between steady-state values of relative humidity in the flow chamber and the flow rate per PEM surface area:

$$\frac{P^s - P_{out}^{v,ss}}{P_{out}^{v,ss} - P_{in}^{v}} = m_{LE} \frac{\dot{V}}{A}, \tag{5.16}$$

with a slope

$$m_{LE} = \frac{1}{k_v} + \frac{P^s}{RTD_w^{\text{eff}} c_{\max}^{liq}} l_{PEM}. \tag{5.17}$$

The inverse water flux as a function of inverse flow rate also prescribes a straight line

$$\frac{1}{J^{ss}} = \frac{RT}{P^s - P_{in}^v} \left[m_{LE} + \frac{A}{\dot{V}}\right]. \tag{5.18}$$

Similarly, the response functions for the VE case are obtained as

$$\frac{P^s - P_{out}^{v,ss}}{P_{out}^{v,ss} - P_{in}^{v}} = m_{VE} \frac{\dot{V}}{A}, \tag{5.19}$$

with a slope

$$m_{VE} = \frac{2}{k_v} + \frac{P^s}{RTD_w^{\text{eff}} c_{\max}^{vap}} l_{PEM} \tag{5.20}$$

and

$$\frac{1}{J^{ss}} = \frac{RT}{P^s - P_{in}^v} \left[m_{VE} + \frac{A}{\dot{V}}\right]. \tag{5.21}$$

These expressions show that the main difference between LE and VE cases is a doubling of the interfacial resistance, due to vaporization exchange, as seen in the first term on the right-hand side of $m_{VE}$. Moreover, the maximal water concentration in the denominator of the second term in $m_{VE}$ could assume a different value.

The linear response functions, presented above, are the result of linearization of the mathematical formalism, by assuming constant transport coefficients of water, $k_v$ and $D_w^{\text{eff}}$, and using a linear approximation for the vapor sorption isotherm. The slopes $m_{VE}$ and $m_{LE}$ represent linear effective resistances, analogous to ohmic resistances in electrical networks.

## Comparison to Experiment

Figure 5.2a shows experimental data for the steady-state RH of Nafion-type PEMs with different thicknesses as a function of the flow rate per surface area. Under both LE and VE conditions, experimental data are accurately represented by the linear relationships that the model predicts. From the linear fits, the slopes $m_{LE}$ and $m_{VE}$ have been extracted and plotted in Figure 5.2b as a function of PEM thickness $l_{PEM}$. Linear fits of the data points for $m_{LE}$ and $m_{VE}$, as functions of $l_{PEM}$, provide values of the vaporization exchange rate $k_v$ and the effective permeability $D_w^{\text{eff}} c_{\max}^{liq/vap}$.

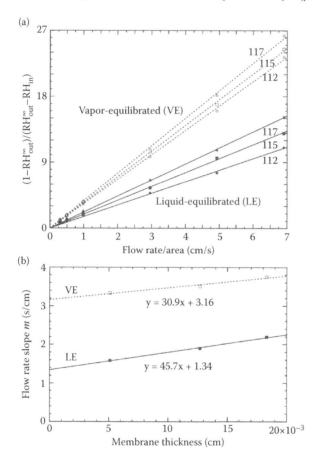

**FIGURE 5.2** Experimental evaluation of the PEM water flux model. (a) The steady-state RH at the outlet of the gas chamber as a function of the flow rate per surface area for PEMs, with different thicknesses (Nafion 117, 115, and 112) at 50°C under VE and LE conditions. The inlet gas was dry helium $RH_{in} = 0$. Lines represent best fits of the data with the model. (b) Slopes of the straight lines in (a), as a function of PEM thickness. The y-axis intercepts give the vaporization exchange resistance and the slopes give the bulk resistance to water perme-ation. (Reprinted from *J. Membr. Sci.*, **324**, Monroe, C. et al. A vaporization-exchange model for water sorption and flux in Nafion, 1–6, Figures 1,2,3, Copyright (2008) Elsevier. With permission.)

For the data shown in Figure 5.2, the procedure gives values $k_v = 0.75$ cm s$^{-1}$ and $D_w^{\mathrm{eff}} c_{\max}^{liq} = 1.0$ µmol dm$^{-1}$ s$^{-1}$ for the LE case and $k_v = 0.60$ cm s$^{-1}$ and $D_w^{\mathrm{eff}} c_{\max}^{vap} = 1.5$ µmol dm$^{-1}$ s$^{-1}$ for the VE case. The last result is somewhat surprising since the water permeability in the VE case is expected to be smaller than in the LE case. The estimates obtained are relatively inaccurate because of the lack of more systematic data at this point; they should be considered as accurate within an order of magnitude. The values of $k_v$ and $D_w^{\mathrm{eff}} c_{\max}^{liq/vap}$ can be used to assess the parameter $\kappa_{PEM}$ that was defined in Equation 5.10. The value obtained is $\kappa_{PEM} = 0.003$ µm$^{-1} \cdot l_{PEM}$.

Correspondingly, a characteristic thickness can be defined:

$$l_{PEM}^C = \frac{v_v RT D^{\mathrm{eff}} c_{\max}^{vap/liq}}{k_v P^s} \tag{5.22}$$

with $v_v = 2$ for VE and $v_v = 1$ for LE conditions. For $l_{PEM} < l_{PEM}^C$, water transport through the PEM is dominated by interfacial vaporization exchange, whereas for $l_{PEM} > l_{PEM}^C$, bulk permeation of water prevails. The data obtained in Monroe et al. (2008) yielded $l_{PEM}^C \simeq 100$–$300$ µm. This indicates that the interfacial vaporization resistance exceeds the resistance due to bulk transport in the membrane when the membrane thickness is $l_{PEM} < 100$ µm.

## Analysis of Transient Water Flux Data

The vaporization exchange model for water sorption and flux in Nafion-type PEMs has been modified and applied to treat transient water flux data (Rinaldo et al., 2011). A decisive modification was the inclusion of transport coefficients that depend on the water concentration or water content in the PEM. A simple form of this dependence is a step function change from slow to fast transport at a given transition concentration $c_*$. For this purpose, a hyperbolic tangent function was introduced:

$$D_w^{\mathrm{eff}}(c) = D_{\mathrm{fast}} \left\{ \frac{D_{slow}}{D_{\mathrm{fast}}} + \left( \frac{1}{2} - \frac{D_{slow}}{D_{\mathrm{fast}}} \right) [\tanh(\omega(c - c_*)) + 1] \right\}, \tag{5.23}$$

where $D_{slow}$ and $D_{\mathrm{fast}}$ are water transport coefficients at low and high water contents, respectively, and $\omega$ determines the "width" of the transition region. Implementing this nonlinear representation of the diffusion coefficient renders the model analytically unsolvable.

The dimensionless form of the equation is

$$\bar{D}(\theta_m) = \beta + \left( \frac{1}{2} - \beta \right) \left[ \tanh\left( \sigma\left( \theta_m - \theta_m^* \right) \right) + 1 \right] \tag{5.24}$$

with $\bar{D}(\theta_m) = D_{\mathrm{eff}}(c)/D_{\mathrm{fast}}$, $\beta = D_{slow}/D_{\mathrm{fast}}$, and $\sigma = \omega c_{\max}^{liq}$. For large values of $\sigma$, the hyperbolic tangent function approaches an ideal step function. Lastly, $\theta_m^*$ is the dimensionless transition concentration. Transport within the PEM can now be described by

$$\frac{\partial \theta_m}{\partial \tau} = \bar{D}(\theta_m) \frac{\partial^2 \theta_m}{\partial \xi^2} + \frac{d\bar{D}(\theta_m)}{d\theta_m} \left(\frac{\partial \theta_m}{\partial \xi}\right)^2. \tag{5.25}$$

The flux boundary condition at the interface to the flow chamber is

$$\left.\frac{\partial \theta_m}{\partial \xi}\right|_{1,\tau} = \kappa_{PEM} [\theta_v(\tau) - \theta_m(1,\tau) - \psi(1 - \theta_m(1,\tau))] \tag{5.26}$$

and $\theta_m(0,\xi) = 0$, that is, the membrane is initially dry. Moreover, it is assumed that the flux at the membrane/chamber interface depends only on $D_{\text{fast}}$.

For a PEM in contact with liquid water as the stagnant fluid (LE), the boundary condition is $\theta_m(\tau,0) = 1$, and for a PEM in contact with water vapor (VE), the boundary condition is

$$\left.\frac{\partial \theta_m}{\partial \xi}\right|_{\tau,0} = -\alpha(1 + \psi)[1 - \theta_m(\tau,0)]. \tag{5.27}$$

The molar concentration of sulfonic acid groups in Nafion 117 is approximately $c_f = 1.2$ kmol m$^{-3}$. Maximal water content, defined as the number of moles of water molecules per moles of ion exchange sites, can be taken to be $\lambda_{\max} = 14$. This corresponds to a maximal water concentration of $c_{\max}^{liq} = \lambda_{\max} c_f = 16.8$ kmol m$^{-3}$.

Figure 5.3a shows a plot of transient water flux data for Nafion 117 under LE conditions along with numerical simulations using constant, as well as concentration-dependent, bulk transport coefficients. Experimental details for the data shown in Figure 5.3 are given in Rinaldo et al. (2011). The analysis of transient sorption data with the model yielded the vaporization exchange rate constant, the bulk transport coefficient at low water content, the bulk transport coefficient at high water content, and a critical water content of the transition from slow to fast transport regime. Values of the diffusion coefficients at low and high water contents are $D_{\text{slow}} = 0.29 \cdot 10^{-5}$ cm$^2$ s$^{-1}$ and $D_{\text{fast}} = 4.5 \cdot 10^{-5}$ cm$^2$ s$^{-1}$. The value $D_{\text{fast}}$ exceeds the self-diffusion coefficient of free bulk water within the porous medium by about a factor 10, suggesting that a dominant contribution to water transport is related to hydraulic permeation. The value of the critical water content ($\lambda_s \approx 3$) for the transition from slow to fast diffusion agrees with the percolation threshold obtained from independent data on structure and transport properties of Nafion. The vaporization exchange rate constant used in the simulation was $k_v = 0.48$ cm s$^{-1}$, which is slightly smaller than the values found under steady-state conditions. Using these parameters, water concentration profiles in Nafion 117 were simulated at different times, as depicted in Figure 5.3b. These profiles show that liquid water needs approximately 25 s to break through the PEM and approach the steady-state distribution.

The extracted parameters are vital for rationalizing mechanisms and amounts of water fluxes in PEFCs. The model could be applied for the analysis of sorption data at varying PEM thickness and equilibrium water content. Experiments running at varying $T$ would provide activation energies of the vaporization-exchange rate constant and bulk transport coefficients. Similar modeling tools can be developed for the study of water sorption and fluxes in catalyst layers. They can be extended, furthermore,

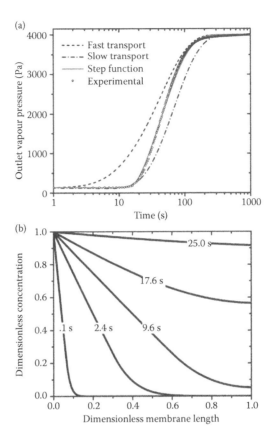

**FIGURE 5.3** Modeling of transient water flux data for Nafion 117. (a) The relaxation of the experimental outlet vapor pressure (open circle) for Nafion 117 in LE mode at 50°C, flow chamber volume $V = 0.125$ L, flow rate $\dot{V} = 0.1$ L min$^{-1}$, membrane area $A = 2$ cm$^2$, and saturation vapor pressure $P^s = 12336.7$ Pa. Plotted for comparison are model simulations for a slow transport coefficient (dash dot), fast transport coefficient (dash), and a concentration-dependent transport coefficient (gray). (b) Water concentration profiles calculated in the model at different time. (Reprinted from *Electrochem. Commun.* **13**, Rinaldo, S. G. et al. Vaporization exchange model for dynamic water sorption in Nafion: Transient solution, 5–7, Figures 1 and 2, Copyright (2011) Elsevier. With permission.)

to complete membrane electrode assemblies that contain various porous layers for water transport and transformation.

## MEMBRANE IN FUEL CELL PERFORMANCE MODELING

### MEMBRANE OPERATION UNDER IDEAL CONDITIONS

Under ideal operation of PEFC, the membrane should retain a uniform saturation level at the optimal water content, providing the highest proton conductivity, $\sigma_p^s$. The optimal water content could be determined from Equation 2.1 or a refinement thereof.

In this case, the PEM operates like a linear ohmic resistance, with irreversible voltage losses $\eta_{PEM} = j_0 l_{PEM}/\sigma_p^s$, where $j_0$ is the fuel cell current density. In reality, this behavior is only observed in the limit of small $j_0$. At normal current densities of fuel cell operation, $j_0 \sim 1$ A cm$^{-2}$, the electro-osmotic coupling between proton and water fluxes causes nonuniform water distributions, which lead to nonlinear effects in $\eta_{PEM}$. These deviations result in a critical current density $j_{pc}$, at which the increase in $\eta_{PEM}$ incurs a dramatic decrease of the cell voltage. It is, thus, crucial to develop membrane models that could predict the value of $j_{pc}$ on the basis of primary experimental data on structure and transport properties.

## MACROSCOPIC MODELING OF PEM OPERATION: GENERAL CONCEPTS

In order to scrutinize the role that the PEM plays in regulating the water fluxes through the PEFC, it is vital to understand spatial distributions of water and water fluxes. Macroscopic or macrohomogeneous modeling approaches that focus on voltage losses in the PEM and on the role of PEM for water management in PEFC have been reviewed in Eikerling et al. (2007a), Weber and Newman (2004b), Eikerling et al. (2008), Eikerling and Malek (2010), and Weber and Newman (2009).

The physical principles of membrane operation and the mathematical structure of modeling approaches, required to incorporate them, are straightforward to determine. Under stationary operation, the inevitable electro-osmotic flux is compensated, at least partly, by a backflux of water from cathode to anode, which ought to be driven by gradients in the chemical potential or liquid pressure of water. Thermal gradients in PEMs are negligible. The driving force for the water backflux is related to a nonuniform water distribution in the PEM, which decreases, under normal operation, from cathode to anode. With increasing $j_0$, the water distribution becomes more nonuniform. At $j_{pc}$, the water content near the anode falls below the percolation threshold of proton conduction, $\lambda < \lambda_c$. This leaves only a small residual proton conductivity in surface water. Since the residual conductivity is usually small, $\eta_{PEM}$ increases dramatically in this regime and the fuel cell voltage decreases to zero, causing the cell to fail.

Models of PEM performance can be classified by the degree of structural complexity that they incorporate. So far, none of the existing performance models has been able to integrate the genuine structural picture of the PEM as a self-organized, phase-segregated polymer, accounting for the complicated intercations of water with the ionomer. The earliest approaches were based on a single phase approximation. They considered the PEM as a continuous nonporous phase, which could be studied with concentrated solution theory (Fuller and Newman, 1989; Springer et al., 1991). In this approach, the thermodynamic state variable to specify the local membrane state, is the chemical potential of water, and water backtransport is by chemical diffusion in a gradient of the chemical potential. The pure diffusion models represent, however, an incomplete description of the membrane response to changing external conditions. They could not predict the net water flux across a saturated membrane, that results from applying a difference in total gas pressures between flow fields on cathode and anode sides. Moreover, owing to the neglect of a one state variable (pressure), diffusion models incompletely describe the state of water in PEM and, therefore, fail in describing the phenomenon that is notoriously known as "Schröder's paradox."

Structural models of membrane operation, on the other hand, treat the membrane as a heterogeneous porous medium in a two-phase approximation. They consider percolating water networks responsible for proton and water transport. This structural concept, thus, involves hydraulic permeation (Dullien, 1979) as a mechanism of water transport that supersedes diffusion at sufficiently large water content. Analogous to water flux in porous rocks, the capillary pressure controls the water saturation in the membrane, and gradients in hydraulic pressure are responsible for the hydraulic water flux from cathode to anode. This approach integrates the previously discussed understanding of the phase-separated membrane structure, water sorption, membrane swelling, and proton transport mechanisms. External gradients in relative humidity or gas pressure between cathode and anode sides may be superimposed on the internal gradient in liquid pressure, providing the means to control water distribution and fluxes in the operating PEFC. This picture is indeed the basis of the hydraulic permeation model of membrane operation.

Bernardi and Verbrugge (1992) built a model of membrane water management that incorporated hydraulic permeation as the counterflux to electro-osmotic water transport, using Schlögl equation. However, they assumed a uniform saturation of the PEM. The first hydraulic permeation model that incorporated the coupling among (i) pore network morphology, (ii) nonuniform spatial distributions of water and pressure, (iii) water transport by electro-osmotic drag and hydraulic flux, and (iv) water management in operating PEFCs, was proposed by Eikerling et al. (1998), and it was refined in Eikerling et al. (2007a).

It should be emphasized again that hydraulic permeation models do not rule out water transport by diffusion. Both mechanisms contribute concurrently. The water content in the PEM determines relative contributions of diffusion and hydraulic permeation to the total backflux of water. Hydraulic permeation prevails at high water contents, that is, under conditions for which water uptake is controlled by capillary condensation. Diffusion prevails at low water contents, that is, under conditions for which water strongly interacts with the polymeric host (chemisorption). The critical water content that marks the transition from diffusion-dominated to hydraulic permeation-dominated transport depends on water-polymer interactions and porous network morphology. Sorption experiments and water flux experiments suggest that this transition occurs at $\lambda \simeq 3$ for Nafion with equivalent weight 1100.

This transition was also discussed in Weber and Newman (2009). However, the authors suggest that diffusion prevails in VE PEM ($\lambda < 14$), while hydraulic permeation prevails in LE PEM ($\lambda \geq 14$). The proposed value of the transition water content is clearly too high. Moreover, the concept of identifying the state of water and corresponding mechanism of water transport inside of the PEM, with the state of water in the adjacent reservoir, is spurious. As discussed above, it is more instructive and consistent with experiments to distinguish bulk-like and surface water, which account for hydraulic and diffusive flux, respectively. The conclusion reached in Weber and Newman (2009) is, however, valid. It states that diffusion and hydraulic permeation in the PEM occur in parallel.

Notwithstanding any particular structural model, water transport in PEMs should be driven by diffusion in gradients of activity or concentration and hydraulic permeation in gradients of liquid pressure. Combination models that consider a superposition of chemical diffusion and hydraulic permeation have been discussed in Eikerling et al. (1998, 2007a, 2008) and Weber and Newman (2009).

The total molar flux of liquid water in the membrane $\mathbf{N}_l$, is given by

$$\mathbf{N}_l = n_d(\lambda)\frac{\mathbf{j}}{F} - D^m(\lambda)\nabla c - \frac{k_p^m(\lambda)}{\mu}\nabla P^l, \qquad (5.28)$$

where $c$, $P^l$, and $\mu$ are the concentration (relative to a unit volume of the membrane), the pressure, and the viscosity of liquid water in pores, $D^m(\lambda)$, is the membrane diffusivity, $k_p^m(\lambda)$ the hydraulic permeability, and $n_d(\lambda)$ the electro-osmotic drag coefficient. It is indicated that these transport parameters are functions of $\lambda$. The hydraulic permeability, $k_p^m(\lambda)$, exhibits a strong dependence on $\lambda$, since larger water contents result in an increased number of pores used for water transport, better connectivity of the porous network, as well as larger mean radii of these pores. A modification of the Hagen–Poiseuille–Kozeny equation was considered in Eikerling et al. (1998, 2007a) to account for these effects

$$k_p^m(\lambda) = \xi\frac{(\lambda - \lambda_c)\rho(r^c)}{8}\Theta(\lambda - \lambda_c), \qquad (5.29)$$

where $\xi$ is the inverse tortuosity factor ($\xi = 1/3$ for isotropic tortuosity in 3D) and $\lambda_c$ is the percolation water content. The Heaviside step function $\Theta$ accounts for the percolation threshold at $\lambda_c$. The mean square radius of pores in a volume element with water content $\lambda$ is

$$\rho(r^c) = \frac{1}{\lambda}\int_0^{r^c}\frac{d\lambda(r)}{dr}r^2 dr, \qquad (5.30)$$

with capillary radius $r^c$. The differential pore size distribution, $d\lambda(r)/dr$, is a measurable property that can be obtained by standard porosimetry (Divisek et al., 1998). Mass conservation of water corresponds to

$$\nabla \cdot \mathbf{N}_l = 0. \qquad (5.31)$$

The proton current is determined by Ohm's law and by the conservation of charge $\mathbf{j} = -\sigma_p(\lambda)\nabla\Phi$ and $\nabla \cdot \mathbf{j} = 0$, where $\sigma_p(\lambda)$ is the PEM conductivity. Since the proton current is predominantly in through-plane direction, it is expedient to consider the problem as one-dimensional with scalar properties $j$ and $N_l$.

The model requires input of the functions $d\lambda(r)/dr$, $D^m(\lambda)$, $k_p^m(\lambda)$, $n_d(\lambda)$, and $\sigma_p(\lambda)$. These relations must be obtained from experiment. Parameterizations have been discussed in Eikerling et al. (1998, 2007a). Moreover, boundary conditions must be given for vapor pressures, gas pressures, and water and proton fluxes.

Logarithmic normal distributions are widely employed to represent experimental porosimetry data of ultrafiltration membranes. In the case of PFSA-type ionomer membranes, functions of the form

$$\frac{d\lambda\,(r)}{dr} = \frac{\lambda_s}{\Lambda}\left[\exp\left\{-\left(\frac{\log\,(r/r_m)}{\log s}\right)^2\right\} - \frac{r}{r_{\max}}\exp\left\{-\left(\frac{\log\,(r_{\max}/r_m)}{\log s}\right)^2\right\}\right]$$

(5.32)

have been shown to provide good results, where $\Lambda$ is a normalization factor, $\lambda_s$ the saturation water content, $r_m$ a parameter that determines the location of the maximum of the PSD, $s$ a measure of the width of the PSD, and $r_{\max}$ the maximal pore radius. A PSD of the form in Equation 40 with $r_m = 1$ nm, $s = 0.15$, and $r_{\max} = 100$ nm closely resembles standard porosimetry data for Nafion 117 (Divisek et al., 1998).

## RESULTS OF THE HYDRAULIC PERMEATION MODEL

The hydraulic permeation model in Eikerling et al. (1998, 2007a) assumed a negligible contribution of diffusional water flux. It is valid at sufficiently high water content. The model helped rationalizing main dependences of the critical current density on membrane parameters. A $\delta$-function-like pore size distribution $d\lambda\,(r)/dr = \lambda_{\max}\delta\,(r - r_1)$, which is completely determined by the maximum water uptake $\lambda_{\max}$, and by the first moment of the pore size distribution (i.e., the average pore size), $r_1$, provided an explicit expression for $j_{pc}$:

$$j_{pc} = \frac{1}{n_d}\left\{j_w + J_m\left(1 - \frac{\lambda_c}{\lambda_{\max}}\right)\right\},$$

(5.33)

where $n_d$ is assumed to be constant and $j_w$ is the net water flux through the PEM. The membrane parameter $J_m$, defined as

$$J_m = \frac{F\sigma\xi c_w\lambda_{\max}}{4\mu}\frac{r_1}{l_{PEM}},$$

(5.34)

depends on $r_1$ and the PEM thickness, $l_{PEM}$. Equations 5.33 and 5.34 suggest that PEMs with higher water uptake, larger pore radii, reduced thickness, and suppressed electro-osmotic drag are less prone to dehydration.

The theoretical analysis of the hydraulic permeation model provided an expression for the current density $j_{ps}$ at which membrane dehydration commences:

$$j_{ps} = \frac{j_w}{n_d} + \frac{Fc_w k_p^m\,(\lambda_{\max})}{n_d\mu}\frac{\Delta P^g}{l_{PEM}}.$$

(5.35)

Below $j_{ps}$, the PEM remains in a fully saturated state and it performs like a linear ohmic resistance. According to Equation 5.35, two modes of water management can be applied to compensate for electro-osmotic drag and, thus, keep the PEM well-hydrated. Sufficient replenishment of water in the membrane can be accomplished either by providing a steady external water supply, $j_w \geq n_d j$, from the anode side (imposed by higher RH at anode compared to cathode side), or by applying an external gas pressure gradient that enforces a sufficiently high internal hydraulic flux in the PEM from cathode to anode, $\Delta P^g = (n_d + v)\dfrac{\mu l_{PEM}}{F c_w k_p^m (\lambda_{\max})} j$. Here, the coefficient $v$ is dependent upon whether product water is removed predominantly via cathode ($v \simeq 0$) or anode ($v = 1/2$).

## DIFFUSION VERSUS HYDRAULIC PERMEATION: COMPARISON

The diffusion model and the hydraulic permeation model differ in their predictions of water content profiles and critical current densities. The origin of this discrepancy is the difference in the functions $D^m(\lambda)$ and $k_p^m(\lambda)$. This point was illustrated in the original publication on this model (Eikerling et al., 1998), where both flux terms, occurring in Equation 5.28, were converted into flux terms with gradients in water content, $\nabla\lambda$, as the driving force and with effective transport coefficients for diffusion, $D_{diff}^{\mathrm{eff}}(\lambda)$, and hydraulic permeation, $D_{hydr}^{\mathrm{eff}}(\lambda)$,

$$\mathbf{N}_\lambda = n_d(\lambda)\frac{\mathbf{j}}{F} - \left\{ D_{diff}^{\mathrm{eff}}(\lambda) + D_{hydr}^{\mathrm{eff}}(\lambda) \right\} \nabla\lambda. \tag{5.36}$$

Direct comparison of $D_{diff}^{\mathrm{eff}}(\lambda)$ and $D_{hydr}^{\mathrm{eff}}(\lambda)$ confirmed that hydraulic permeation dominates at high $\lambda$, whereas diffusion prevails at low $\lambda$.

The hydraulic permeation model predicts highly nonlinear water content profiles, with strong dehydration arising only in the interfacial regions close to the anode. Moreover, severe dehydration occurs only at current densities close to $j_{pc}$. The hydraulic permeation model is consistent with experimental data on water content profiles and differential membrane resistance (Buechi and Scherer, 2001; Mosdale et al., 1996; Xu et al., 2007; Zhang et al., 2008). Bare diffusion models exhibit marked discrepancies in comparison with these data.

Recently, it was shown that the hydraulic permeation model could explain why the membrane performance responds to variations in external gas pressures, observed in operating fuel cells by Renganathan et al. (2006). These authors analyzed the uniformity of water content distributions in operating PEMs by plotting the normalized PEM resistance

$$\frac{R_{PEM}}{R_s} = \frac{\sigma_p (\lambda_{\max})}{l_{PEM}} \int_0^{l_{PEM}} \frac{dz}{\sigma_p (\lambda (z))} \tag{5.37}$$

as a function of the current density of PEFC operation for various applied gas pressures in anode and cathode compartments, similar to what had been done earlier by Buechi and Scherer (2001). In Equation 5.37, $R_s = l_{PEM}/\sigma_p (\lambda_{\max})$ is the resistance of the uniformly saturated membrane (at open circuit conditions). Experimental data for the PEM resistance are reproduced very well within the hydraulic permeation model, as depicted in Figure 5.4. For the basic case with zero gas pressure gradient between cathode and anode side, $\Delta P^g = 0$, the model predicts uniform water distribution and constant membrane resistance at $j < 1$ A cm$^{-2}$ with a steep increase in

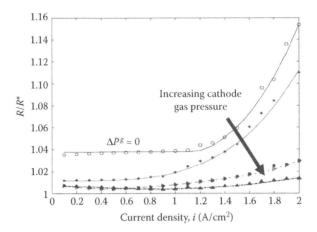

**FIGURE 5.4**   Resistance of a Nafion 112 PEM in an operating PEFC (70°C), as a function of the fuel cell current density. Experimental data (dots) are compared to the hydraulic permeation model for various applied gas pressure differences between cathode and anode. (Reprinted from *J. Power Sources*, Renganathan, S. et al., 2006. Polymer electrolyte membrane resistance model. **160**, 386–397, Figure 5, Copyright (2006), with permission from Elsevier.)

$R_{PEM}/R_s$ beyond this point. These trends are in excellent agreement with experimental data for Nafion 112. A finite positive gas pressure gradient, $\Delta P^g = P_c^g - P_a^g > 0$, improves the internal humidification of the membrane, leading to a more uniform water distribution and a significantly reduced dependence of membrane resistance on $\lambda$. The latter trends are consistent with predictions of the hydraulic permeation model.

The previous discussion suggests that hydraulic permeation is the dominant mode of water transport in Nafion at sufficiently large $\lambda$, whereas a diffusive contribution to water transport will dominate at low $\lambda$. This change in the prevailing mechanism of water transport with $\lambda$ could explain the peculiar transition in water concentration profiles through operating PEM observed in recent neutron scattering experiments.

So far, water management models have assumed a controlled net water flux $j_w$ through the PEM. The basic case evaluated in Eikerling et al. (1998) assumed $j_w = 0$. This approach is incomplete since it does not consider the coupling of water fluxes in the PEM to water fluxes in other components. In the simplest coupling scenario, continuity of water fluxes at the PEM interfaces with adjacent media requires a condition of interfacial water exchange, as suggested in the section "Dynamic Water Sorption and Flux in PEMs":

$$j_w = \frac{F}{2RT} k_v \xi_a P_a^s \left\{ \frac{P_a^v}{P_a^s} - \exp\left( \frac{\Delta G^s(\lambda_a)}{RT} \right) \right\} \tag{5.38}$$

on the anode side and

$$j_w = \frac{F}{2RT} k_v \xi_c P_c^s \left\{ \exp\left( \frac{\Delta G^s(\lambda_c)}{RT} \right) - \frac{P_c^v}{P_c^s} \right\} \tag{5.39}$$

on the cathode side. The factors $\xi_a$ and $\xi_c$ account for the heterogeneity of the interface. The interfacial flux conditions Equations 5.38 and 5.39, can be applied at plain interfaces of the PEM with adjacent homogeneous phases of water, either vapor or liquid. However, in PEFCs with ionomer-impregnated catalyst layers, the ionomer interfaces, with vapor and liquid water, are randomly dispersed inside of the porous composite media. This leads to a highly distributed heterogeneous interface. Attempts to incorporate vaporization exchange into models of catalyst layer operation are underway.

## WATER DISTRIBUTION AND FLUXES IN PEM

Imbalanced distribution of water, corresponding to finite gradients in water content $\lambda$ could be caused by applying differences in relative humidity ($\Delta RH$) or in pressure ($\Delta P$) between the opposite faces of the PEM. As well, it could be caused by the electro-osmotic effect that shuffles water molecules across the PEM along with a finite proton current. It is a commonly found misconception to identify the mechanism of water flux in the membrane with the nature of the external driving force, that is, to say for instance that when $\Delta RH \neq 0$ and $\Delta P = 0$, only vapor diffusion contributes to water flux. Local values of $\lambda$ determine the relative contributions of diffusion and hydraulic permeation of water. Hydraulic permeation prevails over diffusion at high $\lambda$ (large amount of bulk-like water). Diffusion is the only mechanism remaining at low $\lambda$. *Ex situ* measurements (at zero proton current), which measure the net water flux caused by applying a controlled difference in $\Delta RH$ or $\Delta P$, can be used to study the water transport or permeation properties of PEM. Models, discussed in this section, can be used to study contributions to the net water flux due to permeation in the bulk PEM via diffusion or hydraulic permeation and arising from interfacial vaporization exchange. Analysis of transient sorption and flux data with the vaporization exchange model allows isolating and extracting parameters for bulk water transport at low $\lambda$ (presumably corresponding to diffusion) and at high $\lambda$ (presumably corresponding to hydraulic permeation), as well as the interfacial vaporization exchange rate constant. Knowledge of these parameters is vital for performance modeling of operating PEMs in the PEFC. Moreover, evolution of the water distribution in the PEM over time can be studied using this model.

## SUMMARY: PEM OPERATION

Two major groups of performance models have been proposed. The first group considers the membrane as a homogeneous mixture of ionomer and water. The second group involves approaches that consider the membrane as a porous medium. Water vapor equilibrates with this medium by means of capillary forces, osmotic forces resulting from solvated protons and fixed ions, hydration forces, and elastic forces. In this scenario, the thermodynamic state of water in the membrane should be specified by (at least) two independent thermodynamic variables, namely, chemical potential and pressure, subdued to independent conditions of chemical and mechanical equilibrium, respectively. The homogeneous mixture model is the basis of the so-called diffusion model of PEM performance. The porous media model is the basis of the hydraulic permeation model of PEM performance. The literature on PEFC modeling has given preference to diffusion models. This prioritization is mainly based on simplicity. It is unjustified in view of the physics of structure and processes in membranes, which is more consistently accounted for in hydraulic permeation models. The difference between diffusion and hydraulic permeation models is not a mere cosmetic effect. Diffusion models fail in predicting effects of pressure variations on membrane water sorption, water transport, and electrochemical performance of PEFCs.

The most complete description of variations in local distributions and fluxes of water is provided by combination models, which allow for concurrent contributions of diffusion and hydraulic permeation to the water backflux that competes with the electro-osmotic drag. The main conclusions from these models for membrane water management under operation are

- *Local distribution of water.* During fuel cell operation, strong dehydration of the PEM arises in the interfacial regions close to the anode, whereas the other membrane regions remain in a well-hydrated state, close to saturation.
- *Critical current density.* A critical current density $j_{pc}$ exists at which the total membrane resistance increases strongly. It is reached when the water content at the anode side drops below the percolation value for proton conductance. For Nation-type PEMs, estimated critical current densities are found in the range of 1-10 A cm$^{-2}$. The critical $j_{pc}$ is proportional to the saturation water content and the first moment of the pore size distribution. As well, it is inversely proportional to the electro-osmotic drag coefficient and the membrane thickness. Large water uptake, larger medium pore sizes, suppressed electro-osmotic drag, and reduced membrane thickness result in improved water-back transport and more uniform distributions of water, thereby, reducing voltage losses due to nonuniform water distribution. All of these predicted trends are in agreement with experimental findings.
- *Nonlinear corrections.* Nonlinear corrections in current-voltage relations are only relevant in the proximity of the critical current density. Below this value, the ohmic resistance of the saturated state determines the membrane performance. Above the critical current density, the residual conductivity in dehydrated PEM parts, determines the performance. Ohmic resistance as well as nonohmic corrections in the proximity of the critical current density are smaller for thinner membranes.
- *Water management.* Water management in PEM can be enhanced by letting a large net water-flux pass from anode to cathode (high RH on anode side and low RH on cathode side) or by applying an external gas pressure gradient that pushes water hydraulically from cathode to anode. These measures, however, cannot be considered out of relation with the problems of flooding at the cathode. The optimum set of conditions may involve a combination of $\Delta RH$ and $\Delta P$. For thinner PEMs and PEMs with larger permeability, a smaller gas pressure gradient is sufficient to provide optimum humidification at a given current density. These suggestions still need further experimental verification.

In summary, design, integration and performance optimization of advanced PEM needs systematic experimental-theoretical efforts to focus on studying the effects of chemical modifications of the base ionomer, thereby, increasing the ion exchange capacity (to improve transport properties without sacrificing stability) and reducing thickness.

## PERFORMANCE MODELING OF A FUEL CELL

### INTRODUCTORY REMARKS

The polarization curve is a "portrait" of a fuel cell. Measuring this curve is a routine procedure performed in numerous labs worldwide. However, modeling of the polarization curve is one of the biggest challenges for fuel cell theory.

Generally, any electrochemical half-cell conversion requires protons, electrons, and neutral molecules. These species must be transported to catalyst sites, where the reaction events take place. Electrons that are transported typically do not pose any problems, as electron conductivity of fuel cell layers could be made sufficiently high. However, transport of protons and neutral species is not that simple. In a fuel cell, any hurdle in the transport of reaction participants transforms into potential loss, that is, any transport costs some potential. The transport expenses sum up with the kinetic overpotential required to activate the conversion reaction. Metaphorically speaking, while the open-circuit potential is the total capital, the cell polarization curve displays the money (potential) left at our disposal for a given value of current in the external load.

In many applications (for example, in automatic control systems), an analytical equation for the cell polarization curve is highly desirable. Such an equation is also useful in cell performance and aging studies: fitting an analytical equation to the measured curve may help to understand the contribution of kinetic and transport processes to the total potential loss in the cell.

A number of attempts have been made to suggest semiempirical analytical equations for the cell polarization curve (Boyer et al., 2000; Kim et al., 1995; Squadrito et al., 1999). However, the empirical equations usually do not follow from the conservation laws, which makes them unreliable. The predicting capability of these equations is limited.

In this section, conservation laws are used to derive analytical solutions for the polarization curve of the cathode side at finite oxygen stoichiometry, when either oxygen or proton transport in the CCL is poor. These equations help in understanding the type of transport loss in the CCL by fitting the cell polarization curve. Furthermore, the results of this chapter could be used as MEA submodels in CFD models of cells and stacks. Last but not least, the solutions below are simple enough to be used in real-time control systems.

### OXYGEN TRANSPORT LOSS IN THE GAS DIFFUSION LAYER

The equations in the section "CCL Polarization Curves" of Chapter 4 contain the oxygen concentration $c_1$ at the CCL/GDL interface. This concentration is, itself, related to the cell current density. In the GDL, the oxygen diffusion flux is proportional to the cell current density $j_0$,

$$D_b^c \frac{\partial c}{\partial x} = \frac{j_0}{4F}, \tag{5.40}$$

where $D_b^c$ is the oxygen diffusion coefficient in the GDL of a thickness $l_b^c$, and $c$ is the local oxygen concentration in the GDL. The superscript $c$ marks the values at the cathode side.

The right-hand side of Equation 5.40 is independent of $x$. Thus, $c$ is linear in $x$ and may be written (Figure 5.5) as

$$D_b^c \frac{c_h - c_1}{l_b^c} = \frac{j_0}{4F}.$$

Solving this relation for $c_1$ yields

$$c_1 = c_h - \frac{l_b^c j_0}{4F D_b^c}.$$

Dividing both sides of this relation by $c_h^0$, finally results in

$$\tilde{c}_1 = \tilde{c}_h - \frac{\tilde{j}_0}{\tilde{j}_{\lim}^{c0}}, \tag{5.41}$$

where

$$\tilde{j}_{\lim}^{c0} = \frac{4F D_b^c c_h^0}{l_b^c} \tag{5.42}$$

is the limiting current density resulting from oxygen transport in the GDL.

The polarization curve, which takes into account the oxygen transport loss in the GDL, results from the substitution of Equation 5.41 into the CL polarization curves obtained in the previous sections. Consider first the simplest Tafel equation, Equation

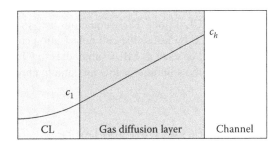

**FIGURE 5.5** A schematic of the oxygen concentration profile through the GDL and CCL thicknesses.

4.130. Substituting Equation 5.41 into Equation 4.130 yields

$$\tilde{\eta}_0 = \ln\left(2\varepsilon^2\tilde{j}_0\right) - \ln\left(\tilde{c}_h - \frac{\tilde{j}_0}{\tilde{j}_{lim}^{c0}}\right). \tag{5.43}$$

The first term on the right describes the activation potential loss (the overpotential needed to activate the ORR). The second term represents the potential loss resulting from the oxygen transport in the GDL. As can be seen, when $\tilde{c}_h = \tilde{j}_0/\tilde{j}_{lim}^{c0}$, the second logarithm tends to minus infinity and the cell potential tends to zero. Physically, this condition means that the oxygen concentration $c_1$ at the CCL/GDL interface tends to zero (Figure 5.5), and the cell is not able to produce a higher current. In dimensional variables, Equation 5.43 reads

$$\eta_0 = b\ln\left(\frac{j_0}{i_* l_{CL}}\right) - b\ln\left(\frac{c_h}{c_h^0} - \frac{j_0}{j_{lim}^{c0}}\right). \tag{5.44}$$

Quite analogous, substituting Equation 5.41 into other expressions for $\tilde{\eta}_0$, derived in Chapter 4, gives the *local polarization curve of the cathode side*, which takes into account the oxygen transport loss in the GDL. For example, substituting Equation 5.41 into Equation 4.141 gives the general local polarization curve of the cell with ideal oxygen transport in the CCL

$$\tilde{\eta}_0 = \text{arcsinh}\left(\frac{\varepsilon^2\tilde{j}_0^2}{2\left(1 - \exp(-\tilde{j}_0/2)\right)}\right) - \ln\left(\tilde{c}_h - \frac{\tilde{j}_0}{\tilde{j}_{lim}^{c0}}\right). \tag{5.45}$$

Again, the second logarithm describes the potential loss due to oxygen transport in the GDL. In the dimensional form, this equation reads

$$\eta_0 = b\,\text{arcsinh}\left(\frac{j_0^2/(2j_{\sigma 0}^2)}{1 - \exp(-j_0/(2j_{ref}))}\right) - b\ln\left(\frac{c_h}{c_h^0} - \frac{j_0}{j_{lim}^{c0}}\right), \tag{5.46}$$

where

$$j_{\sigma 0} = \sqrt{2i_*\sigma_p b}. \tag{5.47}$$

Polarization curves (5.46) corresponding to the two values of $D_b^c$ are depicted in Figure 5.6a. As can be seen, both curves exhibit the effect of limiting current density, which is proportional to $D_b^c$.

## VOLTAGE LOSS DUE TO OXYGEN TRANSPORT IN THE CHANNEL

### Oxygen Mass Conservation in the Channel

The equations of the previous section allow us to take a next step by taking into account the potential loss resulting from the oxygen transport in the channel. Suppose that the cell has a straight oxygen channel. Let the coordinate $z$ be directed along the

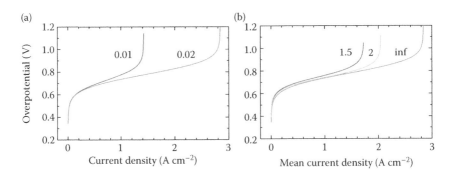

**FIGURE 5.6** (a) A local polarization curve of the cathode side of a PEFC (5.46) for the indicated values of the oxygen diffusion coefficient in the cathode GDL (cm$^2$ s$^{-1}$). Both the curves correspond to infinite oxygen stoichiometry $\lambda = \infty$. (b) A polarization curve with the account of finite oxygen stoichiometry (5.59) for the indicated $\lambda$ and $D_b^c = 0.02$ cm$^2$ s$^{-1}$. The other parameters are listed in Table 5.7.

channel from inlet to outlet (Figure 5.7). Assume that the flow in the cathode channel is a well-mixed one with constant velocity (plug flow). This assumption is reasonable for air cathodes (Kulikovsky, 2001).

The oxygen mass conservation equation in the air channel is

$$v^0 \frac{\partial c_h}{\partial z} = -\frac{j_0}{4Fh}, \tag{5.48}$$

where $v^0$ is the flow velocity and $h$ is the channel depth. Equation 5.48 implies that the oxygen concentration $c_h$ decays along $z$ at a rate proportional to the local current density $j_0$.

Introducing dimensionless $\tilde{z}$ according to

$$\tilde{z} = \frac{z}{L}, \tag{5.49}$$

Equation 5.48 takes the form

$$\lambda \tilde{J} \frac{\partial \tilde{c}_h}{\partial \tilde{z}} = -\tilde{j}_0, \quad \tilde{c}_h(0) = 1. \tag{5.50}$$

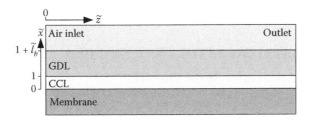

**FIGURE 5.7** Schematic of a cathode side of the cell with the linear oxygen channel.

Here,

$$\lambda = \frac{4Fhv^0 c_h^0}{LJ} \tag{5.51}$$

is the oxygen stoichiometry and $J$ is the mean current density in the cell. Integrating Equation 5.50 over $\tilde{z}$ from 0 to 1, and taking into account that

$$\int_0^1 \tilde{j}_0 \, d\tilde{z} = \tilde{J}, \tag{5.52}$$

a useful relation is obtained for the oxygen concentration $\tilde{c}_1$ at the channel outlet

$$\tilde{c}_1 = 1 - \frac{1}{\lambda}. \tag{5.53}$$

## Low-Current Polarization Curve

To rationalize the effect of oxygen stoichiometry $\lambda$ on the potential loss, consider first the case of a small cell current. In this case, the local polarization curve is given by Equation 5.43. To calculate the polarization curve as a function of *mean* cell current density $\tilde{\eta}_0(\tilde{J})$, it is advisable to divide the terms under the logarithms in Equation 5.43 by $\tilde{c}_h$. This yields

$$\tilde{\eta}_0 = \ln\left(\frac{2\varepsilon^2 \tilde{j}_0}{\tilde{c}_h}\right) - \ln\left(1 - \frac{\tilde{j}_0}{\tilde{j}_{lim}^{c0} \tilde{c}_h}\right). \tag{5.54}$$

This procedure does not change $\tilde{\eta}_0$, as the logarithms in Equation 5.54 appear with different signs. Suppose that $\tilde{\eta}_0$ is independent of $\tilde{z}$ (the rationale for this assumption is discussed below). From Equation 5.54, one can see that $\tilde{\eta}_0$ is constant along $\tilde{z}$, provided that the ratio $\tilde{j}_0/\tilde{c}_h$ is independent of $\tilde{z}$. Note that in that case, both logarithms in Equation 5.54 are constant.

Denoting $\tilde{j}_0 = \gamma \tilde{c}_h$, where $\gamma > 0$ is constant, substituting this relation into Equation 5.50, and solving the resulting equation yields $\tilde{c}_h = \exp\left(-\gamma \tilde{z}/(\lambda \tilde{J})\right)$. Using Equation 5.53 to calculate $\gamma$, one finally obtain (Kulikovsky, 2004)

$$\tilde{c}_h = \left(1 - \frac{1}{\lambda}\right)^{\tilde{z}} \tag{5.55}$$

and

$$\tilde{j}_0 = \tilde{J} f_\lambda \left(1 - \frac{1}{\lambda}\right)^{\tilde{z}}, \tag{5.56}$$

where

$$f_\lambda = -\lambda \ln\left(1 - \frac{1}{\lambda}\right) \tag{5.57}$$

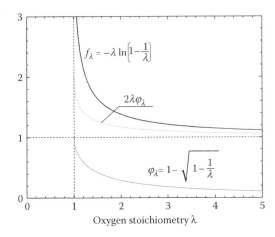

**FIGURE 5.8**  Functions of oxygen stoichiometry $\lambda$ appearing in the analysis.

is a function of the oxygen stoichiometry only. Equations 5.55 and 5.56 reveal that the oxygen concentration and the local current density vary exponentially with $\tilde{z}$, with the characteristic scale of the exponent dependent on the oxygen stoichiometry, $\lambda$.

From Equations 5.55 and 5.56, it is clear that $\tilde{j}_0/\tilde{c}_h = f_\lambda \tilde{J}$. Using this in Equation 5.54, gives the integral polarization curve of the cathode side

$$\tilde{\eta}_0 = \ln\left(2\varepsilon^2 f_\lambda \tilde{J}\right) - \ln\left(1 - \frac{f_\lambda \tilde{J}}{\tilde{j}_{\text{lim}}^{c0}}\right). \tag{5.58}$$

In dimensional form, this equation reads

$$\eta_0 = b\ln\left(\frac{f_\lambda J}{i_* l_{CL}}\right) - b\ln\left(1 - \frac{f_\lambda J}{j_{\text{lim}}^{c0}}\right). \tag{5.59}$$

As can be seen, the function $f_\lambda$ simply rescales the mean cell current density $J$. In particular, finite $\lambda$ lowers the effective limiting current by a factor $f_\lambda$. The plot of $f_\lambda$ is given in Figure 5.8. This function tends to infinity as $\lambda \to 1$ and it tends to 1 as $\lambda \to \infty$. The plot of the cathode side overpotential, in Equation 5.58 for the three values of $\lambda$, is shown in Figure 5.6b.

## Why $\eta_0$ is Nearly Constant along the Channel?

In PEFCs, the constancy of $\tilde{\eta}_0$ along $\tilde{z}$ results from the following argumentation. Generally, the local cell voltage $E_{cell}$ is

$$E_{cell} = E_{oc} - \eta_0 - R_\Omega j_0. \tag{5.60}$$

Here, $E_{oc}$ is the cell open-circuit potential (independent of $z$)* and $R_\Omega$ is the cell ohmic resistivity, which includes the membrane and contact resistivities. Typically, the variation of the ohmic term in Equation 5.60 along $\tilde{z}$ is not large, and this term can be approximated by $R_\Omega J$.

Owing to high electric conductivity, the cell electrodes are equipotential. Thus, in Equation 5.60, one has $E_{cell}$, $E_{oc}$, and $R_\Omega J$, which are independent of $z$. Therefore, $\eta_0$ also must be independent of $z$.

Note that this is correct if the cell resistivity $R_\Omega$ is small, or the variation of the local cell current with $\tilde{z}$ is not large. If, for example, membrane humidification is strongly nonuniform along $z$, the variation of the ohmic term is large, and this variation must be compensated by the respective variation in $\eta_0$. Writing Equation 5.60 in the form

$$\eta_0 + R_\Omega j_0 = E_{oc} - E_{cell}, \tag{5.61}$$

shows that in general, the sum on the left must be constant along $z$.

### Large Current, Poor Oxygen Diffusivity in the CCL

What is the effect of finite $\lambda$ at high currents? In this section, the case will be considered for poor oxygen diffusivity and ideal proton conductivity of the CCL. In this case, the local polarization curve of the CCL is given by Equation 4.87. Substituting Equation 5.41 into Equation 4.87, one obtains the local polarization curve with the term describing oxygen transport in the GDL:

$$\tilde{\eta}_0 = 2\ln\left(\varepsilon\tilde{j}_0\sqrt{\frac{2}{\tilde{D}}}\right) - 2\ln\left(\tilde{c}_h - \frac{\tilde{j}_0}{\tilde{j}_{\lim}^{c0}}\right). \tag{5.62}$$

Dividing the terms under both logarithms by $\tilde{c}_h$ (which does not change $\tilde{\eta}_0$) yields

$$\tilde{\eta}_0 = 2\ln\left(\frac{\varepsilon\tilde{j}_0}{\tilde{c}_h}\sqrt{\frac{2}{\tilde{D}}}\right) - 2\ln\left(1 - \frac{\tilde{j}_0}{\tilde{j}_{\lim}^{c0}\tilde{c}_h}\right). \tag{5.63}$$

Assuming, again, that $\tilde{\eta}_0$ is independent of $\tilde{z}$ (the section "Why $\eta_0$ is Nearly Constant along the Channel?"), this can only be possible if the ratio $\tilde{j}_0/\tilde{c}_h$ is constant. Repeating the calculations in the section "Low-current Polarization Curve" leads to the cell polarization curve (Kulikovsky, 2011a)

$$\tilde{\eta}_0 = 2\ln\left(\varepsilon f_\lambda\tilde{J}\sqrt{\frac{2}{\tilde{D}}}\right) - 2\ln\left(1 - \frac{f_\lambda\tilde{J}}{\tilde{j}_{\lim}^{c0}}\right), \tag{5.64}$$

where $f_\lambda$ is given by Equation 5.57. For the $\tilde{z}$ shapes we still obtain Equations 5.55 and 5.56.

---

* By definition, $E_{oc}$ must be measured at equilibrium, that is, at zero cell current. In this case, the oxygen concentration is constant over the cell surface and, according to the Nernst equation, $E_{oc}$ is independent of coordinates.

In dimensional form, Equation 5.64 reads

$$\eta_0 = 2b \ln \left( \frac{f_\lambda J}{j_{D0}} \right) - 2b \ln \left( 1 - \frac{f_\lambda J}{j_{\lim}^{c0}} \right), \tag{5.65}$$

where

$$j_{D0} = \sqrt{4FDc_h^0 i_*}. \tag{5.66}$$

Equation 5.65 exhibits the Tafel slope doubling (the factor $2b$ in the first term on the right-hand side). Moreover, poor oxygen transport in the CCL doubles the potential loss due to the oxygen transport in the GDL (the factor $2b$ in the second term). Physically, when the oxygen transport in the CCL is poor, a much larger oxygen flux must be transported through the GDL. This effect is one of the main reasons for very poor performance of a cell with a flooded CCL.

## Large Current, Poor Proton Transport in the CCL

In that case, the local polarization curve of the CCL is given by Equation 4.133. Substituting Equation 5.41 into Equation 4.133, we obtain the local polarization curve for the cathode side of the cell with poor ionic transport in the CCL:

$$\tilde{\eta}_0 = 2\ln \left( \varepsilon \tilde{j}_0 \right) - \ln \left( \tilde{c}_h - \frac{\tilde{j}_0}{\tilde{j}_{\lim}^{c0}} \right). \tag{5.67}$$

Adding and subtracting on the right side $\ln \tilde{c}_h$, leads to

$$\tilde{\eta}_0 = 2\ln \left( \frac{\varepsilon \tilde{j}_0}{\sqrt{\tilde{c}_h}} \right) - \ln \left( 1 - \frac{\tilde{j}_0}{\tilde{j}_{\lim}^{c0} \tilde{c}_h} \right). \tag{5.68}$$

The equation $\tilde{\eta}_0 = $ const. has to be solved together with Equation 5.50. The exact analytical solution of this system is hardly achievable.

Before proceeding to numerical results, it is advisable to consider the case of ideal oxygen transport in the GDL. Neglecting the second logarithm in Equation 5.68, the constancy of $\tilde{\eta}_0$ is provided if $\tilde{j}_0 / \sqrt{\tilde{c}_h} = \gamma$, where $\gamma$ is constant. Substituting $\tilde{j}_0 = \gamma \sqrt{\tilde{c}_h}$ into Equation 5.50 and solving the resulting equation yields

$$\tilde{c}_h = \left( 1 - \frac{\gamma \tilde{z}}{2\lambda \tilde{J}} \right)^2.$$

Setting $\tilde{z} = 1$ here, equating the result to the right side of Equation 5.41 and solving the resulting equation for $\gamma$ gives $\gamma = \tilde{J} 2\lambda \phi_\lambda$. With this, one obtains

$$\tilde{c}_h = (1 - \phi_\lambda \tilde{z})^2 \tag{5.69}$$

and

$$\tilde{j}_0 = \tilde{J}\, 2\lambda\phi_\lambda\left(1 - \phi_\lambda \tilde{z}\right), \tag{5.70}$$

where

$$\phi_\lambda = 1 - \sqrt{1 - \frac{1}{\lambda}}. \tag{5.71}$$

Thus, the oxygen concentration decreases parabolically with $\tilde{z}$, while the local current density decays linearly along the channel.

Substituting Equations 5.69 and 5.70 into the first term in Equation 5.68, results in the cell polarization curve under ideal oxygen transport in the GDL:

$$\tilde{\eta}_0 = 2\ln\left(2\lambda\phi_\lambda\varepsilon\tilde{J}\right).$$

To take into account the oxygen transport in the GDL, the system (5.68) and (5.50) has to be solved numerically. Analysis shows that the numerical solution is well approximated by the following relation (Kulikovsky, 2011a):

$$\tilde{\eta}_0 = 2\ln\left(2\lambda\phi_\lambda\varepsilon\tilde{J}\right) - \ln\left(1 - \frac{f_\lambda\tilde{J}}{\tilde{j}^{c0}_{\lim}}\right). \tag{5.72}$$

Note that in Equation 5.72, the transport term is not doubled. In dimensional form, this equation reads

$$\eta_0 = 2b\ln\left(\frac{2\lambda\phi_\lambda J}{j_{\sigma 0}}\right) - b\ln\left(1 - \frac{f_\lambda J}{j^{c0}_{\lim}}\right). \tag{5.73}$$

The functions $\phi_\lambda$ and $2\lambda\phi_\lambda$ are shown in Figure 5.8. The function $\phi_\lambda$ tends to unity as $\lambda \to \infty$, and it tends to 2 as $\lambda \to 1$ (Figure 5.8). From Equation 5.73, it follows that for the activation polarization (the first term on the right-hand side of Equation 5.73), regardless of the cell current, one obtains $\eta^{act}_0|_{\lambda=1} = \eta^{act}_0|_{\lambda=\infty} + 2b\ln 2$. With $b = 50$ mV, lowering of $\lambda$ from infinity to unity increases the cell activation polarization potential by only 30 mV. Physically, in the regime with poor proton transport in the CCL, the local activation overpotential rather weakly depends on the oxygen concentration (see the first term in Equation 5.68). Thus, variation of $\lambda$ is less important in this case, as compared to the case of poor oxygen transport in the CCL (the section "Large Current but Poor Oxygen Diffusivity in the CCL").

The results of the section "Voltage Loss due to the Oxygen Transport in the Channel" are summarized in Tables 5.2 and 5.3. As can be seen, the low-current polarization curves contain neither $\sigma_p$ nor $D$. Thus, in the low-current regime, the cell performance is not affected by the species transport in the CCL. The performance is determined by oxygen transport in channel and GDL and by reaction kinetics (through the parameters $b$ and $i_*$).

In the high-current mode, the local and integral polarization curves depend on the transport parameter, which limits the performance: this parameter is $D$ in the oxygen-limiting regime, and it is $\sigma_p$ in the proton-limiting mode. In all the cases listed in Tables 5.2 and 5.3, the pair of activation and transport terms are unique. Thus, fitting the cell polarization curve to Equation 5.60 with one of the three alternatives for $\eta_0$ might give information on the regime of CCL operation.

Note that in the high-current proton-limiting case, the $\lambda$ dependence of activation polarization is represented by the function $2\lambda\phi_\lambda$, which differs from $f_\lambda$, appearing in the activation terms in the other equations. The asymptotic behavior of $f_\lambda$ and $2\lambda\phi_\lambda$ as $\lambda \to 1$ are different: $f_\lambda$ tends to infinity, while $2\lambda\phi_\lambda$ tends to 2 (Figure 5.8). Also note that the combination of high current and poor oxygen transport is very detrimental for the cell performance: the transport term is twice larger than in the other cases, which results in a much larger polarization potential.

---

## TABLE 5.2
## Local and Integral Dimensionless Polarization Curves of the PEFC Cathode Side

| | Dimensionless Equations | |
|---|---|---|
| **Regime** | **Local Curve** | **Integral Curve** |
| Low current | $\ln\left(2\varepsilon^2\tilde{j}_0\right) - \ln\left(\tilde{c}_h - \dfrac{\tilde{j}_0}{\tilde{j}_{\lim}^{c0}}\right)$ | $\ln\left(2\varepsilon^2 f_\lambda\tilde{J}\right) - \ln\left(1 - \dfrac{f_\lambda\tilde{J}}{\tilde{j}_{\lim}^{c0}}\right)$ |
| High current | | |
| — Poor $O_2$ | $2\ln\left(\varepsilon\tilde{j}_0\sqrt{\dfrac{2}{\tilde{D}}}\right) - 2\ln\left(\tilde{c}_h - \dfrac{\tilde{j}_0}{\tilde{j}_{\lim}^{c0}}\right)$ | $2\ln\left(\varepsilon f_\lambda\tilde{J}\sqrt{\dfrac{2}{\tilde{D}}}\right) - 2\ln\left(1 - \dfrac{f_\lambda\tilde{J}}{\tilde{j}_{\lim}^{c0}}\right)$ |
| — Poor $H^+$ | $2\ln\left(\varepsilon\tilde{j}_0\right) - \ln\left(\tilde{c}_h - \dfrac{\tilde{j}_0}{\tilde{j}_{\lim}^{c0}}\right)$ | $2\ln\left(2\lambda\phi_\lambda\varepsilon\tilde{J}\right) - \ln\left(1 - \dfrac{f_\lambda\tilde{J}}{\tilde{j}_{\lim}^{c0}}\right)$ |
| Any current, | | |
| — Poor $H^+$ | $\text{arcsinh}\left(\dfrac{\varepsilon^2\tilde{j}_0^2}{2\left(1 - \exp(-\tilde{j}_0/2)\right)}\right)$ $-\ln\left(\tilde{c}_h - \dfrac{\tilde{j}_0}{\tilde{j}_{\lim}^{c0}}\right)$ | |
| Medium current, | | |
| — Poor $H^+$ and $O_2$ | $\text{arcsinh}\left(\dfrac{\varepsilon^2\tilde{j}_0^2}{2\tilde{c}_h\left(1 - \exp(-\tilde{j}_0/2)\right)}\right)$ $+\dfrac{1}{\tilde{c}_h\tilde{D}}\left(\tilde{j}_0 - \ln\left(1 + \dfrac{\tilde{j}_0^2}{\gamma^2}\right)\right)$ $\times\left(1 - \dfrac{\tilde{j}_0}{\tilde{j}_{\lim}^{c0}\tilde{c}_h}\right)^{-1}$ $-\ln\left(1 - \dfrac{\tilde{j}_0}{\tilde{j}_{\lim}^{c0}\tilde{c}_h}\right)$ | |

Note that at large $\lambda$, we have $f_\lambda \to 1$, $2\lambda\phi_\lambda \to 1$, $c_h \to c_h^0$, $j_0 \to J$, and the local and integral curves coincide.

## TABLE 5.3
## Local and Integral Dimensional Polarization Curves of the PEFC Cathode Side

| Regime | Dimensional Equations | |
|---|---|---|
| | Local Curve | Integral Curve |
| Low current | $b \ln\left(\dfrac{j_0}{i_* l_{CL}}\right) - b \ln\left(\dfrac{c_h}{c_h^0} - \dfrac{j_0}{j_{lim}^{c0}}\right)$ | $b \ln\left(\dfrac{f_\lambda J}{i_* l_{CL}}\right) - b \ln\left(1 - \dfrac{f_\lambda J}{j_{lim}^{c0}}\right)$ |
| High current | | |
| — Poor $O_2$ | $2b \ln\left(\dfrac{j_0}{j_{D0}}\right) - 2b \ln\left(\dfrac{c_h}{c_h^0} - \dfrac{j_0}{j_{lim}^{c0}}\right)$ | $2b \ln\left(\dfrac{f_\lambda J}{j_{D0}}\right) - 2b \ln\left(1 - \dfrac{f_\lambda J}{j_{lim}^{c0}}\right)$ |
| — Poor $H^+$ | $2b \ln\left(\dfrac{j_0}{j_{\sigma 0}}\right) - b \ln\left(\dfrac{c_h}{c_h^0} - \dfrac{j_0}{j_{lim}^{c0}}\right)$ | $2b \ln\left(\dfrac{2\lambda\phi_\lambda J}{j_{\sigma 0}}\right) - b \ln\left(1 - \dfrac{f_\lambda J}{j_{lim}^{c0}}\right)$ |
| Any current, | | |
| — Poor $H^+$ | $b \,\mathrm{arcsinh}\left(\dfrac{j_0^2/(2j_{\sigma 0}^2)}{1 - \exp(-j_0/(2j_{ref}))}\right)$ | |
| | $-b \ln\left(\dfrac{c_h}{c_h^0} - \dfrac{j_0}{j_{lim}^{c0}}\right)$ | |
| Medium current, | | |
| — Poor $H^+$ and $O_2$ | $b \,\mathrm{arcsinh}\left(\dfrac{j_0^2/(2j_{\sigma 0}^2)}{1 - \exp(-j_0/(2j_{ref}))}\right)$ | |
| | $+\dfrac{\sigma_p b^2}{nFDc_h}\left(\dfrac{j_0}{j_{ref}} - \ln\left(1 + \dfrac{j_0^2}{j_{ref}^2 \gamma^2}\right)\right)$ | |
| | $\times\left(1 - \dfrac{j_0 c_h^0}{j_{lim}^{c0} c_h}\right)^{-1}$ | |
| | $-b \ln\left(1 - \dfrac{j_0 c_h^0}{j_{lim}^{c0} c_h}\right)$ | |

## POLARIZATION CURVE FITTING

The equation in the last row of Table 5.3

$$\eta_0 = b \,\mathrm{arcsinh}\left(\frac{j_0^2/(2j_{\sigma 0}^2)}{1 - \exp(-j_0/(2j_{ref}))}\right)$$

$$+ \frac{\sigma_p b^2}{nFDc_h}\left(\frac{j_0}{j_{ref}} - \ln\left(1 + \frac{j_0^2}{j_{ref}^2 \gamma^2}\right)\right)\left(1 - \frac{j_0 c_h^0}{j_{lim}^{c0} c_h}\right)^{-1}$$

$$- b \ln\left(1 - \frac{j_0 c_h^0}{j_{lim}^{c0} c_h}\right) \quad (5.74)$$

## TABLE 5.4
### Common Parameters for the Curves in Figure 5.9

| | |
|---|---|
| GDL thickness $l_b$ (cm) | 0.025* (250 μm) |
| CCL thickness $l_{CL}$ (cm) | 0.001* (10 μm) |
| Membrane thickness $l_m$ (cm) | 0.0025* (25 μm) |
| Oxygen concentration $c_h^*$ ($p = 1$ bar) (mol cm$^{-3}$) | $7.36 \cdot 10^{-6}$ |
| Cell open-circuit potential $V_{oc}$ (V) | 1.145* |
| **Fitting parameters** | |
| CCL proton conductivity $\sigma_p$ ($\Omega^{-1}$ cm$^{-1}$) | 0.03 |
| Tafel slope $b$ (V) | 0.03 |

*Note:* The data marked with the symbol * are taken from Dobson, P. et al. 2012. *J. Electrochem. Soc.*, **159**, B514–B523.

is obtained by substitution of Equation 5.41 into Equation 4.141. This equation can be used for fitting PEFC polarization curves, provided that the two conditions are met: (i) oxygen stoichiometry is at large ($\lambda \gtrsim 5$) and (ii) the cell is not flooded locally.

The cell potential is given by Equation 5.60. A set of PEFC polarization curves corresponding to different operating temperatures, pressures, and relative humidities of the inlet gases has been published by Dobson et al. (2012). Equation 5.60 with $\eta_0$ given by Equation 5.74, has been fitted to the data of Dobson et al. (2012) obtained at normal pressure of 1 bar on either side of the cell and temperature of 353 K. Dobson et al. (2012) reported two polarization curves corresponding to different humidities of the inlet gases. Fitting has been performed using intrinsic Maple procedure *NonlinearFit* (for details, see Kulikovsky (2014)).

The common parameters for curve fitting are listed in Table 5.4, while the curve-specific parameters are gathered in Table 5.5. The results of fitting are shown in Figure 5.9.

## TABLE 5.5
### Fitting Parameters for the Model Polarization Curves in Figure 5.9

| Curve | 1 | 2 |
|---|---|---|
| **RHA:RHC** | **0.7:0.7** | **0.5:0.5** |
| $i_*$ (A cm$^{-3}$) | $0.817 \cdot 10^{-3}$ | $0.942 \cdot 10^{-3}$ |
| $R_\Omega$ ($\Omega$ cm$^2$) | 0.126 | 0.207 |
| $D_b$ (cm$^2$ s$^{-1}$) | 0.0259 | 0.0227 |
| $D$ (cm$^2$ s$^{-1}$) | $1.36 \cdot 10^{-4}$ | $2.13 \cdot 10^{-4}$ |

*Note:* The second row shows relative humidities of the anode and cathode in the format RHA:RHC. Cell temperature is 353 K and pressure is 1 bar.

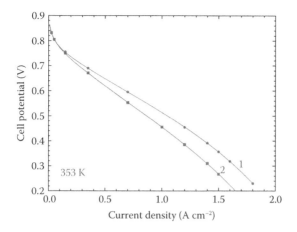

**FIGURE 5.9**   Points: experiment (Dobson et al. 2012). Lines: Equation 5.60 with $\eta_0$ given by Equation 5.74. The experimental conditions and oxygen diffusion coefficients are listed in Table 5.5. Common for all the curves, fitting parameters are collected in Table 5.4.

For both curves, fitting yields very close values of the exchange current density of $i_* \simeq 10^{-3}$ A cm$^{-3}$ (Table 5.5). This value of $i_*$ is nearly two orders of magnitude less than the one obtained in Dobson et al. (2012) from the same set of curves using a flooded agglomerate model. Note that, as discussed in Dobson et al. (2012), their value is an order of magnitude higher than expected; thus, the values of $i_*$ indicated in Table 5.5 could be closer to reality.

The ohmic cell resistivities $R_\Omega$ are 0.13 and 0.21 $\Omega$ cm$^2$ for curves 1 and 2, respectively. This variation can be attributed to lower membrane conductivity because of the lower amount of water in the inlet gases for curve 2.

The GDL oxygen diffusivity is fairly constant: 0.026 and 0.023 cm$^2$ s$^{-1}$ for curves 1 and 2, respectively (Table 5.5). Taking the oxygen diffusivity in air in the order of $D_{free} = 0.2$ cm$^2$ s$^{-1}$, and using the equation for Bruggemann correction $D_b = D_{free}\epsilon_{GDL}^{3/2}$, for the GDL porosity, one obtains $\epsilon_{GDL} \simeq 0.26$. This low value suggests that the GDL is partially flooded.

The CCL oxygen diffusivities resulting from fitting are $1.4 \cdot 10^{-4}$ and $2.1 \cdot 10^{-4}$ cm$^2$ s$^{-1}$ for curves 1 and 2, respectively (Table 5.5). Direct measurement of $D$, in a similar Nafion-based catalyst layer, gave an order of magnitude larger $D = 1.37 \cdot 10^{-3}$ cm$^2$ s$^{-1}$ (Shen et al., 2011). Note, however, that $D$ strongly depends on the amount of liquid water in the CCL: lower liquid saturation leads to higher $D$, the trend correctly captured by the model. Overall, the results of fitting suggest significant flooding of the CCL.

Figure 5.10 shows the overpotentials corresponding to the terms in Equation 5.74 for curve 2. At a typical current of 1 A cm$^{-2}$, the ratio of activation:resistive:CCL transport losses is about 2:1:0.1. The transport loss in the GDL is small. Note that the potential loss due to oxygen transport in the CCL markedly increases with the growth of the cell current density.

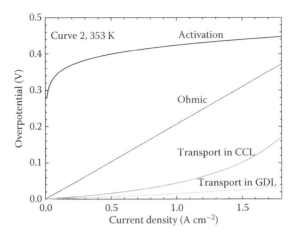

**FIGURE 5.10**   Overpotentials for curve 2 in Figure 5.9.

## PHYSICAL MODELING OF CATALYST LAYER IMPEDANCE

### INTRODUCTORY REMARKS

Electrochemical impedance spectroscopy (EIS) is a powerful tool for studying electrochemical systems (Bard and Faulkner, 2000; Orazem and Tribollet, 2008). Response of an electrochemical system to a small-amplitude sinusoidal electric perturbation gives more information than the steady-state polarization curve. The key aspect in impedance measurements is the possibility of changing the frequency $\omega$ of the exciting signal. This brings additional "dimensionality" to measurements: varying $\omega$, one may determine eigenfrequencies and impedances of the system that correspond to processes with different characteristic timescales.

The impedance spectrum of an electrochemical system is most often represented in Nyquist coordinates $(\Re(Z), \Im(Z))$, where $Z$ is the system impedance. In this representation, the spectrum intersections with the real axis $\Re(Z)$ give the main resistivities in the system. Fitting model equations to experimental impedance spectra enables determining the kinetic and transport parameters of the electrochemical system. In recent years, this technique has been widely used to study fuel cells and cell components (Orazem and Tribollet, 2008; Yuan et al., 2009).*

A key element of PEM fuel cell is the CCL, which contains a double layer capacitance connected to "resistivities" due to proton and oxygen transport. To understand measured CCL spectra, many works employ the *equivalent circuit method* (ECM). This method is based on the construction of an equivalent transmission line, which gives an EIS spectrum close to the spectrum of the system of interest. The components of the resulting circuit are then attributed to CL physical parameters. For example, in a recent work, Nara et al. (2011) used the ECM to study the CCL degradation mechanisms in the PEFC.

---

* Relevant literature accounts of several hundred items. For this reason, the two recent books, which cover the field, are referred to.

Unfortunately, being simple and fast, the ECM is not reliable, since the equivalent circuit is not unique, and it usually ignores important minor features of the spectrum. More reliable information provides the physical modeling of the CL impedance. Following the classical work of de Levie (1967), Lasia (1995, 1997) obtained several fundamental solutions for impedance of the cylindrical porous electrode. Eikerling and Kornyshev (1999) published numerical and analytical studies of the CCL impedance, based on the macrohomogeneous model for electrode performance. Recently, Makharia et al. (2005) used a similar model for processing the experimental spectra. Jaouen and Lindbergh (2003) derived equations for the CCL impedance based on a flooded agglomerate model. An overview of CCL impedance studies has been published by Gomadam and Weidner (2005).

In this section, the physical modeling of impedance of a CCL is discussed. The model is based on a macrohomogeneous model for the CCL performance discussed in Chapter 4. To demonstrate the idea of EIS, the impedance of a simple parallel RC-circuit is considered first.

## IMPEDANCE OF A PARALLEL RC-CIRCUIT

The idea of impedance spectroscopy can be best understood with the following example. Consider a parallel RC-circuit and suppose that the values of $R$ and $C$ are unknown (Figure 5.11). To determine these values, a sinusoidal potential $\phi(t)$ is applied to the circuit and a response current is measured between points $a$ and $b$ (Figure 5.11).

To calculate the total current between $a$ and $b$, it is noted that the current in the resistor $R$ is simply $\phi/R$, while the AC current in the capacitance $C$ is $\partial q/\partial t = C\partial\phi/\partial t$, where $q$ is the instant charge of the capacitor. Summing these contributions gives the total current between $a$ and $b$:

$$I = \frac{\phi}{R} + C\frac{\partial\phi}{\partial t}. \tag{5.75}$$

The potential $\phi(t)$ is harmonic in time. Since our circuit is linear, the total current induced in the system is also harmonic. Thus, $\phi$ and $I$ can be represented as

$$\phi(t) = \hat{\phi}(\omega)\exp(i\omega t),$$

$$I(t) = \hat{I}(\omega)\exp(i\omega t),$$

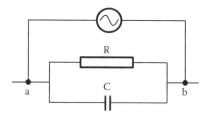

**FIGURE 5.11** The calculation of RC-circuit impedance.

where $\omega = 2\pi f$ is the circular frequency (rad$^{-1}$), $f$ is the regular frequency (Hz), and $\hat{\phi}(\omega)$ and $\hat{I}(\omega)$ are the complex amplitudes, with the exponents describing harmonic variations in time. Note that the phase of the current is included in the amplitude function $\hat{I}(\omega)$.

Substituting these Fourier transforms into Equation 5.75, one obtains

$$\hat{I} = \frac{\hat{\phi}}{R} + i\omega C\hat{\phi}. \tag{5.76}$$

By definition, the impedance $\hat{Z}$ of a linear system is

$$\hat{Z} = \frac{\hat{\phi}}{\hat{I}} \tag{5.77}$$

and from Equation 5.76,

$$\frac{1}{\hat{Z}} = \frac{1}{R} + i\omega C. \tag{5.78}$$

This shows that $R$ and $1/(i\omega C)$ are parallel impedances. Solving Equation 5.78 for $\hat{Z}$, results in

$$\hat{Z} = \frac{R}{1 + (\omega RC)^2} - i\frac{\omega R^2 C}{1 + (\omega RC)^2} \tag{5.79}$$

or, introducing the real $\hat{Z}_{re}$ and imaginary $\hat{Z}_{im}$ parts

$$\hat{Z}_{re} = \frac{R}{1 + (\omega RC)^2}$$

$$\hat{Z}_{im} = -\frac{\omega R^2 C}{1 + (\omega RC)^2}.$$

A *Nyquist plot* of the system is obtained plotting the $\hat{Z}_{im}$ versus $\hat{Z}_{re}$ (Figure 5.12). It is easy to verify that the following relation holds:

$$\left(\hat{Z}_{re} - \frac{R}{2}\right)^2 + \hat{Z}_{im}^2 = \left(\frac{R}{2}\right)^2$$

and, hence, the Nyquist plot of the RC-circuit is an ideal semicircle with radius $R/2$ centered at the point $R/2$ on the real axis.

From Equation 5.79, it is seen that at $\omega = 0$, one obtains $\hat{Z}_{re} = R$. Thus, the right intercept of the impedance spectrum with the real axis gives the resistivity of the circuit. The same is true for any system, which includes capacitances and resistivities: at $\omega = 0$, the spectrum gives the total static resistivity of the system. Indeed, $\omega = 0$ is a direct current (DC) limit. In this limit, all capacitances behave simply as a circuit break and the remaining network of resistances determines the total system resistance.

Further, it is easy to show that the maximal value of $-\hat{Z}_{im}$ corresponds to $\omega_{max} = 1/(RC)$, that is, to the inverse time constant of the system. Knowing $R$ and $\omega_{max}$, $C$

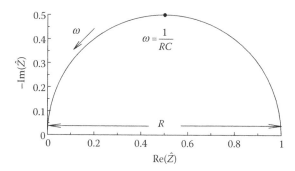

**FIGURE 5.12** Nyquist plot of a parallel RC-circuit.

is obtained. In the section "Finite but Small Current: An Analytical Solution," it will be shown that under certain constrains, the imaginary part of the CCL impedance spectrum also gives the time constant of the system.

A fuel cell electrode contains a charged double layer of capacitance $C_{dl}$ and a number of transport resistances connected to $C_{dl}$. The way these resistances are connected to $C_{dl}$ could be understood by analyzing the spectrum of the electrode. The next section deals with the model for the CCL impedance, as this model is a key component of the complete electrode impedance model.

## IMPEDANCE OF A CATHODE CATALYST LAYER

The physical models for the catalyst layer impedance stem from the theory of impedance of a porous electrode developed in the pioneering work of de Levie (1967). de Levie formulated model equations for the impedance of a single cylindrical pore filled with electrolyte. In the limiting case of either ideal transport of reactants or constant electrolyte potential, he derived analytical solutions. Inspired by the work of de Levie, two subsequent branches of impedance modeling consist of more sophisticated descriptions for a single pore and for a macrohomogeneous electrode.

Lasia obtained the impedance of a single pore, neglecting oxygen transport (Lasia, 1995) and taking it into account (Lasia, 1997). Importantly, Lasia derived solutions using the nonlinear Butler–Volmer equation for the rate of electrochemical conversion, which makes his results valid at arbitrary currents in the system. However, the features of cylindrical geometry and the boundary conditions, used in the cylindrical pore problem, do not allow direct usage of these results for the macrohomogeneous porous electrode. On the other hand, most of the models for macrohomogeneous electrodes utilize the linearized Butler–Volmer equation, which limits the validity of these models to the impedance evaluation close to open-circuit conditions (Devan et al., 2004; Paasch et al., 1993; Rangarajan, 1969).

The analytical model for the impedance of the CCL in PEM fuel cells has been developed by Eikerling and Kornyshev (1999). Their model is based on a nonstationary version of the performance model discussed in the section "MHM with Constant Coefficients: Analytical Solutions" in Chapter 4. In the limiting cases of either fast oxygen transport or small currents, Eikerling and Kornyshev provided analytical solutions. However, Eikerling and Kornyshev neglected the nonstationary term in the oxygen transport equation.

Makharia et al. (2005) developed a physical model for the CCL impedance, neglecting the oxygen transport losses in the CCL. The model was used for fitting the experimental impedance spectra acquired at several cell currents.

Generally, impedance measurements in low-temperature fuel cells are performed at finite cell current. This provides information on electrode performance at practically relevant currents. At high currents, the concentration polarization and the potential loss resulting from proton transport in the electrode could be significant and, hence, the realistic impedance model of the CCL should take these processes into account.

In this section, the analysis is based on a nonstationary model for the CCL performance with the Butler–Volmer conversion function. The steady-state version of this model has been discussed in Chapter 4.

## Basic Equations

The time-dependent CCL performance model is obtained if time derivatives are added to the system of Equations 4.7 through 4.9. The nonstationary version of this system is

$$C_{dl}\frac{\partial \eta}{\partial t} + \frac{\partial j}{\partial x} = -2i_* \left(\frac{c}{c_h^0}\right) \sinh\left(\frac{\eta}{b}\right), \tag{5.80}$$

$$j = -\sigma_p \frac{\partial \eta}{\partial x}, \tag{5.81}$$

$$\frac{\partial c}{\partial t} - D\frac{\partial^2 c}{\partial x^2} = -\frac{2i_*}{nF}\left(\frac{c}{c_h^0}\right) \sinh\left(\frac{\eta}{b}\right). \tag{5.82}$$

Here, $C_{dl}$ is the double layer volumetric capacitance (F cm$^{-3}$) and $t$ is time.

Equation 5.80 describes charge conservation in the CCL. It shows that the proton current decays toward the GDL because of the charging of the double layer (the term with $C_{dl}$) and due to the proton conversion in the ORR (the right side of this equation). Equation 5.80 follows from a general charge conservation equation

$$\frac{\partial \rho}{\partial t} + \nabla \cdot \mathbf{j} = R_{ORR}.$$

With $\rho = C_{dl}\eta$ and a Butler–Volmer rate of the ORR, this equation reduces to Equation 5.80. Note that $\rho = C_{dl}\eta$ means that only the relative variation of the double layer charge, associated with the current transport, needs to be taken into account. The constant space charge associated with the open-circuit potential is independent of time and, hence, it does not appear in Equation 5.80.

To simplify calculations, in addition to Equation 4.51, one can introduce dimensionless time and impedance

$$\tilde{t} = \frac{t}{t_*}, \quad \tilde{Z} = \frac{Z\sigma_p}{l_{CL}}, \tag{5.83}$$

where

$$t_* = \frac{C_{dl}b}{2i_*} \tag{5.84}$$

is the scaling parameter for time.

With these variables, Equations 5.80 through 5.82 take the form

$$\frac{\partial\tilde{\eta}}{\partial\tilde{t}} + \varepsilon^2\frac{\partial\tilde{j}}{\partial\tilde{x}} = -\tilde{c}\sinh\tilde{\eta}, \tag{5.85}$$

$$\tilde{j} = -\frac{\partial\tilde{\eta}}{\partial\tilde{x}} \tag{5.86}$$

and

$$\mu^2\frac{\partial\tilde{c}}{\partial\tilde{t}} - \varepsilon^2\tilde{D}\frac{\partial^2\tilde{c}}{\partial\tilde{x}^2} = -\tilde{c}\sinh\tilde{\eta}. \tag{5.87}$$

Here,

$$\mu^2 = \frac{nFc_h^0}{C_{dl}b} \tag{5.88}$$

determines the characteristic time of oxygen concentration variation in the CL. The typical set of physical parameters, together with the resulting parameters $\varepsilon$ and $\mu$, are listed in Table 5.6.

The proton current can be eliminated from the system by substituting Equation 5.86 into Equation 5.85. With this, is obtained, an equivalent system of two equations

$$\frac{\partial\tilde{\eta}}{\partial\tilde{t}} - \varepsilon^2\frac{\partial^2\tilde{\eta}}{\partial\tilde{x}^2} = -\tilde{c}\sinh\tilde{\eta}, \tag{5.89}$$

$$\mu^2\frac{\partial\tilde{c}}{\partial\tilde{t}} - \varepsilon^2\tilde{D}\frac{\partial^2\tilde{c}}{\partial\tilde{x}^2} = -\tilde{c}\sinh\tilde{\eta}. \tag{5.90}$$

Let $\tilde{\eta}^0$ and $\tilde{c}^0$ be the steady-state solutions to the system (5.89) and (5.90). Substituting

$$\tilde{\eta} = \tilde{\eta}^0 + \tilde{\eta}^1\exp(i\tilde{\omega}\tilde{t}), \quad \tilde{\eta}^1 \ll 1$$

$$\tilde{c} = \tilde{c}^0 + \tilde{c}^1\exp(i\tilde{\omega}\tilde{t}), \quad \tilde{c}^1 \ll 1,$$

into Equations 5.89 and 5.90, neglecting the terms with products of the disturbances, and subtracting the steady-state equations for $\tilde{\eta}^0$ and $\tilde{c}^0$, a system of linear equations

**TABLE 5.6**
**The Physical and Dimensionless Parameters**

| Parameter | PEFC | DMFC |
|---|---|---|
| Tafel slope $b$ (V) | 0.05 | 0.05 |
| Proton conductivity $\sigma_p$ ($\Omega$ cm$^{-1}$) | 0.03 | 0.03 |
| Exchange current density $i_*$ (A cm$^{-3}$) | $10^{-3}$ | 0.1 |
| Effective oxygen diffusion coefficient D (cm$^2$ s$^{-1}$) | $10^{-3}$ | $10^{-3}$ |
| CL capacitance $C_{dl}$ (F cm$^{-3}$) | 20 | 200 |
| CL thickness $l_{CL}$ (cm) | 0.001 | 0.01 |
| $j_*$ (A cm$^{-2}$) | 1.5 | 0.15 |
| $t_*$ (s) | 500 | 50 |
| $D_*$ (cm$^2$ s$^{-1}$) | $5.28 \cdot 10^{-4}$ | $5.28 \cdot 10^{-4}$ |
| $\varepsilon^2$ | $7.5 \cdot 10^5$ | 75 |
| $\mu^2$ | 2.86 | 0.426 |
| $\tilde{D}$ | 1.894 | 1.894 |

*Note:* Catalyst layer volumetric capacitance is estimated with the data (Adapted from Makharia, R., Mathias, M. F., and Baker, D. R. 2005. *J. Electrochem. Soc.*, **152**, A970–A977).

is obtained for the small complex perturbation amplitudes $\tilde{\eta}^1(\tilde{\omega}, \tilde{x})$, $\tilde{c}^1(\tilde{\omega}, \tilde{x})$,

$$\varepsilon^2 \frac{\partial^2 \tilde{\eta}^1}{\partial \tilde{x}^2} = \sinh(\tilde{\eta}^0)\tilde{c}^1 + \left(\tilde{c}^0 \cosh \tilde{\eta}^0 + i\tilde{\omega}\right) \tilde{\eta}^1 \tag{5.91}$$

and

$$\varepsilon^2 \tilde{D} \frac{\partial^2 \tilde{c}^1}{\partial \tilde{x}^2} = \left(\sinh \tilde{\eta}^0 + i\tilde{\omega}\mu^2\right) \tilde{c}^1 + \tilde{c}^0 \cosh(\tilde{\eta}^0)\tilde{\eta}^1, \tag{5.92}$$

where $\tilde{\omega} = \omega t_*$ is the dimensionless frequency of the disturbance. The key assumption is the smallness of $\tilde{\eta}^1$ and $\tilde{c}^1$, which allows arriving at the linear system of Equations 5.91 and 5.92.

The perturbation of the total potential loss in the CCL is $\tilde{\eta}^1|_{\tilde{x}=0}$. By definition, the impedance of a (generally nonlinear) system is the ratio of small perturbations of potential and current density,

$$\tilde{Z} = \frac{\tilde{\eta}^1}{\tilde{j}^1}\bigg|_{\tilde{x}=0} = -\frac{\tilde{\eta}^1}{\partial \tilde{\eta}^1/\partial \tilde{x}}\bigg|_0. \tag{5.93}$$

Solution to the system (5.91) and (5.92) is subjected to the following boundary conditions:

$$\tilde{\eta}^1(1) = \tilde{\eta}_1^1, \qquad \left.\frac{\partial \tilde{\eta}^1}{\partial \tilde{x}}\right|_1 = 0 \tag{5.94}$$

and

$$\left.\frac{\partial \tilde{c}^1}{\partial \tilde{x}}\right|_0 = 0, \quad \tilde{c}^1(1) = 0. \tag{5.95}$$

In this section, the impedance of the GDL is ignored. The first part of Equation 5.94 establishes a small-amplitude perturbation at the CCL/GDL interface (at $\tilde{x} = 1$). The second part of Equation 5.94 means zero proton current at $\tilde{x} = 1$. Equations 5.95 express zero oxygen flux through the membrane and zero concentration perturbation at the CCL/GDL interface, respectively. Note that the disturbance of the overpotential, can be applied on either side of the CCL: the linearity of the system (5.91) and (5.92) guarantees that the solution would be the same in both cases.

The only parameter, which appears in the boundary conditions, is the applied perturbation of potential $\tilde{\eta}_1^1$. It is easy to show that $\tilde{Z}$ does not depend on this parameter. Indeed, the system (5.91) and (5.92) is linear and homogeneous. The solution to this system can be represented in the form

$$\tilde{\eta}^1 = \tilde{\eta}_1^1 \phi(\tilde{\omega}, \tilde{x})$$
$$\tilde{c}^1 = \tilde{\eta}_1^1 \psi(\tilde{\omega}, \tilde{x}). \tag{5.96}$$

Evidently, Equations 5.91 through 5.95 for $\phi$ and $\psi$ do not contain $\tilde{\eta}_1^1$ (the first of Equations 5.94 takes the form $\phi(1) = 1$). Thus, $\tilde{\eta}_1^1$ appears as a scaling factor in the expression for $\tilde{\eta}^1$ and, hence, the impedance

$$\tilde{Z} = -\left.\frac{\phi}{\partial\phi/\partial\tilde{x}}\right|_0$$

is independent of $\tilde{\eta}_1^1$.

## High-Frequency Limit

At large $\tilde{\omega}$, all the terms on the right-hand side of Equation 5.91, except $i\tilde{\omega}\tilde{\eta}^1$, can be omitted. This equation decouples from the system and simplifies to

$$\varepsilon^2 \frac{\partial^2 \tilde{\eta}^1}{\partial \tilde{x}^2} = i\tilde{\omega}\tilde{\eta}^1. \tag{5.97}$$

Below, it will be shown that the solution to Equation 5.97 yields

$$\tilde{Z}_{re} = -\tilde{Z}_{im} = \frac{\varepsilon}{\sqrt{2\tilde{\omega}}}. \tag{5.98}$$

Equation 5.98 means that in the Nyquist plot ($-\tilde{Z}_{im}$ versus $\tilde{Z}_{re}$), the high-frequency branch of the spectrum is a straight line with the slope of 45°. This is the general property of all CCL impedance curves (Eikerling and Kornyshev, 1999).

In the coordinates $\tilde{Z}_{im}$ versus $\tilde{\omega}^{-1/2}$, this linear branch has a slope proportional to $\varepsilon$. In dimensional form, Equation 5.98 is

$$Z_{re} = -Z_{im} = \frac{1}{\sqrt{2\sigma_p C_{dl}\omega}}. \tag{5.99}$$

Thus, if one of the parameters $\sigma_p$ or $C_{dl}$ is known, the slope of the high-frequency line $Z_{im}$ versus $\omega^{-1/2}$ yields the other parameter. Physically, in this limit, the electric field changes so fast that the inertial electrochemical processes do not respond to the field changes, and the impedance is determined solely by the double layer charging. Since this charging involves protons, Equation 5.99 contains $\sigma_p$.

In the general case, a full analytical solution to the system (5.91) and (5.92) can hardly be obtained. Part of the problem is that the general analytical steady-state solutions $\tilde{\eta}^0$ and $\tilde{c}^0$ have not been found yet. However, in the limiting case of ideal oxygen transport, an explicit steady-state function $\tilde{\eta}^0(\tilde{x})$ has been derived in the section "The Case of $\varepsilon_* \ll 1$ and $\varepsilon_*^2 \tilde{j}_0^2 \ll 1$." This allows analyzing the system (5.91) and (5.92), and constructing a hybrid solution to this system, as discussed below (Kulikovsky, 2012g).

### Ideal Oxygen Transport

In the case of ideal oxygen transport in the CCL, one may set $\tilde{c}^1 = 0$, $\tilde{c}^0 = 1$ in Equation 5.91, and omit Equation 5.92. Equation 5.91 takes the form

$$\varepsilon^2 \frac{\partial^2 \tilde{\eta}^1}{\partial \tilde{x}^2} = \left( \cosh \tilde{\eta}^0 + i\tilde{\omega} \right) \tilde{\eta}^1. \tag{5.100}$$

### Impedance at an Open-Circuit Potential

Close to the OCP, $\tilde{\eta}^0 \to 0$, and in Equation 5.100, one may set $\cosh \tilde{\eta}^0 = 1$. With this, Equation 5.100 simplifies to*

$$\varepsilon^2 \frac{\partial^2 \tilde{\eta}^1}{\partial \tilde{x}^2} = (1 + i\tilde{\omega}) \tilde{\eta}^1. \tag{5.101}$$

---

* It can be shown that this approximation works well if the inequality $\varepsilon^2 \tilde{j}_0^2 \ll 1$ holds. In PEM fuel cells, this inequality is fulfilled at vanishingly small currents $j_0 \lesssim 100\ \mu A\ cm^{-2}$.

Solution to Equation 5.101 with the boundary conditions (5.94) is

$$\frac{\tilde{\eta}^1}{\tilde{\eta}^1_1} = \cos\left(\varphi\left(1 - \tilde{x}\right)\right), \quad \varphi = \sqrt{\frac{-1 - i\tilde{\omega}}{\varepsilon^2}}. \tag{5.102}$$

Differentiating Equation 5.102 over $\tilde{x}$, one obtains the disturbance of the cell current

$$\frac{\tilde{j}^1}{\tilde{\eta}^1_1} = -\varphi \sin\left(\varphi\left(1 - \tilde{x}\right)\right). \tag{5.103}$$

Setting $\tilde{x} = 0$ in Equations 5.102 and 5.103 and calculating the impedance according to Equation (5.93), yields

$$\tilde{Z} = -\frac{1}{\varphi \tan \varphi}. \tag{5.104}$$

Separating the real and imaginary parts, one finally finds

$$\tilde{Z}_{re} = \frac{\varepsilon^2}{\sqrt{1 + \tilde{\omega}^2}} \left(\frac{\gamma \sinh(2\gamma) - \alpha \sin(2\alpha)}{\cosh(2\gamma) - \cos(2\alpha)}\right) \tag{5.105}$$

and

$$\tilde{Z}_{im} = -\frac{\varepsilon^2}{\sqrt{1 + \tilde{\omega}^2}} \left(\frac{\gamma \sin(2\alpha) + \alpha \sinh(2\gamma)}{\cosh(2\gamma) - \cos(2\alpha)}\right), \tag{5.106}$$

where

$$\alpha = \sqrt{\frac{\sqrt{1 + \tilde{\omega}^2} - 1}{2\varepsilon^2}}, \quad \gamma = \sqrt{\frac{\sqrt{1 + \tilde{\omega}^2} + 1}{2\varepsilon^2}}. \tag{5.107}$$

The Nyquist plot is a semicircle linked to the straight line in the high-frequency domain (Figure 5.13).* The right intersection of the semicircle with the real axis corresponds to $\tilde{\omega} = 0$, and it gives the static CCL differential resistivity $\tilde{R}_{ccl}$. Setting in Equation 5.105 $\tilde{\omega} = 0$, results in

$$\tilde{R}_{ccl} = \frac{\varepsilon \sinh(2/\varepsilon)}{\cosh(2/\varepsilon) - 1}. \tag{5.108}$$

For small and large $\varepsilon$, this equation further simplifies to

$$\tilde{R}_{ccl} = \begin{cases} \varepsilon, & \varepsilon \ll 1 \\ \dfrac{1}{3} + \varepsilon^2, & \varepsilon \gg 1 \end{cases}. \tag{5.109}$$

---

* Note that the coordinates $(\tilde{Z}_{re}/\varepsilon^2, \tilde{Z}_{im}/\varepsilon^2)$ in Figure 5.13 (impedance close to OCP) differ from the standard impedance coordinates $(\tilde{Z}_{re}, \tilde{Z}_{im})$. When plotted in standard coordinates, the curves in Figure 5.13 exhibit rapid growth to "infinity," on the right side of the linear high-frequency domain. This growth is, in fact, to a large value $\tilde{Z}_{im} \simeq \varepsilon^2$. In the stretched coordinates $(\tilde{Z}_{re}/\varepsilon^2, \tilde{Z}_{im}/\varepsilon^2)$, the curve is seen as a normal semicircle.

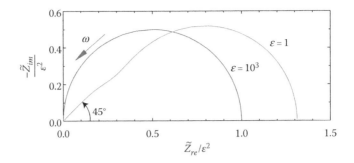

**FIGURE 5.13** The Nyquist plot of a CCL, close to OCP, for the indicated values of parameter $\varepsilon$. The curve for $\varepsilon = 1$ clearly shows the high-frequency $45°$ linear branch. When $\varepsilon = 10^3$, in the scale of this plot, the linear branch is not seen.

In dimensional form, these equations are

$$R_{ccl} = \begin{cases} \sqrt{b/(2i_*\sigma_p)}, & \varepsilon \ll 1 \\ \dfrac{l_{CL}}{3\sigma_p} + b/(2i_*l_{CL}), & \varepsilon \gg 1 \end{cases} . \qquad (5.110)$$

The term $l_{CL}/\sigma_p$ in Equation 5.110 is the proton resistivity of the CCL. This will be discussed in the section "Finite but Small Current: An Analytical Solution." Interestingly, if $\varepsilon$ is small, $R_{ccl}$ does not contain this term. Physically, if the reaction penetration depth is small (hydrogen electrode), the reaction runs close to the membrane and, therefore, protons do not need to be transported deep into the catalyst layer. Thus, the contribution of the proton transport to the cell resistivity does not appear as a separate term. The proton transport represented by $\sigma_p$ is included into the total CL resistivity $\sqrt{b/(2i_*\sigma_p)}$.

So far in this section, the system (5.85) through (5.87) has been discussed in the context of the CCL. However, exactly the same system of equations describes the impedance of the anode catalyst layer. The case of $\varepsilon \ll 1$ is characteristic of the PEFC anode, while $\varepsilon \gg 1$ is typical of the PEFC cathode. Small $\varepsilon$ is typical for the PEFC anode because of the high exchange current density. Thus, the solutions above allow estimating the differential resistivity of the hydrogen anode CL, $R_{acl}$.

With the data from Table 5.7, for anode and the cathode resistivities at the OCP, one finds

$$R_{acl} \simeq 0.01 \quad \Omega \, cm^2,$$

$$R_{ccl} \simeq 2.5 \cdot 10^4 \quad \Omega \, cm^2.$$

Note that $R_{acl}$ and $R_{ccl}$ are the *differential* CL resistivities close to the OCP. As discussed in the section "x Shapes" in Chapter 4, small $\varepsilon$ means that the reaction runs close to the membrane interface. As the amplitude of the perturbing AC potential is small, the electrode (anode) works in the linear regime. The parameter $\varepsilon$ represents

**TABLE 5.7**

Typical Electrode Parameters and the Resulting Newman's Dimensionless Reaction Penetration Depth $\varepsilon$, the Typical Dimensionless Current Density $\tilde{j}_0$, and the Non-Newman Dimensionless RPD $\tilde{l}_\sigma$ for Various Electrodes.

| | PEFC Anode | PEFC Cathode | DMFC Anode | DMFC Cathode |
|---|---|---|---|---|
| $i_*$ (A cm$^{-3}$) | $10^4$ | 0.001–0.1 | 0.1 | 0.1 |
| $l_{CL}$ (cm) | $10^{-3}$ | $10^{-3}$ | $10^{-2}$ | $10^{-2}$ |
| $\sigma_p$ ($\Omega^{-1}$ cm$^{-1}$) | 0.01–0.04 | 0.01–0.04 | $10^{-3}$ | $10^{-3}$ |
| $b$, (V) | 0.015 | 0.05 | 0.05 | 0.05 |
| $j_0$, (A cm$^{-2}$) | 1.0 | 1.0 | 0.1 | 0.1 |
| $\varepsilon$ | 0.1 | $10^2$-$10^3$ | 1 | 1 |
| $\tilde{j}_0$ | 10 | 1 | 10 | 10 |
| $\tilde{l}_\sigma$ | 0.15 | 0.5 | 0.05 | 0.05 |

*Note:* The data on $i_*$ are estimated from Kucernak, A. R. and Toyoda, E. 2008. *Electrochem. Commun.*, **10**, 1728–1731.

an internal space scale of the problem, and the open-circuit resistivity of the electrode is simply proportional to $\varepsilon$. Owing to that, $R_{acl}$ in dimensional form is independent of $l_{CL}$, as seen in Equation 5.110. On the cathode side, large $\varepsilon$ means that the ORR runs uniformly over the CCL thickness and the dimensional charge transfer resistivity of the layer appears to be inversely proportional to $l_{CL}$ (Equation 5.110). Indeed, the thicker the layer, the larger the current that can be converted and vice versa. This is the main reason for the tremendous difference (six orders of magnitude) in $R_{acl}$ and $R_{ccl}$ at the OCP.

In the case of $\varepsilon \gg 1$, the term 1/3 in Equation 5.110 can be neglected and we obtain $R_{ccl} \simeq b/(2i_* l_{CL})$. As $l_{CL}$ is usually known, the right intercept of the semicircle with the real axis gives the ratio $b/i_*$. If one of the parameters $b$, $i_*$ is known (see the section "Finite but Small Current: An Analytical Solution" on how to measure $b$ using EIS), the other can be directly calculated. Note that this result holds only for EIS evaluated close to the OCP, when currents in the system are vanishingly small.

The opposite (left) side of the spectrum represents a high-frequency branch $\tilde{\omega} \to \infty$. In that case, one has

$$\alpha \simeq \gamma \simeq \sqrt{\frac{\tilde{\omega}}{2\varepsilon^2}} \gg 1,$$

and the trigonometric terms in Equations 5.105 and 5.106 can be neglected. After simple algebraic transformations, one obtains the straight line (Equation 5.98).

Note that at large $\varepsilon$, the length of the linear branch in the scale of Figure 5.13 is very small. This branch is clearly seen at $\varepsilon \lesssim 1$ only. At $\varepsilon \gg 1$, in the coordinates $(\tilde{Z}_{re}/\varepsilon^2, \tilde{Z}_{im}/\varepsilon^2)$, the spectrum is represented by almost an ideal semicircle with unit diameter (Figure 5.13).

It is concluded that, close to the OCP, the semicircle radius is independent of the working current density (as long as this current density is small). In dimensionless coordinates (5.83), the radius of the semicircle only depends on the reaction penetration depth $\varepsilon$ (Equation 5.108). In typical PEFC cathodes $\varepsilon \gg 1$, the right intercept of the spectrum with the real axis gives either the Tafel slope or the exchange current density provided that one of these parameters is known.

### Finite Cell Current, Numerical Solution

In this section, it is assumed that $\varepsilon^2 \tilde{j}_0^2 \gg 1$. As discussed in the section "Large $\varepsilon^2 \tilde{j}_0^2$" in Chapter 4, owing to large $\varepsilon$ in PEFCs, this condition is fulfilled already at very small current densities.

The steady-state distribution of overpotential $\tilde{\eta}^0$ is given by Equation 4.121. With this, after simple manipulations, Equation 5.100 takes the form

$$\varepsilon^2 \frac{\partial^2 \tilde{\eta}^1}{\partial \tilde{x}^2} = \left( i\tilde{\omega} + \sqrt{1 + \frac{\varepsilon^4 \beta^4}{4\cos^4(\beta(1-\tilde{x})/2)}} \right) \tilde{\eta}^1. \tag{5.111}$$

An analytical solution to Equation 5.111 is hopelessly cumbersome. A numerical solution to the complex Equation 5.111 can be obtained if one substitutes $\tilde{\eta}^1 = \tilde{\eta}_{re}^1 + i\tilde{\eta}_{im}^1$ into this equation and separates equations for the real $\tilde{\eta}_{re}^1$ and imaginary $\tilde{\eta}_{im}^1$ parts. The resulting system of real equations can be solved numerically using mathematical software, for example, Maple. The impedance is calculated according to Equation 5.93.

The resulting Nyquist spectra are shown in Figure 5.14. The high-frequency branch of all the spectra is again the straight line, described by Equation 5.98 (Figure 5.14). At high $\tilde{\omega}$, the contribution of the $\tilde{x}$-dependent term in Equation 5.111 is small, and this equation reduces to Equation 5.97 with the solution given by Equation 5.98. Note that the spectra in Figure 5.14 do not intersect each other; the spectra corresponding to a higher cell current are of a smaller radius.

It is advisable to calculate the static differential charge transfer resistivity $\tilde{R}_{ct}$ of the CCL. Differentiating Equations 4.130 and 4.133 over $\tilde{j}_0$ yields[*]:

$$\tilde{R}_{ct} = \begin{cases} 1/\tilde{j}_0, & \text{normal Tafel regime} \\ 2/\tilde{j}_0, & \text{double Tafel regime} \end{cases} \tag{5.112}$$

Thus, in both cases, the charge transfer resistivity is inversely proportional to the cell current density. However, the numerator in Equation 5.112 changes from 1 to 2 as the current increases.

In accordance with Equation 5.112, the radii of the semicircles decrease with the growth of the cell current (Figure 5.14). However, at high currents in the double Tafel mode, $\tilde{R}_{ct}$ grows with the decrease in $\tilde{j}_0$ twice faster, which leads to the elongation of the spectra along the real axis (cf. Figure 5.14a and b). To clarify this effect, it is

---

[*] The same equations are obtained by differentiating more general relations (4.129) and (4.132), under conditions of large $\varepsilon^2 \tilde{j}_0$ and $\varepsilon^2 \tilde{j}_0^2$. In this section, it is assumed that these conditions are fulfilled.

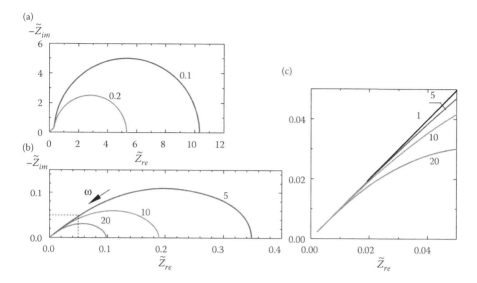

**FIGURE 5.14**   The impedance spectra of the CCL with an ideal oxygen transport in (a) normal Tafel, (b) and (c) double Tafel regimes for the indicated values of the dimensionless cell current density $\tilde{j}_0$, with (c) the high-frequency domain (linear branch) of the spectra in (b). Parameters for the calculations are listed in the second (PEFC) column of Table 5.6.

advisable to multiply the axes in Figure 5.14b by $\tilde{j}_0$ (Jaouen and Lindbergh, 2003). According to Equation 5.112, it can be expected that in these coordinates, the spectra at small and large currents would cross the real axis close to unity and 2, respectively. Figure 5.15 shows that these expectations are correct.

Numerical calculations show that in the regime with ideal oxygen transport, the impedance spectra are independent of $\varepsilon$. To understand the parametric dependencies, it is insightful to analytically solve a low-current version of Equation 5.111.

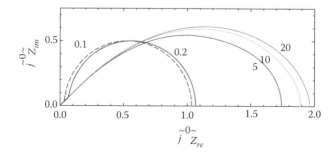

**FIGURE 5.15**   The spectra from Figure 5.14a and b, in the stretched coordinates $(\tilde{j}_0\tilde{Z}_{re}, \tilde{j}_0\tilde{Z}_{im})$. Note that the diameter of the low-current and high-current spectra tend to 1 and 2, respectively.

## Finite but Small Current: An Analytical Solution

In this section, the superscript 0 at $\tilde{j}_0^0$ will be omitted, that is, $\tilde{j}_0 \equiv \tilde{j}_0^0$ is the undisturbed cell current density. The results of this section are valid if the relations $\varepsilon^2 \tilde{j}_0^2 \gg 1$ and $\tilde{j}_0 \ll 1$ hold simultaneously. These relations can be combined in one to yield*

$$\frac{1}{\varepsilon} \ll \tilde{j}_0 \ll 1. \tag{5.113}$$

In dimensional form, this equation reads

$$\sqrt{2i_* \sigma_p b} \ll j_0 \ll \sigma_p b / l_{CL}. \tag{5.114}$$

With the typical values for PEFCs (Table 5.7), the left-hand side of Equation 5.114 is in the order of 1 mA cm$^{-2}$, while the right-hand side is about 1 A cm$^{-2}$. Thus, for PEFCs, the model of this section works for impedance curves measured under the cell currents between 10 and 100 mA cm$^{-2}$.

Approximate solutions to Equation 5.111 can be obtained using the following arguments (Kulikovsky and Eikerling, 2013). As $\varepsilon \gg 1$ (Table 5.6), it can be assumed that $\varepsilon \tilde{j}_0 \gg 1$, therefore, unity may be neglected under the square root in Equation 5.111. Small $\tilde{j}_0$ means that the parameter $\beta \simeq \sqrt{2 \tilde{j}_0}$ (see Equation 4.126) is also small and $\cos^2$ in Equation 5.111 may be replaced by unity. With this, Equation 5.111 takes the form

$$\varepsilon^2 \frac{\partial^2 \tilde{\eta}^1}{\partial \tilde{x}^2} = \left( \varepsilon^2 \tilde{j}_0 + i\tilde{\omega} \right) \tilde{\eta}^1. \tag{5.115}$$

The solution to Equation 5.115 is

$$\tilde{\eta}^1 = \tilde{\eta}_1^1 \cos \left( \sqrt{-\tilde{j}_0 - \frac{i\tilde{\omega}}{\varepsilon^2}} (1 - \tilde{x}) \right). \tag{5.116}$$

Differentiating Equation 5.116 over $\tilde{x}$ and calculating the impedance (Equation 5.93), gives

$$\tilde{Z} = -\frac{1}{\varphi' \tan \varphi'}, \quad \text{where} \quad \varphi' = \sqrt{-\tilde{j}_0 - \frac{i\tilde{\omega}}{\varepsilon^2}}. \tag{5.117}$$

---

* In fact, the right-hand side of inequality (Equation 5.113) can be mitigated. Indeed, expansion of the tan-function in Equation 4.122 is justified if $\beta \ll \pi$. From Equation 4.123, for the upper limit of cell current, it then yields $\tilde{j}_0 \ll \pi^2/2$. Thus, Equation 5.113 transforms to

$$1/\varepsilon \ll \tilde{j}_0 \ll \pi^2/2.$$

In the dimensional form, this extends the upper limit of cell current densities to $\simeq 500$ mA cm$^{-2}$. However, at this current, oxygen transport in the CCL can contribute to the impedance.

Separating the real and imaginary parts, here, one arrives at

$$\tilde{Z}_{re} = \frac{1}{2\sqrt{\tilde{j}_0^2 + (\tilde{\omega}/\varepsilon^2)^2}} \left( \frac{\gamma_* \sinh(\gamma_*) - \alpha_* \sin(\alpha_*)}{\cosh(\gamma_*) - \cos(\alpha_*)} \right) \tag{5.118}$$

and

$$\tilde{Z}_{im} = -\frac{1}{2\sqrt{\tilde{j}_0^2 + (\tilde{\omega}/\varepsilon^2)^2}} \left( \frac{\alpha_* \sinh(\gamma_*) + \gamma_* \sin(\alpha_*)}{\cosh(\gamma_*) - \cos(\alpha_*)} \right), \tag{5.119}$$

where

$$\alpha_* = \sqrt{2\sqrt{\tilde{j}_0^2 + (\tilde{\omega}/\varepsilon^2)^2} - 2\tilde{j}_0}, \quad \gamma_* = \sqrt{2\sqrt{\tilde{j}_0^2 + (\tilde{\omega}/\varepsilon^2)^2} + 2\tilde{j}_0}.$$

Note that in all equations of this section, $\varepsilon$ appears in a combination $\tilde{\omega}/\varepsilon^2$ only. This is immediately seen if Equation 5.115 is divided by $\varepsilon^2$. Thus, $\varepsilon$ simply rescales the frequency, and the shape of the Nyquist spectrum does not depend on $\varepsilon$.

Setting in Equation 5.118 $\tilde{\omega} = 0$, one obtains the steady-state differential CCL resistivity

$$\tilde{R}_{ccl} = \frac{\sinh\left(2\sqrt{\tilde{j}_0}\right)}{\sqrt{\tilde{j}_0}\left(\cosh\left(2\sqrt{\tilde{j}_0}\right) - 1\right)}.$$

Since $\tilde{j}_0$ is small, the last equation transforms to

$$\tilde{R}_{ccl} = \frac{1}{3} + \frac{1}{\tilde{j}_0}. \tag{5.120}$$

The last term on the right-hand side coincides with the low-current limit in Equation 5.112. Thus, this term is the charge transfer (activation) resistivity of the CCL, $\tilde{R}_{ct}$. The first term represents the CCL ionic resistivity (see below).

Equations 5.118 and 5.119 determine the small-current impedance spectrum as a function of the dimensionless cell current density. Examples of these spectra are shown in Figure 5.16.

An important characteristic point of the spectrum is the maximal value of $-\tilde{Z}_{im}$. Parameter $p \equiv \tilde{\omega}/\varepsilon^2$ corresponding to this point is obtained from the solution of the equation $\partial \tilde{Z}_{im}/\partial \tilde{Z}_{re} = 0$. This equation is equivalent to

$$\frac{\partial \tilde{Z}_{im}/\partial p}{\partial \tilde{Z}_{re}/\partial p} = 0, \quad \text{or} \quad \frac{\partial \tilde{Z}_{im}}{\partial p} = 0. \tag{5.121}$$

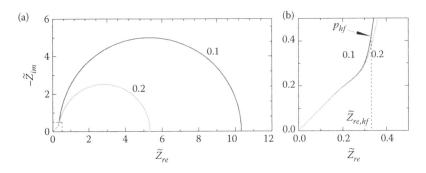

**FIGURE 5.16** (a) Analytical spectra for $\tilde{j}_0 = 0.1$ and 0.2 (Equations 5.118 and 5.119). The spectra in this figure are indistinguishable with the exact numerical spectra in Figure 5.14a. (b) The high-frequency part of the spectra showing the transition from the semicircle to the straight HF line. The point where $\partial^2 \tilde{Z}_{im}/\partial \tilde{Z}_{re}^2 = 0$ is indicated by the arrow.

The numerical solution of Equation 5.121 shows that to a high accuracy, the following relation holds:

$$\frac{\tilde{\omega}_{max}}{\varepsilon^2} = \tilde{j}_0, \tag{5.122}$$

where $\tilde{\omega}_{max}$ is the frequency corresponding to $\max\{-\tilde{Z}_{im}\}$.

For $\tilde{j}_0 \ll 1$ (Equation 5.113), the ionic resistivity term 1/3 in Equation 5.120 can be ignored. In dimensional variables, Equations 5.120 and 5.122 then take the form $R_{ccl} \simeq b/j_0$ and $j_0 = \omega_{max} b C_{dl} l_{CL}$, respectively. Combining these equations yields

$$b \simeq R_{ccl} j_0 \tag{5.123}$$

and

$$\omega_{max} R_{ccl} C_{dl} l_{CL} = 1. \tag{5.124}$$

Equation 5.124 means that in the frequency domain up to $\omega_{max}$, the CCL behaves as a parallel RC-circuit with the resistive element $R_{ccl}$ and the capacitance $C_{dl} l_{CL}$ (cf. the section "Impedance of a Parallel RC Circuit").

Thus, measuring a single small-current spectrum, the Tafel slope can immediately be obtained from Equation 5.123 and the double layer capacitance from Equation 5.124. Note that the ratio $\tilde{\omega}/\varepsilon^2$ does not contain the exchange current density and, hence, this parameter cannot be determined from the spectrum. To determine $i_*$, a spectrum close to OCP has to be measured (the section "Impedance at Open-Circuit Potential").

With $C_{dl}$ in hand, the CCL proton conductivity $\sigma_p$ can be determined from the slope of the high-frequency straight line $Z_{re}$ versus $\omega^{-1/2}$, as discussed in the section "High-Frequency Limit." However, $\sigma_p$ can also be determined directly from the Nyquist spectrum. The point of interest is the junction of the semicircle with the straight high-frequency line (Figure 5.16). At this point, the second derivative is

$\partial^2 \tilde{Z}_{im}/\partial \tilde{Z}_{re}^2 = 0$ (Figure 5.16b).* Expressing this derivative in terms of derivatives over $p$, using Equation 5.121 for the parameter $p_{hf}$ corresponding to the junction, one obtains an equation

$$\left(\frac{\partial^2 \tilde{Z}_{re}}{\partial p^2}\right) \frac{\partial \tilde{Z}_{im}}{\partial p} - \left(\frac{\partial^2 \tilde{Z}_{im}}{\partial p^2}\right) \frac{\partial \tilde{Z}_{re}}{\partial p} = 0. \tag{5.125}$$

The numerical solution to this equation shows that the following relation holds:

$$p_{hf} = \frac{7}{2} \tilde{j}_0^{1/4}, \text{ or } \tilde{j}_0 = \left(\frac{2\tilde{\omega}_{hf}}{7\varepsilon^2}\right)^4. \tag{5.126}$$

The exact numerical solution and the fitting Equation (5.126) are depicted in Figure 5.17a. As can be seen, in a wide range of dimensionless current densities, the accuracy of the approximate Equation (5.126) is high.

Transforming Equation 5.126 into dimensional form and solving the resulting equation for $\sigma_p$ yields

$$\sigma_p = \left(\frac{4\pi}{7}\right)^{4/3} \left(\frac{f_{hf}^4 C_{dl}^4 l_{CL}^7 b}{j_0}\right)^{1/3} \simeq 2.18 \left(\frac{f_{hf}^4 C_{dl}^4 l_{CL}^7 b}{j_0}\right)^{1/3}, \tag{5.127}$$

where $f_{hf} = \omega_{hf}/(2\pi)$ is the regular frequency (Hz), corresponding to the circular frequency $\omega_{hf}$.

A simpler though somewhat less accurate equation for $\sigma_p$ is obtained if one substitutes the first of Equations 5.126 into $\tilde{Z}_{re}$ (Equation 5.118). The resulting function $\tilde{Z}_{re,hf}(\tilde{j}_0)$ is depicted in Figure 5.17b. As can be seen, at the junction of the semicircle

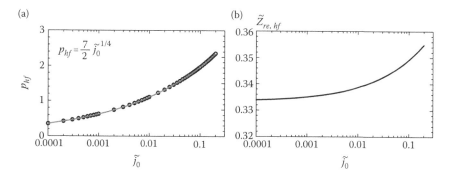

**FIGURE 5.17** (a) Points: the exact numerical solution to Equation 5.125. Line: the fitting equation (Equation 5.126). (b) Real part of the CCL impedance at the point corresponding to parameter $p_{hf} = \tilde{\omega}_{hf}/\varepsilon^2$ (the junction of the semicircle with the straight HF line, see Figure 5.16b).

* This point corresponds to the minimal root of equation $\partial^2 \tilde{Z}_{im}/\partial \tilde{Z}_{re}^2 = 0$.

with the straight HF line, the real part of the CCL impedance is almost independent of $\tilde{j}_0$, and is approximately equal to $1/3$.* From $\tilde{Z}_{re,hf} = 1/3$, one obtains the following dimensional relation:

$$\sigma_p = \frac{l_{CL}}{3Z_{re,hf}}. \tag{5.128}$$

Equation 5.128 gives the last undefined parameter, the CCL proton conductivity $\sigma_p$. Thus, Equations 5.123 and 5.124 and one of Equations 5.127 and 5.128 allow characterizing the CCL from a single low-current impedance curve without fitting.

Unfortunately, parameter $\omega_{max}$ is usually not reported in the literature. An exclusion is the work of Makharia et al. (2005). Figure 5.18, reprinted from this work, shows that the impedance curve for the cell current density of $j_0 = 0.03$ A cm$^{-2}$ (the largest semicircle) reaches a maximum of $-Z_{im}$ at the frequency $f_{max} \simeq 6$ Hz. Further, for 0.03 A cm$^{-2}$, the charge transfer resistivity of the cell is $R_{ct} \simeq 1.1$ $\Omega$ cm$^2$ (Figure 5.18). With these data, from Equation 5.123, one obtains $b \simeq 0.033$ V, which is a typical value. Taking into account that $f = 2\pi\omega$ and $l_{CL} = 0.0013$ cm, from Equation 5.124 one obtains $C_{dl} \simeq 18.6$ F cm$^{-3}$. This value is close to 15.4 F cm$^{-3}$, reported in Makharia et al. (2005) from an equivalent circuit analysis. Figure 10 in Makharia et al. (2005) (not shown) displays an enlarged junction of the semicircle and the straight HF line. From this figure, it is seen that $Z_{re,hf} \simeq 0.04$ $\Omega$ cm$^2$. With $l_{CL} = 0.0013$ cm, from Equation 5.128 one obtains $\sigma_p \simeq 0.011$ $\Omega^{-1}$ cm$^{-1}$, which agrees with literature data. With the parameters obtained from this analysis and in

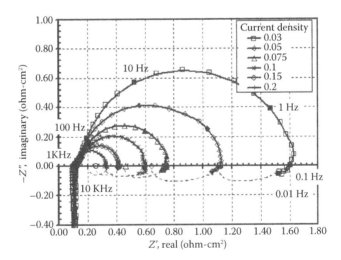

**FIGURE 5.18** Complex-plane impedance plots for H$_2$/O$_2$ operation at constant current density (A cm$^{-2}$). (Reprinted from Makharia, R. et al. 2005. *J. Electrochem. Soc.*, **152**, A970–A977, Figure 3. Copyright (2005), the Electrochemical Society. With permission.)

---

* Calculation shows that $1/3$ is the exact limit of $\tilde{Z}_{re,hf}$ as $\tilde{j}_0 \to 0$. This result has been obtained by Makharia et al. (2005), for the case of zero cell current from a slightly different model. It also correlates with Equation 5.120.

the exchange current density from Table 5.7, it is easy to verify that the conditions in Equation 5.113 are fulfilled.

An experiment (Makharia et al., 2005) has been performed with pure oxygen under large stoichiometry of the cathode flow. This, together with the low cell current, makes the resistivity due to oxygen transport in the channel and in the GDL vanishingly small. These conditions correspond to the model assumptions and they can be recommended for CCL characterization by means of EIS.

It is worth mentioning that the equations of the section "Impedance at an Open-Circuit Potential" (impedance at OCP) *cannot* be derived from equations of this section by passing to the limit $\tilde{j}_0 \to 0$. This is because the results of this section are obtained assuming that the cell current is small but finite (Equation 5.113).

## GENERAL CASE OF MIXED PROTON AND OXYGEN TRANSPORT LIMITATIONS

In the general case of mixed activation, oxygen and proton transport losses, the steady-state system of performance Equations (5.89 and 5.90) has to be solved first in order to determine the undisturbed profiles $\tilde{c}^0(\tilde{x})$ and $\tilde{\eta}^0(\tilde{x})$. Note that in some cases, numerical solution to the steady-state version of the system (5.89) and (5.90) is difficult to obtain. To work around the problem, the equivalent system (4.156) and (4.55) can be solved. With $\tilde{c}^0(\tilde{x})$ and $\tilde{\eta}^0(\tilde{x})$, the system for perturbations (5.91) and (5.92) can be solved and the CL impedance can be calculated.

In PEFCs, the CCL is thin and at current densities below several hundred mA cm$^{-2}$, the oxygen transport in the CCL is typically fast enough to neglect the respective potential loss. In direct methanol fuel cells, the CCL is typically 10 times thicker and oxygen transport may result in significant potential loss.

Figure 5.19a shows the impedance spectra corresponding to the DMFC cathode (the last column in Table 5.6), in which both oxygen and proton transport contribute to potential loss. For simplicity, the effect of fuel (methanol) crossover peculiar to

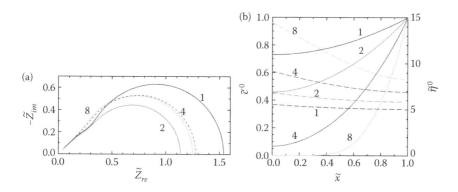

**FIGURE 5.19** (a) The impedance spectra of the CCL in the general case of mixed oxygen and proton transport losses. Indicated is the dimensionless cell current density $\tilde{j}_0$. The other parameters are listed in the last column of Table 5.6. (b) The oxygen concentration (solid lines) and the local overpotential (dashed lines) through the CCL for the same currents.

DMFC cathodes is ignored (this effect will be considered in the section "Impedance of DMFC Cathode"). Figure 5.19b depicts the respective shapes of the oxygen concentration and local overpotential through the CCL.

At lower currents (the curve for $\tilde{j}_0 = 1$), the spectrum is almost an ideal semicircle, linked to the straight HF line. With the growth of the current, the distortion (elongation along the real axis) is seen (cf. Figure 5.14a,b). This elongation manifests the transition to the double Tafel regime, caused by poor oxygen transport (Figure 5.19b).

However, in contrast to the case of ideal oxygen transport, the semicircle radius is no longer a monotonic function of the current density (Figure 5.19a). For smaller currents ($\tilde{j}_0 = 1$ and 2), oxygen depletion across the CCL is not large, and the respective semicircles exhibit regular behavior: they do not intersect, and the spectrum for larger current has a lower radius (Figures 5.19a,b). However, for the currents $\tilde{j}_0 = 4$ and 8, oxygen depletion through the CCL is significant (Figure 5.19b). The radii of the respective spectra exhibit nonmonotonic behavior, and the spectra intersect each other (Figure 5.19a).

In the case of proton transport limitations and fast oxygen diffusion, the CCL resistivity increases monotonically from $1/\tilde{j}_0$ to $2/\tilde{j}_0$ (Equation 5.112 and Figure 5.15). In other words, the general expression is $\tilde{R}_{ccl} = k(\tilde{j}_0)/\tilde{j}_0$, where $k$ is a monotonic function varying from 1 to 2 as $\tilde{j}_0$ increases. In the mixed case of oxygen- and proton transport limitations, $k$ is no longer a monotonic function of $\tilde{j}_0$. Moreover, the limiting value of $k$ is unknown. Jaouen and Lindbergh (2003) reported that in the mixed case, $k$ tends to 4, as $\tilde{j}_0 \to \infty$ (quadrupling of Tafel slope). However, they have used a flooded agglomerate model for the CCL performance. Generally, the curves in Figure 5.19a do not confirm this trend.

## IMPEDANCE OF DMFC CATHODE

The feature of DMFC cathode is a large flux of methanol permeated through the polymer membrane from the anode. The methanol crossover may be expected to affect DMFC cathode impedance spectra. These spectra have been measured by Müller and Urban (1998), Diard et al. (2003), Furukawa et al. (2005), and Piela et al. (2006). However, up to now, the effect of crossover on DMFC cathode spectra has not been fully understood: in none of these works, the features of the spectra, due to methanol crossover, have been clearly identified. This suggests that the effect of crossover is either small or it is masked by other effects.

Measuring DMFC cathode impedance is difficult for two reasons. First, it is hard to achieve true steady-state DMFC operation. Typically, the cell potential slightly varies with time (drift). Another problem is that in a cell without a reference electrode, the cathode impedance cannot be measured directly. Usually, the impedance of the whole cell is measured first. The oxygen on the cathode is then replaced by hydrogen, and the impedance of this quasi-half cell (anode impedance) is subtracted from the whole-cell curve to obtain the cathode spectrum.

As discussed above, experimental spectra do not allow to unambiguously understand the effect of crossover. In this situation, modeling may give valuable hints. Du et al. (2007b) developed a kinetic model of simultaneous MOR and ORR in the

DMFC cathode. The model was used to fit experimental impedance spectra; the contributions due to catalyst poisoning by MOR intermediates and due to parasitic current produced in the MOR have been calculated. However, Du et al. (2007b) assumed that the ORR and MOR are running uniformly over the cathode thickness. In that case, the above effects only weakly distort the shape of the cathode semicircle.

In another work, Du et al. (2007a) measured the impedance spectra of a half-cell cathode, with and without 0.5M methanol solution on the anode side of the membrane. The resulting spectra show an increase in the charge transfer resistance due to crossover, and a somewhat larger inductive low-frequency loop. Using the equivalent circuit approach, the increase in the loop radius was attributed to the poisoning of the Pt surface by MOR intermediates.

An impedance model of the DMFC cathode, with a detailed description of the MOR and ORR reaction mechanisms, has been reported by Chen et al. (2009). However, in this work, the cathode was treated as an infinitely thin interface and no space-resolved transport processes within the CCL were taken into account.

Below, the model for DMFC cathode impedance is presented, assuming the electrochemical mechanism of MOR on the cathode side (Kulikovsky, 2012b). In this section, the nonstationary version of the DMFC cathode performance model (the section "Cathode Catalyst Layer in a DMFC") is used to calculate the cathode impedance. As discussed in the section "Cathode Catalyst Layer in a DMFC," the model takes into account spatial distribution of the MOR and ORR, through the cathode thickness. It is shown below that the spatial separation of MOR and ORR, discussed in the section "Cathode Catalyst Layer in a DMFC," leads to the formation of a separate semicircle in the impedance spectrum.

## Basic Equations

As before, $x$ is the coordinate through the CCL with the origin at the membrane interface. The nonstationary version of the DMFC cathode performance Equations 4.192 through 4.195 (the section "Basic Equations"), reads

$$C_{dl}\frac{\partial \eta_{ox}}{\partial t} + \frac{\partial j}{\partial x} = -R_{ORR} + R_{MOR}, \tag{5.129}$$

$$\frac{\partial c_{ox}}{\partial t} - D_{ox}\frac{\partial^2 c_{ox}}{\partial x^2} = -\frac{R_{ORR}}{4F}, \tag{5.130}$$

$$\frac{\partial c_{mt}}{\partial t} - D_{mt}\frac{\partial^2 c_{mt}}{\partial x^2} = -\frac{R_{MOR}}{6F}, \tag{5.131}$$

and

$$j = -\sigma_p\frac{\partial \eta_{ox}}{\partial x}. \tag{5.132}$$

Equation 5.129 describes the proton current conservation: protons are consumed in the ORR (the first term on the right side), produced in the MOR (the second term), and charge the double layer (the first term on the left side). Equations (5.130) and (5.131) express the mass balance of oxygen and methanol, respectively. Equation (5.132) is Ohm's law. From Equation 4.201, it follows that $\nabla \eta_{ox} = -\nabla \eta_{mt}$, where $\eta_{ox}$ and $\eta_{mt}$ are the ORR and MOR polarization potentials (overpotentials). Thus, any of these gradients, with the proper sign, can be used in Equation 5.132.

In this section, the ORR and MOR rates are assumed to follow the Butler–Volmer kinetics:

$$R_{ORR} = -2i_{ox} \left( \frac{c_{ox}}{c_{ox}^{h0}} \right) \sinh \left( \frac{\eta_{ox}}{b_{ox}} \right) \tag{5.133}$$

and

$$R_{MOR} = 2i_{mt} \left( \frac{c_{mt}}{c_{mt}^{h0}} \right) \sinh \left( \frac{\eta_{mt}}{b_{mt}} \right). \tag{5.134}$$

Here, (i) $i_{ox}$ and $i_{mt}$ are the ORR and MOR volumetric exchange current densities, (ii) $c_{ox}$ and $c_{ox}^{h0}$ are the local and reference (inlet) oxygen concentrations, (iii) $c_{mt}$ and $c_{mt}^{h0}$ are the local and reference methanol concentrations, (iv) $b_{ox}$ and $b_{mt}$ are the ORR and MOR Tafel slopes, and (v) $D_{ox}$ and $D_{mt}$ are the oxygen and methanol diffusion coefficients, respectively.

Dimensionless variables, used in this section, are given by Equation 4.196. Dimensionless diffusion coefficients are those listed in Equation 4.200. Substituting Equation 5.132 into Equation 5.129, using the variables (Equation 4.196) and taking into account Equations 5.133 and 5.134, Equations 5.129 through 5.131 take the form

$$\frac{\partial \tilde{\eta}_{ox}}{\partial \tilde{t}} - \varepsilon^2 \frac{\partial^2 \tilde{\eta}_{ox}}{\partial \tilde{x}^2} = -\tilde{R}_{ORR} + \tilde{R}_{MOR}, \tag{5.135}$$

$$\mu_{ox}^2 \frac{\partial \tilde{c}_{ox}}{\partial \tilde{t}} - \varepsilon^2 \tilde{D}_{ox} \frac{\partial^2 \tilde{c}_{ox}}{\partial \tilde{x}^2} = -\tilde{R}_{ORR} \tag{5.136}$$

and

$$\mu_{mt}^2 \frac{\partial \tilde{c}_{mt}}{\partial \tilde{t}} - \varepsilon^2 \tilde{D}_{mt} \frac{\partial^2 \tilde{c}_{mt}}{\partial \tilde{x}^2} = -\tilde{R}_{MOR}, \tag{5.137}$$

where

$$\tilde{R}_{ORR} = \tilde{c}_{ox} \sinh(\tilde{\eta}_{ox}), \tag{5.138}$$

$$\tilde{R}_{MOR} = r_* \tilde{c}_{mt} \sinh(\tilde{\eta}_{mt}/p) \tag{5.139}$$

and

$$\mu_{ox} = \sqrt{\frac{4Fc_{ox}^{h0}}{C_{dl}b_{ox}}}, \quad \mu_{mt} = \sqrt{\frac{6Fc_{mt}^{h0}}{C_{dl}b_{ox}}}. \tag{5.140}$$

The relation of the MOR overpotential $\tilde{\eta}_{mt}$ to the ORR overpotential $\tilde{\eta}_{ox}$ is given by Equation 4.201. Solutions to Equations 5.136 and 5.137 are subject to the boundary conditions (4.202) and (4.203) (the section "Boundary Conditions" in Chapter 4). The boundary conditions to Equation 5.135 are given by Equation 4.202.

### Linearization and Fourier Transform

The applied potential perturbation is small; therefore, one can write

$$\tilde{\eta}_{ox} = \tilde{\eta}_{ox}^0(\tilde{x}) + \tilde{\eta}_{ox}^1(\tilde{x}, \tilde{t}), \quad \tilde{\eta}_{mt} = \tilde{\eta}_{mt}^0(\tilde{x}) + \tilde{\eta}_{mt}^1(\tilde{x}, \tilde{t}), \quad \tilde{\eta}_{ox}^1, \tilde{\eta}_{mt}^1 \ll 1$$
$$\tilde{c}_{ox} = \tilde{c}_{ox}^0(\tilde{x}) + \tilde{c}_{ox}^1(\tilde{x}, \tilde{t}), \quad \tilde{c}_{mt} = \tilde{c}_{mt}^0(\tilde{x}) + \tilde{c}_{mt}^1(\tilde{x}, \tilde{t}), \quad \tilde{c}_{ox}^1, \tilde{c}_{mt}^1 \ll 1. \tag{5.141}$$

As in the previous sections, the superscript 0 marks the steady-state solution to the system (5.135) through (5.137), and the superscript 1 marks the small-amplitude perturbations.

Substituting Equations 5.141 into Equations 5.135 through 5.137, neglecting the terms with the product of perturbations, and subtracting the steady-state equations, results in a system of linear equations for the perturbations:

$$\frac{\partial \tilde{\eta}_{ox}^1}{\partial \tilde{t}} - \varepsilon^2 \frac{\partial^2 \tilde{\eta}_{ox}^1}{\partial \tilde{x}^2} = -\tilde{R}_{ORR}^1 + \tilde{R}_{MOR}^1, \tag{5.142}$$

$$\mu_{ox}^2 \frac{\partial \tilde{c}_{ox}^1}{\partial \tilde{t}} - \varepsilon^2 \tilde{D}_{ox} \frac{\partial^2 \tilde{c}_{ox}^1}{\partial \tilde{x}^2} = -\tilde{R}_{ORR}^1, \tag{5.143}$$

$$\mu_{mt}^2 \frac{\partial \tilde{c}_{mt}^1}{\partial \tilde{t}} - \varepsilon^2 \tilde{D}_{mt} \frac{\partial^2 \tilde{c}_{mt}^1}{\partial \tilde{x}^2} = -\tilde{R}_{MOR}^1, \tag{5.144}$$

where

$$\tilde{R}_{ORR}^1 = \sinh(\tilde{\eta}_{ox}^0) \tilde{c}_{ox}^1 + \tilde{c}_{ox}^0 \cosh(\tilde{\eta}_{ox}^0) \tilde{\eta}_{ox}^1 \tag{5.145}$$

and

$$\tilde{R}_{MOR}^1 = r_* \sinh(\tilde{\eta}_{mt}^0/p) \tilde{c}_{mt}^1 - \left(\frac{r_*}{p}\right) \tilde{c}_{mt}^0 \cosh(\tilde{\eta}_{mt}^0/p) \tilde{\eta}_{ox}^1 \tag{5.146}$$

are the reaction rate perturbations and $\tilde{\eta}_{ox}^1 = -\tilde{\eta}_{mt}^1$ was used.

Substituting Fourier transforms

$$\tilde{\eta}_{ox}^1(\tilde{x}, \tilde{t}) = \tilde{\eta}_{ox}^1(\tilde{x}, \tilde{\omega}) \exp(i\tilde{\omega}\tilde{t})$$

$$\tilde{c}_{ox}^1(\tilde{x}, \tilde{t}) = \tilde{c}_{ox}^1(\tilde{x}, \tilde{\omega}) \exp(i\tilde{\omega}\tilde{t})$$

$$\tilde{c}_{mt}^1(\tilde{x}, \tilde{t}) = \tilde{c}_{mt}^1(\tilde{x}, \tilde{\omega}) \exp(i\tilde{\omega}\tilde{t})$$

into Equations 5.142 through 5.144, gives equations for the complex perturbation amplitudes $\tilde{\eta}_{ox}^1(\tilde{x}, \tilde{\omega})$, $\tilde{c}_{ox}^1(\tilde{x}, \tilde{\omega})$, and $\tilde{c}_{mt}^1(\tilde{x}, \tilde{\omega})$:

$$\varepsilon^2 \frac{\partial^2 \tilde{\eta}_{ox}^1}{\partial \tilde{x}^2} = \tilde{R}_{ORR}^1 - \tilde{R}_{MOR}^1 + i\tilde{\omega}\tilde{\eta}_{ox}^1, \tag{5.147}$$

$$\varepsilon^2 \tilde{D}_{ox} \frac{\partial^2 \tilde{c}_{ox}^1}{\partial \tilde{x}^2} = \tilde{R}_{ORR}^1 + i\tilde{\omega}\mu_{ox}^2 \tilde{c}_{ox}^1, \tag{5.148}$$

$$\varepsilon^2 \tilde{D}_{mt} \frac{\partial^2 \tilde{c}_{mt}^1}{\partial \tilde{x}^2} = \tilde{R}_{MOR}^1 + i\tilde{\omega}\mu_{mt}^2 \tilde{c}_{mt}^1. \tag{5.149}$$

Note that $\tilde{\eta}_{ox}^1$, $\tilde{c}_{ox}^1$, and $\tilde{c}_{mt}^1$ in Equations 5.145 and 5.146 are now the complex amplitudes in $\tilde{\omega}$ space. Also note that $\tilde{\eta}_{mt}^0$ can be eliminated from this system using Equation 4.201.

The boundary conditions to Equations 5.147 through 5.149 are as follows:

$$\tilde{\eta}_{ox}^1(0) = \tilde{\eta}_0^1, \quad \left.\frac{\partial \tilde{\eta}_{ox}^1}{\partial \tilde{x}}\right|_1 = 0, \tag{5.150}$$

$$\left.\frac{\partial \tilde{c}_{ox}^1}{\partial \tilde{x}}\right|_0 = 0, \quad \tilde{c}_{ox}^1(1) = 0, \tag{5.151}$$

and

$$\tilde{D}_{mt} \left.\frac{\partial \tilde{c}_{mt}^1}{\partial \tilde{x}}\right|_0 = -\beta_* \left.\frac{\partial \tilde{\eta}^1}{\partial \tilde{x}}\right|_0, \quad \tilde{c}_{mt}^1(1) = 0. \tag{5.152}$$

Equations 5.150 fix the applied perturbation $\tilde{\eta}_0^1$ at $\tilde{x} = 0$ and zero proton current at the CCL/GDL interface. The conditions (5.151) and (5.152) follow from linearization of the boundary conditions in Equation 4.203 and Equation 4.204.

## Impedance Spectra and the Virtual Anode

The definition of the CCL impedance is given by Equation 5.93. However, before proceeding to impedance spectra, it is advisable to discuss the steady-state shapes, corresponding to different methanol concentrations. Parameters for calculations are listed in Table 5.8. Figure 5.20 shows the MOR and ORR rates, and the local proton current density, through the half of the CCL, adjacent to the membrane for the 3 M concentrations of methanol feed (0.5, 1.0, and 1.5 M).

In all three cases, the cathode is a two-layer structure: close to the membrane interface, the MOR rate exceeds the ORR rate, while in the rest part of the CCL, the ORR rate dominates (cf. Figure 4.25, the section "Polarization Curves and $x$ Shapes" in Chapter 4). As a result, the local proton current first increases with $\tilde{x}$ and then decreases.

The useful current density entering the CCL is kept at 50 mA cm$^{-2}$, whereas the peak proton current in the layer exceeds 100 mA cm$^{-2}$ (Figure 5.20). The excessive

## TABLE 5.8
## DMFC Physical and Operation Parameters Used in the Calculations

| | |
|---|---|
| Reference oxygen molar concentration $c_{ox}^{h0}$ (mol cm$^{-3}$) | $7.36 \cdot 10^{-6}$ |
| Reference methanol molar concentration $c_{mt}^{h0}$ (mol cm$^{-3}$) | $10^{-3}$ (1 M) |
| Methanol diffusion coefficient in the ABL and CCL $D_{mt}$ (cm$^2$ s$^{-1}$) | $2 \cdot 10^{-5}$ |
| Oxygen diffusion coefficient in the CCL $D_{ox}$ (cm$^2$ s$^{-1}$) | $10^{-3}$ |
| Oxygen diffusion coefficient in the GDL $D_b^c$ (cm$^2$ s$^{-1}$) | $10^{-2}$ |
| Limiting current density due to oxygen transport in the GDL $j_{lim}^{c0}$ (A cm$^{-2}$) | 1.42 |
| Limiting current density due to methanol transport in the ABL $j_{lim}^{a}$ (A cm$^{-2}$) | 0.579 |
| Crossover parameter $\beta_*$ | 0.4 |
| Catalyst layer thickness $l_{CL}$ (cm) | 0.01 |
| Backing layer thickness $l_b$ (cm) | 0.02 |
| Catalyst layer proton conductivity $\sigma_p$ ($\Omega^{-1}$cm$^{-1}$) | 0.001 |
| Double layer volumetric capacitance $C_{dl}$ (F cm$^{-3}$) | 100 |
| ORR Tafel slope $b_{ox}$ (V) | 0.05 |
| MOR Tafel slope $b_{mt}$ (V) | 0.05 |
| ORR exchange current density $i_{ox}$ (A cm$^{-3}$) | 0.1 |
| MRR exchange current density $i_{mt}$ (A cm$^{-3}$) | 0.1 |
| ORR equilibrium potential $E_{ox}^{eq}$ (V) | 1.23 |
| MOR equilibrium potential $E_{mt}^{eq}$ (V) | 0.028 |
| Reference current density $j_* = \sigma_p b / l_{CL}$ (A cm$^{-2}$) | 0.005 |
| Parameter $\varepsilon$ | 1.58 |
| Parameter $t_*$ | 25.0 |
| Dimensionless oxygen diffusivity in the CCL $\tilde{D}_{ox}$ | 56.82 |
| Dimensionless methanol diffusivity in the ABL and CCL $\tilde{D}_{mt}$ | 115.8 |

50 mA cm$^{-2}$ arise due to conversion of the permeated methanol close to the membrane. In other words, the MOR–dominating part serves as a "virtual anode" within the CCL, which converts the incoming methanol flux into a proton current. This proton current is then converted to water flux in the ORR-dominating part of the CCL, adjacent to the GDL.

With 0.5 M methanol feed, the virtual anode is very thin (Figure 5.20a), and it does not disturb the cathode impedance. The cathode spectrum is a single elongated semicircle, representing the combined effect of faradaic (ORR), double layer charging and oxygen transport processes in the CCL (Figure 5.21). However, at the 1 M methanol concentration, the rate of crossover increases, and the virtual anode covers 6% of the CCL thickness (Figure 5.20b). The respective impedance spectrum exhibits a "shoulder" in the high-frequency range (Figure 5.21). At the 1.5 M feed, the virtual anode width increases to up to 30% of the CCL thickness, and the virtual anode appears as a well-separated high-frequency arc (Figure 5.21).

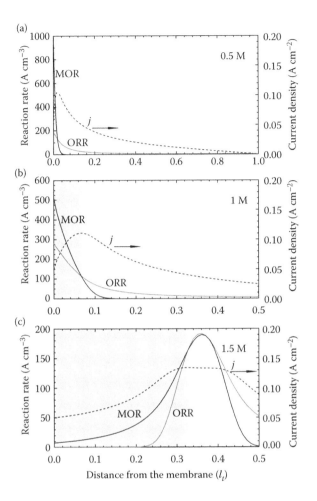

**FIGURE 5.20** The MOR and ORR rates and the local proton current density in the half of the CCL, adjacent to the membrane, for the indicated molar concentrations of methanol feed. The shaded region indicates the virtual anode. The cell current density is fixed at 50 mA cm$^{-2}$. Each frame shows the ORR and MOR reaction rates and the proton current density (dashed line) at the methanol concentration of (a) 0.5 M, (b) 1 M, and (c) 1.5 M.

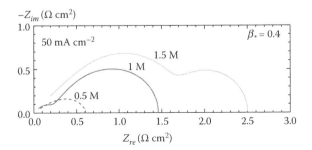

**FIGURE 5.21** Calculated impedance spectra of the CCL for the indicated methanol molar concentrations. The semicircles intercept is at 12 Hz for 1 M curve, and at 0.153 Hz for 1.5 M curve.

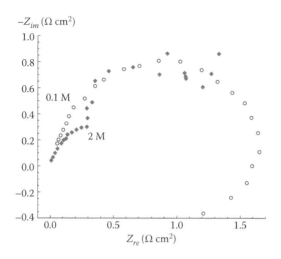

**FIGURE 5.22** Measured impedance spectra of the DMFC cathode for the indicated methanol molar concentrations. (Data from Piela, P., Fields, R., and Zelenay, P. 2006. *J. Electrochem. Soc.*, **153**, A1902–A1913.)

Figure 5.22 shows two experimental spectra of the DMFC cathode, measured by Piela et al. (2006). As can be seen, the spectrum for 0.1 M methanol concentration is a single semicircle. The inductive low-frequency loop has been attributed to the inertial effect of CO coverage on the rate of ORR in the cathode (Piela et al., 2006). However, the spectrum corresponding to the 2 M methanol feed clearly indicates the shoulder (the second semicircle) in the high-frequency range. This correlates with the modeling results above.

## IMPEDANCE OF THE CATHODE SIDE OF A PEM FUEL CELL

In low-temperature cells, oxygen required for the ORR is transported to the CCL through the gas channels and the GDL. Understanding the potential losses resulting from oxygen transport in the cell components is one of the key issues for better cell design.

The EIS has been widely used to study this problem. In general, EIS can give separate contributions of each transport and kinetic process to the total cell resistance. However, unambiguous interpretations of EIS spectra require modeling.

A widely used approach toward the interpretation of cell impedance is a method of equivalent circuit (Orazem and Tribollet, 2008). The great advantage of the method is simplicity. However, as discussed in the section "Introductory Remarks," the equivalent circuit is not unique, that is, similar spectra could be generated by different circuits. In recent years, there has been a growing interest in the direct physical modeling of the cell impedance.

Springer et al. (1996) developed a physical model for the impedance of the cathode side of a PEFC, taking into account oxygen transport in the GDL. They fitted the model to measured spectra and calculated the transport resistivities and the double layer capacitance of the CCL. Similar cell impedance models, based on the flooded agglomerate model for the active layer, has been developed by Guo and White (2004).

A physical-based impedance model of the PEFC cathode, which included oxygen transport in the GDL, has been developed by Bultel et al. (2005). The authors reported qualitative similarity of measured and calculated impedance spectra and analyzed the effect of GDL diffusion resistivity on the cell performance.

A typical impedance spectrum of the PEFC consists of two or three partly over-lapping semicircles. Following Springer et al. (1996), it has been agreed that the lowest-frequency arc is attributed to the oxygen transport in the GDL. However, Schneider et al. (2007a,b) measured local impedance in a single-channel PEFC with segmented electrodes, and showed that this arc is, in fact, due to the oxygen transport in the channel. Experiments of Schneider et al. (2007a,b) clearly show that the position and radius of the LF arc depends on the stoichiometry of oxygen flow. Similar results have been obtained by Brett et al. (2003), who were the first to measure local impedance in a straight channel cell with segmented electrodes.

In this section, a model for the PEFC cathode impedance is discussed, including oxygen transport in the channel (Kulikovsky, 2012d). The model is based on the transient CCL performance model from the section "Basic Equations" linked to the nonstationary extensions of the models for oxygen transport in the GDL and in the channel, discussed in the section "Performance Modeling of a Fuel Cell."

## MODEL ASSUMPTIONS

Consider a linear PEM fuel cell with the straight cathode channel and segmented electrodes (Figure 5.23). The cell of this design has been used in experiments of Brett et al. (2003) and Schneider et al. (2007a,b). The coordinates $\tilde{x}$ and $\tilde{z}$ are normalized according to

$$\tilde{x} = \frac{x}{l_{CL}}, \quad \tilde{z} = \frac{z}{L}, \tag{5.153}$$

where $L$ is the channel length. Note the different scalings for $\tilde{x}$ and $\tilde{z}$.

The goal is to formulate the nonstationary performance model for the cathode side of such a cell. The main assumptions are as follows:

- Current in the cell flows only in the through-plane direction (along the $x$-axis, Figure 5.23).
- In the CCL and GDL, oxygen transport is directed along the $x$-axis. The $z$ fluxes in the porous layers are negligible.
- A well-mixed oxygen flow with the constant velocity in the channel (plug flow) is directed along the $z$-axis.
- The ORR rate is described by the Butler–Volmer equation.
- The cell is well humidified. Liquid water in porous layers may only uni-formly reduce the transport coefficients. The effects from local flooding or membrane drying are not considered.

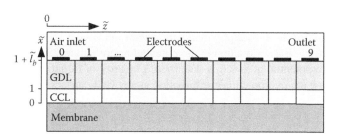

**FIGURE 5.23** A schematic of the cathode side of a PEM fuel cell with a single straight air channel and segmented electrodes. The $x$ model is formulated for the single cell segment consisting of a part of the MEA under the electrode. The $z$ model for the flow in the channel links the segments.

The segmented electrodes enable measuring the local impedance spectra (Figure 5.23). Owing to the absence of $z$-directed fluxes and currents in the porous media, the cell can be considered as a number of elementary segments connected in parallel. Each segment contains a single electrode with the GDL and CCL domains underneath (Figure 5.23).

The segments are "linked" by the flow in the air channel, which transports oxygen from one segment to another. Thus, the oxygen transport in the channel provides an upper boundary condition for the $x$ problem in each segment. Our next goal is the formulation of the $x$ problem for the segment impedance, beginning with the problem of the CCL impedance.

### IMPEDANCE OF THE CATHODE CATALYST LAYER

Equations describing the CCL impedance are Equations 5.91 and 5.92 (the section "Basic Equations"). However, boundary conditions to these equations now differ from those discussed in the section "Basic Equations." Here, the solution to the system (5.91) and (5.92) is subjected to more general boundary conditions:

$$\tilde{\eta}^1(0) = \tilde{\eta}_0^1, \quad \left.\frac{\partial \tilde{\eta}^1}{\partial \tilde{x}}\right|_1 = 0 \tag{5.154}$$

and

$$\left.\frac{\partial \tilde{c}^1}{\partial \tilde{x}}\right|_0 = 0, \quad \tilde{c}^1(1) = \tilde{c}_b^1(1). \tag{5.155}$$

The conditions for $\tilde{\eta}^1$ mean applied perturbation amplitude $\tilde{\eta}_0^1$ at the membrane interface (perturbation of the total potential loss in the system) and zero perturbation of proton current at the CCL/GDL interface. The first of the conditions (5.155) means a zero oxygen flux in the membrane, while the second one expresses continuity of the oxygen concentration at the CCL/GDL interface. Here, $\tilde{c}_b^1$ is the perturbation of the oxygen concentration in the GDL (the section "Impedance of the Cathode Catalyst Layer"). Note that the subscript $b$ marks the values in the GDL.

In the section "Basic Equations," the potential perturbation was applied at $\tilde{x} = 1$. Here, the overpotential at $\tilde{x} = 0$ is disturbed. Linearity of Equations 5.91 and 5.92 allow applying the overpotential perturbation at any CCL surface; the results would be the same.

### Steady-State Problems in the CCL

Equations for the steady-state profiles $\tilde{\eta}^0$ and $\tilde{c}^0$ can be obtained from the system (5.89) and (5.90) by chalking out the time derivatives. This approach, however, has the following drawback: when the cell current is fixed, the boundary condition for $\tilde{\eta}^0$ at $\tilde{x} = 0$ is not known and it should be obtained as a result of iterations.

The alternative approach is to solve a simpler equivalent system (4.156) and (4.55), which does not require iterations. The boundary conditions for Equations 4.156 and 4.55 are straightforward:

$$\tilde{j}^0|_{\tilde{x}=0} = \tilde{j}_0(\tilde{z}), \quad \tilde{j}^0|_{\tilde{x}=1} = 0, \quad \tilde{c}^0|_{\tilde{x}=1} = \tilde{c}_h^0(\tilde{z}) - \frac{\tilde{j}_0(\tilde{z})}{\tilde{j}_{\lim}^{c0}}, \tag{5.156}$$

where $\tilde{j}_0(\tilde{z})$ is the local cell current density Equation 5.172, $\tilde{c}_h^0(\tilde{z})$ is the local oxygen concentration in the channel Equation 5.171, and $j_{\lim}^{c0}$ is the limiting current density resulting from the oxygen transport in the GDL, Equation 5.42. Note that the subscript $h$ marks the oxygen concentration in the channel. The boundary condition for $\tilde{c}^0$ at $\tilde{x} = 1$ follows from the linear diffusion equation for the oxygen transport in the GDL (Section 5).

The functions $\tilde{c}_h^0(\tilde{z})$ and $\tilde{j}_0(\tilde{z})$ in Equation 5.156 are the steady-state oxygen concentration in the channel and the local cell current density, respectively. These functions result from the solution of the steady-state problem for oxygen transport in the cathode channel (the section "Oxygen Transport in the Channel").

### Oxygen Transport in the GDL

The oxygen diffusion equation in the GDL is

$$\frac{\partial c_b}{\partial t} - D_b\frac{\partial^2 c_b}{\partial x^2} = 0, \quad D_b\frac{\partial c_b}{\partial x}\bigg|_{x=l_{CL}+} = D\frac{\partial c}{\partial x}\bigg|_{x=l_{CL}-}, \quad c_b(l_{CL} + l_b) = c_h(z),$$
$$\tag{5.157}$$

where $c_b$ is the oxygen concentration and $D_b$ is the effective oxygen diffusion coefficient in the GDL of a thickness $l_b$. The boundary conditions to this equation mean continuity of the oxygen flux at the GDL/CCL interface and equality of the oxygen concentration at the GDL/channel interface to the channel concentration $c_h$.

After nondimensionalization with Equations 4.51, 5.83, and 5.153, Equation 5.157 takes the form

$$\mu^2 \frac{\partial \tilde{c}_b}{\partial \tilde{t}} - \varepsilon^2 \tilde{D}_b \frac{\partial^2 \tilde{c}_b}{\partial \tilde{x}^2} = 0, \tag{5.158}$$

where $\mu$ is given by Equation 5.88. Equation 5.158 is linear and may directly be written as the frequency domain equation for the perturbation amplitude $\tilde{c}_b^1$ in the GDL

$$\varepsilon^2 \tilde{D}_b \frac{\partial^2 \tilde{c}_b^1}{\partial \tilde{x}^2} = i\omega\mu^2 \tilde{c}_b^1, \quad \tilde{D}_b \frac{\partial \tilde{c}_b^1}{\partial \tilde{x}}\bigg|_{\tilde{x}=1+} = \tilde{D} \frac{\partial \tilde{c}^1}{\partial \tilde{x}}\bigg|_{\tilde{x}=1-}, \quad \tilde{c}_b^1(1+\tilde{l}_b) = \tilde{c}_h^1, \tag{5.159}$$

where $\tilde{c}_h^1$ is the perturbation of the oxygen concentration in the channel (the section "Oxygen Transport in the Channel"). The problem (5.159) links the channel and the CCL problems for the oxygen concentration perturbations.

Equation 5.159 has an analytical solution

$$\tilde{c}_b^1 = \frac{(1+i)\varepsilon f_1^1 \sin\left(a\left(1+\tilde{l}_b - \tilde{x}\right)\right)}{\mu\sqrt{2\tilde{\omega}\tilde{D}_b}\cos\left(a\tilde{l}_b\right)} + \frac{\tilde{c}_h^1 \cos\left(a\left(1-\tilde{x}\right)\right)}{\cos\left(a\tilde{l}_b\right)}, \tag{5.160}$$

where

$$a = \frac{(i-1)\mu}{\varepsilon}\sqrt{\frac{\tilde{\omega}}{2\tilde{D}_b}}$$

and

$$f_1^1 \equiv \tilde{D} \frac{\partial \tilde{c}^1}{\partial \tilde{x}}\bigg|_{\tilde{x}=1-} \tag{5.161}$$

is the perturbation of the oxygen flux at the catalyst layer side of the CCL/GDL interface. At $\tilde{x} = 1$, from Equation 5.160 one obtains

$$\tilde{c}_b^1(1) = \frac{(1+i)\varepsilon f_1^1}{\mu\sqrt{2\tilde{\omega}\tilde{D}_b}} \tan\left(a\tilde{l}_b\right) + \frac{\tilde{c}_h^1}{\cos\left(a\tilde{l}_b\right)}, \tag{5.162}$$

which is required in the boundary condition (5.155) for the CCL problem.

For the channel problem, the perturbation of the oxygen $x$ flux will be needed at the channel/GDL interface. Differentiating Equation 5.160 over $\tilde{x}$, multiplying the result by $\tilde{D}_b$, and substituting $\tilde{x} = 1 + \tilde{l}_b$, one finds

$$\tilde{D}_b \frac{\partial \tilde{c}_b^1}{\partial \tilde{x}}\bigg|_{\tilde{x}=1+\tilde{l}_b} = -\frac{(1+i)a\varepsilon f_1^1}{\mu\sqrt{2\tilde{\omega}\tilde{D}_b}\cos(a\tilde{l}_b)} - a\tilde{c}_h^1 \tan(a\tilde{l}_b). \tag{5.163}$$

## OXYGEN TRANSPORT IN THE CHANNEL

Voltage perturbation disturbs the oxygen concentration in the channel. Propagation of this perturbation is a time-dependent process, which affects the cell impedance spectrum. To take into account this effect, a time-dependent mass balance equation is written for the oxygen concentration in the channel $c_h$

$$\frac{\partial c_h}{\partial t} + v\frac{\partial c_h}{\partial z} = \left(\frac{D_b}{h}\right)\frac{\partial c_b}{\partial x}\bigg|_{x=l_{CL}+l_b}, \tag{5.164}$$

where $v$ is the flow velocity and $h$ is the channel height. On the right-hand side of Equation 5.164 stands the diffusion flux of oxygen in the GDL at the channel/GDL interface. The mass conservation prescribes that this flux must be divided by $h$.

With dimensionless variables (4.51), (5.83), and (5.153), Equation 5.164 takes the form

$$\xi^2\frac{\partial \tilde{c}_h}{\partial \tilde{t}} + \lambda \tilde{J}\frac{\partial \tilde{c}_h}{\partial \tilde{z}} = \tilde{D}_b\frac{\partial \tilde{c}_b}{\partial \tilde{x}}\bigg|_{\tilde{x}=1+\tilde{l}_b}, \tag{5.165}$$

where $J$ is the mean current density in the cell,

$$\lambda = \frac{4Fhvc_h^0}{LJ}$$

is the oxygen stoichiometry, and

$$\xi = \sqrt{\frac{8Fhc_h^0 i_* l_{CL}}{C_{dl}\sigma_p b^2}} \tag{5.166}$$

is the constant parameter. $\xi^2/(\lambda \tilde{J})$ determines the characteristic time of the oxygen concentration variation in the channel.

Equation 5.165 is linear; thus, an equation for the small-amplitude perturbation $\tilde{c}_h^1$ is

$$\xi^2\frac{\partial \tilde{c}_h^1}{\partial \tilde{t}} + \lambda \tilde{J}\frac{\partial \tilde{c}_h^1}{\partial \tilde{z}} = \tilde{D}_b\frac{\partial \tilde{c}_b^1}{\partial \tilde{x}}\bigg|_{\tilde{x}=1+\tilde{l}_b}. \tag{5.167}$$

Substituting here $\tilde{c}_h^1(\tilde{z},\tilde{t}) = \tilde{c}_h^1(\tilde{z},\tilde{\omega})\exp(i\tilde{\omega}\tilde{t})$, gives an equation for the complex perturbation amplitude $\tilde{c}_h^1(\tilde{z},\tilde{\omega})$:

$$\lambda \tilde{J}\frac{\partial \tilde{c}_h^1}{\partial \tilde{z}} = -i\tilde{\omega}\xi^2\tilde{c}_h^1 + \tilde{D}_b\frac{\partial \tilde{c}_b^1}{\partial \tilde{x}}\bigg|_{\tilde{x}=1+\tilde{l}_b}, \quad \tilde{c}_h^1(0) = 0. \tag{5.168}$$

The boundary condition at $\tilde{z} = 0$ means undisturbed inlet oxygen concentration. The diffusion term on the right-hand side of Equation 5.168 is given by Equation 5.163.

Solution to Equation 5.168 is

$$\tilde{c}_h^1 = \frac{1}{\lambda \tilde{J}} \exp\left(-\frac{i\tilde{\omega}\xi^2 \tilde{z}}{\lambda \tilde{J}}\right) \int_0^{\tilde{z}} f_D(\tilde{z}') \exp\left(\frac{i\tilde{\omega}\xi^2 \tilde{z}'}{\lambda \tilde{J}}\right) d\tilde{z}', \qquad (5.169)$$

where

$$f_D(\tilde{z}) \equiv \tilde{D}_b \left. \frac{\partial \tilde{c}_b^1}{\partial \tilde{x}} \right|_{\tilde{x}=1+\tilde{l}_b}.$$

The expression for $f_D$ contains the perturbation of cell current density $\tilde{j}_0^1$. Ohm's law

$$\tilde{j}_0^1 = - \left. \frac{\partial \tilde{\eta}^1}{\partial \tilde{x}} \right|_0 \qquad (5.170)$$

and the boundary condition (5.159) link the result (5.169) to the CCL and GDL problems.

The solution of the steady-state version of the problem (5.165), under the condition of equipotential cell electrodes, gives the undisturbed shapes of the oxygen concentration and local current density along the channel given by Equations 5.55 and 5.56, respectively. In the notations of this section, Equations 5.55 and 5.56 take the form

$$\tilde{c}_h^0(\tilde{z}) = \left(1 - \frac{1}{\lambda}\right)^{\tilde{z}} \qquad (5.171)$$

and

$$\tilde{j}_0(\tilde{z}) = -\tilde{J}\lambda \ln\left(1 - \frac{1}{\lambda}\right)\left(1 - \frac{1}{\lambda}\right)^{\tilde{z}}. \qquad (5.172)$$

## NUMERICAL SOLUTION AND IMPEDANCE

The whole problem is solved in the following way. At $\tilde{z} = 0$, the perturbation of oxygen concentration in the channel is set to zero and the local $\tilde{x}$ problem is solved with $\tilde{c}_h^1 = 0$. In the next segment, $\tilde{c}_h^1$ is calculated, using the simplest rectangle rule for numerical approximation of integral (5.169) and the local $\tilde{x}$-problem is solved at $\tilde{z} = 0 + d\tilde{z}$ (here, $d\tilde{z}$ is the segment length). Typically, the channel length was divided into 100 segments, and the results were printed for each tenth segment. At low oxygen stoichiometries, the number of numerical segments must be larger. This procedure is repeated until the end of the channel is reached.

The local impedance is calculated according to Equation 5.93. Since the cell electrode is equipotential, the local perturbations of current density are summing up and the total cell impedance $\tilde{Z}_{cell}$ is given by

$$\frac{1}{\tilde{Z}_{cell}} = \int_0^1 \frac{d\tilde{z}}{\tilde{Z}}. \qquad (5.173)$$

Equation 5.173 means that the individual segments are parallel impedances.

### Local and Total Spectra

Generally, in each segment, three time-dependent processes run simultaneously: oxygen transport in the GDL, oxygen transport in the CCL, and the double layer charging. The fourth time-dependent process is oxygen transport in the channel, which "links" the segments. To understand the contribution of each process to the local spectra, the respective time derivatives will be "switched on" one by one.

Owing to a small CCL thickness, oxygen transport in the CCL is fast. The time constant of this process is in the order of the time constant of the double layer charging. This means that the contribution of oxygen transport in the CCL to the cell impedance cannot be distinguished from double layer charging. In the following discussion, it will be assumed that oxygen transport in the CCL is infinitely fast.

Figure 5.24 shows the impedance spectra for the zeroth, third, sixth, and ninth segments (the segments are numerated from the oxygen inlet, Figure 5.23). The last frame in Figure 5.24 represents the average (the cell) spectrum calculated according to Equation 5.173. Parameters for the calculation are listed in Table 5.9, the oxygen stoichiometry is $\lambda = 2$.

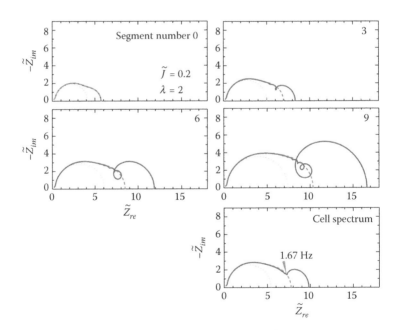

**FIGURE 5.24** A local and average (the bottom frame) impedance spectra for the oxygen stoichiometry $\lambda = 2$. The dotted (innermost) spectra are calculated assuming the infinitely fast relaxation of oxygen transport in GDL and channel. The dashed curve takes into account unsteady oxygen transport in the GDL, but it ignores the propagation of oxygen concentration perturbations along the channel. These spectra correspond to the excitation of separate segments in the cell, provided that all the other segments are not excited. The solid lines are the full spectra, taking into account unsteady oxygen transport in the channel. These spectra account for the synchronous excitation of all segments.

**TABLE 5.9**
**The Physical and Dimensionless Parameters**

| | |
|---|---|
| Tafel slope $b$ (V) | 0.05 |
| CCL proton conductivity $\sigma_p$ ($\Omega$ cm$^{-1}$) | 0.03 |
| Exchange current density $i_*$ (A cm$^{-3}$) | $10^{-3}$ |
| Effective oxygen diffusion coefficient in the CCL, $D$ (cm$^2$ s$^{-1}$) | $1.37 \cdot 10^{-3}$ |
| Effective oxygen diffusion coefficient in the GDL, $D_b$ (cm$^2$ s$^{-1}$) | 0.01 |
| CL capacitance $C_{dl}$ (F cm$^{-3}$) | 20 |
| CL thickness $l_{CL}$ (cm) | 0.001 |
| GDL thickness $l_b$ (cm) | 0.02 |
| Channel depth $h$ (cm) | 0.1 |
| Working current density $J$ (A cm$^{-2}$) | 0.3 |
| Limiting current density $j_{\lim}$ (A cm$^{-2}$) | 1.42 |
| $j_*$ (A cm$^{-2}$) | 1.5 |
| $t_*$ (s) | 500 |
| $D_*$ (cm$^2$ s$^{-1}$) | $5.28 \cdot 10^{-4}$ |
| $\tilde{J}$ | 0.2 |
| $\tilde{j}_{\lim}$ | 0.947 |
| $\tilde{D}$ | 2.59 |
| $\tilde{D}_b$ | 18.94 |
| $\varepsilon$ | 866 |
| $\mu$ | 1.686 |
| $\xi$ | 0.01946 |

*Note:* Catalyst layer volumetric capacitance is estimated with the data (Adapted from Makharia, R., Mathias, M. F., and Baker, D. R. 2005. *J. Electrochem. Soc.*, **152**, A970–A977).

In each frame, the innermost (dotted) semicircle is the local (average in the last frame) impedance spectrum, calculated assuming infinitely fast relaxation of oxygen transport in the channel and in the GDL.* In other words, these spectra represent processes in the CCL only. Note that the local current density decays toward the channel outlet, which increases the radius of the local CCL spectrum (Figure 5.24).

The dashed semicircle in each frame represents the local (average in the last frame) spectrum with nonstationary oxygen transport in the GDL and steady-state (infinitely fast relaxation) transport in the channel. These spectra correspond to separate measurements of the impedance of every individual segment, provided that all the other segments are not excited. Under these conditions, the oxygen concentration in the upstream segments is not perturbed, that is, the resistivity resulting from oxygen transport in the channel is not seen.

---

* Fast relaxation means that the respective steady-state shape of the oxygen concentration is established at a negligibly short time.

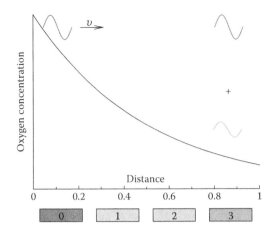

**FIGURE 5.25**  Schematic of the interference of the local and transported perturbations.

These local spectra consist of two overlapped semicircles, corresponding to the CCL (HF domain) and GDL (LF domain). However, a sophisticated fitting algorithm is necessary to separate the contribution of the transport in the GDL.

Solid curves in Figure 5.24 show the local spectra measured when all the segments are excited synchronously, that is, the potential perturbation is applied to the segments in the same phase. The resulting average spectrum (the solid curve in the last frame in Figure 5.24) represents the spectrum of the whole cell. Note that this spectrum coincides with the spectrum of a nonsegmented cell.

As can be seen, the local CCL+GDL+channel spectrum of the zeroth module coincides with the local CCL+GDL spectrum. However, the full spectra of the segments 3, 6, and 9, as well as the average spectrum, exhibit an additional low-frequency arc. This arc is due to oxygen transport in the channel, as suggested by Schneider et al. (2007a). The channel-transport resistance increases with the distance along the channel (Figure 5.24). Physically, this resistance increases with the average rate of oxygen consumption in the upstream segments (in the zeroth segment this rate is zero).

Another feature of the spectra in segments 6 and 9 is a number of closed loops in the frequency range between 1 and 100 Hz (Figure 5.24). These loops are resulting from the interference of the local concentration perturbation and concentration perturbations transported along the channel. The effect is illustrated in Figure 5.25. On all the electrodes, the initial phase of the potential perturbation is the same. Owing to oxygen consumption in the ORR, the sinusoidal potential perturbation generates a periodic perturbation of the oxygen concentration in the channel.

The concentration perturbation, generated at the zeroth segment, is transported with the flow down the channel. The segments located downstream, thus, "feel" the sum of the locally disturbed concentration, plus the concentration perturbation arriving with the flow from the zeroth segment. These two waves interfere: at certain frequencies of the exciting signal, the phases are the same and the amplitudes of the local signal and signals arriving with the flow sum up (Figure 5.25). At other frequencies, the phases are different and the concentration waves partly cancel out

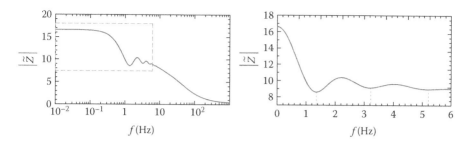

**FIGURE 5.26** Bode plot for the ninth segment in Figure 5.24. Note that the minima of $|\tilde{Z}|$ are equidistant along the frequency axis.

each other. In the remote segments, this leads to oscillation of the channel-transport resistance, which is seen as closed loops threaded on the local CCL+GDL spectrum (Figure 5.24).

One may expect this effect to be periodical along the frequency axis: the perturbation from the zeroth segment arrives at the last segment in the same phase as the local perturbation, if an even number of half-waves fits into the channel length: $L = 2\pi n v/\omega$, where $n = 1, 2, \ldots$. Figure 5.26 confirms this suggestion: the local minima in the Bode plot of $|\tilde{Z}|$ versus $\omega$ are equidistant in the frequency domain.

Figure 5.26 also shows that the amplitude of oscillations $|\tilde{Z}|$ decreases with the frequency. This is seen as a lowering of the loop radius as $\tilde{\omega}$ rises (Figure 5.24). At low frequencies, the sum of the local and transported perturbations penetrates deep into the CCL and generates a well-resolved, closed loop in the spectrum (Figure 5.24). However, at higher frequencies, the perturbation transported with the flow weakly affects the inertial oxygen transport in the GDL and the effect discussed disappears (the loops degrade, Figure 5.24). Thus, the variation of the loop radius with the frequency is another sign of the quality of oxygen transport in the GDL. The larger the GDL oxygen diffusivity, the slower the loop radius decays with $\tilde{\omega}$.

The effect of the formation of loops is best seen at small oxygen stoichiometries, when the shape of the oxygen concentration is strongly nonuniform and the concentration perturbation at zeroth segment significantly exceeds the perturbations at the downstream segments. At higher $\lambda$, the perturbations arriving from all the upstream segments have nearly the same amplitude and different phases, related to various distances traveled along the channel. Thus, the transported perturbations partly cancel out at all frequencies and the loops degenerate into small bumps.

Recently, the problem discussed has been considered by Maranzana et al. (2012). They neglected oxygen transport loss in the CCL and solved the reduced problem analytically. The impedance spectra of individual segments derived in their work also exhibit closed loops. Unfortunately, the spectra measured by Schneider et al. do not show the loops. This, perhaps, is a consequence of poor spectra resolution in their experiments. The measurements of Schneider et al. (2007a,b) have been performed with 10 points per decade, while the resolution of the loops in Figure 5.24 requires about a hundred points per decade.

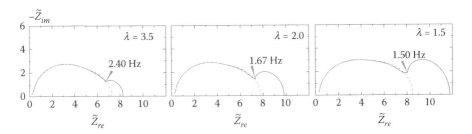

**FIGURE 5.27** Average impedance spectra of the cell for the indicated oxygen stoichiometries. Dashed curves show the spectra calculated assuming steady-state oxygen transport in the channel. The intersection of low- and high-frequency arcs occurs at frequencies between 1 and 3 Hz.

The impedance spectra of the whole cell for the three stoichiometries of the oxygen flow are shown in Figure 5.27. As expected, the resistivity due to oxygen transport in the channel increases with the decrease in $\lambda$. Note that for all $\lambda$, the low-frequency ("channel") arc intersects the CCL+GDL arc at a frequency $f$ in the range of 1–3 Hz (Figure 5.27). The measurements of Schneider et al. (2007a) show that the intersection is independent of $\lambda$ and, in their experiments, it occurs at $f \simeq 7.9$ Hz.

## CONSTANT STOICHIOMETRY VERSUS CONSTANT OXYGEN FLOW

The results of the section "Impedance of the Cathode Side of a PEM Fuel Cell," have been obtained assuming constant oxygen stoichiometry $\lambda$ of the flow. However, impedance experiments are usually performed at a constant oxygen flow rate, rather than at a constant $\lambda$. Indeed, keeping constant $\lambda$ means that the inlet flow rate must be changed in phase with the mean current density perturbation, which is hardly possible.

Qualitatively, keeping $\lambda = \text{const}$ in the equations above means that the regimes with negative slopes of the local static polarization curves, of remote segments, are not considered. These regimes arise under fixed inlet oxygen flow, when this flow is below the critical value (the section "Negative Local Resistance")

$$hwv^0 c_h^0 < Lw\frac{D_b c_h^0}{l_b}, \qquad (5.174)$$

where $w$ is the in-plane width of the oxygen channel. Equation 5.174 means that the slope of the local polarization curves becomes negative if the inlet total oxygen molar flow $hwv^0 c_h^0$ (mol s$^{-1}$) is less than the total maximal limiting oxygen flow in the GDL $LwD_b c_h^0/l_b$. Physically, if oxygen transport through the GDL is fast, the inlet segments consume too much oxygen, and the remote segments experience oxygen starvation. Thus, to homogenize the oxygen distribution along the channel, the GDL diffusivity *should not be large*. This is one of the main reasons for using GDLs in fuel cell design.

At low inlet fluxes, Equation 5.174 holds and the local polarization curves of remote segments exhibit negative slopes (Kulikovsky et al., 2004). In this case, the local impedance spectra of remote segments turn into quadrant $\Re(\tilde{Z}) < 0$, that is, they exhibit negative resistivity.

Note that the *total* cell resistivity is positive, though it depends on the regime of cell feeding, while the local (segment) resistivity may be negative. Static differential resistivity of the cell $\tilde{R}_{cell}$, which takes into account oxygen transport in the channel, can be calculated using the expression for the half-cell overpotential (5.58). However, before proceeding to this equation, it is advisable to relate oxygen stoichiometry and oxygen flow. By definition, $\lambda = 4Fhv^0c_h^0/(LJ)$. Multiplying numerator and denominator by the channel width $w$, and taking into account that the molar flow is $M_{ox} = hwv^0c_h^0$ yields

$$\lambda = \frac{4FM_{ox}}{LwJ}, \quad M_{ox} = hwv^0c_h^0. \tag{5.175}$$

Introducing a dimensionless molar flow of oxygen

$$\tilde{M}_{ox} = \frac{M_{ox}}{M_*}, \quad \text{where } M_* = \frac{Lwj_{ref}}{4F} = \frac{Lw\sigma_p b}{4Fl_{CL}}. \tag{5.176}$$

Equation 5.175 can be transformed to the following dimensionless form:

$$\lambda \tilde{J} = \tilde{M}_{ox}. \tag{5.177}$$

## Constant Oxygen Stoichiometry

In the case of $\lambda = \text{const}$, Equation 5.58 can be directly differentiated, keeping $f_\lambda = \text{const}$. Differentiating Equation 5.58 over $\tilde{J}$, one obtains the static differential cell resistivity $\tilde{R}_{cell} = \partial\tilde{\eta}_0/\partial\tilde{J}$,

$$\tilde{R}_{cell} = \frac{1}{\tilde{J}} + \frac{f_\lambda}{\tilde{j}_{lim}^{c0} - f_\lambda\tilde{J}}. \tag{5.178}$$

Here, $f_\lambda$ is given by Equation 5.57. Comparing this to Equation 5.120, it is seen that the term $1/\tilde{J}$ describes the CCL charge-transfer resistivity, and the last term $f_\lambda/(\tilde{j}_{lim}^{c0} - f_\lambda\tilde{J})$, accounts for the combined resistivity of oxygen transport in the GDL and in the channel.

Note that in the $\lambda = \text{const}$. case, the charge transfer resistivity (averaged over the channel length) is independent of $\lambda$. This means that the local charge transfer resistivities of individual segments are connected in parallel. Indeed, with the local current along the channel given by Equation 5.56, the inverse local charge transfer resistivity is $1/\tilde{R}_{ct}(\tilde{z}) = \tilde{j}_0(\tilde{z})$. Under the assumption of parallel local resistivities, the total charge transfer resistivity of the cell is

$$\frac{1}{\tilde{R}_{ct}^{tot}} = \int_0^1 \frac{d\tilde{z}}{\tilde{R}_{ct}} = \int_0^1 \tilde{j}_0(\tilde{z}) \, d\tilde{z} = \tilde{J},$$

that is, $\tilde{R}_{ct}^{tot} = 1/\tilde{J}$, regardless of $\lambda$.

In the limit of infinite oxygen stoichiometry, $f_\lambda \to 1$, and Equation 5.178 takes the form

$$\tilde{R}_{cell}^{lim} = \frac{1}{\tilde{J}} + \frac{1}{\tilde{j}_{lim}^{c0} - \tilde{J}}. \tag{5.179}$$

In this limit, the resistivity resulting from oxygen transport in the channel is zero and, hence, the last term in this equation is the resistivity because of the oxygen diffusion in the GDL. As it should be, this resistivity tends to infinity, as the cell current density $\tilde{J}$ approaches the limiting current density $\tilde{j}_{lim}^{c0}$.

## Constant Oxygen Inlet Flow

In the case of $\tilde{M}_{ox} = $ const, Equation 5.58 should be reformulated in terms of $\tilde{M}_{ox}$. Using Equation 5.177, and the definition of $f_\lambda$, Equations 5.57 and 5.58 transform to

$$\tilde{\eta}_0 = \ln\left(-2\varepsilon^2 \tilde{M}_{ox} \ln\left(1 - \frac{\tilde{J}}{\tilde{M}_{ox}}\right)\right) - \ln\left(1 + \frac{\tilde{M}_{ox}}{\tilde{j}_{lim}^{c0}} \ln\left(1 - \frac{\tilde{J}}{\tilde{M}_{ox}}\right)\right). \tag{5.180}$$

Differentiating this equation over $\tilde{J}$ yields

$$\tilde{R}_{cell} = -\frac{1}{\tilde{M}_{ox}\left(1 - \dfrac{\tilde{J}}{\tilde{M}_{ox}}\right)\ln\left(1 - \dfrac{\tilde{J}}{\tilde{M}_{ox}}\right)}$$
$$+ \frac{1}{\left(1 - \dfrac{\tilde{J}}{\tilde{M}_{ox}}\right)\left(\tilde{j}_{lim}^{c0} + \tilde{M}_{ox}\ln\left(1 - \dfrac{\tilde{J}}{\tilde{M}_{ox}}\right)\right)}. \tag{5.181}$$

It is easy to verify that in the limit of $\tilde{M}_{ox} \to \infty$, the first term on the right-hand side of Equation 5.181 tends to $1/\tilde{J}$, while the third term tends to $1/(\tilde{j}_{lim}^{c0} - \tilde{J})$. Thus, the first term represents the total CCL charge transfer resistivity at a constant $\tilde{M}_{ox}$. By analogy to Equation 5.178, the last term in Equation 5.181 gives the combined resistivity, due to oxygen transport in the GDL and in the channel.

Note that in the constant-flow case, the charge transfer resistivity is a strong function of $\tilde{M}_{ox}$ (Figure 5.28). Substituting Equation 5.177 into Equation 5.56, gives the local current in the form

$$\tilde{j}_0(\tilde{z}) = -\tilde{M}_{ox}\ln\left(1 - \frac{\tilde{J}}{\tilde{M}_{ox}}\right)\left(1 - \frac{\tilde{J}}{\tilde{M}_{ox}}\right)^{\tilde{z}}.$$

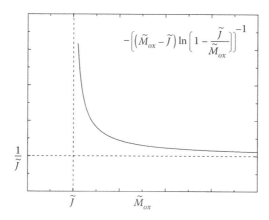

**FIGURE 5.28** The CCL charge transfer resistivity at a constant oxygen flow rate (the first term on the right-hand side of Equation 5.181).

Comparing this to the first term in Equation 5.181, one can see that the following relation holds:

$$\tilde{R}_{ct}^{tot} = \frac{1}{\tilde{j}_0(1)}, \quad \tilde{M}_{ox} = \text{const.} \tag{5.182}$$

Thus, in the constant-flow case, the total charge transfer resistivity is given by the inverse local current at the channel outlet. Clearly, as $\tilde{j}_0(1) < \tilde{J}$, the cell charge transfer resistivity, at a constant flow, always exceeds this resistivity at a constant $\lambda$ (provided that the flow corresponds to $\lambda$ at the mean cell current). In other words, the charge transfer resistivity from EIS measurements with constant oxygen flow rate is always larger than this resistivity calculated from the $\lambda = \text{const.}$ polarization curve.

### Negative Local Resistance

In this section, the work Kulikovsky et al. (2005) is followed. Suppose that the segmented cell in Figure 5.23 is fed under constant oxygen flow rate $\tilde{M}_{ox}$. Experiment shows that under certain conditions, the remote segments exhibit negative local resistance (Brett et al., 2003). Physically, at low inlet flow rate, with the growth of overpotential, more oxygen is consumed close to the channel inlet, less oxygen is left for the remote segments, and the local current there decreases.

To rationalize the effect, suppose that the local polarization curve of an individual segment is given by Equation 5.43, which contains Tafel activation overpotential (the first term) and the potential loss resulting from the oxygen transport in the GDL (the second term).

Taking into account Equation 5.177, Equation 5.50 for the oxygen transport along the channel reads

$$\tilde{M}_{ox}\frac{\partial \tilde{c}_h}{\partial \tilde{z}} = -\tilde{j}_0, \quad \tilde{c}_h(0) = 1. \tag{5.183}$$

Solving Equation 5.43 for $\tilde{j}_0$ yields

$$\tilde{j}_0 = \alpha(\tilde{\eta}_0)\tilde{c}_h(\tilde{z}), \tag{5.184}$$

where

$$\alpha(\tilde{\eta}_0) = \frac{\tilde{j}_{\lim}^{c0} \exp \tilde{\eta}_0}{2\varepsilon^2 \tilde{j}_{\lim}^{c0} + \exp \tilde{\eta}_0}. \tag{5.185}$$

Note that because of equipotential electrode, $\tilde{\eta}_0$ and $\alpha$ are independent of $\tilde{z}$.

Using Equation 5.184 in Equation 5.183 and solving the resulting equation, one obtains the $\tilde{z}$ shape of the oxygen concentration:

$$\tilde{c}_h = \exp\left(-\frac{\alpha\tilde{z}}{\tilde{M}_{ox}}\right), \tag{5.186}$$

where $\alpha$ is given by Equation 5.185. Substituting this result into Equation 5.184, one finds the local polarization curve of the segment,

$$\tilde{j}_0 = \alpha \exp\left(-\frac{\alpha\tilde{z}}{\tilde{M}_{ox}}\right). \tag{5.187}$$

Local curves are shown in Figure 5.29. The remote segments at $\tilde{z} = 0.7$ and $1.0$ exhibit negative resistance. The critical potential and local current, where these curves "fold back," are given by equation $\partial \tilde{j}_0 / \partial \tilde{\eta}_0 = 0$. Differentiating Equation 5.187 over $\tilde{\eta}_0$ and solving this equation, one obtains the critical parameters

$$\tilde{\eta}_0^{crit} = \ln\left(2\varepsilon^2 \tilde{M}_{ox}\right) - \ln\left(\tilde{z} - \frac{\tilde{M}_{ox}}{\tilde{j}_{\lim}^{c0}}\right) \tag{5.188}$$

and

$$\tilde{j}_0^{crit} = \frac{\tilde{M}_{ox}}{\tilde{z}\exp(1)}. \tag{5.189}$$

Note that $\tilde{j}_0^{crit}$ is independent of $\tilde{j}_{\lim}^{c0}$, while $\tilde{\eta}_0^{crit}$ depends strongly on this parameter. As can be seen, the position of the segment closest-to-the-inlet, which exhibits negative resistance, is determined by

$$\tilde{z}^{crit} = \frac{\tilde{M}_{ox}}{\tilde{j}_{\lim}^{c0}}. \tag{5.190}$$

At $\tilde{z} > \tilde{z}^{crit}$, the second logarithm in Equation 5.188 is real, which means that $\tilde{\eta}_0^{crit}$ exists. Closer to the inlet, this logarithm is imaginary, that is, the local polarization curve has no turning point. With the data in Figure 5.29, $\tilde{z}^{crit} = 0.5$.

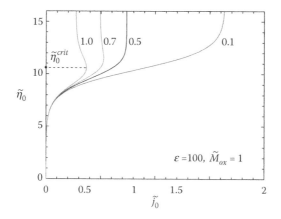

**FIGURE 5.29** Local polarization curves at the indicated normalized distances $\tilde{z}$ from the channel inlet (Equation 5.187). Parameters $\varepsilon = 100$, $\tilde{M}_{ox} = 1$, $\tilde{j}_{\lim}^{c0} = 2$. The critical overpotential $\tilde{\eta}_*^{crit}$ is where the curve turns (infinite differential resistivity).

In the dimensional form, Equations 5.190 and 5.188 read

$$z^{crit} = \frac{4FM_{ox}}{Lwj_{\lim}^{c0}} \tag{5.191}$$

and

$$\eta_0^{crit} = b \ln\left(\frac{4FM_{ox}}{Lwi_*l_{CL}}\right) - b \ln\left(\frac{z}{L} - \frac{4FM_{ox}}{Lwj_{\lim}^{c0}}\right). \tag{5.192}$$

If local impedance spectra are available, Equation 5.191 can be used to evaluate $j_{\lim}^{c0}$. If, in addition, local polarization curves are measured, the curve for any segment with $z > z_{crit}$ can be used to determine one of the parameters $b$ or $i_*$, provided that the other parameter is known from independent measurements. Indeed, with $j_{\lim}^{c0}$ and $\eta_0^{crit}$ at hand, Equation 5.192 gives $b$ or $i_*$. Note that no fitting is necessary. Also note that the model of this section does not take into account the local flooding or membrane drying effects; thus, care should be taken to avoid these effects in using the equations above for the interpretation of measurements.

Suppose that $\tilde{j}_{\lim}^{c0}$ is fixed. The minimal oxygen flow $\tilde{M}_{ox}^{crit}$, which provides positive local resistance of all segments is obtained if $\tilde{z} = 1$ is set in the second logarithm in Equation 5.188. This gives $\tilde{M}_{ox}^{crit} = \tilde{j}_{\lim}^{c0}$. At smaller flow rates, part of the cell exhibits negative resistance. In dimensional form, this critical flow is given by the right-hand side of Equation 5.174.

The local differential segment resistivity is given by $\tilde{R}_{seg} = \partial\tilde{\eta}_0/\partial\tilde{j}_0$. By differentiating Equation 5.187, one finds

$$\tilde{R}_{seg} = \frac{1}{\tilde{j}_0}\left(\left(1 - \frac{\alpha\tilde{z}}{\tilde{M}_{ox}}\right)\left(1 - \frac{\alpha}{\tilde{j}_{\lim}^{c0}}\right)\right)^{-1}. \tag{5.193}$$

From Equation 5.185, it follows that $\alpha/\tilde{j}_{\lim}^{c0} < 1$, so that the change in the resistivity sign is due to the factor $\left(1 - \alpha\tilde{z}/\tilde{M}_{ox}\right)$, which depends on the oxygen flow rate.

The results of this section can be expressed in terms of the mean current density in the cell $\bar{J}$. Integrating Equation 5.187 over $\tilde{z}$ gives the relation of $\alpha$ and $\bar{J}$, which can be used for this transformation

$$\alpha = -\tilde{M}_{ox}\ln\left(1 - \frac{\bar{J}}{\tilde{M}_{ox}}\right). \tag{5.194}$$

## CARBON CORROSION DUE TO FEED MALDISTRIBUTION

### CARBON CORROSION IN PEFCs FROM HYDROGEN DEPLETION

#### Preliminary Remarks

A vast majority of fuel cell aging studies consider cell degradation as uniform over the cell surface process (Borup et al., 2007). However, nonuniformities dramatically accelerate the rate of *local* aging. In a fuel cell, any local inhomogeneity of transport or kinetic parameters induces inhomogeneity in the distribution of the membrane potential over the cell surface. This, in turn, inevitably leads to nonuniform overpotentials of parasitic electrochemical reactions running in the cell. Domains where the parasitic overpotential increases "die" much faster. Thus, modeling of nonuniformities in cells is important in the context of cell durability.

One of the most detrimental processes for PEM fuel cell longevity is carbon corrosion. Under a lack of hydrogen, oxygen penetrates to the anode side and reverses the respective domain of the cell. In this domain, carbon support in the CCL is utilized as a fuel, while oxygen on the anode side is reduced.

The reversal may happen under local hydrogen starvation conditions, when oxygen penetrates to the anode side of the starving domain, through the membrane (Patterson and Darling, 2006). The reversal also occurs during the start-stop cycle, when part of the anode channel is filled with hydrogen and the other part with oxygen (Figure 5.30) (Reiser et al., 2005). Last but not least, the reversal happens if the hydrogen stoichiometry is less than one.

Following Reiser et al. (2005) in this section, the situation during the start-stop cycle is considered. Let the first half of the anode channel be filled with hydrogen, and the second half with air (Figure 5.30). Part of the cell with hydrogen in the anode channel operates as a normal hydrogen-oxygen cell. Current produced in this direct domain (referred to as the D-domain below), is counterbalanced by the negative current in the reverse domain (R-domain). In the R-domain, carbon support in the cathode is utilized as a fuel, while the ORR runs on the anode side and converts protons produced in the carbon corrosion reaction (Figure 5.30).

Functioning of the R-domain is provided by the jump of the membrane phase potential $\Phi$ at the D/R interface (Figure 5.30). A quasi-2D model for the shape of $\Phi$ (Reiser et al., 2005), leads to the Poisson equation with the small parameter (see below). Accurate numerical solution of this equation poses significant difficulties. This, perhaps, was the reason why in Reiser et al. (2005) only a qualitative shape of $\Phi$ has been reported not detailing the distribution of current densities resulting from this shape. The distribution of current densities along the hydrogen channel, partially filled with air, has been modeled by Meyers and Darling (2006). Their model takes into account the oxygen crossover through the membrane and current limitation resulting from oxygen transport in the GDL. However, the model ignores the in-plane current in the membrane and, hence, it fails to describe the physics of the D/R interface.

A fully 2D model of the process has been developed by Ohs et al. (2011). The model takes into account water transport and gas permeation through the membrane and it resolves the catalyst layers. However, neither the details of the transition region between the D- and R-domains nor the distribution of local currents along the electrodes have been reported.

The structure of the transition region at the D/R domains interface for the case of local hydrogen starvation in a small spot (strictly speaking, stripe) has been modeled by Takeuchi and Fuller (2008). Their model includes a 2D equation for the membrane phase potential with fixed pressures of reactants in both electrodes. To simulate crossover, oxygen pressure in the hydrogen-depleted zone was assumed to be five orders of magnitude lower than in air. However, the simulated stripe was only 200 $\mu$m wide and, hence, the large-scale distribution of current densities along the electrode surface is beyond the scope of this model.

In this section, a model is developed for the reversal of the cell with 10-cm electrodes (Kulikovsky, 2011b). The approach of Reiser et al. (2005) is employed and it is shown that the model can be reduced to an algebraic equation, which expresses the quasi-2D balance of currents in the membrane.

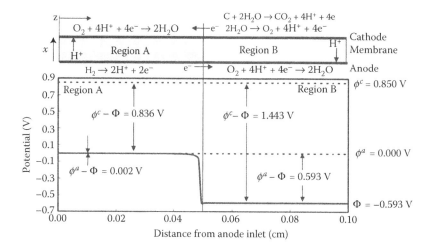

**FIGURE 5.30** The potentials distribution in the cell with the anode channel partially filled with oxygen. For the parameters, see Reiser et al. (2005). In this section, regions A and B are referred to as direct and reverse domains. Note that in the text, the anode and cathode potentials of the carbon phase are denoted as $\phi^a$ and $\phi^c$, respectively. (Reprinted from Reiser, C. A. et al. 2005. *Electrochem. Solid State Lett.*, **8**, A273–A276, Figure 1. Copyright (2005), The Electrochemical Society.)

## Basic Equations

Since no charges are produced or consumed in the membrane, the membrane potential $\Phi$ obeys the Laplace equation

$$\frac{\partial^2 \Phi}{\partial x^2} + \frac{\partial^2 \Phi}{\partial z^2} = 0. \tag{5.195}$$

As in the previous sections, the $z$-axis is directed along the channel and $x$ is the through-plane coordinate (Figure 5.30).

Generally, Equation 5.195 should be supplemented with the boundary conditions specifying current densities at the membrane–catalyst layers interface. This leads to a 2D problem. However, Equation 5.195 can be simplified. The membrane thickness $l_{PEM}$ is four orders of magnitude smaller than the channel length $L$, while our primary interest is the shape of $\Phi$ along $z$. Thus, it is reasonable to approximate the $x$ derivative in Equation 5.195 as

$$\frac{\partial^2 \Phi}{\partial x^2} = \frac{1}{\sigma_{PEM}} \frac{\partial}{\partial x} \left( \sigma_{PEM} \frac{\partial \Phi}{\partial x} \right) \simeq \frac{\sigma_{PEM} \left.\frac{\partial \Phi}{\partial x}\right|^a - \sigma_{PEM} \left.\frac{\partial \Phi}{\partial x}\right|^c}{\sigma_{PEM} l_{PEM}} = \frac{j^c - j^a}{\sigma_{PEM} l_{PEM}}, \tag{5.196}$$

where $\sigma_{PEM}$ and $l_{PEM}$ are the membrane proton conductivity and thickness, respectively, $j$ is the normal ($x$-) component of the proton current density coming in/out of the membrane, and the Ohm's law

$$j = -\sigma_{PEM} \frac{\partial \Phi}{\partial x} \tag{5.197}$$

has been used. Equation 5.196 results from a well-known finite-difference approximation of the derivative $\partial f / \partial x \simeq (f^a - f^c)/l_{PEM}$, where $f^a$ and $f^c$ are the function values at the anode and the cathode sides of the membrane, respectively.

Substituting Equation 5.196 into Equation 5.195, we obtain

$$\frac{\partial^2 \Phi}{\partial z^2} = \frac{j^a - j^c}{\sigma_{PEM} l_{PEM}}. \tag{5.198}$$

The idea to replace the second derivative $\partial^2 \Phi / \partial x^2$ by the difference of local currents has been suggested in a classic work of Newman and Tobias (1962).

Equations 5.196 and 5.198 mean that the second derivative $\partial^2 \Phi / \partial x^2$ is constant along $x$, and it is a function of $z$ only. In other words, the through-plane component of the proton current in the membrane (5.197) is assumed to be linear in $x$.

To simplify calculations, we can introduce dimensionless variables

$$\tilde{z} = \frac{z}{L}, \quad \tilde{\Phi} = \frac{\Phi}{b_{ox}}, \quad \tilde{\eta} = \frac{\eta}{b_{ox}}, \quad \tilde{\phi} = \frac{\phi}{b_{ox}}, \quad \tilde{j} = \frac{j}{j_{ref}^m}, \quad \tilde{c} = \frac{c}{c_h^0}, \tag{5.199}$$

where

$$j_{ref}^m = \frac{\sigma_{PEM} b_{ox}}{l_{PEM}} \tag{5.200}$$

is the characteristic (reference) current density. Note that in the equations below, the reference concentration $c_h^0$ is individual for each type of neutral species, while all potentials are normalized to the ORR Tafel slope $b_{ox}$.

With these variables, Equation 5.198 transforms to

$$\epsilon^2 \frac{\partial^2 \tilde{\Phi}}{\partial \tilde{x}^2} = \tilde{j}^a - \tilde{j}^c, \tag{5.201}$$

where

$$\epsilon = \frac{l_m}{L} \ll 1. \tag{5.202}$$

## Currents

Below, it is assumed that the transport of protons and neutral species in the anode and cathode catalyst layers is ideal. This assumption means that the rates of electrochemical reactions are distributed uniformly through the catalyst layer thickness, and these rates can be calculated simply as a product of the Butler–Volmer rate and the CL thickness. The reaction rates are, thus, determined by the shapes of the membrane phase potential and species concentration along the channel.

### Hydrogen on the Anode

Currents generated on the anode side of the cell are due to the HOR in the D-domain and due to the ORR in the R-domain (Figure 5.30). The HOR current density $\tilde{j}_{hy}^a$ is determined by the Butler–Volmer equation

$$\tilde{j}_{hy}^a = 2\tilde{j}_{hy}^* \left( \frac{c_{hy}}{c_{hy}^{h0}} \right) \sinh \left( \frac{\tilde{\eta}_{hy}}{\tilde{b}_{hy}} \right), \tag{5.203}$$

where $\tilde{j}_{hy}^*$ is the exchange current density, $c_{hy}$ and $c_{hy}^{h0}$ are local and inlet hydrogen molar concentrations, respectively, $\tilde{\eta}_{hy}$ is the half-cell overpotential, and $\tilde{b}_{hy}$ is the HOR Tafel slope.

To take into account hydrogen transport loss in the anode GDL, the hydrogen concentration in the catalyst layer $c_{hy}$ is related to this concentration $c_{hy}^h$ in the anode channel as

$$c_{hy} = c_{hy}^h \left( 1 - \frac{\tilde{j}_{hy}^a}{\tilde{j}_{hy}^{loc}} \right), \tag{5.204}$$

where

$$\tilde{j}_{hy}^{loc} = \frac{2FD_{hy} c_{hy}^h}{l_b^a j_{ref}^m} \tag{5.205}$$

is the local dimensionless limiting current density due to the hydrogen diffusion transport through the anode GDL of a thickness $l_b^a$. Here, $D_{hy}$ is the hydrogen diffusion coefficient in the GDL.

The shape of hydrogen concentration along the anode channel is modeled as

$$c_{hy}^h = c_{hy}^{h0} f_n, \qquad (5.206)$$

where

$$f_n = \frac{1}{2} \left( 1 - \tanh \left( \frac{\tilde{z} - \tilde{z}_0}{\tilde{s}} \right) \right) \qquad (5.207)$$

describes a steep change from unity to zero at $\tilde{z}_0$, with $\tilde{s}$ being the characteristic width of the transition region (Figure 5.31a).

Using Equation 5.206 in Equations 5.204 and 5.205, substituting the result into Equation 5.203, and solving the resulting equation for $\tilde{j}_{hy}^a$, one obtains

$$\frac{f_n}{\tilde{j}_{hy}^a} = \frac{1}{\tilde{j}_{hy}^{BV}} + \frac{1}{\tilde{j}_{hy}^{lim}}, \qquad (5.208)$$

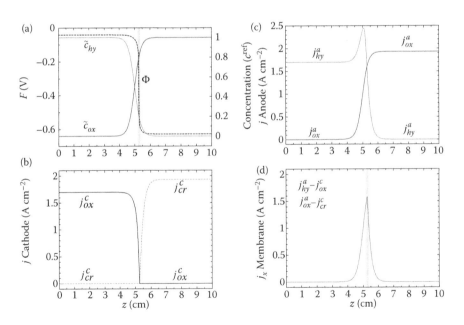

**FIGURE 5.31** (a) Normalized hydrogen and oxygen concentrations, and the shape of the membrane phase potential along the channel. (b) The ORR and *carbon corrosion* (COR) current densities on the cathode side of the cell. (c) The HOR and ORR current densities on the anode side of the cell. (d) The differences between HOR$^a$-ORR$^c$ and ORR$^a$-COR$^c$ are coincidental and equal to the $z$ component of proton current density in the membrane. The shaded region indicates the domain where the approximation of $\epsilon = 0$ fails. For all the curves, $s = 0.04$ and $\phi^c = 0.63329$ V.

where

$$\tilde{j}_{hy}^{BV} = 2\tilde{j}_{hy}^* \sinh\left(\frac{\tilde{\eta}_{hy}}{\tilde{b}_{hy}}\right) \qquad (5.209)$$

is the hydrogen current density at the channel inlet, and

$$\tilde{j}_{hy}^{lim} = \frac{2FD_{hy}c_{hy}^{h0}}{l_b^a j_{ref}^m} \qquad (5.210)$$

is the hydrogen-limiting current density therein.

## Oxygen on the Anode Side

On the anode side, the dimensionless oxygen concentration is zero in the first part of the cell, and it rapidly increases to unity at the D/R interface, $z_0 = 5$ cm (Figure 5.31a). The reasoning quite analogous to that in the previous section, leads to the following equation for the local ORR current density $\tilde{j}_{ox}^a$ on the anode side:

$$\frac{f_p}{\tilde{j}_{ox}^a} = \frac{1}{\tilde{j}_{ox}^{BV}} + \frac{1}{\tilde{j}_{ox}^{lim,a}}, \qquad (5.211)$$

where

$$\tilde{j}_{ox}^{BV} = 2\tilde{j}_{ox}^* \sinh \tilde{\eta}_{ox}^a \qquad (5.212)$$

is the ORR current density at $\tilde{z} = 1$ (at the anode channel outlet), $\tilde{j}_{ox}^{lim,a}$ is the oxygen-limiting current density on the anode side, and $\tilde{j}_{ox}^*$ is the ORR exchange current density.

In Equation 5.211, the function $f_p$ describes steep variations of the oxygen concentration at the D/R interface (Figure 5.31a)

$$f_p = \frac{1}{2}\left(1 + \tanh\left(\frac{\tilde{z} - \tilde{z}_0}{\tilde{s}}\right)\right). \qquad (5.213)$$

## Oxygen and Carbon on the Cathode Side

In the D-domain, the ORR current density on the cathode side, $\tilde{j}_{ox}^c$, is given by

$$\frac{1}{\tilde{j}_{ox}^c} = \frac{1}{\tilde{j}_{ox}^* \exp \tilde{\eta}_{ox}^c} + \frac{1}{\tilde{j}_{ox}^{lim,c}}, \qquad (5.214)$$

where $\tilde{j}_{ox}^{lim,c}$ is the oxygen-limiting current density, due to oxygen transport through the cathode GDL. In Equation 5.214, the reverse exponent in the Butler–Volmer term is neglected, since the cathodic ORR is strongly shifted from equilibrium. It is also assumed that in the cathode channel, the oxygen concentration does not change along $\tilde{z}$ (large stoichiometry of the oxygen flow).

**TABLE 5.10**

**The Equations for the Exchange Current Densities and Overpotentials in Equations 5.208, 5.211, 5.214, 5.215, and parameters**

| | Hydrogen | Oxygen | Carbon |
|---|---|---|---|
| | $\tilde{j}_{hy}^* = \dfrac{i_{hy} L_{pt} A_{pt}}{2 j_{ref}^m}$ | $\tilde{j}_{ox}^* = \dfrac{i_{ox} L_{pt} A_{pt}}{4 j_{ref}^m}$ | $\tilde{j}_{cr}^* = \dfrac{i_{cr} L_{cr} A_{cr}}{4 j_{ref}^m}$ |
| Anode | $\tilde{\eta}_{hy}^a = \tilde{\phi}^a - \tilde{\Phi} - \tilde{E}_{hy}^{eq}$ | $\tilde{\eta}_{ox}^a = -(\tilde{\phi}^a - \tilde{\Phi} - \tilde{E}_{ox}^{eq})$ | |
| Cathode | | $\tilde{\eta}_{ox}^c = -(\tilde{\phi}^c - \tilde{\Phi} - \tilde{E}_{ox}^{eq})$ | $\tilde{\eta}_{cr}^c = \tilde{\phi}^c - \tilde{\Phi} - \tilde{E}_{cr}^{eq}$ |
| $i_*$ (A cm$_*^{-2}$) | $10^{-3}$ | $10^{-9}$ | $6.06 \cdot 10^{-19}$ |
| $L_*$ (g cm$_*^2$) | $4 \cdot 10^{-4}$ | $4 \cdot 10^{-4}$ | $4 \cdot 10^{-4}$ |
| $A_*$ (cm$_*^2$ g$^{-1}$) | $6 \cdot 10^5$ | $6 \cdot 10^5$ | $6 \cdot 10^6$ |
| $E^{eq}$ (V) | 0.0 | 1.23 | 0.207 |
| $\alpha$ | 1 | 0.5 | 0.75 |
| $l_m$ (μm) | | 25 | |
| $\tilde{s}$ | | 0.04 | |
| $\sigma_p$ ($\Omega^{-1}$ cm$^{-1}$) | | 0.1 | |
| $L$ (cm) | | 10 | |
| $T$ (K) | | $273 + 65$ | |
| $j_{ox}^{lim,a}$ (A cm$^{-2}$) | | 2.0 in Figure 5.31; 0.02 in Figure 5.32 | |
| $j_{ox}^{lim,c}$ (A cm$^{-2}$) | | 2.0 | |
| $j_{hy}^{lim}$ (A cm$^{-2}$) | | 8.0 | |
| $E_{cr}$ (V) | | $\phi^a = 0$, $\phi^c$ is a free parameter to fulfill Equation 5.218 | |

*Source:* Reprinted from Reiser, C. A. et al. 2005. *Electrochem. Solid State Lett.*, **8**, A273–A276.

The local current density resulting from carbon corrosion on the cathode side is given by

$$\tilde{j}_{cr}^c = \tilde{j}_{cr}^* \exp\left(\frac{\tilde{\eta}_{cr}^c}{\tilde{b}_{cr}}\right). \tag{5.215}$$

The expressions for the exchange current densities $\tilde{j}_{ox}^*$ and $\tilde{j}_{cr}^*$, overpotentials $\tilde{\eta}_{ox}^c$ and $\tilde{\eta}_{cr}^c$, and the numerical values of parameters are given in Table 5.10 taken from Reiser et al. (2005).

**Final Form of the Basic Equation**

With the expressions above, Equation 5.201 finally takes the form

$$\epsilon^2 \frac{\partial^2 \tilde{\Phi}}{\partial \tilde{z}^2} = \left(\tilde{j}_{hy}^a - \tilde{j}_{ox}^a\right) - \left(\tilde{j}_{ox}^c - \tilde{j}_{cr}^c\right). \tag{5.216}$$

Formally, Equation 5.216 is a Poisson equation with the right-hand side being the difference of normal current densities coming from/to the catalyst layers. Equation

5.216 is strongly nonlinear, as the current densities on the right-hand side exponentially depend on $\tilde{\Phi}$ through the respective overpotentials (Table 5.10). However, the largest problem with this equation is the small parameter $\epsilon^2$ on the left-hand side. Owing to the smallness of this parameter, an accurate numerical solution of Equation 5.216 is a difficult and computationally expensive task.

In the problem discussed, $\epsilon = 2.5 \cdot 10^{-4}$, and, hence, in Equation 5.201, the second derivative $\partial^2 \tilde{\Phi}/\partial \tilde{z}^2$ is multiplied by a very small factor $\epsilon^2 = 6.25 \cdot 10^{-8}$. The smallness of $\epsilon$ allows obtaining the leading-order solution to Equation 5.216, neglecting the left-hand side of this equation. Setting in Equation 5.216 $\epsilon = 0$, one arrives at

$$\tilde{j}_{hy}^a - \tilde{j}_{ox}^a = \tilde{j}_{ox}^c - \tilde{j}_{cr}^c. \tag{5.217}$$

Validity of this approximation can be checked directly using $\Phi$ resulting from Equation 5.217 in Equation 5.216.*

Solution to Equation 5.217 is subject to the following constrain:

$$\int_0^1 \left(\tilde{j}_{hy}^a - \tilde{j}_{ox}^a\right) d\tilde{z} = \int_0^1 \left(\tilde{j}_{ox}^c - \tilde{j}_{cr}^c\right) d\tilde{z} = \tilde{J}, \tag{5.218}$$

where $\tilde{J}$ is the dimensionless mean current density in a cell. In the calculations below, zero current in the external load $\tilde{J} = 0$ is assumed, which is a typical condition during start/stop operation. Equation 5.218, thus, means that the total currents on the anode and cathode sides must be zero.

## Numerical Results

The anode side of the cell is grounded and, hence, $\tilde{\phi}^a = 0$. Solution of Equation 5.217, subject to Equation 5.218, is obtained by varying the only free parameter in this problem, the carbon phase potential of the cathode $\tilde{\phi}^c$. For simplicity, $\tilde{s}$ in Equations 5.207 and 5.213 is taken to be the same.

The shapes of the membrane phase potential $\Phi$, the hydrogen and oxygen concentrations, and the components of the anode and cathode current densities are shown in Figure 5.31. The membrane phase potential exhibits steep gradient at the interface between the direct and reverse domains (Reiser et al., 2005). Because of the symmetry of $H_2$ and $O_2$ concentration profiles, the concentration interface is located at $x_0 = 5$ cm. However, the electric interface (the steep front of $\Phi$) appears to be shifted to $x \simeq 5.27$ cm, as discussed below.

Figure 5.31b shows that the carbon corrosion current density in the R-domain is equal to the oxygen-limiting current density $j_{ox}^{lim}$ (2 A cm$^{-2}$ in this simulation, Table 5.10). Thus, during the start–stop cycle, carbon corrosion runs very fast and even short transients can severely damage the catalyst. The solution to this problem is in lowering the cell potential during the transient (Takeuchi and Fuller, 2008).

On the cathode side, the ORR and carbon corrosion currents are well separated (Figure 5.31b). The feature of the problem is the formation of the HOR peak on the anode side of the cell (Figure 5.31c). This peak manifests the following effect.

---

* Note that the solution to Equation 5.217 is independent of membrane thickness. The smaller the thickness, the smaller the $\epsilon$ and the better the approximation of Equation 5.217 works.

The steep gradient of $\Phi$ induces large proton current in the membrane along the $z$-axis. To support this current, on the anode side of the D/R interface, a "virtual" fuel cell forms.

This is explicitly seen in Figure 5.31d, which shows the differences $j_{hy}^a - j_{ox}^c$ and $j_{ox}^a - j_{cr}^c$. According to Equation 5.217, at $\tilde{J} = 0$, these differences are the same. Physically, both $j_{hy}^a - j_{ox}^c$ and $j_{ox}^a - j_{cr}^c$ represent the uncompensated $x$ *component* of the proton current density in the membrane. Evidently, this uncompensated component *flows along the z-axis*, that is, Figure 5.31d also shows the $z$ component of proton current in the membrane. This component peaks at the D/R interface and supports the large in-plane gradient $\partial\Phi/\partial z$ (Figure 5.31a).

Thus, on the anode side of the D/R interface, a virtual hydrogen–oxygen fuel cell forms, which serves as an electrochemical connector between the D- and R-domains. The D-side of the interface serves as an anode for this cell, while the R-side works as a cathode. Note that the gap of 0.27 cm between the concentration and electric interfaces (Figure 5.31a) is a characteristic width of the virtual cell (Figure 5.31d). It is worth mentioning that the peak of $z$-current in Figure 5.31d is also in the order of the oxygen-limiting current density in the cell. Calculation with the twice larger $j_{ox}^{\lim} = 4$ A cm$^{-2}$ confirms this conclusion.

Figure 5.32 shows the same as in Figure 5.31 curves for the hundred times smaller oxygen-limiting current density on the anode side ($j_{ox}^{\lim,a} = 0.02$ A cm$^{-2}$). This mimics the situation when in the R-domain oxygen penetrates to the anode side through the membrane (crossover).

As can be seen, the carbon corrosion current density again equals the oxygen-limiting current density (Figure 5.32b). At low current, the position of the $\Phi$ gradient, notably (by $\simeq 1$ cm) shifts from the concentration gradient (Figure 5.32a), that is, the virtual cell gets thicker. The reason is that this cell must generate proton current required to support the large gradient of $\Phi$ in the membrane. Since the absolute value of $\Phi$ is lower, a larger thickness of virtual electrodes is needed to generate the required current. The virtual cell has, thus, a self-adjusting width at the D/R interface.

Formation of a virtual fuel cell, at the D/R interface, is shown schematically in Figure 5.33. The jump of $\Phi$ at the D/R interface induces high proton current in the membrane directed along the $z$-axis. To support this current, the hydrogen side of the interface generates excess protons and the oxygen side consumes them in the ORR. The membrane phase potential and the proton trajectory are shown schematically in Figure 5.34. The proton generated on the anode side of the virtual cell moves along the arc trajectory. A similar trajectory has been discussed by Takeuchi and Fuller (2008) based on fully 2D simulations in a 200-μm stripe in the vicinity of D/R interface.[*]

Variation of the thickness $\tilde{s}$ of the concentration transition region in Equations 5.207 and 5.213 does not affect the peak value in Figure 5.31d. Parameter $\tilde{s}$ only scales the width of this peak. At smaller $\tilde{s}$, the peak width is lower, while larger $\tilde{s}$ increases it.

Note that the model above describes the step in $\Phi$ approximately. The calculation of the product $\epsilon^2 \partial^2 \tilde{\Phi}/\partial \tilde{z}^2$ with $\tilde{\Phi}$, resulting from Equation 5.217, shows that the left-

---

[*] Note that the width of the stripe, modeled by Takeuchi and Fuller (2008), is 10 times less than the width of the shaded domain in Figure 5.31.

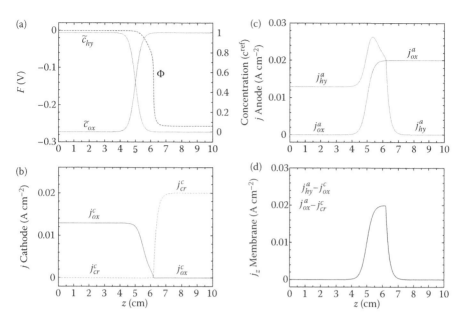

**FIGURE 5.32** (a) Normalized hydrogen and oxygen concentrations, and the shape of the membrane phase potential along the channel. (b) The ORR and *carbon corrosion* (COR) current densities on the cathode side of the cell. (c) The HOR and ORR current densities on the anode side of the cell. (d) The differences between $HOR^a$–$ORR^c$ and $ORR^a$–$COR^c$ are coincidental and equal to the $z$–component of proton current density in the membrane. The shaded region indicates the domain where the approximation of $\epsilon = 0$ fails. For all the curves, $s = 0.04$ and $\phi^c = 0.63329$ V.

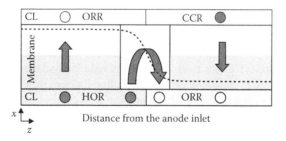

**FIGURE 5.33** The three fuel cells in a PEFC with the oxygen in the second half of the anode channel (cf. Figure 5.30). HOR, ORR, and CCR stand for the hydrogen oxidation, oxygen reduction, and carbon corrosion reactions, while the CL abbreviates the catalyst layer. Arrows indicate trajectories of proton (filled circles) transport. The dashed line is the shape of the membrane phase potential.

side of Equation 5.216 is negligible everywhere along $\tilde{z}$ except the small domain at the D/R interface, indicated in yellow in Figure 5.31. In this domain, the zero-order approximation of Equation 5.217 is not accurate. Note, however, that this domain is very narrow so that the simple model above gives a physically correct picture.

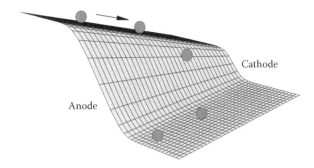

Cathode

Anode

**FIGURE 5.34** A schematic of the proton (ball) trajectory in the membrane at the D/R interface. The surface depicts the membrane phase potential.

The negative effect of cell reversal under local hydrogen starvation can be mitigated by using a thicker membrane. The equivalent current density of oxygen crossover, through the membrane, is inversely proportional to the membrane thickness. Thus, larger $l_{PEM}$ lowers the carbon corrosion current density and increases the cell lifetime.

## CARBON AND RU CORROSION IN A DMFC DUE TO METHANOL DEPLETION

### Problem Description

In a stack environment, state-of-the-art DMFC loses 10–20 $\mu$V per hour of operation at a current density in the order of 100 mA cm$^{-2}$. This limits DMFC lifetime by several thousand hours, which is not sufficient for wide commercialization.

Among the most detrimental processes for DMFC lifetime is Ru corrosion on the anode side. DMFC utilizes PtRu alloy as the anode catalyst to prevent rapid poisoning of the Pt surface by CO molecules produced during methanol oxidation. Unfortunately, when the cell is under load, Ru is dissolved electrochemically and Ru ions are transported through the membrane to the cathode, where they are deposited onto the Pt surface. This process lowers the number of active sites available for methanol oxidation on the anode, and it seemingly reduces ORR activity of the cathode catalyst (Gancs et al., 2006). In addition, $Ru^{2+}$ ions block $SO_3^-$ groups in the membrane, thereby reducing the membrane proton conductivity.

Detailed experimental study of Ru crossover in DMFC has been performed by the Los Alamos group (Piela et al., 2004). Piela et al. studied Ru crossover through the membrane in a DMFC stack under normal operating conditions. They also reported Ru crossover at zero cell current *(currentless crossover)*, most probably caused by diffusion of unalloyed $RuO_2$ through the membrane (Liu et al., 2009). Later, Choi et al. (2006) found that Ru crossover does not depend on the rate of methanol and water crossover in the membrane. This suggests that under long-term operating conditions, Ru travels through the membrane as $Ru^{2+}$ ion (see below).

Recently, Dixon et al. (2011) published a detailed microscopic study of the degradation effects in a DMFC operated in a stack environment. They observed a significant amount of oxidized Ru on the cathode side and in the membrane. Importantly, Ru crossover effects were more significant close to the methanol outlet, where methanol starvation could happen.

In this section, we show that electrochemical Ru corrosion is induced in domains with strong methanol depletion. Such domains may arise in cell/stack environment due to nonuniformity of methanol supply over the cell surface. In these domains, methanol oxidation is substituted by the carbon corrosion reaction. This strongly lowers the membrane phase potential, thereby activating Ru electrochemical dissolution.

## Governing Equation

Consider a fragment of DMFC: the membrane stripe with the infinitely thin catalyst layers clamped between the straight cathode and anode channels (Figure 5.35). Let the oxygen concentration be constant along the channel (this condition is usually fulfilled, as real DMFCs operate at high oxygen stoichiometry). Suppose that on the anode side, the first half of the cell is filled with 1 M methanol–water solution, whereas the second half experiences strong methanol depletion (Figure 5.35). Below, the methanol-rich and methanol-depleted domains are referred to as the MR- and MD-domains.

In the MD-domain, owing to methanol depletion, the negative membrane phase potential $\Phi$ increases by the absolute value (Figure 5.35). This dramatically (by many

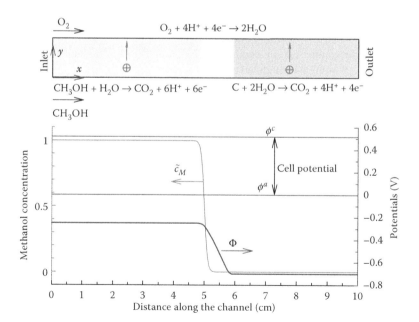

**FIGURE 5.35** Schematic of DMFC with strong methanol depletion in the second half of the channel. The normalized methanol concentration, membrane potential, and carbon phase (electrode) potentials are shown in the bottom.

orders of magnitude) raises the rate of carbon corrosion in the anode catalyst layer. With the new level of $\Phi$ in the MD-domain, the carbon corrosion reaction (CCR) serves as a source of protons. As in the MR-domain, these protons move through the membrane and participate in the ORR on the cathode side (Figure 5.35). In other words, in the MD-domain, carbon on the anode side is utilized as a fuel instead of methanol.

Qualitatively, this picture is analogous to carbon corrosion on the cathode side in the hydrogen-depleted region of a PEM fuel cell discussed in the section "Carbon Corrosion in PEFCs from Hydrogen Depletion." The major difference is that in PEFC, hydrogen depletion causes reversal of the depleted domain due to the presence of oxygen on the anode side. In DMFC, no reversal happens, the anode in the MD-domain continues generating protons, but the source of these protons is CCR.

A model of this process can be constructed using the arguments discussed in the section "Basic Equation." The conductivity of carbon phase in the electrodes is high and the electrode potentials can be safely assumed constant along $\tilde{z}$. The key variable in the methanol-depleted cell function is the membrane potential $\Phi$. Repeating the arguments from the section "Basic Equations," for the dimensionless membrane phase potential $\tilde{\Phi}$, we again come to the Laplace equation (5.201).

Indeed, in the membrane, no current is produced or consumed and hence $\Phi$ obeys the Laplace equation (5.195). Further, the membrane thickness is much smaller than the in-plane size, which allows us to reduce Equation 5.195 to Equation 5.198. With the dimensionless variables (5.199), Equation 5.198 takes the form 5.201.

## Current Densities

All concentrations in this section are normalized to their inlet values; the other dimensionless variables in this section are given by Equation 5.199. The shape of methanol concentration $\tilde{c}_{mt}$ along $z$ is modeled as

$$\tilde{c}_{mt}(\tilde{z}) = \frac{1}{2}\left(1 - \tanh\left(\frac{\tilde{z} - \tilde{z}_0}{\tilde{s}}\right)\right), \tag{5.219}$$

where $\tilde{z}_0 = 0.5$. The function (5.219) describes rapid decay of methanol concentration at the MR/MD interface, with $\tilde{s}$ being the characteristic width of the transition region (Figure 5.35).

In analogy to Equation 5.208, the current density $\tilde{j}_{mt}^a$ due to the methanol oxidation on the anode side is determined by

$$\frac{\tilde{c}_{mt}}{\tilde{j}_{mt}^a} = \frac{1}{\tilde{j}_{mt}^T} + \frac{1}{\tilde{j}_{\lim}^{a0}}. \tag{5.220}$$

Here

$$\tilde{j}_{mt}^T = \tilde{j}_{mt}^* \exp\left(\frac{\tilde{\eta}_{mt}}{\tilde{b}_{mt}}\right) \tag{5.221}$$

is the Tafel methanol current density at the anode channel inlet (where $\tilde{c}_{mt} = 1$), $\tilde{j}_{mt}^*$ is the dimensionless superficial exchange current density of MOR in the anode catalyst layer, $\tilde{\eta}_{mt}$ is the anodic MOR overpotential, and $\tilde{b}_{mt} = b_{mt}/b_{ox}$ is the dimensionless Tafel slope of MOR. Parameter $\tilde{j}_{mt}^{lim}$ is the dimensionless methanol-limiting current density due to the diffusion transport of methanol in the anode backing layer:

$$\tilde{j}_{lim}^{a0} = \frac{6FD_b^a c_{mt}^0}{l_b^a j_{ref}^m}, \tag{5.222}$$

where $c_{mt}^0$ is the inlet methanol concentration.

The carbon corrosion current density on the anode side is given by

$$\tilde{j}_{cr}^a = \tilde{j}_{cr}^* \exp\left(\frac{\tilde{\eta}_{cr}}{\tilde{b}_{cr}}\right), \tag{5.223}$$

where $\tilde{j}_{cr}^*$ is the respective superficial exchange current density, and $\tilde{\eta}_{cr} = \eta_{cr}/b_{ox}$ and $\tilde{b}_{cr} = b_{cr}/b_{ox}$ are the dimensionless CCR overpotential and Tafel slope, respectively.

On the cathode side, the ORR current density $\tilde{j}_{ox}^c$ follows from Equation 5.44 for the cathodic ORR overpotential $\tilde{\eta}_{ox}$, which in the notations of this section reads

$$\tilde{\eta}_{ox} = \ln\left(\frac{\tilde{j}_{ox}^c}{\tilde{j}_{ox}^*}\right) - \ln\left(1 - \frac{(\tilde{j}_{ox}^c + \tilde{j}_{cross})}{\tilde{j}_{ox}^{lim}}\right). \tag{5.224}$$

Here $\tilde{j}_{ox}^*$ is the superficial exchange current density of the ORR and

$$\tilde{j}_{cross} = \beta_* \left(\tilde{j}_{lim}^{a0} \tilde{c}_{mt} - \tilde{j}_{ox}^c\right) \tag{5.225}$$

is the equivalent current density of methanol crossover through the membrane (cf. Equation 4.205).

Solving Equation 5.224 for $\tilde{j}_{ox}^c$ results in

$$\tilde{j}_{ox}^c = \frac{\tilde{j}_{ox}^* \left(\tilde{j}_{ox}^{lim} - \beta_* \tilde{j}_{lim}^{a0} \tilde{c}_{mt}\right) \exp \tilde{\eta}_{ox}}{\tilde{j}_{ox}^*(1 - \beta_*) \exp \tilde{\eta}_{ox} + \tilde{j}_{ox}^{lim}}. \tag{5.226}$$

Equation 5.226 contains the ORR Tafel slope implicitly, as it is used as a normalization constant for $\tilde{\eta}_{ox}$ (see Equation 5.199).

With these currents, Equation 5.201 takes the form

$$\varepsilon^2 \frac{\partial^2 \tilde{\Phi}}{\partial \tilde{z}^2} = \tilde{j}_{mt}^a + \tilde{j}_{cr}^a - \tilde{j}_{ox}^c. \tag{5.227}$$

With the 10 cm channel and the 100 μm membrane, we have $\varepsilon = 10^{-3}$. Thus, the second derivative $\partial^2 \tilde{\Phi}/\partial \tilde{z}^2$ is multiplied by a small number $\varepsilon^2 = 10^{-6}$. On the other hand, estimates show that the currents on the right-hand side of Equation 5.227 are

**TABLE 5.11**

**The Equations for the Exchange Current Densities, Overpotentials, and Parameters**

|  | Methanol | Oxygen | Carbon |
|---|---|---|---|
|  | $\tilde{j}^*_{mt} = \dfrac{j^*_{mt}}{j^m_{ref}}$ | $\tilde{j}^*_{ox} = \dfrac{i_{ox} L_{pt} A_{pt}}{4 j^m_{ref}}$ | $\tilde{j}^*_{cr} = \dfrac{i_{cr} L_{cr} A_{cr}}{4 j^m_{ref}}$ |
| Anode | $\tilde{\eta}_{mt} = \tilde{\phi}^a - \tilde{\Phi} - \tilde{E}^{eq}_{mt}$ |  | $\tilde{\eta}_{cr} = \tilde{\phi}^a - \tilde{\Phi} - \tilde{E}^{eq}_{cr}$ |
| Cathode |  | $\tilde{\eta}_{ox} = -(\tilde{\phi}^c - \tilde{\Phi} - \tilde{E}^{eq}_{ox})$ |  |
| $i_*$ (A cm$_*^{-2}$) |  | $10^{-9}$ | $6.06 \cdot 10^{-19}$ |
| $L_*$ (g cm$_*^2$) |  | $4 \cdot 10^{-4}$ | $4 \cdot 10^{-4}$ |
| $A_*$ (cm$_*^2$ g$^{-1}$) |  | $6 \cdot 10^5$ | $6 \cdot 10^6$ |
| $E^{eq}_*$ (V) | 0.028 | 1.23 | 0.207 |
| $\alpha$ | 0.5 | 1.0 | 0.25 |
| $l_m$ (μm) |  | 100 |  |
| $\tilde{s}$ |  | 0.01 |  |
| $\sigma_p$ (Ω$^{-1}$ cm$^{-1}$) |  | 0.1 |  |
| $L$ (cm) |  | 10 |  |
| $T$ (K) |  | $273 + 65$ |  |
| $j^{lim,c}_{ox}$ (A cm$^{-2}$) |  | 1.0 |  |
| $j^{lim}_{mt}$ (A cm$^{-2}$) |  | 1.0 |  |
| $E_{cr}$ (V) |  | $\phi^a = 0$, $\phi^c$ is a free parameter to fulfill Equation 5.229 |  |

*Source:* Part of the data is taken from Reiser, C. A. et al. 2005. *Electrochem. Solid State Lett.*, **8**, A273–A276.

in the order of unity. Therefore, at leading order, the left-hand side of Equation 5.227 can be ignored and this equation reduces to

$$\tilde{j}^a_{mt} + \tilde{j}^a_{cr} - \tilde{j}^c_{ox} = 0. \tag{5.228}$$

The expressions for overpotentials, the data for the exchange current densities, and the other constants are listed in Table 5.11. Following the convention of Reiser et al. (2005), we set the potential of the carbon phase on the anode side to zero: $\tilde{\phi}^a = 0$; this corresponds to the grounded anode.

From Table 5.11, it can be seen that the expression for the ORR overpotential contains yet undefined parameter, the carbon phase potential of the cathode $\phi^c$ (the cell potential). The solution to Equation 5.228 must obey the following condition:

$$\int_0^1 (\tilde{j}^a_{mt} + \tilde{j}^a_{cr}) \, d\tilde{z} = \int_0^1 \tilde{j}^c_{ox} \, d\tilde{z} = \tilde{J}, \tag{5.229}$$

where $\tilde{J}$ is the dimensionless mean current density in a cell. This condition determines $\tilde{\phi}^c$.

## Numerical Results

The system of Equations 5.228 and 5.229 was solved numerically using Maple software. Figure 5.36a shows the shape of normalized methanol concentration and membrane potential $\Phi$. Equation 5.219 prescribes rapid decrease of $\tilde{c}_{mt}$ from unity to zero at $x = 5$ cm (Figure 5.36a). As can be seen, this decrease induces the drop of membrane potential by 450 mV (from $-0.25$ V in the MR-domain down to $-0.7$ V in the MD-domain, Figure 5.36a).

Physically, the lack of methanol in the second half of the channel forces the anodic reaction to run with carbon as a fuel. Owing to the small exchange current density (Table 5.11), the carbon corrosion reaction requires quite a significant overpotential

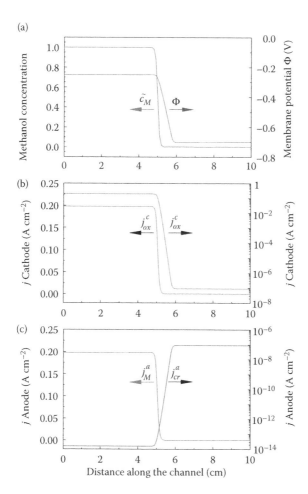

**FIGURE 5.36** (a) The normalized methanol concentration and the shape of the membrane phase potential along the channel. (b) The ORR current density in linear and logarithmic scales. (c) The MOR and carbon corrosion current densities. For all the curves, the mean current density is 100 mA cm$^{-2}$, $s = 0.01$, and the cell potential $\phi^c = 0.5168$ V.

to produce proton current, which can be balanced by the ORR current on the cathode side. Therefore, the drop of $\Phi$ in the second half of the cell solves the two problems: It increases the CCR overpotential and decreases the ORR overpotential to balance the CCR and ORR currents (cf. the ORR current in the MD domain, Figure 5.36b, and the CCR current there, Figure 5.36c).

The high absolute value of $\Phi$ in the methanol-depleted region creates positive overpotential for the Ru dissolution reaction

$$Ru \rightarrow Ru^{2+} + 2e^-. \tag{5.230}$$

An equilibrium potential of this reaction is $E_{Ru}^{eq} = 0.455$ V (Bard, 1976). The overpotential for the reaction (5.230) is given by

$$\eta_{Ru} = \phi^a - \Phi - E_{Ru}^{eq}, \tag{5.231}$$

where $\phi^a = 0$ is the anode carbon phase potential. The overpotential $\eta_{Ru}$ jumps from the negative value in the methanol-rich part of the cell to the positive value in the methanol-depleted domain (Figure 5.37a). Thus, in the MR-domain, the reaction is shifted toward reduction and no Ru corrosion occurs. In the MD-domain, the reaction (5.230) is strongly (by more than 200 mV) shifted to oxidation. Unfortunately, the exchange current density and the Tafel slope of the reaction (5.230) are poorly known, which makes it difficult to estimate the respective Ru dissolution current.

Figure 5.37b shows the dependence of the cell potential and overpotentials for carbon and Ru corrosion on the fraction of the MD-domain. In this figure, the mean current in the cell is fixed at 100 mA cm$^{-2}$. As can be seen, as long as the fraction of MD-domain is below 70%, the variation of the cell potential and overpotentials is not large. Thus, a very substantial methanol-depleted area manifests itself as a

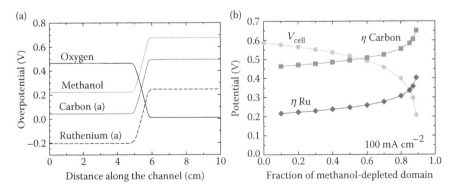

**FIGURE 5.37** (a) Overpotentials along the channel. For all the curves, $s = 0.01$ and $\phi^c = 0.5168$ V. (b) Cell potential and overpotentials for carbon and Ru corrosion versus the fraction of methanol-depleted domain. Note the dramatic drop of $E_{cell}$ and growth of overpotentials when more than 80% of the channel length is methanol-depleted. The mean cell current density is 100 mA cm$^{-2}$.

moderate (below 100 mV) lowering of cell potential. For example, 50% methanol-depleted surface area leads to only 70 mV loss in the cell potential (Figure 5.37b). Thus, a small reduction of cell potential may indicate strong methanol depletion and as a result, fast Ru dissolution.

However, $E_{cell}$ dramatically drops and the corrosion overpotential dramatically rises if more than 80% of the cell surface suffers from methanol depletion (Figure 5.37b). MD-area of 80% or more leads to a catastrophically fast carbon and Ru corrosion (Figure 5.37b).

It should be noted that $E_{Ru}^{eq} = 0.455$ V is the equilibrium potential of pure Ru dissolution in aqueous environment. For PtRu alloy in an acid environment of a fuel cell anode, this potential could differ from the indicated value. The model above, thus, gives a qualitative picture of what we could expect in the cell with methanol-depleted region.

It is worth noting that the mechanisms of Ru dissolution and transport are out of the scope of this calculation. Understanding these mechanisms would require a delicate electrochemical studies. The model above only shows that strong fuel depletion creates an environment for Ru corrosion.

## DEAD SPOTS IN THE PEM FUEL CELL ANODE

So far in this chapter, the problems have been studied assuming that the catalyst layers on both sides of the cell are not destroyed and, therefore, nonuniformity is caused by in-plane gradient of the fuel concentration in the anode. What happens if a spot in the anode catalyst is "dead" for the fuel oxidation reaction? This may happen if the anode catalyst is locally poisoned by CO molecules, or if it suffers from agglomeration of Pt particles. Similar situations arise when, in one of the electrodes, the catalyst is deliberately removed in a small spot to provide a non-Pt window for the transmission *x-ray absorption spectroscopy* (XAS) of another electrode (Roth et al., 2005).

This section considers a circular dead spot in the anode catalyst layer of a PEM fuel cell and solves a problem for the distribution of potentials and currents in and around the spot (Kulikovsky, 2013b). The spot is modeled as a circular domain with many orders of magnitude lower exchange current density of the HOR, which mimics much lower catalyst active surface.

### MODEL

#### Schematic of Potentials

A schematic of potentials in a cell with the dead spot on the anode side is shown in Figure 5.38. The spot is characterized by much lower electrochemical activity. Let $R_s$ be the spot radius and $r$ be the radial coordinate from the spot center (Figure 5.38). Lower electrochemical activity means reduction of the HOR exchange current density $j_{hy}$ inside the spot. To describe this reduction, one can use the smooth tanh function

$$j_{hy} = j_{hy}^{\infty} \left( k_s + \frac{1 - k_s}{2} \left( 1 + \tanh\left( \frac{r - R_s}{s} \right) \right) \right), \qquad (5.232)$$

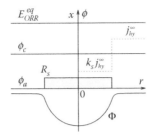

**FIGURE 5.38** A schematic of the potentials in the cell with the dead spot (shaded disk). $\phi^a = 0$ and $\phi^c$ are the anode and cathode carbon phase (electrode) potentials, while $\Phi$ is the membrane potential. The distribution of the HOR exchange current density is shown by the dashed line. Inside the spot, this current density is nine orders of magnitude lower than outside the spot ($k_s = 10^{-9}$).

where $j_{hy}^\infty$ is the HOR exchange current density in the regular domain, $k_s$ is the ratio of the exchange currents in the spot and in the regular domain ($k_s = 10^{-9}$), and $s$ is the thickness of the transition region between the spot and the regular domain. With $s \ll R_s$, the function (5.232) provides a smooth interpolation of the $j_{hy}$ steep change at the spot boundary (Figure 5.38).

The key variable in the problem is the membrane potential $\Phi$. Let $\phi^a$ and $\phi^c$ be the potentials of the electron-conducting phase of the anode and the cathode, respectively. Suppose that the anode is grounded $\phi^a = 0$. The expected distribution of potentials around the spot is depicted in Figure 5.38. Qualitatively, as the anode inside the spot does not produce proton current, the membrane potential would decrease in front of the spot, in order to lower the cathodic overpotential and to reduce ORR current in this domain.

## Equation for the Membrane Potential

In the bulk membrane, potential $\Phi$ obeys the Laplace equation (5.195). In the cylindrical coordinates with the axis $x$, perpendicular to the spot center (Figure 5.38), this equation reads

$$\frac{1}{r}\frac{\partial}{\partial r}\left(r\frac{\partial \Phi}{\partial r}\right) + \frac{\partial^2 \Phi}{\partial x^2} = 0. \tag{5.233}$$

Suppose that the spot radius largely exceeds the membrane thickness $R_s \gg l_m$. In that case, the distribution of $\Phi$ along $x$ is of minor interest, as the key effects show up along the radius. Thus, the $x$-derivative in Equation 5.233 can be approximated using Equation 5.196, and Equation 5.233 takes the form

$$\frac{1}{r}\frac{\partial}{\partial r}\left(r\frac{\partial \Phi}{\partial r}\right) = \frac{j^c - j^a}{\sigma_m l_m}. \tag{5.234}$$

To simplify further calculations, dimensionless variables are introduced according to

$$\tilde{r} = \frac{r}{R_s}, \quad \tilde{j} = \frac{j l_m}{\sigma_m b_{ox}}, \quad \tilde{\Phi} = \frac{\Phi}{b_{ox}}, \quad \tilde{b}_{hy} = \frac{b_{hy}}{b_{ox}}, \qquad (5.235)$$

where

$$b_{hy} = \frac{\alpha_{hy} F}{R_g T}, \quad b_{ox} = \frac{\alpha_{ox} F}{R_g T} \qquad (5.236)$$

are the HOR and ORR Tafel slopes, respectively, and $\alpha_{hy}$ and $\alpha_{ox}$ are the respective transfer coefficients.

With these variables, Equation 5.234 takes the form

$$\epsilon^2 \frac{1}{\tilde{r}} \frac{\partial}{\partial \tilde{r}} \left( \tilde{r} \frac{\partial \tilde{\Phi}}{\partial \tilde{r}} \right) = \tilde{j}^c - \tilde{j}^a, \qquad (5.237)$$

where

$$\epsilon = \frac{l_m}{R_s} \qquad (5.238)$$

is a small parameter ($\epsilon \ll 1$).

With the dimensionless variables (5.235), Equation 5.232 transforms to

$$\tilde{j}_{hy} = \tilde{j}_{hy}^\infty f_s, \qquad (5.239)$$

where

$$f_s = k_s + \frac{1 - k_s}{2} \left( 1 + \tanh \left( \frac{\tilde{r} - 1}{\tilde{s}} \right) \right). \qquad (5.240)$$

## Current Densities

The right-hand side of Equation 5.237 is the proton current density $j^a$, produced in the ACL, minus the proton current density $j^c$ consumed in the CCL. In this section, our goal is to demonstrate the effect due to the dead spot, and all transport losses on either side of the cell will be ignored. If necessary, transport losses in the GDL can be taken into account, as described in the section "Carbon Corrosion in PEFCs from Hydrogen Depletion," while the transport losses in the CLs can be accounted for, using equations of Chapter 4.

For $j^a$ and $j^c$ one can write the Butler–Volmer equations of the form

$$j^a = 2 j_{hy} \sinh \left( \frac{\eta_{hy}}{b_{hy}} \right) \qquad (5.241)$$

and

$$j^c = 2 j_{ox}^\infty \sinh \left( \frac{\eta_{ox}}{b_{ox}} \right), \qquad (5.242)$$

where $j_{ox}^\infty$ is the exchange current density of the ORR, and $\eta_{hy}$ and $\eta_{ox}$ are the HOR and ORR overpotentials, respectively. In Equation 5.241, $j_{hy}$ is given by Equation 5.232. This is where the defect of the anode catalyst appears in the equations. Also note that the constant concentration factors are included into $j_{hy}$ and $j_{ox}^\infty$.

With the dimensionless variables (5.235), Equations 5.241 and (5.242) take the form

$$\tilde{j}^a = 2\tilde{j}_{hy}^\infty f_s \sinh\left(\tilde{\eta}_{hy}/\tilde{b}_{hy}\right), \tag{5.243}$$

$$\tilde{j}^c = 2\tilde{j}_{ox}^\infty \sinh\tilde{\eta}_{ox}, \tag{5.244}$$

and for Equation 5.237, we obtain

$$\epsilon^2 \frac{1}{\tilde{r}} \frac{\partial}{\partial\tilde{r}}\left(\tilde{r}\frac{\partial\tilde{\Phi}}{\partial\tilde{r}}\right) = 2\tilde{j}_{ox}^\infty \sinh\tilde{\eta}_{ox} - 2\tilde{j}_{hy}^\infty f_s\sinh\left(\tilde{\eta}_{hy}/\tilde{b}_{hy}\right). \tag{5.245}$$

By definition, the HOR and ORR overpotentials are

$$\eta_{hy} = \phi^a - \Phi - E_{HOR}^{eq} \tag{5.246}$$

and

$$\eta_{ox}' = \phi^c - \Phi - E_{ORR}^{eq}, \tag{5.247}$$

where $E_{HOR}^{eq} = 0$ and $E_{ORR}^{eq} = 1.23$ V are the HOR and ORR equilibrium potentials. Taking into account that $\phi^a = 0$ and that the ORR Butler–Volmer equation (5.244) is written for the positive ORR overpotential $\eta_{ox} = -\eta_{ox}'$, the last two equations in the dimensionless coordinates transform to

$$\tilde{\eta}_{hy} = -\tilde{\Phi} \tag{5.248}$$

and

$$\tilde{\eta}_{ox} = -\tilde{\phi}^c + \tilde{\Phi} + \tilde{E}_{ORR}^{eq}. \tag{5.249}$$

Substituting these equations into Equation 5.245, finally gives an equation for $\tilde{\Phi}$

$$\epsilon^2 \frac{1}{\tilde{r}} \frac{\partial}{\partial\tilde{r}}\left(\tilde{r}\frac{\partial\tilde{\Phi}}{\partial\tilde{r}}\right) = 2\tilde{j}_{ox}^\infty\sinh\left(-\tilde{\phi}^c + \tilde{\Phi} + \tilde{E}_{ORR}^{eq}\right) - 2\tilde{j}_{hy}^\infty f_s\sinh\left(\frac{-\tilde{\Phi}}{\tilde{b}_{hy}}\right). \tag{5.250}$$

The boundary conditions to Equation 5.250 result from the following arguments. Clearly, at a large distance from the spot, the radial derivative $\partial\tilde{\Phi}/\partial\tilde{r}$ vanishes and, thus, $f_s = 1$. This means that at $\tilde{r} \gg 1$, $\tilde{\Phi}^\infty$ can be found setting $f_s = 1$ and equating the right-hand side of Equation 5.250 to zero

$$2\tilde{j}_{hy}^\infty \sinh\left(\frac{-\tilde{\Phi}^\infty}{\tilde{b}_{hy}}\right) = 2\tilde{j}_{ox}^\infty \sinh\left(-\tilde{\phi}^c + \tilde{\Phi}^\infty + \tilde{E}_{ORR}^{eq}\right). \tag{5.251}$$

At $\tilde{r} = 0$, a symmetry condition holds and, hence, we get

$$\left.\frac{\partial \tilde{\Phi}}{\partial \tilde{r}}\right|_{\tilde{r}=0} = 0, \quad \tilde{\Phi}(\infty) = \tilde{\Phi}^\infty, \tag{5.252}$$

where $\tilde{\Phi}^\infty$ is given by Equation 5.251.

In the section "Carbon Corrosion in PEFCs from Hydrogen Depletion," the second derivative along the channel was neglected. In the Poisson equations (5.201) and (5.250), the characteristic thickness of the domain of the $\Phi$ variation is in the order of $\epsilon$. In the problem with the air/hydrogen front $\epsilon \simeq 10^{-4}$, it allowed a chalking-out, at leading order, the second derivative in Equation 5.201. Here, however, we will consider the case of relatively small dead spots $\epsilon = 10^{-1}$, although with the spot radius still largely exceeding the membrane thickness. In the case of $\epsilon = 0.1$, the second derivative in Equation 5.250 has to be retained.

Equation 5.250 is formally a Poisson equation in radial coordinates. Equations of that type are well known in the theory of space charge effects in electrolytes and plasmas. The right-hand side of the Poisson equation for the self-consistent electric field in the plasma contains a difference of positive and negative charge densities (Chen, 1974). The right-hand side of Equation 5.250 contains the difference of HOR and ORR current densities, that is, $j^a$ and $j^c$ play a role of the positive and negative charge densities in the plasma theory. In plasmas, the Poisson equation describes the formation of a double layer, a structure with separated positive and negative charges at the interface between two medias or in the external electric field because of different charge mobilities. Below, it will be shown that Equation 5.250 describes the formation of a current double layer at the spot boundary.

## CURRENT DOUBLE LAYER

The base-case parameters used in the calculations are listed in Table 5.12. Equation 5.250 has a single free parameter $\tilde{\phi}^c$ a potential of the carbon phase on the cathode side (the cell potential, Figure 5.38). This parameter determines the mean current density $J$ in the regular domain, which is fixed at 1 A cm$^{-2}$ (Table 5.12).

In spite of the steep change in the HOR exchange current at $\tilde{r} = 1$, the radial distributions of $\Phi$, $\eta_{ox}$, and $j^c$ appear to be rather smooth (Figure 5.39a,b). However, the shape of the HOR current density on the anode side $j^a$ exhibits a very steep gradient at $\tilde{r} = 1$ (Figure 5.39b). Figure 5.39c shows the formation of the current double layer at the spot boundary $\tilde{r} = 1$. To the left of this boundary, $j^a$ is negligibly small and only the ORR current flows, while on the right side of this boundary, the HOR

**TABLE 5.12**

**The Base-Case Physical Parameters**

| | |
|---|---|
| ORR transfer coefficient $\alpha_{ox}$ | 0.8 |
| ORR exchange current density $j_{ox}^{\infty}$ (A cm$^{-2}$) | $10^{-6}$ |
| ORR equilibrium potential $E_{ox}^{eq}$ V | 1.23 |
| HOR transfer coefficient $\alpha_{hy}$ | 1.0 |
| HOR exchange current density in the regular domain $j_{hy}^{\infty}$ (A cm$^{-2}$) | 1 |
| HOR equilibrium potential $E_{hy}^{eq}$ V | 0.0 |
| Membrane proton conductivity $\sigma_m$ ($\Omega^{-1}$ cm$^{-1}$) | 0.1 |
| Membrane thickness $l_m$ (cm) | 0.0025 (25 μm) |
| Spot radius $R_s$ (cm) | $10 l_m = 0.025$ |
| Thickness of the transition region $s$ in Equation 5.232 (cm) | $10^{-3} R_s = 2.5 \cdot 10^{-5}$ |
| The ratio of the HOR exchange current density in the spot to that current density outside the spot $k_s$ | $10^{-9}$ |
| Cell potential $\phi_c$ (V) | 0.7131 |
| Mean current density in the regular domain $J$ (A cm$^{-2}$) | 1.0 |
| Cell temperature $T$ (K) | $273 + 65$ |

*Note:* Note that the spot radius is 10 times the membrane thickness.

---

current dominates. This structure forms in order to support the lowering of the ORR current inside the spot. A 3D representation of the current double layer is shown in Figure 5.40.

Physically, lowering of $j^c$ at $\tilde{r} < 1$ occurs, since the opposite domain of the anode (dead spot) does not produce protons. However, lowering of $j^c$ requires quite a significant lowering of $\Phi$ inside the spot, as $\Phi$ determines the ORR overpotential. Therefore, the radial gradient of $\Phi$ forms. This gradient, in turn, induces a radial proton current in the membrane, which must be supported by an enhanced rate of HOR outside the spot. Multiplying the difference $j^a - j^c$ in Figure 5.39c by $2\pi r\, dr$ and integrating over $r$ yields zero, which means that the total ORR current, produced in front of the spot on the cathode side, equals the total excessive HOR current just outside the spot boundary.*

Figure 5.39b also shows the radial proton current density inside the membrane $-j_r = \sigma_m \partial \Phi / \partial r$. Though Equation 5.250 for $\Phi$ has been formulated in a quasi-2D approximation, the $x$ currents $j^a$ and $j^c$, together with the $r$ current $j_r$ in Figure 5.39b,

---

* This result follows from Equation 5.237. Indeed, multiplying both sides of Equation 5.237 by $2\pi \tilde{r}\, d\tilde{r}$ and integrating over $\tilde{r}$ from 0 to $\tilde{R}$, we get

$$\int_0^{\tilde{R}} 2\pi \tilde{r}(\tilde{j}_c - \tilde{j}_a)\, d\tilde{r} = 2\pi \epsilon^2 \tilde{R} \left. \frac{\partial \Phi}{\partial \tilde{r}} \right|_{\tilde{r}=\tilde{R}}.$$

For sufficiently large $\tilde{R}$, the derivative on the right-hand side is zero, and we come to the relation discussed.

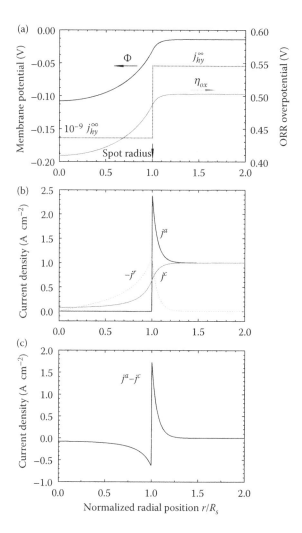

**FIGURE 5.39** The base-case variant. (a) The shapes of the membrane potential $\Phi$ and ORR overpotential $\eta_{ox}$. The step change of the HOR exchange current density is shown schematically by the dotted line. (b) The HOR $j^a$, ORR $j^c$, and radial $\sigma_m \partial\Phi/\partial r$ current densities. (c) The difference of the HOR and ORR current densities (current double layer). The cell potential $\phi^c = 0.7131$ V. The cell current density in the regular domain is fixed at 1 A cm$^{-2}$.

allow reconstructing the 2D trajectory of protons in the membrane. Inside the spot cylinder,[*] both $r$ and $x$ components of proton current increase with $r$. Outside the spot cylinder, near the boundary, the $x$ component is twice larger than the $r$ component, that is, the proton current moves mainly along $x$, with the small curvature directed toward the spot center (Figure 5.39b).

---

[*] A cylinder in the bulk membrane with the base equal to the anode spot.

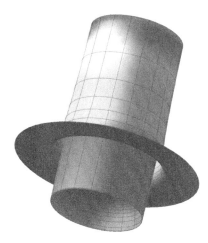

**FIGURE 5.40**   A 3D view of the current double layer in Figure 5.39c.

Figure 5.39b shows an interesting paradox. At $\tilde{r} = 0$, the radial component of proton current in the membrane is zero, though the ORR current density $j^c(0)$ is not zero. There is no electrochemical activity on the anode side at $r = 0$ and, hence, it is not obvious as to what the bridging mechanism is for $j^c(0)$.

The key is that $j^c(0)$ is a *current density*, while to calculate the real current at the spot axis, $j^c$ must be integrated over the surface of a small disk at $r = 0$. Calculations show that inside the spot, for an arbitrary disk radius $r$, the following relation holds:

$$\int_0^r j^c\, 2\pi r'\, dr' = -2\pi r l_m j_r, \qquad (5.253)$$

where $j_r = -\sigma_m \partial \Phi / \partial r$. Thus, in the membrane, the total axial current, crossing the cathode base of the cylinder of an arbitrary radius $r$, enters this cylinder through its side surface moving in the radial direction.

Note that the peak value of $j^a$ is 2.5 times larger than the mean cell current density, that is, a small ring around the spot generates much higher proton current on the anode side (Figure 5.39b). The Joule heat in the membrane is proportional to the square of the proton current density; hence, inside this ring, a risk of local overheat and of membrane drying increases dramatically.

Solutions for the 10 times larger spot radius $R_s = 0.25$ cm ($l_m/R_s = 0.01$) show that the peak of anodic current $j^a$ at the spot boundary remains nearly the same as in the base-case variant (cf. Figures 5.39b and 5.41b). Thus, the structure of the current double layer seemingly does not depend on the spot radius, provided that this radius largely exceeds the membrane thickness. Calculations show that this structure is determined by the mean current density $J$, in the cell, and by the HOR and ORR kinetic parameters. Analytical solution to the problem (Kulikovsky and Berg, 2013) confirms this result.

Note that in larger spots, the drop of $\Phi$ and the respective decrease in $\eta_{ox}$ in the spot center is larger (Figure 5.41a). As a result of that, the ORR current density in the

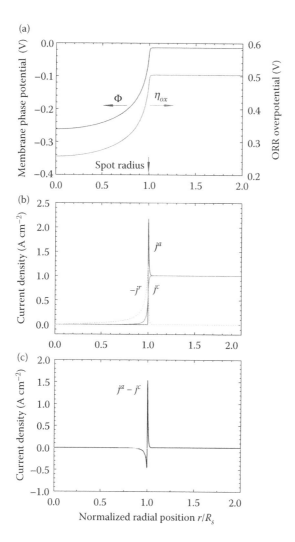

**FIGURE 5.41**  The same as Figure 5.39 for the 10 times larger spot, $l_m/R_s = 0.01$. Shown are the radial shapes of (a) the membrane potential and ORR overpotential, (b) ORR and HOR current densities, and radial component of the proton current density in the membrane, and (c) the difference of the HOR and ORR current densities.

spot center vanishes in Figure 5.41b, while in the smaller spot $j^c(0)$ is quite noticeable (Figure 5.39b). As the HOR current density in the double layer remains the same, in larger spot, a smaller $j^c$ is needed to provide the same total radial current crossing this interface. In other words, in large spots, the regime of the cathode operation in front of the spot center tends to open-circuit conditions.

# References

Adachi, M., Navessin, T., Xie, Z., Frisken, B., and Holdcroft, S. 2009. Correlation of in situ and ex situ measurements of water permeation through Nafion NRE211 proton exchange membranes. *J. Electrochem. Soc.*, **156**, B782–B790.

Adachi, M., Navessin, T., Xie, Z., Li, F. H., Tanaka, S., and Holdcroft, S. 2010. Thickness dependence of water permeation through proton exchange membranes. *J. Membr. Sci.*, **364**(1–2), 183–193.

Adamson, A. W. 1979. *A Textbook of Physical Chemistry*. New York: Academic Press.

Adzic, R. R., Zhang, J., Sasaki, K., Vukmirovic, M. B., Shao, M., Wang, J. X., Nilekar, A. U., Mavrikakis, M., Valerio, J. A., and Uribe, F. 2007. Platinum monolayer fuel cell electrocatalysts. *Top. Catal.*, **46**, 249–262.

Affoune, A. M., Yamada, A., and Umeda, M. 2005. Conductivity and surface morphology of Nafion membrane in water and alcohol environments. *J. Power Sources*, **148**(0), 9–17.

Agmon, N. 1995. The Grotthus mechanism. *Chem. Phys. Lett.*, **244**, 456–462.

Ahmadi, T. S., Wang, Z. L., Green, T. C., Henglein, A., and El-Sayed, M. A. 1996. Shape-controlled synthesis of colloidal platinum nanoparticles. *Science*, **272**, 1924–1926.

Alava, M. J., Nukala, P. K. V. V., and Zapperi, S. 2006. Statistical models of fracture. *Adv. Phys.*, **55**(3-4), 349–476.

Alcañiz-Monge, J., Linares-Solano, A., and Rand, B. 2001. Water adsorption on activated carbons: Study of water adsorption in micro-and mesopores. *J. Phys. Chem. B*, **105**(33), 7998–8006.

Aleksandrova, E., Hiesgen, R., Friedrich, K. A., and Roduner, E. 2007. Electrochemical atomic force microscopy study of proton conductivity in a Nafion membrane. *Phys. Chem. Chem. Phys.*, **9**, 2735–2743.

Allahyarov, E., and Taylor, P. L. 2007. Role of electrostatic forces in cluster formation in a dry ionomer. *J. Chem. Phys.*, **127**(15), 154901–154901.

Allen, M. P. and Tildesley, D. J. 1989. *Computer Simulation of Liquids*. Oxford: Oxford University Press.

Alsabet, M., Grden, M., and Jerkiewicz, G. 2006. Comprehensive study of the growth of thin oxide layers on Pt electrodes under well-defined temperature, potential, and time conditions. *J. Electroanal. Chem.*, **589**, 120–127.

Ambegaokar, V., Halperin, B. I., and Langer, J. S. 1971. Hopping conductivity in disordered systems. *Phys. Rev. B*, **4**(8), 2612–2620.

Andreaus, B. and Eikerling, M. 2007. Active site model for CO adlayer electrooxidation on nanoparticle catalysts. *J. Electroanal. Chem.*, **607**, 121–132.

Andreaus, B. and Scherer, G. G. 2004. Proton-conducting polymer membranes in fuel cells—humidification aspects. *Solid State Ionics*, **168**, 311–320.

Andreaus, B., Maillard, F., Kocylo, J., Savinova, E. R., and Eikerling, M. 2006. Kinetic modeling of COad monolayer oxidation on carbon-supported platinum nanoparticles. *J. Phys. Chem. B*, **110**, 21028–21040.

Angerstein-Kozlowska, H., Conway, B. E., and Sharp, W. B. A. 1973. The real condition of electrochemically oxidized Part I. Resolution of component processes. *J. Electroanal. Chem. Interfacial Electrochem.*, **43**, 9–36.

Ansermet, J. P. 1985. A new approach to the study of surface phenomena. PhD thesis, University of Illinois, Urbana-Champaign.

Antolini, E. 2003. Formation, microstructural characteristics and stability of carbon supported platinum catalysts for low temperature fuel cells. *J. Mater. Sci.*, **38**, 2995–3005.

Antolini, E., Cardellin, F., Giacometti, E., and Squadrito, G. 2002. Study on the formation of Pt/C catalysts by non-oxidized active carbon support and a sulfur-based reducing agent. *J. Mater. Sci.*, **37**(1), 133–139.

Aotani, K., Miyazaki, S., Kubo, N., and Katsuta, M. 2008. An analysis of the water transport properties of polymer electrolyte membrane. *ECS Trans.*, **16**, 341–352.

Aqvist, J. and Warshel, A. 1993. Simulations of enzyme reactions using valence bond force fields and other hybrid quantum/classical approaches. *Chem. Rev.*, **93**, 2523–2544.

Arcella, V., Troglia, C., and Ghielmi, A. 2005. Hyflon ion membranes for fuel cells. *Ind. Eng. Chem. Res.*, **44**(20), 7646–7651.

Arenz, M., Mayrhofer, K. J. J., Stamenkovic, V., Blizanac, B. B., Tomoyuki, T., Ross, P. N., and Markovic, N. M. 2005. The effect of the particle size on the kinetics of CO electrooxidation on high surface area Pt catalysts. *J. Am. Chem. Soc.*, **127**(18), 6819–6829.

Argyropoulos, P., Scott, K., Shukla, A. K., and Jackson, C. 2002. Empirical model equations for the direct methanol fuel cell. *Fuel Cells*, **2**(2), 78–82.

Ashcroft, N. W. and Mermin, N. D. 1976. *Solid State Physics. Science: Physics.* Philadelphia: Saunders College.

Astill, T. 2008. *Factors influencing electrochemical properties and performance of hydrocarbon based ionomer PEMFC catalyst layers.* PhD thesis, Simon Fraser University.

Astill, T., Xie, Z., Shi, Z., Navessin, T., and Holdcroft, S. 2009. Factors influencing electrochemical properties and performance of hydrocarbon-based electrolyte PEMFC catalyst layers. *J. Electrochem. Soc.*, **156**(4), B499–B508.

Babadi, M., Ejtehadi, M. R., and Everaers, R. 2006. Analytical first derivatives of the RE-squared interaction potential. *J. Comput. Phys.*, **219**(2), 770–779.

Baghalha, M., Stumper, J., and Eikerling, M. 2010. Model-based deconvolution of potential losses in a PEM fuel cell. *ECS Trans.*, **28**(23), 159–167.

Bagotsky, V. 2012. *Fuel Cells, Problems and Solutions.* New York: Wiley.

Bagotzky, V. S. and Vasilyev, Yu. B. 1964. Some characteristics of oxidation reactions of organic compounds of platinum electrodes. *Electrochim. Acta*, **9**, 869–882.

Bagotzky, V. S. and Vasilyev, Yu. B. 1967. Mechanism of electrooxidation of methanol on the platinum electrodes. *Electrochim. Acta*, **12**, 1323–1343.

Balandin, A. A. 1969. *Modern state of the multiplet theory of heterogeneous catalysis*, D. D. Eley, H. Pines, P. B. Weisz (Eds), *Advanced Catalysis*, New York: Academic Press, p.1.

Balbuena, P. B., Lamas, E. J., and Wang, Y. 2005. Molecular modeling studies of polymer electrolytes for power sources. *Electrochim. Acta*, **50**(19), 3788–3795.

Baldaus, M. and Preidel, W. 2001. Experimental results on the electrochemical oxidation of methnaol in PEM fuel cells. *J. Appl. Electrochem.*, **31**, 781–786.

Barbir, F. 2012. *PEM Fuel Cells Theory and Practice.* Amsterdam: Elsevier.

Bard, A. J. 1976. *Encyclopedia of Electrochemistry of the Elements*, Vol. 6. New York: Marcel Dekker.

Bard, A. J. and Faulkner, L. R. 2000. *Electrochemical Methods: Fundamentals and Applications.* New York: John Wiley & Sons.

Barsoukov, E. and Macdonald, J. R. (eds). 2005. *Impedance Spectroscopy: Theory, Experiment, and Applications*, 2nd ed. New Jersey: John Wiley & Sons, Inc.

Baschuk, J. J. and Li, X. 2000. Modelling of polymer electrolyte membrane fuel cells with variable degrees of water flooding. *J. Power Sources*, **86**, 181–196.

Baxter, S. F., Battaglia, V. S., and White, R. E. 1999. Methanol fuel cell model: Anode. *J. Electrochem. Soc.*, **146**(2), 437–447.

Bayati, M., Abad, J. M., Bridges, C. A., Rosseinsky, M. J., and Schiffrin, D. J. 2008. Size control and electrocatalytic properties of chemically synthesized platinum nanoparticles grown on functionalised HOPG. *J. Electroanal. Chem.*, **623**, 19–28.

Bazeia, D., Leite, V. B. P., Lima, B. H. B., and Moraes, F. 2001. Soliton model for proton conductivity in Langmuir films. *Chem. Phys. Lett.*, **340**, 205–210.

Becerra, L. R., Klug, C. A., Slichter, C. P., and Sinfelt, J. H. 1993. NMR study of diffusion of carbon monoxide on alumina-supported platinum clusters. *J. Phys. Chem.*, **97**(46), 12014–12019.

Becke, A. D. 1988. Density functional exchange energy approximation with correct asymptotic behavior. *Phys. Rev. A*, **38**(6), 3098–3100.

Becker, J., Flueckiger, R., Reum, M., Buechi, F., Marone, F., and Stampanoni, M. 2009. Determination of material properties of gas diffusion layers: Experiments and simulations using phase contrast tomographic microscopy. *J. Electrochem. Soc.*, **156**, B1175–B1181.

Bellac, M. Le, Mortessagne, F., and Batrouni, G. 2004. Equilibrium and Nonequilibrium Statistical Thermodynamics. Cambridge: Cambridge University Press.

Benziger, J., Kimball, E., Mejia-Ariza, R., and Kevrekidis, I. 2011. Oxygen mass transport limitations at the cathode of polymer electrolyte membrane fuel cells. *AIChE J.*, **57**, 2505–2517.

Berendsen, H. J. C., van der Spoel, D., and van Drunen, R. 1995. GROMACS: A message-passing parallel molecular dynamics implementation. *Comp. Phys. Comm.*, **91**(1–3), 43–56.

Berg, P. and Boland, A. 2013. Analysis of ultimate fossil fuel reserves and associated $CO_2$ emissions in IPCC scenarios. *Natural Resources Res.* **23**, 141–158.

Berg, P. and Ladipo, K. 2009. Exact solution of an electro-osmotic flow problem in a cylindrical channel of polymer electrolyte membranes. *Proc. Roy. Soc. A-Math. Phys. Eng. Sci.*, **465**(2109), 2663–2679.

Berg, P., Promislow, K., Pierre, J. St., Stumper, J., and Wetton, B. 2004. Water management in PEM fuel cells. *J. Electrochem. Soc.*, **151**(3), A341–A353.

Bergelin, M., Herrero, E., Feliu, J. M., and Wasberg, M. 1999. Oxidation of CO adlayers on Pt (111) at low potentials: An impinging jet study in $H_2SO_4$ electrolyte with mathematical modeling of the current transients. *J. Electroanal. Chem.*, **467**(1), 74–84.

Bernardi, D. M. and Verbrugge, M. W. 1992. A mathematical model of the solid-polymer-electrolyte fuel cell. *J. Electrochem. Soc.*, **139**(9), 2477–2491.

Bewick, A., Fleischmann, M., and Thirsk, H. R. 1962. Kinetics of the electrocrystallization of thin films of calomel. *Trans. Faraday Soc.*, **58**, 2200–2216.

Biesheuvel, P. M. and Bazant, M. Z. 2010. Nonlinear dynamics of capacitive charging and desalination by porous electrodes. *Phys. Rev. E*, **81**(3), 031502.

Biesheuvel, P. M., van Soestbergen, M., and Bazant, M. Z. 2009. Imposed currents in galvanic cells. *Electrochim. Acta*, **54**(21), 4857–4871.

Bird, R. B., Stewart, W. E., and Lightfoot, E. N. 1960. *Transport Phenomena*. New York: Wiley.

Birss, V.I., Chang, M., and Segal, J. 1993. Platinum oxide film formation-reduction: An *in situ* mass measurement study. *J. Electroanal. Chem.*, **355**, 181–191.

Björnbom, P. 1987. Modelling of a double-layered PTFE-bonded oxygen electrode. *Electrochim. Acta*, **32**(1), 115–119.

Bolhuis, P. G., Chandler, D., Dellago, C., and Geissler, P. L. 2002. Transition path sampling: Throwing ropes over rough mountain passes, in the dark. *Annu. Rev. Phys. Chem.*, **53**(1), 291–318.

Bonakdarpour, A., Stevens, K., Vernstrom, G. D., Atanasoski, R., Schmoeckel, A.K., Debe, M. K., and Dahn, J. R. 2007. Oxygen reduction activity of Pt and Pt–Mn–Co electrocatalysts sputtered on nano-structured thin film support. *Electrochim. Acta*, **53**(2), 688–694.

Booth, F. 1951. The dielectric constant of water and the saturation effect. *J. Chem. Phys.*, **19**(4), 391–394.

Borup, R., Meyers, J., Pivovar, B., Kim, Y. S., Mukundan, R., Garland, N., Myers, D., Wilson, M., Garzon, F., Wood, D., Zelenay, P., More, K., Stroh, K., Zawodzinski, T., Boncella, J., McGrath, J. E., Inaba, N., Miyatake, K., Hori, M., Ota, K., Ogumi, Z., Miyata, S., Nishikata, A., Siroma, Z., Uchimoto, Y., Yasuda, K., Kimijima, K., and Iwashita, N. 2007. Scientific aspects of polymer electrolyte fuel cell durability and degradation. *Chem. Rev.*, **107**, 3904–3951.

Bossel, U. 2000. *The Birth of the Fuel Cell*. Göttingen, Germany: Jürgen Kinzel.

Boudart, M. 1969. Catalysis by supported metals. *Adv. Catal.*, **20**, 153–166.

Boyer, C. C., Anthony, R. G., and Appleby, A. J. 2000. Design equations for optimized PEM fuel cell electrodes. *J. Appl. Electrochem.*, **30**, 777–786.

Bradley, J. S. (ed). 2007. *The Chemistry of Transition Metal Colloids, in Clusters and Colloids: From Theory to Applications*. Weinheim: Wiley-VCH Verlag GmbH.

Brett, D. J. L., Atkins, S., Brandon, N. P., Vesovic, V., Vasileadis, N., and Kucernak, A. 2003. Localized impedance measurements along a single channel of a solid polymer fuel cell. *Electrochem. Solid State Lett.*, **6**, A63–A66.

Broadbent, S. R., and Hammersley, J. M. 1957. Percolation processes. *Math. Proc. Cambridge*, **53**(6), 629–641.

Brownson, D. A. C., Kampouris, D. K., and Banks, C. E. 2012. Graphene electrochemistry: Fundamental concepts through to prominent applications. *Chem. Soc. Rev.*, **41**, 6944–6976.

Buechi, F. N. and Scherer, G. G. 2001. Investigation of the transversal water profile in Nafion membranes in polymer electrolyte fuel cells. *J. Electrochem. Soc.*, **148**(3), A183–A188.

Bultel, Y., Wiezell, K., Jaouen, F., Ozil, P., and Lindbergh, G. 2005. Investigation of mass transport in gas diffusion layer at the air cathode of a PEMFC. *Electrochim. Acta*, **51**, 474–488.

Bunde, A. and Kantelhardt, J. W. 1998. Introduction to percolation theory. J. Karger, R. Heitjans, R. Haberlandt (Eds), *Diffusion in Condensed Matter*, (Chapter 15), Wiesbaden: Vieweg-Sohn.

Burda, C., Chen, X., Narayanan, R., and El-Sayed, M. A. 2005. Chemistry and properties of nanocrystals of different shapes. *Chem. Rev.*, **105**(4), 1025–1102.

Burnett, D. J., Garcia, A. R., and Thielmann, F. 2006. Measuring moisture sorption and diffusion kinetics on proton exchange membranes using a gravimetric vapor sorption apparatus. *J. Power Sources*, **160**, 426–430.

Cable, K. M., Mauritz, K. A., and Moore, R. B. 1995. Effects of hydrophilic and hydrophobic counterions on the Coulombic interactions in perfluorosulfonate ionomers. *J. Pol. Sci. Part B: Pol. Phys.*, **33**(7), 1065–1072.

Cai, Y., Hu, J., Ma, H., Yi, B., and Zhang, H. 1006. Effect of water transport properties on a PEM fuel cell operating with dry hydrogen. *Electrochim. Acta*, **51**, 6361–6366.

Cappadonia, M., Erning, J. W., and Stimming, U. 1994. Proton conduction of Nafion® 117 membrane between 140 K and room temperature. *J. Electroanal. Chem.*, **376**(1–2), 189–193.

Cappadonia, M., Erning, J. W., Niaki, S. M. S., and Stimming, U. 1995. Conductance of Nafion 117 membranes as a function of temperature and water content. *Solid State Ionics*, **77**(0), 65–69.

Car, R. and Parrinello, M. 1985. Unified approach for molecular dynamics and density-functional theory. *Phys. Rev. Lett.*, **55**, 2471–2474.

Carmo, M., dos Santos, A. R., Poco, J. G. R., and Linardi, M. 2007. Physical and electrochemical evaluation of commercial carbon black as electrocatalysts supports for DMFC applications. *J. Power Sources*, **173**(2), 860–866.

Casalegno, A. and Marchesi, R. 2008. DMFC anode polarization: Experimental analysis and model validation. *J. Power Sources*, **175**, 372–382.

Chan, K. and Eikerling, M. 2011. A pore-scale model of oxygen reduction in ionomer-free catalyst layers of PEFCs. *J. Electrochem. Soc.*, **158**(1), B18–B28.

Chan, K. and Eikerling, M. 2012. Impedance model of oxygen reduction in water-flooded pores of ionomer-free PEFC catalyst layers. *J. Electrochem. Soc.*, **159**, B155–B164.

Chan, K. and Eikerling, M. 2014. Water balance model for polymer electrolyte fuel cells with ultrathin catalyst layers. *Phys. Chem. Chem. Phys.*, **16**, 2106–2117.

Chan, K., Roudgar, A., Wang, L., and Eikerling, M. (eds). 2010. Nanoscale Phenomena in Catalyst Layers for PEM Fuel Cells: From Fundamental Physics to Benign Design. Hoboken: Wiley.

Chen, F. F. 1974. *Introduction to Plasma Physics*. New York: Plenum Press.

Chen, J. and Chan, K. Y. 2005. Size-dependent mobility of platinum cluster on a graphite surface. *Mol. Simulat.*, **31**(6-7), 527–533.

Chen, M., Du, C. Y., Yin, G. P., Shi, P. F., and Zhao, T. S. 2009. Numerical analysis of the electrochemical impedance spectra of the cathode of direct methanol fuel cells. *Int. J. Hydrogen Energy*, **34**, 1522–1530.

Chen, Q. and Schmidt-Rohr, K. 2007. Backbone dynamics of the Nafion ionomer studied by 19F-13C solid-state NMR. *Mac. Chem. Phys.*, **208**(19-20), 2189–2203.

Cheng, X., Yi, B., Han, M., Zhang, J., Qiao, Y., and Yu, J. 1999. Investigation of platinum utilization and morphology in catalyst layer of polymer electrolyte fuel cells. *J. Power Sources*, **79**(1), 75–81.

Cherstiouk, O. V., Simonov, P. A, and Savinova, E. R. 2003. Model approach to evaluate particle size effects in electrocatalysis: Preparation and properties of Pt nanoparticles supported on GC and HOPG. *Electrochim. Acta*, **48**, 3851–3860.

Chizmadzhev, Y. A., Markin, V. S., Tarasevich, M. R., Chirkov, Y. G., and Bikerman, J. J. 1971. Macrokinetics of processes in porous media (fuel cells). Moscow (in Russian): Nanka Publisher.

Cho, C., Kim, Y., and Chang, Y.-S. 2009. Perfromance analysis of direct methanol fuel cell for optimal operation. *J. Thermal Sci. Techn.*, **4**, 414–423.

Choi, J.-H., Kim, Y. S., Bashyam, R., and Zelenay, P. 2006. Ruthenium crossover in DMFCs operating with different proton conducting membranes. *ECS Trans.*, **1**, 437–445.

Choi, K. H., Peck, D. H., Kim, C. S., Shin, D. R., and Lee, T. H. 2000. Water transport in polymer membranes for PEMFC. *J. Power Sources*, **86**, 197–201.

Choi, P., and Datta, R. 2003. Sorption in proton-exchange membranes—an explanation of Schroeder's paradox. *J. Electrochem. Soc.*, **150**(12), E601–E607.

Choi, P., Jalani, N. H., and Datta, R. 2005. Thermodynamics and proton transport in Nafion II. Proton diffusion mechanisms and conductivity. *J. Electrochem. Soc.*, **152**(3), E123–E130.

Clark, J. K., Paddison, S. J., Eikerling, M., Dupuis, M., and Jr., T. A. Zawodzinski. 2012. A comparative ab initio study of the primary hydration and proton dissociation of various imdide and sulfonic acid ionomers. *J. Phys. Chem. A*, **116**, 1801–1813.

Claus, P. and Hofmeister, H. 1999. Electron microscopy and catalytic study of silver catalysts: Structure sensitivity of the hydrogenation of crotonaldehyde. *J. Phys. Chem. B*, **103**, 2766–2775.

Clavilier, J., Rodes, A., Achi, K. El, and Zamakchari, M. A. 1991. Electrochemistry at platinum single crystal surfaces in acidic media:*J. Chim. Phys. Phys.-Chim. Biol*, **88**, 1291–1337.

Clay, C., Haq, S., and Hodgson, A. 2004. Hydrogen bonding in mixed $OH + H_2O$ overlayers on Pt(111). *Phys. Rev. Lett.*, **92**, 046102.

Climent, V., Garcia-Araez, N., Herrero, E., and Feliu, J. 2006. Potential of zero total charge of platinum single crystals: A local approach to stepped surfaces vicinal to Pt (111). *Russ. J. Electrochem.*, **42**(11), 1145–1160.

Coalson, R. D., and Kurnikova, M. G. 2007. Poisson–Nernst–Planck theory of ion permeation through biological channels. S.-H. Chung, O.S. Andersen, V. Krishnamurthy (Eds), *Biological Membrane Ion Channels*. New York: Springer, pp. 449–484.

Collins, M. A., Blumen, A., Currie, J. F., and Ross, J. 1979. Dynamics of domain walls in ferrodistortive materials. I. Theory. *Phys. Rev. B*, **19**, 3630–3644.

Commer, P., Cherstvy, A. G., Spohr, E., and Kornyshev, A. A. 2002. The effect of water content on proton transport in polymer electrolyte membranes. *Fuel Cells*, **2**(3–4), 127–136.

Conway, B. and Jerkiewicz, G. 1992. Surface orientation dependence of oxide film growth at platinum single crystals. *J. Elect. Chem.*, **339**, 123–146.

Conway, B. E. 1995. Electrochemical oxide film formation at noble metals as a surface-chemical process. *Progr. Surf. Sci.*, **49**(4), 331–452.

Conway, B. E. and Gottesfeld, S. 1973. Real condition of oxidized platinum electrodes. Part 2.—Resolution of reversible and irreversible processes by optical and impedance studies. *J. Chem. Soc., Faraday Trans. 1*, **69**, 1090–1107.

Conway, B. E., Barnett, B., Angerstein Kozlowska, H., and Tilak, B. V. 1990. A surface electrochemical basis for the direct logarithmic growth law for initial stages of extension of anodic oxide films formed at noble metals. *J. Electroanal. Chem. Interfacial Electrochem.*, **93**, 8361–8373.

Corry, B., Kuyucak, S., and Chung, S. H. 2003. Dielectric self-energy in Poisson–Boltzmann and Poisson–Nernst–Planck models of ion channels. *Biophys. J.*, **84**(6), 3594–3606.

Cui, S., Liu, J., Selvan, M. E., Keffer, D. J., Edwards, B. J., and Steele, W. V. 2007. A molecular dynamics study of a Nafion polyelectrolyte membrane and the aqueous phase structure for proton transport. *J. Phys. Chem. B*, **111**(9), 2208–2218.

Cwirko, E. H. and Carbonell, R. G. 1992a. Interpretation of transport coefficients in Nafion using a parallel pore model. *J. Membr. Sci.*, **67**(2–3), 227–247.

Cwirko, E. H. and Carbonell, R. G. 1992b. Ionic equilibria in ion-exchange membranes—A comparison of pore model predictions with experimental results. *J. Membr. Sci.*, **67**(2-3), 211–226.

Daiguji, H. 2010. Ion transport in nanofluidic channels. *Chem. Soc. Rev.*, **39**(3), 901–911.

Damjanovic, A. 1992. Progress in the studies of oxygen reduction during the last thirty years. 1992, O. J. Murphy et al. (Eds), *Electrochemistry in Transition*. New York: Plenum Press, pp. 107–126.

Damjanovic, A. and Yeh, L. S. R. 1979. Oxide growth at platinum anodes with emphasis on the pH dependence of growth. *J. Electrochem. Soc.*, **126**, 555–562.

Damjanovic, A., Yeh, L. S. R., and Wolf, J. F. 1980. Temperature study of oxide film growth at platinum anodes in $H_2SO_4$ solutions. *J. Electrochem. Soc.*, **127**, 874–877.

Davydov, A. S. 1985. *Solitons in Molecular Systems*. Boston: Reidel.

de Gennes, P. G. 1979. *Scaling Concepts in Polymer Physics*. Ithaca: Cornell University.

de Grotthuss, C. J. T. 1806. Sur la décomposition de l'eau et des corps qu'elle tient en dissolution à l'aide de l'électricité galvanique. *Ann. Chim.*, **58**, 54–73.

de Levie, R. 1964. On porous electrodes in electrolyte solutions—IV. *Electrochim. Acta*, **9**(9), 1231–1245.

de Levie, R. 1967. Electrochemical response of porous and rough electrodes. Delahay, P. (ed.), *Advances in Electrochemistry and Electrochemical Engineering*, Vol. 6. pp. 329–397. New York: Interscience.

Debe, M. K. 2012. Electrocatalyst approaches and challenges for automotive fuel cells. *Nature*, **486**, 43–51.

Debe, M. K. 2013. Tutorial on the fundamental characteristics and practical properties of nanostructured thin film (NSTF) catalysts. *J. Electrochem. Soc.*, **160**, 522–534.

Debe, M. K., Schmoeckel, A. K., Vernstrom, G. D., and Atanasoski, R. 2006. High voltage stability of nanostructured thin film catalysts for PEM fuel cells. *J. Power Sources*, **161**, 1002–1011.

Dellago, C., Bolhuis, P. G., and Chandler, D. 1998. Efficient transition path sampling: Application to Lennard–Jones cluster rearrangements. *J. Chem. Phys.*, **108**, 9236–9245.

Devan, S., Subramanian, V. R., and White, R. E. 2004. Analytical solution for the impedance of a porous electrode. *J. Electrochem. Soc.*, **151**, A905–A913.

Devanathan, R., Venkatnathan, A., and Dupuis, M. 2007a. Atomistic simulation of Nafion membrane. 2. Dynamics of water molecules and hydronium ions. *J. Phys. Chem. B*, **111**(45), 13006–13013.

Devanathan, R., Venkatnathan, A., and Dupuis, M. 2007b. Atomistic simulation of Nafion membrane: I. Effect of hydration on membrane nanostructure. *J. Phys. Chem. B*, **111**(28), 8069–8079.

Devanathan, R., Venkatnathan, A., and Dupuis, M. 2007c. Atomistic simulation of Nafion membrane: I. Effect of hydration on membrane nanostructure. *J. Phys. Chem. B*, **111**(28), 8069–8079.

Dhathathreyan, K. S., Sridhar, P., Sasikumar, G., Ghosh, K. K., Velayutham, G., Rajalakshmi, N., Subramaniam, C. K., Raja, M., and Ramya, K. 1999. Development of polymer electrolyte membrane fuel cell stack. *Int. J. Hydrogen Energ.*, **24**(11), 1107–1115.

Diard, J.-P., Glandut, N., Landaud, P., Gorrec, B. Le, and Montella, C. 2003. A method for determining anode and cathode impedances of a direct methanol fuel cell running on a load. *Electrochim. Acta*, **48**, 555–562.

Ding, J., Chuy, C., and Holdcroft, S. 2001. A self-organized network of nanochannels enhances ion conductivity through polymer films. *Chem. Mater.*, **13**(7), 2231–2233.

Ding, J., Chuy, C., and Holdcroft, S. 2002. Enhanced conductivity in morphologically controlled proton exchange membranes: Synthesis of macromonomers by SFRP and their incorporation into graft polymers. *Macromolecules*, **35**(4), 1348–1355.

Divisek, J., Eikerling, M., Mazin, V., Schmitz, H., Stimming, U., and Volfkovich, Y. M. 1998. A study of capillary porous structure and sorption properties of Nafion proton exchange membranes swollen in water. *J. Electrochem. Soc.*, **145**(8), 2677–2683.

Dixon, D., Wippermann, K., Mergel, J., Schoekel, A., Zils, S., and Roth, C. 2011. Degradation effects at the methanol inlet, outlet and center region of a stack MEA operated in DMFC. *J. Power Sources*, **196**, 5538–5545.

Dlubek, G., Buchhold, R., Hübner, C., and Nakladal, A. 1999. Water in local free volumes of polyimides: A positron lifetime study. *Macromolecules*, **32**(7), 2348–2355.

Dobrynin, A. V. 2005. Electrostatic persistence length of semiflexible and flexible polyelectrolytes. *Macromolecules*, **38**(22), 9304–9314.

Dobrynin, A. V. and Rubinstein, M. 2005. Theory of polyelectrolytes in solutions and at surfaces. *Progr. Pol. Sci.*, **30**(11), 1049–1118.

Dobson, P., Lei, C., Navessin, T., and Secanell, M. 2012. Characterization of the PEM fuel cell catalyst layer microstructure by nonlinear least-squares parameter estimation. *J. Electrochem. Soc.*, **159**, B514–B523.

Dohle, H., Divisek, J., and Jung, R. 2000. Process engineering of the direct Methanol fuel cell. *J. Power Sources*, **86**, 469–77.

Dohle, H., Divisek, J., Oetjen, H. F., Zingler, C., Mergel, J., and Stolten, D. 2002. Recent developments of the measurement of methanol permeation in a direct methanol fuel cell. *J. Power Sources*, **105**, 274–82.

Du, C. Y., Zhao, T. S., and Wang, W. W. 2007a. Effect of methanol crossover on the cathode behavior of a DMFC: A half-cell investigation. *Electrochim. Acta*, **52**, 5266–5271.

Du, C. Y., Zhao, T. S., and Xu, C. 2007b. Simultaneous oxygen-reduction and methanol-oxidation reactions at the cathode of a DMFC: A model-based electrochemical impedance spectroscopy study. *J. Power Sources*, **167**, 265–271.

Duan, Q., Wang, H., and Benziger, J. B. 2012. Transport of liquid water through Nafion membranes. *J. Membr. Sci.*, **392–393**, 88–94.

Dubbeldam, D. and Snurr, R. Q. 2007. Recent developments in the molecular modeling of diffusion in nanoporous materials. *Mol. Simul.*, **33**(4–5), 305–325.

Dullien, F. A. L. 1979. *Porous Media: Fluid Transport and Pore Structure*. New York: Academic Press.

Easton, E. B. and Pickup, P. G. 2005. An electrochemical impedance spectroscopy study of fuel cell electrodes. *Electrochim. Acta*, **50**(12), 2469–2474.

Eichler, A., Mittendorfer, F., and Hafner, J. 2000. Precursor-mediated adsorption of oxygen on the (111) surfaces of platinum-group metals. *Phys. Rev. B.*, **62**(7), 4744.

Eigen, M. 1964. Proton transfer, acid–base catalysis, and enzymatic hydrolysis. Part I: Elementary processes. *Ang. Chem. Intern. Ed.*, **3**, 1–19.

Eikerling, M. 2006. Water management in cathode catalyst layers of PEM fuel cells: A structure-based model. *J. Electrochem. Soc.*, **153**(3), E58–E70.

Eikerling, M. and Berg, P. 2011. Poroelectroelastic theory of water sorption and swelling in polymer electrolyte membranes. *Soft Matter*, **7**(13), 5976–5990.

Eikerling, M. and Kornyshev, A. A. 1998. Modelling the performance of the cathode catalyst layer of polymer electrolyte fuel cells. *J. Electroanal. Chem.*, **453**, 89–106.

Eikerling, M. and Kornyshev, A. A. 1999. Electrochemical impedance of the cathode catalyst layer in polymer electrolyte fuel cells. *J. Electroanal. Chem.*, **475**(2), 107–123.

Eikerling, M. and Kornyshev, A. A. 2001. Proton transfer in a single pore of a polymer electrolyte membrane. *J. Electroanal. Chem.*, **502**(1-2), 1–14.

Eikerling, M. and Malek, K. 2009. Electrochemical materials for PEM fuel cells: Insights from physical theory and simulation. In: Schlesinger, M. (ed.), *Modern Aspects of Electrochem.* Vol. 43, New York: Springer.

Eikerling, M. and Malek, K. 2010. Physical modeling of materials for PEFC: structure, properties and performance. In: Wilkinson, D. P., Zhang, J., Hui, R., Fergus, J., and Li, X. (eds), *Proton Exchange Membrane Fuel Cells: Materials Properties and Performance*. New York: CRC Press, Taylor & Francis Group.

Eikerling, M., Kornyshev, A. A., and Stimming, U. 1997. Electrophysical properties of polymer electrolyte membranes: A random network model. *J. Phys. Chem. B*, **101**(50), 10807–10820.

Eikerling, M., Kharkats, Yu. I., Kornyshev, A. A., and Volfkovich, Yu. M. 1998. Phenomenological theory of electroâosmotic effect and water management in polymer electrolyte protonâconducting membranes. *J. Electrochem. Soc.*, **145**(8), 2684–2699.

Eikerling, M., Kornyshev, A. A., Kuznetsov, A. M., Ulstrup, J., and Walbran, S. 2001. Mechanisms of proton conductance in polymer electrolyte membranes. *J. Phys. Chem. B*, **105**(1-2), 3646–3662.

Eikerling, M., Paddison, S. J., and Zawodzinski, T. A. 2002. Molecular orbital calculations of proton dissociation and hydration of various acidic moieties for fuel cell polymers. *J. New Mater. Electrochem. Sys.*, **5**, 15–24.

Eikerling, M., Paddison, S. J., Pratt, L. R., and Zawodzinski, Jr., T. A. 2003. Defect structure for proton transport in a triflic acid monohydrate solid. *Chem. Phys. Lett.*, **368**, 108–114.

Eikerling, M., Ioselevich, A. A., and Kornyshev, A. A. 2004. How good are the electrodes we use in PEFC? *Fuel Cells*, **4**(3), 131–140.

Eikerling, M., Kornyshev, A. A., and Kucernak, A. R. 2006. Water in polymer electrolyte fuel cells: Friend and foe. *Phys. Today*, **59**, 38–44.

Eikerling, M., Kornyshev, A., and Kucernak, A. 2007a. Driving the hydrogen economy. *Phys. World*, **20**, 32–36.

Eikerling, M., Kornyshev, A. A., and Kulikovsky, A. A. 2007b. Physical modeling of fuel cells and their components. Bard, A.J., Stratmann, M., Macdonald, D., and Schmuki, P. (eds), *Encyclopaedia of Electrochemistry*, Vol. 5, pp. 447–543, New York: Wiley, ISBN: 978-3-527-30397-7.

Eikerling, M., Kornyshev, A. A., and Spohr, E. 2008. Proton-conducting polymer electrolyte membranes: Water and structure in charge. G. G. Scherer (Ed), *Fuel Cells I*, New York: Springer, pp. 15–54.

Eisenberg, A. 1970. Clustering of ions in organic polymers. A theoretical approach. *Macromolecules*, **3**(2), 147–154.

Elfring, G. J. and Struchtrup, H. 2008. Thermodynamics of pore wetting and swelling in Nafion. *J. Membr. Sci.*, **315**(1), 125–132.

Elliott, J. A. and Paddison, S. J. 2007. Modeling of morphology and proton transport in PFSA membranes. *Phys. Chem. Chem. Phys.*, **9**, 2602–2618.

Elliott, J. A., Elliott, A. M. S., and Cooley, G. E. 1999. Atomistic simulation and molecular dynamics of model systems for perfluorinated ionomer membranes. *Phys. Chem. Chem. Phys.*, **1**(20), 4855–4863.

Elliott, J. A., Hanna, S., Elliott, A. M. S., and Cooley, G. E. 2000. Interpretation of the small-angle x-ray scattering from swollen and oriented perfluorinated ionomer membranes. *Macromolecules*, **33**(11), 4161–4171.

Ensing, B., Laio, A., Parrinello, M., and Klein, M. L. 2005. A recipe for the computation of the free energy barrier and the lowest free energy path of concerted reactions. *J. Phys. Chem. B*, **109**(14), 6676–6687.

Erdey-Gruz, T. 1974. *Transport Phenomena in Electrolyte Solutions*. London: Adam Hilger.

Eucken, A. 1948. Assoziation in flussigkeiten. (Association in liquids). *Z. Elecktrochem.*, **52**, 255–269.

Everaers, R. and Ejtehadi, M. R. 2003. Interaction potentials for soft and hard ellipsoids. *Phys. Rev. E*, **67**(4), 041710.

Falk, M. 1980. An infrared study of water in perfluorosulfonate (Nafion) membranes. *Can. J. Chem.*, **58**(14), 1495–1501.

Farebrother, M., Goledzinowski, M., Thomas, G., and Birss, V.I. 1991. Early stages of growth of hydrous platinum oxide films. *J. Electroanal. Chem. Interfacial Electrochem.*, **297**, 469–488.

Feibelman, P. J. 1997. D-electron frustration and the large fcc versus hcp binding preference in O adsorption on Pt (111). *Phys. Rev. B.*, **56**(16), 10532.

Feibelman, P. J., Hammer, B., Nørskov, J. K., Wagner, F., Scheffler, M., Stumpf, R., Watwe, R., and Dumesic, J. 2001. The CO/Pt (111) puzzle. *J. Phys. Chem. B*, **105**(18), 4018–4025.

Feldberg, S. W., Enke, C. G., and Bricker, C. E. 1963. Formation and dissolution of platinum oxide film: Mechanism and kinetics. *J. Electrochem. Soc.*, **110**, 826–834.

Feng, S. and Voth, G. A. 2011. Proton solvation and transport in hydrated Nafion. *J. Phys. Chem. B*, **115**(19), 5903–5912.

Feng, X. and Wang, Y. 2010. Multi-layer configuration for the cathode electrode of polymer electrolyte fuel cell. *Electrochim. Acta*, **55**, 4579–4586.

Fernandez, R., Ferreira-Aparicio, P., and Daza, L. 2005. PEMFC electrode preparation: Influence of the solvent composition and evaporation rate on the catalytic layer microstructure. *J. Power Sources*, **151**, 18–24.

Ferreira, P. J., O, G. J. Ia, Shao-Horn, Y., Morgan, D., Makharia, R., Kocha, S., and Gasteiger, H. A. 2005. Instability of Pt/C electrocatalysts in proton exchange membrane fuel cells: A mechanistic investigation. *J. Electrochem. Soc.*, **152**(11), A2256–A2271.

Fishman, Z. and Bazylak, A. 2011. Heterogeneous through-plane distributions of tortuosity, effective diffusivity, and permeability for PEMFC GDLs. *J. Electrochem. Soc.*, **158**, B247–B252.

Flory, P. J. 1969. *Statistical Mechanics of Chain Molecules.* New York: Interscience Publishers.

Flory, P. J. and Rehner, J. 1943. Statistical mechanincs of cross-linked polymer networks II. Swelling. *J. Chem. Phys.*, **11**, 521–526.

Forinash, K., Bishop, A. R., and Lomadhl, P. S. 1991. Nonlinear dynamics in a double-chain model of DNA. *Phys. Rev. B*, **43**, 10743–10750.

Franck, E. U., Hartmann, D., and Hensel, F. 1965. Proton mobility in water at high temperatures and pressures. *Discuss. Faraday Soc.*, **39**, 200–206.

Freger, V. 2002. Elastic energy in microscopically phase-separated swollen polymer networks. *Polymer*, **43**, 71–76.

Freger, V. 2009. Hydration of ionomers and Schroeder's paradox in Nafion. *J. Phys. Chem. B*, **113**, 24–36.

Frenkel, A. I., Hills, C. W., and Nuzzo, R. G. 2001. A view from the inside: Complexity in the atomic scale ordering of supported metal nanoparticles. *J. Phys. Chem. B*, **105**(51), 12689–12703.

Friedrich, K. A., Henglein, F., Stimming, U., and Unkauf, W. 2000. Size dependence of the CO monolayer oxidation on nanosized Pt particles supported on gold. *Electrochim. Acta*, **45**(20), 3283–3293.

Frumkin, A. N. 1949. O raspredelenii korrozionnogo protsessa po dline trubki. (On the distribution of corrosion processes along the length of the tube. *Zh. Fiz. Khim.*, **23**(12), 1477–1482.

Frumkin, A. N. and Petrii, O. A. 1975. Potentials of zero total and zero free charge of platinum group metals. *Electrochim. Acta*, **20**(5), 347–359.

Fujimura, M., Hashimoto, T., and Kawai, H. 1981. Small-angle x-ray scattering study of perfluorinated ionomer membranes. 1. Origin of two scattering maxima. *Macromolecules*, **14**(5), 1309–1315.

Fuller, T. and Newman, J. 1989. Fuel cells. White, R. E., and Appleby, A. J. (eds), *The Electrochemical Society Softbound Proceedings Series*, Vol. PV89-14, p. 25. New Jersey: Pennington.

Fuller, T. F. and Newman, J. 1993. Water and thermal management in solid–polymer–electrolyte fuel cells. *J. Electrochem. Soc.*, **140**, 1218–1225.

Furukawa, K., Okajima, K., and Sudoh, M. 2005. Structural control and impedance analysis of cathode for direct methanol fuel cell. *J. Power Sources*, **139**, 9–14.

Futerko, P. and Hsing, I. 1999. Thermodynamics of water vapor uptake in perfluorosulfonic acid membranes. *J. Electrochem. Soc.*, **146**(6), 2049–2053.

Galperin, D. Y., and Khokhlov, A. R. 2006. Mesoscopic morphology of proton-conducting polyelectrolyte membranes of Nafion® type: A self-consistent mean field simulation. *Macromol. Theory Simul.*, **15**(2), 137–146.

Gancs, L., Hakim, N., Hult, B. N., and Mukerjee, S. 2006. Dissolution of Ru from PtRu electrocatalysts and its consequences in DMFCs. *ECS Trans.*, **3**, 607–618.

Garcia, B. L., Sethuraman, V. A., Weidner, J. W., White, R. E., and Dougal, R. 2004. Mathematical model of a direct methanol fuel cell. *J. Fuel Cell Sci. Techn.*, **1**, 43–48.

Gardel, M. L., Shin, J. H., MacKintosh, F. C., Mahadevan, L., Matsudaira, P., and Weitz, D. A. 2004. Elastic behavior of cross-linked and bundled actin networks. *Science*, **304**(5675), 1301–1305.

Gasteiger, H. A. and Markovic, N. M. 2009. Just a dream—Or future reality? *Science*, **324**(5923), 48–49.

Gasteiger, H. A. and Yan, S. G. 2004. Dependence of PEM fuel cell performance on catalyst loading. *J. Power Sources*, **127**(1), 162–171.

Gasteiger, H. A., Marković, N., Ross, Jr. P. N., and Cairns, E. J. 1993a. Methanol electrooxidation on well-characterized Pt–Ru alloys. *J. Phys. Chem.*, **97**, 12020–12029.

Gasteiger, H. A., Markovic, N., Ross, Jr, P. N., and Cairns, E. J. 1993b. Methanol electrooxidation on well-characterized platinum–ruthenium bulk alloys. *J. Phys. Chem*, **97**(46), 12020–12029.

Gatrell, M. and MacDougall, B. 2003. Reaction mechanisms in the $O_2$ reduction/evolution reaction. *Handbook of Fuel Cells: Fundamentals, Technology, Applications*, W. Vielstich, A. Lamm, H. A. Gasteiger, (eds). Vol. 2. Chichester: John Wiley, pp. 443–464.

Ge, S., Yi, B., and Ming, P. 2006. Experimental determination of electro-osmotic drag coefficient in Nafion membrane for fuel cells. *J. Electrochem. Soc.*, **153**, A1443–A1450.

Gebel, G. 2000. Structural evolution of water swollen perfluorosulfonated ionomers from dry membrane to solution. *Polymer*, **41**(15), 5829–5838.

Gebel, G. and Diat, O. 2005. Neutron and X-ray scattering: Suitable tools for studying ionomer membranes. *Fuel Cells*, **5**(2), 261–276.

Gebel, G. and Lambard, J. 1997. Small-angle scattering study of water-swollen perfluorinated ionomer membranes. *Macromolecules*, **30**(25), 7914–7920.

Gebel, G. and Moore, R. B. 2000. Small-angle scattering study of short pendant chain perfluorosulfonated ionomer membranes. *Macromolecules*, **33**(13), 4850–4855.

Georgievskii, Y., Medvedev, E. S., and Tuchebrukhov, A. A. 2002. Proton transport via the membrane surface. *Biophys. J.*, **82**(6), 2833–2846.

Gierer, A. 1950. Anomale D+- und OD—ionenbeweglichkeit in schwerem wasser. (Anomalous D+- and OD-ion mobility in heavy water). *Z. Naturforsch*, **5**, 581–589.

Gierer, A., and Wirtz, K. 1949. Anomale H-ioncnbeweglichkeit und OH-ionenbeweglichkeit im wasser. (Anomalous mobility of H and OH ions in water). *Annalen Phys.*, **6**, 257–304.

Gierke, T. D., Munn, G. E., and Wilson, F. C. 1981. The morphology in Nafion perfluorinated membrane products, as determined by wide-angle and small-angle X-ray studies. *J. Polym. Sci. Part B: Polym. Phys.*, **19**(11), 1687–1704.

Gileadi, E. 2011. *Physical Electrochemistry: Fundamentals, Techniques and Applications.* Weinheim: Wiley-VCH.

Gillespie, D. T. 1976. A general method for numerically simulating the stochastic time evolution of coupled chemical reactions. *J. Comp. Phys.*, **22**(4), 403–434.

Gilroy, D. 1976. Oxide growth at platinum electrodes in $H_2SO_4$ at potentials below 1.7 V. *J. Electroanal. Chem. Interfacial Electrochem.*, **151**, 257–277.

Gilroy, D. and Conway, B. E. 1968. Surface oxidation and reduction of platinum electrodes: Coverage, kinetic and hysteresis studies. *Can. J. Chem.*, **46**, 875–890.

Giner, J. and Hunter, C. 1969. The mechanism of operation of the Teflon-bonded gas diffusion electrode: A mathematical model. *J. Electrochem. Soc.*, **116**(8), 1124–1130.

Gloaguen, F. and Durand, R. 1997. Simulations of PEFC cathodes: An effectiveness factor approach. *J. Appl. Electrochem.*, **27**(9), 1029–1035.

Goddard, III, W., Merinov, B., Duin, A. Van, Jacob, T., Blanco, M., Molinero, V., Jang, S. S., and Jang, Y. H. 2006. Multi-paradigm multi-scale simulations for fuel cell catalysts and membranes. *Mol. Simul.*, **32**(3-4), 251–268.

Goedecker, S., Teter, M., and Hutter, J. 1996. Separable dual-space Gaussian pseudopotentials. *Phys. Rev. B*, **54**(3), 1703–1710.

Golovnev, A. and Eikerling, M. 2013a. Theoretical calculation of proton mobility for collective surface proton transport. *Phys. Rev. E*, **87**, 062908.

Golovnev, A. and Eikerling, M. 2013b. Theory of collective proton motion at interfaces with densely packed protogenic surface groups. *J. Phys.: Conds. Matter*, **25**(4), 045010.

Gomadam, P. and Weidner, J. W. 2005. Analysis of electrochemical impedance spectroscopy in proton exchange membrane fuel cells. *Int. J. Energy. Res.*, **29**, 1133–1151.

Gomez-Marin, A. M., Clavilier, J., and Feliu, J. M. 2013. Sequential Pt(111) oxide formation in perchloric acid: An electrochemical study of surface species inter-conversion. *J. Electroanal. Chem.*, **688**(0), 360–370.

Gordon, A. 1990. On soliton mechanism for diffusion of ionic defects in hydrogen-bonded solids. *Nuovo Cimento Della Societa Italiana Di Fisica D-Conds. Matter Atomic Molecular and Chem. Phys. Fluids Plasmas Biophys.*, **12**(2), 229–232.

Gore, W.L. and Associates. 2003. *GORE® PRIMEA® MEAs for Transportation*.

Graf, P., Kurnikova, M. G., Coalson, R. D., and Nitzan, A. 2004. Comparison of dynamic lattice Monte Carlo simulations and the dielectric self-energy Poisson–Nernst–Planck continuum theory for model ion channels. *J. Phys. Chem. B*, **108**(6), 2006–2015.

Grassberger, P. 1983. On the critical behavior of the general epidemic process and dynamical percolation. *Math. Biosci.*, **63**(2), 157–172.

Greeley, J., Jaramillo, T. F., Bonde, J., Chorkendorff, I. B., and Nørskov, J. K. 2006. Computational high-throughput screening of electrocatalytic materials for hydrogen evolution. *Nat. Mater.*, **5**, 909–913.

Greeley, J., Stephens, I. E. L., Bondarenko, A. S., Johansson, T. P., Hansen, H. A., Jaramillo, T. F., Rossmeisl, J., Chorkendorff, I. B., and Nørskov, J. K. 2009. Alloys of platinum and early transition metals as oxygen reduction electrocatalysts. *Nat. Chem.*, **1**(7), 552–556.

Groot, R. D. 2003. Electrostatic interactions in dissipative particle dynamics—Simulation of polyelectrolytes and anionic surfactants. *J. Chem. Phys.*, **118**, 11265.

Groot, R. D. and Warren, P. B. 1997. Dissipative particle dynamics: Bridging the gap between atomistic and mesoscopic simulation. *J. Chem. Phys.*, **107**, 4423.

Gross, A. 2006. Reactivity of bimetallic systems studied from first principles. *Top. Catal.*, **37**, 29–39.

Grot, W. 2011. *Fluorinated Ionomers*. Plastics Design Library. Elsevier Science.

Grubb, W. T. 1959. *Fuel Cell*. US Patent 2,913,511.

Grubb, W. T. and Niedrach, L. W. 1960. Batteries with solid ion exchange membrane electrolytes: II. Low temperature hydrogen oxygen fuel cells. *J. Electrochem. Soc.*, **107**, 131–135.

Gruber, D., Ponath, N., Muller, J., and Lindstaedt, F. 2005. Sputter-deposited ultra-low catalyst loadings for PEM fuel cells. *J. Power Sources*, **150**, 67–72.

Gruger, A., Regis, A., Schmatko, T., and Colomban, P. 2001. Nanostructure of Nafion® membranes at different states of hydration: An IR and Raman study. *Vib. Spectrosc.*, **26**(2), 215–225.

Gubler, L., Gursel, S. A., and Scherer, G. G. 2005. Radiation grafted membranes for polymer electrolyte fuel cells. *Fuel Cells*, **5**(3), 317–335.

Guo, Q. and White, R. E. 2004. A steady–state impedance model for a PEMFC cathode. *J. Electrochem. Soc.*, **151**, E133–E149.

Ha, B.-Y. and Liu, A. J. 1999. Counterion-mediated, non-pairwise-additive attractions in bundles of like-charged rods. *Phys. Rev. E.*, **60**(1), 803.

Halperin, W. P. 1986. Quantum size effects in metal particles. *Rev. Mod. Phys.*, **58**, 533–606.

Halsey, T. C. 1987. Stability of a flat interface in electrodeposition without mixing. *Phys. Rev. A*, **36**(7), 3512.

Hambourger, M., Moore, G. F., Kramer, D. A., Gust, D., Moore, A. L., and Moore, T. A. 2009. Biology and technology for photochemical fuel production. *Chem. Soc. Rev.*, **38**, 25–35.

Hamm, U. W., Kramer, D., Zhai, R. S., and Kolb, D. M. 1996. The pzc of Au (111) and Pt (111) in a perchloric acid solution: an ex situ approach to the immersion technique. *J. Electroanal. Chem.*, **414**(1), 85–89.

Hammer, B. and Nørskov, J. K. 1995. Electronic factors determining the reactivity of metal surfaces. *Surf. Sci.*, **343**, 211–211.

Hammer, B. and Nørskov, J. K. 2000. Theoretical surface science and catalysis—Calculations and concepts. *Adv. Catal.*, **45**, 71–129.

Hammer, B., Nielsen, O. H., and Nørskov, J. K. 1997. Structure sensitivity in adsorption: CO interaction with stepped and reconstructed Pt surfaces. *Catal. Lett.*, **46**, 31–35.

Han, B. C., Miranda, C. R., and Ceder, G. 2008. Effect of particle size and surface structure on adsorption of O and OH on platinum nanoparticles: A first-principles study. *Phys. Rev. B*, **77**(7), 75410–75410.

Hansen, J. P. and McDonald, I. R. 2006. *Theory of Simple Liquids*. Elsevier Science.

Hansen, L. B., Stoltze, P., and Nørskov, J. K. 1990. Is there a contraction of the interatomic distance in small metal particles? *Phys. Rev. Lett.*, **64**, 3155–3158.

Hao, X., Spieker, W. A., and Regalbuto, J. R. 2003. A further simplification of the revised physical adsorption (RPA) model. *J. Coll. Interface Sci.*, **267**(2), 259–264.

Harrington, D. 1997. Simulation of anodic Pt oxide growth. *J. ELectroanal Chem.*, **420**, 101–109.

Harris, L. B. and Damjanovic, A. 1975. Initial anodic growth of oxide film on platinum in $2NH_2SO_4$ under galvanostatic, potentiostatic, and potentiodynamic conditions: The question of mechanism. *J. Electrochem. Soc.*, **122**, 593–600.

Harvey, D., Pharoah, J. G., and Karan, K. 2008. A comparison of different approaches to modelling the PEMFC catalyst layer. *J. Power Sources*, **179**, 209–219.

Havranek, A. and Wippermann, K. 2004. Determination of proton conductivity in anode catalyst layers of the direct methanol fuel cell (DMFC). *J. Electroanal. Chem.*, **567**, 305–315.

Hayashi, H., Yamamoto, S., and Hyodo, S. A. 2003. Lattice-Boltzmann simulations of flow through NAFION polymer membranes. *Int. J. Mod. Phys. B*, **17**(01n02), 135–138.

Hayes, R. L., Paddison, S. J., and Tuckerman, M. E. 2009. Proton transport in triflic acid hydrates studied via path integral Car–Parrinello molecular dynamics. *J. Phys. Chem. B*, **113**, 16574–16589.

Hayes, R. L., Paddison, S. J., and Tuckerman, M. E. 2011. Proton transport in triflic acid pentahydrate studied via ab initio path integral molecular dynamics. *J. Phys. Chem. A*, **115**, 6112–6124.

Heberle, J., Riesle, J., Thiedemann, G., Oesterhelt, D., and Dencher, N. A. 1994. Proton migration along the membrane surface and retarded surface to bulk transfer. *Nature*, **370**, 379–381.

Henle, M. L. and Pincus, P. A. 2005. Equilibrium bundle size of rodlike polyelectrolytes with counterion-induced attractive interactions. *Phys. Rev. E*, **71**(6), 060801.

Herz, H. G., Kreuer, K. D., Maier, J., Scharfenberger, G., Schuster, M. F. H., and Meyer, W. H. 2003. New fully polymeric proton solvents with high proton mobility. *Electrochim. Acta*, **48**(14–16), 2165–2171.

Heyd, D. V. and Harrington, D. A. 1992. Platinum oxide growth kinetics for cyclic voltammetry. *J. Electroanal. Chem.*, **335**, 19–31.

Hickner, M. A., Ghassemi, H., Kim, Y. S., Einsla, B. R., and McGrath, J. E. 2004. Alternative polymer systems for proton exchange membranes (PEMs). *Chem. Rev.*, **104**, 4587.

Hickner, M. A. and Pivovar, B. S. 2005. The chemical and structural nature of proton exchange membrane fuel cell properties. *Fuel Cells*, **5**(2), 213–229.

Hickner, M. A., Siegel, P. N., Chen, K. S., McBrayer, D. N., Hussey, D. S., Jacobson, D. L., and Arif, M. 2006. Real-time imaging of liquid water in an operating proton exchange membrane fuel cell. *J. Electrochem. Soc.*, **153**, A902–A908.

Hiesgen, R., Wehl, I., leksandrova, A. E., Roduner, E., Bauder, A., and Friedrich, K. A. 2010. Nanoscale properties of polymer fuel cell materials—A selected review. *Int. J. Energ. Res.*, **34**(14), 1223–1238.

Hiesgen, R., Helmly, S., Galm, I., Morawietz, T., Handl, M., and Friedrich, K. A. 2012. Microscopic analysis of current and mechanical properties of Nafion studied by atomic force microscopy. *Membranes*, **2**(4), 783–803.

Hinebaugh, J., Bazylak, A., and Mukherjee, P. P. 2012. Multi-scale modeling of two-phase transport in polymer electrolyte membrane fuel cells. Ch. Hartnig, Ch. Roth (Eds), *Polymer Electrolyte Membrane and Direct Methanol Fuel Cell Technology, Volume 1*. Oxford: Woodhead Publishing, pp. 254–290.

Holdcroft, S. 2014. Fuel cell catalyst layers: A polymer science perspective. *Chem. Mater.*, **26**, 381–393.

Hoogerbrugge, P. J. and Koelman, J. M. V. A. 1992. Simulating microscopic hydrodynamic phenomena with dissipative particle dynamics. *Europhys. Lett.*, **19**(3), 155.

Housmans, T. H. M., Wonders, A. H., and Koper, M. T. M. 2006. Structure sensitivity of methanol electrooxidation pathways on platinum: An on-line electrochemical mass spectrometry study. *J. Phys. Chem. B*, **110**, 10021–10031.

Hsu, W. Y., Barkley, J. R., and Meakin, P. 1980. Ion percolation and insulator-to-conductor transition in Nafion perfluorosulfonic acid membranes. *Macromolecules*, **13**(1), 198–200.

Hsu, W. Y. and Gierke, T. D. 1982. Elastic theory for ionic clustering in perfluorinated ionomers. *Macromolecules*, **15**(1), 101–105.

Hsu, W. Y. and Gierke, T. D. 1983. Ion-transport and clustering in Nafion perfluorinated membranes. *J. Membr. Sci.*, **13**(3), 307–326.

Humphrey, W., Dalke, A., and Schulten, K. 1996. VMD: Visual molecular dynamics. *J. Mol. Graph.*, **14**(1), 33–38.

Hunt, A. G. 2005. *Percolation Theory for Flow in Porous Media*. Berlin, Heidelberg: Springer.

Hunt, A. G. and Ewing, R. P. 2003. On the vanishing of solute diffusion in porous media at a threshold moisture content. *Soil Sci. Soc. Am. J.*, **67**(6), 1701–1702.

Hyman, M. P. and Medlin, J. W. 2005. Theoretical study of the adsorption and dissociation of oxygen on Pt(111) in the presence of homogeneous electric fields. *J. Phys. Chem. B.*, **109**, 6304–6310.

Iczkowski, R. P. and Cutlip, M. B. 1980. Voltage losses in fuel cell cathodes. *J. Electrochem. Soc.*, **127**(7), 1433–1440.

Ihonen, J., Jaouen, F., Lindbergh, G., Lundblad, A., and Sundholm, G. 2002. Investigation of mass-transport limitations in the solid polymer fuel cell cathode: II. Experimental. *J. Electrochem. Soc.*, **149**, 448–454.

Ihonen, J., Mikkola, M., and Lindbergha, G. 2004. Flooding of gas diffusion backing in PEFCs. *J. Electrochem. Soc.*, **151**(8), A1152–A1161.

Iijima, S. and Ichihashi, T. 1986. Structural instability of ultrafine particles of metals. *Phys. Rev. Lett.*, **56**, 616–619.

Ioselevich, A. S. and Kornyshev, A. A. 2001. Phenomenological theory of solid oxide fuel cell anode. *Fuel Cells*, **1**(1), 40–65.

Ioselevich, A. S. and Kornyshev, A. A. 2002. Approximate symmetry laws for percolation in complex systems: Percolation in polydisperse composites. *Phys. Rev. E*, **65**(2), 021301.

Ioselevich, A. S., Kornyshev, A. A., and Steinke, J. H. G. 2004. Fine morphology of proton-conducting ionomers. *J. Phys. Chem. B*, **108**(32), 11953–11963.

Intergovernmental Panel on Climate Change (IPCC). *Special Report on Renewable Energy Sources and Climate Change Mitigation.* Technical report. 2011, http://srren.ipcc-wg3.de/report.

Ise, M., Kreuer, K. D., and Maier, J. 1999. Electroosmotic drag in polymer electrolyte membranes: An electrophoretic NMR study. *Solid State Ionics*, **125**, 213–223.

Isichenko, M. B. 1992. Percolation, statistical topography, and transport in random media. *Rev. Mod. Phys.*, **64**(Oct), 961–1043.

Iwasita, T. and Xia, X. 1996. Adsorption of water at Pt(111) electrode in $HClO_4$ solutions. The potential of zero charge. *J. Electroanal. Chem.*, **411**, 95–102.

Izvekov, S. and Violi, A. 2006. A coarse-grained molecular dynamics study of carbon nanoparticle aggregation. *J. Chem. Theory Comput.*, **2**, 504–512.

Izvekov, S. and Voth, G. A. 2005. A multiscale coarse-graining method for biomolecular systems. *J. Phys. Chem. B*, **109**, 2469–2473.

Izvekov, S., Violi, A., and Voth, G. A. 2005. Systematic coarse-graining of nanoparticle interactions in molecular dynamics simulation. *J. Phys. Chem. B*, **109**(36), 17019–17024.

Jacob, T. 2006. The mechanism of forming $H_2O$ from $H_2$ and $O_2$ over a Pt catalyst via direct oxygen reduction. *Fuel Cells*, **6**, 159–181.

Jacob, T., Merinov, B. V., and Goddard, III, W. A. 2004. Chemisorption of atomic oxygen on Pt (111) and Pt/Ni (111) surfaces. *Chem. Phys. Lett.*, **385**(5), 374–377.

Jalani, N. H., and Datta, R. 2005. The effect of equivalent weight, temperature, cationic forms, sorbates, and nanoinorganic additives on the sorption behavior of Nafion. *J. Membr. Sci.*, **264**, 167–175.

James, P. J., Elliott, J. A., McMaster, T. J., Newton, J. M., Elliott, A. M. S., Hanna, S., and Miles, M. J. 2000. Hydration of Nafion® studied by AFM and x-ray scattering. *J. Mater. Sci.*, **35**(20), 5111–5119.

Jang, S. S., Molinero, V., Cagin, T., and Goddard, W. A. 2004. Nanophase-segregation and transport in Nafion 117 from molecular dynamics simulations: Effect of monomeric sequence. *J. Phys. Chem. B*, **108**(10), 3149–3157.

Janik, M. J., Wasileski, S. A., Taylor, C. D., and Neurock, M. 2008. First-principles simulation of the active sites and reaction environment in electrocatalysis, M. T. M. Koper (Ed). *Fuel Cell Catalysis: A Surface Science Approach.* John Wiley & Sons, Inc. pp. 93–128.

Janssen, G. J. M., and Overvelde, M. L. J. 2001. Water transport in the proton-exchange-membrane fuel cell: Measurements of the effective drag coefficient. *J. Power Sources*, **101**, 117–125.

Jaouen, F. and Lindbergh, G. 2003. Transient techniques for investigating mass-transport limitations in gas diffusion electrode. *J. Electrochem. Soc.*, **150**, A1699–A1710.

Jaouen, F., Lindbergh, G., and Sundholm, G. 2002. Investigation of mass-transport limitations in the solid polymer fuel cell cathode I. Mathematical model. *J. Electrochem. Soc.*, **149**(4), A437–A447.

Jeng, K. T. and Chen, C. W. 2002. Modeling and simulation of a direct methanol fuel cell anode. *J. Power Sources*, **112**, 367–375.

Jerkiewicz, G., Vatankhah, G., Lessard, J., Soriaga, M. P., and Park, Y.-S. 2004. Surface-oxide growth at platinum electrodes in aqueous $H_2SO_4$: Reexamination of its mechanism through combined cyclic-voltammetry, electrochemical quartz-crystal nanobalance, and Auger electron spectroscopy measurements. *Electrochim. Acta*, **49**(9), 1451–1459.

Jiang, J. and Kucernak, A. 2005. Solid polymer electrolyte membrane composite microelectrode investigations of fuel cell reactions. II: Voltammetric study of methanol oxidation at the nanostructured platinum microelectrode|Nafion® membrane interface. *J. Electroanal. Chem.*, **576**, 223–236.

Jiang, J., Oberdoerster, G., Elder, A., Gelein, R., Mercer, P., and Biswas, P. 2008. Does nanoparticle activity depend upon size and crystal phase? *Nanotoxicology*, **2**, 33–42.

Jiang, Q., Liang, L. H., and Zhao, D. S. 2001. Lattice contraction and surface stress of fcc nanocrystals. *J. Phys. Chem. B.*, **105**, 6275–6277.

Jiang, R. and Chu, D. 2004. Comparative studies of methanol crossover and cell performance for a DMFC. *J. Electrochem. Soc.*, **151**, A69–A76.

Jinnouchi, R. and Anderson, A.B. 2008. Structure calculations of liquid–solid interfaces: A combination of density functional theory and modified Poisson–boltzmann theory. *Phys. Rev. B.*, **77**, 245417–245435.

Jusys, Z. and Behm, R. J. 2004. Simultaneous oxygen reduction and methanol oxidation on a carbon-supported Pt catalyst and mixed potential formation-revisited. *Electrochim. Acta*, **49**, 3891–3900.

Kang, M. S. and Martin, C. R. 2001. Investigations of potential-dependent fluxes of ionic permeates in gold nanotubule membranes prepared via the template method. *Langmuir*, **17**(9), 2753–2759.

Kaplan, T., Gray, L. J., and Liu, S. H. 1987. Self-affine fractal model for a metal-electrolyte interface. *Phys. Rev. B*, **35**(10), 5379.

Karan, K. 2007. Assessment of transport-limited catalyst utilization for engineering of ultra-low Pt loading polymer electrolyte fuel cell anode. *Electrochem. Commun.*, **9**(4), 747–753.

Kasai, N. and Kakudo, M. 2005. *X-Ray Diffraction by Macromolecules*. Springer Series in Chemical Physics. Kodansha Limited and Springer-Verlag, Berlin, Heidelberg.

Kast, W. and Hohenthanner, C. R. 2000. Mass transfer within the gas-phase of porous media. *Int. J. Heat Mass Transfer*, **43**(5), 807–823.

Katsounaros, I., Auinger, M., Cherevko, S., Meier, J. C., Klemm, S. O., and Mayrhofer, K. J. J. 2012. Dissolution of platinum: Limits for the deployment of electrochemical energy conversion? *Angew. Chem. Int. Ed.*, **51**, 12613–12615.

Kavitha, L., Jayanthi, S., Muniyappan, A., and Gopi, D. 2011. Protonic transport through solitons in hydrogen-bonded systems. *Phys. Scr.*, **84**, 035803.

Keener, J. P. and Sneyd, J. 1998. *Mathematical Physiology*, Vol. 8. New York: Springer.

Kelly, M. J., Egger, B., Fafilek, G., Besenhard, J. O., Kronberger, H., and Nauer, G. E. 2005. Conductivity of polymer electrolyte membranes by impedance spectroscopy with microelectrodes. *Solid State Ionics*, **176**(25–28), 2111–2114.

Khalatur, P. G., Talitskikh, S. K., and Khokhlov, A. R. 2002. Structural organization of water-containing Nafion: The integral equation theory. *Macromol. Theory Simul.*, **11**(5), 566–586.

Khandelwal, M. and Mench, M. 2006. Direct measurement of through-plane thermal conductivity and contact resistance in fuel cell materials. *J. Power Sources*, **161**, 1106–1115.

Kim, J., Lee, S.-M., Srinivasan, S., and Chamberlin, Ch. E. 1995. Modeling of proton exchange membrane fuel cell performance with an empirical equation. *J. Electrochem. Soc.*, **142**(8), 2670–2674.

Kim, M. H., Glinka, C. J., Grot, S. A., and Grot, W. G. 2006. SANS study of the effects of water vapor sorption on the nanoscale structure of perfluorinated sulfonic acid (NAFION) membranes. *Macromolecules*, **39**(14), 4775–4787.

Kinkead, B., van Drunen, J., Paul, M. T. Y., Dowling, K., Jerkiewicz, G., and Gates, B. D. 2013. Platinum ordered porous electrodes: Developing a platform for fundamental electro-chemical characterization. *Electrocatalysis*, **4**, 179–186.

Kinoshita, K. 1988. *Carbon: Electrochemical and Physicochemical Properties*. New York: John Wiley Sons.

Kinoshita, K. 1990. Particle-size effects for oxygen reduction on highly dispersed platinum in acid electrolytes. *J. Electrochem. Soc.*, **137**, 845–848.

Kinoshita, K. 1992. *Electrochemical Oxygen Technology, the Electrochemical Society Series.* New York: Wiley.

Kirkpatrick, S. 1973. Percolation and conduction. *Rev. Mod. Phys.*, **45**(4), 574–588.

Kisljuk, O. S., Kachalova, G. S., and Lanina, N. P. 1994. An algorithm to find channels and cavities within protein crystals. *J. Mol. Graph.*, **12**(4), 305–307.

Kobayashi, T., Babu, P. K., Gancs, L., Chung, J. H., Oldfield, E., and Wieckowski, A. 2005. An NMR determination of CO diffusion on platinum electrocatalysts. *J. Am. Chem. Soc.*, **127**(41), 14164–14165.

Kolb, D. M., Engelmann, G. E., and Ziegler, J. C. 2000. On the unusual electrochemical stability of nanofabricated copper clusters. *Angew. Chem. Int. Edit.*, **39**, 1123–1125.

Koper, M. T. M. 2011. Thermodynamic theory of multi-electron transfer reactions: Implications for electrocatalysis. *J. Electroanal. Chem.*, **660**, 254–260.

Koper, M. T. M., and Heering, H. A. 2010. Comparison of electrocatalysis and bio-electrocatalysis of hydrogen and oxygen redox reactions, A. Wiecleowski and A. H. Heering, (Eds), *Fuel Cell Science: Theory, Fundamentals and Biocatalysis.* pp. 71–110. Hoboken, NJ: John Wiley & Sons, Inc.

Koper, M. T. M., Jansen, A. P. J., Santen, R. A. Van, Lukkien, J. J., and Hilbers, P. A. J. 1998. Monte Carlo simulations of a simple model for the electrocatalytic CO oxidation on platinum. *J. Chem. Phys.*, **109**, 6051.

Koper, M. T. M., Lebedeva, N. P., and Hermse, C. G. M. 2002. Dynamics of CO at the solid/liquid interface studied by modeling and simulation of CO electro-oxidation on Pt and PtRu electrodes. *Faraday Discuss*, **121**, 301–311.

Kornyshev, A. A. and Leikin, S. 1997. Theory of interaction between helical molecules. *J. Chem. Phys.*, **107**, 3656.

Kornyshev, A. A. and Leikin, S. 2000. Electrostatic interaction between long, rigid helical macromolecules at all interaxial angles. *Phys. Rev. E.*, **62**(2), 2576.

Kresse, G. and Furthmüller, J. 1996a. Efficiency of ab-initio total energy calculations for metals and semiconductors using a plane-wave basis set. *Comput. Mat. Sci.*, **6**, 15–50.

Kresse, G. and Furthmüller, J. 1996b. Efficient iterative schemes for ab initio total-energy calculations using a plane-wave basis set. *Phys. Rev. B*, **54**, 11169–11186.

Kresse, G. and Hafner, J. 1993. Ab initio molecular dynamics for liquid metals. *Phys. Rev. B*, **47**, 558–561.

Kresse, G. and Hafner, J. 1994a. Ab initio molecular-dynamics simulation of the liquid–metal–amorphous–semiconductor transition in germanium. *Phys. Rev. B.*, **49**, 14251–14269.

Kresse, G. and Hafner, J. 1994b. Norm-conserving and ultrasoft pseudopotentials for first-row and transition elements. *J. Phys.: Condens. Mat.*, **6**, 8245–8258.

Kreuer, K. D. 1997. On the development of proton conducting materials for technological applications. *Solid State Ionics*, **97**(1), 1–15.

Kreuer, K. D., Paddison, S. J., Spohr, E., and Schuster, M. 2004. Transport in proton conductors for fuel-cell applications: Simulations, elementary reactions, and phenomenology. *Chem. Rev.*, **104**, 4637–4678.

Kreuer, K. D., Schuster, M., Obliers, B., Diat, O., Traub, U., Fuchs, A., Klock, U., Paddison, S. J., and Maier, J. 2008. Short-side-chain proton conducting perfluorosulfonic acid ionomers: Why they perform better in PEM fuel cells. *J. of Power Sources*, **178**(2), 499–509.

Krewer, U., Christov, M., Vidakovic, T., and Sundmacher, K. 2006. Impedance spectroscopic analysis of the electrochemical methanol oxidation kinetics. *J. Electroanal. Chem.*, **589**, 148–159.

Krishtalik, L. I. 1986. *Charge Transfer Reactions in Electrochemical and Chemical Processes.* New York: Plenum.

Krtil, A. T. P. and Samec, Z. 2001. Kinetics of water sorption in Nafion thin films—Quartz Crystal Microbalance study. *J. Phys. Chem. B*, **105**, 7979–7983.

Kucernak, A. R. and Toyoda, E. 2008. Studying the oxygen reduction and hydrogen oxidation reactions under realistic fuel cell conditions. *Electrochem. Commun.*, **10**, 1728–1731.

Kulikovsky, A. A. 2001. Gas dynamics in channels of a gas-feed direct methanol fuel cell: Exact solutions. *Electrochem. Comm.*, **3**(10), 572–79.

Kulikovsky, A. A. 2002a. The voltage–current curve of a polymer electrolyte fuel cell: "exact" and fitting equations. *Electrochem. Commun.*, **4**(11), 845–852.

Kulikovsky, A. A. 2002b. The voltage current curve of a direct methanol fuel cell: "Exact" and fitting equations. *Electrochem. Comm.*, **4**, 939–946.

Kulikovsky, A. A. 2004. The effect of stoichiometric ratio $\lambda$ on the performance of a polymer electrolyte fuel cell. *Electrochim. Acta*, **49**(4), 617–625.

Kulikovsky, A. A. 2005. Active layer of variable thickness: The limiting regime of anode catalyst layer operation in a DMFC. *Electrochem. Commun.*, **7**, 969–975.

Kulikovsky, A. A. 2006. Heat balance in the catalyst layer and the boundary condition for heat transport equation in a low-temperature fuel cell. *J. Power Sources*, **162**, 1236–1240.

Kulikovsky, A. A. 2007. Heat transport in a PEFC: Exact solutions and a novel method for measuring thermal conductivities of the catalyst layers and membrane. *Electrochem. Commun.*, **9**, 6–12.

Kulikovsky, A. A. 2009a. A model of SOFC anode performance. *Electrochim. Acta*, **54**, 6686–6695.

Kulikovsky, A. A. 2009b. Optimal effective diffusion coefficient of oxygen in the cathode catalyst layer of polymer electrode membrane fuel cells. *Electrochem. Solid State Lett.*, **12**, B53–B56.

Kulikovsky, A. A. 2009c. Optimal shape of catalyst loading across the active layer of a fuel cell. *Electrochem. Commun.*, **11**, 1951–1955.

Kulikovsky, A. A. 2010a. *Analytical Modelling of Fuel Cells*. Amsterdam: Elsevier.

Kulikovsky, A. A. 2010b. The regimes of catalyst layer operation in a fuel cell. *Electrochim. Acta*, **55**, 6391–6401.

Kulikovsky, A. A. 2011a. Polarization curve of a PEM fuel cell with poor oxygen or proton transport in the cathode catalyst layer. *Electrochem. Commun.*, **13**, 1395–1399.

Kulikovsky, A. A. 2011b. A simple model for carbon corrosion in PEM fuel cell. *J. Electrochem. Soc.*, **158**, B957–B962.

Kulikovsky, A. A. 2012a. Catalyst layer performance in PEM fuel cell: Analytical solutions. *Electrocatalysis*, **3**, 132–138.

Kulikovsky, A. A. 2012b. A model for DMFC cathode impedance: The effect of methanol crossover. *Electrochem. Commun.*, **24**, 65–68.

Kulikovsky, A. A. 2012c. A model for DMFC cathode performance. *J. Electrochem. Soc.*, **159**, F644–F649.

Kulikovsky, A. A. 2012d. A model for local impedance of the cathode side of PEM fuel cell with segmented electrodes. *J. Electrochem. Soc.*, **159**, F294–F300.

Kulikovsky, A. A. 2012e. A model for mixed potential in direct methanol fuel cell cathode and a novel cell design. *Electrochim. Acta*, **79**, 52–56.

Kulikovsky, A. A. 2012f. A model for optimal catalyst layer in a fuel cell. *Electrochim. Acta*, **79**, 31–36.

Kulikovsky, A. A. 2012g. A physical model for the catalyst layer impedance. *J. Electroanal. Chem.*, **669**, 28–34.

Kulikovsky, A. A. 2013a. Analytical polarization curve of DMFC anode. *Adv. Energy Res.*, **1**, 35–52.

Kulikovsky, A. A. 2013b. Dead spot in the PEM fuel cell anode. *J. Electrochem. Soc.*, **160**, F401–F405.

Kulikovsky, A. A. 2014. A physically-based analytical polarization curve of a PEM fuel cell. *J. Electrochem. Soc.*, **161**, F263–F270.

Kulikovsky, A. A. and Berg, P. 2013. Analytical description of a dead spot in a PEMFC anode. *ECS Electrochem. Lett.*, **9**, F64–F67.

Kulikovsky, A. A. and Eikerling, M. 2013. Analytical solutions for impedance of the cathode catalyst layer in PEM fuel cell: Layer parameters from impedance spectrum without fitting. *J. Electroanal. Chem.*, **691**, 13–17.

Kulikovsky, A. A., Kučernak, A., and Kornyshev, A. 2005. Feeding PEM fuel cells. *Electrochim. Acta*, **50**, 1323–1333.

Kulikovsky, A. A., and McIntyre, J. 2011. Heat flux from the catalyst layer of a fuel cell. *Electrochim. Acta*, **56**, 9172–9179.

Kulikovsky, A. A., Scharmann, H., and Wippermann, K. 2004. On the origin of voltage oscillations of a polymer electrolyte fuel cell in galvanostatic regime. *Electrochem. Commun.*, **6**, 729–736.

Kümmel, R. 2011. *The Second Law of Economics: Energy, Entropy, and the Origins of Wealth.* New York: Springer.

Kuntova, Z., Chvoj, Z., Sima, V., and Tringides, M. C. 2005. Limitations of the thermodynamic, Gibbs–Thompson analysis of nanoisland decay. *Phys. Rev. B*, **71**, 1165–1174.

Kurzynski, M. 2006. *The Thermodynamic Machinery of Life.* New York: Springer.

Kusoglu, A., Kwong, A., Clark, K., Gunterman, H. P., and Weber, A. Z. 2012. Water uptake of fuel-cell catalyst layers. *J. Electrochem. Soc.*, **159**, F530–F535.

Kuznetsov, A. M. and Ulstrup, J. 1999. *Electron Transfer in Chemistry and Biology: An Introduction to the Theory.* Chichester: John Wiley & Sons, Ltd.

Laasonen, K., Sprik, M., Parrinello, M., and Car, R. 1993. Ab initio liquid water. *J. Chem. Phys.*, **99**, 9080–9090.

Lam, A., Wetton, B., and Wilkinson, D. P. 2011. One-dimensional model for a direct methanol fuel cell with a 3D anode structure. *J. Electrochem. Soc.*, **158**, B29–B35.

Lamas, E. J. and Balbuena, P. B. 2003. Adsorbate effects on structure and shape of supported nanoclusters: A molecular dynamics study. *J. Phys. Chem. B*, **107**(42), 11682–11689.

Lamas, E. J. and Balbuena, P. B. 2006. Molecular dynamics studies of a model polymer-catalyst-carbon interface. *Electrochim. Acta*, **51**(26), 5904–5911.

Lampinen, M. J., and Fomino, M. 1993. Analysis of free energy and entropy changes for half-cell reactions. *J. Electrochem. Soc.*, **140**, 3537–46.

Lamy, C. and Leger, J.-M. 1991. Electrocatalytic oxidation of small organic molecules at platinum single crystals. *J. Chim. Phys.*, **88**, 1649–1671.

Lasia, A. 1995. Impedance of porous electrodes. *J. Electroanal. Chem.*, **397**, 27–33.

Lasia, A. 1997. Porous electrodes in the presence of a concentration gradient. *J. Electroanal. Chem.*, **428**, 155–164.

Lebedeva, N. P., Koper, M. T. M., Feliu, J. M., and Santen, R. A. Van. 2002. Role of crystalline defects in electrocatalysis: Mechanism and kinetics of CO adlayer oxidation on stepped platinum electrodes. *J. Phys. Chem. B*, **106**(50), 12938–12947.

Leberle, K., Kempf, I., and Zundel, G. 1989. An intramolecular hydrogen bond with large proton polarizability within the head group of phosphatidylserine. *Biophys. J.*, **55**, 637–648.

Lee, I., Morales, R., Albiter, M. A., and Zaera, F. 2008. Synthesis of heterogeneous catalysts with well shaped platinum particles to control reaction selectivity. *Proc. Natl. Acad. Sci. USA*, **105**, 15241–15246.

Lee, M., Uchida, M., Yano, H., Tryk, D. A., Uchida, H., and Watanabe, M. 2010. New evaluation method for the effectiveness of platinum/carbon electrocatalysts under operating conditions. *Electrochim. Acta*, **55**(28), 8504–8512.

Lee, S. J., Mukerjee, S., McBreen, J., Rho, Y. W., Kho, Y. T., and Lee, T. H. 1998. Effects of Nafion impregnation on performances of PEMFC electrodes. *Electrochim. Acta*, **43**(24), 3693–3701.

Lehmani, A., Bernard, O., and Turq, P. 1997. Transport of ions and solvent in confined media. *J. Stat. Phys.*, **89**(1-2), 379–402.

Lehmani, A., Durand-Vidal, S., and Turq, P. 1998. Surface morphology of Nafion 117 membrane by tapping mode atomic force microscope. *J. Appl. Polym. Sci.*, **68**(3), 503–508.

Leite, V. B. P., Cavalli, A., and Oliveira, O. N. 1998. Hydrogen-bond control of structure and conductivity of Langmuir films. *Phys. Rev. E*, **57**, 6835–6839.

Levie, R. De. 1963. On porous electrodes in electrolyte solutions: I. Capacitance effects. *Electrochim. Acta*, **8**(10), 751–780.

Levie, R. De. 1967. Electrochemical response of porous and rough electrodes. *Adv. Electroch. El. Eng.*, **6**, 329–397.

Li, G. and Pickup, P. 2003. Ionic conductivity of PEMFC electrodes: Effect of Nafion loading. *J. Electrochem. Soc.*, **150**(11), C745–C752.

Li, W. and Wang, C.-Y. 2007. Three-dimensional simulations of liquid feed direct methanol fuel cells. *J. Electrochem. Soc.*, **154**(3), B352–B361.

Lide, D.R. (ed). 1990. *CSIR Handbook of Chemistry and Physics*. Boca Raton, FL: CRC Press.

Lin, W. F., Jin, J. M., Christensen, P. A., Zhu, F., and Shao, Z. G. 2008. In-situ FT-IR spectroscopic studies of fuel cell electro-catalysis: From single-crystal to nanoparticle surfaces. *Chem. Eng. Commun.*, **195**, 147–166.

Lin, X., Ramer, N. J., Rappe, A. M., Hass, K. C., Schneider, W. F., and Trout, B. L. 2001. Effect of particle size on the adsorption of O and S atoms on Pt: A density-functional theory study. *J. Phys. Chem. B*, **105**, 7739–7747.

Lindahl, E., Hess, B., and Spoel, D. Van Der. 2001. GROMACS 3.0: A package for molecular simulation and trajectory analysis. *Mol. Model. Annu.*, **7**(8), 306–317.

Linford, R. G. (ed.). 1973. *Solid State Surface Science*. New York: Marcel Dekker.

Lisiecki, I. 2005. Size, shape, and structural control of metallic nanocrystals. *J. Phys. Chem. B*, **109**, 12231–12244.

Liu, F., Lu, G., and Wang, C. Y. 2007. Water transport coefficient distribution through the membrane in a polymer electrolyte fuel cell. *J. Membr. Sci.*, **287**, 126–131.

Liu, J. and Eikerling, M. 2008. Model of cathode catalyst layers for polymer electrolyte fuel cells: The role of porous structure and water accumulation. *Electrochim. Acta*, **53**(13), 4435–4446.

Liu, L., Zhang, L., Cheng, X., and Zhang, Y. 2009. On-time Determination of Ru crossover in DMFC. *ECS Trans.*, **19**, 43–51.

Liu, P. and Nørskov, J. K. 2001. Kinetics of the anode processes in PEM fuel cells—The promoting effect of Ru in PtRu anodes. *Fuel Cells*, **1**(3–4), 192–201.

Liu, W. J., Wu, B. L., and Cha, C. S. 1999. Surface diffusion and the spillover of H-adatoms and oxygen-containing surface species on the surface of carbon black and Pt/C porous electrodes. *J. Electroanal. Chem.*, **476**, 101–108.

Longworth, R. and Vaughan, D. J. 1968a. *Polym. Prepr. Am. Chem. Soc. Div. Polym. Chem.*, **9**, 525.

Longworth, R. and Vaughan, D. J. 1968b. Physical structure of ionomers. *Nature*, **218**(0), 85–87.

Lopez, N., Janssens, T. V. W., Clausen, B. S., Xu, Y., Mavrikakis, M., Bligaard, T., and Nørskov, J. K. 2004. On the origin of the catalytic activity of gold nanoparticles for low-temperature CO oxidation. *J. Catal.*, **223**, 232–235.

Loppinet, B. and Gebel, G. 1998. Rodlike colloidal structure of short pendant chain perfluorinated ionomer solutions. *Langmuir*, **14**(8), 1977–1983.

Lota, G., Fic, K., and Frackowiak, E. 2011. Carbon nanotubes and their composites in electrochemical applications. *Energy Environ. Sci.*, **4**(5), 1592–1605.

MacKay, G. and Jan, N. 1984. Forest fires as critical phenomena. *J. Phys. A: Math. Gen.*, **17**(14), L757–760.

MacMillan, B., Sharp, A. R., and Armstrong, R. L. 1999. N.m.r. relaxation in Nafion—The low temperature regime. *Polymer*, **40**(10), 2481–2485.

Maillard, F., Savinova, E. R., Simonov, P. A., Zaikovskii, V. I., and Stimming, U. 2004. Infrared spectroscopic study of CO adsorption and electro-oxidation on carbon-supported Pt nanoparticles: Interparticle versus intraparticle heterogeneity. *J. Phys. Chem. B*, **108**(46), 17893–17904.

Maillard, F., Savinova, E. R., and Stimming, U. 2007. CO monolayer oxidation on Pt nanoparticles: Further insights into the particle size effects. *J. Electroanal. Chem.*, **599**, 221–232.

Maillard, F., Schreier, S., Hanzlik, M., Savinova, E. R., Weinkauf, S., and Stimming, U. 2005. Influence of particle agglomeration on the catalytic activity of carbon-supported Pt nanoparticles in CO monolayer oxidation. *Phys. Chem. Chem. Phys*, **7**, 385–393.

Majsztrik, P. W., Satterfield, M. B., Bocarsly, A. B., and Benziger, J. B. 2007. Water sorption, desorption and transport in Nafion membranes. *J. Membr. Sci.*, **301**, 93–106.

Makharia, R., Mathias, M. F., and Baker, D. R. 2005. Measurement of catalyst layer electrolyte resistance in PEFCs using electrochemical impedance spectroscopy. *J. Electrochem. Soc.*, **152**, A970–A977.

Maldonado, L., Perrin, J-C., Dillet, J., and Lottin, O. 2012. Characterization of polymer electrolyte Nafion membranes: Influence of temperature, heat treatment and drying protocol on sorption and transport properties. *J. Membr. Sci.*, **389**, 43–56.

Malek, K., Eikerling, M., Wang, Q., Navessin, T., and Liu, Z. 2007. Self-organization in catalyst layers of polymer electrolyte fuel cells. *J. Phys. Chem. C*, **111**(36), 13627–13634.

Malek, K., Eikerling, M., Wang, Q., Liu, Z., Otsuka, S., Akizuki, K., and Abe, M. 2008. Nanophase segregation and water dynamics in hydrated Nafion: Molecular modeling and experimental validation. *J. Chem. Phys.*, **129**, 204702.

Malek, K., Mashio, T., and Eikerling, M. 2011. Microstructure of catalyst layers in PEM fuel cells redefined: A computational approach. *Electrocatalysis*, **2**(2), 141–157.

Manning, G. S. 1969. Limiting laws and counterion condensation in polyelectrolyte solutions I. colligative properties. *J. Chem. Phys.*, **51**(3), 924–933.

Manning, G. S. 2011. Counterion condensation theory of attraction between like charges in the absence of multivalent counterions. *Eur. Phys. J. E.*, **34**(12), 1–18.

Maranzana, G., Mainka, J., Lottin, O., Dillet, J., Lamibrac, A., Thomas, A., and Didierjean, S. 2012. A proton exchange membrane fuel cell impedance model taking into account convection along the air channel: On the bias between the low frequency limit of the impedance and the slope of the polarization curve. *Electrochim. Acta*, **83**, 13–27.

Markovic, N. M., Grgur, B. N., and Ross, P. N. 1997. Temperature-dependent hydrogen electrochemistry on platinum low-index single-crystal surfaces in acid solutions. *J. Phys. Chem. B*, **101**, 5405–5413.

Marković, N. M., and Ross, Jr., P. N. 2002. Surface science studies of model fuel cell electrocatalysts. *Surf. Sci. Rep.*, **45**(4), 117–229.

Markovic, N. M., Schmidt, T. J., Grgur, B. N., Gasteiger, H. A., Behm, R. J., and P. N. Ross, Jr. 1999. The effect of temperature on the surface processes at the Pt(111)-liquid interface: Hydrogen adsorption, oxide formation and CO-oxidation. *J. Phys. Chem. B*, **103**, 8568–8577.

Markvoort, A. J. 2010. Coarse-grained molecular dynamics. In: R. A. van Santen, P. Sautet (Eds), *Computational Methods in Catalysis and Materials Science*. Wiley-VCH.

Marrink, S. J., Risselada, H. J., Yefimov, S., Tieleman, D. P., and de Vries, A. H. 2007. The MARTINI force field: coarse grained model for biomolecular simulations. *J. Phys. Chem. B*, **111**(27), 7812–7824.

Marx, D. 2006. Proton transfer 200 years after von Grotthus: Insights from ab initio simulations. *Chem. Phys. Phys. Chem.*, **7**, 1848–1870.

Marx, D. and Hutter, J. 2009. *Ab Initio Molecular dynamics: Basic Theory and Advanced Methods*. Cambridge: Cambridge University Press.

Marx, D., Tuckerman, M. E., Hutter, J., and Parrinello, M. 1999. The nature of the hydrated excess proton in water. *Nature*, **397**, 601–604.

Mathias, M. F., Makharia, R., Gasteiger, H. A., Conley, J. J., Fuller, T. J., Gittleman, C. J., Kocha, S. S., Miller, D. P., Mittelsteadt, C. K., Xie, T., Yan, S. G., and Yu, P. T. 2005. Two fuel cell cars in every garage? *Electrochem. Soc. Interface*, **14**, 24–35.

Matsui, J., Miyata, H., Hanaoka, Y., and Miyashita, T. 2011. Layered ultrathin proton conductive film based on polymer nanosheet assembly. *ACS Appl. Mater. Interfaces*, **3**, 1394–1397.

Mauritz, K.A. and Moore, R.B. 2004. State of understanding Nafion®. *Chem. Rev.*, **104**, 4535–4585.

Mavrikakis, M., Hammer, B., and Nørskov, J. K. 1998. Effect of strain on the reactivity of metal surfaces. *Phys. Rev. Lett.*, **81**, 2819–2822.

Mayo, S. L., Olafson, B. D., and Goddard, W. A. 1990. DREIDING: A generic force field for molecular simulations. *J. Phys. Chem.*, **94**(26), 8897–8909.

Mayrhofer, K. J. J., Blizanac, B. B., Arenz, M., Stamenkovic, V. R., Ross, P. N., and Markovic, N. M. 2005. The impact of geometric and surface electronic properties of Pt-catalysts on the particle size effect in electrocatalysis. *J. Phys. Chem. B*, **109**, 14433–14440.

Mayrhofer, K. J. J., Strmcnik, D., Blizanac, B. B., Stamenkovic, V., Arenz, M., and Markovic, N. M. 2008. Measurement of oxygen reduction activities via the rotating disc electrode method: From Pt model surfaces to carbon-supported high surface area catalysts. *Electrochim. Acta*, **53**(7), 3181–3188.

McCallum, C. and Pletcher, D. 1976. An investigation of the mechanism of the oxidation of carbon monoxide adsorbed onto a smooth Pt electrode in aqueous acid. *J. Electroanal. Chem.*, **70**(3), 277–290.

Meiboom, S. 1961. Nuclear magnetic resonance study of the proton transfer in water. *J. Chem. Phys.*, **34**(2), 375.

Meier, J., Friedrich, K. A., and Stimming, U. 2002. Novel method for the investigation of single nanoparticle reactivity. *Faraday Discuss.*, **121**, 365–372.

Meier, J. C., Katsounaros, I., Galeano, C., Bongard, H. J., Topalov, A. A., Kostka, A., Karschin, A., Schüth, F., and Mayrhofer, K. J. J. 2012. Stability investigations of electrocatalysts on the nanoscale. *Energy Environ. Sci.*, **5**, 9319–9330.

Melchy, A. M. and Eikerling, M. 2014. Physical theory of ionomer aggregation in water. *Phys. Rev. E*, **89**, 032603.

Mench, M. M., Dong, Q. L., and Wang, C. Y. 2003. In situ water distribution measurements in a polymer electrolyte fuel cell. *J. Power Sources*, **124**, 90–98.

Meulenkamp, E. A. 1998. Size dependence of the dissolution of ZnO nanoparticles. *J. Phys. Chem. B*, **102**, 7764–7769.

Meyers, J. and Newman, J. 2002a. Simulation of the direct methanol fuel cell. II. Modeling and data analysis of transport and kinetic phenomena. *J. Electrochem. Soc.*, **149**, A718–A728.

Meyers, J. P. and Darling, R. M. 2006. Model of carbon corrosion in PEM fuel cells. *J. Electrochem. Soc.*, **153**, A1432–A1442.

Meyers, J. P. and Newman, J. 2002b. Simulation of the direct methanol fuel cell II. Modeling and data analysis of transport and kinetic phenomena. *J. Electrochem. Soc.*, **149**(6), A718–A728.

Mezedur, M.M., Kaviany, M., and Moore, W. 2002. Effect of pore structure, randomness sand size on effective mass diffusivity. *AICHE J.*, **48**, 15–24.

Miao, Zh., He, Y.-L., Li, X.-L., and Zou, J.-Q. 2008. A two-dimensional two-phase mass transport model for direct methanol fuel cells adopting a modified agglomerate approach. *J. Power Sources*, **185**, 1233–1246.

Mills, G., Jonsson, H., and Schenter, G. K. 1995. Reversible work transition state theory: Application to dissociative adsorption of hydrogen. *Surf. Sci.*, **324**, 305–337.

Milton, G. W. 2002. *The Theory of Composites*. Cambridge, MA: Cambridge University Press.

Mirkin, M. V. 1996. Peer reviewed: Recent advances in scanning electrochemical microscopy. *Anal. Chem.*, **68**(5), 177A–182A.

Mitchell, P. 1961. Coupling of phosphorylation to electron and hydrogen transfer by a chemiosmotic type of mechanism. *Nature*, **191**, 144–148.

Mogensen, M. 2012. Private communication.

Moldrup, P., Olesen, T., Komatsu, T., Schjønning, P., and Rolston, D. E. 2001. Tortuosity, diffusivity, and permeability in the soil liquid and gaseous phases. *Soil Sci. Soc. Am. J.*, **65**(3), 613–623.

Mologin, D. A., Khalatur, P. G., and Khokhlov, A. R. 2002. Structural organization of water-containing Nafion: A cellular-automaton-based simulation. *Macromol. Theory Simul.*, **11**(5), 587–607.

Mond, L. and Langer, C. 1889. A new form of gas battery. *Proc. Royal Soc. London*, **46**, 296–304.

Monroe, C., Romero, T., Merida, W., and Eikerling, M. 2008. A vaporization-exchange model for water sorption and flux in Nafion. *J. Membr. Sci.*, **324**, 1–6.

Moore, R. B., and Martin, C. R. 1988. Chemical and morphological properties of solution-cast perfluorosulfonate ionomers. *Macromolecules*, **21**(5), 1334–1339.

Morgan, H., Taylor, D. M., and Oliveira, O. N. 1991. Proton transport at the monolayer–water interface. *Biochim. Biophys. Acta*, **1062**, 149–156.

Moriguchi, I., Nakahara, F., Furukawa, H., Yamada, H., and Kudo, T. 2004. Colloidal crystal-templated porous carbon as a high performance electrical double-layer capacitor material. *Electrochem. Solid-State Lett.*, **7**(8), A221–A223.

Morris, D. R. and Sun, X. 1993. Water-sorption and transport properties of Nafion 117 H. *J. Appl. Pol. Sci.*, **50**(8), 1445–1452.

Morrow, B. H. and Striolo, A. 2007. Morphology and diffusion mechanism of platinum nanoparticles on carbon nanotube bundles. *J. Phys. Chem. C*, **111**(48), 17905–17913.

Mosdale, R. and Srinivasan, S. 1995. Analysis of performance and of water and thermal management in proton exchange membrane fuel cells. *Electrochim. Acta*, **40**(4), 413–421.

Mosdale, R., Gebel, G., and Pineri, M. 1996. Water profile determination in a running proton exchange membrane fuel cell using small-angle neutron scattering. *J. Membr. Sci.*, **118**(2), 269–277.

Motupally, S., Becker, A. J., and Weidner, J. W. 2000. Diffusion of water in Nafion 115 membranes. *J. Electrochem. Soc.*, **147**, 3171–3177.

Mukerjee, S. 1990. Particle-size and structural effects in platinum electrocatalysis. *J. Appl. Electrochem.*, **20**, 537–548.

Mukerjee, S. 2003. *Catalysis and Electrocatalysis at Nanoparticle Surfaces*. New York: Marcel Dekker.

Mukerjee, S. and McBreen, J. 1998. Effect of particle size on the electrocatalysis by carbon-supported Pt electrocatalysts: An in situ XAS investigation. *J. Electroanal. Chem.*, **448**, 163–171.

Mukherjee, P. P., and Wang, C.-Y. 2007. Direct numerical simulation modeling of bilayer cathode catalyst layers in polymer electrolyte fuel cells. *J. Electrochem. Soc.*, **154**, B1121–B1131.

Mulla, S. S., Chen, N., Cumaranatunge, L., Blau, G. E., Zemlyanov, D. Y., Delgass, W. N., Epling, W. S., and Ribeiro, F. H. 2006. Reaction of NO and $O_2$ to $NO_2$ on Pt: Kinetics and catalyst deactivation. *J. Catal.*, **241**, 389–399.

Müller, J. T., and Urban, P. M. 1998. Characterization of direct methanol fuel cells by AC impedance spectroscopy. *J. Power Sources*, **75**, 139–143.

Mund, K. and Sturm, F. V. 1975. Degree of utilization and specific effective surface area of electrocatalyst in porous electrodes. *Electrochim. Acta*, **20**(6-7), 463–467.

Murahashi, T., Naiki, M., and Nishiyama, E. 2006. Water transport in the proton exchange-membrane fuel cell: Comparison of model computation and measurements of effective drag. *J. Power Sources*, **162**, 1130–1136.

Murgia, G., Pisani, L., Shukla, A. K., and Scott, K. 2003. A numerical model of a liquid-feed solid polymer electrolyte DMFC and its experimental validation. *J. Electrochem. Soc.*, **150**, A1231–A1245.

Murtola, T., Bunker, A., Vattulainen, I., Deserno, M., and Karttunen, M. 2009. Multiscale modeling of emergent materials: Biological and soft matter. *Phys. Chem. Chem. Phys.*, **11**(12), 1869–1892.

Nadler, B., Hollerbach, U., and Eisenberg, R. S. 2003. Dielectric boundary force and its crucial role in gramicidin. *Phys. Rev. E*, **68**(2), 021905.

Nagle, J. F. and Morowitz, H. J. 1978. Molecular mechanisms for proton transport in membranes. *Proc. Natl. Acad. Sci.*, **75**, 298–302.

Nagle, J. F. and Tristam-Nagle, S. 1983. Hydrogen bonded chain mechanisms for proton conduction and proton pumping. *J. Membr. Biol.*, **74**, 1–14.

Nakamura, K., Hatakeyama, T., and Hatakeyama, H. 1983. Relationship between hydrogen bonding and bound water in polyhydroxystyrene derivatives. *Polymer*, **24**(7), 871–876.

Nam, J. H. and Kaviany, M. 2003. Effective diffusivity and water-saturation distribution in single-and two-layer PEMFC diffusion medium. *Int. J. Heat Mass Trans.*, **46**(24), 4595–4611.

Nara, H., Tominaka, S., Momma, T., and Osaka, T. 2011. Impedance analysis counting reaction distribution on degradation of cathode catalyst layer in PEFCs. *J. Electrochem. Soc.*, **158**, B1184–B1191.

Narasimachary, S. P., Roudgar, A., and Eikerling, M. H. 2008. Ab initio study of interfacial correlations in polymer electrolyte membranes for fuel cells at low hydration. *Electrochim. Acta*, **53**(23), 6920–6927.

Narayanan, R. and El-Sayed, M. A. 2008. Some aspects of colloidal nanoparticle stability, catalytic activity, and recycling potential. *Top. Catal.*, **47**(1-2), 15–21.

Narayanan, R., Tabor, C., and El-Sayed, M. A. 2008. Can the observed changes in the size or shape of a colloidal nanocatalyst reveal the nanocatalysis mechanism type: Homogeneous or heterogeneous? *Top. Catal.*, **48**(1-4), 60–74.

Nashawi, I.S., Malallah, A., and Al-Bisharah, M. 2010. Forecasting world crude oil production using multicyclic Hubbert model. *Energy Fuels*, **24**, 1788–1800.

Natarajan, D. and Nguyen, T. V. 2001. A two-dimensional, two-phase, multicomponent, transient model for the cathode of a proton exchange membrane fuel cell using conventional gas distributors. *J. Electrochem. Soc.*, **148**(12), A1324–A1335.

Nazarov, I. and Promislow, K. 2007. The impact of membrane constraint on PEM fuel cell water management. *J. Electrochem. Soc.*, **154**(7), B623–B630.

Newman, J. S. and Tobias, C. W. 1962. Theoretical analysis of current distribution in porous electrodes. *J. Electrochem. Soc.*, **109**, 1183–1191.

Neyerlin, K. C., Gu, W., Jorne, J., and Gasteiger, H. 2006. Determination of catalyst unique parameters for the oxygen reduction reaction in a PEMFC. *J. Electrochem. Soc.*, **153**, A1955–A1963.

Nie, S., Feibelman, P. J., Bartelt, N. C., and Thürmer, K. 2010. Pentagons and heptagons in the first water layer on Pt(111). *Phys. Rev. Lett.*, **105**, 026102.

Nishizawa, M., Menon, V. P., and Martin, C. R. 1995. Metal nanotubule membranes with electrochemically switchable ion-transport selectivity. *Science*, **268**, 700–702.

Nocera, D. G. 2009. Living healthy on a dying planet. *Chem. Soc. Rev.*, **38**, 13–15.

Nordlund, J. and Lindbergh, G. 2004. Temperature-dependent kinetics of the anode in the DMFC. *J. Electrochem. Soc.*, **151**(9), A1357–A1362.

Nørskov, J. K., Rossmeisl, J., Logadottir, A., Lindqvist, L., Kitchin, J. R., Bligaard, T., and Jonsson, H. 2004. Origin of the overpotential for oxygen reduction at a fuel-cell cathode. *J. Phys. Chem. B.*, **108**(46), 17886–17892.

Noskov, S. Y., Im, W., and Roux, B. 2004. Ion permeation through the $\alpha$-hemolysin channel: Theoretical studies based on Brownian dynamics and Poisson–Nernst–Plank electrodiffusion theory. *Biophys. J.*, **87**(4), 2299.

Noyes, A. A. 1910. *J. Chim. Phys.*, **6**, 505.

Noyes, A. A. and Johnston, J. 1909. The conductivity and ionization of polyionic salts. *J. Am. Chem. Soc.*, **31**, 987–1010.

Ogasawara, H., Brena, B., Nordlund, D., Nyberg, M., Pelmenschikov, A., Pettersson, L. G. M., and Nilsson, A. 2002. Structure and bonding of water on Pt(111). *Phys. Rev. Lett.*, **89**, 276102.

O'Hayre, R., Lee, S. J., Cha, S. W., and Prinz, F. B. 2002. A sharp peak in the performance of sputtered platinum fuel cells at ultra-low platinum loading. *J. Power Sources*, **109**(2), 483–493.

Ohs, J. H., Sauter, U., Maas, S., and Stolten, D. 2011. Modeling hydrogen starvation conditions in proton-excahnge membrane fuel cells. *J. Power Sources*, **196**, 255–263.

Oliveira, O. N., Leite, A., and Riuland V. B. P. 2004. Water at interfaces and its influence on the electrical properties of adsorbed films. *Braz. J. Phys.*, **34**, 73–83.

Onishi, L. M., Prausnitz, J. M., and Newman, J. 2007. Water–nafion equilibria. Absence of Schroeder's paradox. *J. Phys. Chem. B*, **111**(34), 10166–10173.

Oosawa, Fumio. 1968. A theory on the effect of low molecular salts on the dissociation of linear polyacids. *Biopolymers*, **6**(1), 135–144.

Orazem, M. E., and Tribollet, B. 2008. *Electrochemical Impedance Spectroscopy*. New York: Wiley.

Orilall, M. C., Matsumoto, F., Zhou, Q., Sai, H., Abruna, H. D., DiSalvo, F. J., and Wiesner, U. 2009. One-pot synthesis of platinum-based nanoparticles incorporated into mesoporous niobium oxide-carbon composites for fuel cell electrodes. *J. Am. Chem. Soc.*, **131**(26), 9389–9395.

Ostwald, F. W. 1894. Die Wissenschaftliche Elektrochemie der Gegenwart und die Technische der Zukunft. (Scientific electrochemistry of today and technical electrochemistry of tomorrow). *Z. für Elektrotechnik und Elektrochemie*, **1**, 122–125.

Paasch, G., Micka, K., and Gersdorf, P. 1993. Theory of the electrochemical impedance of macrohomogeneous porous electrode. *Electrochim. Acta*, **38**, 2653–2662.

Paddison, S. J. 2001. The modeling of molecular structure and ion transport in sulfonic acid based ionomer membranes. *J. New Mater. Electrochem. Sys.*, **4**, 197–208.

Paddison, S. J., and Elliott, J. A. 2006. On the consequences of side chain flexibility and backbone conformation on hydration and proton dissociation in perfluorosulfonic acid membranes. *Phys. Chem. Chem. Phys.*, **8**, 2193–2203.

Paganin, V. A., Ticianelli, E. A., and Gonzalez, E. R. 1996. Development and electrochemical studies of gas diffusion electrodes for polymer electrolyte fuel cells. *J. Appl. Electrochem.*, **26**(3), 297–304.

Pajkossy, T. and Nyikos, L. 1990. Scaling-law analysis to describe the impedance behavior of fractal electrodes. *Phys. Rev. B*, **42**(1), 709.

Parsons, R. 2011. *Volacano Curves in Electrochemistry.* Hoboken, NJ: Wiley & Sons.

Parthasarathy, A., Srinivasan, S., Appleby, A. J., and Martin, C. R. 1992. Temperature dependence of the electrode kinetics of oxygen reduction at the platinum/Nafion® interface—A microelectrode investigation. *J. Electrochem. Soc.*, **139**(9), 2530–2537.

Passalacqua, E., Lufrano, F., Squadrito, G., Patti, A., and Giorgi, L. 2001. Nafion content in the catalyst layer of polymer electrolyte fuel cells: Effects on structure and performance. *Electrochim. Acta*, **46**(6), 799–805.

Patterson, T. W. and Darling, R. M. 2006. Damage to the cathode catalyst of a PEM fuel cell caused by localized fuel starvation. *Electrochem. Solid State Lett.*, **9**, A183–A185.

Paul, R. and Paddison, S. J. 2001. A statistical mechanical model for the calculation of the permittivity of water in hydrated polymer electrolyte membrane pores. *J. Chem. Phys.*, **115**(16), 7762–7771.

Paulus, U. A., Wokaun, A., Scherer, G. G., Schmidt, T. J., Stamenkovic, V., Radmilovic, V., Markovic, N. M., and Ross, P. N. 2002. Oxygen reduction on carbon-supported Pt–Ni and Pt–Co alloy catalysts. *J. Phys. Chem. B*, **106**(16), 4181–4191.

Peckham, T. J. and Holdcroft, S. 2010. Structure–morphology–property relationships of non-perfluorinated proton-conducting membranes. *Adv. Mater.*, **22**(42), 4667–4690.

Peckham, T. J., Yang, Y., and Holdcroft, S. 2010. Proton exchange membranes, D. P. Willkinson, J. J. Zhang, R. Hui, J. Fergus, and X. Li (Eds), *Proton Exchange Membrane Fuel Cells: Materials, Properties and Performance.* pp. 107–190. Boca Raton: CRC Press.

Peron, J., Mani, A., Zhao, X., Edwards, D., Adachi, M., Soboleva, T., Shi, Z., Xie, Z., Navessin, T., and Holdcroft, S. 2010. Properties of Nafion® NR-211 membranes for PEMFCs. *J. Membr. Sci.*, **356**(1–2), 44–51.

Perrin, J. C., Lyonnard, S., Guillermo, A., and Levitz, P. 2006. Water dynamics in ionomer membranes by field-cycling NMR relaxometry. *J. Phys. Chem. B*, **110**(11), 5439–5444.

Perrin, J. C., Lyonnard, S., and Volino, F. 2007. Quasielastic neutron scattering study of water dynamics in hydrated Nafion membranes. *J. Phys. Chem. C.*, **111**(8), 3393–3404.

Perry, M. L., Newman, J., and Cairns, E. J. 1998. Mass transport in gas-diffusion electrodes: A diagnostic tool for fuel-cell cathodes. *J. Electrochem. Soc.*, **145**(1), 5–15.

Peter, C. and Kremer, K. 2009. Multiscale simulation of soft matter systems—From the atomistic to the coarse-grained level and back. *Soft Matter*, **5**(22), 4357–4366.

Petersen, M. K. and Voth, G. A. 2006. Characterization of the solvation and transport of the hydrated proton in the perfluorosulfonic acid membrane Nafion. *J. Phys. Chem. B*, **110**(37), 18594–18600.

Petersen, M. K., Wang, F., Blake, N. P., Metiu, H., and Voth, G. A. 2005. Excess proton solvation and delocalization in a hydrophilic pocket of the proton conducting polymer membrane Nafion. *J. Phys. Chem. B*, **109**(9), 3727–3730.

Petrii, O. A. 1996. Surface electrochemistry of oxides: Thermodynamic and model approaches. *Electrochim. Acta*, **41**(14), 2307–2312.

Petry, O. A., Podlovchenko, B. I., Frumkin, A. N., and Lal, H. 1965. The behaviour of platinized-platinum and platinum-ruthenium electrodes in methanol solutions. *J. Electroanal. Chem.*, **10**, 253–269.

Petukhov, A. V. 1997. Effect of molecular mobility on kinetics of an electrochemical Langmuir–Hinshelwood reaction. *Chem. Phys. Lett.*, **277**(5), 539–544.

Peyrard, M. and Flytzanis, N. 1987. Dynamics of two-component solitary waves in hydrogen-bonded chains. *Phys. Rev. A*, **36**, 903–914.

Piela, P., Eickes, Ch., Brosha, E., Garzon, F., and Zelenay, P. 2004. Ruthenium crossover in direct methanol fuel cell with Pt–Ru black anode. *J. Electrochem. Soc.*, **151**, A2053–A2059.

Piela, P., Fields, R., and Zelenay, P. 2006. Electrochemical impedance spectroscopy for direct methanol fuel cell diagnostics. *J. Electrochem. Soc.*, **153**, A1902–A1913.

Pisani, L., Valentini, M., and Murgia, G. 2003. Analytical pore scale modeling of the reactive regions of polymer electrolyte fuel cells. *J. Electrochem. Soc.*, **150**, A1549–A1559.

Pivovar, A. M., and Pivovar, B. S. 2005. Dynamic behavior of water within a polymer electrolyte fuel cell membrane at low hydration levels. *J. Phys. Chem. B*, **109**(2), 785–793.

Pivovar, B. S. 2006. An overview of electro-osmosis in fuel cell polymer electrolytes. *Polymer*, **47**(11), 4194–4202.

Pnevmatikos, S. 1988. Soliton dynamics of hydrogen-bonded networks: A mechanism for proton conductivity. *Phys. Rev. Lett.*, **60**, 1534–1537.

Polle, A. and Junge, W. 1989. Proton diffusion along the membrane surface of thylakoids is not enhanced over that in bulk water. *Biophys. J.*, **56**, 27–31.

Porod, G. 1982. Chapter 2. *Small Angle X-Ray Scattering*. O. Glatter and O. Kratley (Eds), London: Academic Press, pp. 17–51.

Proton transport. 2011. Special section papers on transport phenomena in proton conducting media. *J. Phys.: Condens. Matter*, **23**, 234101–234111.

Pushpa, K. K., Nandan, D., and Iyer, R. M. 1988. Thermodynamics of water sorption by perfluorosulphonate (Nafion-117) and polystyrene–divinylbenzene sulphonate (Dowex 50W) ion-exchange resins at 298pm 1K. *J. Chem. Soc., Faraday Trans. 1*, **84**(6), 2047–2056.

Qi, Z. and Kaufman, A. 2002. open-circuit voltage and methanol crossover in DMFC. *J. Power Sources*, **110**, 177–185.

Raistrick, I. D. 1986. Diaphragms, separators, and ion-exchange membranes. *Proceedings of the Symposium*, Boston, MA. *Electrochem. Soc. Proc. Series*, **86**(13), 172.

Raistrick, I. D. 1989 (Oct. 24). *Electrode assembly for use in a solid polymer electrolyte fuel cell*. US Patent 4,876,115.

Raistrick, I. D. 1990. Impedance studies of porous electrodes. *Electrochim. Acta*, **35**(10), 1579–1586.

Ramesh, P., Itkis, M. E., Tang, J. M., and Haddon, R. C. 2008. SWNT-MWNT hybrid architecture for proton exchange membrane fuel cell cathodes. *J. Phys. Chem. C*, **112**(24), 9089–9094.

Rangarajan, S. K. 1969. Theory of flooded porous electrodes. *J. Electroanal. Chem.*, **22**, 89–104.

Ravikumar, M. K. and Shukla, A. K. 1996. Effect of methanol crossover in a liquid-feed polymer-electrolyte direct methanol fuel cell. *J. Electrochem. Soc.*, **143**(8), 2601–2606.

Reiser, C. A., Bregoli, L., Patterson, T. W., Yi, J. S., Yang, J. D., Perry, M. L., and Jarvi, Th. D. 2005. A reverse-current decay mechanism for fuel cells. *Electrochem. Solid State Lett.*, **8**, A273–A276.

Reith, D., Pütz, M., and Müller-Plathe, F. 2003. Deriving effective mesoscale potentials from atomistic simulations. *J. Comput. Chem.*, **24**(13), 1624–1636.

Ren, X. and Gottefeld, S. 2001. Electro-osmotic drag of water in poly(perfluorosulfonic acid) membranes. *J. Electrochem. Soc.*, **148**, A87–A93.

Ren, X., Springer, T., Zawodzinski, T. A., and Gottefeld, S. 2000a. Methanol transport through Nafion membranes: Electro-osmotic drag effects on potential step measurements. *J. Electrochem. Soc.*, **147**, 466–474.

Ren, X., Springer, T. E., and Gottesfeld, S. 2000b. Water and methanol uptakes in Nafion membranes and membrane effects in direct methanol cell performance. *J. Electrochem. Soc.*, **147**(1), 92–8.

Renganathan, S., Guo, Q., Sethuraman, V. A., Weidner, J. W., and White, R. E. 2006. Polymer electrolyte membrane resistance model. *J. Power Sources*, **160**, 386–397.

Rice, C. L. and Whitehead, R. 1965. Electrokinetic flow in a narrow cylindrical capillary. *J. Phys. Chem.*, **69**(11), 4017–4024.

Rieberer, S. and Norian, K. H. 1992. Analytical electron microscopy of Nafion ion exchange membranes. *Ultramicroscopy*, **41**(1–3), 225–233.

Rigsby, M. A., Zhou, W. P., Lewera, A., Duong, H. T., Bagus, P. S., Jaegermann, W., Hunger, R., and Wieckowski, A. 2008. Experiment and theory of fuel cell catalysis: Methanol and formic acid decomposition on nanoparticle Pt/Ru. *J. Phys. Chem. C*, **112**, 15595–15601.

Rinaldo, S. G., Stumper, J., and Eikerling, M. 2010. Physical theory of platinum nanoparticle dissolution in polymer electrolyte fuel cells. *J. Phys. Chem. C*, **114**(13), 5773–5785.

Rinaldo, S. G., Monroe, C. W., Romero, T., Merida, W., and Eikerling, M. 2011. Vaporization-exchange model for dynamic water sorption in Nafion: Transient solution. *Electrochem. Commun.*, **13**, 5–7.

Rinaldo, S. G., Lee, W., Stumper, J., and Eikerling, M. 2012. Nonmonotonic dynamics in Lifshitz–Slyozov–Wagner theory: Ostwald ripening in nanoparticle catalysts. *Phys. Rev. E*, **86**, 041601.

Rinaldo, S. G., Lee, W., Stumper, J., and Eikerling, M. 2014. Mechanistic principles of platinum oxide formation and reduction. *Electrocatalysis*, 1–11, DOI 10.1007/S12678-014-0189-y.

Rioux, R. M., Song, H., Grass, M., Habas, S., Niesz, K., Hoefelmeyer, J. D., Yang, P., and Somorjai, G. A. 2006. Monodisperse platinum nanoparticles of well-defined shape: Synthesis, characterization, catalytic properties and future prospects. *Top. Catal.*, **39**(3-4), 167–174.

Rivin, D., Kendrick, C. E., Gibson, P. W., and Schneider, N. S. 2001. Solubility and transport behavior of water and alcohols in Nafion™. *Polymer*, **42**(2), 623–635.

Roche, E. J., Pineri, M., Duplessix, R., and Levelut, A. M. 1981. Small-angle scattering studies of Nafion membranes. *J. Polym. Sci. Pol. Phys.*, **19**(1), 1–11.

Roduner, E. (ed). 2006. *Nanoscopic Materials: Size-Dependent Phenomena*. Cambridge: Royal Society of Chemistry.

Rollet, A. L., Diat, O., and Gebel, G. 2002. A new insight into Nafion structure. *J. Phys. Chem. B*, **106**(12), 3033–3036.

Romero, T. and Merida, W. 2009. Water transport in liquid and vapor equilibrated Nafion membranes. *J. Membr. Sci.*, **338**, 135–144.

Rosi-Schwartz, B., and Mitchell, G. R. 1996. Extracting force fields for disordered polymeric materials from neutron scattering data. *Polymer*, **37**(10), 1857–1870.

Ross, P. N. (ed.). 2003. *Handbook of Fuel Cells*. Chichester: Wiley.

Rossmeisl, J., Logadottir, A., and Nørskov, J. K. 2005. Electrolysis of water on (oxidized) metal surfaces. *Chem. Phys.*, **319**(1), 178–184.

Rossmeisl, J., Greeley, J., and Karlberg, G. S. 2009. *Electrocatalysis and Catalyst Screening from Density Functional Theory Calculations*, pp. 57–92. John Wiley & Sons, Inc.

Rossmeisl, J., Chan, K., Ahmed, R., Tripkovic, V., and Bjerketun, M. E. 2013. pH in atomic scale simulations of electrochemical interfaces. *Phys. Chem. Chem. Phys.*, **15**, 10321–10325.

Roth, C., Benker, N., Mazurek, M., Fuess, F., and Scheiba H. 2005. Development of an in-situ cell for X-ray absorption measurements during fuel cell operation. *Adv. Eng. Mater.*, **7**(10), 952–956.

Roudgar, A., and Groß, A. 2004. Local reactivity of supported metal clusters: Pd-n on Au(111). *Surf. Sci.*, **559**, L180–L186.

Roudgar, A., Eikerling, M., and van Santen, R. 2010. Ab initio study of oxygen reduction mechanism at Pt$_4$ cluster. *Phys. Chem. Chem. Phys.*, **12**, 614–620.

Roudgar, A., Narasimachary, S. P., and Eikerling, M. 2006. Hydrated arrays of acidic surface groups as model systems for interfacial structure and mechanisms in PEMs. *J. Phys. Chem. B*, **110**(41), 20469–20477.

Roudgar, A., Narasimachary, S. P., and Eikerling, M. 2008. Ab initio study of surface-mediated proton transfer in polymer electrolyte membranes. *Chem. Phys. Lett.*, **457**(4), 337–341.

Rouquerol, J., Baron, G., Denoyel, R., Giesche, H., Groen, J., Klobes, P., Levitz, P., Neimark, A. V., Rigby, S., Skudas, S. R., Sing, K., Thommes, M., and Unger, K. 2011. Liquid intrusion and alternative methods for the characterization of macroporous materials (IUPAC Technical Report). *Pure Appl. Chem.*, **84**(1), 107–136.

Rouzina, I. and Bloomfield, V. A. 1996. Macroion attraction due to electrostatic correlation between screening counterions. 1. Mobile surface-adsorbed ions and diffuse ion cloud. *J. Phys. Chem.*, **100**(23), 9977–9989.

Rubatat, L., Gebel, G., and Diat, O. 2004. Fibrillar structure of Nafion: Matching Fourier and real space studies of corresponding films and solutions. *Macromolecules*, **37**(20), 7772–7783.

Rubatat, L., Rollet, A. L., Gebel, G., and Diat, O. 2002. Evidence of elongated polymeric aggregates in Nafion. *Macromolecules*, **35**(10), 4050–4055.

Rubinstein, M. and Colby, R. H. 2003. *Polymer Physics*. Oxford: Oxford University Press.

Rudi, S., Tuaev, X., and Strasser, P. 2012. Electrocatalytic oxygen reduction on dealloyed Pt1-xNix alloy nanoparticle electrocatalysts. *Electrocatalysis*, **3**, 265–273.

Rupprechter, G. (ed). 2007. *Catalysis on Well-Defined Surfaces: From Single Crystals to Regular Nanoparticles*. New York: Kluwer Academic/Plenum.

Sabatier, F. 1920. *La Catalyse en chimie organique*. Berauge, Paris.

Sadeghi, E., Djilali, N., and Bahrami, M. 2008. Analytic determination of the effective thermal conductivity of PEM fuel cell gas diffusion layers. *J. Power Sources*, **179**, 200–208.

Sadeghi, E., Djilali, N., and Bahrami, M. 2011. A novel approach to determine the in-plane thermal conductivity of gas diffusion layers in Proton exchange membrane fuel cells. *J. Power Sources*, **196**, 3565–3571.

Sadeghi, E., Putz, A., and Eikerling, M. 2013a. Effects of ionomer coverage on agglomerate effectiveness in catalyst layers of polymer electrolyte fuel cells. *J. Solid State Electrochem.*, F1159–F1169.

Sadeghi, E., Putz, A., and Eikerling, M. 2013b. Hierarchical model of reaction rate distribuiton and effectiveness factors in catalyst layers of polymer electrolyte fuel cells. *J. Electrochem. Soc.*, **160**, F1159–F1169.

Saha, M. S., Gullá, A. F., Allen, R. J., and Mukerjee, S. 2006. High performance polymer electrolyte fuel cells with ultra-low Pt loading electrodes prepared by dual ion-beam assisted deposition. *Electrochi. Acta*, **51**(22), 4680–4692.

Sahimi, M. 1993. Flow phenomena in rocks: from continuum models to fractals, percolation, cellular automata, and simulated annealing. *Rev. Mod. Phys.*, **65**(4), 1393–1534.

Sahimi, M. 1994. *Applications of Percolation Theory*. Boca Raton, FL: Taylor & Francis.

Sahimi, M. 2003. *Heterogeneous Materials I: Linear Transport and Optical Properties*. Berlin, Germany: Springer.

Sahimi, M., Gavalas, G. R., and Tsotsis, T. T. 1990. Statistical and continuum models of fluid solid reactions in porous-media. *Chem. Eng. Sci.*, **45**, 1443–1502.

Sakurai, I., and Kawamura, Y. 1987. Lateral electrical conduction along a phosphatidylcholine monolayer. *Biochim. Biophys. Acta*, **904**, 405–409.

Sapoval, B. 1987. Fractal electrodes and constant phase angle response: Exact examples and counter examples. *Solid State Ionics*, **23**(4), 253–259.

Sapoval, B., Chazalviel, J. N., and Peyriere, J. 1988. Electrical response of fractal and porous interfaces. *Phys. Rev. A*, **38**(11), 5867.

Sarapuu, A., Kallip, S., Kasikov, A., Matisen, L., and Tammeveski, K. 2008. Electroreduction of oxygen on gold-supported thin Pt films in acid solutions. *J. Electroanal. Chem.*, **624**(1), 144–150.

Sasaki, K., Zhang, L., and Adzic, R. R. 2008. Niobium oxide-supported platinum ultra-low amount electrocatalysts for oxygen reduction. *Phys. Chem. Chem. Phys.*, **10**(1), 159–167.

Sasikumar, G., Ihm, J. W., and Ryu, H. 2004. Dependence of optimum Nafion content in catalyst layer on platinum loading. *J. Power Sources*, **132**(1), 11–17.

Satterfield, M. B. and Benziger, J. B. 2008. Non-Fickian water vapor sorption dynamics by Nafion membranes. *J. Phys. Chem. B*, **112**, 3693–3704.

Sattler, M. L. and Ross, P. N. 1986. The surface structure of Pt crystallites supported on carbon black. *Ultramicroscopy*, **20**, 21–28.

Schlick, S., Gebel, G., Pineri, M., and Volino, F. 1991. Fluorine-19 NMR spectroscopy of acid Nafion membranes and solutions. *Macromolecules*, **24**(12), 3517–3521.

Schmickler, W. 1996. *Interfacial Electrochemistry*. New York: Oxford University Press.

Schmickler, W. and Santos, E. 2010. *Interfacial Electrochemistry*, 2nd ed. Berlin: Springer.

Schmidt, T. J., Gasteiger, H. A., Stäb, G. D., Urban, P. M., Kolb, D. M., and Behm, R. J. 1998. Characterization of high-surface-area electrocatalysts using a rotating disk electrode configuration. *J. Electrochem. Soc.*, **145**(7), 2354–2358.

Schmidt-Rohr, K. and Chen, Q. 2008. Parallel cylindrical water nanochannels in Nafion fuel-cell membranes. *Nat. Mat.*, **7**, 75–83.

Schmuck, M. and Bazant, M. Z. 2012. Homogenization of the Poisson–Nernst–Planck equations for ion transport in charged porous media. *arXiv*, **32**, 168–174.

Schneider, I. A., Kramer, D., Wokaun, A., and Scherer, G. G. 2007a. Oscillations in gas channels. II. unraveling the characteristics of the low-frequency loop in air-fed PEFC impedance spectra. *J. Electrochem. Soc.*, **154**, B770–B3782.

Schneider, I. A., Freunberger, S. A., Kramer, D., Wokaun, A., and Scherer, G. G. 2007b. Oscillations in gas channels. Part I. The forgotten player in impedance spectroscopy in PEFCs. *J. Electrochem. Soc.*, **154**, B383–B388.

Schneider, I. A., Kuhn, H., Wokaun, A., and Scherer, G. G. 2005. Study of water balance in a polymer electrolyte fuel cell by locally resolved impedance spectroscopy. *J. Electrochem. Soc.*, **152**, A2383–A2389.

Schoenbein, Prof. 1839. On the voltaic polarization of certain solid and fluid substances. *Philos. Mag.*, **XIV**, 43–45.

Schuster, M., Rager, T., Noda, A., Kreuer, K. D., and Maier, J. 2005. About the choice of the protogenic group in PEM separator materials for intermediate temperature, low humidity operation: A critical comparison of sulfonic acid, phosphonic acid and imidazole functionalized model compounds. *Fuel Cells*, **5**(3), 355–365.

Schwarz, D. H., and Djilali, N. 2007. 3D modeling of catalyst layers in PEM fuel cells. *J. Electrochem. Soc.*, **154**, B1167–B1178.

Scott, K., Taama, W. M., Kramer, S., Argyropoulos, P., and Sundmacher, K. 1999. Limiting current behaviour of the direct methanol fuel cell. *Electrochim. Acta*, **45**, 945–57.

Seeliger, D., Hartnig, C., and Spohr, E. 2005. Aqueous pore structure and proton dynamics in solvated Nafion membranes. *Electrochim. Acta*, **50**(21), 4234–4240.

Senn, S. M. and Poulikakos, D. 2004. Tree network channels as fluid distributors constructing double-staircase polymer electrolyte fuel cells. *J. Appl. Phys.*, **96**(1), 842–852.

Senn, S. M. and Poulikakos, D. 2006. Pyramidal direct methanol fuel cell. *Int. J. Heat Mass Transfer*, **49**, 1516–1528.

Sepa, D. B., Vojnovic, M. V., and Damjanovic, A. 1981. Reaction intermediates as a controlling factor in the kinetics and mechanism of oxygen reduction at platinum electrodes. *Electrochim. Acta*, **26**(6), 781–793.

Sepa, D. B., Vojnovic, M. V., Vracar, L. J. M., and Damjanovic, A. 1987. Different views regarding the kinetics and mechanisms of oxygen reduction at Pt and Pd electrodes. *Electrochim. Acta*, **32**(1), 129–134.

Serowy, S., Saparov, S. M., Antonenko, Y. N., Kozlovsky, W., Hagen, V., and Pohl, P. 2003. Structural proton diffusion along lipid bilayers. *Biophys. J.*, **84**, 1031–1037.

Seung, Y. S., Dong, L., Hickner, M. A., Glass, T. E., Webb, V., and McGrath, J. E. 2003. State of water in disulfonated poly (arylene ether sulfone) copolymers and a perfluorosulfonic acid copolymer (Nafion) and its effect on physical and electrochemical properties. *Macromolecules*, **36**(17), 6281–6285.

Shan, J. and Pickup, P. G. 2000. Characterization of polymer supported catalysts by cyclic voltammetry and rotating disk voltammetry. *Electrochim. Acta*, **46**(1), 119–125.

Shao, Y., Zhang, S., Engelhard, M. H., Li, G., Shao, G., Wang, Yong, Liu, Jun, Aksay, I. A., and Lin, Y. 2010. Nitrogen-doped graphene and its electrochemical applications. *J. Mater. Chem.*, **20**, 7491–7496.

Shao-Horn, Y., Sheng, W. C., Chen, S., Ferreira, P. J., Holby, E. F., and Morgan, D. 2007. Instability of supported platinum nanoparticles in low-temperature fuel cells. *Top. Catal.*, **46**, 285–305.

Shekhawat, A., Zapperi, S., and Sethna, J. P. 2013. Viewpoint: The breaking of brittle materials. *Phys. Rev. Lett*, **110**, 185505.

Shen, J., Zhou, J., Astrath, N. G. C., Navessin, T., Liu, Z.-S. (Simon), Lei, C., Rohling, J. H., Bessarabov, D., Knights, S., and Ye, S. 2011. Measurement of effective gas diffusion coefficients of catalyst layers of PEM fuel cells with a Loschmidt diffusion cell. *J. Power Sources*, **96**, 674–78.

Shklovskii, B. I., and Efros, A. L. 1982. *Electronic Properties of Doped Semiconductors*. New York: Springer.

Silberstein, N. 2008. *Mechanics of proton exchange membranes: Time, temperature, and hydration dependence of the stress–strain behaviour of persulfonated polytetrafluorethylene*. Massachusetts Institue of Technology, Department of Mechanical Engineering.

Silva, R. F., Francesco, M. De, and Pozio, A. 2004. Tangential and normal conductivities of Nafion membranes used in polymer electrolyte fuel cells. *J. Power Sources*, **134**, 1826.

Siu, A., Schmeisser, J. and Holdcroft, S. 2006. Effect of water on the low temperature conductivity of polymer electrolytes. *J. Phys. Chem. B*, **110**(12), 6072–6080.

Smalley, R. E. 2005. Future global energy properity: The terawatt challenge. *MRS Bull.*, **30**, 413–417.

Smil, V. 2010. *Visions of Discovery: New Light on Physics, Cosmology, and Consciousness*. Cambridge: Cambridge University Press.

Smitha, B., Sridhar, S., and Khan, A.A. 2005. Solid polymer electrolyte membranes for fuel cell applications—A review. *J. Membr. Sci.*, **259**(1–2), 10–26.

Soboleva, T., Zhao, X., Malek, K., Xie, Z., Navessin, T., and Holdcroft, S. 2010. On the micro-, meso-, and macroporous structures of polymer electrolyte membrane fuel cell catalyst layers. *ACS Appl. Mater. Interfaces*, **2**(2), 375–384.

Soboleva, T., Malek, K., Xie, Z., Navessin, T., and Holdcroft, S. 2011. PEMFC catalyst layers: The role of micropores and mesopores on water sorption and fuel cell activity. *ACS Appl. Mater. Interfaces*, **3**, 1827–1837.

Soin, N., Roy, S. S., Karlsson, L., and McLaughlin, J. A. 2010. Sputter deposition of highly dispersed platinum nanoparticles on carbon nanotube arrays for fuel cell electrode material. *Diamond Relat. Mater.*, **19**, 595–598.

Solla-Gullón, J, Vidal-Iglesias, FJ, Herrero, E, Feliu, JM, and Aldaz, A. 2006. CO monolayer oxidation on semi-spherical and preferentially oriented (100) and (111) platinum nanoparticles. *Electrochem. Comm.*, **8**(1), 189–194.

Somorjai, G. A. (ed). 1994. *Introduction to Surface Chemistry and Catalysis*. New York: Wiley.

Song, C., Ge, Q., and Wang, L. 2005. DFT studies of Pt/Au bimetallic clusters and their interactions with the CO molecule. *J. Phys. Chem. B*, **109**, 22341–22350.

Spencer, J. B., and Lundgren, J-O. 1973. Hydrogen bond studies. LXXIII. The crystal structure of trifluoromethanesulphonic acid monohydrate, $H3O+CF3SO3-$, at 298 and 83 K. *Acta Cryst. B*, **29**, 1923–1928.

Spohr, E. 2004. Molecular dynamics simulations of proton transfer in a model Nafion pore. *Mol. Simul.*, **30**, 107–115.

Spohr, E., Commer, P., and Kornyshev, A. A. 2002. Enhancing proton mobility in polymer electrolyte membranes: Lessons from molecular dynamics simulations. *J. Phys. Chem. B*, **106**, 10560–10569.

Springer, T. E., Zawodzinski, T. A., and Gottesfeld, S. 1991. Polymer electrolyte fuel cell model. *J. Electrochem. Soc.*, **138**(8), 2334–2342.

Springer, T. E., Zawodzinski, T. A., Wilson, M. S., and Gottefeld, S. 1996. Characterization of polymer electrolyte fuel cells using AC impedance spectroscopy. *J. Electrochem. Soc.*, **143**, 587–599.

Squadrito, G., Maggio, G., Passalacqua, E., Lufrano, F., and Patti, A. 1999. An empirical equation for polymer electrolyte fuel cell (PEFC) behaviour. *J. Appl. Electrochem.*, **29**, 1449–1455.

Srinivasan, S. and Hurwitz, H. D. 1967. Theory of a thin film model of porous gas-diffusion electrodes. *Electrochim. Acta*, **12**, 495–512.

Srinivasan, S., Hurwitz, H. D., and Bockris, J. O'.M. 1967. Fundamental equations of electrochemical kinetics at porous gas-diffusion electrodes. *J. Chem. Phys*, **46**, 3108.

Srinivasan, S., Mosdale, R., Stevens, Ph., and Yang, Ch. 1999. Fuel Cells: Reaching the era of clean and efficient power generation in the twenty–first century. *Annu. Rev. Energy Environ.*, **24**, 281–328.

Stamenkovic, V., Mun, B. S., Mayrhofer, K. J. J., Ross, P. N., Markovic, N. M., Rossmeisl, J., Greeley, J., and Nørskov, J. K. 2006. Changing the activity of electrocatalysts for oxygen reduction by tuning the surface electronic structure. *Angew. Chem. Int. Edit.*, **45**, 2897–2901.

Stamenkovic, V. R., Fowler, B., Mun, B. S., Wang, G., Ross, P. N., Lucas, C. A., and Marković, N. M. 2007a. Improved oxygen rduction activity on Pt3Ni(111) via increased surface site availability. *Science*, **315**(5811), 493–497.

Stamenkovic, V. R., Mun, B. S., Arenz, M., Mayrhofer, K. J. J., Lucas, C. A., Wang, G., Ross, P. N., and Markovic, N. M. 2007b. Trends in electrocatalysis on extended and nanoscale Pt-bimetallic alloy surfaces. *Nat. Mat.*, **6**(3), 241–247.

Stauffer, D., and Aharony, A. 1994. *Introduction to Percolation Theory*, 2nd ed. London: Taylor & Francis.

Stein, D., Kruithof, M., and Dekker, C. 2004. Surface-charge-governed ion transport in nanofluidic channels. *Phys. Rev. Lett.*, **93**, 035901.

Stejskal, E. O. 1965. Use of spin echoes in a pulsed magnetic-field gradient to study restricted diffusion and flow. *J. Chem. Phys.*, **43**, 3597.

Stephens, I. L., Bondarenko, A. S., Perez-Alonso, F. J., Calle-Vallejo, F., Bech, L., Johansson, T. P., Jepsen, A. K., Frydendal, R., Knudsen, B. P., Rossmeisl, J., et al. 2011. Tuning the activity of Pt (111) for oxygen electroreduction by subsurface alloying. *J. Am. Chem. Soc.*, **133**(14), 5485–5491.

Tanimura, S. and Matsuoka, T. 2004. Proton transfer in nafion membrane by quantum chemistry calculation. *J. Polym. Sci. Pol. Phys.*, **42**(10), 1905–1914.

Stilbs, P. 1987. Fourier transform pulsed-gradient spin-echo studies of molecular diffusion. *Prog. Nuc. Mag. Res. Spec.*, **19**(1), 1–45.

Stonehart, P. and Ross, Jr, P. N. 1976. Use of porous electrodes to obtain kinetic rate constants for rapid reactions and adsorption isotherms of poisons. *Electrochim. Acta*, **21**(6), 441–445.

Studebaker, M. L., and Snow, C. W. 1955. The influence of ultimate composition upon the wettability of carbon blacks. *J. Phys. Chem.*, **59**(9), 973–976.

Su, X., Lianos, L., Shen, Y. R., and Somorjai, G. A. 1998. Surface-induced ferroelectric ice on Pt(111). *Phys. Rev. Lett.*, **80**, 1533–1536.

Sun, A., Franc, J., and Macdonald, D. D. 153. Growth and properties of oxide films on platinum: I. EIS and X-Ray photoelectron spectroscopy studies. *J. Electrochem. Soc.*, **2006**, B260–B277.

Sun, W., Peppley, B. A., and Karan, K. 2005. An improved two-dimensional agglomerate cathode model to study the influence of catalyst layer structural parameters. *Electrochim. Acta*, **50**(16), 3359–3374.

Suntivich, J., Gasteiger, H. A., Yabuuchi, N., Nakanishi, H., Goodenough, J. B., and Shao-Horn, Y. 2011. Design principles for oxygen-reduction activity on perovskite oxide catalysts for fuel cells and metal-air batteries. *Nat. Chem.*, **3**(7), 546–550.

Susut, C., Nguyen, T. D., Chapman, G. B., and Tong, Y. 2008. Shape and size stability of Pt nanoparticles for MeOH electro-oxidation. *Electrochim. Acta*, **53**, 6135–6142.

Sutton, A.P. and Chen, J. 1990. Long-range Finnis–Sinclair Potentials. *Philos. Mag. Lett.*, **61**(3), 139–146.

Suzuki, T., Tsushima, S., and Hirai, S. 2011. Effects of Nafion® ionomer and carbon particles on structure formation in a proton-exchange membrane fuel cell catalyst layer fabricated by the decal-transfer method. *Int. J. Hydrogen Energy*, **36**, 12361–12369.

Tabe, Y., Nishino, M., Takamatsu, H., and Chikahisa, T. 2011. Effects of cathode catalyst layer structure and properties dominating polymer electrolyte fuel cell performance. *J. Electrochem. Soc.*, **158**, B1246–B1254.

Takaichi, S., Uchida, H., and Watanabe, M. 2007. Response of specific resistance distribution in electrolyte membrane to load change at PEFC operation. *J. Electrochem. Soc.*, **154**, B1373–B1377.

Takeuchi, N., and Fuller, T. F. 2008. Modeling and investigation of design factors and their impact on carbon corrosion of PEMFC electrodes. *J. Electrochem. Soc.*, **155**, B770–B775.

Tamayol, A., Wong, K. W., and Bahrami, M. 2012. Effects of microstructure on flow properties of fibrous porous media at moderate reynolds numbers. *Phys. Rev. E*, **85**, 026318.

Tang, J. M., Jensen, K., Waje, M., Li, W., Larsen, P., Pauley, K., Chen, Z., Ramesh, P., Itkis, M. E., Yan, Y., and Haddon, R. C. 2007. High performance hydrogen fuel cells with ultralow pt loading carbon nanotube thin film catalysts. *J. Phys. Chem.*, **111**, 17901–17904.

Tantram, A.D.S. and Tseung, A. C. 1969. Structure and performance of hydrophobic gas electrodes. *Nature*, **221**, 167–168.

Tarasevich, M. R., Sadkowski, A., and Yeager, E., 1983. Comprehensive Treatise of Electrochemistry, Vol. 7: Kinetics and Mechanisms of Electrode Processes. New York: Plenum Press.

Tarasevich, M. R., Sadkowski, A., and Yeager, E. 1983. Oxygen electrochemistry. Coway, B. E., Bockris, J. O'M., Yeager, E., Khan, S. U. M., and White, R. E. (eds), *Comprehensive Treatise of Electrochemistry*, Vol. 7, pp. 310–398. New York: Plenum Press.

Taylor, C. D., Wasileski, S. A., Filhol, J. S., and Neurock, M. 2006. First principles reaction modeling of the electrochemical interface: Consideration and calculation of a tunable surface potential from atomic and electronic structure. *Phys. Rev. B.*, **73**, 165402–165418.

Teissie, J., Prats, M., Soucaille, P., and Tocanne, J. F. 1985. Evidence for conduction of protons along the interface between water and a polar lipid monolayer. *Proc. Natl. Acad. Sci.*, **82**, 3217–3221.

Teranishi, K., Tsushima, S., and Hirai, S. 2006. Analysis of water transport in PEFCs by magnetic resonance imaging measurement. *J. Electrochem. Soc.*, **153**, A664–A668.

Springer, T. E., Wilson, M. S., and Gottesfeld, S. 1993. Modeling and experimental diagnostics in polymer electrolyte fuel cells. *J. Electrochem. Soc.*, **140**(12), 3513–3526.

Thampan, T., Malhotra, S., Tang, H., and Datta, R. 2000. Modeling of conductive transport in proton-exchange membranes for fuel cells. *J. Electrochem. Soc.*, **147**(9), 3242–3250.

Thiedmann, R., Gaiselmann, G., Lehnert, W., and Schmidt, V. 2012. Stochastic modeling of fuel cell components. Stolten, D., and Emonts, B. (eds), *Fuel Cell Science and Engineering: Materials, Processes, Systems and Technology*, Vol. 2, pp. 669–702. Weinheim: Wiley–VCH.

Thiele, E. W. 1939. Relation between catalytic activity and size of particle. *Ind. Eng. Chem.*, **31**(7), 916–920.

Thiele, S., Furstenhaupt, T., Banham, D., Hutzenlaub, T., Birss, V., and Zengerle, R. 2013. Multiscale tomography of nanoporous carbon-supported noble metal catalyst layers. *J. Power Sources*, **228**, 185–192.

Thomas, S. C., Ren, X., Gottesfeld, S., and Zelenay, P. 2002. Direct methanol fuel cells: Progress in cell performance and cathode research. *Electrochim. Acta*, **47**, 3741–3748.

Tian, F., Jinnouchi, R., and Anderson, A. B. 2009. How potentials of zero charge and potentials for water oxidation to OH (ads) on Pt (111) electrodes vary with coverage. *J. Phys. Chem. C*, **113**(40), 17484–17492.

Ticianelli, E. A., Derouin, Ch. R., and Srinivasan, S. 1988. Localization of platinum in low catalyst loading electrodes to attain high power densities in SPE fuel cells. *J. Electroanal. Chem.*, **251**, 275–295.

Topalov, A. A., Katsounaros, I., Auinger, M., Cherevko, S., Meier, J. C., Klemm, S. O., and Mayrhofer, K. J. J. 2012. Dissolution of platinum: Limits for the deployment of electrochemical energy conversion? *Angew. Chem. Int. Ed.*, **51**(50), 12613–12615.

Torquato, S. 2002. *Random Heterogeneous Materials: Microstructure and Macroscopic Properties*. Interdisciplinary Applied Mathematics. New York: Springer.

Torrie, G. M., and Valleau, J. P. 1974. Monte Carlo free energy estimates using non-Boltzmann sampling: Application to the sub-critical Lennard–Jones fluid. *Chem. Phys. Lett.*, **28**, 578–581.

Tsang, E. M. W., Zhang, Z., Shi, Z., Soboleva, T., and Holdcroft, S. 2007. Considerations of macromolecular structure in the design of proton conducting polymer membranes: Graft versus diblock polyelectrolytes. *J. Am. Chem. Soc.*, **129**(49), 15106–15107.

Tsang, E. M. W., Zhang, Z., Yang, A. C. C., Shi, Z., Peckham, T. J., Narimani, R., Frisken, B. J., and Holdcroft, S. 2009. Nanostructure, morphology, and properties of fluorous copolymers bearing ionic grafts. *Macromolecules*, **42**(24), 9467–9480.

Tsironis, G. P. and Pnevmatikos, S. 1989. Proton conductivity in quasi-one-dimensional hydrogen-bonded systems—nonlinear approach. *Phys. Rev. B*, **39**(10), 7161–7173.

Tuckerman, M. E., Laasonen, L., Sprik, M. and Parrinello, M. 1994. Ab initio simulations of water and water ions. *J. Phys.: Condens. Matter*, **6**, A93–A100.

Tuckerman, M. E.., Laasonen, K., Sprik, M., and Parrinello, M. 1995. Ab initio molecular dynamics simulation of the solvation and transport of $H_3O^+$ and $OH^-$ ions in water. *J. Phys. Chem.*, **99**, 5749–5752.

Tuckerman, M. E., Marx, D., and Parrinello, M. 2002. The nature and transport mechanism of hydrated hydroxide in aqueous solution. *Nature*, **417**, 925–929.

Vreven, T., Morokuma, K., Farkas, Ö., Schlegel, H. B., and Frisch, M. J. 2003. Geometry optimization with QM/MM, ONIOM, and other combined methods. I. Microiterations and constraints. *J. Comp. Chem.*, **24**(6), 760–769.

Uchida, M., Aoyama, Y., Eda, N., and Ohta, A. 1995a. Investigation of the microstructure in the catalyst layer and effects of both perfluorosulfonate ionomer and PTFE-loaded carbon on the catalyst layer of polymer electrolyte fuel cells. *J. Electrochem. Soc.*, **142**(12), 4143–4149.

Uchida, M., Aoyama, Y., Eda, N., and Ohta, A. 1995b. New preparation method for polymer electrolyte fuel cells. *J. Electrochem. Soc.*, **142**(2), 463–468.

Uchida, M., Fukuoka, Y., Sugawara, Y., Eda, N., and Ohta, A. 1996. Effects of microstructure of carbon support in the catalyst layer on the performance of polymer-electrolyte fuel cells. *J. Electrochem. Soc.*, **143**(7), 2245–2252.

United Nations Environment Program (UNEP) and Bloomberg New Energy Finance, *Global Trends in Renewable Energy Investment*. Technical report. 2011.

Urata, S., Irisawa, J., Takada, A., Shinoda, W., Tsuzuki, S., and Mikami, M. 2005. Molecular dynamics simulation of swollen membrane of perfluorinated ionomer. *J. Phys. Chem. B*, **109**(9), 4269–4278.

van der Geest, M. E., Dangerfield, N. J., and Harrington, D. A. 1997. An ac voltammetry study of Pt oxide growth. *J. Electroanal. Chem.*, **420**, 89–100.

van der Vliet, D. F., Wang, C., Tripkovic, D., Strmcnik, D., Zhang, X. F., Debe, M. K., Atanasoski, R. T., Markovic, N. M., and Stamenkovic, V. R. 2012. Mesostructured thin films as electrocatalysts with tunable composition and surface morphology. *Nat. Mat.*, **11**, 1051–1058.

Van deVondele, J. and Hutter, J. 2007. Gaussian basis sets for accurate calculations on molecular systems in gas and condensed phases. *J. Chem. Phys.*, **127**(11), 114105.

Van deVondele, J., Krack, M., Mohamed, F., Parrinello, M., Chassaing, T. and Hutter, J. 2005. Quickstep: Fast and accurate density functional calculations using a mixed Gaussian and plane waves approach. *Comput. Phys. Commun.*, **167**(2), 103–128.

van Santen, R. A., and Neurock, M. (eds). 2006. Molecular Heterogeneous Catalysis: A Conceptual and Computational Approach. Weinheim: Wiley-VCH.

van Soestbergen, M., Biesheuvel, P. M., and Bazant, M. Z. 2010. Diffuse-charge effects on the transient response of electrochemical cells. *Phys. Rev. E*, **81**(2), 021503.

Vartak, S., Roudgar, A., Golovnev, A., and Eikerling, M. 2013. Collective proton dynamics at highly charged interfaces studied by ab initio metadynamics. *J. Phys. Chem. B*, **117**(2), 583–588.

Venkatnathan, A., Devanathan, R., and Dupuis, M. 2007. Atomistic simulations of hydrated Nafion and temperature effects on hydronium ion mobility. *J. Phys. Chem. B*, **111**(25), 7234–7244.

Vetter, K. J., and Schultze, J. W. 1972a. The kinetics of the electrochemical formation and reduction of monomolecular oxide layers on platinum in 0.5 M $H_2SO_4$: Part I. Potentiostatic pulse measurements. *J. Electroanal. Chem. Interfacial Electrochem.*, **34**, 131–139.

Vetter, K. J., and Schultze, J. W. 1972b. The kinetics of the electrochemical formation and reduction of monomolecular oxide layers on platinum in 0.5 M $H_2SO_4$: Part I. Potentiostatic pulse measurements. *J. Electroanal. Chem. Interfacial Electrochem.*, **34**, 141–158.

Vielstich, W., Lamm, A., and Gasteiger, H. A. (eds). 2003. *Handbook of Fuel Cells: Fundamentals, Technology, Applications, 4-Volume Set*. Chichester: Wiley & Sons.

Vielstich, W., Paganin, V. A., Lima, F. H. B., and Ticianelli, E. A. 2001. Non-electrochemical pathway of methanol oxidation at a platinum-catalyzed oxygen gas diffusion electrode. *J. Electrochem. Soc.*, **148**, A502–A505.

Vishnyakov, A. and Neimark, A. V. 2000. Molecular simulation study of Nafion membrane solvation in water and methanol. *J. Phys. Chem. B*, **104**(18), 4471–4478.

Vishnyakov, A. and Neimark, A. V. 2001. Molecular dynamics simulation of microstructure and molecular mobilities in swollen Nafion membranes. *J. Phys. Chem. B*, **105**(39), 9586–9594.

Vishnyakov, A. and Neimark, A. V. 2005. Final report for US Army Research Office. *DAAD190110545, March*, **430**.

Vol'fkovich, Y. M. and Bagotsky, V. S. 1994. The method of standard porosimetry: 1. Principles and possibilities. *J. Power Sources*, **48**(3), 327–338.

Vol'fkovich, Y. M., Bagotsky, V. S., Sosenkin, V. E., and Shkolnikov, E. I. 1980. Techniques of standard porosimery, and possible areas of their use in electrochemistry. *Soviet Electrochem.*, **16**(11), 1325–1353.

Vol'fkovich, Y. M., Sosenkin, V. E., and Nikol'skaya, N. F. 2010. Hydrophilic–hydrophobic and sorption properties of the catalyst layers of electrodes in a proton-exchange membrane fuel cell: A stage-by-stage study. *Russ. J. Electrochem.*, **46**, 438–449.

Volino, F., Perrin, J. C., and Lyonnard, S. 2006. Gaussian model for localized translational motion: Application to incoherent neutron scattering. *J. Phys. Chem. B*, **110**(23), 11217–11223.

Voth, G. A. 2008. *Coarse-Graining of Condensed Phase and Biomolecular Systems*. Taylor & Francis.

Wakisaka, M., Asizawa, S., Uchida, H., and Watanabe, M. 2010. *In situ* STM observation of morphological changes of the Pt (111) electrode surface during potential cycling in 10 mM HF solution. *Phys. Chem. Chem. Phys.*, **12**, 4184–4190.

Walbran, S. and Kornyshev, A. A. 2001. Proton transport in polarizable water. *J. Chem. Phys.*, **114**(22), 10039–10048.

Wang, C.-Y. 2004. Fundamental models for fuel cell engineering. *Chem. Rev.*, **104**, 4727–4766.

Wang, Q., Eikerling, M., Song, D., Liu, Zh., Navessin, T., Xie, Zh., and Holdcroft, S. 2004. Functionality graded cathode catalyst layers for polymer electrolyte fuel cell. *J. Electrochem. Soc.*, **151**, A950–A957.

Wang, J. C. 1988. Impedance of a fractal electrolyte–electrode interface. *Electrochim. Acta*, **33**(5), 707–711.

Wang, J. X., Zhang, J., and Adzic, R. R. 2007. Double-trap kinetic equation for the oxygen reduction reaction on Pt(111) in acidic media. *J. Phys. Chem. A*, **111**, 12702–12710.

Wang, L., Roudgar, A. and Eikerling, M. 2009. Ab initio study of stability and site-specific oxygen adsorption energies of Pt nanoparticles. *J. Phys. Chem. C*, **113**(42), 17989–17996.

Wang, L., Stimming, U., and Eikerling, M. 2010. Kinetic model of hydrogen evolution at an array of Au-supported catalyst nanoparticles. *Electrocatalysis*, **1**, 60–71.

Wang, X., Kumar, R., and Myers, D. J. 2006. Effect of voltage on platinum dissolution: Relevance to polymer electrolyte fuel cells. *Electrochem. Solid State Lett.*, **9**, A225–A227.

Wang, Y., and Feng, X. 2009. Analysis of reaction rates in the cathode electrode of polymer electrolyte fuel cell. II. Dual-layer electrodes. *J. Electrochem. Soc.*, **156**, B403–B409.

Wang, Y., Kawano, Y., Aubuchon, S. R., and Palmer, R. A. 2003. TGA and time-dependent FTIR study of dehydrating Nafion Na membrane. *Macromolecules*, **36**(4), 1138–1146.

Wang, Z. H., and Wang, C. Y. 2003. Mathematical modeling of liquid-feed direct methanol fuel cells. *J. Electrochem. Soc.*, **150**(4), A508–A519.

Ward, A., Damjanovic, A., Gray, E., and O'Jea, M. 1976. Kinetics of the extended growth of anodic oxide films at platinum in $H_2SO_4$ solution. *J. Electrochem. Soc.*, **123**, 1599–1604.

Warshel, A. 1991. Computer Modeling of Chemical Reactions in Enzymes and in Solutions. New York: Wiley.

Warshel, A., and Weiss, R. M. 1980. An empirical valence bond approach for comparing reactions in solutions and in enzymes. *J. Am. Chem. Soc.*, **102**, 6218–6226.

Wasserman, H.J., and Vermaak, J.S. 1972. On the determination of the surface stress of copper and platinum. *J. Surf. Sci.*, **32**, 168–174.

Watanabe, M., and Motoo, S. 1975. Electrocatalysis by ad-atoms: Part III. Enhancement of the oxidation of carbon monoxide on platinum by ruthenium ad-atoms. *J. Electroanal. Chem.*, **60**, 275–283.

Weaver, M. J. 1998. Potentials of zero charge for platinum (111)-aqueous interfaces: A combined assessment from in-situ and ultrahigh-vacuum measurements. *Langmuir*, **14**(14), 3932–3936.

Weber, A., and Newman, J. 2004a. Modeling transport in polymer–electrolyte fuel cells. *Chem. Rev.*, **104**, 4679–4726.

Weber, A. Z., and Newman, J. 2003. Transport in polymer–electrolyte membranes. *J. Electrochem. Soc*, **150**, 1008.

Weber, A. Z., and Newman, J. 2004b. A theoretical study of membrane constraint in polymer–electrolyte fuel cells. *AIChE J.*, **50**(12), 3215–3226.

Weber, A. Z., and Newman, J. 2009. A combination model for macroscopic transport in polymer electrolyte membranes. In: Paddison, S. J., and Promislow, K. S. (eds), *Device and Materials Modeling in PEM Fuel Cells*. Springer.

Weiner, J. H., and Askar, A. 1970. Proton migration in hydrogen-bonded chains. *Nature*, **226**(5248), 842–844.

Wen, Z. H., Wang, Q., and Li, J. H. 2008. Template synthesis of aligned carbon nanotube arrays using glucose as a carbon source: Pt decoration of inner and outer nanotube surfaces for fuel cell catalysts. *Adv. Funct. Mater.*, **18**(6), 959–964.

Wescott, J. T., Qi, Y., Subramanian, L., and Capehart, T. W. 2006. Mesoscale simulation of morphology in hydrated perfluorosulfonic acid membranes. *J. Chem. Phys.*, **124**, 134702.

Whitaker, S. 1998. *The Method of Volume Averaging. Theory and Applications of Transport in Porous Media*. Dordrecht, the Netherlands: Kluwer Academic Publ.

Whitesides, R. W., and Crabtree, R. W. 2007. Don't forget long-term fundamental research in energy. *Science*, **315**, 796–798.

Wieckowski, A. Savinova, E. R., and Vayenas, C. G. 2003. *Catalysis and Electrocatalysis at Nanoparticle Surfaces*. New York: Marcel Dekker.

Wilson, M. S. and Gottesfeld, S. 1992. High-performance catalyzed membranes of ultra-low Pt loadings for polymer electrolyte fuel cells. *J. Electrochem. Soc.*, **139**(2), L28–L30.

Wu, D. S., Paddison, S. J., and Elliott, J. A. 2008. A comparative study of the hydrated morphologies of perfluorosulfonic acid fuel cell membranes with mesoscopic simulations. *Energy Environ. Sci.*, **1**, 284–293.

Wu, X., Wang, X., He, G., and Benziger, J. 2011. Differences in water sorption and proton conductivity between Nafion and SPEEK. *J. Polym. Sci. B: Polym. Phys.*, **49**, 1437–1445.

Wu, Y., Tepper, H. L., and Voth, G. A. 2006. Flexible simple point-charge water model with improved liquid-state properties. *J. Chem. Phys.*, **124**, 024503.

Wynblatt, P., and Gjostein, N. A. 1976. Particle growth in model supported metal catalysts—I. Theory. *Acta Metall.*, **24**, 1165–1174.

Xia, Z., Wang, Q., Eikerling, M., and Liu, Z. 2008. Effectiveness factor of Pt utilization in cathode catalyst layer of polymer electrolyte fuel cells. *Can. J. Chem.*, **86**(7), 657–667.

Xiao, L., and Wang, L. C. 2004. Structures of platinum clusters: Planar or spherical? *J. Phys. Chem. A*, **108**, 8605–8614.

Xiao, L., Zhang, H., Scanlon, E., Ramanathan, L. S., Choe, E. W., Rogers, D., Apple, T., and Benicewicz, B. C. 2005. High-temperature polybenzimidazole fuel cell membranes via a sol–gel process. *Chem. Mater.*, **17**, 5328–5333.

Xie, G., and Okada, T. 1995. Water transport behavior in Nafion 117 membranes. *J. Electrochem. Soc.*, **142**, 3057–3062.

Xie, J., More, K. L., Zawodzinski, T. A., and H. Smith, Wayne. 2004. Porosimetry of MEAs made by "thin film decal" method and its effect on performance of PEFCs. *J. Electrochem. Soc.*, **151**(11), A1841–A1846.

Xie, J., Wood, D. L., More, K. L., Atanassov, P., and Borup, R. L. 2005. Microstructural changes of membrane electrode assemblies during PEFC durability testing at high humidity conditions. *J. Electrochem. Soc.*, **152**(5), A1011–A1020.

Xie, Z., Zhao, X., Adachi, M., Shi, Z., Mashio, T., Ohma, A., Shinohara, K., Holdcroft, S., and Navessin, T. 2008. Fuel cell cathode catalyst layers from "green" catalyst inks. *Energy Environ. Sci.*, **1**, 184–193.

Xing, L., Hossain, M. A., Tian, M., Beauchemin, D., Adjemian, K. T., and Jerkiewicz, G. 2014. *Electrocatalysis*, **5**, 96–112.

Xiong, Y., Wiley, B. J., and Xia, Y. 2007. Nanocrystals with unconventional shapes—A class of promising catalysts. *Angew. Chem. Int. Edit.*, **46**, 7157–7159.

Xu, C., He, Y. L., Zhao, T. S., Chen, R., and Ye, Q. 2006. Analysis of mass transport of methanol at the anode of a direct methanol fuel cell. *J. Electrochem. Soc.*, **153**, A1358–A1364.

Xu, F., Diat, O., Gebel, G., and Morin, A. 2007. Determination of transverse water concentration profile through MEA in a fuel cell using neutron scattering. *J. Electrochem. Soc.*, **154**(12), B1389–B1398.

Yacaman, M. J., Ascencio, J. A., Liu, H. B., and Gardea-Torresdey, J. 2001. Structure shape and stability of nanometric sized particles. *J. Vac. Sci. Technol. B*, 1091–1103.

Yamada, H., Nakamura, H., Nakahara, F., Moriguchi, I., and Kudo, T. 2007. Electrochemical study of high electrochemical double layer capacitance of ordered porous carbons with both meso/macropores and micropores. *J. Phys. Chem. C*, **111**(1), 227–233.

Yamamoto, K., Kolb, D. M., Ktz, R., and Lehmpfuhl, G. 1979. Hydrogen adsorption and oxide formation on platinum single crystal electrodes. *J. Electroanal. Chem. Interfacial Electrochem.*, **96**, 233–239.

Yamamoto, S., and Hyodo, S. 2003. A computer simulation study of the mesoscopic structure of the polyelectrolyte membrane Nafion. *Polymer J.*, **35**(6), 519–527.

Yan, Q., Toghiani, H., and Wu, J. 2006. Investigation of water transport through membrane in a PEM fuel cell by water balance experiments. *J. Power Sources*, **158**, 316–325.

Yan, T. Z., and Jen, T.-C. 2008. Two-phase flow modeling of liquid-feed direct methanol fuel cell. *Int. J. Heat Mass Transfer*, **51**, 1192–1204.

Yang, C., Srinivasan, S., Bocarsly, A. B., Tulyani, S., and Benziger, J. B. 2004. A comparison of physical properties and fuel cell performance of Nafion and zirconium phosphate/Nafion composite membranes. *J. Membr. Sci.*, **237**(1-2), 145–161.

Yang, W. W., and Zhao, T. S. 2007. A two-dimensional, two-phase mass transport model for liquid feed DMFCs. *Electrochim. Acta*, **52**, 6125–6140.

Yang, W. W., and Zhao, T. S. 2008. A transient two-phase mass transport model for liquid feed direct methanol fuel cells. *J. Power Sources*, **185**, 1131–1140.

Yang, Y., and Liang, Y. C. 2009. Modelling and analysis of a direct methanol fuel cell with under–rib mass transport and two-phase flow at the anode. *J. Power Sources*, **194**, 712–729.

Yang, Y. and Holdcroft, S. 2005. Synthetic strategies for controlling the morphology of proton conducting polymer membranes. *Fuel Cells*, **5**(2), 171–186.

Yang, Y., Siu, A., Peckham, T. J., and Holdcroft, S. 2008. Structural and morphological features of acid-bearing polymers for PEM fuel cells. Scherer, Günther, G. (ed.), *Fuel Cells I*, pp. 55–126. Adv. in Polym. Sci., Vol. 215. Berlin, Heidelberg: Springer.

Yao, K. Z., Karan, K., McAuley, K. B., Osthuizen, P., Peppley, B., and Xie, T. 2004. A review of mathematical models for hydrogen and direct methanol polymer electrolyte membrane fuel cells. *Fuel Cells*, **4**, 3–29.

Ye, Q., Yang, X.-G., and Cheng, P. 2012. Modeling of spontaneous hydrogen evolution in a direct methanol fuel cell. *Electrochim. Acta*, **69**, 230–238.

Ye, X., and Wang, C. Y. 2007. Measurement of water transport properties through membrane–electrode assemblies. *J. Electrochem. Soc.*, **154**, B676–B682.

Yeager, E., O'Grady, W. E., Woo, M. Y. C., and Hagans, P. 1978. Hydrogen adsorption on single crystal platinum. *J. Electrochem. Soc.*, **125**, 348–349.

Yoshida, H. and Miura, Y. 1992. Behavior of water in perfluorinated ionomer membranes containing various monovalent cations. *J. Membr. Sci.*, **68**(1), 1–10.

Yoshioka, N., Kun, F., and Ito, N. 2010. Kertész line of thermally activated breakdown phenomena. *Phys. Rev. E.*, **82**(5), 055102.

Yoshitake, M. and Watakabe, A. 2008. Perfluorinated ionic polymers for PEFCs (including supported PFSA). Scherer, Günther, G. (ed.), *Fuel Cells I*, pp. 127–155. Adv. in Polym. Sci., Vol. 215. Berlin, Heidelberg: Springer.

Yuan, X.-Z., Song, C., Wang, H., and Zhang, J. (JJ). 2009. *Electrochemical Impedance Spectroscopy in PEM Fuel Cells: Fundamentals and Applications*. Berlin: Springer.

Zalitis, C. M., Kramer, D., and Kucernak, A. R. 2013. Electrocatalytic performance of fuel cell reactions at low catalyst loading and high mass transport. *Phys. Chem. Chem. Phys.*, **15**(12), 4329–4340.

Zallen, R. 1983. *The Physics of Amorphous Solids*. A Wiley-Interscience Publication, Weinheim: Wiley.

Zamel, N., and Li, X. 2013. Effective transport properties for polymer electrolyte membrane fuel cells with a focus on the gas diffusion layer. *Progr. Energy Comb. Sci.*, **39**, 111–146.

Zawodzinski, Jr., T. A., Neeman, M., Sillerud, L. O., and Gottesfeld, S. 1991. Determination of water diffusion coefficients in perfluorosulfonate ionomeric membranes. *J. Phys. Chem.*, **95**(15), 6040–6044.

Zawodzinski, Jr., T. A., Springer, T. E., Uribe, F., and Gottesfeld, S. 1993. Characterization of polymer electrolytes for fuel cell applications. *Solid State Ionics*, **60**, 199–211.

Zawodzinski, T. A., Springer, T. E., Davey, J., Jestel, R., Lopez, C., Valerio, J., and Gottesfeld, S. 1993a. A comparative study of water uptake by and transport through ionomeric fuel cell membranes. *J. Electrochem. Soc.*, **140**(7), 1981–1985.

Zawodzinski, T. A., Gottesfeld, S., Shoichet, S., and McCarthy, T. J. 1993b. The contact angle between water and the surface of perfluorosulphonic acid membranes. *J. Appl. Electrochem.*, **23**(1), 86–88.

Zawodzinski, T. A., Derouin, C., Radzinski, S., Sherman, R. J., Smith, V. T., Springer, T. E., and Gottefeld, S. 1993c. Water uptake by and transport through Nafion 117 membranes. *J. Electrochem. Soc.*, **140**, 1041–1047.

Zeis, R., Mathur, A., Fritz, G., Lee, J., and Erlebacher, J. 2007. Platinum-plated nanoporous gold: An efficient, low Pt loading electrocatalyst for PEM fuel cells. *J. Power Sources*, **165**(1), 65–72.

Zeitler, S., Wendler-Kalsch, E., Preidel, W., and Tegeder, V. 1997. Corrosion of platinum electrodes in phosphate buffered saline solution. *Mater. Corros.*, **48**, 303–310.

Zhang, J., and Unwin, P. R. 2002. Proton diffusion at phospholipid assemblies. *J. Am. Chem. Soc.*, **124**, 2379–2383.

Zhang, J., Vukmirovic, M. B., Xu, Y., Mavrikakis, M., and Adzic, R. R. 2005. Controlling the catalytic activity of platinum-monolayer electrocatalysts for oxygen reduction with different substrates. *Angew. Chem. Int. Ed.*, **44**(14), 2132–2135.

Zhang, J. J. (ed). 2008. *PEM Fuel Cell Electrocatalysts and Catalyst Layers*. London: Springer.

Zhang, L., Wang, L., Holt, C. M. B., Navessin, T., Malek, K., Eikerling, M. H., and Mitlin, D. 2010. Oxygen reduction reaction activity and electrochemical stability of thin-film bilayer systems of platinum on niobium oxide. *J. Phys. Chem. C*, **114**(39), 16463–16474.

Zhang, L., Wang, L. Y., Holt, C. M. B., Zahiri, B., Li, Z., Navessin, T., Malek, K., Eikerling, M., and Mitlin, D. 2012. Highly corrosion resistant platinum–niobium oxide–carbon nanotube electrodes for the oxygen reduction in PEM fuel cells. *Energy Environ. Sci.*, **5**(3), 6156–6172.

Zhang, Z. H., Marble, A. E., MacMillan, B., Promislow, K., Martin, J., Wang, H. J., and Balcom, B. J. 2008. Spatial and temporal mapping of water content across Nafion membranes under wetting and drying conditions. *J. Magn. Res.*, **194**(2), 245–253.

Zhao, T. S., Xu, C., Chen, R., and Yang, W. W. 2009. Mass Transport Phenomena in direct methanol fuel cells. *Progr. Energy Combust. Sci.*, **35**, 275–292.

Zhdanov, V. P. and Kasemo, B. 1997. Kinetics of rapid reactions on nanometer catalyst particles. *Phys. Rev. B*, **55**(7), 4105.

Zhdanov, V. P. and Kasemo, B. 2000. Simulations of the reaction kinetics on nanometer supported catalyst particles. *Surf. Sci. Rep.*, **39**(2), 25–104.

Zhdanov, V. P. and Kasemo, B. 2003. One of the scenarios of electrochemical oxidation of CO on single-crystal Pt surfaces. *Surf. Sci.*, **545**(1), 109–121.

Zhou, X., Chen, Z., Delgado, F., Brenner, D., and Srivastava, R. 2007. Atomistic simulation of conduction and diffusion processes in Nafion polymer electrolyte and experimental validation. *J. Electrochem. Soc.*, **154**(1), B82–B87.

Zolfaghari, A., and Jerkiewicz, G. 1999. Temperature-dependent research on Pt(111) and Pt(100) electrodes in aqueous $H_2SO_4$. *J. Electroanal. Chem.*, **467**, 177–185.

Zolfaghari, A., Chayer, M., and Jerkiewicz, G. 1997. Energetics of the underpotential deposition of hydrogenelectrodes: I. Absence of coadsorbed species. *J. Electrochem. Soc.*, **144**, 3034–3041.

Zolotaryuk, A. V., Pnevmatikos, St., and Savin, A. V. 1991. Charge transport by solitons in hydrogen-bonded materials. *Phys. Rev. Lett.*, **67**, 707–710.

Zundel, G., and Fritsch, J. 1986. *The Chemical Physics of Solvation*, Vol. 2. Amsterdam: Elsevier.

# Abbreviations

| | |
|---|---|
| ABL | Anode backing layer |
| ACL | Anode catalyst layer |
| AIMD | *Ab initio* molecular dynamics |
| CCL | Cathode catalyst layer |
| CCR | Carbon corrosion reaction |
| CGMD | Coarse-grained molecular dynamics |
| CPMD | Car-Parrinello molecular dynamics |
| CL | Catalyst layer |
| DMFC | Direct methanol fuel cell |
| DPD | Dissipative particle dynamics |
| EIS | Electrochemical impedance spectroscopy |
| EMF | Electromotive force |
| EVB | Empirical valence bond method |
| EW | Equivalent weight |
| FAM | Flooded agglomerate model |
| GDE | Gas diffusion electrode |
| GDL | Gas–diffusion layer |
| HOR | Hydrogen oxidation reaction |
| IEC | Ion exchange capacity |
| LE | Liquid equilibrated |
| MD | Methanol–depleted (domain) |
| MEA | Membrane–electrode assembly |
| MHM | Macrohomogeneous model |
| MOR | Methanol oxidation reaction |
| MPL | Microporous layer |
| MR | Methanol–rich (domain) |
| ORR | Oxygen reduction reaction |
| PEFC | Polymer electrolyte fuel cell |
| PEM | Polymer electrolyte membrane |
| PNP | Poisson-Nernst-Planck (theory) |
| PRD | Particle (or pore) radius distribution function |
| PSD | Pore size distribution function |
| PRD | Particle radius distribution function |
| pzc | potential of zero charge |
| RDF | Radial distribution function |
| REV | Representative elementary volume |
| SG | Surface group |
| SHE | Standard hydrogen electrode |

TAM     Triflic acid monohydrate crystal
TEM     Transmission electron microscopy
TLM     Transmission line model
UTCL    Ultrathin catalyst layer
VASP    Vienna *ab initio* simulation package
VE      Vapor equilibrated

# Nomenclature

## Greek Letters

$\alpha$    Dimensionless constant, Equation 4.276

$\alpha$    Coefficient of surface group reorganization in PEM (dimensionless), Equation 2.36

$\alpha$    Electron transfer coefficient

$\alpha(\eta_0)$    Dimensionless function of overpotential, Equation 5.185

$\alpha_{\text{eff}}$    Effective transfer coefficient of electrode reaction (dimensionless)

$\alpha_c$    Effective transfer coefficient of electrode reaction at the cathode side (dimensionless)

$\beta$    Ratio of the net water flux to proton flux

$\beta_*$    Crossover parameter, Equation 4.207

$\beta_{cross}$    Crossover parameter, Equation 4.208

$\gamma$    Dimensionless parameter, Equation 4.127

$\gamma$    Dimensionless solution to Equation 4.163

$\gamma$    Surface tension (J cm$^{-2}$)

$\gamma_{\text{H}}^{+}$    Reaction order of protons (dimensionless)

$\gamma_{O_2}$    Reaction order of oxygen (dimensionless)

$\Gamma_{\text{agg}}$    Effectiveness factor of agglomerates (dimensionless)

$\bar{\Gamma}_{\text{agg}}$    Average effectiveness factor of agglomerates (dimensionless)

$\Gamma_{CL}$    Effectiveness factor of Pt utilization (dimensionless)

$\Gamma_{np}$    Surface-to-volume atom ratio of Pt nanoparticles (dimensionless)

$\delta_{CL}$    Reaction penetration depth of the catalyst layer due to ineffective mass transport (cm)

$\delta_{pt}$    Mass density of platinum (kg m$^{-3}$)

$\delta_C$    Mass density of carbon (kg m$^{-3}$)

$\delta_I$    Mass density of ionomer (kg m$^{-3}$)

$\Delta E_o$    Oxygen adsorption energy (eV), Equation 3.36

$\Delta G$    Gibbs free energy change in the reaction (J mol$^{-1}$)

$\Delta G^s$    Gibbs energy of water sorption (kj mol$^{-1}$)

$\Delta G^w$    Gibbs energy of water sorption at free water surface (kj mol$^{-1}$)

$\Delta H$    Enthalpy change in the reaction (J mol$^{-1}$)

$\Delta H_{vap}$    Evaporation enthalpy (J mol$^{-1}$), Equation 1.47

$\Delta S$    Entropy change in the reaction (J mol$^{-1}$ K$^{-1}$)

$\Delta S_{ORR}$    Entropy change in the half-cell ORR (J mol$^{-1}$ K$^{-1}$), Equation 1.45

$\Delta V$    Volume expansion of membrane due to water uptake in (cm$^3$), Equation 2.52

$\epsilon$    Dimensionless parameter, Equation 5.202

$\epsilon$    Small parameter, Equation 4.166

$\epsilon_v$    Voltage efficiency of a single fuel cell (dimensionless)

$\epsilon_{\text{cell}}$    Fuel cell efficiency

$\epsilon_{fuel}$    Total fuel efficiency of a fuel cell

$\epsilon_{rev}$    Theoretical thermodynamic efficiency of a fuel cell

$\epsilon_{rev}^{heat}$    Efficiency of a Carnot heat machine

$\varepsilon$    Dimensionless reaction penetration depth, Equation 4.56

$\varepsilon$    Potential well depth (eV), Equation 2.60

$\varepsilon_*$    $\varepsilon/\sqrt{c_1}$, Equations 4.100, 4.56

$\varepsilon_F$    Fermi level (energy) of electrons in catalyst particle (J)

$\varepsilon_F^{el}$    Pseudo–Fermi energy of electrolyte (J)

$\varepsilon_{CL}$    Catalyst layer porosity

$\varepsilon_{ij}$    Interaction strength in the Lennard–Jones potential (J mol$^{-1}$)

$\varepsilon_r$    Relative permittivity of water (dimensionless)

$\sigma_\mu$    Surface charge density of metal (Ccm$^{-2}$)

$\zeta_0$    Dimensionless parameter, Equation 4.72

$\eta$    Local overpotential in the catalyst layer (V)

$\eta$    Pore swelling function (dimensionless), Chapter 2

$\eta^1$    Small perturbation of overpotential (V)

$\eta_0$    Overpotential at the membrane/CL interface (V)

$\eta_0^{crit}$    Critical overpotential for negative slope of the local polarization curve (V), Equation 5.188

$\eta_1$    Overpotential at the CL/GDL interface (V)

$\eta_1^{lim}$    Limiting overpotential at the CL/GDL interface (V), Equation 4.136

$\eta_{cr}^c$    Carbon corrosion overpotential on the cathode side (V)

$\eta_{HOR}$    HOR overpotential (V), Equation 1.37

$\eta_{hy}$    HOR overpotential (V)

$\eta_{mt}$    MOR overpotential (V)

$\eta_{mt}^1$    Small perturbation of the MOR overpotential (V)

$\eta_{ORR}$    ORR overpotential (V), Equation 1.35

$\eta_{ox,0}$    ORR overpotential at the membrane/CCL interface (V)

$\eta_{ox,1}$    ORR overpotential at the CCL/GDL interface (V)

$\eta_{ox}$    ORR overpotential (V)

$\eta_{ox}^1$    Small perturbation of the ORR overpotential (V)

$\eta_{ox}^a$    ORR overpotential on the anode side (V)

$\eta_{ox}^c$    ORR overpotential on the cathode side (V)

$\eta_{PEM}$    Potential loss in the bulk membrane (V), Equation 1.36

$\eta_{Ru}$    Ruthenium oxidation overpotential (V), Equation 5.231

$\eta_{tot}$    Total cell overpotential (V), Equation 1.38

$\Theta(x)$    Heaviside step function

$\kappa$    Composite parameter determining the regime of catalyst layer operation, Equation 1.88

$\lambda$    Membrane water content, a number of water molecules per sulfonic group

$\lambda_b$    Bulk water contribution to water content in PEM

$\lambda_p$    Characteristic pore parameter (cm), Equation 2.83

$\lambda_s$    Surface water contribution to water content in PEM

$\lambda_D$    Debye length (cm), Equation 2.81

$\lambda$    Stoichiometry of the oxygen flow, Equation 5.51

$\tilde{\mu}$    Electrochemical potential (J mol$^{-1}$)

| | |
|---|---|
| $\mu$ | Dimensionless parameter, Equation 5.88 |
| $\mu$ | Dynamic viscosity $(m^2(s^{-1}))$, Equation 2.85 |
| $\mu$ | Soliton mobility $(m(Ns^{-1}))$, Equation 2.63 |
| $\mu_{e^-}^{0,M}$ | Chemical potential of electron in metal $(J\ mol^{-1})$, Equation 1.3 |
| $\mu_{H^+}^{b}$ | Proton mobility in bulk water $(cm^2\ V^{-1}\ s^{-1})$ |
| $\mu_{H^+}^{0,s}$ | Chemical potential of proton in electrolyte in standard state $(J\ mol^{-1})$, Equation 1.3 |
| $\mu_{H_2}^{0,g}$ | Chemical potential of gaseous hydrogen in standard state $(J\ mol^{-1})$, Equation 1.3 |
| $\mu_{mt}$ | Dimensionless parameter, Equation 5.140 |
| $\mu_{O_2}^{0,g}$ | Chemical potential of gaseous oxygen in standard state $(J\ mol^{-1})$, Equation 1.6 |
| $\mu_{O_2}^{0,g}$ | Chemical potential of liquid water in standard state $(J\ mol^{-1})$, Equation 1.6 |
| $\mu_{ox}$ | Dimensionless parameter, Equation 5.140 |
| $\mu_{sol}$ | Soliton mobility $(cm^2\ V^{-1}\ s^{-1})$, Equation 2.63 |
| $v$ | Reaction order, or small parameter |
| $v_{net}$ | Net overall rate of oxidation/reduction $(A\ cm^{-3})$ |
| $\xi$ | Dimensionless parameter, Equation 5.166 |
| $\xi$ | Dimensionless parameter, Equation 2.39 |
| $\xi$ | Inverse dimensionless characteristic rate of methanol adsorption, Equation 4.235 |
| $\xi_k$ | Molar fraction of component $k$ |
| $\xi_{CL}^{lv}$ | Total liquid/vapor interfacial area per total geometric electrode surface area, Equation 1.47 |
| $\xi_{pt}$ | Surface area enhancement factor of pt catalyst (heterogeneity factor, dimensionless) |
| $\rho$ | Space charge density $(C\ cm^{-3})$ |
| $\rho_p^{dry}$ | Mass density of dry polymer $(kg\ m^{-3})$ |
| $\rho_w$ | Liquid water density $(kg\ m^{-3})$ |
| $\rho_{H^+}$ | Proton charge density $(C\ cm^{-3})$ |
| $\sigma$ | Surface charge density of bundle or pore in membrane $(C\ cm^{-2})$ |
| $\sigma_m$ | Surface charge density of metal $(C\ cm^{-2})$ |
| $\sigma_R$ | Ideal surface charge density of the rod in membrane $(C\ cm^{-2})$, Equation 2.8 |
| $\sigma_t$ | Proton conductivity of the catalyst layer $(s\ cm^{-1})$ |
| $\sigma_{ij}$ | Effective bead radius (cm), Equation 2.14) |
| $\sigma_{PEM}$ | Membrane proton conductivity $(s\ cm^{-1})$ |
| $\sigma_{SG}$ | Scattering cross section $(cm^2)$ |
| $\tau$ | GDL tortuosity |
| $\phi$ | Potential of electronic conductor phase (V) |
| $\phi$ | Potential (V); auxiliary variable |
| $\phi^{pzc}$ | Potential of zero change (V), Equation 3.77 |
| $\phi^{eq}$ | Equilibrium potential of electronic conductor phase (V) |
| $\phi_\lambda$ | Dimensionless function, Equation 5.71 |

$\Phi$    Dimensionless function, Equation 4.274

$\Phi$    Electrolyte (membrane) phase potential (V)

$\Phi^{\infty}$    Undisturbed membrane potential (V), Equation 5.251

$\Phi^{eq}$    Electrolyte (membrane) phase equilibrium potential (V)

$\Gamma_{stat}$    Pt utilization factor (dimensionless)

$\omega$    Circular frequency (rad s$^{-1}$)

$\Omega$    Dimensionless parameter, Equation 2.60

## Roman Letters

$a$    Dimensionless parameter (see equation below Equation 5.160)

$a_{H^+}$    Proton activity, Equation 1.2

$b$    Tafel slope (V)

$b_{cr}$    Tafel slope of the carbon corrosion reaction (V)

$b_{hy}$    HOR Tafel slope (V)

$b_{mt}$    MOR Tafel slope (V)

$b_{ox}$    ORR Tafel slope (V)

$C$    Capacitance (F)

$C_{dl}$    Double layer volumetric capacitance ($F$ cm$^{-3}$)

$c$    Molar concentration (mol cm$^{-3}$)

$c^0$    Molar concentration at the channel inlet (mol cm$^{-3}$)

$c^0$    Steady–state molar concentration (mol cm$^{-3}$)

$c^1$    Molar concentration at the channel outlet (mol cm$^{-3}$)

$c^1$    Small perturbation of the molar concentration (mol cm$^{-3}$)

$c_1$    Molar concentration at the CCL/GDL interface (mol cm$^{-3}$)

$c_b$    Oxygen concentration in the GDL (mol cm$^{-3}$)

$c_1$    Oxygen concentration in the cathode channel (mol cm$^{-3}$)

$c_h$    Oxygen molar concentration in the air channel (mol cm$^{-3}$)

$c_h^0$    Inlet oxygen molar concentration in the air channel (mol cm$^{-3}$)

$c_h^0$    Reference (inlet) molar concentration (mol cm$^{-3}$)

$c_O$    Molar concentration of oxidized species (mol cm$^{-3}$)

$c_R$    Molar concentration of reduced species (mol cm$^{-3}$)

$c_t$    Oxygen molar concentration in the CCL (mol cm$^{-3}$)

$c_w$    Water molar concentration (mol cm$^{-3}$)

$c_{\text{tot}}$    Total molar concentration of a mixture of gases (mol cm$^{-3}$)

$c_{hy}$    Hydrogen molar concentration (mol cm$^{-3}$)

$c_{hy}^h$    Hydrogen molar concentration in the anode channel (mol cm$^{-3}$)

$c_{hy}^{h0}$    Reference hydrogen molar concentration (mol cm$^{-3}$)

$c_{mt}$    Methanol molar concentration (mol cm$^{-3}$)

$c_{mt}^1$    Small perturbation of the methanol concentration (mol cm$^{-3}$)

$c_{ox,1}$    Oxygen concentration at the CCL/GDL interface (mol cm$^{-3}$), Equation 4.209

$D$    Effective diffusion coefficient of feed molecules in the CL (cm$^2$ s$^{-1}$)

$D_b$    Oxygen diffusion coefficient in the GDL (cm$^2$ s$^{-1}$)

$D_K$    Knudsen diffusion coefficient (cm$^2$ s$^{-1}$), Equation 1.32

$D_s$     Self-diffusion coefficient of water in membrane (cm$^2$ s$^{-1}$)

$D_{ik}$     Binary diffusion coefficient of the component $i$ in a mixture with the component $k$ (cm$^2$ s$^{-1}$), Equation 1.31

$D_{mt}$     Methanol diffusion coefficient in the CCL (cm$^2$ s$^{-1}$), Equation 4.200

$D_{ox}$     Oxygen diffusion coefficient in the CCL (cm$^2$ s$^{-1}$), Equation 4.200

$D_{lr}$     Long-range diffusion coefficient of water in membrane (cm$^2$ s$^{-1}$)

$D_t$     Local diffusion coefficient of water in membrane (cm$^2$ s$^{-1}$)

$E$     Soliton energy (eV), Equation 2.61

$E_0$     Soliton energy (g), Equation 2.61

$E_{coh}$     Cohesive energy of Pt nanoparticles (eV), Equation 3.30

$E^{eq}$     Equilibrium potential of a half–cell reaction (V)

$E^{eq}_{1/2}$     Half–cell equilibrium potential (V)

$E_{cell}$     Fuel cell potential (V)

$E_{cell}$     Cell potential (V)

$E^{0,M}_{H^+,H_2}$     Equilibrium potential of the hydrogen electrode, at standard conditions (V), Equation 1.2

$E^{eq}_{H^+,H_2}$     Equilibrium potential of the hydrogen electrode (V), Equation 1.2

$E^{eq}_{mt}$     MOR equilibrium potential (V)

$E^{0,M}_{O_2,H^+}$     Equilibrium potential of the oxygen electrode at standard conditions (V), Equation 1.5

$E^{eq}_{O_2,H^+}$     Equilibrium potential of the oxygen electrode (V), Equation 1.5

$E^{0}_{O_2,H_2}$     Equilibrium potential of the hydrogen/oxygen fuel cell at standard conditions (V), Equation 1.8

$E^{eq}_{O_2,H_2}$     Equilibrium cell voltage of the hydrogen/oxygen fuel cell (V), Equation 1.8

$E^{eq}_{ox}$     ORR equilibrium potential (V)

$E^{eq}_{Ru}$     Equilibrium potential of ruthenium oxidation (V)

$E_{SHE}$     Standard potential of the hydrogen electrode. By convention, $E_{SHE} = 0$

$E_{solv}$     Solvation energy (J)

$F$     Faraday constant

$F_H$     Helmholtz energy (J)

$f$     Frequency (Hz)

$f^1_1$     Dimensionless perturbation of the oxygen flux at the catalyst layer side of the CCL/GDL interface, Equation 5.161

$f_c$     Correction factor, Equation 4.171

$f_D$     Dimensionless perturbation of the oxygen flux at the channel/GDL interface, Equation 5.163

$f_n$     Dimensionless profile function, Equation 5.207

$f_p$     Dimensionless profile function, Equation 5.213

$f_s$     Dimensionless shaping function, Equation 5.240

$f_\lambda$     Dimensionless function, Equation 5.57

$G$     Shear modulus (Pa)

$g$     Relative variation of the exchange current density through the CCL depth, Equation 4.266

$g$     Dimensionless parameter, Equation 4.25

$h$  Channel height (depth) (cm)

$I$  Total current (A)

IEC  Ion exchange capacity $(\text{mol g}^{-1})$

$i$  Imaginary unit

$i_*$  Volumetric exchange current density $(\text{A cm}^{-3})$

$i_{mt}$  MOR volumetric exchange current density $(\text{A cm}^{-3})$

$i_{ox}$  ORR volumetric exchange current density $(\text{A cm}^{-3})$

$i_{ads}$  Characteristic rate of methanol adsorption on the catalyst surface $(\text{A cm}^{-3})$

$J$  Mean current density in a cell $(\text{A cm}^{-2})$

$j$  Local proton current density in the catalyst layer $(\text{A cm}^{-2})$

$j^0$  Exchange current density per unit geometrical electrode surface area $(\text{A cm}^{-2})$

$j^a$  Current density on the anode side $(\text{A cm}^{-2})$

$j^c$  Current density on the cathode side $(\text{A cm}^{-2})$

$j_*^0$  Exchange current density per unit catalyst active surface area (intrinsic activity) $(\text{A cm}^{-2})$

$j_0$  Cell current density $(\text{A cm}^{-2})$

$j_0^{crit}$  Critical current density for negative slope of the local polarization curve $(\text{A cm}^{-2})$, Equation 5.189

$j_D$  Oxygen-transport current density in the CCL, $(\text{A cm}^{-2})$, Equation 4.77

$j_w$  Net water flux through the membrane

$j_{\lim}^a$  Limiting current density due to methanol transport in the ABL $(\text{A cm}^{-2})$, Equation 4.206

$j_{\lim}^R$  Resistive limiting current density $(\text{A cm}^{-2})$, Equation 4.221

$j_{\lim}^{a0}$  Methanol-limiting current density $(\text{A cm}^{-2})$, Equation 5.222

$j_{\lim}^{c0}$  Limiting current density due to oxygen transport in the GDL at the channel inlet $(\text{mol cm}^{-3})$, Equation 4.210

$j^l$  Liquid water flux density $(\text{A cm}^{-2})$

$j_m$  Dimensionless membrane parameter, Equation 5.34

$j_{\sigma,0}$  Characteristic current density $(\text{A cm}^{-2})$, Equation 5.47

$j_\sigma$  Characteristic current density $(\text{A cm}^{-2})$, (4.105)

$j_{crit}^0$  Critical current density for the linear regime onset $(\text{A cm}^{-2})$, Equation 4.220

$j_{cross}$  Equivalent current density of methanol crossover $(\text{A cm}^{-2})$, Equation 4.205

$j_{cr}^*$  Carbon corrosion superficial exchange current density $(\text{A cm}^{-2})$

$j_{cr}^a$  Carbon corrosion current density on the anode $(\text{A cm}^{-2})$, Equation 5.223

$j_{cr}^c$  Carbon corrosion current density on the cathode side $(\text{A cm}^{-2})$, Equation 5.215

$j_{D0}$  Characteristic current density $(\text{A cm}^{-2})$, Equation 5.66

$j_{hy}^*$  HOR superficial exchange current density $(\text{A cm}^{-2})$

$j_{hy}^a$  HOR current density on the anode side $(\text{A cm}^{-2})$, Equation 5.203

$j_{hy}^{c}$    HOR current density on the cathode side (A cm$^{-2}$)

$j_{hy}^{\infty}$    HOR superficial exchange current density (A cm$^{-2}$)

$j_{hy}^{loc}$    Limiting current density resulting from hydrogen diffusion through the anode GDL (A cm$^{-2}$), Equation 5.205

$j_{mt}^{*}$    MOR superficial exchange current density (A cm$^{-2}$)

$j_{mt}^{a}$    Total MOR current density on the anode side (A cm$^{-2}$)

$j_{mt}^{T}$    MOR current density on the anode side (A cm$^{-2}$), Equation 5.221

$j_{ox}^{*}$    ORR superficial exchange current density (A cm$^{-2}$)

$j_{ox}^{a}$    ORR current density on the anode side (A cm$^{-2}$), Equation 5.211

$j_{ox}^{c}$    ORR current density on the cathode side (A cm$^{-2}$), Equation 5.214

$j_{ox}^{\infty}$    ORR superficial exchange current density (A cm$^{-2}$)

$j_{ox}^{\lim,a}$    Oxygen-limiting current density on the anode side (A cm$^{-2}$)

$j_{ox}^{\lim,c}$    Oxygen-limiting current density on the cathode side (A cm$^{-2}$)

$j_{ox}^{BV}$    Butler–Volmer ORR current density (A cm$^{-2}$), Equation 5.212

$j_{ref}$    Reference (characteristic) current density (A cm$^{-2}$), Equation 4.52

$j_{ref}^{m}$    Reference current density (A cm$^{-2}$), Equation 5.199

$j^{v}$    Vapor flux density (A cm$^{-2}$)

$k$    Bundle aggregation number in membrane

$k$    Hydronium–hydronium coupling constant, Equation 2.57

$k_{b}$    Boltzmann constant, $k_{b} = 1.3806\,jk^{-1}$

$k_{s}$    Dimensionless parameter, Equation 5.232

$K_{vap}$    Evaporation rate constant (atm$^{-1}$ s$^{-1}$), Equation 1.47

$L$    Channel length (cm)

$L$    Length of cylindrical pore in PEM (cm), Chapter 2

$L_{0}$    Length of cylindrical reference pore in PEM (cm), Chapter 2

$L_{cell}$    Single cell thickness (cm)

$l_{b}$    GDL thickness (cm)

$l_{CL}$    Catalyst layer thickness (cm)

$l_{D}$    Reaction penetration depth, Equation 4.91

$l_{b}^{a}$    Thickness of the Anode GDL (cm)

$l_{b}^{c}$    Thickness of the Cathode GDL (cm)

$l_{D}$    Characteristic length of the conversion domain under poor oxygen transport (cm), Equation 1.87

$L_{h}$    Oxygen channel length (cm)

$l_{m}$    Membrane thickness (cm)

$l_{N}$    Reaction penetration depth (cm)

$l_{s}$    Scattering length (cm)

$l_{free}$    Mean free path of molecules in atmospheric pressure air (cm)

$l_{\sigma}$    Characteristic length of the conversion domain under poor proton transport (cm), Equation 1.86

$l_{PEM}$    Membrane thickness

$\tilde{l}_{D}$    Dimensionless reaction penetration depth, Equation 4.77

$m$    H$_3$O$^{+}$ ion mass (g), Equation 2.57

$m_{C}$    Mass loading of carbon (mg cm$^{-2}$)

$m_{I}$    Mass loading of ionomer (mg cm$^{-2}$)

$m_{Pt}$    Mass loading platinum (mg cm$^{-2}$)

$M_w$    Molecular weight of water (kg mol$^{-1}$)

$M_{ox}^{crit}$    Minimal oxygen flow, which provides positive local resistance of all segments in a segmented cell (mol s$^{-1}$), Equation 5.174

$N$    Molar flux (mol cm$^{-2}$ s$^{-1}$)

$N_o$    Number of H$_3$O$^+$ ions per solution dimensionless, Equation 2.62

$n$    Number of electrons transferred in reaction (V)

$n_H$    Proton number density in membrane (cm$^{-3}$)

$N_{uc}$    Number of unit cells, Equation 2.47

$p$    Relative variation of the Nafion content through the CCL depth, Equation 4.267

$p$    Site occupation probability in percolation theory (dimensionless)

$p_c$    Percolation threshold of site occupation probability (dimensionless)

Pe    Péclet number, Equation 1.30

PRD    Particle radius distribution function

$P^c$    Capillary pressure (Pa)

$P^g$    Gaseous pressure (Pa)

$P^l$    Liquid pressure (Pa)

$P^s$    Saturated vapor pressure (Pa)

$P_{cell}^s$    Specific power of a single fuel cell (W g$_{Pt}^{-1}$)

$P^v$    Water vapor pressure (Pa)

$P^{el}$    Elastic pressure (Pa)

$P^{osm}$    Osmotic pressure (Pa)

$P^v$    Partial pressure of water vapor (Pa)

$P_{cell}$    Fuel cell power density (W cm$^{-2}$)

$P_{cell}$    Volumetric power density of a single fuel cell (W L$^{-1}$)

$P_{H_2}$    Partial pressure of hydrogen normalized to the standard pressure Equation 1.2

$P_{O_2}$    Oxygen partial pressure normalized to the standard pressure Equation 1.5

$q$    Wave number (cm$^{-1}$)

$q_{tot}$    Heat flux from the catalyst layer (W cm$^{-2}$), Equation 4.298

$q_{tot}^{high}$    Heat flux from the CL in the high-current regime (W cm$^{-2}$), Equation 4.294

$q_{tot}^{low}$    Heat flux from the CL in the low-current regime (W cm$^{-2}$), Equation 4.290

$Q_{PEM}$    Volumetric rate of Joule heating in membrane

$Q_{rev}$    Reversible heat of the fuel cell reaction (J mol$^{-1}$)

$Q_J$    Rate of the Joule heat production (W cm$^{-3}$), Equation 1.46

$Q_{ORR}$    Volumetric rate of heat production in the ORR (W cm$^{-3}$), Equation 1.45

$Q_{vap}$    Rate of heat consumption due to liquid water evaporation (W cm$^{-3}$), Equation 1.47

$R$    Pore radius (cm)

$R$    Radius of cylindrical pore in PEM (cm), Chapter 2

$R$    Resistance ($\Omega$)

$R_g$   Gas constant

$R_0$   Radius of cylindrical reference pore in PEM (cm), Chapter 2

$R_{lv}$   Source term of water vapor (A cm$^{-3}$), Equation 4.17

$R_{reac}$   Electrochemical source term (A cm$^{-3}$), Equation 4.16

$R_\Omega$   Cell resistance ($\Omega$ cm$^2$)

$r$   Radial coordinate (cm)

$r_B$   Radius of ionomer bundle (cm)

$R_c$   Capillary radius (cm)

$r_C$   Radius of cylindrical unit cell that contains single ionomer bundle (cm)

$R_R$   Rod length in membrane (cm)

$R_s$   Spot radius (cm)

$R_{acl}$   Anode catalyst layer resistance ($\Omega$ cm$^2$)

$R_{act}$   Activation resistance of the CL ($\Omega$ cm$^2$), Equations 1.79 and 4.115

$R_{ccl}$   CCL differential resistance ($\Omega$ cm$^2$)

$R_{ct}$   Charge transfer resistance ($\Omega$ cm$^2$)

$R_{exp}$   Exponential approximation of the ORR rate (A cm$^{-3}$), Equation 4.153

$R_{MOR}$   MOR rate (A cm$^{-3}$)

$R_{MOR}^1$   Small perturbation of the MOR rate (A cm$^{-3}$)

$R_{ORR}$   Volumetric rate of the ORR (A cm$^{-3}$)

$R_{ORR}^1$   Small perturbation of the ORR rate (A cm$^{-3}$)

$R_{PEM}$   Electrolyte (bulk membrane) resistance ($\Omega$ cm$^{-2}$)

$R_{reac}$   Reaction rate (A cm$^{-3}$)

$R_{seg}$   Segment differential resistance ($\Omega$ cm$^2$), Equation 5.193

$s$   Thickness of transition region (cm), Equation 5.232

$S$   Liquid saturation

$S_r$   Liquid water saturation in the cathode catalyst layer (dimensionless), Equation 3.104

$S_{ECSA}$   Electrochemically active surface area of catalyst (cm$^2$)

$T$   Cell temperature (K)

$t_*$   Characteristic time (s), Equation 5.84

$u$   Oxygen or fuel utilization

$u_i$   Displacement of the $i$th $H_3O^+$ ion (cm), Equation 2.57

$U_{LJ}$   Lennard–Jones (LJ) potential (J mol$^{-1}$)

$v$   Flow velocity in the channel (cm s$^{-1}$)

$V(\mathbf{r})$   Potential energy of the system (J)

$V(u_i)$   Effective substrate potential (eV), Equation 2.57

$v^0$   Inlet flow velocity in the channel (cm s$^{-1}$)

$v_0$   Maximal possible soliton velocity (cm s$^{-1}$), Equation 2.60

$\bar{V}_p$   Molar volume of ionomer (per backbone repeat unit that includes one sidechain) (L mol$^{-1}$)

$\bar{V}_w$   Molar volume of water (L mol$^{-1}$)

$V_p^{dry}$   Volume of dry polymer (L)

$V_w$   Water volume (L)

$w$   In-plane width of the channel (cm)

$v_0$   Unit cell volume in dry PEM (cm$^3$), Chapter 2

$v_p$   volume of cylindrical pore in PEM ($cm^3$), Chapter 2

$W_{cell}$   Volumetric energy density of a single fuel cell ($J\,L^{-1}$)

$W_{out}$   Real electrical work of a fuel cell ($J\,mol^{-1}$)

$W_{rev,el}$   Maximal amount of electrical work in a reversible process ($J\,mol^{-1}$)

$x$   Coordinate through the catalyst layer, or through the MEA (cm)

$X_c$   Volume fraction at percolation threshold (dimensionless)

$X_P$   Total pore volume fraction in catalyst layer (dimensionless)

$X_w$   Water volume fraction (dimensionless)

$X_{el}$   Volume fraction of ionomer in catalyst layer (dimensionless)

$X_{PtC}$   Volume fraction of solid matrix of Pt and C in catalyst layer (dimensionless)

$X_P$   Gas porosity of the catalyst layer

$Z$   Impedance ($\Omega\,cm^2$)

$z$   Coordinate along the channel (cm)

$Z_{im}$   Imaginary part of impedance ($\Omega\,cm^2$)

$Z_{re}$   Real part of impedance ($\Omega\,cm^2$)

# Index

Printed and bound by CPI Group (UK) Ltd, Croydon, CR0 4YY

18/10/2024

01776261-0018